Mechanical Engineering Series

Frederick F. Ling
Editor-in-Chief

Mechanical Engineering Series

G. Genta, **Vibration Dynamics and Control**

R. Firoozian, **Servomotors and Industrial Control Theory**

G. Genta and L. Morello, **The Automotive Chassis, Volumes 1 & 2**

F. A. Leckie and D. J. Dal Bello, **Strength and Stiffness of Engineering Systems**

Wodek Gawronski, **Modeling and Control of Antennas and Telescopes**

Makoto Ohsaki and Kiyohiro Ikeda, **Stability and Optimization of Structures: Generalized Sensitivity Analysis**

A.C. Fischer-Cripps, **Introduction to Contact Mechanics, 2nd ed.**

W. Cheng and I. Finnie, **Residual Stress Measurement and the Slitting Method**

J. Angeles, **Fundamentals of Robotic Mechanical Systems: Theory Methods and Algorithms, 3rd ed.**

J. Angeles, **Fundamentals of Robotic Mechanical Systems: Theory, Methods, and Algorithms, 2nd ed.**

P. Basu, C. Kefa, and L. Jestin, **Boilers and Burners: Design and Theory**

J.M. Berthelot, **Composite Materials: Mechanical Behavior and Structural Analysis**

I.J. Busch-Vishniac, **Electromechanical Sensors and Actuators**

J. Chakrabarty, **Applied Plasticity**

K.K. Choi and N.H. Kim, **Structural Sensitivity Analysis an Optimization 1: Linear Systems**

K.K. Choi and N.H. Kim, **Structural Sensitivity Analysis and Optimization 2: Nonlinear Systems and Applications**

G. Chryssolouris, **Laser Machining: Theory and Practice**

V.N. Constantinescu, **Laminar Viscous Flow**

G.A. Costello, **Theory of Wire Rope, 2nd ed.**

K. Czolczynski, **Rotordynamics of Gas-Lubricated Journal Bearing Systems**

M.S. Darlow, **Balancing of High-Speed Machinery**

W. R. DeVries, **Analysis of Material Removal Processes**

J.F. Doyle, **Nonlinear Analysis of Thin-Walled Structures: Statics, Dynamics, and Stability**

J.F. Doyle, **Wave Propagation in Structures: Spectral Analysis Using Fast Discrete Fourier Transforms, 2nd Edition**

P.A. Engel, **Structural Analysis of Printed Circuit Board Systems**

A.C. Fischer-Cripps, **Introduction to Contact Mechanics**

A.C. Fischer-Cripps, **Nanoindentation, 2nd ed.**

J. García de Jalón and E. Bayo, **Kinematic and Dynamic Simulation of Multibody Systems: The Real-Time Challenge**

W.K. Gawronski, **Advanced Structural Dynamics and Active Control of Structures**

W.K. Gawronski, **Dynamics and Control of Structures: A Modal Approach**

G. Genta, **Dynamics of Rotating Systems**

D. Gross and T. Seelig, **Fracture Mechanics with Introduction to Micro-mechanics**

K.C. Gupta, **Mechanics and Control of Robots**

(continued after index)

Mechanical Engineering Series

Frederick F. Ling
Editor-in-Chief

Mechanical Engineering Series

(continued after index)

Giancarlo Genta

Vibration Dynamics
and Control

 Springer

Giancarlo Genta
Politecnico di Torino
Torino, Italy
giancarlo.genta@polito.it

ISBN: 978-1-4899-7784-7   ISBN: 978-0-387-79580-5 (eBook)
DOI: 10.1007/978-0-387-79580-5

springer.com

Mechanical Engineering Series

Frederick F. Ling
Editor-in-Chief

The Mechanical Engineering Series features graduate texts and research monographs to address the need for information in contemporary mechanical engineering, including areas of concentration of applied mechanics, biomechanics, computational mechanics, dynamical systems and control, energetics, mechanics of materials, processing, production systems, thermal science, and tribology.

Mechanical Engineering Series

Frederick F. Ling
Editor-in-Chief

The Mechanical Engineering Series features graduate texts and research monographs to address the need for information in contemporary mechanical engineering, including areas of concentration of applied mechanics, biomechanics, computational mechanics, dynamical systems and control, energetics, mechanics of materials, processing, production systems, thermal science, and tribology.

Advisory Board/Series Editors

Applied Mechanics	F.A. Leckie
	University of California,
	Santa Barbara
	D. Gross
	Technical University of Darmstadt
Biomechanics	V.C. Mow
	Columbia University
Computational Mechanics	H.T. Yang
	University of California,
	Santa Barbara
Dynamic Systems and Control/	D. Bryant
Mechatronics	University of Texas at Austin
Energetics	J.R. Welty
	University of Oregon, Eugene
Mechanics of Materials	I. Finnie
	University of California, Berkeley
Processing	K.K. Wang
	Cornell University
Production Systems	G.-A. Klutke
	Texas A&M University
Thermal Science	A.E. Bergles
	Rensselaer Polytechnic Institute
Tribology	W.O. Winer
	Georgia Institute of Technology

Series Preface

Mechanical engineering, and engineering discipline born of the needs of the industrial revolution, is once again asked to do its substantial share in the call for industrial renewal. The general call is urgent as we face profound issues of productivity and competitiveness that require engineering solutions, among others. The Mechanical Engineering Series is a series featuring graduate texts and research monographs intended to address the need for information in contemporary areas of mechanical engineering.

The series is conceived as a comprehensive one that covers a broad range of concentrations important to mechanical engineering graduate education and research. We are fortunate to have a distinguished roster of series editors, each an expert in one of the areas of concentration. The names of the series editors are listed on page vi of this volume. The areas of concentration are applied mechanics, biomechanics, computational mechanics, dynamic systems and control, energetics, mechanics of materials, processing, thermal science, and tribology.

Preface

After 15 years since the publication of *Vibration of Structures and Machines* and three subsequent editions a deep reorganization and updating of the material was felt necessary. This new book on the subject of *Vibration dynamics and control* is organized in a larger number of shorter chapters, hoping that this can be helpful to the reader. New material has been added and many points have been updated. A larger number of examples and of exercises have been included.

Since the first edition, these books originate from the need felt by the author to give a systematic form to the contents of the lectures he gives to mechanical, aeronautical, and then mechatronic engineering students of the Technical University (Politecnico) of Torino, within the frames of the courses of Principles and Methodologies of Mechanical Design, Construction of Aircraft Engines, and Dynamic Design of Machines. Their main aim is to summarize the fundamentals of mechanics of vibrations to give the needed theoretical background to the engineer who has to deal with vibration analysis and to show a number of design applications of the theory. Because the emphasis is mostly on the practical aspects, the theoretical aspects are not dealt with in detail, particularly in areas in which a long and complex study would be needed.

The book is structured in 30 Chapters, subdivided into three Parts.

The first part deals with the dynamics of linear, time invariant, systems. The basic concepts of linear dynamics of discrete systems are summarized in the first 10 Chapters. Following the lines just described, some specialized topics, such as random vibrations, are just touched on, more to remind the reader that they exist and to stimulate him to undertake a deeper study of these aspects than to supply detailed information.

Chapter 11 constitutes an introduction to the dynamics of controlled structural systems, which are increasingly entering design practice and will unquestionably be used more often in the future.

The dynamics of continuous systems is the subject of the following two chapters. As the analysis of the dynamic behavior of continuous systems is now mostly performed using discretization techniques, only the basic concepts are dealt with. Discretization techniques are described in a general way in Chapter 14, while Chapter 15 deals more in detail with the finite element method, with the aim of supplying the users of commercial computer codes with the theoretical background needed to build adequate mathematical models and critically evaluate the results obtained from the computer.

The following two Chapters are devoted to the study of multibody modeling and of the vibration dynamics of systems in motion with respect to an inertial reference frame. These subjects are seldom included in books on vibration, but the increasing use of multibody codes and the inclusion in them of flexible bodies modeled through the finite element method well justifies the presence of these two chapters.

Part II (including Chapters from 18 to 22 is devoted to the study of nonlinear and non time-invariant systems. The subject is dealt with stressing the aspects of these subjects that are of interest to engineers more than to theoretical mechanicists. The recent advances in all fields of technology often result in an increased nonlinearity of machines and structural elements and design engineers must increasingly face nonlinear problems: This part is meant to be of help in this instance.

Part III deals with more applied aspects of vibration mechanics. Chapters from 23 to 28 are devoted to the study of the dynamics of rotating machines, while Chapters 29 and 30 deal with reciprocating machines. They are meant as specific applications of the more general topics studied before and intend to be more application oriented than the previous ones. However, methods and mathematical models that have not yet entered everyday design practice and are still regarded as research topics are dealt with herein.

Two appendices related to solution methods and Laplace transform are then added.

The subjects studied in this book (particularly in the last part) are usually considered different fields of applied mechanics or mechanical design. Specialists in rotor dynamics, torsional vibration, modal analysis, nonlinear mechanics, and controlled systems often speak different languages, and it is difficult for students to be aware of the unifying ideas that are at the base of all these different specialized fields. The inconsistency of the symbols used in the different fields can be particularly confusing. In order to use a consistent symbol system throughout the book, some deviation from the common practice is unavoidable.

The author believes that it is possible to explain all the aspects related to mechanical vibrations (actually not only mechanical) using a unified approach. The current book is an effort in this direction.

S.I. units are used in the whole book, with few exceptions. The first exception is the measure of angles, for which in some cases the old unit degree is preferred to the S.I. unit radian, particularly where phase angles are concerned. Frequencies and angular velocities should be measured in rad/s. Sometimes the older units (Hz for frequencies and revolutions per minute [rpm]) are used, when the author feels that this makes things more intuitive or where normal engineering practice suggests it. In most formulas, at any rate, consistent units are used. In very few cases this rule is not followed, but the reader is explicitly warned in the text.

For frequencies, no distinction is generally made between frequency in Hz and circular frequency in rad/s. Although the author is aware of the subtle differences between the two quantities (or better, between the two different ways of seeing the same quantity), which are subtended by the use of two different names, he chose to regard the two concepts as equivalent. A single symbol (ω) is used for both, and the symbol f is never used for a frequency in Hz. The period is then always equal to $T = 2\pi/\omega$ because consistent units (in this case, rad/s) must be used in all formulas. A similar rule holds for angular velocities, which are always referred to with the symbol Ω.[1] No different symbol is used for angular velocities in rpm, which in some texts are referred to by n.

In rotor dynamics, the speed at which the whirling motion takes place is regarded as a whirl frequency and not a whirl angular velocity (even if the expression *whirl speed* is often used in opposition to spin speed), and symbols are used accordingly. It can be said that the concept of angular velocity is used only for the rotation of material objects, and the rotational speed of a vector in the complex plane or of the deformed shape of a rotor (which does not involve actual rotation of a material object) is considered a frequency.

The author is grateful to colleagues and students in the Mechanics Department of the Politecnico di Torino for their suggestions, criticism, and general exchange of ideas and, in particular, to the postgraduate students working in the dynamics field at the department for reading the whole manuscript and checking most of the equations. Particular thanks are due to my wife, Franca, both for her encouragement and for doing the tedious work of revising the manuscript.

G. Genta
Torino, July 2008

---

[1] In *Vibration of Structures and Machines* λ was used instead of ω for frequencies to avoid using Ω for angular velocities. In the present text a more standard notation is adopted.

Contents

Symbols

c	viscous damping coefficient, clearance
c^*	complex viscous damping coefficient
c_{cr}	critical value of c
c_{eq}	equivalent damping coefficient
d	distance (between axis of cylinder and center of crank)
e	base of natural logarithms
$\mathbf{f}$	force vector
$\overline{f}_i$	ith modal force
f_0	amplitude of the force $F(t)$
g	acceleration of gravity
$g(t)$	response to a unit-step input
h	thickness of oil film, relaxation factor
$h(t)$	response to a unit-impulse function
i	imaginary unit ($i = \sqrt{-1}$)
k	stiffness, gain
k^*	complex stiffness ($k^* = k' + ik''$)
l	length, length of the connecting rod
l_0	length at rest, length in a reference condition
m	mass, number of modes taken into account, number of outputs
n	number of degrees of freedom
p	pressure
q_i	ith generalized coordinate
$\mathbf{q}$	vector of the (complex) coordinates
$\mathbf{q}_i$	ith eigenvector
$q_i(xyz)$	ith eigenfunction
r	radius, number of inputs

$\mathbf{r}$	Ritz vector, vector of the complex coordinates (rotating frame), vector of the command inputs, vector of modal participation factors
s	Laplace variable
$\mathbf{s}$	state vector (transfer-matrices method)
t	time, thickness
u	displacement
$\mathbf{u}$	vector of the inputs, displacement vector (FEM)
$u(t)$	unit-step function
v	velocity
$\mathbf{v}$	vector containing the derivatives of the generalized coordinates
v_s	velocity of sound
$\mathbf{w}$	vector of the generalized velocities
$\mathbf{x}$	vector of the coordinates
x_0	amplitude of $x(t)$
x_m	maximum value of periodic law $x(t)$
xyz	(fixed) reference frame
$\mathbf{y}$	vector of the outputs
z	complex coordinate $(z = x + iy)$
$\mathbf{z}$	state vector
A	area of the cross-section
$\mathbf{A}$	matrix linking vectors $\mathbf{w}$ and $\mathbf{v}$
$\mathcal{A}$	dynamic matrix (state-space approach)
$\mathbf{B}$	matrix linking vectors $\mathbf{w}$ and $\mathbf{v}$
$\mathcal{B}$	input gain matrix
$\mathbf{C}$	damping matrix
$\mathcal{C}$	output gain matrix
$\overline{\mathbf{C}}$	modal-damping matrix
$\mathbf{D}$	dynamic matrix (configuration space approach)
$\mathcal{D}$	direct link matrix
E	Young's modulus
$\mathbf{E}$	stiffness matrix of the material (FEM)
F	force
$\mathcal{F}$	Rayleigh dissipation function
G	shear modulus, balance-quality grade
$G(s)$	transfer function
$\mathbf{G}$	gyroscopic matrix
$\mathbf{G}(s)$	matrix of the transfer functions
$H(\omega)$	frequency response
$\mathbf{H}$	controllability matrix
$\mathbf{H}(\omega)$	dynamic compliance matrix
I	area moment of inertia
$\mathbf{I}$	identity matrix
J	mass moment of inertia
$\mathcal{L}$	work
$\mathcal{L}(f)$	Laplace transform of function f
$\mathbf{K}$	stiffness matrix.
$\mathbf{K}_{dyn}$	dynamic stiffness matrix
$\mathbf{K}_g$	geometric stiffness matrix

$\mathbf{K}''$ imaginary part of the stiffness matrix

$\overline{K}_i$ ith modal stiffness

M moment

$\mathbf{M}$ mass matrix

$\overline{M}_i$ ith modal mass

$\mathbf{M}_g$ geometric stiffness matrix

$\mathbf{N}$ matrix of the shape functions

O load factor (Ocvirk number)

$\mathbf{O}$ observability matrix

Q quality factor

Q_i ith generalized force

R radius

$\mathbf{R}$ rotation matrix

S Sommerfeld number

$\mathbf{S}$ Jacobian matrix

$S(\lambda)$ power spectral density

T period of the free oscillations

$\mathcal{T}$ kinetic energy

$\mathbf{T}$ transfer matrix, matrix linking the forces to the inputs

$\mathcal{U}$ potential energy

V velocity, volume

w power

α slenderness of a beam, phase of static unbalance, nondimensional parameter

β attitude angle, phase of couple unbalance, nondimensional parameter

γ shear strain

δ logarithmic decrement, phase in phase-angle diagrams

$\delta(t)$ unit-impulse function (Dirac delta)

$\delta\mathcal{L}$ virtual work

$\delta\mathbf{x}$ virtual displacement

ϵ strain, eccentricity

$\boldsymbol{\epsilon}$ strain vector

ζ damping factor ($\zeta = c/c_{cr}$); nondimensional coordinate ($\zeta = z/l$), complex coordinate ($\zeta = \xi + i\eta$)

η loss factor

$\boldsymbol{\eta}$ modal coordinates

θ angular coordinate, pitch angle

μ coefficient of the nonlinear term of stiffness, viscosity

ν Poisson's ratio, complex frequency

$\xi\eta\zeta$ rotating reference frame

ρ density

σ decay rate, stress

$\boldsymbol{\sigma}$ stress vector

σ_y yield stress

τ shear stress

ϕ angular displacement, roll angle, complex coordinate ($\phi = \phi_y - i\phi_x$)

χ Torsional stiffness, shear factor, angular error for couple unbalance.

ψ specific damping capacity, yaw angle

ω frequency, complex frequency, whirl speed

ω'	whirl speed in the rotating frame
$[\omega^2]$	eigenvalue matrix
ω_n	natural frequency of the undamped system
ω_p	frequency of the resonance peak in damped systems
$\mathbf{B}$	compliance matrix
Γ	torsional damping coefficient
Φ	phase angle
$\mathbf{\Phi}$	eigenvector matrix
$\mathbf{\Phi}^*$	eigenvector matrix reduced to m modes
Ω	angular velocity (spin speed)
$\mathbf{\Omega}$	angular velocity vector
Ω_{cr}	critical speed
$\Im$	imaginary part
$\Re$	real part
—	complex conjugate ($\bar{a}$ is the conjugate of a)
~	Laplace transform ($\tilde{f}$ is the Laplace transform of f).

SUBSCRIPTS

d	deviatoric
m	mean
n	nonrotating
r	rotating
I	imaginary part
R	real part.

Introduction

Vibration

Vibration is one of the most common aspects of life. Many natural phenomena, as well as man-made devices, involve periodic motion of some sort. Our own bodies include many organs that perform periodic motion, with a wide spectrum of frequencies, from the relatively slow motion of the lungs or heart to the high-frequency vibration of the eardrums. When we shiver, hear, or speak, even when we snore, we directly experience vibration.

Vibration is often associated with dreadful events; indeed one of the most impressive and catastrophic natural phenomena is the earthquake, a manifestation of vibration. In man-made devices vibration is often less impressive, but it can be a symptom of malfunctioning and is often a signal of danger. When traveling by vehicle, particularly driving or flying, any increase of the vibration level makes us feel uncomfortable. Vibration is also what causes sound, from the most unpleasant noise to the most delightful music.

Vibration can be put to work for many useful purposes: Vibrating sieves, mixers, and tools are the most obvious examples. Vibrating machines also find applications in medicine, curing human diseases. Another useful aspect of vibration is that it conveys a quantity of useful information about the machine producing it.

Vibration produced by natural phenomena and, increasingly, by man-made devices is also a particular type of pollution, which can be heard as noise if the frequencies that characterize the phenomenon lie within the audible range, spanning from about 18 Hz to 20 kHz, or felt directly as vibration. This type of pollution can cause severe discomfort. The discomfort due to noise depends on the intensity of the noise and its frequency, but many other features are also of great importance. The sound of a bell and the noise from some machine may have the same intensity and frequency but create very different sensations. Although even the psychological disposition of the subject can be important in assessing how much discomfort a certain sound creates, some standards must be assessed in order to evaluate the acceptability of noise sources.

Generally speaking, there is growing awareness of the problem and designers are asked, sometimes forced by standards and laws, to reduce the noise produced by all sorts of machinery.

When vibration is transmitted to the human body by a solid surface, different effects are likely to be felt. Generally speaking, what causes discomfort is not the amplitude of the vibration but the peak value (or better, the root mean square value) of the acceleration. The level of acceleration that causes discomfort depends on the frequency and the time of exposure, but other factors like the position of the human body and the part that is in contact with the source are also important. Also, for this case, some standards have been stated. The maximum r.m.s. (root mean square) values of acceleration that cause reduced proficiency when applied for a stated time in a vertical direction to a sitting subject are plotted as a function of frequency in Fig. 1. The figure, taken from the ISO 2631-1978 standard, deals with a field from 1 to 80 Hz and with daily exposure times from 1 min to 24 h.

The exposure limits can be obtained by multiplying the values reported in Fig. 1 by 2, while the reduced comfort boundary is obtained by dividing the same values by 3.15 (i.e., by decreasing the r.m.s. value by 10 dB). From the plot, it is clear that the frequency field in which humans are more affected by vibration lies between 4 and 8 Hz.

Frequencies lower than 1 Hz produce sensations similar to motion sickness. They depend on many parameters other than acceleration and are variable from individual to individual.

At frequencies greater than 80 Hz, the effect of vibration is also dependent on the part of the body involved and on the skin conditions and it is impossible to give general guidelines.

An attempt to classify the effects of vibration with different frequencies on the human body is shown in Fig. 2. Note that there are resonance fields at which some parts of the body vibrate with particularly large amplitudes.

As an example, the thorax–abdomen system has a resonant frequency of about 3–6 Hz, although all resonant frequency values are dependent on individual characteristics. The head–neck–shoulder system has a resonant frequency of about 20–30 Hz, and many other organs have more or less pronounced resonances at other frequencies (e.g., the eyeball at 60–90 Hz, the lower jaw-skull system at 100–220 Hz).

In English, as in many other languages, there are two terms used to designate oscillatory motion: oscillation and vibration. The two terms are

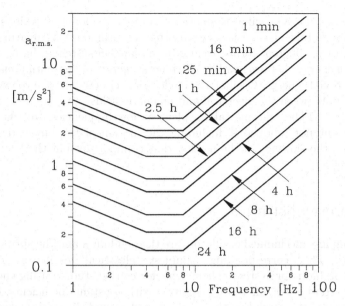

FIGURE 1. Vertical vibration exposure criteria curves defining the 'fatigue-decreased proficiency boundary' (ISO 2631-1978 standard).

used almost interchangeably; however, if a difference can be found, *oscillation* is more often used to emphasize the kinematic aspects of the phenomenon (i.e., the time history of the motion in itself), while *vibration* implies dynamic considerations (i.e., considerations on the relationships between the motion and the causes from which it originates).

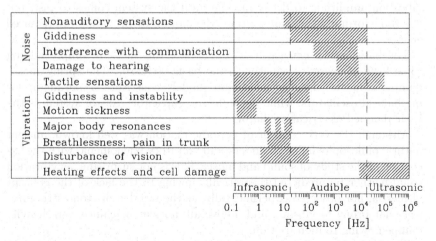

FIGURE 2. Effects of vibration and noise (intended as airborne vibration) on the human body as functions of frequency (R.E.D. Bishop, *Vibration*, Cambridge Univ. Press, Cambridge, 1979).

Actually, not all oscillatory motions can be considered vibrations: For a vibration to take place, it is necessary that a continuous exchange of energy between two different forms occurs. In mechanical systems, the particular forms of energy that are involved are kinetic energy and potential (elastic or gravitational) energy. Oscillations in electrical circuits are due to exchange of energy between the electrical and magnetic fields.

Many periodic motions taking place at low frequencies are thus oscillations but not vibrations, including the motion of the lungs. It is not, however, the slowness of the motion that is important but the lack of dynamic effects.

Theoretical studies

The simplest mechanical oscillators are the pendulum and the spring–mass system. The corresponding simplest electrical oscillator is the capacitor–inductor system. Their behavior can be studied using the same linear second-order differential equation with constant coefficients, even if in the case of the pendulum the application of a simple linear model requires the assumption that the amplitude of the oscillation is small.

For centuries, the pendulum, and later the spring–mass system (later still the capacitor–inductor system), has been more than a model. It constituted a paradigm through which the oscillatory behavior of actual systems has been interpreted. All oscillatory phenomena in real life are more complex than that, at least for the presence of dissipative mechanisms causing some of the energy of the system is dissipated, usually being transformed into heat, at each vibration cycle, i.e., each time the energy is transformed forward and backward between the two main energy forms. This causes the vibration amplitude to decay in time until the system comes to rest, unless some form of excitation sustains the motion by providing the required energy.

The basic model can easily accommodate this fact, by simply adding some form of energy dissipator to the basic oscillator. The spring–mass–damper and the damped-pendulum models constitute a paradigm for mechanical oscillators, while the inductor–capacitor–resistor system is the basic damped electrical oscillator.

Although the very concept of periodic motion was well known, ancient natural philosophy failed to understand vibratory phenomena, with the exception of the study of sound and music. This is not surprising, as vibration could neither be predicted theoretically, owing to the lack of the concept of inertia, nor observed experimentally, as the wooden or stone structures were not prone to vibrate, and, above all, ancient machines were heavily damped owing to high friction.

The beginnings of the theoretical study of vibrating systems are traced back to observations made by Galileo Galilei in 1583 regarding the motions of one of the lamps hanging from the ceiling of the cathedral of Pisa. It is said that he timed the period of oscillations using the beat of his heart as a time standard to conclude that the period of the oscillations is independent from the amplitude.

Whether or not this is true, he described in detail the motion of the pendulum in his *Dialogo sopra i due massimi sistemi del mondo*, published in 1638, and stated clearly that its oscillations are isochronous. It is not surprising that the beginning of the studies of vibratory mechanics occurred at the same time as the formulation of the law of inertia.

The idea that a mechanical oscillator could be used to measure time, due to the property of moving with a fixed period, clearly stimulated the theoretical research in this field. While Galileo seems to have believed that the oscillations of a pendulum have a fixed period even if the amplitude is large (he quotes a displacement from the vertical as high as 50°), certainly Huygens knew that this is true only in linear systems and around 1656 introduced a modified pendulum whose oscillations would have been truly isochronous even at large amplitudes. He published his results in his *Horologium Oscillatorium* in 1673.

The great development of theoretical mechanics in the eighteenth and nineteenth centuries gave the theory of vibration very deep and solid roots. When it seemed that theoretical mechanics could not offer anything new, the introduction of computers, with the possibility of performing very complex numerical experiments, revealed completely new phenomena and disclosed unexpected perspectives.

The study of chaotic motion in general and of chaotic vibrations of nonlinear systems in particular will hopefully clarify some phenomena that have been beyond the possibility of scientific study and shed new light on known aspects of mechanics of vibration.

Vibration analysis in design

Mechanics of vibration is not just a field for theoretical study. Design engineers had to deal with vibration for a long time, but recently the current tendencies of technology have made the dynamic analysis of machines and structures more important.

The load conditions the designer has to take into account in the structural analysis of any member of a machine or a structure can be conventionally considered as static, quasi-static, or dynamic. A load condition belongs to the first category if it is constant and is applied to the structure for all or most of its life. A typical example is the self-weight of a building. The task of the structural analyst is usually limited to determining whether the stresses static loads produce are within the allowable limits of the material, taking into account all possible environmental effects (creep, corrosion, etc.). Sometimes the analyst must check that the deformations of the structure are consistent with the regular working of the machine.

Also, loads that are repeatedly exerted on the structure, but that are applied and removed slowly and stay at a constant value for a long enough time, are assimilated to static loads. An example of these static load conditions is the pressure loading on the structure of the pressurized fuselage of an airliner and the thermal loading of many pressure vessels. In this case, the designer also has to take into account the fatigue phenomena that can be caused by repeated application of the load. Because the number of stress cycles is usually low, low-cycle fatigue is encountered.

Quasi-static load conditions are those conditions that, although due to dynamic phenomena, share with static loads the characteristics of being applied slowly and remaining for comparatively long periods at more or less constant values. Examples are the centrifugal loading of rotors and the loads on the structures of space vehicles due to inertia forces during launch or re-entry. Also, in this case, fatigue phenomena can be very important in the structural analysis.

Dynamic load conditions are those in which the loads are rapidly varying and cause strong dynamic effects. The distinction is due mainly to the speed at which loads vary in time. Because it is necessary to state in some way a time scale to assess whether a certain load is applied slowly, it is possible to say that a load condition is static or quasi-static if the characteristic times of its variation are far longer than the longest period of the free vibrations of the structure.

A given load can thus be considered static if it is applied to a structure whose first natural frequency is high or dynamic if it is applied to a structure that vibrates at low frequency.

Dynamic loads may cause the structure to vibrate and can sometimes produce a resonant response. Causes of dynamic loading can be the motion of what supports the structure (as in the case of seismic loading of buildings or the stressing of the structure of ships due to wave motion), the motion of the structure (as in the case of ground vehicles moving on uneven roads), or the interaction of the two motions (as in the case of aircraft flying in gusty air). Other sources of dynamic loads are unbalanced rotating or reciprocating machinery and aero- or gas-dynamic phenomena in jet and rocket engines.

The task the structural analyst must perform in these cases is much more demanding. To check that the structure can withstand the dynamic loading for the required time and that the amplitude of the vibration does not affect the ability of the machine to perform its tasks, the analyst must acquire a knowledge of its dynamic behavior that is often quite detailed. The natural frequencies of the structure and the corresponding mode shapes must first be obtained, and then its motion under the action of the dynamic loads and the resulting stresses in the material must be computed. Fatigue must generally be taken into account, and often the methods based on fracture mechanics must be applied.

Fatigue is not necessarily due to vibration; it can be defined more generally as the possibility that a structural member fails under repeated loading at stress levels lower than those that could cause failure if applied only once. However, the most common way in which this repeated loading takes place is linked with vibration. If a part of a machine or structure vibrates, particularly if the frequency of the vibration is high, it can be called on to withstand a high number of stress cycles in a comparatively short time, and this is usually the mechanism triggering fatigue damage.

Another source of difficulty is the fact that, while static loads are usually defined in deterministic terms, often only a statistical knowledge of dynamic loads can be reached.

Progress causes machines to be lighter, faster, and, generally speaking, more sophisticated. All these trends make the tasks of the structural analyst more complex and demanding. Increasing the speed of machines is often a goal in itself, like in the transportation field. This is sometimes useful in increasing production and lowering costs (as in machine tools) or causing more power to be produced, transmitted, or converted (as in energy-related devices). Faster machines, however, are likely to be the cause of more intense vibrations, and, often, they are prone to suffer damages due to vibrations.

Speed is just one of the aspects. Machines tend to be lighter, and materials with higher strength are constantly being developed. Better design procedures allow the exploitation of these characteristics with higher stress levels, and all these efforts often result in less stiff structures, which are more prone to vibrate. All these aspects compel designers to deal in more detail with the dynamic behavior of machines.

Dynamic problems, which in the past were accounted for by simple overdesign of the relevant elements, must now be studied in detail, and dynamic design is increasingly the most important part of the design of many machines.

Most of the methods used nowadays in dynamic structural analysis were first developed for nuclear or aerospace applications, where safety and, in the latter case, lightness are of utmost importance. These methods are spreading to other fields of industry, and the number of engineers working in the design area, particularly those involved in dynamic analysis, is growing. A good technical background in this field, at least enough to understand the existence and importance of these problems, is increasingly important for persons not directly involved in structural analysis, such as production engineers, managers, and users of machinery.

It is now almost commonplace to state that about half of the engineers working in mechanical industries, and particularly in the motor-vehicle industry, are employed in tasks directly related to design. A detailed analysis of the tasks in which engineers are engaged in an industrial group working in the field of energy systems is reported in Fig. 3a. An increasing number of engineers are engaged in design and the relative economic weight of design activities on total production costs is rapidly increasing. An increase of 300% in the period from 1950 to 1990 has been recorded.

Within design activities, the relative importance of structural analysis, mainly dynamic analysis, is increasing, while that of activities generally indicated as drafting is greatly reduced (Fig. 3b).

Economic reasons advocate the use of predictive methods for the study of the dynamic behavior of machines from the earliest stages of design,

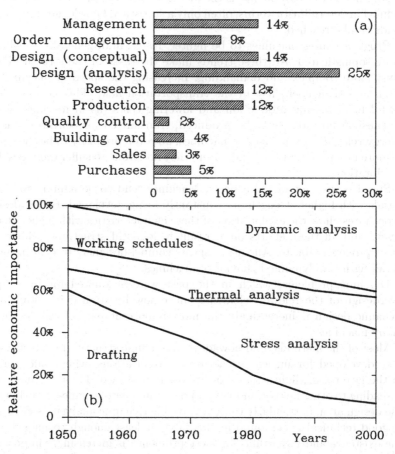

FIGURE 3. (a) Tasks in which engineers are employed in an Italian industrial group working in the field of energy systems; (b) relative economic weight of the various activities linked with structural design (P.G. Avanzini, *La formazione universitaria nel campo delle grandi costruzioni meccaniche*, Giornata di studio sull'insegnamento della costruzione delle macchine, Pisa, March 31, 1989).

without having to wait until prototypes are built and experimental data are available. The cost of design changes increases rapidly during the progress of the development of a machine, from the very low cost of changes introduced very early in the design stage to the dreadful costs (mostly in terms of loss of image) that occur when a product already on the market has to be recalled to the factory to be modified. On the other hand, the effectiveness of the changes decreases while new constraints due to the progress of the design process are stated.

This situation is summarized in the plot of Fig. 4. Because many design changes can be necessary as a result of dynamic structural analysis, it must be started as early as possible in the design process, at least in the form of first-approximation studies. The analysis must then be refined and detailed when the machine takes a more definite form.

The quantitative prediction, and not only the qualitative understanding, of the dynamic behavior of structures is then increasingly important. To understand and, even more, to predict quantitatively the behavior of any system, it is necessary to resort to models that can be analyzed using mathematical tools. Such analysis work is unavoidable, even if in some of its aspects it can seem that the physical nature of the problem is lost within the mathematical intricacy of the analytical work.

After the analysis has been performed it is necessary to extract results and interpret them to obtain a synthetic picture of the relevant phenomena. The analytical work is necessary to ensure a correct interpretation of the relevant phenomena, but if it is not followed by a synthesis, it remains only a sterile mathematical exercise. The tasks designers are facing in modern

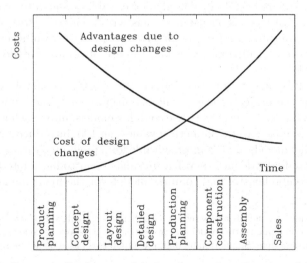

FIGURE 4. Cost and effectiveness of design changes as a function of the stage at which the changes are introduced.

technology force them to understand increasingly complex analytical techniques. They must, however, retain the physical insight and engineering common sense without which no sound synthesis can be performed.

Mathematical modeling

The computational predictions of the characteristics and the performance of a physical system are based on the construction of a *mathematical model*,[1] constructed from a number of equations, whose behavior is similar to that of the physical system it replaces. In the case of discrete dynamic models, such as those used to predict the dynamic behavior of discrete mechanical systems, the model usually is made by a number of ordinary differential equations[2] (ODE).

The complexity of the model depends on many factors that are the first choice the analyst has to make. The model must be complex enough to allow a realistic simulation of the system's characteristics of interest, but no more. The more complex the model, the more data it requires, and the more complicated are the solution and the interpretation of results. Today it is possible to built very complex models, but overly complex models yield results from which it is difficult to extract useful insights into the behavior of the system.

Before building the model, the analyst must be certain about what he wants to obtain from it. If the goal is a good physical understanding of the underlying phenomena, without the need for numerically precise results, simple models are best. Skilled analysts were able to simulate even complex phenomena with precision using models with a single degree of freedom. If, on the contrary, the aim is precise quantitative results, even at the price of more difficult interpretation, the use of complex models becomes unavoidable.

Finally, it is important to take into account the data available at the stage reached by the project: Early in the definition phase, when most data are not yet available, it is useless to use complex models, into which more or less arbitrary estimates of the numerical values must be introduced. Simplified, or synthetic, models are the most suitable for a preliminary analysis. As the design is gradually defined, new features may be introduced into the model, reaching comprehensive and complex models for the final simulations.

---

[1]Simulations are not always based on a mathematical models in a strict sense. In the case of analog computers, the model was an electric circuit whose behavior simulated that of the physical system. Simulation on digital computers is based on actual mathematical models.

[2]A dynamic model, or a dynamic system, is a model expressed by one or more differential equations containing derivatives with respect to time.

Such complex models, useful for simulating many characteristics of the machine, may be considered as true *virtual prototypes*. Virtual reality techniques allow these models to yield a large quantity of information, not only on performance and the dynamic behavior of the machine, but also on the space taken by the various components, the adequacy of details, and their esthetic qualities, that is comparable to what was once obtainable only from physical prototypes.

The models of a given machine thus evolve initially toward a greater complexity, from synthetic models to virtual prototypes, to return later to simpler models.

Models are useful not only to the designer but also to the test engineer in interpreting the results of testing and performing all adjustments. Simplified models allow the test engineer to understand the effect of adjustments and reduce the number of tests required, provided they are simple enough to give an immediate idea of the effect of the relevant parameters. Here the final goal is to adjust the virtual prototype on the computer, transferring the results to the physical machine and hoping that at the end of this process only a few physical validation tests are required.

Simplified models that can be integrated in real time on relatively low-power hardware are also useful in control systems. A mathematical model of the controlled system (*plant*, in control jargon) may constitute an *observer* (always in the sense of the term used in control theory) and be a part of the control architecture.

The analyst has the duty not only of building, implementing, and using the models correctly but also of updating and maintaining them. The need to build a mathematical model of some complexity is often felt at a certain stage of the design process, but the model is then used much less than needed, and above all is not updated with subsequent design changes, with the result that it becomes useless or must be updated when the need for it arises again.

There are usually two different approaches to mathematical modeling: models made by equations describing the physics of the relevant phenomena, – these may be defined as *analytical models* – and empirical models, often called *black box* models.

In analytical models the equations approximating the behavior of the various parts of the system, along with the required approximations and simplifications, are written. Even if no real-world spring behaves exactly like the linear spring, producing a force proportional to the relative displacement of its ends through a constant called stiffness, and even if no device dissipating energy is a true linear damper, the dynamics of a mass–spring–damper system (see Chapter 1) can be described, often to a very good approximation, by the usual ordinary differential equation (ODE)

$$m\ddot{x} + c\dot{x} + kx = f(t) \, .$$

The behavior of some systems, on the other hand, is so complex that writing equations to describe it starting from the physical and geometrical characteristics of their structure is forbiddingly difficult. Their behavior is studied experimentally and then a mathematical expression able to describe it is sought, identifying the various parameters from the experimental data. While each of the parameters m, c, and k included in the equation of motion of the mass–spring–damper system refers to one of the parts of the system and has a true physical meaning, the many coefficients appearing in empirical models usually have no direct physical meaning and refer to the system as a whole.

Among the many ways to build black box models, that based on neural networks must be mentioned.[3] Such networks can simulate complex and highly nonlinear systems, adapting their parameters (the *weights* of the network) to produce an output with a relationship to the input that simulates the input–output relationship of the actual system.

Actually, the difference between analytical and black box models is not as clear-cut as it may seem. The complexity of the system is often such that it is difficult to write equations precisely describing the behavior of its parts, while the values of the parameters cannot always be known with the required precision. In such cases the model is built by writing equations approximating the general pattern of the response of the system, with the parameters identified to make the response of the model as close as possible to that of the actual system. In this case, the identified parameters lose a good deal of their physical meaning related to the various parts of the system they are conceptually linked to and become global parameters.

In this book primarily analytical models will be described and an attempt will be made to link the various parameters to the components of the system.

Once the model has been built and the equations of motion written, there is no difficulty in studying the response to any input, assuming the initial conditions are stated. A general approach is to numerically integrate the ordinary differential equation constituting the model, using one of the many available numerical integration algorithms. In this way, the *time history* of the generalized coordinates (or of the state variables) is obtained from any given time history of the inputs (or of the forcing functions)

This approach, usually referred to as simulation or numerical experimentation, is equivalent to physical experimentation, where the system is subjected to given conditions and its response measured.

---

[3]Strictly speaking, neural networks are not sets of equations and thus do not belong to the mathematical models here described. However, at present neural networks are usually simulated on digital computers, in which case their model is made of a set of equations.

This method is broadly applicable, because it

- may be used on models of any type and complexity
- allows the response to any type of input to be computed

Its limitations are also clear:

- it does not allow the general behavior of the system to be known, but only its response to given experimental conditions,
- it may require long computation time (and thus high costs) if the model is complex, or has characteristics that make numerical integration difficult, and
- it allows the effects of changes of the values of the parameters to be predicted only at the cost of a number of different simulations.

If the model can be reduced to a set of linear differential equations with constant coefficients, it is possible to obtain a general solution of the equations of motion. The free behavior of the system can be studied independently from its forced behavior, and it is possible to use mathematical instruments such as Fourier or Laplace transforms to obtain solutions in the *frequency domain* or in the *Laplace domain*. These solutions are often much more expedient than solutions in the *time domain* that are in general the only type of solution available for nonlinear systems.

The possibility of obtaining general results makes it convenient to start the study by writing a linear model through suitable linearization techniques. Only after a good insight of the behavior of the linearized models is obtained will the study of the nonlinear model be undertaken. When dealing with nonlinear systems it is also expedient to begin with simplified methods, based on techniques like harmonic balance, or to look for series solutions before starting to integrate the equations numerically.

Computational vibration analysis

If technological advances force the designer to perform increasingly complex tasks, it also provides the instruments for the fulfillment of the new duties with powerful means of theoretical and experimental analysis.

The availability of computers of increasing power has deeply changed the methods, the mathematical means, and even the language of structural analysis, while extending the mathematical study to problems that previously could be tackled only through experiments. However, the basic concepts and theories of structural dynamics have not changed: Its roots are very deep and strong and can doubtless sustain the new rapid growth. Moreover, only the recent increase of computational power enabled a deeper utilization of the body of knowledge that accumulated in the last two centuries and often remained unexploited owing to the impossibility of handling the exceedingly complex computations. The numerical solution of problems that, until a few years ago, required an experimental approach can only be attempted by applying the aforementioned methods of theoretical mechanics.

At the same time, together with computational instruments, there was a striking progress in test machines and techniques. Designers can now base their choices on large quantities of experimental data obtained on machines similar to those being studied, which are often not only more plentiful but also more detailed and less linked with the ability and experience of the experimenter than those that were available in the past. Tests on prototypes or on physical models of the machine (even if numerical experimentation is increasingly replacing physical experimentation) not only yield a large amount of information on the actual behavior of machines but also allow validation of theoretical and computational techniques.

Modern instruments are increasingly used to monitor more or less continuously machines in operating conditions. This allows designers to collect a great deal of data on how machines work in their actual service conditions and to reduce safety margins without endangering, but actually increasing, safety.

As already said, designers can now rely on very powerful computational instruments that are widely used in structural analysis. Their use is not, however, free of dangers. A sort of disease, called *number crunching syndrome*, has been identified as affecting those who deal with computational mechanics. Oden and Bathe[4] defined it as 'blatant overconfidence, indeed the arrogance, of many working in the field [of computational mechanics] ... that is becoming a disease of epidemic proportions in the computational mechanics community. Acute symptoms are the naive viewpoint that because gargantuan computers are now available, one can code all the complicated equations of physics, grind out some numbers, and thereby describe every physical phenomena of interest to mankind'.

Methods and instruments that give the user a feeling of omnipotence, because they supply numerical results on problems that can be of astounding complexity, without allowing the user to control the various stages of the computation, are clearly potentially dangerous. They give the user a feeling of confidence and objectivity, because the computer cannot be wrong or have its own subjective bias.

The finite element method, perhaps the most powerful computational method used for many tasks, among which the solution of problems of

---

[4]Oden T.J., Bathe K.J., *A Commentary on Computational Mechanics*, Applied Mechanics Review, 31, p. 1053, 1978.

structural dynamics is one of the most important is, without doubt, the most dangerous from this viewpoint.

In the beginning, computers entered the field of structural analysis in a quiet and reserved way. From the beginning of the 1950s computers were used to automatically perform those computational procedures that required long and tedious work, for which electromechanical calculators were widely used. Because the computations required for the solution of many problems (like the evaluation of the critical speeds of complex rotors or the torsional vibration analysis of crankshafts) were very long, the use of automatic computing machines was an obvious improvement.

At the end of the 1950s computations that nobody could even think of performing without using computers became routine work. Programs of increasing complexity were often prepared by specialists, and analysts started to concentrate their attention on the preparation of data and the interpretation of results more than on how the computation was performed. In the 1960s the situation evolved further, and the first commercial finite element codes appeared on the market. Soon they had some sort of preprocessors and postprocessors to help the user handle the large amounts of data and results.

In the 1970s general-purpose codes that can tackle a wide variety of different problems were commonly used. These codes, which are often prepared by specialists who have little knowledge of the specific problems for which the code can be used, are generally considered by the users to be tools to use without bothering to find out how they work and the assumptions on which the work is based. Often the designer who must use these commercial codes tends to accept noncritically any result that comes out of the computer.

Moreover, these codes allow a specialist in a single field to design a complex system without seeking the cooperation of other specialists in the relevant matters in the belief that the code can act as a most reliable and unbiased consultant.

On the contrary, the user must know well what the code can do and the assumptions at its foundation. He must have a good physical perception of the meaning of the data being introduced and the results obtained in order to be able to give a critical evaluation.

There are two main possible sources of errors in the results obtained from a code. First there can be errors (*bugs*, in the jargon of computer users) in the code itself. This may even happen in well-known commercial codes, particularly if the problem being studied requires the use of parts of the code that are seldom used or insufficiently tested. The user may try to solve problems the programmer never imagined the code could be asked to tackle and may thus follow (without having the least suspicion of doing so) paths that were never imagined and thus never tested.

More often, it is the modeling of the physical problem that is to blame for poor results. The user must always be aware that even the most sophisticated code always deals with a simplified model of the real world, and it is a part of the user's task to ascertain that the model retains the relevant features of the actual problem.

Generally speaking, a model is acceptable only if it yields predictions close to the actual behavior of the physical system. Other than this, only its internal consistency can be unquestionable, but internal consistency alone has little interest for the applications of a model.

The availability of programs that automatically prepare data (preprocessors) can make things worse. Together with the advantages of reducing the work required from the user and avoiding the errors linked with the manual preparation and introduction of a large amount of data, there is the drawback of giving a false confidence. The mathematical model prepared by the machine is neither better nor more objective than a handmade one, and it is always the operator who must use engineering knowledge and common sense to reach a satisfactory model.

The use of general-purpose codes requires the designer to have a knowledge of the physical features of the actual systems and of the modeling methods not much less than that required to prepare the code. The designer must also be familiar with the older simplified methods through which approximate, or at least order-of-magnitude, results can be quickly obtained, allowing the designer to keep a close control over a process in which he has little influence.

The use of sophisticated computational methods must not decrease the skill of building very simple models that retain the basic feature of the actual system with a minimum of complexity. Some very ingenious analysts can create models, often with only one, or very few, degrees of freedom, which can simulate the actual behavior of a complicated physical system. The need for this skill is actually increasing, and such models often constitute a base for a physical insight that cannot be reached using complex numerical procedures. The latter are then mandatory for the collection of quantitative information, whose interpretation is made easier by the insight already gained.

Concern about vibration and dynamic analysis is not restricted to designers. No matter how good the dynamic design of a machine is, if it is not properly maintained, the level of vibration it produces can increase to a point at which it becomes dangerous or causes discomfort. The balance conditions of a rotor, for example, may change in time, and periodic rebalancing may be required.

Maintenance engineers must be aware of vibration-related problems to the same extent as design engineers. The analysis of the vibration produced by a machine can be a very powerful tool for the engineer who has to maintain a machine in working condition. It has the same importance that the study of the symptoms of disease has for medical doctors.

In the past, the experimental study of the vibration characteristics of a machine was a matter of experience and was more an art than a science: Some maintenance engineers could immediately recognize problems developed by machines and sometimes even foretell future problems just by pressing an ear against the back of a screwdriver whose blade is in contact with carefully chosen parts of the outside of the machine. The study of the motion of water in a transparent bag put on the machine or of a white powder distributed on a dark vibrating panel could give other important indications. Modern instrumentation, particularly electronic computer-controlled instruments, gives a scientific basis to this aspect of the mechanics of machines.

The ultimate goal of *preventive maintenance* is that of continuously obtaining a complete picture of the working conditions of a machine in such a way as to plan the required maintenance operations in advance, without having to wait for malfunctions to actually take place.

In some more advanced fields of technology, such as aerospace or nuclear engineering, this approach has already entered everyday practice. In other fields, these are more indications for future developments than current reality.

Unfortunately, the subject of vibration analysis is complex and the use of modern instrumentation requires a theoretical background beyond the knowledge of many maintenance or practical engineers.

Active vibration control

The revolution in all fields of technology, and increasingly in everyday life, due to the introduction of computers, microprocessors, and other electronic devices did not only change the way machines and structures are designed, built, and monitored but also had an increasingly important impact on how they work and will deeply change the very idea of machines. A typical example is the expression *intelligent machines*, which until a few decades ago would have been considered an intrinsically contradictory statement, but now is commonly accepted.

The recent developments in the fields of electronics, information, and control systems made it possible to tackle dynamic problems of structures in a new and often more effective way. While the traditional approach for reducing dynamic stressing has always been that of changing (usually increasing) the stiffness of the structure or adding damping, now control systems that can either adapt the behavior of the structure to the changing dynamic requirements or fight vibrations directly by applying adequate dynamic forces to the structure are increasingly common. This trend is widespread in all fields of structural mechanics, with civil, mechanical, and aeronautical engineering applications. For example, structural control has been successfully attempted in tall buildings and bridges, machine tools, aircraft, bearing systems for rotating machinery, robots, space structures, and ground vehicles. In the latter case, the term *active suspensions* has even become popular among the general public.

The advantages of this approach over the conventional one are clear and can be easily evidenced by the example of a large lightweight structure designed to be deployed in space in a microgravity environment. The absence or the low value of static forces allows the design of very light structures, and lightness is a fundamental prerequisite for any structure that has to be brought into orbit. This leads to very low natural frequencies and corresponding vibration modes that can easily be excited and are very lightly damped. Any attempt to maintain the dynamic stresses and displacements within reasonable limits with conventional techniques, i.e., by stiffening the structure and adding damping, would lead to large increases in the mass and, hence, the cost of the structure. The application of suitable control devices can achieve the same goals in a far lighter and cheaper way.

A structure provided with actuators that can adapt its geometric shape or modify its mechanical characteristics to stabilize a number of working parameters (e.g., displacements, stresses, and temperatures) is said to be an *adaptive structure*. An adaptive structure can be better defined as a structure with actuators allowing controlled alterations of the system states and characteristics.

If there are sensors, the structure can be defined a *sensory structure*. The two things need not go together, as in the case of a structure provided with embedded optical fibers that supply information about the structural integrity of selected components or in the case of a machine with a built-in diagnostic system. If, however, the structure is both adaptive and sensory, it is a *controlled structure*.

Active structures are a subset of controlled structures in which there is an external source of power, aimed at supplying the control energy and modulated by the control system using the information supplied by the sensors. Another typical characteristic is that the integration between the structure and the control system is so strong that the distinction between structural functionality and control functionality is blurred and no separate optimization of the parts is possible.

Intelligent structures can be tentatively differentiated from active structures by the presence of a highly distributed control system that takes care of most of the functions. Most biological structures fall in this category; a good example of the operation of an intelligent structure is the way the wing of a bird regulates the aerodynamic forces needed to fly. Not only is the shape constantly adapted, but the dynamic behavior of the structure is also controlled. Although a central control system coordinates all this, most of the control action is committed to peripheral subsystems, distributed on the whole structure.

A tentative classification of adaptive and sensory structures is shown in Fig. 5.

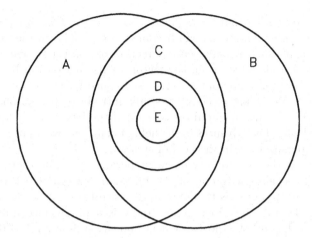

FIGURE 5. Tentative classification of adaptive and sensory structures. A: adaptive structures; B: sensory structures; C: controlled structures; D: active structures; and E: intelligent structures.

In all types of controlled structures the control system may need to perform different tasks, with widely different requirements. For example, it can be used to change some critical parameter to adapt the characteristics of the system to the working conditions, like a device that varies the stiffness of the supports of a rotor with the aim of changing its critical speed during start-up to allow a shift from subcritical to supercritical conditions without having to actually pass a critical speed. In this case, there is no need to have a very complicated control system, and even a manual control can be used, if a slow start-up is predicted. Other examples requiring a slow control system are the suspension systems for ground vehicles that are able to maintain the vehicle body in a prescribed attitude even when variations of static or quasi-static forces (e.g., centrifugal forces in road bends) occur.

In the case where the control system has to supply forces to control vibrations, its response has to be faster. If only a few modes of a large and possibly soft structure are to be controlled, as in the case of tall buildings, bridges, and some space structures, the requirements for the control system may not be severe, but they become tougher when the characteristic time of the phenomena to be kept under control gets shorter since the relevant frequencies are high.

In other cases, when the structural elements are movable and a control system is already present to control the rigid-body motions, the control of the dynamic behavior of the system can be achieved by suitably modulating the inputs to the devices that operate the machine. This is the case of robot arms or deployable space structures in which the dynamic behavior is strongly affected by the way the actuators perform their task of driving the structural elements to the required positions.

It is easy to predict that the application of structural control, particularly using active control systems, will become more popular in the future. The advances in performance and cost reduction of control systems are going to make it cost-effective, but a key factor for its success will be the incorporation of complex microprocessor-based control systems into machines of different kinds. They, although basically introduced for reasons different from structural control, can also take care of the latter in an effective and economical way. The advances in the field of neural networks may also open promising perspectives in the field of structural control.

However, if the control system must perform the vibration control of a structure, a malfunctioning of the first can cause a structural failure. The reliability required is that typical of control systems performing vital functions, like in fly-by-wire systems, and this requirement can have heavy effects on costs, on both the component and the system level, and can slow down the application of structural control in low-cost, mass-production applications.

As already stated, the trend is toward an increasing integration between the structural and control functions, and this leads to the need for a unified approach at the design and analysis stages. The control subsystem must no longer be seen as something added to an already existing structural subsystem that has been designed independently.

There is a trend toward a unified approach to many aspects of structural dynamics and control, from both the theoretical viewpoint and its practical applications. A further interdisciplinary effort must also include those aspects that are more strictly linked with the electrical and electronic components that are increasingly found in all kinds of machinery.

This interdisciplinary approach to the design of complex machines is increasingly referred to as *mechatronics.*

Although there are many definitions of what mechatronics is, it can be safely stated that it deals with the integration of mechanics, electronics, and control science to design products that reach their specifications mainly through a deep integration of their structural and control subsystems. A tentative graphical definition is shown in Fig. 6, which must be regarded to as an approximation.[5] First, the sets defining the various component technologies are not crisply defined; they are fuzzy sets. Second, it is questionable whether computer technology is to be so much stressed, as in this way analogic devices seem to be ruled out.

But what is actually lacking in Fig. 6 are the economic aspects, which must enter such an interdisciplinary approach from the onset of any practical application. The very need for an integrated approach allowing a true simultaneous engineering of the various components of any machine comes

---

[5]S. Ashley, "Getting a hold on mechatronics", *Mechanical Engineering*, 119 (5), May 1997.

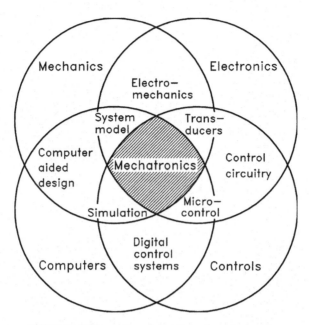

FIGURE 6. Tentative definition of mechatronics.

from economic consideration, even before thinking of the performance or other technical aspects. No wonder that among the first applications of mechatronics were consumer goods like cameras and accessories for personal computers.

It is the integration of a sound mechanical design, which includes static and dynamic analysis and simulation, with electronic and control design which allows construction of machines that offer better performance with increased safety levels at potentially lower costs.

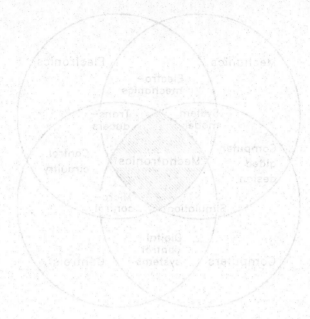

Part I

Dynamics of Linear, Time Invariant, Systems

1
Conservative Discrete Vibrating Systems

The equations of motion of single- and multi-degrees-of-freedom undamped vibrating systems are obtained both by writing the dynamic equilibrium equations and by resorting to Lagrange equations. The vibrating system is assumed to be constrained to either an inertial reference frame or a body moving with a known time history with respect to an inertial frame and whose motion is translational. The equations of motion are obtained both in the configuration and in the state space.

1.1 Oscillator with a single degree of freedom

The simplest system studied by structural dynamics is the linear mechanical oscillator with a single degree of freedom. It consists of a point mass suspended by a massless linear spring (Fig. 1.1a). Historically, however, the mathematical pendulum (Fig. 1.1c) represented for centuries the most common paradigm of an oscillator, which could be assumed to be linear, at least within adequate limitations.

The study of the simple linear oscillators of Fig. 1.1 is important for more than just historical reasons. In the first instance it is customary to start the study of mechanics of vibration with a model that is very simple but demonstrates, at least qualitatively, the behavior of more complex systems.

The arrangements shown in Fig. 1.1 also have a great practical importance: They constitute models that can often be used to study, with good approximation, the behavior of systems of greater complexity. Moreover,

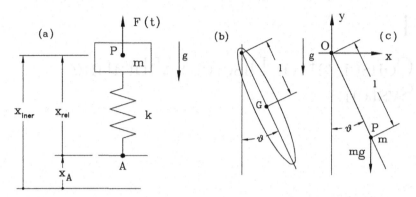

FIGURE 1.1. Linear oscillators with one degree of freedom: (a) Spring–mass system; the coordinate x for the study of the motion of point P can expressed in an inertial reference frame or be a relative displacement; (b) physical pendulum; (c) mathematical pendulum.

systems with many degrees of freedom, and even continuous systems, can be reduced, under fairly wide simplifying assumptions, to a set of independent systems with a single degree of freedom.

A linear spring is an element that, when stretched of the quantity $l - l_0$, reacts with a force

$$F_s = -k(l - l_0), \tag{1.1}$$

where l_0 is the length at rest of the spring and k is a constant, usually referred to as the stiffness of the spring, expressing the ratio between the force and the elongation. In SI units, it is measured in N/m. If constant k is positive, the force is a restoring force, opposing the displacement of point P. The system is then statically stable, in the sense that, when displaced from its equilibrium position, it tends to return to it.[1]

A force function of time $F(t)$ can act on point P and the supporting point A can move in x-direction with a known time history $x_A(t)$.

The dynamic equilibrium equation states that the inertia force is, at any time, in equilibrium with the elastic reaction of the spring added to the external forces. Written with reference to the inertial x-coordinate, it is simply

$$m\ddot{x} = -k\left[x - l_0 - x_A(t)\right] + F(t) - mg . \tag{1.2}$$

Owing to the linearity of the system of Fig. 1.1a, the length at rest of the spring l_0 and all constant forces such as those due to the gravitational acceleration g, affect the static equilibrium position but not its dynamic behavior. The dynamic problem can thus be separated from the

---

[1] For a more detailed definition of stability see Chapter 20. Only stable systems will be dealt with in this chapter.

static problem by neglecting all constant forces and writing the dynamic equilibrium equation

$$m\ddot{x} + kx = kx_A(t) + F(t).$$ (1.3)

Equation (1.3) expresses the motion of point P in terms of its displacement from the position of static equilibrium $x = 0$ characterized by $F = 0$ and $x_A = 0$. The excitation provided by the motion of the supporting point and that provided by an external force can be dealt with in exactly the same way.

When the excitation to the system is provided by the motion of the supporting point, it may be expedient to express the position of point P with reference to point A: coordinate x_{rel} in Fig. 1.1a. The absolute acceleration of point P is now expressed as

$$\ddot{x}_{iner} = \ddot{x}_{rel} + \ddot{x}_A,$$

and, neglecting the terms that are constant and whose effect is just displacing the static position of equilibrium, the equation of motion becomes

$$m\ddot{x} + kx = -m\ddot{x}_A + F(t),$$ (1.4)

where subscript rel has been dropped.

It is very similar to Eq. (1.3), the only difference being the way the displacement of the supporting point is taken into account.

Remark 1.1 *Because the equation of motion is a second-order differential equation, two conditions on the initial values must be stated to obtain a unique solution.*

Instead of the translational oscillator of Fig. 1.1a, a torsional oscillator can be devised. It consists of a rigid body free to rotate about an axis passing through its center of mass, constrained by a torsional spring. Equation (1.3) still holds, provided the parameters involved are changed according to Table 1.1.

TABLE 1.1. Formal equivalence between mechanical oscillators with translational and rotational motion. Quantities entering the equation of motion, common symbols, and SI units.

	Displacement x [m]	Mass m [kg]	Stiffness k [N/m]	Force F [N]
	Rotation θ [rad]	Moment of inertia J [kgm^2]	Torsional stiffness χ [Nm/rad]	Moment M [Nm]

Example 1.1 *Consider the pendulum shown in Fig. 1.1b. It can be considered as a rotational oscillator, with moment of inertia J and restoring generalized force, due to the gravitational field, equal to*

$$mgl\sin(\theta).$$

In the case of the mathematical pendulum of Fig. 1.1c, the rigid body reduces to a point mass suspended to a massless rod, and the moment of inertia reduces to

$$J = ml^2.$$

The equation of motion for the free oscillations can be easily computed from Eq. (1.3):

$$J\ddot{\theta} + mgl\sin(\theta) = 0.$$

If the amplitude of the oscillations is small enough, the equation of motion can be linearized by substituting θ for $\sin(\theta)$:

$$J\ddot{\theta} + mgl\theta = 0.$$

1.2   Systems with many degrees of freedom

Consider a discrete system consisting of two point masses connected to point A through a number of springs (Fig. 1.2). A force F_i acts on each mass. By introducing the same inertial coordinates seen for the case of systems with a single degree of freedom, the following dynamic equilibrium equations can be written:

$$\begin{cases} m_1\ddot{x}_1 + k_1(x_1 - x_A) - k_{12}(x_1 - x_2) = F_1(t) \\ m_2\ddot{x}_2 + k_2(x_2 - x_A) - k_{12}(x_2 - x_1) = F_2(t), \end{cases} \tag{1.5}$$

or in matrix form

$$\begin{bmatrix} m_1 & 0 \\ 0 & m_2 \end{bmatrix} \begin{Bmatrix} \ddot{x}_1 \\ \ddot{x}_2 \end{Bmatrix} + \begin{bmatrix} k_1 + k_{12} & -k_{12} \\ -k_{12} & k_2 + k_{12} \end{bmatrix} \begin{Bmatrix} x_1 \\ x_2 \end{Bmatrix} = \\ = \begin{Bmatrix} k_1 x_A + F_1(t) \\ k_2 x_A + F_2(t) \end{Bmatrix}. \tag{1.6}$$

The structure of Eq. (1.6) holds for conservative linear discrete systems with any number of degrees of freedom. The dynamic equilibrium equations of a system made by a number of masses connected with each other and to a supporting frame by linear springs can thus be written in the compact form

$$\mathbf{M\ddot{x} + Kx = f}(t). \tag{1.7}$$

The matrices and vectors included in Eq. (1.7) are

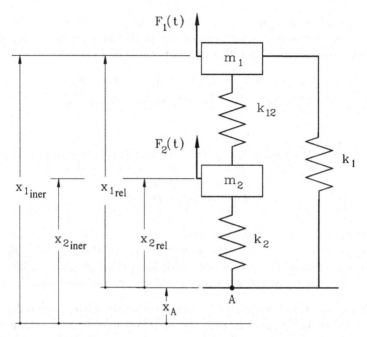

FIGURE 1.2. Sketch of a system with two degrees of freedom, made of two masses connected to the supporting point A by linear springs. The position of the two masses may be expressed by inertial coordinates or by the displacements relative to point A.

- **M** is the mass matrix of the system. It is diagonal if all coordinates x_i are related to translational degrees of freedom and measured with reference to an inertial frame.

- **K** is the stiffness matrix. Generally it is not a diagonal matrix, although it usually has a band structure. In some cases, it is possible to resort to a set of generalized coordinates for which the stiffness matrix is diagonal (e.g., using as coordinates the length of the various springs), but such a choice results in a non-diagonal mass matrix. The only exception is that of the modal coordinates that allow the use of mass and stiffness matrices, which are both diagonal.

- **x** is a vector[2] in which the generalized coordinates are listed.

- **f** is a time-dependent vector containing the forcing functions due to external forces or to the motion of the supporting points.

---

[2]Here the term *vector* is used with the meaning of column matrix: it is not a vector in three-dimensional space.

If, instead of using inertial coordinates, the positions of the two masses m_1 and m_2 are expressed in terms of the displacements relative to point A (coordinates $x_{i_{rel}}$ in Fig. 1.2), the equations of motion are

$$\begin{cases} m_1\,(\ddot{x}_1 + \ddot{x}_A) + k_1 x_1 - k_{12}(x_1 - x_2) = F_1(t) \\ m_2\,(\ddot{x}_2 + \ddot{x}_A) + k_2 x_2 - k_{12}(x_2 - x_1) = F_2(t)\,, \end{cases} \qquad (1.8)$$

i.e.,

$$\begin{bmatrix} m_1 & 0 \\ 0 & m_2 \end{bmatrix} \begin{Bmatrix} \ddot{x}_1 \\ \ddot{x}_2 \end{Bmatrix} + \begin{bmatrix} k_1 + k_{12} & -k_{12} \\ -k_{12} & k_2 + k_{12} \end{bmatrix} \begin{Bmatrix} x_1 \\ x_2 \end{Bmatrix} =$$
$$= \begin{Bmatrix} -m_1 \ddot{x}_A + F_1(t) \\ -m_2 \ddot{x}_A + F_2(t) \end{Bmatrix}. \qquad (1.9)$$

Remark 1.2 *The homogeneous part of the equation is not affected by the use of relative coordinates. The excitation due to the motion of the constraints is expressed in terms of the acceleration of the latter while in the case of inertial coordinates it is expressed in terms of their displacements.*

This is consistent with what is seen for systems with a single degree of freedom, but in the present case the situation may be more complicated because, even if the generalized coordinates x_i are all displacements, they may occur in different directions.

Consider for instance a discrete system where all degrees of freedom are translational, constrained to a rigid frame that can move in the directions of the axes of the inertial reference frame xyz. Let the components of the displacement of the rigid frame be x_A, y_A, and z_A and express the generalized coordinates x_i with reference to the moving frame. A two-dimensional example is shown in Fig. 1.3.

The vector containing the absolute accelerations can be obtained from that containing the second derivatives of the coordinates x_i by the relationship

$$\ddot{\mathbf{x}}_{iner} = \ddot{\mathbf{x}} + \boldsymbol{\delta}_x \ddot{x}_A + \boldsymbol{\delta}_y \ddot{y}_A + \boldsymbol{\delta}_z \ddot{z}_A\,, \qquad (1.10)$$

where the terms δ_{x_i}, δ_{y_i}, and δ_{z_i} are simply the direction cosines of the displacement x_i in the system of reference xyz. The equation of motion written with reference to the relative coordinates is thus

$$\mathbf{M}\ddot{\mathbf{x}} + \mathbf{K}\mathbf{x} = -\mathbf{M}\boldsymbol{\delta}_x \ddot{x}_A - \mathbf{M}\boldsymbol{\delta}_y \ddot{y}_A - \mathbf{M}\boldsymbol{\delta}_z \ddot{z}_A + \mathbf{f}(t)\,. \qquad (1.11)$$

If a simple discrete system consists of point masses connected with each other and to the ground by springs, the generalized coordinates x_i can be simply the components of the displacements along the directions of the reference axes, in exactly the same way as for the system with a single degree of freedom.

Generally speaking, matrices $\mathbf{M}$ and $\mathbf{K}$ are symmetrical matrices of order n, where n is the number of degrees of freedom of the system.

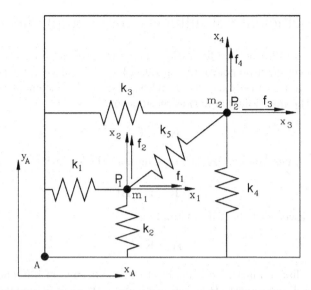

FIGURE 1.3. Example of a two-dimensional system excited by the motion of the constraints. In this case, $\boldsymbol{\delta}_x = \begin{bmatrix} 1 & 0 & 1 & 0 \end{bmatrix}^T$ and $\boldsymbol{\delta}_y = \begin{bmatrix} 0 & 1 & 0 & 1 \end{bmatrix}^T$. Displacements $\mathbf{x}_i$ are referred to the static equilibrium conditions and are considered to be small displacements.

Remark 1.3 *The symmetry of the matrices can be destroyed if some equations are substituted by linear combinations of the equations or are just multiplied by a constant. The equations of motion can thus be written in forms in which the relevant matrices are not symmetrical.*

Generally matrices $\mathbf{M}$ and $\mathbf{K}$ are positive semidefinite, but in many cases they are positive definite. The mass matrix is positive defined when a non-vanishing mass is associated to all degrees of freedom. The stiffness matrix is positive defined when no rigid body motion is allowed. Sometimes a system in which the constraints prevent all rigid body motions is said to be a *structure*, and the term *mechanism* is used for the opposite case.

Many devices, such as spacecraft, aircraft, or drivelines, are actually unconstrained and, if modeled as a whole, are characterized by singular stiffness matrices.

Sometimes the difficulties linked with the presence of a singular stiffness matrix can be circumvented by adding very soft constraints, which cause low-frequency rigid body oscillatory motions, but this can be done only when the vibrational behavior of the structure is studied as uncoupled with the rigid body (or attitude) dynamics of the system. If their coupling is accounted for, there is no way of removing the singularity of the stiffness matrix.

1.3   Coefficients of influence and compliance matrix

The static equilibrium configuration the system takes under the action of constant forces can be computed through Eq. (1.7). If vector $\mathbf{f}$ is assumed to be constant and only a solution with constant displacement vector $\mathbf{x}$ is searched, the mentioned equation reduces to

$$\mathbf{Kx} = \mathbf{f} \ . \tag{1.12}$$

If the stiffness matrix $\mathbf{K}$ is not singular, the equation can be solved obtaining

$$\mathbf{x} = \mathbf{K}^{-1}\mathbf{f} = \mathbf{Bf} \ , \tag{1.13}$$

where the inverse of the stiffness matrix

$$\mathbf{B} = \mathbf{K}^{-1}$$

is the compliance matrix[3] or matrix of the coefficients of influence.

The compliance matrix $\mathbf{B}$ is symmetrical, like the stiffness matrix $\mathbf{K}$, but while the latter has usually a band structure, the former is generally full.

Remark 1.4 *The generic element β_{ij} of matrix $\mathbf{B}$ has an obvious physical meaning: It is the ith generalized displacement due to a unit jth generalized force, i.e., it is what is commonly called an* influence coefficient.

Remark 1.5 *Matrix $\mathbf{B}$ exists only if the stiffness matrix is not singular.*

1.4   Lagrange equations

The generalized coordinates appearing in Eq. (1.7) are directly the coordinates x, y, and z of the various point masses. The number of degrees of freedom of the system has been assumed to coincide with the number of coordinates of points $\mathbf{P}_i$.

If a number of constraints are located between the point masses, the number of degrees of freedom of the system is smaller than the number of coordinates and the displacement vectors $\vec{r}_i$ can be expressed as functions of a number n of parameters x_i

$$\vec{r}_i = \vec{r}_i(x_1, x_2, \ldots, x_n) \ . \tag{1.14}$$

Because the number of parameters needed to state the configuration of the system is n, it has n degrees of freedom. Vector $\mathbf{x}$ is thus the vector

---

[3]In the literature, the compliance matrix is often referred to with the symbol $\mathbf{C}$. Here, an alternative symbol ($\mathbf{B}$, i.e., capital β) had to be used to avoid confusion with the viscous damping matrix $\mathbf{C}$.

of the generalized coordinates, and the corresponding elements of vector **f** are the generalized forces. Some of the x_i can be true displacements or rotations, but they can also have a less direct meaning, as in the case where they are coefficients of a series expansion. Correspondingly, the generalized forces are true forces, moments, or just mathematical expressions linked to the forces and moments acting on the system in a less direct way.

Remark 1.6 *The choice of the generalized coordinates is in a way arbitrary, and different sets of generalized coordinates can be devised for a given system. However, this choice is not immaterial and the complexity of the mathematical model can strongly depend on it.*

In this way, what has been seen for a system made of point masses can be extended to any mechanical system, provided that a finite number of generalized coordinates can express its configuration.

The equations of motion can be obtained directly by writing the dynamic equilibrium equations for each of the masses m_i, i.e., by imposing that the sum of all forces acting on each mass is equal to zero. These forces must include those due to the springs, external forces as well as inertia forces due to the motion of the reference frame. Although this approach is straightforward if the system is simple enough, if the number of degrees of freedom is high or if some of the generalized coordinates are not easily linked with the displacements and rotations of masses m_i, it is convenient to resort to the methods of analytical mechanics like the principle of virtual works, Hamilton's principle, or Lagrange equations in order to write the equations of motion. In this book, Lagrange equations

$$\frac{d}{dt}\left(\frac{\partial T}{\partial \dot{x}_i}\right) - \frac{\partial T}{\partial x_i} + \frac{\partial U}{\partial x_i} = Q_i \qquad (1.15)$$

will be used extensively, although the choice of one of these techniques is often just a matter of personal preference.

To understand the equivalence of two approaches (Lagrange equations and dynamic equilibrium equations), it is sufficient to observe that the first two terms of Eq. (1.15) are the expression of inertia forces as functions of the kinetic energy T, the third term expresses conservative forces obtainable from the potential energy U, and that on the right-hand side is a generic expression of forces that, although being functions of time, cannot be obtained from the potential energy. Their expression can be obtained from the virtual work δL performed by the forces applied to the system when the virtual displacement $\delta\mathbf{x}$ is given:

$$Q_i = \frac{\partial \delta L}{\partial \delta x_i}. \qquad (1.16)$$

In the case of linear systems the potential energy is a quadratic form in the displacements and, apart from constant terms which do not affect the equation of motion, can be expressed as

$$\mathcal{U} = \frac{1}{2}\mathbf{x}^T\mathbf{K}\mathbf{x} + \mathbf{x}^T\mathbf{f}_0 \ , \tag{1.17}$$

where $\mathbf{K}$ is a symmetric matrix.

Even in the case of nonlinear systems, the potential energy does not depend on the generalized velocities: its derivatives with respect to the generalized velocities $\dot{x}_i$ vanish. Equation (1.15) may thus be written by resorting to the *Lagrangian function* or *Lagrangian* $(\mathcal{T} - \mathcal{U})$

$$\frac{d}{dt}\left[\frac{\partial(\mathcal{T}-\mathcal{U})}{\partial\dot{x}_i}\right] - \frac{\partial(\mathcal{T}-\mathcal{U})}{\partial x_i} = Q_i \ . \tag{1.18}$$

The kinetic energy is usually assumed to be a quadratic function of the generalized velocities

$$\mathcal{T} = \mathcal{T}_0 + \mathcal{T}_1 + \mathcal{T}_2 \ , \tag{1.19}$$

where $\mathcal{T}_0$ does not depend on the generalized velocities, $\mathcal{T}_1$ is linear, and $\mathcal{T}_2$ is quadratic.

In the case of linear systems, the kinetic energy must contain terms in which no power greater than two of the displacements and velocities is present. As a consequence, $\mathcal{T}_2$ cannot contain the displacements, i.e.,

$$\mathcal{T}_2 = \frac{1}{2}\sum_{i=1}^{n}\sum_{j=1}^{n}m_{ij}x_i x_j = \frac{1}{2}\dot{\mathbf{x}}^T\mathbf{M}\dot{\mathbf{x}} \ , \tag{1.20}$$

where $\mathbf{M}$ is a symmetric matrix whose elements m_{ij} do not depend on either $\mathbf{x}$ or $\dot{\mathbf{x}}$. In the present chapter only systems with constant parameters will be considered, and hence $\mathbf{M}$ will be assumed to be constant.

$\mathcal{T}_1$ is linear in the velocities and then can contain powers not greater than the first one in the generalized displacements:

$$\mathcal{T}_1 = \frac{1}{2}\dot{\mathbf{x}}^T\left(\mathbf{M}_1\mathbf{x} + \mathbf{f}_1\right) \ , \tag{1.21}$$

where matrix $\mathbf{M}_1$ and vector $\mathbf{f}_1$ do not contain the generalized coordinates, although $\mathbf{f}_1$ may be a function of time.

$\mathcal{T}_0$ does not contain the generalized velocities but only terms of order not higher than two in the displacements:

$$\mathcal{T}_o = \frac{1}{2}\mathbf{x}^T\mathbf{M}_g\mathbf{x} + \mathbf{x}^T\mathbf{f}_2 + e \ , \tag{1.22}$$

where matrix $\mathbf{M}_g$, vector $\mathbf{f}_2$, and the scalar e are constant. $\mathcal{T}_0$ has a structure similar to that of the potential energy: The term $\mathcal{U} - \mathcal{T}_0$ is usually referred to as *dynamic potential*.

By performing the derivatives appearing in the Lagrange equations it follows that

$$\frac{\partial(\mathcal{T}-\mathcal{U})}{\partial\dot{x}_i} = \mathbf{M}\dot{\mathbf{x}} + \frac{1}{2}\left(\mathbf{M}_1\mathbf{x} + \mathbf{f}_1\right) \ , \tag{1.23}$$

$$\frac{d}{dt}\left[\frac{\partial(\mathcal{T}-\mathcal{U})}{\partial \dot{x}_i}\right] = \mathbf{M}\ddot{\mathbf{x}} + \frac{1}{2}\mathbf{M}_1\dot{\mathbf{x}} + \dot{\mathbf{f}}_1, \tag{1.24}$$

$$\frac{\partial(\mathcal{T}-\mathcal{U})}{\partial x_i} = \frac{1}{2}\mathbf{M}_1^T\dot{\mathbf{x}}^T + \mathbf{M}_g\mathbf{x} - \mathbf{K}\mathbf{x} + \mathbf{f}_2 - \mathbf{f}_0. \tag{1.25}$$

The equation of motion can thus be written in the form

$$\mathbf{M}\ddot{\mathbf{x}} + \frac{1}{2}\left(\mathbf{M}_1 - \mathbf{M}_1^T\right)\dot{\mathbf{x}} + (\mathbf{K} - \mathbf{M}_g)\,\mathbf{x} = -\dot{\mathbf{f}}_1 + \mathbf{f}_2 - \mathbf{f}_0 + \mathbf{Q}. \tag{1.26}$$

Matrix $\mathbf{M}_1$ is usually skew-symmetric. However, even if it is not so, it can be considered as the sum of a symmetric and a skew-symmetric part

$$\mathbf{M}_1 = \mathbf{M}_{1sy} + \mathbf{M}_{1sk}\ . \tag{1.27}$$

When it is introduced into Eq. (1.26), the term

$$\mathbf{M}_1 - \mathbf{M}_1^T$$

becomes

$$\mathbf{M}_{1sy} + \mathbf{M}_{1sk} - \mathbf{M}_{1sy} + \mathbf{M}_{1sk} = 2\mathbf{M}_{1sk}\ .$$

Only the skew-symmetric part of $\mathbf{M}_1$ appears in the equation of motion. Let $2\mathbf{M}_{1sk}$ be indicated as $\mathbf{G}$ and vectors $\mathbf{f}_0$, $\dot{\mathbf{f}}_1$, and $\mathbf{f}_2$ be included into the external forces vector $\mathbf{Q}$. The equation of motion then becomes

$$\mathbf{M}\ddot{\mathbf{x}} + \mathbf{G}\dot{\mathbf{x}} + (\mathbf{K} - \mathbf{M}_g)\,\mathbf{x} = \mathbf{Q}, \tag{1.28}$$

The mass and stiffness matrices $\mathbf{M}$ and $\mathbf{K}$ have already been defined. The skew-symmetric matrix $\mathbf{G}$ is usually referred to as the *gyroscopic matrix* and the symmetric matrix $\mathbf{M}_g$ is usually called the *geometric stiffness matrix*.[4]

A system in which $\mathcal{T}_1$ vanishes is said to be a *natural system* and no gyroscopic matrix is present. In many cases also $\mathcal{T}_0$ is not present and the kinetic energy is expressed by Eq. (1.20); such is the case for example of linear nonrotating structures.

While in the case of linear systems the Lagrangian is a quadratic form in the generalized coordinates and their derivatives, for general nonlinear systems it may have a different expression.

Remark 1.7 *When writing the linearized equations of motion of a nonlinear system, two alternatives are possible: either the nonlinear equations are written first and then linearization is performed directly on the final equations or the expressions of the energies are reduced to quadratic forms by expanding them in series and truncating the series after the quadratic terms. The two approaches yield the same results, but the first one is generally far heavier from a computational viewpoint.*

---

[4]Symbol $\mathbf{M}_g$ has been used here for the geometric stiffness matrix instead of $\mathbf{K}_g$ to stress that it derives from the kinetic energy.

Example 1.2 *Write the equation of motion of the system sketched in Fig. 1.4. It consists of three discs linked with each other by shafts that are torsionally deformable; the first shaft is clamped in point A to a fixed frame. The system can be modeled as a lumped-parameter system, with three rigid inertias and three massless springs. Note that the numerical values reported in the figure are unrealistic for a system of that type and were chosen only in order to work with simple numbers. Consider the rotations θ_1, θ_2, and θ_3 as generalized coordinates.*
By remembering the equivalences in Table 1.1, the equation of motion of the third disc is

$$J_3\ddot{\theta}_3 + k_{T3}(\theta_3 - \theta_2) = M_3.$$

The equations for the other two discs can be written in a similar way, obtaining a set of three second-order differential equations which, after introducing the numerical values of the parameters, is

$$\begin{bmatrix} 1 & 0 & 0 \\ 0 & 4 & 0 \\ 0 & 0 & 0.5 \end{bmatrix} \begin{Bmatrix} \ddot{\theta}_1 \\ \ddot{\theta}_2 \\ \ddot{\theta}_3 \end{Bmatrix} + \begin{bmatrix} 20 & -10 & 0 \\ -10 & 14 & -4 \\ 0 & -4 & 4 \end{bmatrix} \begin{Bmatrix} \theta_1 \\ \theta_2 \\ \theta_3 \end{Bmatrix} = \begin{Bmatrix} M_1 \\ M_2 \\ M_3 \end{Bmatrix}.$$

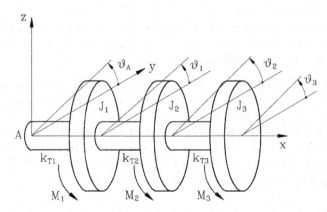

FIGURE 1.4. System with three degrees of freedom. $J_1 = 1$ kg m^2; $J_2 = 4$ kg m^2; $J_3 = 0.5$ kg m^2; $k_{T1} = 10$ N m/rad; $k_{T2} = 10$ N m/rad; $k_{T3} = 4$ N m/rad.

Alternatively, the equations of motion can be obtained from Lagrange equations. The kinetic and potential energies and the virtual work of the external moments due to a virtual displacement $[\delta\theta_1, \delta\theta_2, \delta\theta_3]^T$ are

$$2T = J_1\dot{\theta_1}^2 + J_2\dot{\theta_2}^2 + J_3\dot{\theta_3}^2,$$
$$2\mathcal{U} = k_{T_1}\theta_1^2 + k_{T_2}(\theta_2 - \theta_1)^2 + k_{T_3}(\theta_3 - \theta_2)^2,$$
$$\delta\mathcal{L} = M_1\delta\theta_1 + M_2\delta\theta_2 + M_3\delta\theta_3.$$

By performing the relevant derivatives, the same equation seen above is obtained.

Example 1.3 *Write the equation of motion of the mathematical pendulum of Fig. 1.1c using Lagrange equation. The position of point P is*

$$(\overline{P - O}) = \left\{ \begin{array}{c} l\sin(\theta) \\ -l\cos(\theta) \end{array} \right\}.$$

Angle θ can be taken as generalized coordinate. By differentiating the coordinates of P with respect to time, the velocity and then the kinetic energy are readily obtained:

$$V_P = (\overline{\dot{P - O}}) = l\dot{\theta} \left\{ \begin{array}{c} \cos(\theta) \\ \sin(\theta) \end{array} \right\},$$

$$T = \frac{1}{2}m|V|_P^2 = \frac{1}{2}ml^2\dot{\theta}^2.$$

The gravitational potential energy is simply

$$\mathcal{U} = mgy = -mgl\cos(\theta).$$

The derivatives included in the Lagrange equation are

$$\frac{\partial T}{\partial\dot{\theta}} = ml^2\dot{\theta}, \qquad \frac{d}{dt}\left(\frac{\partial T}{\partial\dot{\theta}}\right) = ml^2\ddot{\theta},$$

$$\frac{\partial T}{\partial\theta} = 0, \qquad \frac{\partial\mathcal{U}}{\partial\theta} = mgl\sin(\theta).$$

The equation of motion thus coincides with that already obtained:

$$ml^2\ddot{\theta} + mgl\sin(\theta) = 0.$$

The linearization can be performed in two ways: either by linearizing directly the equation of motion or by expressing the potential energy as a series in θ and truncating it after the quadratic term

$$\mathcal{U} \approx -mgl \left(1 - \frac{\theta^2}{2}\right).$$

The kinetic energy is already a quadratic form in the generalized coordinate and its derivative, so the inertial term is already linear. Since

$$\frac{\partial \mathcal{U}}{\partial \theta} \approx mgl\theta,$$

the linearized equation of motion is readily obtained:

$$ml^2 \ddot{\theta} + mgl\theta = 0.$$

Remark 1.8 *If the expression of the potential energy in Example 1.4 were linearized*

$$\mathcal{U} = -mgl \cos(\theta) \approx -mgl,$$

its derivative with respect to θ would have vanished and a wrong expression of the equation of motion would have been obtained. To obtain a linearized equation of motion, the expressions of the kinetic and potential energies must be quadratic and not linear.

Example 1.4 *Consider the two identical pendulums connected by a massless spring shown in Fig. 1.5. The length of the spring at rest is equal to the distance d between the suspension points.*

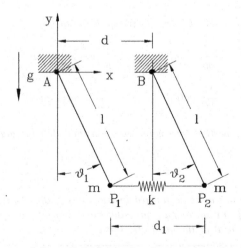

FIGURE 1.5. Two pendulums linked together by a spring: sketch of the system.

ven¶

Write the kinetic and potential energies of the system and obtain the equation of motion through Lagrange equations. Finally, linearize the equations of motion.

There are four dynamic equilibrium equations; they state that all forces, including inertia forces, that act on the two point masses in x- and y-directions balance each other. Two constraint equations, stating that the distances $(\overline{P_1 - A})$ and $(\overline{P_2 - B})$ are equal to the lengths l of the two pendulums, must be added to the dynamic equilibrium equations.

The system has thus two degrees of freedom, and angles θ_1 and θ_2 can be chosen as generalized coordinates. Equations (1.14) linking the positions of the point masses with the generalized coordinates are

$$\vec{r}_1 = (\overline{P_1 - A}) = \left\{ \begin{array}{c} l\sin(\theta_1) \\ -l\cos(\theta_1) \end{array} \right\} , \ \vec{r}_2 = (\overline{P_2 - A}) = \left\{ \begin{array}{c} d + l\sin(\theta_2) \\ -l\cos(\theta_2) \end{array} \right\} .$$

Note that the relationship between the positions of the point masses and the generalized coordinates is nonlinear. The kinetic energy is thus

$$\mathcal{T} = \frac{1}{2}m\left(\dot{x}_1^2 + \dot{y}_1^2 + \dot{x}_2^2 + \dot{y}_2^2\right) = \frac{1}{2}ml^2(\dot{\theta_1}^2 + \dot{\theta_2}^2) .$$

The gravitational potential energy can be defined with reference to any zero level, for example, that of point A. The potential energy is

$$\mathcal{U} = -mgl[\cos(\theta_1) + \cos(\theta_2)] + \frac{1}{2}k(d_1 - d)^2 ,$$

where the distance d_1 between points P_1 and P_2 can be easily shown to be

$$d_1 = \sqrt{d^2 + 2l^2[1 - \cos(\theta_1 - \theta_2)] - 2dl[\sin(\theta_2) - \sin(\theta_1)]} .$$

By performing all the relevant derivatives of the Lagrangian function, $\mathcal{T} - \mathcal{U}$, the equations of motion of the system are obtained:

$$\left\{ \begin{array}{l} ml^2\ddot{\theta}_1 + mgl\sin(\theta_1) + k(d_1 - d)\dfrac{\partial d_1}{\partial \theta_1} = 0 , \\[3mm] ml^2\ddot{\theta}_2 + mgl\sin(\theta_2) + k(d_1 - d)\dfrac{\partial d_1}{\partial \theta_2} = 0 . \end{array} \right.$$

The derivatives of d_1 with respect to θ_1 and θ_2 can be easily computed and the equations can be written in explicit form. They are clearly nonlinear, but as angles θ_1 and θ_2 are assumed to be small, they can be linearized:

$$d_1 \approx \sqrt{d^2 - 2dl(\theta_2 - \theta_1)} \approx d - l(\theta_2 - \theta_1) , \ \partial d_1/\partial\theta_1 \approx l , \ \partial d_1/\partial\theta_2 \approx -l .$$

The linearized equation of motion can be written in the form

$$ml\left[\begin{array}{cc} 1 & 0 \\ 0 & 1 \end{array} \right]\left\{ \begin{array}{c} \ddot{\theta}_1 \\ \ddot{\theta}_2 \end{array} \right\} + \left[\begin{array}{cc} mg + kl & -kl \\ -kl & mg + kl \end{array} \right]\left\{ \begin{array}{c} \theta_1 \\ \theta_2 \end{array} \right\} = \left\{ \begin{array}{c} 0 \\ 0 \end{array} \right\} .$$

The same result could be obtained by approximating the expressions of the kinetic and potential energies using quadratic forms in the generalized coordinates and velocities. The kinetic energy is already such, while the potential energy can be simplified by developing the cosines in series and truncating them after the second term and using the simplified expression for d_1

$$\mathcal{U} = -mgl\left[\left(1 - \frac{\theta_1^2}{2}\right) + \left(1 - \frac{\theta_1^2}{2}\right)\right] + \frac{1}{2}kl^2(\theta_1 - \theta_2)^2 ,$$

i.e.,

$$\mathcal{U} = \frac{l}{2}\left\{\begin{array}{c} \theta_1 \\ \theta_2 \end{array}\right\}^T \left[\begin{array}{cc} mg + kl & -kl \\ -kl & mg + kl \end{array}\right] \left\{\begin{array}{c} \theta_1 \\ \theta_2 \end{array}\right\} + 2mgl .$$

The same linearized equations of motion are thus obtained.

1.5   Configuration space

Vector **x** in which the generalized coordinates are listed is a vector in the sense it is column matrix. However, any set of n numbers may be interpreted as a vector in an n-dimensional space. This space containing vector **x** is usually referred to as the *configuration space*, since any point in this space can be associated to a configuration of the system.

Actually, not all points of the configuration space, intended as an infinite n-dimensional space, correspond to configurations that are physically possible for the system: A subset of possible configurations may thus be defined. Moreover, even systems that are dealt with using linear equations of motion are linear only for configurations not much displaced from a reference configuration (usually the equilibrium configuration) and then the linear Eq. (1.28) applies only in an even smaller subset of the configuration space.

A simple system with two degrees of freedom is shown in Fig. 1.6a; it consists of two masses and two springs whose behavior is linear in a zone around the equilibrium configuration with $x_1 = x_2 = 0$ but then behave in a nonlinear way to fail at a certain elongation (Fig. 1.6b). In the configuration space, that in the case of a two-degrees-of-freedom system has two dimensions and thus is a plane, there is a linearity zone, surrounded by a zone where the system behaves in a nonlinear way. Around the latter there is another zone where the system loses its structural integrity.

During motion, the point representing the system's configuration moves in the configuration space and its trajectory is referred to as the *dynamical path*. The dynamical paths corresponding to different time histories of the system can intersect each other, and a given configuration can be instantaneously taken during different motions.

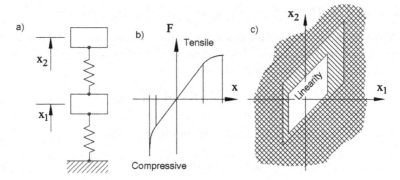

FIGURE 1.6. Sketch of a system with two degrees of freedom (**a**) made of two masses and two springs, whose characteristics (**b**) are linear only in a zone about the equilibrium position. Three zones can be identified in the configuration space (**c**) in the inner one the system behaves linearly, in another one the system is nonlinear. The latter zone is surrounded by a 'forbidden' zone.

1.6   State space

Knowledge of the system's configuration at a given time and of the time history of the forcing function does not allow one to predict its future evolution or to know its past time history. If, on the contrary, the generalized velocities are also known, the state of motion of the system is completely known at any time. Positions and velocities, taken together, are thus the state variables of the system, even if this choice is not unique and other pairs of variables correlated with them can be used (e.g., positions and momenta).

A state vector

$$ \mathbf{z} = \left\{ \begin{array}{c} \mathbf{v} \\ \mathbf{x} \end{array} \right\}, $$

where

$$ \mathbf{v} = \dot{\mathbf{x}} , $$

containing the displacements and velocities can thus be defined.[5] It has $2n$ components and defines a point in a space with $2n$ dimensions, the *state space*, defined by a reference frame whose coordinates are the state variables of the system. In the case of systems with a single degree of freedom, the state space has only two dimensions and is called the *state plane*.

---

[5]The state vector can be alternatively defined as $\mathbf{z} = \left\{ \begin{array}{c} \mathbf{x} \\ \mathbf{v} \end{array} \right\}$. There is no difficulty in modifying all relevant matrices to cope with this definition.

Remark 1.9 *The configuration space is a subspace of the state space.*

With reference to the state space, the equation of motion (1.28) of a linear system can be transformed into a set of $2n$ first-order linear differential equations, the state equations of the system

$$\begin{cases} \mathbf{M}\dot{\mathbf{v}} + \mathbf{G}\mathbf{v} + (\mathbf{K} - \mathbf{M}_g)\mathbf{x} = \mathbf{Q} \\ \dot{\mathbf{x}} = \mathbf{v} \ . \end{cases} \tag{1.29}$$

The state equations are usually written in the form

$$\dot{z}(t) = \mathcal{A}z(t) + \mathcal{B}u(t) \ , \tag{1.30}$$

where

$$\mathcal{A} = \begin{bmatrix} -\mathbf{M}^{-1}\mathbf{G} & -\mathbf{M}^{-1}\left(\mathbf{K} - \mathbf{M}_g\right) \\ \mathbf{I} & 0 \end{bmatrix}$$

is the dynamic matrix of the system. It is neither symmetrical nor positive defined.

Vector $\mathbf{u}(t)$, whose size need not be equal to the number of degrees of freedom of the system, is the vector in which the inputs affecting the behavior of the system are listed. $\mathcal{B}$ is the input gain matrix; if the number of inputs is r, it has $2n$ rows and r columns.

If the inputs $\mathbf{u}(t)$ are linked with the generalized forces $\mathbf{Q}(t)$ acting on the various degrees of freedom by the relationship

$$\mathbf{Q}(t) = \mathbf{T}\mathbf{u}(t) \ , \tag{1.31}$$

then the expression of the input gain matrix is

$$\mathcal{B} = \begin{bmatrix} \mathbf{M}^{-1}\mathbf{T} \\ 0 \end{bmatrix} \ . \tag{1.32}$$

If the output of the system consists of a linear combination of the state variables, to which a linear combination of the inputs can be added, a second equation can be added to Eq. (1.30)

$$y(t) = \mathcal{C}z(t) + \mathcal{D}u(t) \ , \tag{1.33}$$

where

- $\mathbf{y}$ is the output vector, i.e., a vector in which the m outputs of the system are listed.

- $\mathcal{C}$ is a matrix with m rows and n columns, often referred to as the *output gain matrix*.[6] If all generalized displacements are taken as outputs of the system, matrix $\mathcal{C}$ is simply $\mathcal{C} = [\mathbf{0}, \mathbf{I}]$.

---

[6]The output gain matrix is usually referred to as $\mathcal{C}$. This symbol is used here even though it is similar to that used for the damping matrix $\mathbf{C}$ because the author thinks no confusion between them is possible.

- $\mathcal{D}$ is a matrix with m rows and r columns, expressing the direct influence of the inputs on the outputs; it is therefore referred to as the *direct link matrix*.

The set of four matrices $\mathcal{A}$, $\mathcal{B}$, $\mathcal{C}$, and $\mathcal{D}$ is usually referred to as the *quadruple* of the dynamic system.

Summarizing, the equations that define the dynamic behavior of the system, from input to output, are

$$\begin{cases} \dot{\mathbf{z}} = \mathcal{A}\mathbf{z} + \mathcal{B}\mathbf{u} \\ \mathbf{y} = \mathcal{C}\mathbf{z} + \mathcal{D}\mathbf{u} \end{cases} . \tag{1.34}$$

The input–output relationship described by Eq. (1.34) may be described by the block diagram shown in Fig. 1.7.

If $r = 1$, i.e., there is a single input $u(t)$, and $m = 1$, i.e., there is a single output $y(t)$, the system is referred to as a *single-input, single-output* (SISO) *system*. Otherwise, if there are several inputs and outputs, the system is a *multiple-input, multiple-output* (MIMO) one. This distinction has nothing to do with the number of degrees of freedom or of state variables. A single-degree-of-freedom system has two state variables (position and velocity) and may be a MIMO system, where the input–output relationship is concerned.

Remark 1.10 *The state equation is a differential equation, but the output equation is simply algebraic.*

The points representing the state of the system in subsequent instants describe a trajectory in the state space. This trajectory defines the motion. The various trajectories obtained with different initial conditions constitute the state portrait of the system. In the case of autonomous systems, i.e., systems modeled by an equation of motion not containing explicitly the independent variable time, it is possible to demonstrate that, with the exception of possible singular points, only one trajectory can pass through any given point of the state space.

Equation (1.30) is non-autonomous, as time appears explicitly in the input vector. In this case, one more dimension, namely time, is added to the

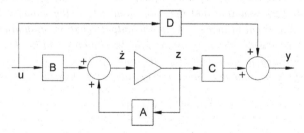

FIGURE 1.7. Block diagram corresponding to Eq. (1.34).

state space to prevent the trajectories from crossing each other, as would happen in $(\mathbf{x}, \dot{\mathbf{x}})$ space. The state space for a non-autonomous system with a single degree of freedom is consequently a tridimensional space $(x, \dot{x}, t)$. Even in this case only the state projection in the $(x, \dot{x})$ plane is often represented to reduce the complexity of the state portrait. Another technique is that of representing only some selected points of the trajectories, chosen at fixed time intervals, usually the period of the forcing function when the latter is periodic, as if a strobe were used. This strobed map is usually referred to as a *Poincaré section* or *Poincaré map*.

A point in the state space such that

$$\mathcal{A}\mathbf{z} + \mathcal{B}\mathbf{u} = \mathbf{0}$$

for any value of time is an equilibrium point. Because it is a static solution, it can be defined only if the input vector $\mathbf{u}$ is constant in time. All generalized velocities are identically equal to zero and thus the equilibrium point lies in the configuration space, thought as a subspace of the state space. Although a nonlinear system can have a number of equilibrium points, a single equilibrium point exists if the system is linear. If $\mathbf{u}$ is equal to zero, the equilibrium point is the solution of the homogeneous algebraic equation

$$\mathcal{A}\mathbf{z} = \mathbf{0} \ ,$$

i.e., the trivial solution $\mathbf{z} = \mathbf{0}$, except when the dynamic matrix is singular.

In the case of nonlinear systems, the equations of motion can often be linearized about any given equilibrium points. The motion of the linearized system about an equilibrium point is usually referred to as *motion in the small*.

Remark 1.11 *The equation of motion in the state space can be written in many different forms, but the current formulation is standard for the study of dynamic systems in general. When the generalized momenta are used instead of the generalized velocities, the term* phase *is used instead of* state.

Example 1.5 *Write the equation of motion in the state space of the system of Example 1.2, assuming that the only input $u(t)$ is the moment M_3 acting on the third moment of inertia. Write the output equation, assuming that only one output is considered, the rotation of the third moment of inertia. Introduce three auxiliary variables $v_1 = \dot{\theta}_1$, $v_2 = \dot{\theta}_2$, and $v_3 = \dot{\theta}_3$.*

Because there is only one input ($r=1$), matrix T has three rows and one column:

$$T = [0, 0, 1]^T \ .$$

The state equation is thus

$$\begin{Bmatrix} \dot{v}_1 \\ \dot{v}_2 \\ \dot{v}_3 \\ \dot{\theta}_1 \\ \dot{\theta}_2 \\ \dot{\theta}_3 \end{Bmatrix} = \begin{bmatrix} 0 & 0 & 0 & -20 & 10 & 0 \\ 0 & 0 & 0 & 2.5 & -3.5 & 1 \\ 0 & 0 & 0 & 0 & 8 & -8 \\ 1 & 0 & 0 & 0 & 0 & 0 \\ 0 & 1 & 0 & 0 & 0 & 0 \\ 0 & 0 & 1 & 0 & 0 & 0 \end{bmatrix} \begin{Bmatrix} v_1 \\ v_2 \\ v_3 \\ \theta_1 \\ \theta_2 \\ \theta_3 \end{Bmatrix} + \begin{Bmatrix} 0 \\ 0 \\ 2 \\ 0 \\ 0 \\ 0 \end{Bmatrix} M_3 . \quad (1.35)$$

Because the output is only θ_3, matrix $\mathbf{D}$ vanishes while the output gain matrix has one row and six columns:

$$\mathcal{C} = \begin{bmatrix} 0 & 0 & 0 & 0 & 0 & 1 \end{bmatrix} .$$

If matrix $\mathbf{M}$ is singular, it is impossible to write the dynamic matrix in the usual way. Usually this occurs because a vanishingly small inertia is associated to some degrees of freedom and the problem may be circumvented by associating a very small mass to them. However, it has little sense to resort to tricks of this kind when it is possible to overcome the problem in a more correct and essentially simple way.

Consider the system described by Eq. (1.28) and assume that matrices $\mathbf{G}$ and $\mathbf{M}_g$ are zero (the system is natural). Moreover, assume that matrix $\mathbf{M}$ is diagonal, which is not a lack of generality, since it is always possible to write the system in this form.

The degrees of freedom can be subdivided into two sets: a vector $\mathbf{x}_1$, containing those to which a non-vanishing inertia is associated, and a vector $\mathbf{x}_2$, containing all other ones. In a similar way all matrices and forcing functions may be split:

$$\mathbf{M} = \begin{bmatrix} \mathbf{M}_{11} & \mathbf{M}_{12} \\ \mathbf{M}_{21} & \mathbf{M}_{22} \end{bmatrix}, \quad \mathbf{K} = \begin{bmatrix} \mathbf{K}_{11} & \mathbf{K}_{12} \\ \mathbf{K}_{21} & \mathbf{K}_{22} \end{bmatrix}, \quad \mathbf{Q} = \begin{Bmatrix} \mathbf{Q}_1 \\ \mathbf{Q}_2 \end{Bmatrix} .$$

The mass matrix $\mathbf{M}_{22}$ vanishes and, since the mass matrix is diagonal, also $\mathbf{M}_{12}$ and $\mathbf{M}_{21}$ vanish.

The equations of motion can be written in the form

$$\begin{cases} \mathbf{M}_{11}\ddot{\mathbf{x}}_1 + \mathbf{K}_{11}\mathbf{x}_1 + \mathbf{K}_{12}\mathbf{x}_2 = \mathbf{Q}_1(t) \\ \mathbf{K}_{21}\mathbf{x}_1 + \mathbf{K}_{22}\mathbf{x}_2 = \mathbf{Q}_2(t) . \end{cases} \quad (1.36)$$

The second set of equations can be readily solved in $\mathbf{x}_2$:

$$\mathbf{x}_2 = -\mathbf{K}_{22}^{-1}\mathbf{K}_{21}\mathbf{x}_1 + \mathbf{K}_{22}^{-1}\mathbf{Q}_2(t) . \quad (1.37)$$

It is thus possible to write an equation of motion containing only the generalized coordinates $\mathbf{x}_1$:

$$\mathbf{M}_{11}\ddot{\mathbf{x}}_1 + \left(\mathbf{K}_{11}\mathbf{x}_1 - \mathbf{K}_{22}^{-1}\mathbf{K}_{21} \right) \mathbf{x}_1 = \mathbf{Q}_1(t) + \mathbf{K}_{22}^{-1}\mathbf{Q}_2(t) \quad (1.38)$$

whose mass matrix is not singular. This procedure is essentially what in Chapter 10 will be defined as *static reduction*.

1.7   Exercises

Exercise 1.1 *Write the equations of motion for the system of Fig. 1.3, assuming that the displacements of points P_1 and P_2 are small. Obtain the explicit expressions of all matrices.*

Exercise 1.2 *Write the equations of motion of the system of Fig. 1.8. Eliminate the generalized coordinate x_C, and consider the system as a system with two degrees of freedom only.*

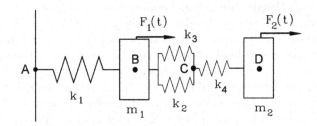

FIGURE 1.8. System with two degrees of freedom.

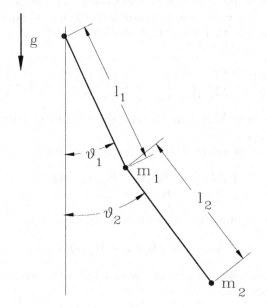

FIGURE 1.9. Double pendulum.

Exercise 1.3 *Consider the double pendulum of Fig. 1.9. Write the kinetic and potential energies of the system and obtain the equation of motion through Lagrange equations. Linearize the equation of motion in order to study the small oscillations about the static equilibrium position.*

Exercise 1.4 *Write the quadruple of a system governed by Eq. (1.11) in which forces $\mathbf{f}(t)$ have been neglected, assuming the accelerations of the constraints as inputs and the generalized coordinates as outputs.*

2

Equations in the Time, Frequency, and Laplace Domains

The equations of motion of a discrete system are ordinary differential equations containing the derivatives of the generalized coordinates with respect to time, usually up to the second order. If the time history of the response is assumed, e.g. if it is stated that the time history is harmonic or poly-harmonic, the equations of motion can be transformed into algebraic equations containing the frequency but not the time. Another alternative to transform the ordinary differential equations into algebraic equations is to use Laplace transforms. In this case the equations contain the Laplace variable, usually indicated with symbol s, instead of time.

2.1 Equations in the time domain

The equations of motion written in the previous chapter (Eq. (1.3) or (1.4)) contain the time history of both the excitation $F(t)$ (or $\mathbf{f}(t)$) and the response $x(t)$ (or $\mathbf{x}(t)$). If the system is not time invariant, also the parameters (m, k, etc.) are functions of time.

The equation of motion is thus said to be written in the *time domain*.

The inputs and the outputs are thus time histories and to solve a dynamic problem means to obtain the time history of the output knowing that of the input. Sometimes the input is derived from the output and the problem is said to be an inverse problem.

This approach is fairly straightforward both in case of linear systems (and then analytical solutions are possible, at least if the time history of the input is not too complicated) and in case of nonlinear systems. In the latter case a general solution cannot usually be obtained and the only possible approach is to resort to the numerical integration or to some approximate methods whose results are increasingly unreliable with increasing nonlinearity.

Also the equation written with reference to the state space (Eq. (1.30)) is referred to the time domain. Their solution can usually be obtained in closed form only in case of linear systems, excited by a not-too-complicated law $u(t)$.

2.2   Equations in the frequency domain

2.2.1   Harmonic motion

Assume that the system is linear and that the equations are time invariant. As it will be demonstrated later, when the excitation either is not present (free behavior) or has an harmonic time history, the response of the system is harmonic in time.

This means that if

$$\mathbf{f} = \mathbf{f}_1 \cos(\omega t) + \mathbf{f}_2 \sin(\omega t) \tag{2.1}$$

(with the particular case $\mathbf{f}_1 = \mathbf{f}_2 = \mathbf{0}$ describing free behavior), the response can be written in the form

$$\mathbf{x} = \mathbf{x}_1 \cos(\omega t) + \mathbf{x}_2 \sin(\omega t) \ . \tag{2.2}$$

Because

$$x_0 \cos(\omega t + \Phi) = x_0 \left[\cos(\Phi) \cos(\omega t) - \sin(\Phi) \sin(\omega t) \right] \ ,$$

the harmonic time history of the ith response can be written as

$$x_i = x_{0i} \cos(\omega t + \Phi_i), \tag{2.3}$$

where x_{0i} and Φ_i are, respectively, its amplitude and phase. Clearly, the amplitudes of the cosine and sine components are

$$\mathbf{x}_1 = \{x_{0i} \cos(\Phi_i)\} \ , \quad \mathbf{x}_2 = \{-x_{0i} \sin(\Phi_i)\} \ . \tag{2.4}$$

The same holds for the excitation.

It is common to use exponential functions to express a harmonic time history. Instead of writing the ith time history using Eq. (2.3), the following expression is very common:

$$x_i = x_{0i} e^{i\omega t} , \tag{2.5}$$

where x_{0i} is a complex number, the *complex amplitude* of the response and ω is a real constant, the *circular frequency* or simply the *frequency*.

By expanding Eq. (2.5),

$$x_i = \Re\left(x_{0i}\right)\cos\left(\omega t\right) - \Im\left(x_{0i}\right)\sin\left(\omega t\right) + i\left[\Re\left(x_{0i}\right)\sin\left(\omega t\right) + \Im\left(x_{0i}\right)\cos\left(\omega t\right)\right] , \tag{2.6}$$

it is clear that it does not coincide with Eq. (2.3).

In particular, while Eq. (2.3) yields a real result, Eq. (2.5) yields a complex displacement.

Remark 2.1 *In the real world the displacement and the excitation are both real quantities. Equation (2.5) as such is thus unsatisfactory.*

To avoid this problem, Eq. (2.5) can be written in the form

$$x_i = \Re\left(x_{0i}e^{i\omega t}\right) \tag{2.7}$$

that yields Eq. (2.3), provided

$$x_{1i} = \Re\left(x_{0i}\right), \quad x_{2i} = -\Im\left(x_{0i}\right) . \tag{2.8}$$

This amounts to represent the force f and the displacements x as the projections on the real axis of vectors

$$f^* = f_0 e^{i\omega t} , \quad x^* = x_0 e^{i\omega t} \tag{2.9}$$

rotating in the complex plane with an angular velocity ω (Fig. 2.1a). Angle Φ in the figure is the phase difference between the two vectors.

In practice Eq. (2.5) is directly used instead of Eq. (2.7). Neglecting the $\Re$ symbol amounts to writing a relationship between f^* and x^* instead of between f and x.

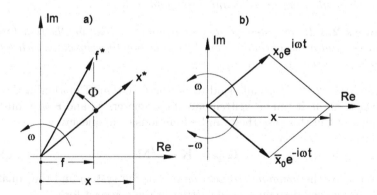

FIGURE 2.1. (a) Force f and displacement x as projections of the complex quantities f^* and x^* represented as rotating vectors in the complex plane. Situation at time t. (b) Displacement thought as the sum of two counter-rotating vectors in the complex plane.

Another way to solve the inconsistency between Eq. (2.5) and Eq. (2.3) is that of assuming implicitly that there are two solutions of the type of Eq. (2.5), one with positive ω and one with negative ω, and that the time history is given by the linear combination

$$x_i = x_{0i}e^{i\omega t} + \overline{x}_{0i}e^{-i\omega t} , \qquad (2.10)$$

where $\overline{x}_{0i}$ is the conjugate of x_{0i}.

A quick check shows that Eq. (2.10) is completely equivalent to Eq. (2.3). The vector x with harmonic time history can be thought as the sum of two vectors counter-rotating in the complex plane (Fig. 2.1b).

2.2.2 Frequency domain

By introducing the time history (2.5) for both excitation and response into the equation of motion (1.7), an algebraic equation is obtained:

$$\left(-\omega^2 \mathbf{M} + \mathbf{K}\right) \mathbf{x}_0 e^{i\omega t} = \mathbf{f}_0 e^{i\omega t} . \qquad (2.11)$$

This equation holds for any value of time t. Expression $e^{i\omega t}$ never goes to zero in the complex plane and its projection on the real axis vanishes only for selected values of time. Equation (2.11) can thus be simplified as

$$\left(-\omega^2 \mathbf{M} + \mathbf{K}\right) \mathbf{x}_0 = \mathbf{f}_0 . \qquad (2.12)$$

Remark 2.2 *By introducing a time history of both response and excitation into the equation of motion, the latter transforms from a differential equation of order 2n (where n is the number of degrees of freedom) into an algebraic equation containing the frequency but not the time. The matrix of the coefficients contains constant terms and terms in* ω^2.

Remark 2.3 *If the system performs harmonic motion and the time history of the excitation is harmonic, Eq. (2.12) is exactly equivalent to the time domain equation (1.7).*

Equation (2.12) is usually said to be the equation of motion in the *frequency domain*. It can be written only if parameters $\mathbf{M}$ and $\mathbf{K}$ are constant in time, but in this case they can be functions of frequency.

Matrix

$$\mathbf{K}_{dyn} = \mathbf{K} - \omega^2 \mathbf{M} \qquad (2.13)$$

is said to be the *dynamic stiffness matrix* of the system and the equation in the frequency domain can be written in the compact form

$$\mathbf{K}_{dyn}\mathbf{x}_0 = \mathbf{f}_0 . \qquad (2.14)$$

The dynamic stiffness matrix is a function of the frequency ω.

The inverse of the dynamic stiffness matrix is the *dynamic compliance matrix* or the *frequency response* of the system $\mathbf{H}(\omega)$

$$\mathbf{H}\left(\omega\right) = \mathbf{K}_{dyn}^{-1} . \tag{2.15}$$

In the case of single-degree-of-freedom systems, its value is

$$H\left(\omega\right) = \frac{1}{k - \omega^2 m} , \tag{2.16}$$

The matrix of the frequency responses for a system with n degrees of freedom contains a total of n^2 functions of the frequency. However, since both the dynamic stiffness and the frequency response matrices are symmetrical, only $n\left(n+1\right)/2$ are different from each other.

Also the state equation (2.18) can be written in the frequency domain. If both the state vector $\mathbf{z}$ and the input vector $\mathbf{u}$ are harmonic in time and are represented through Eq. (2.5)

$$\mathbf{z} = \mathbf{z}_0 e^{i\omega t}, \qquad \mathbf{u} = \mathbf{u}_0 e^{i\omega t}, \tag{2.17}$$

it follows that

$$(i\omega\mathbf{I} - \mathcal{A})\,\mathbf{z}_0 = \mathcal{B}\mathbf{u}_0, \tag{2.18}$$

where matrices $\mathcal{A}$ and $\mathcal{B}$ must be constant in time, but may be functions of the frequency. Vectors $\mathbf{z}_0$ and $\mathbf{u}_0$ are complex, and their meaning is exactly the same already seen for $\mathbf{x}_0$ and $\mathbf{f}_0$.

Example 2.1 *Write the frequency domain equation in both the configuration and state space for the pendulum shown in Fig. 1.1b (Example 1.1).*
The frequency domain equation is applicable only to linear systems (as it will be seen later, the response of nonlinear systems is not harmonic) and then only the linearized equation

$$J\ddot{\theta} + mgl\theta = 0$$

will be considered.
By introducing a solution of the type of Eq. (2.5), it yields the frequency domain equation

$$\left(-\omega^2 J + mgl\right)\theta_0 = 0 .$$

To write a state space equation, an auxiliary state variable, such as the velocity

$$v_\theta = \dot{\theta},$$

must be introduced. The time domain state equation is thus

$$\left\{\begin{array}{c} \dot{v}_\theta \\ \dot{\theta} \end{array}\right\} = \left[\begin{array}{cc} 0 & -\dfrac{mgl}{J} \\ 1 & 0 \end{array}\right]\left\{\begin{array}{c} v_\theta \\ \theta \end{array}\right\}.$$

The equation in the state space is obtained from Eq. (2.18):

$$\left[\begin{array}{cc} i\omega & \dfrac{mgl}{J} \\ -1 & i\omega \end{array}\right]\left\{\begin{array}{c} v_\theta \\ \theta \end{array}\right\} = \mathbf{0}.$$

2.3 Equations in the Laplace domain

2.3.1 Laplace transforms

Consider a function of time $f(t)$ defined for $t \geq 0$. Its Laplace transform $\mathcal{L}[f(t)] = \tilde{f}(s)$ is defined as

$$\mathcal{L}[f(t)] = \tilde{f}(s) = \int_0^\infty f(t)e^{-st}dt, \tag{2.19}$$

where s is a complex variable. For the mathematical details on Laplace transforms and the conditions on function $f(t)$ which make the transform possible, see one of the many textbooks on the subject.[1]

Laplace transform is a linear transform, i.e., the transform of a linear combination of functions is equal to the linear combination of the transforms of the various functions.

The main property that makes the Laplace transform useful in structural dynamics is that regarding the transform of the derivatives of function $f(t)$:

$$\mathcal{L}[\dot{f}(t)] = s\mathcal{L}[f(t)] - f(0), \qquad \mathcal{L}[\ddot{f}(t)] = s^2\mathcal{L}[f(t)] - sf(0) - \dot{f}(0). \tag{2.20}$$

The transform thus enables changing a differential equation into an algebraic equation without actually assuming the time histories of the excitation and the response, as was seen for the equations written in the frequency domain.

Given the equation of motion (1.3) of a conservative linear system with a single degree of freedom,

$$m\ddot{x} + kx = f(t),$$

by transforming both functions $f(t)$ and $x(t)$ into $\tilde{f}(s)$ and $\tilde{x}(s)$, the following equation in the Laplace domain is obtained:

---

[1] For example, W.T. Thompson, *Laplace transformation*, Prentice Hall, Englewood Cliffs, 1960.

$$(s^2 m + k)\tilde{x}(s) - msx(0) - m\dot{x}(0) = \tilde{f}(s) . \tag{2.21}$$

Since the Laplace transforms of the most common functions $f(t)$ are tabulated (see Appendix B), Eq. (2.21) can be used to compute the Laplace transform of the response of the system. The time history $x(t)$ can thus be obtained through the inverse transformation or, more simply, by using Laplace transform tables.

Equation (2.21) holds also for systems with many degrees of freedom:

$$(s^2 \mathbf{M} + \mathbf{K})\tilde{\mathbf{x}}(s) - \mathbf{M}sx(0) - \mathbf{M}\dot{\mathbf{x}}(0) = \tilde{\mathbf{f}}(s) . \tag{2.22}$$

If at time $t = 0$ both $\mathbf{x}(0)$ and $\dot{\mathbf{x}}(0)$ are equal to zero, it follows that

$$(s^2 \mathbf{M} + \mathbf{K})\tilde{\mathbf{x}}(s) = \tilde{\mathbf{f}}(s) . \tag{2.23}$$

The parameters of the system $\mathbf{M}$ and $\mathbf{K}$ must be constant in time, but may be functions of the Laplace variable s.

Remark 2.4 *The main limitation of the Laplace transform approach is that of being restricted to the solution of linear differential equations with constant coefficients.*

2.3.2 Transfer functions

Since Eq. (2.23) is a very simple algebraic equation, it is easily solved in $\tilde{\mathbf{x}}(s)$:

$$\tilde{\mathbf{x}}(s) = (s^2 \mathbf{M} + \mathbf{K})^{-1}\tilde{\mathbf{f}}(s) . \tag{2.24}$$

The function of s that, once multiplied by the transform of the excitation $\tilde{\mathbf{f}}(s)$, yields the transform of the response $\tilde{\mathbf{x}}(s)$ is the *transfer function* $\mathbf{G}(s)$ of the system

$$\mathbf{G}(s) = (s^2 \mathbf{M} + \mathbf{K})^{-1} , \tag{2.25}$$

or, in case of single-degree-of-freedom systems,

$$G(s) = \frac{1}{ms^2 + k} . \tag{2.26}$$

It coincides with the frequency response of the system $H(\omega)$ once s has been substituted for $i\omega$.

The block diagram of the system can be drawn with reference to the Laplace domain using the transfer function as shown in Fig. 2.2. The transfer function and the frequency response are strictly related to each other: The second can be obtained from the first by substituting the frequency multiplied by the imaginary unit $i\omega$ for the Laplace variable s.

A number n^2 of transfer functions is included in matrix $\mathbf{G}(s)$, which is often referred to as a *transfer matrix*, or matrix of the transfer functions. It is symmetrical like matrix $\mathbf{H}(\omega)$.

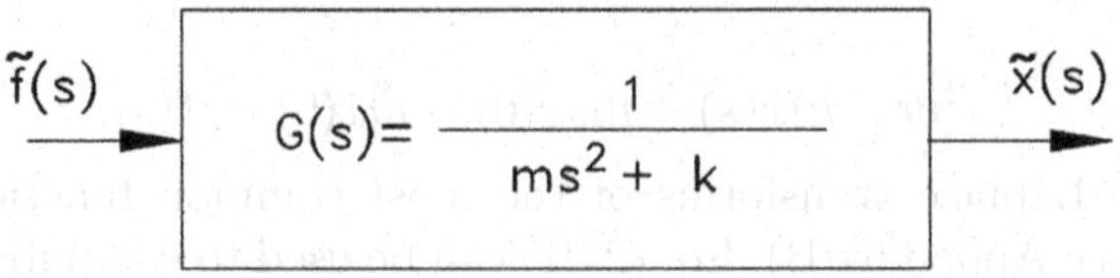

$$\tilde{f}(s) \longrightarrow \boxed{\quad G(s)= \dfrac{1}{ms^2 + k} \quad} \longrightarrow \tilde{x}(s)$$

FIGURE 2.2. Block diagram of a conservative linear single-degree-of-freedom system in the Laplace domain.

Remark 2.5 *When operating in the frequency domain, the motion is assumed to be harmonic, and hence the frequency ω is expressed by a real number. When operating in the Laplace domain, no limitation is set on the type of time history and hence s is complex.*

Example 2.2 *Assume that a torque $M(t)$ is applied to the pendulum of Fig. 1.1c (Example 1.1). Write the Laplace domain equation in the configuration space and the transfer function.*
Again, only the linearized equation will be dealt with. By adding the driving torque $M(t)$, the equation of motion becomes

$$J\ddot{\theta} + mgl\theta = M(t).$$

By introducing the Laplace transforms $\tilde{\theta}(s)$ and $\tilde{M}(s)$ of the response and of the excitation, and assuming that for $t = 0$ the pendulum is at rest in its central position ($\theta(0) = 0$, $\dot{\theta}(0) = 0$), the Laplace domain equation is readily obtained:

$$\left(s^2 J + mgl\right) \tilde{\theta}(s) = \tilde{M}(s).$$

The transfer function is thus

$$G(s) = \frac{1}{s^2 J + mgl}.$$

2.3.3 State space equations

Also the state equation (2.18) can be transformed into an algebraic equation through Laplace transform. Assuming that at time $t = 0$ the value of all state variables is zero, the state and output equations of the system in the Laplace domain are

$$\begin{cases} s\tilde{\mathbf{z}}(s) = \mathcal{A}\tilde{\mathbf{z}}(s) + \mathcal{B}\tilde{\mathbf{u}}(s) \\ \tilde{\mathbf{y}}(s) = \mathcal{C}\tilde{\mathbf{z}}(s) + \mathcal{D}\tilde{\mathbf{u}}(s), \end{cases} \tag{2.27}$$

where $\tilde{\mathbf{z}}(s)$, $\tilde{\mathbf{u}}(s)$, and $\tilde{\mathbf{y}}(s)$ are the Laplace transforms of the state, input, and output vectors, respectively.

The first equation can be solved in the state vector, obtaining

$$\tilde{\mathbf{z}}\,(s) = (s\mathbf{I} - \mathcal{A})^{-1}\,\mathcal{B}\tilde{\mathbf{u}}(s)\,. \tag{2.28}$$

By introducing the state vector into the output equation, the following input–output transfer function is obtained:

$$\mathbf{G}(s) = \frac{\tilde{\mathbf{y}}(s)}{\tilde{\mathbf{u}}(s)} = \mathcal{C}\,(s\mathbf{I} - \mathcal{A})^{-1}\,\mathcal{B} + \mathcal{D}\,. \tag{2.29}$$

The generic transfer function $\mathbf{G}_{ij}(s)$, which links the ith output with the jth input, can be written as the ratio of two polynomials:

$$\mathbf{G}_{ij}(s) = \frac{\beta_m s^m + \beta_{m-1}s^{m-1} + \cdots + \beta_1 s + \beta_0}{s^n + \alpha_{n-1}s^{n-1} + \cdots + \alpha_1 s + \alpha_0}\,, \tag{2.30}$$

where n is the order of the system and m (with $m \le n$) is the order of the numerator of the transfer function. The difference $n - m$ is referred to as the *pole excess* or *relative order* of the system.

Remark 2.6 *The roots of the denominator of the transfer function are the poles of the system, i.e., the eigenvalues of the dynamic matrix $\mathbf{A}$. The roots of the numerator are the zeros. Note that the poles are characteristics of the system, and the zeros are typical of each transfer function.*

The transfer functions can be written in the form

$$\mathbf{G}_{ij}(s) = k\frac{(s + z_1)(s + z_2)\cdots}{(s + p_1)(s + p_2)\cdots}\,, \tag{2.31}$$

where z_i and p_i are the zeros and poles, respectively.

Remark 2.7 *The poles and zeros are either real or complex-conjugate pairs, at least if the quadruple is real.*

2.4   Exercises

Exercise 2.1 *Plot the dynamic compliance of the system sketched in Fig. 1.4 and already studied in Example 1.2.*

Exercise 2.2 *A tubular cantilever beam has a length l and inner and outer diameters d_i and d_o. At the end of the beam a mass m has been attached. Compute the dynamic compliance of the system, neglecting the mass of the beam. Data: $l = 1$ m, $d_i = 60$ mm, $d_o = 80$ mm, $m = 30$ kg, $E = 2.1 \times 10^{11}$ N/m^2.*

Exercise 2.3 *Write the transfer functions of the system of Fig. 1.8 and already studied in Exercise 1.2.*

Exercise 2.4 *Consider the system of Fig. 1.4 and already studied in Examples 1.2 and 1.5. Assume that the input into the system is the torque M_3 applied at point 3 and the output is the rotation θ_2. Compute the input–output transfer function.*

3

Damped Discrete Vibrating Systems

Damping is a feature of all real-world systems, but is usually not easily modeled. In the present chapter linearized models, namely viscous, hysteretic (or structural), and general nonviscous damping are discussed in detail. The equations of motion, in both the time and the frequency domain, for linear damped systems are introduced, and the important issue of how including nonviscous damping in time-domain equations is tackled.

3.1 Linear viscous damping

3.1.1 Definition of viscous damping

During vibration the energy stored in the system in different forms is continually exchanged between them. Mechanical vibration entails the transformation of energy from the kinetic to the potential form and vice versa. In a similar way, electrical oscillations are characterized by the energy exchange between the magnetic and the electric field.

However, each time energy is transformed from one form to another in a real-world system, some energy is lost, or better is transformed, through some irreversible process, into a form (usually heat) from which it cannot be transformed back. In electrical systems this occurs due to the resistance of the conductors (superconductors are an important exception), while in mechanical system there is always some sort of friction or damping causing energy losses. In mechanical vibration, for instance, damping causes some

energy to be lost each time it is transformed from potential to kinetic energy
and back, causing a decrease in time of the amplitude of free oscillations.

The actual mechanisms causing energy losses are complex, and usually
lead to nonlinearities. However, particularly when damping is not large,
the exact way in which the damping force is applied is far less important
than the energy it dissipates in each vibration cycle. In this case the simplest way to introduce energy losses into the system is applying a force
whose direction is opposite to that of the velocity and whose amplitude is
proportional to the speed.

A device producing a force whose amplitude is proportional to the relative velocity of its end points $\dot{l}$ through the damping coefficient c and whose
direction is opposite to that of the relative velocity

$$F_d = -c\dot{l} \tag{3.1}$$

is usually referred to as a linear viscous damper or linear dashpot. It can
be added (in parallel to the spring) to the linear mechanical oscillator with
a single degree of freedom consisting of a point mass suspended by a linear
spring (Fig. 1.1a), obtaining a spring–mass–damper system (Fig. 3.1a).

If the damping coefficient (in S.I. units expressed in Ns/m) is positive,
the damper is a device that dissipates energy, and the amplitude of the free
oscillations of the system decays in time. If the system is statically stable,
it is also dynamically stable because it actually returns to the equilibrium
position, at least asymptotically.

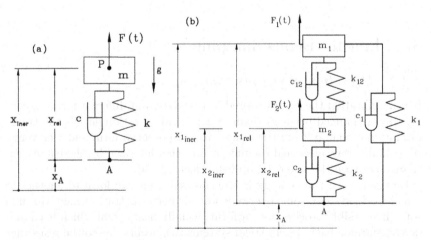

FIGURE 3.1. (a): Damped linear oscillator with one degree of freedom (spring–
mass–damper system, with the spring and the damper in parallel); (b) linear
damped system with 2 degrees of freedom.

3.1.2   Time-domain equation of motion

The dynamic equilibrium equation of a spring–mass–damper system, with the spring and the damper in parallel (Fig. 3.1a), written with reference to the inertial x-coordinate, becomes

$$m\ddot{x} = -c\left[\dot{x} - \dot{x}_A(t)\right] - k\left[x - l_0 - x_A(t)\right] + F(t) - mg\,. \qquad (3.2)$$

The dynamic problem can be separated from the static problem by neglecting all constant forces and the dynamic equilibrium equation can be written in the form

$$m\ddot{x} + c\dot{x} + kx = c\dot{x}_A(t) + kx_A(t) + F(t)\,. \qquad (3.3)$$

The corresponding equation of motion, written in terms of relative coordinates, is

$$m\ddot{x} + c\dot{x} + kx = -m\ddot{x}_A + F(t)\,, \qquad (3.4)$$

A 2 degrees of freedom systems is shown in Fig. 3.1b. The relevant equation of motion can be written in matrix form as

$$
\begin{bmatrix} m_1 & 0 \\ 0 & m_2 \end{bmatrix} \begin{Bmatrix} \ddot{x}_1 \\ \ddot{x}_2 \end{Bmatrix} + \begin{bmatrix} c_1 + c_{12} & -c_{12} \\ -c_{12} & c_2 + c_{12} \end{bmatrix} \begin{Bmatrix} \dot{x}_1 \\ \dot{x}_2 \end{Bmatrix} +
$$

$$
+ \begin{bmatrix} k_1 + k_{12} & -k_{12} \\ -k_{12} & k_2 + k_{12} \end{bmatrix} \begin{Bmatrix} x_1 \\ x_2 \end{Bmatrix} = \begin{Bmatrix} c_1\dot{x}_A + k_1 x_A + F_1(t) \\ c_2\dot{x}_A + k_2 x_A + F_2(t) \end{Bmatrix}\,.
$$

$$(3.5)$$

In the case of a general natural system with any number n of degrees of freedom, the equation of motion can be written in the compact form

$$\mathbf{M}\ddot{\mathbf{x}} + \mathbf{C}\dot{\mathbf{x}} + \mathbf{K}\mathbf{x} = \mathbf{f}(t)\,, \qquad (3.6)$$

where the symbols have the same meaning seen for conservative system and $\mathbf{C}$ is the viscous damping matrix. Generally it is not a diagonal matrix (although it usually has a band structure) and is symmetrical and positive semidefinite.

3.1.3   Dynamic stiffness

If both $\mathbf{x}(t)$ and $\mathbf{f}(t)$ are harmonic in time, and the parameters of the system are constant, Eq. (3.6) can be written in the frequency domain by stating that both the response and the excitation can be expressed by the time history of Eq. (2.5). The algebraic equation so obtained is

$$\left(-\omega^2\mathbf{M} + i\omega\mathbf{C} + \mathbf{K}\right)\mathbf{x}_0 = \mathbf{f}_0\,. \qquad (3.7)$$

The dynamic stiffness

$$\mathbf{K}_{dyn} = \mathbf{K} - \omega^2\mathbf{M} + i\omega\mathbf{C} \qquad (3.8)$$

is thus a complex quantity. Also its inverse, the dynamic compliance or the frequency response, is complex.

Remark 3.1 *The time history of the free oscillations of a damped system is not harmonic. The frequency-domain equation (3.7) implies that the function $x(t)$ is harmonic, and thus cannot be used for free motion.*[1]

Remark 3.2 *The dynamic stiffness of a system made by a spring and a damper in parallel is*

$$k_{dyn} = k + i\omega c. \tag{3.9}$$

Its real part is constant, while its imaginary part is proportional to the frequency. The ratio between the imaginary and the real parts of the dynamic stiffness is usually referred to as the loss factor

$$\eta = \frac{\omega c}{k} . \tag{3.10}$$

It grows linearly with the frequency.

3.1.4 Energy dissipated in harmonic motion

Consider a system with a single degree of freedom. The power dissipated by the damper is simply given by the product of the force it exerts by the speed:

$$W = F_d \dot{x} = -c\dot{x}^2. \tag{3.11}$$

Since the velocity in harmonic motion

$$x = x_0 \cos(\omega t)$$

is

$$\dot{x} = x_0 \omega \sin(\omega t) ,$$

the energy dissipated in a cycle is

$$e_d = \int_0^T W \, dt = -c x_0^2 \omega^2 \int_0^T \sin^2(\omega t) \, dt , \tag{3.12}$$

where T is the period.

The integral is easily solved, yielding

$$e_d = -c x_0^2 \omega \int_0^{2\pi} \sin^2(\omega t) \, d(\omega t) = -\pi c x_0^2 \omega . \tag{3.13}$$

The energy the damper dissipates in a cycle (negative since it is dissipated energy) is thus proportional to the frequency and to the square of the amplitude of the motion.

---

[1] The frequency ω is here assumed to be a real quantity.

The ratio between the energy dissipated in a cycle and the potential energy stored by the spring at the maximum elongation

$$e_s = \frac{1}{2}kx_0^2 \tag{3.14}$$

is the *specific damping capacity* of the system, usually indicated by symbol ψ:

$$\psi = \left|\frac{e_d}{e_s}\right| = 2\pi\omega\frac{c}{k}. \tag{3.15}$$

The specific damping capacity of a viscously damped system is proportional to the frequency.

3.1.5 Transfer function of a system with viscous damping

The equation of motion of a system with viscous damping can be written in the Laplace domain by transforming both functions $f(t)$ and $x(t)$ into $\tilde{f}(s)$ and $\tilde{x}(s)$. Equation (3.6) thus becomes

$$(s^2\mathbf{M} + s\mathbf{C} + \mathbf{K})\tilde{\mathbf{x}}(s) - \mathbf{M}sx(0) - \mathbf{M}\dot{x}(0) - \mathbf{C}x(0) = \tilde{\mathbf{f}}(s). \tag{3.16}$$

If at time $t = 0$ both $\mathbf{x}(0)$ and $\dot{\mathbf{x}}(0)$ are equal to zero, it follows that

$$(s^2\mathbf{M} + s\mathbf{C} + \mathbf{K})\tilde{\mathbf{x}}(s) = \tilde{\mathbf{f}}(s). \tag{3.17}$$

Also the damping matrix $\mathbf{C}$ must be constant in time, but may be a function of the Laplace variable s.

Equation (3.17) is easily solved in $\tilde{\mathbf{x}}(s)$

$$\tilde{\mathbf{x}}(s) = (s^2\mathbf{M} + s\mathbf{C} + \mathbf{K})^{-1}\tilde{\mathbf{f}}(s). \tag{3.18}$$

The transfer function of the damped system is thus

$$\mathbf{G}(s) = (s^2\mathbf{M} + s\mathbf{C} + \mathbf{K})^{-1}, \tag{3.19}$$

or, in case of single degrees of freedom systems,

$$G(s) = \frac{1}{ms^2 + cs + k}. \tag{3.20}$$

3.1.6 Dynamic stiffness of a spring–damper series.

Consider a mass–spring–damper system in which the spring and the damper are in series instead of being in parallel (Fig. 3.2a).

The system has 2 degrees of freedom: the displacement x of mass m and the displacement x_B of point B.

The equations of motion are easily obtained by stating the dynamic equilibrium conditions of mass m and point B

$$\begin{cases} m\ddot{x} + c(\dot{x} - \dot{x}_B) = F(t) \\ c(\dot{x}_B - \dot{x}) + kx_B = kx_A. \end{cases} \tag{3.21}$$

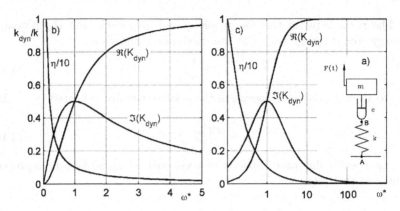

FIGURE 3.2. Spring–mass–damper system with the spring and the damper in series. (a): sketch of the system; (b) and (c): nondimensional dynamic stiffness and loss factor as functions of the nondimensional frequency in linear and logarithmic scales.

Remark 3.3 *This set of differential equations is not a fourth order set, as expected for a system with 2 degrees of freedom, but a third order set, because the acceleration $\ddot{x}_B$ is not present. This is due to the fact that no mass is located in B.*

If all time histories are harmonic, the equation can be written in the frequency domain

$$
\begin{cases}
\left(-\omega^2 m + i\omega c\right) x_0 - i\omega c x_{B_0} = F_0 \\
-i\omega c x_0 + \left(k + i\omega c\right) x_{B_0} = k x_{A_0} .
\end{cases}
\tag{3.22}
$$

The amplitude x_{B_0} can be obtained from the second equation

$$
x_{B_0} = \frac{i\omega c x_0}{k + i\omega c} + \frac{k x_{A_0}}{k + i\omega c}
\tag{3.23}
$$

and substituted into the first, obtaining

$$
\left(-\omega^2 m + \frac{k\omega^2 c^2}{k^2 + \omega^2 c^2} + i\omega \frac{k^2 c}{k^2 + \omega^2 c^2}\right) x_0 = F_0 + k x_{A_0} \frac{k - i\omega c}{k^2 + \omega^2 c^2} .
\tag{3.24}
$$

The dynamic stiffness of the spring–damper series (without the mass m) is thus

$$
k_{dyn} = \frac{k\omega^2 c^2}{k^2 + \omega^2 c^2} + i\omega \frac{k^2 c}{k^2 + \omega^2 c^2} .
\tag{3.25}
$$

Equation (3.25) can be written in nondimensional form as

$$
\frac{k_{dyn}}{k} = \frac{\omega^{*2}}{1 + \omega^{*2}} + i\frac{\omega^*}{1 + \omega^{*2}} ,
\tag{3.26}
$$

where the nondimensional frequency is

$$\omega^* = \omega \frac{c}{k} \ . \tag{3.27}$$

The loss factor is thus

$$\eta = \frac{1}{\omega^*} \tag{3.28}$$

and decreases with increasing frequency.

The real and imaginary parts of the nondimensional dynamic stiffness and the loss factor are plotted as functions of the nondimensional frequency in Fig. 3.2b and c.

When the frequency tends to zero, both the real and the imaginary parts of the dynamic stiffness vanish and the loss factor tends to infinity. At very high frequency, on the contrary, only the imaginary part of the dynamic stiffness tends to zero, together with the loss factor, while the real part tends to the constant value k.

The value of the frequency at which the imaginary part has a peak is easily computed by searching the frequency at which its derivative with respect to ω vanishes:

$$\omega_{peak} = \frac{k}{c} \ , \quad \text{i.e.} \quad \omega^*_{peak} = 1 \ . \tag{3.29}$$

The spring–damper series system acts as a damper at low frequency, while at high frequency it acts as a spring:

$$\lim_{\omega^* \to 0} (k_{dyn}) = ik\omega^* = ic\omega \ ,$$

$$(k_{dyn})_{\omega^*=1} = \tfrac{k}{2} (1+i) \ , \tag{3.30}$$

$$\lim_{\omega^* \to \infty} (k_{dyn}) = k \ .$$

3.2   State-space approach

When using the state-space approach, the presence of damping does not change much the relevant equations with the exception of the dynamic matrix which is now

$$\mathcal{A} = \begin{bmatrix} -\mathbf{M}^{-1}\mathbf{C} & -\mathbf{M}^{-1}\mathbf{K} \\ \mathbf{I} & \mathbf{0} \end{bmatrix} \ . \tag{3.31}$$

Such a formulation holds only if the mass matrix is non-singular. This can easily occur when one of the masses has a null value, like when a spring and a damper are connected in series without any mass in between (see, for instance, Fig. 3.2a).

As already seen for undamped systems, this problem may be circumvented by associating a very small mass to the relevant degrees of freedom. Also here, however, it has little sense to resort to tricks of this kind when it is possible to overcome the problem in a more correct way.

The degrees of freedom can be subdivided into two sets: a vector $\mathbf{x}_1$ containing the generalized coordinates to which a nonvanishing inertia is associated, and a vector $\mathbf{x}_2$ containing all other ones. In a similar way all matrices and forcing functions may be split. The mass matrix $\mathbf{M}_{22}$ vanishes and, if the mass matrix is diagonal, also $\mathbf{M}_{12}$ and $\mathbf{M}_{21} = \mathbf{M}_{12}^T$ vanish.

Assuming that $\mathbf{M}_{12}$ is zero, the equations of motion become

$$\begin{cases} \mathbf{M}_{11}\ddot{\mathbf{x}}_1 + \mathbf{C}_{11}\dot{\mathbf{x}}_1 + \mathbf{C}_{12}\dot{\mathbf{x}}_2 + \mathbf{K}_{11}\mathbf{x}_1 + \mathbf{K}_{12}\mathbf{x}_2 = \mathbf{f}_1(t) \\ \mathbf{C}_{21}\dot{\mathbf{x}}_1 + \mathbf{C}_{22}\dot{\mathbf{x}}_2 + \mathbf{K}_{21}\mathbf{x}_1 + \mathbf{K}_{22}\mathbf{x}_2 = \mathbf{f}_2(t) \ . \end{cases} \tag{3.32}$$

To simplify the equations of motion the gyroscopic and circulatory matrices were not explicitly written, but in what follows no assumption on the symmetry of the matrices will be done. Equation (3.32) thus holds also for gyroscopic and circulatory systems.

By introducing the velocities $\mathbf{v}_1$, together with the generalized coordinates $\mathbf{x}_1$ and $\mathbf{x}_2$, as state variables, the state equation is

$$\mathbf{M}^* \left\{ \begin{array}{c} \dot{\mathbf{v}}_1 \\ \dot{\mathbf{x}}_1 \\ \dot{\mathbf{x}}_2 \end{array} \right\} = \mathcal{A}^* \left\{ \begin{array}{c} \mathbf{v}_1 \\ \mathbf{x}_1 \\ \mathbf{x}_2 \end{array} \right\} + \left[\begin{array}{cc} \mathbf{I} & \mathbf{0} \\ \mathbf{0} & \mathbf{I} \\ \mathbf{0} & \mathbf{0} \end{array} \right] \left\{ \begin{array}{c} \mathbf{f}_1(t) \\ \mathbf{f}_2(t) \end{array} \right\} , \tag{3.33}$$

where

$$\mathbf{M}^* = \left[\begin{array}{ccc} \mathbf{M}_{11} & \mathbf{0} & \mathbf{C}_{12} \\ \mathbf{0} & \mathbf{0} & \mathbf{C}_{22} \\ \mathbf{0} & \mathbf{I} & \mathbf{0} \end{array} \right] , \quad \mathcal{A}^* = - \left[\begin{array}{ccc} \mathbf{C}_{11} & \mathbf{K}_{11} & \mathbf{K}_{12} \\ \mathbf{C}_{21} & \mathbf{K}_{21} & \mathbf{K}_{22} \\ -\mathbf{I} & \mathbf{0} & \mathbf{0} \end{array} \right] . \tag{3.34}$$

The dynamic matrix and the input gain matrix are

$$\mathcal{A} = \mathbf{M}^{*-1}\mathcal{A}^* , \quad \mathcal{B} = \mathbf{M}^{*-1} \left[\begin{array}{cc} \mathbf{I} & \mathbf{0} \\ \mathbf{0} & \mathbf{I} \\ \mathbf{0} & \mathbf{0} \end{array} \right] . \tag{3.35}$$

Alternatively, the expressions of $\mathbf{M}^*$ and $\mathcal{A}^*$ may be

$$\mathbf{M}^* = \left[\begin{array}{ccc} \mathbf{M}_{11} & \mathbf{C}_{11} & \mathbf{C}_{12} \\ \mathbf{0} & \mathbf{C}_{21} & \mathbf{C}_{22} \\ \mathbf{0} & \mathbf{I} & \mathbf{0} \end{array} \right] , \quad \mathcal{A}^* = - \left[\begin{array}{ccc} \mathbf{0} & \mathbf{K}_{11} & \mathbf{K}_{12} \\ \mathbf{0} & \mathbf{K}_{21} & \mathbf{K}_{22} \\ -\mathbf{I} & \mathbf{0} & \mathbf{0} \end{array} \right] . \tag{3.36}$$

If vector $\mathbf{x}_1$ contains n_1 elements and $\mathbf{x}_2$ contains n_2 elements, the order of the set of differential equations and the size of the dynamic matrix $\mathcal{A}$ are $2n_1 + n_2$.

3.3   Rayleigh dissipation function

When the equations of motion are written through Lagrange equations, non-conservative forces can be included into the force vector $\mathbf{Q}$. Alternatively, for a class of damping forces, it is possible to introduce a function of the generalized velocities, usually referred to as the Rayleigh dissipation function $\mathcal{F}$.

The Lagrange equations thus become

$$\frac{d}{dt}\left(\frac{\partial \mathcal{T}}{\partial \dot{x}_i}\right) - \frac{\partial \mathcal{T}}{\partial x_i} + \frac{\partial \mathcal{U}}{\partial x_i} + \frac{\partial \mathcal{F}}{\partial \dot{x}_i} = Q_i \;. \tag{3.37}$$

The Rayleigh dissipation function for the viscous damper of Fig. 3.1a is simply

$$\mathcal{F} = \frac{1}{2}c\left[\dot{x} - \dot{x}_A(t)\right]^2 \;. \tag{3.38}$$

In general, for a linear system it is a quadratic function of the velocity. If it can be written in the simple form

$$\mathcal{F} = \frac{1}{2}\dot{\mathbf{x}}^T\mathbf{C}\dot{\mathbf{x}} \;, \tag{3.39}$$

where $\mathbf{C}$ is the symmetric damping matrix, the damping term appearing in the equation of motion is simply $\mathbf{C}\dot{\mathbf{x}}$, as in equation (3.6).

This is not always the case, and the Rayleigh dissipation function may contain also terms in which the products of the displacements by the velocities are present

$$\mathcal{F} = \frac{1}{2}\dot{\mathbf{x}}^T\mathbf{C}\dot{\mathbf{x}} + \dot{\mathbf{x}}^T\mathbf{H}\mathbf{x} \;, \tag{3.40}$$

where $\mathbf{H}$ is a skew-symmetric matrix, referred to as the *circulatory matrix*.

The equation of motion of a linear, discrete, non-conservative system can thus be written by adding the terms in $\mathbf{C}$ and $\mathbf{H}$ to Eq. (1.28)

$$\mathbf{M}\ddot{\mathbf{x}} + (\mathbf{C}+\mathbf{G})\dot{\mathbf{x}} + (\mathbf{K}-\mathbf{M}_g+\mathbf{H})\mathbf{x} = \mathbf{f}(t) \;, \tag{3.41}$$

Remark 3.4 *The Rayleigh dissipation function is a measure of the power dissipated by non-conservative forces.*

Example 3.1 *Consider the torsional system shown in Fig. 1.4 and already studied in Example 1.2, adding three viscous torsional dampers with damping coefficients* $\Gamma_1 = 0.1$, $\Gamma_2 = 1$ *Ns/m, and* $\Gamma_3 = 0.4$ *Ns/m in parallel to the springs. Write the damping matrix and the dynamic matrix.*
The equation of motion of the third disc becomes

$$J_3\ddot{\theta}_3 + \Gamma_3(\dot{\theta}_3 - \dot{\theta}_2) + k_{T3}(\theta_3 - \theta_2) = M_3.$$

The equations for the other two discs can be written in a similar way, obtaining a set of three second-order differential equations.

The Rayleigh dissipation function is

$$2\mathcal{F} = \Gamma_1\dot{\theta}_1{}^2 + \Gamma_2(\dot{\theta}_2 - \dot{\theta}_1)^2 + \Gamma_3(\dot{\theta}_3 - \dot{\theta}_2)^2 \,,$$

i.e.,

$$\mathcal{F} = \frac{1}{2}\left\{ \begin{array}{c} \dot{\theta}_1 \\ \dot{\theta}_2 \\ \dot{\theta}_3 \end{array} \right\}^T \left[\begin{array}{ccc} \Gamma_1 + \Gamma_2 & -\Gamma_2 & 0 \\ -\Gamma_2 & \Gamma_2 + \Gamma_3 & -\Gamma_3 \\ 0 & -\Gamma_3 & \Gamma_3 \end{array} \right] \left\{ \begin{array}{c} \dot{\theta}_1 \\ \dot{\theta}_2 \\ \dot{\theta}_3 \end{array} \right\} .$$

The damping matrix is thus

$$\mathbf{C} = \left[\begin{array}{ccc} 1.1 & -1 & 0 \\ -1 & 1.4 & -0.4 \\ 0 & -0.4 & 0.4 \end{array} \right] .$$

The dynamic matrix is

$$\mathcal{A} = \left[\begin{array}{cc} -\mathbf{M}^{-1}\mathbf{C} & -\mathbf{M}^{-1}\mathbf{K} \\ \mathbf{I} & \mathbf{0} \end{array} \right] = \left[\begin{array}{cccccc} -1.1 & 1 & 0 & -20 & 10 & 0 \\ 0.25 & -0.35 & 0.1 & 2.5 & -3.5 & 1 \\ 0 & 0.8 & -0.8 & 0 & 8 & -8 \\ 1 & 0 & 0 & 0 & 0 & 0 \\ 0 & 1 & 0 & 0 & 0 & 0 \\ 0 & 0 & 1 & 0 & 0 & 0 \end{array} \right] .$$

Example 3.2 *Consider the system of Fig. 3.1b, but with the spring and damper connecting mass m_2 to the ground in series instead of being in parallel (Fig. 3.3). Write the equations of motion in both the configuration and the state space.*
The system has now 3 degrees of freedom: the displacement of point B must be added to the displacements of point masses m_1 and m_2 The equations of motion are thus

$$\left[\begin{array}{ccc} m_1 & 0 & 0 \\ 0 & m_2 & 0 \\ 0 & 0 & 0 \end{array} \right] \left\{ \begin{array}{c} \ddot{x}_1 \\ \ddot{x}_2 \\ \ddot{x}_B \end{array} \right\} + \left[\begin{array}{ccc} c_1 + c_{12} & -c_{12} & 0 \\ -c_{12} & c_2 + c_{12} & -c_2 \\ 0 & -c_2 & c_2 \end{array} \right] \left\{ \begin{array}{c} \dot{x}_1 \\ \dot{x}_2 \\ \dot{x}_B \end{array} \right\} +$$

$$+ \left[\begin{array}{ccc} k_1 + k_{12} & -k_{12} & 0 \\ -k_{12} & k_{12} & 0 \\ 0 & 0 & k_2 \end{array} \right] \left\{ \begin{array}{c} x_1 \\ x_2 \\ x_B \end{array} \right\} = \left\{ \begin{array}{c} k_1 x_A + F_1(t) \\ F_2(t) \\ k_2 x_A \end{array} \right\} .$$

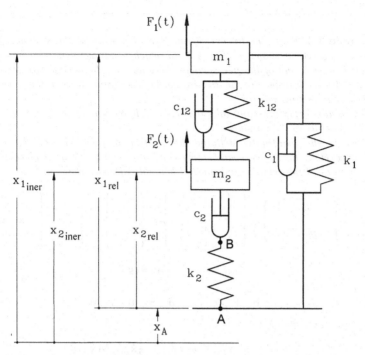

FIGURE 3.3. System of Fig. 3.1b, but with the spring and damper connecting mass m_2 to the ground in series instead of being in parallel.

The state variables are just 5 and, using the first formulation, the state-space equation is

$$
\begin{bmatrix}
m_1 & 0 & 0 & 0 & 0 \\
0 & m_2 & 0 & 0 & -c_2 \\
0 & 0 & 0 & -c_2 & c_2 \\
0 & 0 & 1 & 0 & 0 \\
0 & 0 & 0 & 1 & 0
\end{bmatrix}
\begin{Bmatrix}
\dot{v}_1 \\
\dot{v}_2 \\
\dot{x}_1 \\
\dot{x}_2 \\
\dot{x}_B
\end{Bmatrix} =
$$

$$
=
\begin{bmatrix}
-c_1 - c_{12} & c_{12} & -k_1 - k_{12} & k_{12} & 0 \\
c_{12} & -c_2 - c_{12} & k_{12} & -k_{12} & 0 \\
0 & c_2 & 0 & 0 & -k_{12} \\
1 & 0 & 0 & 0 & 0 \\
0 & 1 & 0 & 0 & 0
\end{bmatrix}
\begin{Bmatrix}
v_1 \\
v_2 \\
x_1 \\
x_2 \\
x_B
\end{Bmatrix} +
$$

$$
+
\begin{bmatrix}
1 & 0 & 0 \\
0 & 1 & 0 \\
0 & 0 & 1 \\
0 & 0 & 0 \\
0 & 0 & 0
\end{bmatrix}
\begin{Bmatrix}
k_1 x_A + F_1(t) \\
F_2(t) \\
k_2 x_A
\end{Bmatrix} .
$$

Example 3.3 *Write the linearized equation of motion of the system with 4 degrees of freedom shown in Fig. 1.3, to which a damper has been added in parallel to each one of the five springs. Assume that the spring with stiffness k_5 is at 45° with respect to the direction of the coordinate axes in the reference (static equilibrium) position.*
Use Lagrange equations and take the relative displacements of points P_1 and P_2 as generalized coordinates.
Let x_{0i} be the coordinates defining the position of points P_1 and P_2 in the reference (i.e., the static equilibrium) condition and x_i their displacements. The vectors of the relative and inertial generalized coordinates are

$$\mathbf{x}_{rel}=\left\{\begin{array}{c}(P_1-A)\\(P_2-A)\end{array}\right\}=\left\{\begin{array}{c}\left\{\begin{array}{c}x_{A_{P_1}}\\y_{A_{P_1}}\end{array}\right\}\\\left\{\begin{array}{c}x_{A_{P_2}}\\y_{A_{P_2}}\end{array}\right\}\end{array}\right\}=\left\{\begin{array}{c}x_1\\x_2\\x_3\\x_4\end{array}\right\}+\left\{\begin{array}{c}x_{01}\\x_{02}\\x_{03}\\x_{04}\end{array}\right\}=\mathbf{x}+\mathbf{x}_0 \ ,$$

$$\mathbf{x}_{iner}=\mathbf{x}_{rel}+\boldsymbol{\delta}_x x_A+\boldsymbol{\delta}_y y_A \ ,$$

where
$$\boldsymbol{\delta}_x = \begin{bmatrix} 1 & 0 & 1 & 0 \end{bmatrix}^T \ , \quad \boldsymbol{\delta}_y = \begin{bmatrix} 0 & 1 & 0 & 1 \end{bmatrix}^T \ .$$
The kinetic energy of the system is
$$\mathcal{T} = \tfrac{1}{2}\left(m_1 V_{P_1}^2 + m_2 V_{P_2}^2\right) = \tfrac{1}{2}\dot{\mathbf{x}}_{iner}^T \mathbf{M}\dot{\mathbf{x}}_{iner}$$

where
$$\mathbf{M} = \begin{bmatrix} m_1 & 0 & 0 & 0 \\ 0 & m_1 & 0 & 0 \\ 0 & 0 & m_2 & 0 \\ 0 & 0 & 0 & m_2 \end{bmatrix} \ .$$

Assuming that the length at rest of the springs is l_{0i}, the potential energy due to the springs is

$$\mathcal{U}=\tfrac{1}{2}k_1\left[\sqrt{(x_1+x_{01})^2+x_2^2}-l_{01}\right]^2+\tfrac{1}{2}k_2\left[\sqrt{(x_2+x_{02})^2+x_1^2}-l_{02}\right]^2+$$
$$\tfrac{1}{2}k_3\left[\sqrt{(x_3+x_{03})^2+x_4^2}-l_{03}\right]^2+\tfrac{1}{2}k_4\left[\sqrt{(x_4+x_{04})^2+x_3^2}-l_{04}\right]^2+$$
$$+\tfrac{1}{2}k_5\left[\sqrt{(x_3+x_{03}-x_1-x_{01})^2+(x_4+x_{04}-x_2-x_{02})^2}-l_{05}\right]^2 \ .$$

Since the displacements x_i are referred to the static equilibrium conditions, the length at rest of the springs is

$$l_i = x_{0i} \ for \ i=1...4 \ ; \quad l_{05} = \sqrt{(x_{03}-x_{01})^2+(x_{04}-x_{02})^2} \ .$$

To linearize the equation of motion, the expression of the potential energy (or better, the increase of the potential energy with respect to the potential energy in static equilibrium conditions) can be simplified by neglecting the squares of the displacements x_i with respect to the products $x_i x_{0j}$. The expression of the potential energy reduces to

$$\mathcal{U} = \tfrac{1}{2}k_1 \left[\sqrt{2x_1 l_{01} + l_{01}^2} - l_{01}\right]^2 + \tfrac{1}{2}k_2\left[\sqrt{2x_2 l_{02} + l_{02}^2} - l_{02}\right]^2 +$$
$$\tfrac{1}{2}k_3\left[\sqrt{2x_3 l_{03} + l_{03}^2} - l_{03}\right]^2 + \tfrac{1}{2}k_4\left[\sqrt{2x_4 l_{04} + l_{04}^2} - l_{04}\right]^2 +$$
$$+\tfrac{1}{2}k_5\left[\sqrt{l_{05}^2 + 2\Delta x\,(x_3 - x_1) + 2\Delta y\,(x_4 - x_2)} - l_{05}\right]^2 ,$$

where

$$\Delta x = x_{03} - x_{01} \quad , \quad \Delta y = x_{04} - x_{02} .$$

Finally, the square root can be substituted by its series truncated at its second term, obtaining

$$\mathcal{U} = \frac{1}{2}\left\{ k_1 x_1^2 + k_2 x_2^2 + k_3 x_3^2 + k_4 x_4^2 + k_5 \left[\frac{\Delta x}{l_{05}}(x_3 - x_1) + \frac{\Delta y}{l_{05}}(x_4 - x_2)\right]^2 \right\} ,$$

i.e., in matrix form,

$$\mathcal{U} = \tfrac{1}{2}\mathbf{x}_{rel}^T \mathbf{K}\mathbf{x}_{rel} =$$

$$= \frac{1}{2}\left\{\begin{array}{c} x_1 \\ x_2 \\ x_3 \\ x_4 \end{array}\right\}^T \left[\begin{array}{cccc} k_1 + c^2 k_5 & csk_5 & -c^2 k_5 & -csk_5 \\ csk_5 & k_2 + s^2 k_5 & -csk_5 & -s^2 k_5 \\ -c^2 k_5 & -csk_5 & k_3 + c^2 k_5 & csk_5 \\ -csk_5 & -s^2 k_5 & csk_5 & k_4 + s^2 k_5 \end{array}\right]\left\{\begin{array}{c} x_1 \\ x_2 \\ x_3 \\ x_4 \end{array}\right\} ,$$

where $c = \Delta x/l_{05}$ and $s = \Delta y/l_{05}$. Since the spring with stiffness k_5 is at 45°, both c and s are equal to $\sqrt{2}/2$.

The Rayleigh dissipation function is the sum of the dissipation functions of all dampers. It is easily expressed as a function of the relative velocities

$$\mathcal{F} = \frac{1}{2}\left\{ c_1 \dot{x}_1^2 + c_2 \dot{x}_2^2 + c_3 \dot{x}_3^2 + c_4 \dot{x}_4^2 + c_5\left[c\,(\dot{x}_3 - \dot{x}_1)^2 + s\,(\dot{x}_4 - \dot{x}_2)^2\right]\right\}$$

i.e., in matrix form,

$$\mathcal{F} = \frac{1}{2}\left\{\begin{array}{c} \dot{x}_1 \\ \dot{x}_2 \\ \dot{x}_3 \\ \dot{x}_4 \end{array}\right\}\left[\begin{array}{cccc} c_1 + c^2 c_5 & csc_5 & -c^2 c_5 & -csc_5 \\ csc_5 & c_2 + s^2 c_5 & -csc_5 & -s^2 c_5 \\ -c^2 c_5 & -csc_5 & c_3 + c^2 c_5 & csc_5 \\ -csc_5 & -s^2 c_5 & csc_5 & c_4 + s^2 c_5 \end{array}\right]\left\{\begin{array}{c} \dot{x}_1 \\ \dot{x}_2 \\ \dot{x}_3 \\ \dot{x}_4 \end{array}\right\} .$$

> *By using Lagrange Equations, it follows*
>
> $$\mathbf{M\ddot{x}}_{rel} + \mathbf{C\dot{x}_{rel}} + \mathbf{Kx}_{rel} = -\mathbf{M}\delta_x\ddot{x}_A - \mathbf{M}\delta_y\ddot{y}_A + \mathbf{f}(t) \ .$$
>
> *It coincides with the equation obtained by introducing the damping matrix into Eq. (1.11).*

3.4   Structural or hysteretic damping

3.4.1   Hysteresis cycle

An elastic material is a material that does not dissipate energy when deformed; if its stress–strain characteristic is linear it is a linear, elastic material. A structural element made of a material of this kind can be modeled as a linear spring (Fig. 3.4a). This model is sometimes referred to as Hooke's model and the relevant stress–strain relationship is[2]

$$\sigma = E\epsilon \ . \tag{3.42}$$

The proportionality constant E is the *Young's modulus* or *modulus of elasticity*.

If instead of reasoning at the material level, (stresses, strains, moduli), one works at the level of structural elements (forces, displacements,

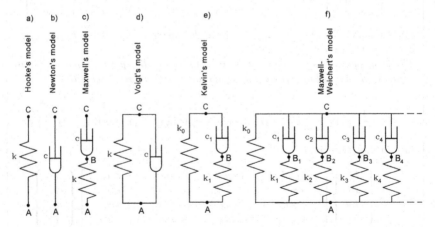

FIGURE 3.4. Models for linear materials: (a) elastic, (b) viscous, (c), (d), (e) and (f): visco-elastic.

---

[2]The sign ($-$) is omitted for consistency with the usual convention for stresses and strains: here the force is positive when the spring is stretched and negative when it is compressed.

stiffness), the force–displacement characteristics is

$$F = -kx \ . \tag{3.43}$$

However, as already stated, no actual material is exactly elastic: the simplest way to model material damping is to assume that the material reacts with a force that depends only on the strain rate. If it is linear it can be modeled as a linear viscous damper (Fig. 3.4b), a model sometimes referred to as Newton's model. Its stress–strain relationship is

$$\sigma = -C\dot{\epsilon} \ , \tag{3.44}$$

or, in terms of forces and displacements

$$F = -c\dot{x} \ . \tag{3.45}$$

Actual materials react with both a restoring and a damping force; to model a linear visco-elastic material at least one spring and one damper are needed. They may be arranged in series (Maxwell's model, Fig. 3.4c) or in parallel (Voigt's model[3], Fig. 3.4d). The stress–strain and force–displacement relationships for the latter model are

$$\sigma = -E\epsilon - C\dot{\epsilon} \ , \qquad F = -kx - c\dot{x} \ . \tag{3.46}$$

A model of this kind does not simulate satisfactorily actual engineering materials. Since the 1920s experiments showed that many materials, when subjected to cyclic loading, exhibit a type of internal damping causing energy losses per cycle that are proportional to the square of the amplitude and independent of the frequency. This behavior is usually described as structural or hysteretic damping.[4] Although subsequent studies showed that there is a certain dependence on the frequency, hysteretic damping is still considered an adequate model for energy dissipation in structural materials in many applications.

Structural damping is thus defined assuming that the time history of the stress cycles is harmonic and, since the material is linear, also the time history of the deformation follows a similar pattern. The time histories of the stress and of the strain are slightly out of phase and an elliptical hysteresis cycle results in the $(\sigma\epsilon)$ plane (Fig. 3.5a): the strain lags the stress by a phase-angle Φ.

Since the internal damping of most engineering materials is small, the two time histories are only slightly out of phase and the elliptical hysteresis cycle is small, usually much smaller than that shown in Fig. 3.5a.

---

[3]These models are often referred to with different names. See Banks H.T., Pinter G.A., *Hysteretic Damping*, in S. Braun (ed.), *Encyclopedia of Vibration*, Academic Press, London, 2001, and D. Roylance, *Engineering Viscoelasticity*, MIT, Cambridge, 2001.

[4]See, for instance, N.O. Myklestad, The concept of complex damping, Jour. of Applied Mechanics, Vol. 19, 1952, p. 284.

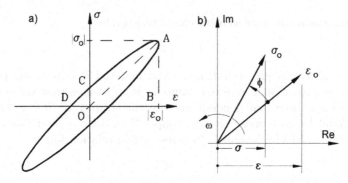

FIGURE 3.5. (a) Hysteresis cycle in $(\sigma \; \epsilon)$ plane. (b) Stresses and strains represented as rotating vectors in the complex plane at time $t = 0$.

Assuming that at time $t = 0$ the strain reaches its maximum value, it follows

$$
\begin{cases}
\sigma = \sigma_0 \cos\left(\omega t + \Phi\right) \\
\epsilon = \epsilon_0 \cos\left(\omega t\right)
\end{cases}
\tag{3.47}
$$

or, using the complex notation

$$
\begin{cases}
\sigma = \sigma_0 e^{i(\omega t + \Phi)} \\
\epsilon = \epsilon_0 e^{i\omega t} \; ,
\end{cases}
\tag{3.48}
$$

where the amplitudes σ_0 and ϵ_0 have been assumed to be real numbers.

The ratio between the stress and the strain, that in an elastic material is the Young's modulus, is now expressed by a complex number, the complex modulus

$$
E^* = \frac{\sigma}{\epsilon} = \frac{\sigma_0}{\epsilon_0} e^{i\Phi} = E\left[\cos\left(\Phi\right) + i \sin\left(\Phi\right)\right].
\tag{3.49}
$$

Its real part

$$
E' = E \cos\left(\Phi\right)
\tag{3.50}
$$

gives the measure of the elastic stiffness of the material and is often referred to as *storage* or *in-phase modulus*. The imaginary part

$$
E'' = E \sin\left(\Phi\right).
\tag{3.51}
$$

is linked with damping and is said to be the *loss* or *in-quadrature stiffness*. Their ratio is the loss factor or loss ratio η

$$
\eta = \frac{E''}{E'} = \arctan(\Phi).
\tag{3.52}
$$

As already stated, the phase-angle Φ by which the strain lags the stress is assumed to be independent from the frequency. This causes the complex modulus to be independent from the frequency: it can be considered as a characteristic of the material.

In the same way also the stiffness of a structural member can be expressed by a complex number, the complex stiffness

$$k^* = k' + ik'',$$
(3.53)

whose real part, the in-phase or storage stiffness k', and imaginary part, the in-quadrature or loss stiffness k'', are independent from the frequency. The loss factor of a structural element is thus defined as

$$\eta = \frac{k''}{k'} = \arctan(\Phi).$$
(3.54)

Another parameter that is sometimes used to quantify the internal damping of materials is the specific damping capacity ψ. It is defined as the ratio between the energy dissipated in a cycle (area of the ellipse in Fig. 3.5) and the elastic energy stored in the system in the condition of maximum amplitude (area of the OAB triangle in the same figure)

$$\psi = \frac{\pi \sigma_0 \epsilon_0 \sin(\Phi)}{\frac{1}{2}\sigma_0 \epsilon_0} = 2\pi \sin(\Phi).$$
(3.55)

Contrarily to what seen for viscous damping, the specific damping capacity of a system with hysteretic damping does not depend on the frequency.

The damping of most engineering materials (except for some elastomers) is quite small (see Table 3.1), and the trigonometric functions of the phase-angle Φ can be linearized. The expressions of the quantities defined earlier can, consequently, be simplified:

$$\begin{aligned} k' &\approx k, & \eta &\approx \Phi, \\ k^* &\approx k(1 + i\eta), & \Psi &\approx 2\pi\Phi \approx 2\pi\eta. \end{aligned}$$
(3.56)

The loss factor of a structural member can be equal to that of the material (as in the case of a homogeneous monolithic spring) or greater, if some damping mechanisms other than material hysteresis are present (as in built-up members, with rivets or threaded joints, elements in viscous fluids, and so on).

TABLE 3.1. Order of magnitude of the loss factor of some engineering materials. The high values typical of some elastomeric materials prevents from using the simplified formulae reported above.

Material	η
Aluminium alloy	$0.0001 - 0.001$
Copper and copper alloys	$0.001 - 0.005$
Cast iron	$0.001 - 0.08$
Steel	$0.01 - 0.06$
Rubber	$0.01 - 3$

3.4.2 Equation of motion in the frequency domain

The complex stiffness and the complex modulus have been introduced in connection with harmonic loading and, as a consequence, they are well suited to equations of motion written in the frequency domain.

The equation of motion of a single degree of freedom system with hysteretic damping can thus be obtained by introducing the complex stiffness expressed by Eq. (3.56) into the frequency-domain equation (2.12), obtaining

$$\left[-\omega^2 m + k(1 + i\eta)\right] x_0 = f_0 \ . \tag{3.57}$$

This equation can be generalized to multi degrees of freedom systems by introducing an in-phase and an in-qudrature stiffness matrix

$$\left(-\omega^2 \mathbf{M} + \mathbf{K}' + i\mathbf{K}''\right) \mathbf{x}_0 = \mathbf{f}_0 \ . \tag{3.58}$$

If the loss factor is constant throughout the system, matrices $\mathbf{K}''$ and $\mathbf{K}'$ are proportional and the complex stiffness matrix reduces to

$$\mathbf{K}^* = (1 + i\eta)\mathbf{K} \ .$$

Remark 3.5 *A time-domain equation of motion of the kind*

$$m\ddot{x} + k(1 + i\eta)x = f(t) \tag{3.59}$$

has no meaning: Both the force f and the displacement x are generic functions of time, while the complex stiffness is defined only if their time histories are both harmonic. Moreover, functions $x(t)$ and $f(t)$ are both real quantities and the complex equation linking them has no meaning.

Although it can be demonstrated that the definition of structural damping can be extended to the more general case of periodic loading, because any periodic time history can be expressed as the sum of harmonic terms, its extension to nonperiodic time histories through Fourier transform is impossible, since it can lead to non-causal results.

The dynamic stiffness of a system with structural damping is thus

$$k_{dyn} = -m\omega^2 + k(1 + i\eta) \ , \tag{3.60}$$

for single degree of freedom systems, or in general,

$$\mathbf{K}_{dyn} = -\omega^2 \mathbf{M} + \mathbf{K}' + i\mathbf{K}'' \ . \tag{3.61}$$

3.4.3 Equivalent viscous damping

Structural damping is a form of linear damping that does not differ much from viscous damping. From the frequency-domain equation of motion (3.7) it is clear that also in the latter case it is possible to define a complex stiffness, with an imaginary part equal to

$$ic\omega$$

instead of

$$ik\eta \ .$$

By equating the two expressions of the in-quadrature stiffness, it is possible to define an equivalent viscous damping

$$c_{eq} = \frac{\eta k}{\omega} \ , \tag{3.62}$$

through which structural damping can be assimilated to viscous damping with a coefficient inversely proportional to the frequency at which the hysteresis cycle is gone through.[5]

In the case of systems with many degrees of freedom, it is also possible to define an equivalent viscous damping matrix

$$\mathbf{C}_{eq} = \frac{1}{\omega}\mathbf{K}'' \ . \tag{3.63}$$

Remark 3.6 *Equation* (3.62) *shows clearly the inconsistence of using the hysteretic damping model at very low frequency, since the equivalent damping tends to infinity when $\omega \rightarrow 0$. This is due to the assumption that the shape of the hysteresis cycle is not affected by the frequency, while in a static test (i.e., for $\omega \rightarrow 0$) there is no hysteresis cycle at all and the stress–strain relationship of a linear material is a straight line.*

To remove the dependency of the equivalent damping from the frequency, Eq. (3.62) can be modified as

$$c_{eq} = \frac{\eta k}{\omega_r} \ , \tag{3.64}$$

where ω_r is a reference frequency, at which the equivalent viscous damping dissipates the same energy as its hysteretic counterpart. As it will be seen later, if ω_r is the natural frequency of the system and damping is low, an acceptable approximation may be obtained.

Remark 3.7 *Structural damping is just a linear model that, while allowing modeling many actual systems better than viscous damping, gives only a rough approximation of the behavior of structural members. The Young's*

---

[5]The formula should be written as $c_{eq} = \eta k / |\omega|$, where $|\omega|$ is the frequency at which the hysteresis cycle is gone through. In the present section the frequency is assumed to be always expressed by a positive number.

modulus E and the loss factor η of most engineering materials are independent of the frequency only in an approximated way. Most metals stick to this rule within a fair or even a good approximation, while elastomers often show very strong dependence of their mechanical characteristics on frequency (see Section 3.7).

The loss factor of all materials is a function of many parameters and is particularly influenced by the amplitude of the stress cycle. During the life of a structural member, strong variations of the damping characteristics with the progress of fatigue phenomena are expected. Actually, damping can be used to obtain information on the extent of fatigue damage.

With all the needed caution, it is possible to mention typical values of damping of different engineering materials (Table 3.1). A large quantity of data can be found in the literature.[6]

The behavior of materials is only approximately linear, but while the non-linearities of the stress–strain curve are usually only found at high stresses and a wide linearity field exists, the nonlinearities in the damping characteristics are found at all values of the load. While the dependence of the characteristics of the material on the frequency can be taken into account easily, the last consideration would lead to nonlinear equations and, consequently, is usually neglected.

With all the aforementioned limitations, the structural damping model remains a powerful tool for structural analysis and finds a wide application.

3.5   Non-viscous damping

3.5.1   Systems with a single degree of freedom

However, if a visco-elastic material is abruptly subject to stress, its strain reaches instantly a certain value, to increase slowly in time. This is known under the name of creep. If on the contrary is abruptly strained, its stress reaches instantly a certain value, and then slowly decreases in time. This is known under the name of relaxation.

The stress (force) is thus dependent not only on the instant values of the displacement (strain) and velocity (strain rate), but also on their past history. Some sort of 'memory' must be included in the constitutive law of the material.

Possibly, the most general model for this phenomenon is the relationship

$$F = -kx - \int_{-\infty}^{t} \mathcal{C}\left(t, \tau\right) \dot{x}\left(\tau\right) d\tau . \tag{3.65}$$

---

[6]See, for instance, B.J. Lazan, *Damping of Materials and Members in Structural Mechanics*, Pergamon Press, Oxford, 1968.

where function $\mathcal{C}\left(t, \tau\right)$, which usually has the form $\mathcal{C}\left(t - \tau\right)$, is referred to as *damping kernel* function, or *retardation, heredity, after-effect* or *relaxation* function.

A common expression for the damping kernel function for a single degree of freedom system is a sum of exponential terms

$$\mathcal{C}\left(t - \tau\right) = \sum_{i=1}^{m} c_i \mu_i e^{-\mu_i (t - \tau)} , \tag{3.66}$$

where the m parameters μ_i are said *relaxation parameters*.

If all μ_i tend to infinity, viscous damping is obtaned.

The equation of motion of a system with a single degree of freedom, which includes also nonviscous damping, modeled using Eq. (3.65) together with Eq. (3.66) to express the damping kernel, is

$$m\ddot{x} + c\dot{x} + \sum_{i=1}^{m} c_i \mu_i \int_{-\infty}^{t} e^{-\mu_i (t - \tau)} \dot{x}\left(\tau\right) d\tau + kx = f(t) . \tag{3.67}$$

It is possible to demonstrate that each exponential term in Eq. (3.66) yields a force equivalent to that due to a spring with a damper in series (system of Fig. 3.4c). The restoring force expressed by Eq. (3.65) is thus equivalent to that due to the Maxwell–Weichart's model shown in Fig. 3.4f, with a number m of dampers.

This is easily shown by observing that the force exerted by each exponential term in Eq. (3.66) is

$$F_i = c_i \mu_i \int_{-\infty}^{t} e^{-\mu_i (t - \tau)} \dot{x}\left(\tau\right) d\tau . \tag{3.68}$$

Since the Laplace transform of

$$f(\tau) = e^{-\mu \tau}$$

is

$$\tilde{f}(s) = \frac{1}{s + \mu} ,$$

the Laplace transform of force $F_i(t)$ is

$$\tilde{F}_i\left(s\right) = c_i \mu_i \frac{s}{s + \mu_i} \tilde{x}\left(s\right) . \tag{3.69}$$

The corresponding stiffness in the Laplace domain is thus

$$k\left(s\right) = c_i \mu_i \frac{s}{s + \mu_i} . \tag{3.70}$$

The frequency-domain complex stiffness is obtained by substituting $i\omega$ for s:

$$k^*\left(\omega\right) = c_i \mu_i \frac{i\omega}{i\omega + \mu_i} = c_i \mu_i \frac{\omega^2 + i\mu_i \omega}{\omega^2 + \mu_i^2} . \tag{3.71}$$

If

$$\mu_i = \frac{k}{c_i} \quad \text{and} \quad c = c_i \tag{3.72}$$

the complex stiffness becomes

$$k^*(\omega) = k\left(\frac{\omega^2 c^2}{k^2 + c^2\omega^2} + i\omega\frac{kc}{k^2 + c^2\omega^2}\right), \tag{3.73}$$

which coincides with the complex stiffness expressed by Eq. (3.25).

Since the nonviscous damping can be expressed by the system of Fig. 3.4f, the m degrees of freedom corresponding to the displacements of points B_i must also be considered. They are usually referred to as *internal* or *damping degrees of freedom*. The system, although containing just one mass, has thus $m + 1$ degrees of freedom.

However, since no mass is associated to points B_i, the accelerations of the internal coordinates do not appear in the equations of motion, and the order of the differential equation is not $2(m + 1)$ but only $m + 2$.

The equation of motion of the system of Fig. 3.4f with mass m located in point C and constrained in point A thus

$$\mathbf{M}\ddot{\mathbf{x}} + \mathbf{C}\dot{\mathbf{x}} + \mathbf{K}\mathbf{x} = \mathbf{f}(t), \tag{3.74}$$

where, remembering Eq. (3.65), the relevant matrices and vectors are

$$\mathbf{x} = \begin{Bmatrix} x_C \\ x_{B1} \\ x_{B2} \\ \dots \end{Bmatrix}, \quad \mathbf{M} = \begin{bmatrix} m & 0 & 0 & \dots \\ & 0 & 0 & \dots \\ & & 0 & \dots \\ \text{symm.} & & & \dots \end{bmatrix}, \quad \mathbf{f}(t) = \begin{Bmatrix} F_c \\ 0 \\ 0 \\ 0 \end{Bmatrix},$$

$$\mathbf{C} = \begin{bmatrix} c + \sum_{i=1}^{m} c_i & -c_1 & -c_2 & \dots \\ & c_1 & 0 & \dots \\ & & c_2 & \dots \\ \text{symm.} & & & \dots \end{bmatrix}, \mathbf{K} = \begin{bmatrix} k & 0 & 0 & \dots \\ & \mu_1 c_1 & 0 & \dots \\ & & \mu_2 c_2 & \dots \\ \text{symm.} & & & \dots \end{bmatrix}.$$

Since the states are only $m + 2$, remembering Eq. (3.34), the state equation is,

$$\begin{Bmatrix} \dot{v}_C \\ \dot{x}_C \\ \dot{\mathbf{x}}_B \end{Bmatrix} = \mathbf{M}^{*-1}\mathcal{A}^* \begin{Bmatrix} v_C \\ x_C \\ \mathbf{x}_B \end{Bmatrix} + \mathbf{M}^{*-1} \begin{Bmatrix} F_C \\ 0 \\ \mathbf{0}_{m \times 1} \end{Bmatrix}, \tag{3.75}$$

where

$$\mathbf{M}^* = \begin{bmatrix} m & 0 & \mathbf{C}_{12} \\ \mathbf{0}_{m \times 1} & \mathbf{0}_{m \times 1} & \mathbf{C}_{22} \\ 0 & 1 & \mathbf{0}_{1 \times m} \end{bmatrix}, \quad \mathcal{A}^* = -\begin{bmatrix} C_{11} & K_{11} & \mathbf{0}_{1 \times m} \\ \mathbf{C}_{21} & \mathbf{0}_{m \times 1} & \mathbf{K}_{22} \\ -1 & 0 & \mathbf{0}_{1 \times m} \end{bmatrix},$$

C_{11} and K_{11} are numbers, $\mathbf{C}_{12}$ is a row matric with m columns, and $\mathbf{C}_{22}$ and $\mathbf{K}_{22}$ are $m \times m$ diagonal matrices.

This form of the state equations is not unique, and many different forms, all essentially equivalent, can be found in the literature.[7] Moreover, it is possible to associate also a mass to points B_i, in such a way that the mass matrix is not singular. This approach is followed by the GHM model intrduced by Golla and Hughes.[8]

Since the various approaches are essentially equivalent, only the one shown above will be dealt with here.

A similar result can be obtained by resorting to the stress–strain relationship

$$\sigma + \tau_\epsilon \dot{\sigma} = E_r \left(\epsilon + \tau_\sigma \dot{\epsilon} \right) , \tag{3.76}$$

where E_r, τ_ϵ, and τ_σ are the relaxed modulus of elasticity, and the constant strain and constant stress relaxation times, respectively. It allows to account for creep and relaxation phenomena. If the stress and strain time histories are harmonic in time

$$\begin{cases} \sigma = \sigma_0 e^{i\omega t} \\ \epsilon = \epsilon_0 e^{i\omega t} , \end{cases} \tag{3.77}$$

where the amplitudes σ_0 and ϵ_0 are expressed by complex numbers, the frequency-domain stress–strain relationship becomes:

$$\sigma_0 \left(1 + i\tau_\epsilon \omega \right) = \epsilon_0 E_r \left(1 + i\tau_\sigma \omega \right) . \tag{3.78}$$

The ratio between the (complex) amplitudes of the stress and the strain is

$$\frac{\sigma_0}{\epsilon_0} = E_r \frac{1 + i\tau_\sigma \omega}{1 + i\tau_\epsilon \omega} = E_r \left[\frac{1 + \tau_\epsilon \tau_\sigma \omega^2}{1 + \tau_\epsilon^2 \omega^2} + i \frac{(\tau_\sigma - \tau_\epsilon) \omega}{1 + \tau_\epsilon^2 \omega^2} \right] . \tag{3.79}$$

This is equivalent to defining a complex Young's modulus whose real part (the in-phase or storage modulus) E' and imaginary part (the in-quadrature or loss modulus) E'' are both functions of the frequency ω

$$E' = E_r \frac{1 + \tau_\epsilon \tau_\sigma \omega^2}{1 + \tau_\epsilon^2 \omega^2} , \quad E'' = E_r \frac{(\tau_\sigma - \tau_\epsilon) \omega}{1 + \tau_\epsilon^2 \omega^2} . \tag{3.80}$$

Their ratio is the loss factor

$$\eta = \frac{E''}{E'} = \frac{(\tau_\sigma - \tau_\epsilon) \omega}{1 + \tau_\epsilon \tau_\sigma \omega^2} . \tag{3.81}$$

It is again possible to demonstrate that this frequency-domain expression is the same that can be obtained from the Kelvin's model, (Fig. 3.4d, Eq.

---

[7]For instance, the equation found in N. Wagner, S. Adhicari, *Symmetric State-Space Method for a Class of Nonviscously Damped Systems*, AIAA Journal, Vol. 41, No5, May 2003, p. 951–956, is obtained starting from Eq. (3.36) instead of Eq. (3.34) and multiplying the last equation by m.

[8]D.J. McTavish, P.C. Hughes, *Modelling of Linear Viscoelastic Space Structures*, Journal of Vibration and Acoustics, , Vol. 115, Jan. 1993, p. 103–110.

(3.88)), provided that

$$k_0 = E_r , \quad k_1 = E_r \left(\frac{\tau_\sigma}{\tau_\epsilon} - 1 \right), \quad c_1 = E_r \left(\tau_\sigma - \tau_\epsilon \right) .$$

This model is sometimes also referred to as the standard linear material model.

3.5.2 Systems with many degrees of freedom

Consider a system with many degrees of freedom, which includes both viscous and nonviscous damping, and use Eq. (3.65) to express the latter. Let the damping kernel be expressed by Eq. (3.66).

The resulting equation of motion is

$$\mathbf{M}\ddot{\mathbf{x}} + \mathbf{C}\dot{\mathbf{x}} + \sum_{i=1}^m \mathbf{C}_i \mu_i \int_{-\infty}^t e^{-\mu_i(t-\tau)} \dot{\mathbf{x}}(\tau)\, d\tau + \mathbf{K}\mathbf{x} = \mathbf{f}(t) . \qquad (3.82)$$

Although being usually symmetrical, the matrices $\mathbf{C}_i$ may have different structures. If, for instance, hystereting damping is distributed on the whole structure, having the same characteristics everywhere, they are not rank deficient, and their number m is equal to the number of exponential terms needed to approximate the actual behavior of the material with the required accuracy.

If, on the contrary, the nonviscous damping has different properties in different parts of the structure (e.g., because different materials are used), their rank is much smaller than the number n of degrees of freedom (i.e., they are rank deficient) and they have nonvanishing terms only in the zone interested by the relevant material. Their number m is much larger, since there is a number of matrices equal to the sum of the numbers of exponential terms needed to model each one of the various materials. For instance, if there are five different materials, and each one is modeled using four exponential terms (i.e. there is a total of 20 exponential terms), $m = 20$.

The equation in the configuration space has now a size $n(m+1)$

$$\mathbf{M}^{**}\ddot{\mathbf{x}}^{**} + \mathbf{C}^{**}\dot{\mathbf{x}}^{**} + \mathbf{K}^{**}\mathbf{x}^{**} = \mathbf{f}^{**}(t) , \qquad (3.83)$$

where

$$\mathbf{x}^{**} = \left\{ \begin{array}{c} \mathbf{x} \\ \mathbf{x}_{B1} \\ \mathbf{x}_{B2} \\ \dots \end{array} \right\}, \quad \mathbf{M}^{**} = \begin{bmatrix} \mathbf{M} & 0 & 0 & \dots \\ & 0 & 0 & \dots \\ & & 0 & \dots \\ \text{symm.} & & & \dots \end{bmatrix}, \quad \mathbf{f}^{**}(t) = \left\{ \begin{array}{c} \mathbf{f} \\ 0 \\ 0 \\ 0 \end{array} \right\},$$

$$\mathbf{C}^{**} = \begin{bmatrix} \mathbf{C} + \sum_{i=1}^m \mathbf{C}_i & -\mathbf{C}_1 & -\mathbf{C}_2 & \dots \\ & \mathbf{C}_1 & 0 & \dots \\ & & \mathbf{C}_2 & \dots \\ \text{symm.} & & & \dots \end{bmatrix},$$

$$\mathbf{K}^{**} = \begin{bmatrix} \mathbf{K} & 0 & 0 & \cdots \\ & \mu_1\mathbf{C}_1 & 0 & \cdots \\ & & \mu_2\mathbf{C}_2 & \cdots \\ \text{symm.} & & & \cdots \end{bmatrix} .$$

If matrices $\mathbf{C}_i$ are not rank deficient, the state equations can be written in the same way seen for single degree of freedom systems. Partitioning the degrees of freedom to separate the coordinates of the actual system $\mathbf{x}$ (they are n) from the internal coordinates $\mathbf{x}_\mathrm{B}$ (they are $n \times m$), and using Eq. (3.34), it follows that

$$\left\{ \begin{array}{c} \dot{\mathbf{v}} \\ \dot{\mathbf{x}} \\ \dot{\mathbf{x}}_\mathrm{B} \end{array} \right\} = \mathbf{M}^{*-1}\mathcal{A}^* \left\{ \begin{array}{c} \mathbf{v} \\ \mathbf{x} \\ \mathbf{x}_\mathrm{B} \end{array} \right\} + \mathbf{M}^{*-1} \left\{ \begin{array}{c} \mathbf{f} \\ \mathbf{0}_{n \times 1} \\ \mathbf{0}_{(n \times m) \times 1} \end{array} \right\}, \qquad (3.84)$$

where

$$\mathbf{M}^* = \begin{bmatrix} \mathbf{M} & 0 & \mathbf{C}_{12}^{**} \\ \mathbf{0}_{(n \times m) \times n} & \mathbf{0}_{(n \times m) \times n} & \mathbf{C}_{22}^{**} \\ \mathbf{0}_{n \times n} & \mathbf{I}_{n \times n} & \mathbf{0}_{n \times (n \times m)} \end{bmatrix},$$

$$\mathcal{A}^* = - \begin{bmatrix} \mathbf{C}_{11}^{**} & \mathbf{K}_{11}^{**} & \mathbf{0}_{n \times (n \times m)} \\ \mathbf{C}_{21}^{**} & \mathbf{0}_{(n \times m) \times n} & \mathbf{K}_{22}^{**} \\ -\mathbf{I}_{n \times n} & 0 & \mathbf{0}_{n \times (n \times m)} \end{bmatrix}.$$

Again, the state equation can be written in different, but equivalent, forms.

If matrices $\mathbf{C}_i$ are rank deficient the singularity of matrix $\mathbf{M}^*$ prevents from performing the inversion needed to obtain the dynamic matrix of the system. Removing the singularity has the added advantage of reducing the number of states of the system.

Assuming that all matrices $\mathbf{C}_i$ are symmetrical and following the procedure outlined by S. Adhicari and N. Wagner,[9] the eigenvalues and eigenvectors of the matrices $\mathbf{C}_i$ are first obtained. If the rank of the generic matrix $\mathbf{C}_i$ is r_i, its r_i (with $r_i < n$) nonzero eigenvalues can be collected in the diagonal eigenvalue matrix $\mathbf{d}_i$, whose size is $r_i \times r_i$.

A rectangular transformation matrix $\mathbf{R}_i$ (whos size is $n \times r_i$) can be defined, so that

$$\mathbf{R}_i^T \mathbf{C}_i \mathbf{R}_i = \mathbf{d}_i . \qquad (3.85)$$

The columns of matrices $\mathbf{R}_i$ are the eigenvectors of matrix $\mathbf{C}_i$ corresponding to the r_i nonzero eigenvalues. If matrices $\mathbf{C}_i$ are not symmetrical, this procedure can be modified by using the left and right eigenvectors.[10]

---

[9]N. Wagner, S. Adhicari, *Symmetric State-Space Method for a Class of Nonviscously Damped Systems*, AIAA Journal, Vol. 41, No5, May 2003, p. 951–956. Here a slightly different definition of the internal coordinates has been used.

[10]S. Adhicari, N. Wagner, *Analysis of Asymmetric Nonviscously Damped Linear Dynamic Systems*, Journal of Applied Mechanics, Vo. 70, Nov. 2003, p. 885–893.

The internal coordinates $\mathbf{x}_{Bi}$ can thus be reduced in number through the transformation

$$\mathbf{x}_{Bi} = \mathbf{R}_i \tilde{\mathbf{x}}_{Bi} .\tag{3.86}$$

By introducing the internal coordinates $\tilde{\mathbf{x}}_{Bi}$ instead of $\mathbf{x}_{Bi}$, the matrices $\mathbf{C}^{**}$ and $\mathbf{K}^{**}$ reduce to

$$\mathbf{C}^{**} = \begin{bmatrix} \mathbf{C} + \sum_{i=1}^{m} \mathbf{C}_i & -\mathbf{C}_1\mathbf{R}_1 & -\mathbf{C}_2\mathbf{R}_2 & \cdots \\ & \mathbf{d}_1 & \mathbf{0} & \cdots \\ & \mathbf{0} & \mathbf{d}_2 & \cdots \\ \text{symm} & & & \cdots \end{bmatrix},$$

$$\mathbf{K}^{**} = \begin{bmatrix} \mathbf{K} & 0 & 0 & \cdots \\ & \mu_1 \mathbf{d}_1 & 0 & \cdots \\ & & \mu_2 \mathbf{d}_2 & \cdots \\ \text{symm.} & & & \cdots \end{bmatrix}$$

and the computation proceeds as before.

3.6   Structural damping as nonviscous damping

3.6.1   Ideal Maxwell–Weichert model

As already stated, the formulation of the structural damping seen in Section 3.4.1, with the exception of Eq. (3.64), can be used only in the frequency domain. This is a severe limitation, particularly nowadays, since it precludes the possibility of integrating the equations of motion numerically in time.

Many attempts have been made, particularly in the 1950s and 1960s, to overcome this limitation. A model that can be applied in the time domain was introduced by Voigt and studied further by Biot[11] and then by Weichert. It consists of a spring, with in parallel a large number of spring and damper system in series (Fig. 3.4f).

If there are n dampers, like in the case seen for non-viscous damping, the system has $m + 2$ degrees of freedom: The displacements of points A and C and of the m points B_i.

Consider the simplified case with $m = 1$ (Kelvin's model, Fig. 3.4e). The dynamic stiffness matrix of the system can be obtained by adding the stiffness of the spring k_0 with that of the parallel of spring k_1 and damper

---

[11]M.A. Biot, *Linear Thermodynamics and the Mechanics of solids*, Proc. Third U.S. National Congress of Applied Mechanics, 1958, p.1, T.K. Caughey, *Vibration of Dynamic Systems with Linear Hysteretic Damping, Linear Theory*, Proceedings of the Fourth US National Congress of Applied Mechanics, 1962, pp. 87–97. In his paper Biot refers to the Maxwell–Weichert model as the Voigt model.

c_1 obtained from Eq. (3.25):

$$k_{dyn} = k_0 + \frac{\omega^2 c_1^2 k_1}{k_1^2 + \omega^2 c_1^2} + i \frac{\omega c_1 k_1^2}{k_1^2 + \omega^2 c_1^2} . \qquad (3.87)$$

The complex stiffness of the system of Fig. 3.4a is thus

$$k_{dyn} = k_0 + \sum_{i=1}^{m} \frac{\omega^2 c_i^2 k_i}{k_i^2 + \omega^2 c_i^2} + i \sum_{i=1}^{m} \frac{\omega c_i k_i^2}{k_i^2 + \omega^2 c_i^2} . \qquad (3.88)$$

Caughey, in the mentioned paper based on the Maxwell–Weichert model of Fig. 3.4f, introduced a ratio

$$\beta_i = \frac{k_i}{c_i} \qquad (3.89)$$

between the stiffness and the damping of the ith branch.[12] Its dimensions are $1/s$ and then has the same dimensions of a frequency. Equation (3.88) thus reduces to

$$k_{dyn} = k_0 + \sum_{i=1}^{m} k_i \left(\frac{\omega^2}{\beta_i^2 + \omega^2} + i \frac{\omega \beta_i}{\beta_i^2 + \omega^2} \right) . \qquad (3.90)$$

By allowing the number of dampers to tend to infinity, the stiffness k_i becomes a function of a parameter identifying the infinity of infinitesimal dampers. Biot proposed to use β as a parameter and to assume that function $k(\beta)$ is

$$k(\beta) = k_0 \frac{g}{\beta} , \qquad (3.91)$$

where g is a constant linked with the damping of the system.

The sums in Eq. (3.88) are thus transformed into integrals

$$k_{dyn} = k_0 \left[1 + g\omega^2 \int_{\epsilon}^{\infty} \frac{d\beta}{\beta(\beta^2 + \omega^2)} + ig\omega \int_{\epsilon}^{\infty} \frac{d\beta}{(\beta^2 + \omega^2)} \right] , \qquad (3.92)$$

where ϵ is the minimum value taken by β.

By integrating and assuming that g is linked to the loss factor by the relationship

$$g = \eta \frac{2}{\pi} ,$$

it follows that

$$k_{dyn} = k_0 \left\{ 1 + \eta \frac{1}{\pi} \ln \left[1 + \frac{\omega^2}{\epsilon^2} \right] + i\eta \left[1 - \frac{2}{\pi} artg \left(\frac{\epsilon}{\omega} \right) \right] \right\} . \qquad (3.93)$$

---

[12]Ratio β_i coincides with the exponent μ_i introduced in the nonviscous damping model.

The complex stiffness of an hysteretic damper should be independent from the frequency, while that expressed by Eq. (3.93) is not. However, in a suitably chosen frequency range, the approximation is quite good. The ratio between the complex stiffness obtained from Eq. (3.93) and that obtained from the hysteretic damping model is shown in Fig. 3.6 as a function of the nondimensional frequency ω/ϵ.

Biot suggested that the frequency should be larger than 10ϵ, i.e., that the nondimensional frequency ω/ϵ should be larger than 10. From the figure it is clear that the error on the imaginary part is smaller than 6% and reduces quickly with increasing frequency. The error on the real part depends on the value of the loss factor: if the latter is small enough an error smaller than a few percent is obtained in a wide frequency range.

Remark 3.8 *The Maxwell–Weichert model is quite satisfactory, considering also that*

(a)   the hysteretic damping model is at any rate an approximation,

(b)   the hysteretic damping model does not hold at low frequency (ϵ must then be chosen accordingly)

(c)   the hysteretic damping model in the form of Eq. (3.56) holds only for low values of the loss factor η.

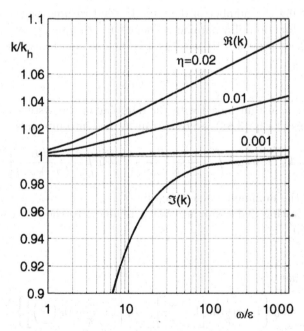

FIGURE 3.6. Ratio between the real and the imaginary parts of the complex stiffness k obtained from Eq. (3.93) and the ones obtained from the hysteretic damping model k_h as a function of the nondimensional frequency ω/ϵ.

Remark 3.9 *The Maxwell–Weichert model is not linked with the frequency-domain formulation and can be used also in time-domain equations. It introduces however a large (theoretical infinite) number of additional degrees of freedom, the displacements of points B_i.*

3.6.2 Practical Maxwell–Weichert model

The Maxwell–Weichert model seen in the previous section is just a theoretical model and cannot be directly applied to numerical integration of the equations of motion, owing to the infinity of spring–damper systems it includes. It may be approximated by using a finite (small) number m of dampers.

The contribution of each damper to the imaginary part of the complex stiffness is expressed by Eq. (3.90) and peaks at a frequency (Eq. (3.29))

$$\omega = \frac{k_i}{c_i} = \beta_i \ . \tag{3.94}$$

It is then possible to identify a number m of frequencies ω_i and stating that at each one of them one of the dampers works at its maximum damping conditions $(\beta_i = \omega_i)$ and that the sum of all terms at that speed is equal to ηk:

$$\sum_{i=1}^{n} k_i \frac{\beta_j \beta_i}{\beta_i^2 + \beta_j^2} = \eta k \ \text{ for } j = 1, \ ..., \ m \ . \tag{3.95}$$

A set of n equations allowing to compute the various k_i is thus obtained

$$\mathbf{Ak} = \eta k \mathbf{e} \ , \tag{3.96}$$

where

$$A_{ij} = \frac{\beta_j \beta_i}{\beta_i^2 + \beta_j^2} \ , \tag{3.97}$$

$\mathbf{k}$ is the vector containing the unknowns k_i and $\mathbf{e}$ is a suitable unit vector.

Since the various β_i are known, all the parameters of the system are known once that the values of k_i have been obtained.

The simplest model is that with a single damper ($m = 1$, Fig. 3.4e): it adds just a single degree of freedom to the system and is made by two springs and one damper. Matrix $\mathbf{A}$ reduces to a number:

$$\mathbf{A} = 0.5 \ . \tag{3.98}$$

If the damper is tuned at the reference frequency ω_r, it follows

$$k_1 = 2\eta k_0 \ , \quad c_1 = 2\frac{\eta k_0}{\omega_r} \ . \tag{3.99}$$

A comparison between the equivalent damping computed using Eq. (3.64) and that computed using the present model is reported in Fig. 3.7a and b.

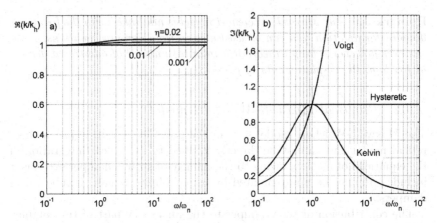

FIGURE 3.7. Maxwell–Weichert damper with $n = 1$ (Kelvin model). Comparison between the equivalent damping computed using Eq. (3.64), curve labeled 'Voight', and that computed using the present model, curve labeled 'Kelvin'. The real part depends also on the value of η: The curves for $\eta = 0.001, 0.01$, and 0.02 are reported.

The figure shows that the real part of the complex stiffness, that in the model of Eq. (3.64) is the same as that for hysteretic damping, is now slightly increasing with the frequency. How much it increases depends on the value of the loss factor and three curves for $\eta = 0.001, 0.01$, and 0.02 are reported. The imaginary part coincides with that of the hysteretic damping at the reference frequency. Both below and above the Maxwell–Weichert model gives a better approximation than the 'parallel' (Voigt's) model of Fig. 3.4d. Above all, it prevents the imaginary part from rising drastically with increasing frequency. In both cases, the behavior at very low frequency is better, since it prevents the equivalent damping from growing without bounds at decreasing frequency.

To obtain a better approximation, a larger number of spring–damper series can be used. Assume that the frequencies ω_i, with $i = 1, ..., m$ at which the dampers are tuned are in geometric progression:

$$\omega_{i+1} = a^i \omega_1 , \quad \text{for } i = 1, ..., m - 1 . \tag{3.100}$$

Since $\beta_i = \omega_i$, the elements of matrix $\mathbf{A}$ are

$$A_{ij} = \frac{a^{(j-i)}}{1 + a^{2(j-i)}} . \tag{3.101}$$

If a is large enough (typically, it can be assumed $a = 10$),

$$a^{2j} >> a_1^{2i} , \quad \text{for } j > i. \tag{3.102}$$

In this case,

$$A_{ij} = \frac{1}{2} \quad \text{for } j = i , \tag{3.103}$$

$$A_{ij} \approx a^{i-j} \quad \text{for } j > i. \tag{3.104}$$

The values of the stiffness and the damping coefficients of the various dampers are thus

$$k_i = \gamma_i \eta k , \quad c_i = \frac{\gamma_i \eta k}{\omega_1 a^{(i-1)}} ,$$

where γ_i are the solutions of the equation

$$\mathbf{A}\gamma = \mathbf{e} \tag{3.105}$$

and are functions only of ratio a. The values for $a = 10$, computed using the exact values for $\mathbf{A}$, are

$n = 1$	2								
$n = 2$	1.6694	1.6694							
$n = 3$	1.7035	1.3254	1.7035						
$n = 4$	1.7001	1.3600	1.3600	1.7001					
$n = 5$	1.7004	1.3566	1.3947	1.3566	1.7004				
$n = 6$	1.7004	1.3569	1.3913	1.3913	1.3569	1.7004			
$n = 7$	1.7004	1.3569	1.3916	1.3878	1.3916	1.3569	1.7004		
$n = 8$	1.7004	1.3569	1.3916	1.3881	1.3881	1.3916	1.3569	1.7004	
$n = 9$	1.7004	1.3569	1.3916	1.3881	1.3885	1.3881	1.3916	1.3569	1.7004
$n = 10$	1.7004	1.3569	1.3916	1.3881	1.3885	1.3885	1.3881	1.3916	1.3569	1.7004

If the range in which the Maxwell–Weichert damper is tuned is centered (in a logarithmic scale) on the reference frequency ω_r, the value of ω_1 is

$$\omega_1 = \omega_r a^{\frac{1-m}{2}} . \tag{3.106}$$

To evaluate the precision that can be obtained using this approach, consider the case with $m = 3$ and $a = 10$.

The results are reported in Fig. 3.8. From the figure it is clear that the imaginary part of the complex frequency is almost constant, and close to that obtained for hysteretic damping, in a range of frequencies spanning from less than 0.1 to more than 10 times the reference frequency, to drop out outside this range. The real part is close to that characterizing hysteretic damping, with an error growing with growing η.

Remark 3.10 *The model here shown coincides with that previously described for nonviscous damping. It allows to write equations of motion in the time domain starting from the hysteretic damping formulation that are a very good approximation in a wide frequency range, at the cost of introducing a number of additional degrees of freedom.*

Remark 3.11 *The larger the number of additional degrees of freedom, the wider is this frequency range or the more precisely the model simulates hysteretic damping.*

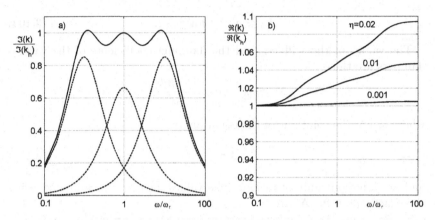

FIGURE 3.8. Ratio between the imaginary (**a**) and the real (**b**) parts of the complex stiffness k obtained from a simplified Maxwell–Weichert model with three dampers and the ones obtained from the hysteretic damping model k_h as a function of the frequency ω. *Full lines*: complex stiffness of the system and *dashed line*: imaginary part of the stiffness of each one of the branches.

3.7   Systems with frequency-dependent parameters

A complex viscous damping

$$c^* = c' + ic''$$ (3.107)

can be introduced in the same way the complex stiffness was defined. The real part c' is coincident with the damping coefficient and the imaginary part c'' is actually a stiffness in the sense that it does not involve any dissipation of energy. As an equivalent damping was defined for the imaginary part of the complex stiffness, an equivalent stiffness can be defined for the imaginary part of the complex damping:

$$k_{eq} = -\omega c''.$$

Consequently, the complex damping model allows to introduce a stiffness that grows linearly with increasing frequency into the linear equation of motion (if c'' is expressed by a negative number).

More generally, it is possible to define coefficients c and k which are general functions of the frequency. However, as already stated for the complex stiffness approach, a time-domain equation of motion like

$$m\ddot{x} + c(\omega)\dot{x} + k(\omega)x = F(t)$$ (3.108)

is conceptually inconsistent, since it is written partly in the time domain, with the time histories $x(t)$ (and its time derivatives) and $F(t)$, and partly in the frequency domain, because the laws $c(\omega)$ and $k(\omega)$ enter explicitly the equation. The frequency ω is, however, defined only when the time history $x(t)$ is harmonic. As a consequence, equation (3.108) should not be used.

The corresponding frequency-domain relationship

$$\left[-m\omega^2 + ic(\omega)\omega + k(\omega) \right] x_0 = f_0 , \qquad (3.109)$$

on the contrary, is correct and useful for the study of harmonic motion. For a more detailed discussion of this issue, see, for example, the well-known paper by S.H. Crandall.[13]

A case in which the dependence of the elastic and damping characteristics of the system on the frequency is very important is that of structural members made of elastomeric materials. The in-phase and in-quadrature stiffness k' and k'' of an elastomeric element are reported as functions of the frequency in Fig. 3.9. The loss factor

$$\eta = \tan(\Phi)$$

has also been plotted in the figure. Three regions are usually defined: At low frequency (rubbery region), the material shows a very low stiffness; at

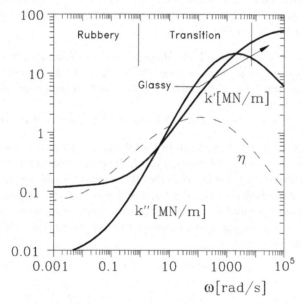

FIGURE 3.9. In-phase and in-quadrature stiffness and loss factor of an elastomeric spring as a function of the frequency.

---

[13]S.H.Crandall, "The role of damping in vibration theory", *J. of Sound and Vibration*, 11(1), (1970), 3–18.

high frequency (glassy region), its stiffness is substantially higher. Between the two regions there is a zone (transition region) in which the loss factor has a maximum.

The stiffness and damping of elastomeric materials is also strongly influenced by temperature, and the effects of an increase of frequency or a decrease of temperature are so similar that it is possible to obtain the curves related to changes in temperature at constant frequency from those related to frequency at constant temperature and vice versa. Note that in the case of Fig. 3.9, the value of the loss factor is quite high, at least in the transition region, and no small-damping assumption can be made.

3.8   Exercises

Exercise 3.1 *Consider the system of Fig. 1.8, already studied in Exercises 1.2, 2.1 and 2.3. Write the transfer function of the system, adding dampers c_1, c_2, c_3, and c_4 in parallel to all springs.*

Numerical data: $m_1 = 10$, $m_2 = 5$, $k_1 = 10$, $k_2 = 8$, $k_3 = 4$, $k_4 = 5$, $c_1 = 2$, $c_2 = 1$, $c_3 = 1$, $c_4 = 1.5$.

Exercise 3.2 *Plot the dynamic compliance of the system sketched in Fig. 1.4 and already studied in Example 1.2 and Exercise 2.4, with the values of the damping coefficients given in Example 3.1.*

Exercise 3.3 *Consider the beam already studied in Exercise 2.2, and assume that the loss factor of the material is $\eta = 0.01$. Compute the dynamic compliance of the system. Since it is a complex number, plot its amplitude and phase as functions of the frequency.*

Data: $l = 1$ m, $d_i = 60$ mm, $d_o = 80$ mm, $m = 30$ kg, $E = 2.1 \times 10^{11}$ N/m^2.

Exercise 3.4 *Repeat the computation of the previous exercise, by substituting the structural damping model with a Maxwell–Weichert model with three dampers. Chose the parameters of the model so that the two model yield the same results at the frequency at which the frequency response has a maximum and at frequencies equal to 0.1 and 10 times that value. Compare the results with those of the previous exercise. Plot also the real and imaginary parts of the dynamic stiffness of the beam without mass.*

4

Free Vibration of Conservative Systems

The free motion of undamped vibrating systems is studied. The properties of the modes of free vibration of multi-degrees-of-freedom systems are studied and it is shown that they allow to perform a coordinate transformation uncoupling the equations of motion.

4.1 Systems with a single degree of freedom

The solution of Eq. (1.3) (or of Eq. (1.4)) can be obtained by adding the *complementary function*, i.e., the general solution of the homogeneous equation, to a *particular integral* of the complete equation. The first describes the behavior of the system when no external excitation acts on it (*free behavior*) and, consequently, it is influenced only by its internal parameters. The homogeneous equation of motion is a second-order autonomous differential equation because the independent variable, time, does not appear explicitly.

Remark 4.1 *When studying the free behavior, it is immaterial whether Eq. (1.3) or (1.4) is used, because their homogeneous parts are identical.*

Remark 4.2 *Contrary to the complementary function, the particular integral describes the motion under the effect of an external excitation and thus is influenced by both the characteristics of the system and the time history of the excitation.*

The homogeneous equation of motion of a conservative system with a single degree of freedom is thus

$$m\ddot{x} + kx = 0 .$$ (4.1)

Its solution is harmonic in time, and thus can be expressed by either a sine or cosine function or an exponential function with imaginary exponent, as discussed in detail in Section 2.2.1.

Here a solution of the type

$$x = x_0 e^{st}$$ (4.2)

is assumed, which amounts to searching for a solution in the Laplace domain.

The acceleration is easily obtained from the time history (4.2):

$$\ddot{x} = x_0 s^2 e^{st} .$$

By introducing the solution (4.2) into the equation of motion, the latter becomes

$$x_0 \left(ms^2 + k \right) e^{st} = 0 .$$ (4.3)

Equation (4.3) holds for any value of time and hence in particular when $e^{st} \neq 0$.[1] By dividing Eq. (4.3) by e^{st}, the following algebraic equation is obtained

$$x_0 (ms^2 + k) = 0 .$$ (4.4)

The condition for the existence of a solution other than the trivial solution

$$x_0 = 0$$

leads to the following characteristic equation

$$ms^2 + k = 0 ,$$ (4.5)

whose two solutions, s_1 and s_2, are

$$s_{1,2} = \pm i\sqrt{\frac{k}{m}} = \pm i\omega_n .$$ (4.6)

The two solutions of the characteristic equation are imaginary, yielding a harmonic oscillation whose circular frequency is the natural frequency ω_n of the system

$$\omega_n = \sqrt{\frac{k}{m}} .$$ (4.7)

---

[1]Since s is in general a complex number, there may be an infinity of values of t for which $e^{st} = 0$.

The complementary function is the sum of two terms of the type shown in Eq. (4.2) with two constants, x_{01} and x_{02}, that depend on the initial conditions and are usually expressed by complex numbers

$$x = x_{01}e^{s_1 t} + x_{02}e^{s_2 t} = \left\{ \left[\Re(x_{01} + x_{02}) + i\Im(x_{01} + x_{02}) \right] \cos(\omega_n t) + \right.$$
$$\left. + \left[\Im(x_{02} - x_{01}) + i\Re(x_{01} - x_{02}) \right] \sin(\omega_n t) \right\} .$$

(4.8)

Remark 4.3 *As already stated in Chapter 2, the displacement x is a real quantity and then the two constants, x_{01} and x_{02}, must be complex conjugate. Equation (4.8) thus reduces to*

$$x = 2\Re(x_{01}) \cos(\omega_n t) - 2\Im(x_{01}) \sin(\omega_n t) .$$

(4.9)

If at time $t = 0$ the position $x(0)$ and the velocity $\dot{x}(0)$ are known, the values of the real and imaginary parts of constant x_{01} are

$$\Re(x_{01}) = \frac{x(0)}{2}$$
$$\Im(x_{01}) = -\frac{1}{2\omega_n}\dot{x}(0) .$$

(4.10)

The equation describing the oscillations of the system is then

$$x = \left[x(0) \cos(\omega_n t) + \frac{\dot{x}(0)}{\omega_n} \sin(\omega_n t) \right] .$$

(4.11)

The trajectories of the free oscillations of a conservative linear system in the state plane are circles, or ellipses, depending on the scales used for displacements and velocities (Fig. 4.1).

Example 4.1 *An instrument whose mass is 20 kg must be mounted on a space vehicle through a cantilever arm of annular cross-section made of light alloy (Young's modulus $E = 72 \times 10^9$ N/m², density $\rho = 2,800$ kg/m³), 600 mm long. Choose the dimensions of the cross-section in such a way that the first natural frequency is higher than 50 Hz.*
If the mass of the beam is neglected and a model with a single degree of freedom is used, in order to obtain a natural frequency higher than 50 Hz (314 rad/s) the stiffness of the beam must be

$$k \geq m\omega_n^2 = 1.97 \times 10^6 \ N/m .$$

(4.12)

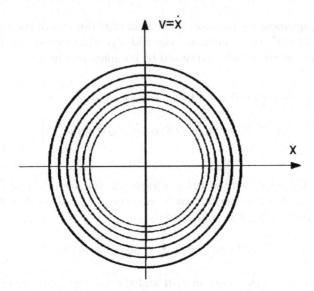

FIGURE 4.1. State-space trajectories for a system with a single degree of freedom. The different trajectories correspond to different initial conditions.

The arm can be modeled as a cantilever beam clamped at one end and loaded by the inertia force of the instrument at the other end. The well-known formula for its stiffness is

$$k = \frac{3EI}{l^3},$$

(4.13)

where l and I are the length of the beam and the area moment of inertia of the cross-section, respectively.
The minimum value of the moment of inertia I can be easily computed:

$$I = \frac{kl^3}{3E} = 1.97 \times 10^{-6} \ m^4.$$

(4.14)

By using a beam with annular cross-section with inner and outer diameters of $100 \ mm$ and $110 \ mm$, respectively, the value of the moment of inertia is $I = 2.28 \times 10^{-6} \ m^4$, yielding the following values of the stiffness and natural frequency:

$$k = 2.28 \times 10^6 \ N/m, \quad \omega_n = 337 \ rad/s = 53.7 \ Hz.$$

(4.15)

The mass of the beam is 2.77 kg, which is only slightly larger than 1/10 of the concentrated mass at the free end. The single-degree-of-freedom model in which the mass of the beam has been neglected can then be used with confidence.

4.2   Systems with many degrees of freedom

The solution can be performed directly with reference to the configuration space. The homogeneous equation describing free motion is

$$\mathbf{M\ddot{x} + Kx = 0}.$$ (4.16)

Equation (4.16) is a set of linear homogeneous second-order differential equations. Such equations are coupled, because at least one of the matrices $\mathbf{M}$ or $\mathbf{K}$ is usually not diagonal. If matrix $\mathbf{M}$ is not diagonal, the system is said to have an *inertial coupling*; if $\mathbf{K}$ is not diagonal, the coupling is said to be *elastic*. Again, a Laplace domain solution similar to Eq. (4.2)

$$\mathbf{x = x_0}e^{st}$$ (4.17)

can be assumed, and an eigenproblem of the same type as Eq. (4.4) can be obtained. Because the system is undamped, all solutions s are imaginary and the use of a frequency-domain solution with the form

$$\mathbf{x = x_0}e^{i\omega t},$$ (4.18)

in which the frequency of oscillation ω is explicitly included, is expedient. Since the acceleration is

$$\mathbf{\ddot{x}} = -\omega^2 \mathbf{x_0}e^{i\omega t},$$ (4.19)

the following algebraic homogeneous equation is obtained:

$$\left(\mathbf{K} - \omega^2 \mathbf{M}\right)\mathbf{x_0} = 0.$$ (4.20)

The characteristic equation of the relevant eigenproblem can be obtained by noting that to obtain a solution different from the trivial solution $\mathbf{x_0 = 0}$, the determinant of the matrix of the coefficients must vanish:

$$\det\left(\mathbf{K} - \omega^2 \mathbf{M}\right) = 0.$$ (4.21)

This eigenproblem can be reduced in standard form in one of the two following ways:

$$\det\left(\mathbf{K}^{-1}\mathbf{M} - \frac{1}{\omega^2}\mathbf{I}\right) = 0, \qquad \det\left(\mathbf{M}^{-1}\mathbf{K} - \omega^2\mathbf{I}\right) = 0.$$ (4.22)

Both matrices $\mathbf{K}^{-1}\mathbf{M}$ and $\mathbf{M}^{-1}\mathbf{K}$ are often referred to as dynamic matrices and symbol $\mathbf{D}$ is used for them. They should not be confused with either the dynamic matrix $\mathcal{A}$ or the direct link matrix $\mathcal{D}$ defined with reference to the state space.

Equations (4.22) are algebraic equations of degree n in ω^2 (or in $1/\omega^2$) that yield the n values of the natural frequencies of the system.

Remark 4.4 *The dynamic matrices* **D** *are not symmetrical, even if both* **M** *and* **K** *are.*

Remark 4.5 *The solutions in terms of natural frequencies are actually $2n$, corresponding to $\pm\omega_{n_i}$. In the following pages only n solutions, corresponding to the n eigenvalues and eigenvectors of the eigenproblem in ω^2 (or in $1/\omega^2$), will be considered. If the solution in the state space is considered, $2n$ conjugate eigenvalues in s are found.*

The eigenvectors give the mode shapes, i.e., the amplitudes of oscillation of the various masses at the corresponding natural frequency. All eigenvalues in ω^2 (or in $1/\omega^2$) are real and positive; the natural frequencies then are real, and undamped oscillations of the system are obtained. Also, the eigenvectors $\mathbf{q}_i$ are real, which means that all masses move in phase, or with a phase lag of $180°$.

Because there are n eigenvectors, a square matrix, the matrix of the eigenvectors

$$\mathbf{\Phi} = [\mathbf{q}_1, \mathbf{q}_2, \dots, \mathbf{q}_n],$$

can be written. Each one of its columns is one of the eigenvectors.

The complete solution of the equation of motion can be transformed in the same way as Eq. (4.9):

$$\mathbf{x} = \sum_{i=1}^{n} \left[\Re(K_i^*)\mathbf{q}_i \cos(\omega_i t) - \Im(K_i^*)\mathbf{q}_i \sin(\omega_i t) \right], \qquad (4.23)$$

where the n complex constants K_i^* can be determined from the $2n$ initial conditions. If at time $t = 0$ the positions $\mathbf{x}_0$ and the velocities $\dot{\mathbf{x}}_0$ are known, it follows that

$$\Re\{K_i^*\} = \mathbf{\Phi}^{-1}\mathbf{x}_0 , \qquad \Im\{\omega_i K_i^*\} = -\mathbf{\Phi}^{-1}\dot{\mathbf{x}}_0 . \qquad (4.24)$$

Remark 4.6 *Since the system is conservative, free vibration does not decay in time, i.e., its amplitude remains constant. This is clearly just an academic result, since all real-world systems have some damping.*

The free motion of a vibrating system can be studied in the state space by assuming a solution of the type

$$\mathbf{z} = e^{\mathcal{A}t}\mathbf{z}_0 , \qquad (4.25)$$

where $\mathbf{z}_0$ is the state vector at time t_0 and $e^{\mathcal{A}t}$ is the so-called transition matrix at time t. It can be expressed by the series

$$e^{\mathcal{A}t} = \mathbf{I} + t\mathcal{A} + \frac{t^2}{2!}\mathcal{A}^2 + \frac{t^3}{3!}\mathcal{A}^3 + \cdots , \qquad (4.26)$$

which converges for any value of t. The computation of the transition matrix can become impractical for large-order systems, and the number of terms in the series (4.26) to be considered grows rapidly with increasing time t.

Example 4.2 *Compute the time history of the free motion of a single-degree-of-freedom conservative system through the series (4.26) for the transition matrix.*

For an undamped single-degree-of-freedom system, the state vector and the dynamic matrix are

$$\mathbf{z} = \left\{ \begin{array}{c} v \\ x \end{array} \right\}, \qquad A = \left[\begin{array}{cc} 0 & -\frac{k}{m} \\ 1 & 0 \end{array} \right] = \left[\begin{array}{cc} 0 & -\omega_n^2 \\ 1 & 0 \end{array} \right].$$

Equation (4.26) yields the transition matrix

$$e^{\left[\begin{array}{cc} 0 & -\omega_n^2 \\ 1 & 0 \end{array} \right] t} = \left[\begin{array}{cc} 1 & 0 \\ 0 & 1 \end{array} \right] + t \left[\begin{array}{cc} 0 & -\omega_n^2 \\ 1 & 0 \end{array} \right] + \frac{t^2}{2} \left[\begin{array}{cc} -\omega_n^2 & 0 \\ 0 & -\omega_n^2 \end{array} \right] +$$

$$+ \frac{t^3}{3!} \left[\begin{array}{cc} 0 & \omega_n^4 \\ -\omega_n^2 & 0 \end{array} \right] + \frac{t^4}{4!} \left[\begin{array}{cc} -\omega_n^4 & 0 \\ 0 & -\omega_n^4 \end{array} \right] + \cdots =$$

$$= \left[\begin{array}{cc} 1 + \frac{t^2}{2}\omega_n^2 + \frac{t^4}{4!}\omega_n^4 + \ldots & -\omega_n^2 \left(t\omega_n - \frac{t^3}{3!}\omega_n^3 + \ldots \right) \\ \frac{1}{\omega_n} \left(t\omega_n - \frac{t^3}{3!}\omega_n^3 + \ldots \right) & 1 + \frac{t^2}{2}\omega_n^2 + \frac{t^4}{4!}\omega_n^4 + \ldots \end{array} \right] =$$

$$= \left[\begin{array}{cc} \cos(\omega_n t) & -\omega_n \sin(\omega_n t) \\ \frac{1}{\omega_n} \sin(\omega_n t) & \cos(\omega_n t) \end{array} \right].$$

After obtaining the product $e^{At}\mathbf{z}_0$, the following time histories for the displacement and the velocity are obtained:

$$\begin{cases} v = v_0 \cos(\omega_n t) - \omega_n x_0 \sin(\omega_n t) \\ \\ x = \dfrac{v_0}{\omega_n} \sin(\omega_n t) + x_0 \cos(\omega_n t) . \end{cases}$$

This solution coincides with that expressed by Eq. (4.9).

4.3   Properties of the eigenvectors

Consider a linear natural system and refer to the space of the configurations. The eigenvectors are orthogonal with respect to both the stiffness and mass matrices. This propriety can be demonstrated simply by writing the dynamic equilibrium equation in harmonic oscillations for the ith mode:

$$\mathbf{K}\mathbf{q}_i = \omega_i^2 \mathbf{M}\mathbf{q}_i . \qquad (4.27)$$

Equation (4.27) can be premultiplied by the transpose of the jth eigenvector

$$\mathbf{q}_j^T \mathbf{K}\mathbf{q}_i = \omega_i^2 \mathbf{q}_j^T \mathbf{M}\mathbf{q}_i . \qquad (4.28)$$

Products $\mathbf{q}_j^T \mathbf{K} \mathbf{q}_i$ and $\mathbf{q}_j^T \mathbf{M} \mathbf{q}_i$ are scalar quantities.

The same can be done for the equation written for the jth mode and premultiplied by the transpose of the ith eigenvector:

$$\mathbf{q}_i^T \mathbf{K} \mathbf{q}_j = \omega_j^2 \mathbf{q}_i^T \mathbf{M} \mathbf{q}_j . \tag{4.29}$$

By subtracting Eq. (4.29) from Eq. (4.28) it follows that

$$\mathbf{q}_j^T \mathbf{K} \mathbf{q}_i - \mathbf{q}_i^T \mathbf{K} \mathbf{q}_j = \omega_i^2 \mathbf{q}_j^T \mathbf{M} \mathbf{q}_i - \omega_j^2 \, \mathbf{q}_i^T \mathbf{M} \mathbf{q}_j . \tag{4.30}$$

Remembering that, owing to the symmetry of matrices $\mathbf{K}$ and $\mathbf{M}$,

$$\mathbf{q}_j^T \mathbf{K} \mathbf{q}_i = \mathbf{q}_i^T \mathbf{K} \mathbf{q}_j$$

and

$$\mathbf{q}_j^T \mathbf{M} \mathbf{q}_i = \mathbf{q}_i^T \mathbf{M} \mathbf{q}_j ,$$

it follows that

$$\left(\omega_i^2 - \omega_j^2 \right) \mathbf{q}_j^T \mathbf{M} \mathbf{q}_i = 0 . \tag{4.31}$$

In the same way, it can be shown that

$$\left(\frac{1}{\omega_i^2} - \frac{1}{\omega_j^2} \right) \mathbf{q}_j^T \mathbf{K} \mathbf{q}_i = 0 . \tag{4.32}$$

From Eqs. (4.31) and (4.32), assuming that all natural frequencies are different from each other, it follows that, if $i \neq j$,

$$\mathbf{q}_i^T \mathbf{M} \mathbf{q}_j = 0 , \qquad \mathbf{q}_i^T \mathbf{K} \mathbf{q}_j = 0 , \tag{4.33}$$

which are the relationships defining the orthogonality properties of the eigenvectors with respect to the mass and stiffness matrices, respectively.

If $i = j$, the results of the same products are not zero:

$$\mathbf{q}_i^T \mathbf{M} \mathbf{q}_i = \overline{M_i} , \qquad \mathbf{q}_i^T \mathbf{K} \mathbf{q}_i = \overline{K_i} . \tag{4.34}$$

Constants $\overline{M_i}$ and $\overline{K_i}$ are the modal mass and modal stiffness of the ith mode, respectively. They are linked with the natural frequencies by the relationship

$$\omega_i = \sqrt{\frac{\overline{K_i}}{\overline{M_i}}} , \tag{4.35}$$

stating that the ith natural frequency coincides with the natural frequency of a system with a single degree of freedom whose mass is the ith modal mass and whose stiffness is the ith modal stiffness.

The modal mass matrix and the modal stiffness matrix can be obtained from the following relationships based on the matrix of the eigenvectors $\boldsymbol{\Phi}$:

$$\begin{cases} \boldsymbol{\Phi}^T \mathbf{M} \boldsymbol{\Phi} = \mathrm{diag}[\overline{M_i}] = \overline{\mathbf{M}}, \\[2mm] \boldsymbol{\Phi}^T \mathbf{K} \boldsymbol{\Phi} = \mathrm{diag}[\overline{K_i}] = \overline{\mathbf{K}}. \end{cases} \tag{4.36}$$

An interpretation of the modal stiffnesses and of the modal masses is straightforward: If the system is deformed following the ith mode shape, the potential energy is

$$\mathcal{U} = \frac{1}{2}\mathbf{q}_i^T \mathbf{K} \mathbf{q}_i = \frac{1}{2}\overline{K_i},$$

i.e., it is equal to half the modal stiffness. In a similar way, the kinetic energy the system stores at its maximum speed while vibrating following a mode shape is equal to the corresponding modal mass, apart from the constant $\omega_i^2/2$.

4.4   Uncoupling of the equations of motion

The matrix of the eigenvectors can be used to perform a coordinate transformation that is particularly useful:

$$\mathbf{x} = \boldsymbol{\Phi}\boldsymbol{\eta}, \qquad \boldsymbol{\eta} = \boldsymbol{\Phi}^{-1}\mathbf{x}. \tag{4.37}$$

This amounts to expressing the generic n-dimensional vector $\mathbf{x}$, which states the configuration of the system, as a linear combination of the eigenvectors using n coefficients of proportionality η_i. This is possible because the eigenvectors are linearly independent and define a reference frame in the space of the configurations of the system.

Remark 4.7 *It must be explicitly stated that the eigenvectors are orthogonal with respect to the mass and stiffness matrices (they are said to be m-orthogonal and k-orthogonal), but they are not orthogonal to each other. The product $\boldsymbol{\Phi}^T \boldsymbol{\Phi}$ does not yield a diagonal matrix, and the inverse $\boldsymbol{\Phi}^{-1}$ of matrix $\boldsymbol{\Phi}$ does not coincide with its transpose $\boldsymbol{\Phi}^T$.*

The eigenvectors are, however, orthogonal if the dynamic matrix is symmetrical, like when a system with a diagonal mass matrix has all masses equal to each other (i.e., the mass matrix is made by an identity matrix multiplied by a constant).

In the space of the configurations, the eigenvectors are n vectors that can be taken as a reference frame. However, as already stated, in general they are not orthogonal with respect to each other.

Remark 4.8 *Equation (4.37) is nothing else than a coordinate transformation in the space of the configurations, as the n values η_i are the n coordinates of the point representing the configuration of the system, with reference to the system of the eigenvectors. They are said to be principal, modal, or normal coordinates.*

Although the eigenvectors are not orthogonal to each other, it is possible to perform a coordinate transformation that yields a system whose dynamic matrix ($\mathbf{D}$, referred to the configuration space) is symmetrical and has orthogonal eigenvectors. By performing a Cholesky decomposition of the mass matrix

$$\mathbf{M} = \mathbf{L}\mathbf{L}^T , \tag{4.38}$$

where $\mathbf{L}$ is a lower triangular non-singular matrix, a new set of generalized coordinates $\mathbf{x}^*$ can be defined by the relationship

$$\mathbf{x}^* = \mathbf{L}^T \mathbf{x}. \tag{4.39}$$

By introducing the generalized coordinates (4.39) into the equation of motion and premultiplying it by $\mathbf{L}^{-1}$, the mass matrix transforms into an identity matrix, while the stiffness matrix and the force vector reduce to[2]

$$\mathbf{K}^* = \mathbf{L}^{-1}\mathbf{K}\mathbf{L}^{-T} , \qquad \mathbf{f}^* = \mathbf{L}^{-1}\mathbf{f} . \tag{4.40}$$

The system so obtained has a unit mass matrix, and its eigenvectors are orthogonal. The eigenvalues are not changed by the transformation, and the eigenvectors are obtained from those of the original system using a simple linear combination. The modal mass matrix can be shown to be an identity matrix, and the modal stiffness matrix is a diagonal matrix with all elements equal to the squares of the natural frequencies:

$$\overline{\mathbf{K}} = [\omega^2] .$$

If the modal coordinates are introduced into the equation of motion (1.7), it follows that

$$\mathbf{M}\mathbf{\Phi}\ddot{\eta} + \mathbf{K}\mathbf{\Phi}\eta = \mathbf{f} . \tag{4.41}$$

By premultiplying the equation so obtained by $\mathbf{\Phi}^T$,

$$\mathbf{\Phi}^T\mathbf{M}\mathbf{\Phi}\ddot{\eta} + \mathbf{\Phi}^T\mathbf{K}\mathbf{\Phi}\eta = \mathbf{\Phi}^T\mathbf{f} , \tag{4.42}$$

it follows that

$$\overline{\mathbf{M}}\ddot{\eta} + \overline{\mathbf{K}}\eta = \overline{\mathbf{f}} , \tag{4.43}$$

---

[2]Symbol $\mathbf{A}^{-T}$ is here used for the transpose of the inverse ($\mathbf{A}^{-T} = (\mathbf{A}^{-1})^T$).

where $\overline{\mathbf{M}}$ and $\overline{\mathbf{K}}$ are the modal mass and modal stiffness matrices, defined by Eq. (4.34), and $\overline{\mathbf{f}}(t)$ is the modal force vector:

$$\overline{f}_i(t) = \mathbf{q}_i^T \mathbf{f}(t) \,. \tag{4.44}$$

Since the modal matrices are diagonal, Eq. (4.43) is a set of n uncoupled second-order differential equations. Each of them is

$$\overline{M}_i \ddot{\eta}_i + \overline{K}_i \eta_i = \overline{f}_i \,,$$

and the system with n degrees of freedom is broken down into a set of n uncoupled systems, each with a single degree of freedom (Fig. 4.2).

The eigenvectors are the solutions of a linear set of homogeneous equations and, thus, are not unique: For each mode, an infinity of eigenvectors exists, all proportional to each other. Because the eigenvectors can be seen as a set of n vectors in the n-dimensional space providing a reference frame, the length of such vectors is not determined, but their directions are known. In other words, the scales of the axes are arbitrary.

There are many ways to normalize the eigenvectors. The simplest is by stating that the value of one particular element or of the largest one is

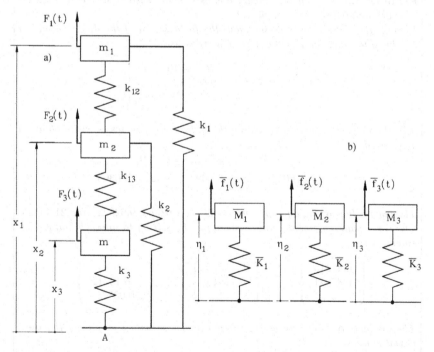

FIGURE 4.2. Modal uncoupling. The coupled system (a) and the uncoupled modal systems (b) are exactly equivalent.

set to unity. Each eigenvector can also be divided by its Euclidean norm, obtaining unit vectors in the space of the configurations.

Another way is to normalize the eigenvectors in such a way that the modal masses are equal to unity. This can be done by dividing each eigenvector by the square root of the corresponding modal mass. In the latter case, each modal stiffness coincides with the corresponding eigenvalue, i.e., with the square of the natural frequency. Equation (4.43) reduces to

$$\ddot{\boldsymbol{\eta}} + [\omega^2]\boldsymbol{\eta} = \overline{\mathbf{f}'} , \tag{4.45}$$

where

$$[\omega^2] = \text{diag}\{\omega_i^2\}$$

is the matrix of the eigenvalues and the modal forces $\overline{\mathbf{f}'}(t)$ are

$$\overline{f_i'} = \frac{\overline{f_i}}{\overline{M_i}} = \frac{\mathbf{q}_i^T \mathbf{f}}{\mathbf{q}_i \mathbf{M} \mathbf{q}_i} . \tag{4.46}$$

Example 4.3 *Perform the modal analysis of the system shown in Fig. 1.4 and already studied in Example 1.2.*
Because the mass matrix is diagonal, the formulation of the dynamic matrix involving the inversion of the mass matrix is used:

$$\mathbf{D} = \mathbf{M}^{-1}\mathbf{K} = \begin{bmatrix} 20 & -10 & 0 \\ -2.5 & 3.5 & -1 \\ 0 & -8 & 8 \end{bmatrix} .$$

The characteristic equation yielding the natural frequencies is easily obtained and solved:

$$\omega^6 - 31.5\omega^4 + 225\omega^2 - 200 = 0 ;$$

$$\omega_1 = 1.0166 ; \qquad \omega_2 = 3.0042 ; \qquad \omega_3 = 4.6305 .$$

The eigenvalues of the dynamic matrix $\mathbf{D}$ *are 1.03353, 9.02522, and 21.44125. The corresponding eigenvectors, normalized by setting to unity the largest element, are*

$$\begin{Bmatrix} 0.45913 \\ 0.87081 \\ 1 \end{Bmatrix} ; \qquad \begin{Bmatrix} 0.11677 \\ 0.12815 \\ -1 \end{Bmatrix} ; \qquad \begin{Bmatrix} 1 \\ -0.14412 \\ 0.08578 \end{Bmatrix} .$$

The products $\mathbf{q}_i^T \mathbf{M} \mathbf{q}_j$ *are easily computed. Those with* $i = j$ *yield the three modal masses*

$$\overline{M_1} = 3.74404 , \qquad \overline{M_2} = 0.57933 , \qquad \overline{M_3} = 1.08616 ,$$

and those with different subscripts yield the values -1.02×10^{-5}, 1.54×10^{-5}, and 4.09×10^{-6}. The values obtained are very small, compared to the modal masses; they represent the deviations from orthogonality (with respect to the mass matrix) due to computational approximations.

The eigenvectors can be normalized by dividing each one of them by the square root of the corresponding modal mass. The matrix of the eigenvectors is

$$\boldsymbol{\Phi} = \begin{bmatrix} 0.23728 & 0.15342 & 0.95925 \\ 0.45004 & 0.16837 & -0.13825 \\ 0.51681 & -1.31383 & 0.08228 \end{bmatrix} .$$

The eigenvectors are represented in the space of the configurations $\theta_1 \theta_2 \theta_3$ in Fig. 4.3a. They are clearly not orthogonal.

The modal mass matrix, then, is the identity matrix, while the modal stiffness matrix is a diagonal matrix containing the squares of the natural frequencies computed earlier:

$$\overline{\mathbf{M}} = \begin{bmatrix} 1 & 0 & 0 \\ 0 & 1 & 0 \\ 0 & 0 & 1 \end{bmatrix}, \qquad \overline{\mathbf{K}} = \begin{bmatrix} 1.03353 & 0 & 0 \\ 0 & 9.02522 & 0 \\ 0 & 0 & 21.44125 \end{bmatrix} .$$

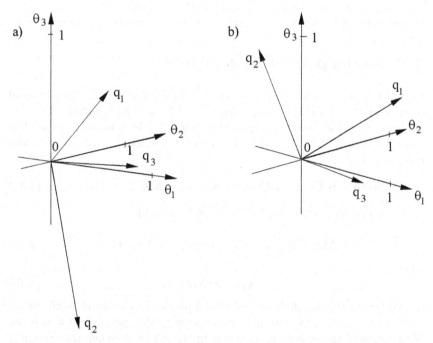

FIGURE 4.3. Eigenvectors represented in the space of the configurations $\theta_1 \theta_2 \theta_3$: (a) eigenvectors normalized in such a way that the modal masses have a unit value; (b) orthogonal eigenvectors obtained by premultiplying the vectors in (a) by matrix $\mathbf{L}^T$.

> *Note that, although the mass and stiffness matrices are the same as obtained by transformation (4.39), the eigenvectors are not orthogonal.*
> *In this case the Cholesky transformation of the mass matrix is very simple, because the mass matrix is diagonal. It yields*
>
> $$\mathbf{L} = \begin{bmatrix} 1 & 0 & 0 \\ 0 & 2 & 0 \\ 0 & 0 & 0.7071 \end{bmatrix}.$$
>
> *By using the transformation (4.40) of the stiffness matrix, it follows that*
>
> $$\mathbf{K}^* = \begin{bmatrix} 20 & -5 & 0 \\ -5 & 3.5 & -2.8284 \\ 0 & -2.8284 & 8 \end{bmatrix}.$$
>
> *In this case, because the mass matrix is an identity matrix, the dynamic matrix* $\mathbf{D} = \mathbf{M}^{-1}\mathbf{K}^*$ *coincides with the stiffness matrix* $\mathbf{K}^*$ *and is symmetrical. The eigenvectors*
>
> $$\mathbf{L}^T\mathbf{\Phi} = \begin{bmatrix} 0.23728 & -0.15342 & 0.95925 \\ 0.90008 & -0.33674 & -0.27650 \\ 0.36544 & 0.92901 & 0.05818 \end{bmatrix}$$
>
> *are orthogonal, as it can be easily verified (Fig. 4.3b).*

4.5   Modal participation factors

Consider a multi-degrees-of-freedom system excited by the translational motion of the constraints and chose the relative displacements between the various masses and the supporting points to describe its motion. The equation of motion is Eq. (1.11); it can be rewritten in terms of modal coordinates as

$$\mathbf{M}\mathbf{\Phi}\ddot{\eta} + \mathbf{K}\mathbf{\Phi}\eta = -\mathbf{M}\delta_x\ddot{x}_A - \mathbf{M}\delta_y\ddot{y}_A - \mathbf{M}\delta_z\ddot{z}_A + \mathbf{f}(t). \tag{4.47}$$

By premultiplying all terms by matrix $\mathbf{\Phi}^T$, it yields

$$\overline{\mathbf{M}}\ddot{\eta} + \overline{\mathbf{K}}\eta = -\mathbf{r}_x\ddot{x}_A - \mathbf{r}_y\ddot{y}_A - \mathbf{r}_z\ddot{z}_A + \overline{\mathbf{f}}(t), \tag{4.48}$$

where

$$\mathbf{r}_j = \mathbf{\Phi}^T\mathbf{M}\delta_j \tag{4.49}$$

is a vector containing the so-called modal participation factors in the direction j ($j = x, y, z$). Each term of this vector gives a measure of how much of the mass of the system participates in the ith mode when the system is excited by a motion of the supporting frame in the relevant direction.

This statement is easily justified. Consider a rigid-body translation of the system in the direction j ($j = x, y, z$) with velocity V. The vector containing the generalized velocities (the derivatives of the generalized coordinates) is

$$\mathbf{v} = V\boldsymbol{\delta}_j . \tag{4.50}$$

The kinetic energy of the system is thus

$$\mathcal{T} = \frac{1}{2}V^2\boldsymbol{\delta}_j^T\mathbf{M}\boldsymbol{\delta}_j . \tag{4.51}$$

Since the system is performing a rigid-body motion, the expression

$$m_T = \boldsymbol{\delta}_j^T\mathbf{M}\boldsymbol{\delta}_j \tag{4.52}$$

is nothing else than the total mass m_T of the system, or at least the mass that can be associated to a motion along the direction j.

Equation (4.49) can be solved in $\boldsymbol{\delta}_j$ by premultiplying both sides by $\mathbf{M}^{-1}\boldsymbol{\Phi}^{-T}$:

$$\boldsymbol{\delta}_j = \mathbf{M}^{-1}\boldsymbol{\Phi}^{-T}\mathbf{r}_j . \tag{4.53}$$

By remembering that the mass matrix $\mathbf{M}$ is symmetrical, the following expression for the total mass of the system is readily obtained:

$$m_T = \mathbf{r}_j^T\boldsymbol{\Phi}^{-1}\mathbf{M}^{-1}\mathbf{M}\mathbf{M}^{-1}\boldsymbol{\Phi}^{-T}\mathbf{r}_j = \mathbf{r}_j^T\boldsymbol{\Phi}^{-1}\mathbf{M}^{-1}\boldsymbol{\Phi}^{-T}\mathbf{r}_j . \tag{4.54}$$

Since

$$\boldsymbol{\Phi}^{-1}\mathbf{M}^{-1}\boldsymbol{\Phi}^{-T} = \left(\boldsymbol{\Phi}^T\mathbf{M}\boldsymbol{\Phi}\right)^{-1} = \overline{\mathbf{M}}^{-1} ,$$

the total mass of the system can be expressed as

$$m_T = \mathbf{r}_j^T\overline{\mathbf{M}}^{-1}\mathbf{r}_j . \tag{4.55}$$

The modal mass matrix is diagonal, and thus Eq. (4.55) can be written in the form

$$m_T = \sum_{i=1}^{n} \left(\frac{r_{j_i}^2}{\overline{M_i}}\right) . \tag{4.56}$$

Often ratios $r_{j_i}^2/\overline{M_i}$, expressed as percentages of mass m_T, are used instead of the modal participation factors. The higher the value of the modal participation factor in a certain direction, the more that mode is excited by a motion of the supports in that direction.

Example 4.4 *Compute the modal participation factors in x- and y-directions for the system of Fig.* 1.3 *(Example 3.3), assuming the following data:*

- *masses: $m_1 = 2$; $m_2 = 3$;*
- *stiffness: $k_1 = 3$; $k_2 = 2$; $k_3 = 5$; $k_4 = 4$; $k_5 = 7$;*
- *geometry: $\Delta x/l_{o5} = 1/\sqrt{2}$; $\Delta y/l_{o5} = 1/\sqrt{2}$ (the fifth spring is at $45°$).*

The relevant matrices are

$$\mathbf{M} = \begin{bmatrix} 2 & 0 & 0 & 0 \\ 0 & 2 & 0 & 0 \\ 0 & 0 & 3 & 0 \\ 0 & 0 & 0 & 3 \end{bmatrix}, \quad \mathbf{K} = \begin{bmatrix} 6.5 & 3.5 & -3.5 & -3.5 \\ 3.5 & 5.5 & -3.5 & -3.5 \\ -3.5 & -3.5 & 8.5 & 3.5 \\ -3.5 & -3.5 & 3.5 & 7.5 \end{bmatrix}.$$

The natural frequencies and the eigenvectors, normalized in such a way that the modal masses have a unit value, are

$$\omega_{n1} = 1.0624 , \quad \omega_{n2} = 1.1823 , \quad \omega_{n3} = 1.2699 , \quad \omega_{n4} = 2.6822 ,$$

$$\Phi = \begin{bmatrix} -0.2057 & -0.4202 & 0.3521 & 0.3964 \\ 0.5928 & 0.1079 & 0.0647 & 0.3644 \\ 0.0946 & 0.1064 & 0.4888 & -0.2722 \\ 0.2488 & -0.4433 & -0.0946 & -0.2568 \end{bmatrix}.$$

The modal participation factors are

$$\mathbf{r}_x = \begin{Bmatrix} -0.1275 \\ -0.5211 \\ 2.1706 \\ -0.0239 \end{Bmatrix} , \quad \mathbf{r}_y = \begin{Bmatrix} 1.9321 \\ -1.1142 \\ -0.1545 \\ -0.0415 \end{Bmatrix}.$$

It is easy to verify that, since the modal masses have a unit value, products

$$\mathbf{r}_x^T \mathbf{r}_x = \mathbf{r}_y^T \mathbf{r}_y = 5$$

are equal to the total mass of the system. The modal participation factors, normalized to the total mass of the system, are thus

Mode #	$\mathbf{r}_x^2/m_t$	$\mathbf{r}_y^2/m_t$
1	0.0032	0.7466
2	0.0543	0.2483
3	0.9423	0.0048
4	0.0001	0.0003

A motion of the supports in x-direction excites practically only the third mode, and only marginally the second one, while a motion in y-direction excites both the first and, to a lesser extent, the second modes. The fourth mode is almost not excited by any motion of the supporting frame.

Example 4.5 *Two identical pendulums connected by a spring*
Consider the two identical pendulums connected by a spring shown in Fig. 1.5 and studied in Example 1.4. Compute the natural frequencies of the linearized system and the time history of the free oscillations when the system is released from a standstill, with the first pendulum displaced at θ_0 and the second in the vertical position. The main data are m=1 kg, l=600 mm, k=2 N/m, and g=9.81 m/s².
By introducing the data, the equation of motion becomes

$$\begin{bmatrix} 0.6 & 0 \\ 0 & 0.6 \end{bmatrix} \begin{Bmatrix} \ddot{\theta}_1 \\ \ddot{\theta}_2 \end{Bmatrix} + \begin{bmatrix} 11.01 & -1.2 \\ -1.2 & 11.01 \end{bmatrix} \begin{Bmatrix} \theta_1 \\ \theta_2 \end{Bmatrix} = \begin{Bmatrix} 0 \\ 0 \end{Bmatrix}.$$

The eigenfrequencies and the corresponding eigenvectors can be easily computed:

$$\omega_1^2 = \frac{g}{l} = 16,35 \qquad\qquad \omega_1 = 4.04\, rad/s,$$
$$\omega_2^2 = \frac{mg + 2kl}{ml} = 20,35 \qquad \omega_2 = 4.51\, rad/s,$$

$$\mathbf{q}_1 = \begin{Bmatrix} 1 \\ 1 \end{Bmatrix}, \qquad \mathbf{q}_2 = \begin{Bmatrix} 1 \\ -1 \end{Bmatrix}.$$

In the first mode the two pendulums move together, without stretching the spring, with the same frequency they would have if they were not connected. This motion is not affected by the characteristics of the spring.
In the second mode the pendulums oscillate in opposition with a frequency affected by the characteristics of both the spring and the pendulums. If the spring is very soft (k/m much smaller than g/l) the two natural frequencies are very close to each other.
The initial conditions are $\theta_1 = \theta_0$ and $\theta_2 = \dot{\theta}_1 = \dot{\theta}_2 = 0$. The time histories of the free oscillations are then easily obtained:

$$\theta_1(t) = \frac{\theta_0}{2} \Big[\cos(\omega_1 t) + \cos(\omega_2 t) \Big], \qquad \theta_2(t) = \frac{\theta_0}{2} \Big[\cos(\omega_1 t) - \cos(\omega_2 t) \Big],$$

or, remembering some trigonometric identities,

$$\begin{cases} \theta_1(t) = \theta_0 \left[\cos\left(\dfrac{\omega_2 - \omega_1}{2}t\right) \cos\left(\dfrac{\omega_2 + \omega_1}{2}t\right) \right], \\[3mm] \theta_2(t) = \theta_0 \left[\sin\left(\dfrac{\omega_2 - \omega_1}{2}t\right) \sin\left(\dfrac{\omega_2 + \omega_1}{2}t\right) \right]. \end{cases}$$

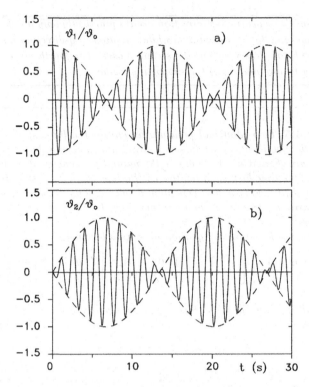

FIGURE 4.4. Two pendulums linked together by a spring: time history of the response.

The motion can then be considered an oscillation with a frequency equal to the average of the natural frequencies of the system $(\omega_2 + \omega_1)/2$ with an amplitude that is modulated with a frequency equal to $(\omega_2 - \omega_1)/2$, as clearly shown in Figs. 4.4a and b.

The system does not include any damping: The energy of the two pendulums is therefore conserved. The initial conditions are such that at time $t = 0$ all energy is concentrated in the first pendulum. The spring slowly transfers energy from the first to the second in such a way that the amplitude of the former decreases in time while the amplitude of the latter increases. This process goes on until the first pendulum stops and all energy is concentrated in the second one.

The initial situation is so reversed and the process of energy transfer starts again in the opposite direction. The frequency of the sine wave that modulates the amplitude is

$$(\omega_2 - \omega_1)/2 = 0.235 \ rad/s = \ 0.037 \ Hz \ ,$$

> *corresponding to a period of 26.72 s. The frequency of the beat is thus twice the frequency computed earlier, i.e., 0.072 Hz, corresponding to a period of 13.37 s. This means that the amplitude increases and decreases with a period that is half of that of the modulating sine wave.*
>
> *The occurrence of the beat is, however, linked with the initial conditions that must be able to excite both modes. If this does not happen, the oscillation is mono-harmonic, and no beat takes place.*

4.6   Structural modification

Many techniques aimed at computing the natural frequencies of a system after some of its characteristics have been modified without solving a new eigenproblem are listed under the general name of *structural modification.* Sometimes, the inverse problem is also considered: to compute the modifications needed to obtain required values of some natural frequencies.

Consider an undamped discrete system and introduce some small modifications in such a way that the mass and stiffness matrices can be written in the form $\mathbf{M} + \Delta\mathbf{M}$ and $\mathbf{K} + \Delta\mathbf{K}$. If the modifications introduced are small enough, the eigenvectors of the new system can be approximated by the eigenvectors of the old one and the ith modal mass, stiffness, and natural frequency of the new system can be approximated as

$$\overline{M}_{i_{mod}} = \mathbf{q}_i^T \left(\mathbf{M} + \Delta\mathbf{M}\right) \mathbf{q}_i = 1 + \mathbf{q}_i^T \Delta\mathbf{M}\mathbf{q}_i \,,$$

$$\overline{K}_{i_{mod}} = \mathbf{q}_i^T \left(\mathbf{K} + \Delta\mathbf{K}\right) \mathbf{q}_i = \omega_i^2 + \mathbf{q}_i^T \Delta\mathbf{K}\mathbf{q}_i \,, \tag{4.57}$$

$$\omega_{i_{mod}}^2 = \frac{\overline{K}_{i_{mod}}}{\overline{M}_{i_{mod}}} = \frac{\omega_i^2 + \mathbf{q}_i^T \Delta\mathbf{K}\mathbf{q}_i}{1 + \mathbf{q}_i^T \Delta\mathbf{M}\mathbf{q}_i} \,,$$

where the eigenvectors have been normalized in such a way that the modal masses of the original system have unit values. The modifications are assumed to be small. In this case, the series expressing the square root of $\omega_{i_{mod}}^2$ can be truncated after the first term, yielding

$$\omega_{i_{mod}} \approx \omega_i \left(1 + \frac{\mathbf{q}_i^T \Delta\mathbf{K}\mathbf{q}_i}{2\omega_i^2} - \frac{\mathbf{q}_i^T \Delta\mathbf{M}\mathbf{q}_i}{2}\right) \,. \tag{4.58}$$

Equation (4.58) can be used to compute the new value of the ith eigenvalue, knowing the modifications that have been introduced into the system. However, the inverse problem can also be solved. If matrices $\Delta\mathbf{M}$ and $\Delta\mathbf{K}$ are functions of a few unknown parameters, a suitable set of equations (4.58) can be used to find the values of the unknowns, which allow the solution for some stated values of the natural frequencies. The procedure described here is approximated and can be used only for small modifications.

Another procedure that can be used to compute the effect of stiffness modifications that do not need to be small is the following. The eigenproblem allowing the computation of the natural frequencies of the modified system expressed in terms of modal coordinates of the original system is

$$\ddot{\boldsymbol{\eta}} + [\omega_i^2]\boldsymbol{\eta} + \boldsymbol{\Phi}^T \Delta \mathbf{K} \boldsymbol{\Phi} \boldsymbol{\eta} = \mathbf{0} \,, \tag{4.59}$$

where matrix $\boldsymbol{\Phi}^T \Delta \mathbf{K} \boldsymbol{\Phi}$ is, in general, not diagonal, because the eigenvectors of the original system do not uncouple the equations of motion of the modified system.

If only one modification is introduced, matrix $\Delta \mathbf{K}$ can be expressed in the form $\Delta \mathbf{K} = \alpha \mathbf{u}\mathbf{u}^T$ where, if the modification consists of the addition of a spring linking degrees of freedom i and j, constant α is nothing other than the stiffness of the spring and all elements of vector $\mathbf{u}$ are zero except elements i and j, which are equal to 1 and -1, respectively. This is actually not a limitation: Because the procedure is not approximated, several modifications can be performed in sequence without losing precision.

The modal matrix linked with the modification can be expressed as

$$\boldsymbol{\Phi}^T \Delta \mathbf{K} \boldsymbol{\Phi} = \alpha \boldsymbol{\Phi}^T \mathbf{u}\mathbf{u}^T \boldsymbol{\Phi} = \alpha \bar{\mathbf{u}}\bar{\mathbf{u}}^T \,, \tag{4.60}$$

where, obviously, $\bar{\mathbf{u}} = \boldsymbol{\Phi}^T \mathbf{u}$.

The eigenproblem

$$(-\omega^2 \mathbf{I} + [\omega_i^2] + \alpha \bar{\mathbf{u}}\bar{\mathbf{u}}^T)\boldsymbol{\eta}_0 = \mathbf{0}$$

linked with Eq. (4.59) yields a set of n equations of the type

$$\left(-\omega^2 + \omega_i^2\right)\frac{\eta_{0_i}}{\bar{u}_i} = -\alpha \sum_{k=1}^{n} \bar{u}_k \eta_{0_k} \qquad (i = 1, 2, \ldots, n)\,. \tag{4.61}$$

Note that ω^2 equals the eigenvalues of the modified system, while ω_i^2 equals those of the original one. The term on the right-hand side of (4.61) is the same in all equations. It then follows that

$$\left(-\omega^2 + \omega_1^2\right)\frac{\eta_{0_1}}{\bar{u}_1} = \left(-\omega^2 + \omega_2^2\right)\frac{\eta_{0_2}}{\bar{u}_2} = \cdots = \left(-\omega^2 + \omega_n^2\right)\frac{\eta_{0_n}}{\bar{u}_n}\,. \tag{4.62}$$

The eigenvector $\boldsymbol{\eta}_0$ can thus be easily computed. By stating that the ith element is equal to unity, the remaining elements can be computed from Eq. (4.62):

$$\eta_{0_k} = \frac{\left(-\omega^2 + \omega_i^2\right)\bar{u}_k}{\left(-\omega^2 + \omega_k^2\right)\bar{u}_i} \qquad (k \neq i)\,. \tag{4.63}$$

By introducing the eigenvector expressed by Eq. (4.63) into Eq. (4.61) and remembering that η_{0_i} was assumed to be equal to one, it follows that

$$
\frac{1}{\alpha} = -\sum_{k=1}^{n} \frac{\bar{u}_k^2}{\left(-\omega^2 + \omega_k^2 \right)} = 0 .
\tag{4.64}
$$

Equation (4.64) can be regarded as a nonlinear equation in ω, yielding the eigenfrequencies of the modified system. The same equation can, however, be used to compute the value of α once a value for the natural frequency of the modified system has been stated. Note that although it is possible to obtain any given value of the eigenfrequency, it is, however, impossible to be sure that a given eigenfrequency is modified as needed.

4.7 Exercises

Exercise 4.1 *A ballistic pendulum of mass m and length l is struck by a bullet, having a mass m_b and velocity v, when it is at rest. Assuming that the bullet remains in the pendulum, compute the frequency and amplitude of the oscillations. Data: $m = 100$ kg; $m_b = 0.05$ kg, $v = 200$ m/s, and $l = 3$ m.*

Exercise 4.2 *Consider the undamped system of Fig. 1.8 (Exercise 1.2). Compute the natural frequencies assuming that $m_1 = 10$ kg, $m_2 = 5$ kg, $k_1 = 10$ kN/m, $k_2 = 8$ kN/m, $k_3 = 4$ kN/m, and $k_4 = 5$ kN/m.*

Exercise 4.3 *Consider the double pendulum of Fig. 1.9 (Exercise 1.3). Using the linearized equations of motion already obtained compute the natural frequencies and the mode shapes. Data: $m_1 = 2$ kg, $m_2 = 4$ kg, $l_1 = 600$ mm, $l = 400$ mm, and $g = 9.81$ m/s^2.*

Exercise 4.4 *Consider a system with 2 degrees of freedom, made by masses m_1 and m_2, connected by a spring k_{12} between each other and springs k_1 and k_2 to point A. Compute the natural frequencies, mode shapes, the modal masses, and stiffnesses. Plot the eigenvectors in the space of the configurations and show that the mode shapes are m-orthogonal and k-orthogonal. Compute the modal participation factor for an excitation due to the motion of point A. Data: $m_1 = 5$ kg, $m_2 = 10$ kg, $k_1 = k_2 = 2$ kN/m, and $k_{12} = 4$ kN/m.*

Exercise 4.5 *Consider the system of the previous exercise and modify the stiffness of the springs so that the first natural frequency is increased by 20%. Check that the second natural frequency is not changed much.*

Exercise 4.6 *The multifilar pendulum is one of the devices used for measuring moments of inertia (Fig. 4.5a). A trifilar pendulum consists of a tray hanging from three wires and the objects to be measured can be simply positioned on the*

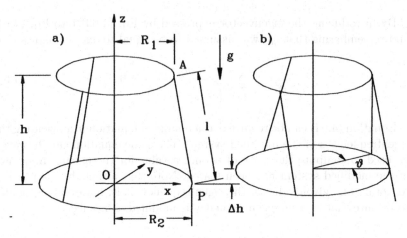

FIGURE 4.5. Trifilar suspension at rest (a) and displaced at angle θ from the static equilibrium position (b).

tray, taking care to put the center of mass of the object over the center of the latter. The period of the torsional oscillations of the tray and of the tray plus the object is measured in two subsequent tests. In Fig. 4.5b the device is shown during the motion, in a position displaced at angle θ from the position at rest.

Write the equation of motion of the system.

Write the equation allowing to compute the moment of inertia of the object being tested from the measurements and the geometrical characteristics of the pendulum (l, R_1, and R_2) and the masses m, m_t, and m_w of the object, the tray and of each one of the suspension wires.

5

Free Vibration of Damped Systems

Damping causes free vibration to decay in time. Moreover, if damping is high enough, some or even all the modes of free vibration may be nonoscillatory and free motion a simple return toward the equilibrium position. Except in a few cases, damping prevents from uncoupling the equations of motion when written in modal coordinates, at least in an exact way. However, if damping is small enough, uncoupling can still be performed in an approximate way, and the concept of modal damping can nonetheless be introduced.

5.1 Systems with a single degree of freedom–viscous damping

The solution of the homogeneous equation associated with the equation of motion of a damped system with a single degree of freedom (3.3) can be assumed to be of the same type already seen for the corresponding undamped (conservative) system

$$x = x_0 e^{st} . \tag{5.1}$$

By introducing solution (5.1) into the equation of motion, the following algebraic equation is obtained:

$$x_0(ms^2 + cs + k) = 0 . \tag{5.2}$$

The condition for the existence of a solution other than the trivial solution $x_0 = 0$ leads to the characteristic equation

$$ms^2 + cs + k = 0 , \qquad (5.3)$$

whose two solutions s_1 and s_2 are

$$s_{1,\,2} = \frac{-c \pm \sqrt{c^2 - 4mk}}{2m} . \qquad (5.4)$$

Generally speaking, s is expressed by a complex number; the time history (5.1) is thus

$$x = x_0 e^{\Re(s)t} e^{i\Im(s)t} . \qquad (5.5)$$

The solution (5.5) of the equation of motion can be regarded as a harmonic oscillation $x_0 e^{i\Im(s)t}$ of the same kind seen for the undamped system, multiplied by a factor $e^{\Re(s)t}$ that decreases in time, if $\Re(s)$ is negative.

The imaginary part of s is the frequency ω of the damped oscillations of the system, while its real part is often referred to as decay rate and symbol σ is used for it.

Remark 5.1 *In some cases the decay rate is defined as $\sigma = -\Re(s)$, so that σ is positive when the motion actually decays in time. The definition used here ($\sigma = \Re(s)$) would be better referred to as* growth rate.

Equation (5.5) can thus be written in the form

$$x = x_0 e^{\sigma t} \left[\cos\left(\omega t\right) + i \sin\left(\omega t\right) \right] . \qquad (5.6)$$

For stability $\sigma = \Re(s)$ must be negative.

If condition

$$c > 2\sqrt{km} \qquad (5.7)$$

is satisfied, the solutions of the characteristic equation (5.3) are real. The motion of the system is not oscillatory, but simply the combination of two terms that decrease monotonically in time, because both roots are negative. The value of the damping expressed by Eq. (5.7) is often referred to as *critical damping*, the highest value of c that allows the system to show an oscillatory free behavior. When condition (5.7) is satisfied, the system is said to be *overdamped*.

Introducing the damping ratio or relative damping ζ, i.e., the ratio between the value of the damping c and its critical value c_{cr}

$$\zeta = \frac{c}{c_{cr}} = \frac{c}{2\sqrt{km}} , \qquad (5.8)$$

the two values of the decay rate σ of an overdamped system ($\zeta > 1$) are

$$\sigma = \Re(s) = -\sqrt{\frac{k}{m}} \left(\zeta \pm \sqrt{\zeta^2 - 1} \right) . \qquad (5.9)$$

If the damping of the system is lower than the critical damping ($\zeta < 1$), the system is said to be *underdamped*. Equation (5.4) still holds but leads to a pair of complex-conjugate solutions for s. The system performs damped harmonic oscillations (Fig. 5.1) with circular frequency ω and decay rate σ

$$\begin{cases} \omega = \Im(s) = \pm\sqrt{\frac{k}{m}}\sqrt{1 - \zeta^2} = \pm\omega_n\sqrt{1 - \zeta^2}, \\[2mm] \sigma = \Re(s) = -\zeta\sqrt{\frac{k}{m}} = -\zeta\omega_n\ , \end{cases} \tag{5.10}$$

where ω_n is the natural frequency of the corresponding undamped system.

As shown in nondimensional form in Figs. 5.2b and c, the decay rate decreases linearly with increasing ζ, while the plot $\omega\left(\zeta\right)$ is an arc of a circle.

The complete expression of the complementary function with its integration constants is similar to that seen in Chapter 4 for the conservative system (Eqs. from (4.8) to (4.11)), with the difference that here a decay rate is present to account for the decrease in time of the amplitude of the free oscillations:

$$x = x_{01}e^{s_1t} + x_{02}e^{s_2t} = e^{\sigma t}\left\{\left[\Re(x_{01} + x_{02}) + i\Im(x_{01} + x_{02})\right]\cos(\omega t) + \right.$$
$$\left. +\left[\Im(x_{02} - x_{01}) + i\Re(x_{01} - x_{02})\right]\sin(\omega t)\right\}.$$

$$\tag{5.11}$$

Displacement x is a real quantity and then the two complex constants, x_{01} and x_{02}, must be conjugate. If at time $t = 0$ the position $x(0)$ and the velocity $\dot{x}(0)$ are known, the values of the real and imaginary parts of

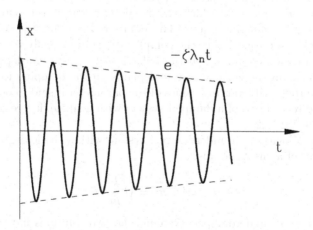

FIGURE 5.1. Time history of the damped oscillations of a lightly damped system.

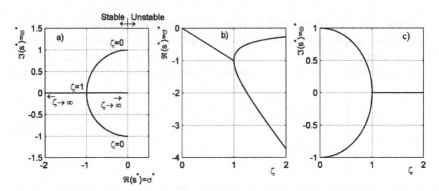

FIGURE 5.2. (a) Nondimensional roots locus for a damped system with a single degree of freedom ($s^* = s/\omega_n$). (b) and (c) Plots of the real and imaginary parts of s^* as functions of the damping ratio ζ.

constant x_{01} are

$$\Re(x_{01}) = \frac{x(0)}{2},$$
$$\Im(x_{01}) = \frac{-1}{2\omega_n\sqrt{1-\zeta^2}}\left[\dot{x}(0) + \zeta\omega_n x(0)\right]. \tag{5.12}$$

The equation describing the damped free oscillations of the system is then

$$x = e^{-\zeta\omega_n t}\left\{ x(0)\cos\left(\omega_n\sqrt{1-\zeta^2}\,t\right) + \right.$$
$$\left. +\frac{1}{\sqrt{1-\zeta^2}}\left[\frac{\dot{x}(0)}{\omega_n} + \zeta x(0)\right]\sin\left(\omega_n\sqrt{1-\zeta^2}\,t\right)\right\}. \tag{5.13}$$

The roots of the characteristic equation can be reported on the complex plane, i.e., on a plane in which the x-axis is taken as the real axis and the y-axis is the imaginary axis. The points representing the solutions of the characteristic equation in the complex plane are usually referred to as the *poles* of the system. When the behavior of the system depends on a parameter, as in the current case it depends on the damping ratio ζ, the plot of the roots with varying values of the parameter is said to be the *roots locus*. The roots locus of a damped system with a single degree of freedom is shown in Fig. 5.2a.

The roots locus of an underdamped system with a single degree of freedom is half of a circle, because

$$|s| = \sqrt{\omega^2 + \sigma^2} = \sqrt{\frac{k}{m}} = \omega_n. \tag{5.14}$$

In the case of underdamped systems, the two solutions have the same real part (decay rate) and imaginary parts equal in modulus but opposite in sign. For overdamped systems, on the contrary, the two solutions, which are real (lying on the real axis), are not coincident. One of them tends to infinity and the other tends to zero when damping tends to infinity.

In the case of critically damped systems ($\zeta = 1$), the two solutions are coincident: The two branches of the locus for underdamped systems meet on the real axis to separate again for overdamped systems. The point where the two branches meet to remain on the real axis is said to be a *break-in point*. When the opposite occurs and two branches lying on the real axis meet to depart from the axis, the point is said to be a *breakaway point*.

Solutions on the left of the complex plane, as in the case of systems with positive damping, are asymptotically stable, while solutions lying on the right part of the plane denote unstable behavior. Solutions located on the imaginary axis are stable but not asymptotically, as the oscillations are not damped, and the system, once set in motion, cannot reach an equilibrium condition although not being unstable.

Remark 5.2 *The systems studied in structural dynamics are usually lightly damped; consequently, factor*

$$e^{-\zeta \omega_n t}$$

decreases monotonically very slowly with time. The frequency of the oscillations is only slightly smaller than the natural frequency of the undamped system ω_n, owing to the presence of factor $\sqrt{1 - \zeta^2}$. In the case of lightly damped systems, this factor is almost equal to unity, and the frequency of the damped free oscillations is almost equal to that of the free oscillations of the undamped system.

Consider the part of the time history of the displacement close to a peak (Fig. 5.3a). The actual peak occurs in point A, which is very close to point B where the harmonic part of the solution $e^{i\omega t}$ has a unit value. The amplitudes at two subsequent peaks, approximated by two subsequent points B, spaced apart by one period of oscillations ($T = 2\pi/\omega$) are

$$x_k = e^{\sigma t_k} , \qquad x_{k+1} = e^{\sigma t_{k+1}} = e^{\sigma(t_k + 2\pi/\omega)} . \tag{5.15}$$

The ratio between the amplitudes of two subsequent peaks is then

$$\frac{x_k}{x_{k+1}} = e^{-2\pi\sigma/\omega}, \tag{5.16}$$

and is constant in time. Its natural logarithm, usually defined as logarithmic decrement, is

$$\delta = \ln\left(\frac{x_k}{x_{k+1}}\right) = 2\pi\frac{\zeta}{\sqrt{1 - \zeta^2}} \approx 2\pi\zeta . \tag{5.17}$$

The two expressions of the logarithmic decrement are reported in Fig. 5.3b.

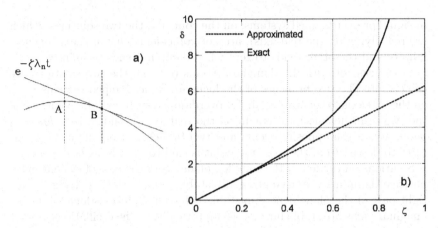

FIGURE 5.3. Damped oscillations of a lightly damped system. (a) Enlargement of the zone close to a peak; (b) logarithmic decrement as a function of ζ.

This expression was obtained by confusing points A and B in the figure, but this approximation is not needed and holds also for large values of ζ. The peaks can be obtained by differentiating the time history of Eq. (5.13) with respect to time and setting to zero the derivative. Since the initial conditions are immaterial, it is possible to state $x(0) = 0$, obtaining

$$\frac{dx}{dt} = e^{\sigma t}\frac{\dot{x}(0)}{\omega}\left[\sigma \sin\left(\omega t\right) - \omega \cos\left(\omega t\right)\right] = 0 \, , \qquad (5.18)$$

which yields

$$\tan\left(\omega t\right) = \frac{\omega}{\sigma} \, , \qquad (5.19)$$

i.e.,

$$\omega t = \operatorname{artg}\left(\frac{\omega}{\sigma} + i\pi\right) \quad \text{for } i = 0, \ 1, \ 2, \ .... \qquad (5.20)$$

The solutions with even values of i are maxima, those with odd values are minima.

The ratio between the amplitude at two subsequent peaks is thus

$$\frac{x_k}{x_{k+1}} = \frac{e^{\frac{\sigma}{\omega}\operatorname{artg}\left(\frac{\omega}{\sigma} + 2k\pi\right)}\frac{\dot{x}(0)}{\omega}\sin\left[\operatorname{artg}\left(\frac{\omega}{\sigma} + 2k\pi\right)\right]}{e^{\sigma\operatorname{artg}\left[\frac{\omega}{\sigma} + 2(k+1)\pi\right]}\frac{\dot{x}(0)}{\omega}\sin\left\{\operatorname{artg}\left[\frac{\omega}{\sigma} + 2\left(k+1\right)\pi\right]\right\}} = e^{-\frac{2\pi\sigma}{\omega}} \, ,$$

$$(5.21)$$

which coincides with the expression obtained earlier.

Remark 5.3 *The logarithmic decrement gives a measure of the damping of the system that is not too difficult to evaluate from the recording of the amplitude versus time (Fig. 5.1). If the first expression of Eq. (5.17) is used, it holds also for large values of ζ.*

Remark 5.4 *The logarithmic decrement is a constant of the system and can be measured in any part of the time history of the free vibrations. Different values measured during the decay of the free motion are usually a symptom of nonlinearity.*

The decay of the amplitude in n oscillation is

$$\frac{x_n}{x_0} = e^{-n\delta} = e^{-\frac{2n\pi\zeta}{\sqrt{1-\zeta^2}}}. \tag{5.22}$$

It is plotted in Fig. 5.4a versus ζ for various values of n.

The trajectories of the free oscillations of an undamped linear system in the state plane were found to be circles (or ellipses, depending on the scales used for displacements and velocities, Fig. 4.1) while those of a damped system are logarithmic spirals. They wind up in a clockwise direction toward the origin, which is a singular point (Fig. 5.4b).

Instead of using Eq. (5.1), the solution of the equation of motion could also be written in the form

$$x = x_0 e^{i\nu t}, \tag{5.23}$$

where $i\nu = s$. The real and imaginary parts of the complex frequency ν are the actual frequency of the motion and the decay rate (changed in sign), respectively,

$$\Re(\nu) = \Im(s) = \omega, \qquad \Im(\nu) = -\Re(s) = -\sigma. \tag{5.24}$$

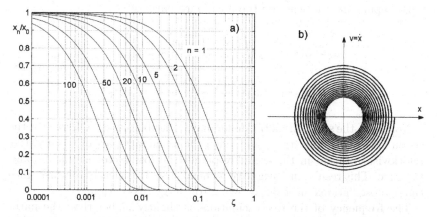

FIGURE 5.4. Damped system with a single degree of freedom. (a) Decrease of the amplitude after a number n of oscillations versus the damping ratio ζ. (b) State-space trajectory. Initial conditions: The system is displaced from the equilibrium position and then let free with zero initial velocity.

5.2   Systems with a single degree of freedom–hysteretic damping

As seen in Section 3.4.2, the dynamic stiffness of a system with structural damping can be obtained by simply introducing the complex stiffness in the expression of the dynamic stiffness of an undamped system

$$k_{dyn} = -m\omega^2 + k(1 + i\eta), \qquad (5.25)$$

or, by remembering that $s = i\omega$,

$$k_{dyn} = ms^2 + k(1 + i\eta). \qquad (5.26)$$

Remark 5.5 *Strictly speaking, this way of proceeding is inconsistent: hysteretic damping was defined with reference to harmonic motion, and the very concept of decay rate implies that the amplitude of the motion changes in time. Equation (5.26) holds only when ν is real (and equal to ω), and thus Eq. (5.26) should not be used for free damped oscillations, i.e., when s has a non-vanishing real part. However, systems with structural damping are usually very lightly damped and the decay rate is small enough, if compared with the frequency of oscillations, to use the concept of hysteretic damping also for damped free oscillations.*

The frequency and the decay rate of the free oscillations of the system can be obtained by equating the dynamic stiffness to zero

$$s = i\sqrt{\frac{k}{m}}\sqrt{1 + i\eta} = i\omega_n\sqrt{1 + i\eta}. \qquad (5.27)$$

By separating the real and the imaginary parts of s[1]

$$\omega = \Im(s) = \omega_n\sqrt{\frac{1 + \sqrt{1 + \eta^2}}{2}} \approx \omega_n,$$

$$\sigma = \Re(s) = -\omega_n\sqrt{\frac{-1 + \sqrt{1 + \eta^2}}{2}} \approx -\omega_n\frac{\eta}{2}. \qquad (5.28)$$

Note that the solution (5.27) should have a double sign ($\pm$). However, the strain always lags the stress, and when the imaginary part of s is negative (clockwise rotation in the complex plane) the loss factor should be also negative. This results in a negative value of σ, as it is clearly the case since the free oscillations must decay owing to energy dissipation.

The frequency of the free oscillations is slightly higher than the natural frequency of the undamped system. This result is different from that

---

[1] $\sqrt{a \pm ib} = \sqrt{\frac{\sqrt{a^2+b^2}+a}{2}} \pm i\sqrt{\frac{\sqrt{a^2+b^2}-a}{2}}.$

found for viscous damping but, also in this case, the frequency shift due to the presence of damping is negligible for lightly damped systems. The logarithmic decrement δ takes the value

$$\delta = \frac{\pi\eta}{\sqrt{1 + (\eta^2/4)}} \approx \pi\eta. \qquad (5.29)$$

5.3 Systems with a single degree of freedom – nonviscous damping

The homogeneous equation associated to Eq. (3.74) describes the free motion of a single degree of freedom spring, mass, damper system in which the damper has nonviscous characteristics.

The natural frequencies are readily computed by finding the eigenvalues of the dynamic matrix

$$\mathcal{A} = \mathbf{M}^{*-1}\mathcal{A}^* . \qquad (5.30)$$

If the nonviscous damper is modeled using m springs and dampers in series, the eigenvalues are $m + 2$. Two of them are either real or complex conjugate while the other m, mostly related to the motion of points B_i, are real. If the two former eigenvalues are complex the system is underdamped, if they are all real the system is overdamped.

To simplify matters, let $m = 1$. By assuming a solution of the type

$$x = x_0 e^{st} ,$$

the equation for free motion of the spring, mass, nonviscous damper system

$$m\ddot{x} + c\mu \int_{-\infty}^{t} e^{-\mu(t-\tau)} \dot{x}(\tau)\, d\tau + kx = 0 \qquad (5.31)$$

becomes

$$\left(ms^2 + c\mu\frac{s}{s+\mu} + k \right) x_0 = 0 . \qquad (5.32)$$

The characteristic equation is thus

$$ms^2 + c\mu\frac{s}{s+\mu} + k = 0 , \qquad (5.33)$$

or, introducing the nondimensional quantities

$$s^* = \frac{s}{\omega_n} = s\sqrt{\frac{m}{k}} , \quad \zeta = \frac{c}{2\sqrt{mk}} , \quad \beta = \frac{\omega_n}{\mu} = \frac{1}{\mu}\sqrt{\frac{k}{m}} ,$$

$$\beta s^{*3} + s^{*2} + (\beta + 2\zeta) s^* + 1 = 0 . \qquad (5.34)$$

The characteristic equation is a cubic equation. Its solutions can easily be computed in closed form, obtaining

$$\begin{cases} s_1^* = -\frac{1}{3\beta} + S + T \,, \\ s_{2,3}^* = -\frac{1}{3\beta} - \frac{1}{2}(S+T) \pm i\frac{\sqrt{3}}{2}(S-T) \,, \end{cases} \tag{5.35}$$

where

$$R = \frac{1}{27\beta^3}\left(9\,\beta\zeta - 9\,\beta^2 - 1\right),$$

$$D = \frac{1}{27\beta^4}\left[8\,\beta\,\zeta^3 + \zeta^2\left(12\,\beta^2 - 1\right) + \zeta\left(6\,\beta^3 - 10\,\beta\right) + \beta^4 + 2\,\beta^2 + 1\right],$$

$$S = \sqrt[3]{R + \sqrt{D}} \,, \quad T = \sqrt[3]{R - \sqrt{D}} \,.$$

This solution holds if $D > 0$ and yields a damped harmonic motion: the system is thus uderdamped.

The condition $D = 0$, i.e.,

$$8\,\beta\,\zeta^3 + \zeta^2\left(12\,\beta^2 - 1\right) + \zeta\left(6\,\beta^3 - 10\,\beta\right) + \beta^4 + 2\,\beta^2 + 1 = 0 \tag{5.36}$$

thus discriminates between under- and overdamped systems.

The behavior of the system depends on two parameters: ζ and β. The former is a nondimensional damping coefficient, the second one is a parameter stating its nonviscosity. If $\beta = 0$ ($\mu \to \infty$) the damping is viscous, and the usual condition $\zeta < 1$ is found for oscillatory free motion. The larger is β, the less viscous is damping.

Once a value of β is stated, Eq. (5.36) can be solved in ζ and a zone in which the system is underdamped can be identified in the parameter plane $\zeta\beta$.

Since it is a cubic equation in ζ, three solutions are found. However, only two are positive and need to be considered

$$\zeta_L = \frac{1}{24\beta}\left[1 - 12\,\beta^2 + 2\sqrt{1 + 216\beta^2}\cos\left(\frac{\theta_c + 4\pi}{3}\right)\right]$$

$$\zeta_U = \frac{1}{24\beta}\left[1 - 12\,\beta^2 + 2\sqrt{1 + 216\beta^2}\cos\left(\frac{\theta_c}{3}\right)\right] \,, \tag{5.37}$$

where

$$\theta_c = \arcos\left[\frac{1 - 5832\beta^4 - 540\beta^2}{\sqrt{\left(216\beta^2 + 1\right)^3}}\right] \,. \tag{5.38}$$

The two lines $\beta(\zeta)$ so obtained are reported in Fig. 5.5. The plot identifies two zones in the parameter plane $\zeta\beta$: The values included in zone A give way to an oscillatory behavior. On the contrary, the values of the parameters in zone B are characterized by nonoscillatory free behavior.

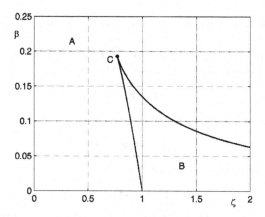

FIGURE 5.5. Free behavior of a simple system with nonviscous damping. The parameters in zone A correspond to an underdamped system, those in zone B to an overdamped system.

The coordinates of point C are

$$\begin{cases} \zeta_C = \frac{4}{3\sqrt{3}} = 0.7698 \\[2mm] \beta_C = \frac{1}{3\sqrt{3}} = 0.1925 \, . \end{cases} \qquad (5.39)$$

If the system is weakly nonviscous, its behavior is not much different from that of a viscously damped system: for low values of ζ it is underdamped while for high values it is overdamped.

With increasing β the critical damping decreases, but an underdamped behavior can be found also for very high values of ζ. The zone in which overdamped behavior is found decreases with increasing β, and when $\beta \geq \beta_c$ it disappears altogether.

5.4   Systems with many degrees of freedom

The solution of the homogeneous equation associated with the equation of motion (3.6) is of the type

$$\mathbf{x} = \mathbf{x}_0 e^{st} \, . \qquad (5.40)$$

By introducing the solution (5.40) into Eq. (3.6) and setting $\mathbf{f} = \mathbf{0}$, the following set of homogeneous linear algebraic equations is obtained:

$$\mathbf{x}_0(\mathbf{M}s^2 + \mathbf{C}s + \mathbf{K}) = \mathbf{0} \, . \qquad (5.41)$$

To obtain solutions other than the trivial solution $\mathbf{x}_0 = \mathbf{0}$, the determinant of the matrix of the coefficients must vanish. The resulting eigenproblem

$$\det(\mathbf{M}s^2 + \mathbf{C}s + \mathbf{K}) = 0 \qquad (5.42)$$

is not reduced in canonical form, but has the form of a so-called *lambda matrix*.[2]

To obtain an eigenproblem in canonical form it is possible to resort to the state space. Assuming a solution of the type

$$z = z_0 e^{st},$$ (5.43)

the homogeneous linear algebraic equations is

$$(\mathcal{A} - s\mathbf{I})z_0 = 0,$$ (5.44)

yielding the eigenproblem of order $2n$

$$\det(\mathcal{A} - s\mathbf{I}) = 0.$$ (5.45)

The eigenvalues s yield the frequencies of oscillation and the decay rates; the eigenvectors yield the complex mode shapes z_0. All considerations on the stability of the system seen in the previous section still hold, with the only difference that now there are n pairs of complex-conjugate solutions. If some of them are real, the corresponding modes are nonoscillatory; if they are imaginary, undamped oscillations may occur. The fact that the eigenvectors are complex[3] causes the time histories related to the various degrees of freedom to be out of phase.

5.5 Uncoupling the equations of motion: space of the configurations

The matrix $\boldsymbol{\Phi}$ of the eigenvectors of the conservative system obtained by neglecting the damping matrix $\mathbf{C}$ can be used to perform a modal transformation of the equation of motion of the damped system. By introducing the modal coordinates

$$\mathbf{x} = \boldsymbol{\Phi}\boldsymbol{\eta}$$ (5.46)

into the equation of motion (3.6) and by premultiplying all its terms by $\boldsymbol{\Phi}^T$, it follows that

$$\overline{\mathbf{M}}\ddot{\boldsymbol{\eta}} + \overline{\mathbf{C}}\dot{\boldsymbol{\eta}} + \overline{\mathbf{K}}\boldsymbol{\eta} = \overline{\mathbf{f}},$$ (5.47)

where $\overline{\mathbf{M}}$ and $\overline{\mathbf{K}}$ are the diagonal modal mass and modal stiffness matrices, respectively, defined by Eq. (4.34) and $\overline{\mathbf{f}}(t)$ is the modal force vector, already defined when dealing with the undamped system.

$\overline{\mathbf{C}}$ is the modal-damping matrix

$$\overline{\mathbf{C}} = \boldsymbol{\Phi}^T \mathbf{C} \boldsymbol{\Phi}.$$ (5.48)

---

[2]The term lambda matrix comes from the habit of using symbol λ for the unknown of the eigenproblem. Here the more modern practice of using symbol s is followed.

[3]An interesting discussion on the meaning of complex modes can be found in G.F. Lang, 'Demystifying complex modes', *Sound and Vibration*, January, 1989.

Remark 5.6 *The modal-damping matrix is generally not diagonal, because the eigenvectors of the undamped system, although orthogonal with respect to the stiffness and mass matrices, are not orthogonal with respect to the damping matrix. The modal-damping matrix is, however, symmetrical, since the original damping matrix is such.*

It has been demonstrated that a condition that is both necessary and sufficient to obtain a diagonal modal-damping matrix is that matrix $\mathbf{M}^{-1}\mathbf{C}$ commutes[4] with matrix $\mathbf{M}^{-1}\mathbf{K}$, or

$$\mathbf{C}\mathbf{M}^{-1}\mathbf{K} = \mathbf{K}\mathbf{M}^{-1}\mathbf{C}. \tag{5.49}$$

In this case, it is possible to define a modal damping $\overline{C_i} = \overline{C_{ii}}$ for each mode and to uncouple the equations of motion.

A particular case that satisfies condition (5.49) is the so-called proportional damping, i.e., the case in which the damping matrix can be expressed as a linear combination of the mass and stiffness matrices:

$$\mathbf{C} = \alpha\mathbf{M} + \beta\mathbf{K}. \tag{5.50}$$

Because condition (5.49) is more general than condition (5.50), a system whose damping satisfies the first will be said to possess generalized proportional damping.

Under these conditions, all matrices are diagonal, and Eq. (5.47) is a set of n uncoupled second-order differential equations. Each of them is

$$\overline{M_i}\ddot{\eta}_i + \overline{C_i}\dot{\eta}_i + \overline{K_i}\eta_i = \overline{f_i}, \tag{5.51}$$

and the system with n degrees of freedom is broken down into a set of n uncoupled systems, each with a single degree of freedom.

By normalizing the eigenvectors in such a way that the modal masses are equal to unity, Eq. (5.47) for a system with generalized proportional damping reduces to

$$\ddot{\boldsymbol{\eta}} + 2[\zeta\omega]\dot{\boldsymbol{\eta}} + [\omega^2]\boldsymbol{\eta} = \overline{\mathbf{f}'}, \tag{5.52}$$

where $[\omega^2] = \text{diag}\{\omega_i^2\}$ is the matrix of the eigenvalues, already defined in Section 1.7, and matrix

$$[\zeta\omega] = \text{diag}\{\zeta_i\omega_i\}$$

contains the damping ratios for the various modes.

Remark 5.7 *The frequencies ω_i are the natural frequencies of the corresponding undamped system.*

Remark 5.8 *Often the modal-damping matrix is not obtained by building matrix $\mathbf{C}$ and then performing the modal transformation, but directly by assuming reasonable values for the modal-damping ratios ζ_i.*

---

[4]Matrices $\mathbf{A}$ and $\mathbf{B}$ commute if $\mathbf{AB} = \mathbf{BA}$.

Although modal uncoupling does not introduce approximations in undamped systems, the equations of motion of damped systems can be uncoupled exactly only in the case of generalized proportional damping. The situation is usually that sketched in Fig. 5.6, in which the various modal systems are coupled to each other by the out-of-diagonal terms of modal damping.

This statement obviously does not exclude the possibility of uncoupling the equations of motion by introducing adequate approximations. The simplest way is by computing the modal-damping matrix using Eq. (5.48) and then neglecting all its terms except those on the diagonal. This corresponds to 'cutting' the dampers connecting the modal systems in Fig. 5.6b and substituting dampers $\overline{C}_i$ with dampers with coefficient $\overline{C}_{ii}$. Another possibility is to compute the eigenvalues related to the damped system solving the eigenproblem (5.44) and then using Eq. (5.10) to compute the modal damping for each mode.

Remark 5.9 *These procedures, which usually produce very similar results, introduce errors that are very small if the system is lightly damped.*

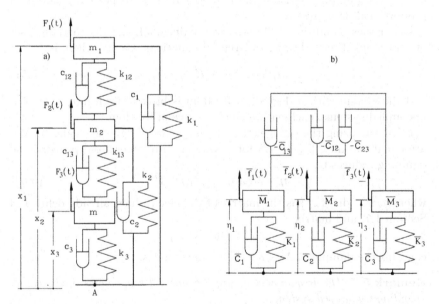

FIGURE 5.6. Modal uncoupling of damped systems. (**a**): Multi-degrees of freedom damped system. The modal systems in (**b**) are coupled by the out-of-diagonal modal-damping terms. The damping coefficients $\overline{C}_i$ are linked to the elements of the modal damping matrix by the relationships $\overline{C}_i = \sum_{j=1}^{3} \overline{C}_{ij}$ for $i = 1, ...,$ 3. Note that the term $\overline{C}_{ij}$ with $i \neq j$ is negative.

There are, however, cases in which neglecting the out-of-diagonal elements of the modal-damping matrix leads to unacceptable results, mainly when the system is highly damped and the damping distribution is far from being proportional. In this case, however, it is still possible to distinguish between a proportional part $\overline{\mathbf{C}}_p$ of the damping matrix, which is a diagonal matrix containing the elements of $\overline{\mathbf{C}}$ on the main diagonal, and a nonproportional part $\overline{\mathbf{C}}_{np}$, containing all other elements of $\overline{\mathbf{C}}$. The latter is symmetrical with all elements on the main diagonal equal to zero.

By applying the inverse modal transformation, it is possible to show also that the original damping matrix $\mathbf{C}$ can be split into a proportional and a nonproportional part,

$$\mathbf{C}_p = \mathbf{\Phi}^{-T}\overline{\mathbf{C}}_p\mathbf{\Phi}^{-1} \text{ and } \mathbf{C}_{np} = \mathbf{\Phi}^{-T}\overline{\mathbf{C}}_{np}\mathbf{\Phi}^{-1} ,$$

respectively. Note that $\mathbf{C}_p$ is not strictly proportional but only proportional in a generalized way.

Once the natural frequencies and mode shapes of the undamped system are known, the equation of motion of the system can be rewritten in modal coordinates in the form

$$\overline{\mathbf{M}}\ddot{\boldsymbol{\eta}} + \overline{\mathbf{C}}_p\dot{\boldsymbol{\eta}} + \overline{\mathbf{K}}\boldsymbol{\eta} = -\overline{\mathbf{C}}_{np}\dot{\boldsymbol{\eta}} + \overline{\mathbf{f}}(t) . \tag{5.53}$$

All matrices on the left-hand side are diagonal and coupling terms are present only on the right-hand side. It is thus possible to devise an iterative procedure allowing the computation of the eigenvalues of the damped system without solving the eigenproblem (5.44), whose size is $2n$. Because a solution of the type of Eq. (5.11) can be assumed for the time history of the response in terms of modal coordinates, the differential homogeneous equation associated with Eq. (5.53) can be transformed into the following algebraic equation in the Laplace domain:

$$\left(s^2\overline{\mathbf{M}} + s\overline{\mathbf{C}}_p + \overline{\mathbf{K}}\right)\boldsymbol{\eta}_0 = -s\overline{\mathbf{C}}_{np}\boldsymbol{\eta}_0 . \tag{5.54}$$

To compute the ith eigenvalue s_i, assume a set of n complex modal coordinates that are all zero, except for the real part of the ith coordinate, which is assumed with unit value. This amounts to assuming a complete uncoupling between the modes, at least where the ith mode is concerned. From the ith equation a first approximation of the complex eigenvalue can be obtained. The remaining $i - 1$ complex equations can be used to obtain new values for the $n - 1$ complex elements of the eigenvector; the ith element is assumed to have unit real part and zero imaginary part. Once the eigenvector has been computed, a new estimate for the ith eigenvalue can be obtained from the ith equation while the other equations yield a new estimate for the eigenvector. This procedure can be repeated until convergence is obtained.

This iterative procedure does not rely on any small-damping assumption. The starting values of the eigenfrequencies are not those of the undamped system, but rather are those of a system that has a generalized proportional distribution of damping. Consequently, it can also be applied to systems with very high damping in which some modes show a nonoscillatory free response. Although no theoretical proof of the convergence of the technique has been attempted, a number of tests did show that convergence is very quick. The reduction of computation time with respect to the conventional state-space approach is particularly remarkable when the root loci are to be obtained and the eigenproblem has to be solved several times.

Example 5.1 *Perform the modal analysis of the system shown in Fig.* 1.4 *(Example* 1.2), *with the damping added in Example* 3.1.
From the state-space dynamic matrix obtained in Example 3.1, *the eigenvalues are readily obtained:*

$$s_1 = -0.0263 \pm i\ 1.017, \quad s_2 = -0.4410 \pm i\ 2.974, \quad s_3 = -0.6577 \pm i\ 4.575.$$

The modal analysis of the undamped system was performed in Example 4.3, *obtaining the following matrix of the eigenvectors, normalized by stating the modal masses have unit values*

$$\Phi = \begin{bmatrix} 0.23728 & 0.15342 & 0.95925 \\ 0.45004 & 0.16837 & -0.13825 \\ 0.51681 & -1.31383 & 0.08228 \end{bmatrix}.$$

The modal-damping matrix can thus be computed. It is not diagonal because the damping of the system is not proportional.

$$\overline{C} = \Phi^T C \Phi = \begin{bmatrix} 0.0527 & -0.0328 & -0.2049 \\ -0.0328 & 0.8813 & -0.1324 \\ -0.2049 & -0.1324 & 1.3160 \end{bmatrix}.$$

Even if the damping matrix is not diagonal, approximate modal uncoupling can be performed by neglecting all elements of the modal-damping matrix lying outside the main diagonal.
The fact that the neglected elements of matrix $\overline{C}$ are of the same order of magnitude as the terms considered should not give the impression of a rough approximation. Actually, the behavior of the system is similar to that of the undamped system, owing to the low value of damping, except when a mode is excited near its resonant frequency. Damping is important only in near-resonant conditions, and even then only in one of the equations that is related to the resonant mode.
In the equation in which damping is important, the element on the diagonal of the modal-damping matrix is multiplied by a generalized coordinate that is far greater than the other ones, i.e., by the modal coordinate of the resonant mode.

From the three uncoupled modal systems, the damped frequencies and decay rates of the free motions can be obtained easily:

$$s_1 = -0.0266 \pm i\, 1.016\,, \quad s_2 = -0.4407 \pm i\, 2.972\,, \quad s_3 = -0.6580 \pm i\, 4.583.$$

They are very close to the values obtained directly as eigenvalues of the dynamic matrix in the state space. The imaginary parts of the eigenvalues can be compared with the natural frequencies of the undamped system.
It is clear that the presence of damping does not greatly affect the frequency of the free oscillations of the system.

5.6 Uncoupling the equations of motion: state space

The eigenvectors of Eq. (5.44) do not uncouple the equations of motion in the state space because matrix $\mathcal{A}$ is not symmetrical.

It is, however, possible to uncouple the state equations by resorting to the eigenvectors of the adjoint eigenproblem, i.e., the eigenvectors of matrix $\mathcal{A}^T$. Let $\mathbf{q}_{Ri}$ be the ith right eigenvector (i.e., the ith eigenvector of matrix $\mathcal{A}$) and let $\mathbf{q}_{Li}$ be the ith left eigenvector (i.e., the ith eigenvector of matrix $\mathcal{A}^T$). They are biorthogonal, i.e., all products $\mathbf{q}_{Lj}^T\mathbf{q}_{Ri}$ are equal to zero, if $i \neq j$.

The eigenvectors can be normalized in such a way that

$$\mathbf{q}_{Li}^T\mathbf{q}_{Ri} = 1\,, \tag{5.55}$$

in which case they are said to be *biorthonormal*.

By introducing into the state equation the modal states $\bar{\mathbf{z}}$ defined by the relationship

$$\mathbf{z} = \mathbf{\Phi}_R\bar{\mathbf{z}}\,, \tag{5.56}$$

where $\mathbf{\Phi}_R$ is the matrix of the right eigenvectors, and premultiplying by the matrix of the left eigenvectors $\mathbf{\Phi}_L$ transposed, the following modal uncoupled set of equations is obtained:

$$\dot{\bar{\mathbf{z}}} = \overline{\mathcal{A}}\bar{\mathbf{z}} + \overline{\mathcal{B}}\mathbf{u}\,, \tag{5.57}$$

where

$$\overline{\mathcal{A}} = \mathbf{\Phi}_L^T\mathcal{A}\mathbf{\Phi}_R \tag{5.58}$$

is a diagonal matrix listing the eigenvalues and

$$\overline{\mathcal{B}} = \mathbf{\Phi}_L^T\mathcal{B} \tag{5.59}$$

is the input gain matrix of the modal system.

Note that this is possible because

$$\overline{\mathbf{\Phi}}_L^T\mathbf{\Phi}_R = \mathbf{I}.$$

Remark 5.10 *The transition matrix $e^{\overline{A}t}$ in this case is a diagonal matrix, which can be easily computed because each of its elements is simply $e^{s_i t}$.*

Example 5.2 *Consider the system shown in Fig. 1.4 and studied in Examples 1.2, 3.1, and 5.1. Write the state-space equations and uncouple them through the right and left eigenvectors.*
The dynamic matrix and the input gain matrix computed in the mentioned examples are

$$
\mathcal{A} = \begin{bmatrix}
-1.1 & 1 & 0 & -20 & 10 & 0 \\
0.25 & -0.35 & 0.1 & 2.5 & -3.5 & 1 \\
0 & 0.8 & -0.8 & 0 & 8 & -8 \\
1 & 0 & 0 & 0 & 0 & 0 \\
0 & 1 & 0 & 0 & 0 & 0 \\
0 & 0 & 1 & 0 & 0 & 0
\end{bmatrix},
$$

$$
\mathcal{B} = \begin{bmatrix} 0 & 0 & 2 & 0 & 0 & 0 \end{bmatrix}^T.
$$

The right and left eigenvectors, suitably normalized, are (since they are conjugate, the six columns of the matrices are reported in synthetic form)

$$
\mathbf{q}_{Ri} = \begin{bmatrix}
0.2028 \mp 0.5513i & 0.1217 \pm 0.0196i & 1.0102 \mp 0.0155i \\
2.2258 \mp 1.1453i & 0.1336 \mp 0.0110i & -0.1494 \mp 0.0313i \\
2.5456 \mp 1.3337i & -1.0418 \pm 0.0201i & 0.0850 \pm 0.0446i \\
-0.5721 \mp 1.1674i & 0.0005 \mp 0.0410i & -0.0344 \mp 0.2158i \\
-1.1815 \mp 2.1572i & -0.0101 \mp 0.0434i & -0.0021 \pm 0.0330i \\
-1.3747 \mp 2.4665i & 0.0575 \pm 0.3418i5 & 0.0069 \mp 0.0196i
\end{bmatrix},
$$

$$
\mathbf{q}_{Li} = \begin{bmatrix}
0.0184 \pm 0.0111i & 0.0932 \pm 0.0342i & 0.4557 \pm 0.0546i \\
0.1425 \pm 0.0775i & 0.4292 \pm 0.0472i & -0.2614 \mp 0.0929i \\
0.0205 \pm 0.0110i & -0.4134 \mp 0.0719i & 0.0178 \pm 0.0126i \\
-0.0272 \pm 0.0112i & -0.1476 \pm 0.2879i & 0.0169 \pm 2.1322i \\
-0.0675 \pm 0.1502i & 0.0581 \pm 1.2954i & 0.0355 \mp 1.2320i \\
-0.0095 \pm 0.0216i & 0.0225 \mp 1.2599i & -0.0291 \pm 0.0924i
\end{bmatrix}.
$$

It can be easily checked that

$$
\overline{\mathbf{\Phi}}_L^T \mathbf{\Phi}_R = \mathbf{I}.
$$

The modal state matrix and the modal input gain matrix are

$$
\overline{A} = \mathbf{\Phi}_L^T \mathcal{A} \mathbf{\Phi}_R =
$$

$$
= diag \begin{bmatrix} -0.0263 \pm 1.0174i & -0.4410 \pm 2.9736i & -0.6577 \pm 4.5753i \end{bmatrix},
$$

$$
\overline{B} = \mathbf{\Phi}_L^T \mathcal{B} =
$$

$$
= \begin{bmatrix} 0.0411 \pm 0.0219i & -0.8269 \pm 0.1438i & 0.0355 \pm 0.0252i \end{bmatrix}^T.
$$

The elements of the modal state matrix coincide with the damped eigenvalues of the system.

5.7 Exercises

Exercise 5.1 *Add a viscous damper with damping coefficient c to the ballistic pendulum of Exercise 4.1. Compute the maximum amplitude, the logarithmic decrement, and the time needed to reduce the amplitude to $1/100$ of the original value. Data: $m = 100$ kg; $m_b = 0.05$ kg, $v = 200$ m/s, $l = 3$ m, $c = 6$ Ns/m.*

Exercise 5.2 *Add a viscous damper $c_{12} = 40$ Ns/m between masses m_1 and m_2 of the system of Exercise 4.4. Compute the modal-damping matrix and say whether damping is proportional or not. In the case where it is not proportional, compute matrices $\mathbf{C}_p$ and $\mathbf{C}_{np}$. Compute the complex frequencies and the complex modes of the system, both in the direct way and through an iterative modal procedure.*

Exercise 5.3 *Consider the system with hysteretic damping studied in Exercise 3.3. Compute the natural frequency of the undamped system and then the frequency of the free oscillations and the decay rate of the damped one. Repeat the computations for the system with Maxwell–Weichert damping of Exercise 3.4.*

Exercise 5.4 *Consider the system with 3 degrees of freedom of Fig. 1.8, already studied in Exercises 1.2, 2.1, and 2.3, with damping added as in Exercise 3.1. Write the state-space equations and uncouple them through the right and left eigenvectors.*

6

Forced Response in the Frequency Domain: Conservative Systems

The response of a linear conservative system excited by a harmonic forcing function is a harmonic time history in phase with the excitation. When the forcing frequency is close to a natural frequency of the system, very large amplitudes can be reached in a short time. This phenomenon is referred to as resonance. Theoretically, resonance leads to infinitely large amplitudes in undamped systems.

6.1 System with a single degree of freedom

6.1.1 Steady-state response

The motion of a system with a single degree of freedom under the effect of an external excitation can be obtained by adding the solution of the homogeneous equation describing the free motion to a particular integral of Eq. (1.3) or (1.4).

Among the different time histories of the excitation $F(t)$ that may be considered, one is of particular interest: Harmonic excitation

$$F(t) = f_1 \cos(\omega t) + f_2 \sin(\omega t) = f_0 \cos(\omega t + \Phi) . \qquad (6.1)$$

As discussed in detail in Section 2.2.1, it is expedient to express quantities that have a harmonic time history as projections on the real axis of vectors that rotate in the complex plane by resorting to the complex notation. The forcing function can thus be expressed in the form

$$F(t) = f_0 e^{i\omega t} . \qquad (6.2)$$

In a similar way, if the excitation is provided by the motion of the constraint, the forcing function is

$$x_A(t) = x_{A_0} e^{i\omega t} \; . \tag{6.3}$$

It is easy to verify that if the forcing function has an harmonic time history, the particular integral of the equation of motion is harmonic in time, oscillating with the same frequency ω. It can be represented by an exponential function as well

$$x(t) = x_0 e^{i\omega t} \; . \tag{6.4}$$

By introducing a harmonic time history for both excitation and response, the differential time-domain equation of motion (1.3) or (1.4) can be transformed into an algebraic, frequency domain, equation yielding the complex amplitude of the response

$$\left(- m\omega^2 + k \right)x_0 = \begin{cases} f_0 \, , \\ k x_{A_0} \, , \\ -m\omega^2 x_{A_0} \, , \end{cases} \tag{6.5}$$

for excitation provided by a force, by the motion of the supporting point A using an inertial coordinate, and by the motion of the supporting point A using a relative coordinate, respectively. The coefficient of the unknown x_0 in Eq. (6.5) is the dynamic stiffness of the system, already defined in Section 2.2.1, with reference to multi-degrees-of-freedom systems

$$k_{dyn} = \left(- m\omega^2 + k \right) = k \left[1 - \left(\frac{\omega}{\omega_n} \right)^2 \right] \; .$$

The dynamic stiffness is a function of the forcing frequency; its reciprocal is usually referred to as *dynamic compliance* or *receptance* and expresses the ratio between the amplitude of the displacement x_0 and the amplitude of the exciting force f_0. When the forcing frequency tends to zero, the dynamic stiffness and compliance tend to their static counterparts, the stiffness k and the compliance $1/k$.

The ratio between the dynamic and static compliance of the system is usually referred to as the *nondimensional frequency response*

$$H(\omega) = \frac{k}{k - m\omega^2} = \frac{1}{1 - \left(\frac{\omega}{\omega_n} \right)^2} \tag{6.6}$$

of the system; it is also referred to as the *magnification factor* (Fig. 6.1a). When plotting the frequency response, logarithmic axes are often used, and the scale of the ordinates may be expressed in decibels (Fig. 6.1b). The value in decibels of the magnification factor $H(\omega)$ is defined as

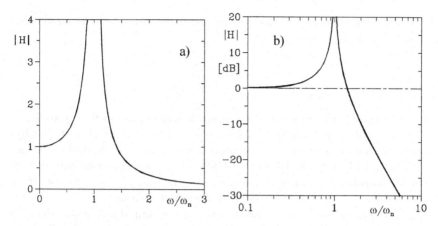

FIGURE 6.1. Magnification factor as a function of the forcing frequency: (a) linear scales, (b) logarithmic scale for frequency and dB scale for amplitudes.

$$H_{dB} = 20 \log_{10}(|H|). \tag{6.7}$$

In the logarithmic plot, the frequency response tends, for very low frequency, to the straight line $H=1$, and for very high frequency to a straight line sloping down with a slope equal to -2. This last situation is often referred to as an attenuation of 12 dB/oct (decibel per octave), even if the actual value is 12.041 dB/oct, or 40 dB/dec (decibel per decade). Where the response follows the first straight line, the system is said to be controlled by the stiffness of the spring because inertia forces are negligible, and the external force $F(t)$ is in equilibrium with the elastic force. In the case of excitation due to the motion of point A, mass m follows the displacement of the support. The response is in phase with the excitation.

When the system follows the sloping straight line, its behavior is said to be controlled by the inertia of the mass m, because the elastic force is negligible if compared to the inertia force, and the latter balances the external force $F(t)$. The response is still in phase with the excitation but its amplitude has an opposite sign: this situation is usually described as a phase angle of $-180°$.

When the excitation frequency ω is close to the natural frequency ω_n, a resonance occurs and the steady-state amplitude tends to be infinitely large. However, in this zone, which is said to be controlled by damping, the damping of the system becomes the governing factor because, at resonance, the inertia force exactly balances the elastic force and, consequently, only the damping force can balance the excitation $F(t)$. In this case, the presence of damping cannot be neglected and the present conservative model loses its accuracy.

If the excitation is provided by the harmonic motion of the supporting point A, a frequency response

$$H(\omega) = \frac{x_0}{x_{A_0}}$$

can be defined. The magnification factor is still expressed by Eq. (6.6).

A very common example of a system excited by the motion of the supports is that of a rigid body supported by compliant mountings whose aim is that of insulating it from vibrations that can be transmitted from the surrounding environment. In this case, the magnification factor $|H|$ is referred to as *transmissibility* of the suspension system.

The transmissibility is the ratio between the amplitude of the absolute displacement of the suspended object and the amplitude of the displacement of the supporting points.

Another problem related to the insulation of mechanical vibrations is that of reducing the excitation exerted on the supporting structure by a rigid body on which a force variable in time is acting. The ratio between the amplitude of the force exerted by the spring on the supporting point kx and the amplitude of the excitation $F(t)$ is also referred to as transmissibility of the suspension.

With simple computations, it is possible to show that the value of the transmissibility so defined is the same as obtained in Eq. (6.6). This explains why the two ratios are referred to by the same name.

Apart from the dynamic compliance and the dynamic stiffness, other frequency responses can be defined. The ratio between the amplitude of the velocity and that of the force F is said to be the mobility; its reciprocal is the mechanical impedance. The ratio between the amplitude of the acceleration and that of the force F is said to be the inertance, and its reciprocal is the dynamic mass. The aforementioned frequency responses are summarized in Table 6.1 and Fig. 6.2. The most widely used are the dynamic compliance and the inertance.

They are all expressed by real numbers, in the case of undamped systems.

TABLE 6.1. Frequency responses.

Frequency response	Definition	S.I. units	$\mathrm{Lim}(\omega \to 0)$
Dynamic compliance	x_0/f_0	m/N	$1/k$
Dynamic stiffness	f_0/x_0	N/m	k
Mobility	$(\dot{x})_0/f_0 = \omega x_0/f_0$	m/sN	0
Mechanical impedance	$f_0/(\dot{x})_0 = f_0/\omega x_0$	Ns/m	∞
Inertance	$(\ddot{x})_0/f_0 = \omega^2 x_0/f_0$	m/s^{2}N	0
Dynamic mass	$f_0/(\ddot{x})_0 = f_0/\omega^2 x_0$	s^{2}N/m	∞

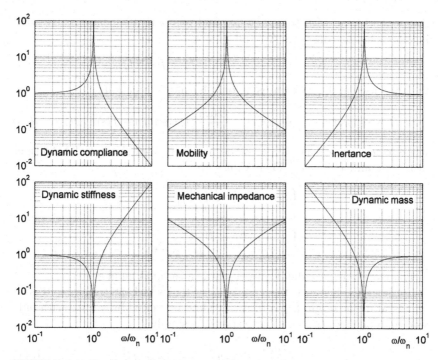

FIGURE 6.2. Nondimensional frequency responses for an undamped single-degree-of-freedom system (logarithmic scales).

6.1.2 Nonstationary response

The complete solution of the equation of motion is obtained by adding the solutions found for the free and forced oscillations (i.e., adding a particular integral to the complementary function)

$$x = K^* e^{i\omega_n t} + H(\omega)\frac{f_0}{k}e^{i\omega t} . \tag{6.8}$$

The complex constant K^* can be determined from the initial conditions. In most actual systems, owing to the presence of damping (see Chapter 5), the first term of Eq. (6.8) tends to zero, often quite quickly, while the second one has a constant amplitude; as a consequence, when studying the response of a damped system to harmonic excitation, usually only the latter is considered. There are, however, cases in which the initial transient cannot be neglected, particularly when dealing with lightly damped systems or when the forcing function is applied to a system that is at rest: in the latter case oscillations with growing amplitude usually result, until the steady-state conditions are reached.

In the present undamped case the free response does not vanish with time, and the solution is a poly-harmonic oscillations made of two components, one with frequency ω_n and one with frequency ω. By writing the complex constants K^* and f_0 as

$$K^* = C_1 - iC_2 \quad , \qquad f_0 = f_1 - if_2 \ ,$$

the response can be written using only harmonic functions

$$x = C_1 \cos(\omega_n t) + C_2 \sin(\omega_n t) + \frac{H(\omega)}{k} \left[f_1 \cos(\omega t) + f_2 \sin(\omega t) \right] \ , \quad (6.9)$$

where constants C_1 and C_2 must be obtained from the initial conditions $x(0) = x_0$ and $\dot{x}(0) = v_0$. Their values are

$$\begin{cases} C_1 = x_0 - \dfrac{H(\omega)}{k} f_1 \\[3mm] C_2 = \dfrac{v_0}{\omega_n} - \dfrac{H(\omega)}{k} \dfrac{\omega}{\omega_n} f_2. \end{cases} \qquad (6.10)$$

The motion is periodic if ratio ω/ω_n is a rational number.

Consider for instance the case of a system which at time $t = 0$ is at standstill in the origin ($x_0 = 0$, $v_0 = 0$) and is excited by a force with only sine components ($f_1 = 0$). The response is simply

$$x = f_2 \frac{H(\omega)}{k} \left[\sin(\omega t) - \frac{\omega}{\omega_n} \sin(\omega_n t) \right] \ . \qquad (6.11)$$

Some nondimensional time histories are reported in Fig. 6.3 for different values of ratio ω/ω_n. In all cases it is assumed to be a rational number so that the motion is periodic, but in the first case the period is fairly long. In cases (b) and (c) the forcing frequency is close to the natural frequency and a beat takes place. The nondimensional period of the beat is $20\pi = 62.8$.

If the frequency of the forcing function coincides with the natural frequency, Eq. (6.11) cannot be used, since $H(\omega)$ is infinitely large while the expression in braces vanishes. The response can be computed as

$$x = \lim_{\omega \to \omega_n} \left\{ f_2 \frac{H(\omega)}{k} \left[\sin(\omega t) - \frac{\omega}{\omega_n} \sin(\omega_n t) \right] \right\} \ , \qquad (6.12)$$

i.e., by introducing the value of $H(\omega)$

$$x = \frac{f_2 \omega_n}{k} \lim_{\omega \to \omega_n} \left[\frac{\omega_n \sin(\omega t) - \omega \sin(\omega_n t)}{\omega_n^2 - \omega^2} \right] \ . \qquad (6.13)$$

Using de L'Hospital's rule, the limit can be computed by differentiating both the numerator and denominator with respect to ω

$$x = \frac{f_2 \omega_n}{k} \lim_{\omega \to \omega_n} \left[\frac{\omega_n t \cos(\omega t) - \sin(\omega_n t)}{-2\omega} \right] = \frac{f_2}{2k} \left[\sin(\omega_n t) - \omega_n t \cos(\omega_n t) \right] \ . \qquad (6.14)$$

The amplitude grows linearly with time, tending to infinity for $t \to \infty$.

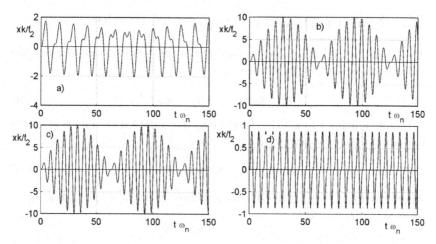

FIGURE 6.3. Nondimensional time history for an undamped system excited by an harmonic force starting from the origin at standstill. (a) $\omega/\omega_n = 0.51$; (b) $\omega/\omega_n = 0.90$; (c) $\omega/\omega_n = 1.1$; (d) $\omega/\omega_n = 2.0$.

Remark 6.1 *To state that when resonance occurs the amplitude of the response becomes infinitely large is an oversimplification, even in the idealized case of undamped systems. The amplitude grows linearly and an infinite time is required to reach an infinite amplitude*

In practice, large values of the amplitude can be reached in a short time, but there are cases where the amplitude buildup is slow.

The time history expressed by Eq. (6.14) is reported in Fig. 6.4.

The equation of the straight line enveloping the response is easily approximated by the line connecting the peaks, which occur at the times when $\sin(\omega_n t)$ vanishes

$$x = \frac{f_2 \omega_n}{2k} t \ . \tag{6.15}$$

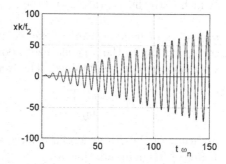

FIGURE 6.4. Resonant response of a conservative system with a single degree of freedom.

6.2   System with many degrees of freedom

The harmonic excitation and response of a general multi-degrees-of-freedom system can be written in the form

$$\mathbf{f}(t) = \mathbf{f}_0 e^{i\omega t} \ , \quad \mathbf{x}(t) = \mathbf{x}_0 e^{i\omega t} \ . \tag{6.16}$$

By introducing the solution (6.16) into the equation of motion (1.7), the frequency-domain equation (2.14) is obtained

$$\mathbf{K}_{dyn}\mathbf{x}_0 = \mathbf{f}_0 \ , \tag{6.17}$$

where the expression of the dynamic stiffness matrix of the system is expressed by Eq. (2.13)

$$\mathbf{K}_{dyn} = -\omega^2 \mathbf{M} + \mathbf{K}.$$

The dynamic stiffness matrix is real and symmetrical but can be non-positive defined.

As stated in Chapter 2, vectors $\mathbf{x}_0$ and $\mathbf{f}_0$ are in general complex vectors. By separating the real and the imaginary parts of Eq. (6.17) it follows

$$\begin{bmatrix} \mathbf{K} - \omega^2 \mathbf{M} & \mathbf{0} \\ \mathbf{0} & \mathbf{K} - \omega^2 \mathbf{M} \end{bmatrix} \left\{ \begin{array}{c} \Re(\mathbf{x}_0) \\ \Im(\mathbf{x}_0) \end{array} \right\} = \left\{ \begin{array}{c} \Re(\mathbf{f}_0) \\ \Im(\mathbf{f}_0) \end{array} \right\} . \tag{6.18}$$

The real part of the response then depends only on the real part of the excitation, and the same holds for the imaginary parts.

Remark 6.2 *The real and imaginary parts of the response can be computed from the real and imaginary parts of the excitation separately. This property, however, holds only for conservative (undamped) systems.*

If all the harmonic exciting forces have the same phasing, i.e., the excitation is said to be monophase or coherently phased, the response of the system is harmonic and is in phase with the excitation. In such a case both $\mathbf{f}_0$ and $\mathbf{x}_0$ can be expressed by real vectors, simply by taking as initial time the instant in which both the excitation and the response are at their maximum.

A system with n degrees of freedom can be excited using n harmonic generalized forces corresponding to the n generalized coordinates, and, for each exciting force, n responses can be obtained. The frequency responses

$$H_{ij}(\omega) = \frac{x_{0i}(\omega)}{f_{0j}} \ , \tag{6.19}$$

where f_{0j} is the amplitude of the jth generalized force and x_{0i} is the response at the ith degree of freedom, are thus n^2.

The static compliance matrix or the matrix of the coefficients of influence is defined as the inverse of the stiffness matrix $\mathbf{K}$. A dynamic compliance matrix can be defined in the same way as the inverse of the dynamic stiffness matrix. It coincides with the matrix of the frequency responses $\mathbf{H}(\omega)$ defined earlier. The compliance matrix is symmetrical, as is the stiffness matrix, but while the latter often has a band structure, the former does not show useful regularities.

A number n^2 of frequency responses can be plotted in the same way seen for systems with a single degree of freedom. Their amplitude has n peaks with infinite height, corresponding to the natural frequencies. Some of the curves $H(\omega)$ can cross the frequency axis, i.e., the amplitude can get vanishingly small at certain frequencies. This condition is usually referred to as antiresonance. It must be noted that while the resonances are the same for all the degrees of freedom, the antiresonances are different and may be absent in some of the responses. The number of antiresonances is $n - 1$ for the transfer functions on the main diagonal and, in the case of in-line systems with the generalized coordinates listed sequentially, decrease by one on each diagonal above or below it. No antiresonance is thus found in $H_{1,n}$ and $H_{n,1}$.

Example 6.1 *Compute the elements H_{11} and H_{13} of the frequency response of the system in Example 1.2.*
The dynamic compliance is easily computed by inverting the dynamic stiffness matrix. The frequency responses are plotted in Fig. 6.5a using logarithmic scales. By multiplying the dynamic compliance by ω^2 the inertance is easily obtained (Fig. 6.5b).

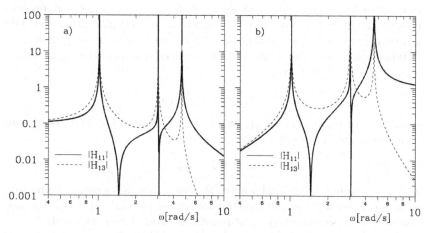

FIGURE 6.5. Two elements of the **(a)** dynamic compliance matrix and **(b)** the inertance matrix of a system with three degrees of freedom. The units are rad/Nm for the dynamic compliance and rad/Nms^2 for the inertance.

6.3   Modal computation of the response

Since the equations of motion of any linear conservative system with n degrees of freedom can be uncoupled into n separate linear equations of the same type of those seen for a single-degree-of-freedom system (Eqs. (4.45)), the time history of the response to an arbitrary excitation can be computed by

1. Computing the eigenvalues and eigenvectors of the system and normalizing the eigenvectors.

2. Computing the modal forces as functions of time.

3. Solving the n Eqs. (4.45), in order to obtain the time histories of the response in terms of modal coordinates $\boldsymbol{\eta}$.

4. Recombining the responses computed through Eq. (4.37), yielding the time history of the system in terms of the physical coordinates $\mathbf{x}$.

This procedure has the notable advantage of dealing with n uncoupled equations, while a direct solution would require the integration of a set of n coupled differential equations.

There is, however, another advantage: Not all modes are equally important in determining the response of the system. If there are many degrees of freedom, a limited number of modes (usually those characterized by the lowest natural frequencies) is sufficient for obtaining the response with good accuracy. If only the first m modes are considered,[1] the savings in terms of computation time, and hence cost, are usually noticeable, because only m eigenvalues and eigenvectors need to be computed and m systems with one degree of freedom have to be studied. Usually the modes that are more difficult to deal with are those characterized by the highest natural frequencies, particularly if the equations of motion are numerically integrated. The advantage of discarding the higher-order modes is, in this case, great.

When some modes are neglected, the reduced matrix of the eigenvectors, which will be referred to as

$$\boldsymbol{\Phi}^* = [\mathbf{q}_1, \mathbf{q}_2, \ldots, \mathbf{q}_m] \ ,$$

is not square because it has n rows and m columns. The first coordinate transformation (4.37) still holds

$$\mathbf{x}_{n \times 1} = \boldsymbol{\Phi}^*_{n \times m} \boldsymbol{\eta}_{m \times 1} \ ,$$

and the m values of the modal mass, stiffness, and force can be computed as usual. However, the inverse transformation (second Eq. (4.37)) is not possible, because the inversion of matrix $\boldsymbol{\Phi}^*$ cannot be performed.

---

[1] In the following pages it is assumed that the modes which are retained are those from the first one to the mth.

The modal coordinates $\boldsymbol{\eta}$ can be computed from the physical coordinates $\mathbf{x}$ by premultiplying by $\boldsymbol{\Phi}^{*T}\mathbf{M}$ both sides of the first Eq. (4.37) computed using the reduced matrix of the eigenvectors $\boldsymbol{\Phi}^*$, obtaining

$$\boldsymbol{\Phi}^{*T}\mathbf{M}\mathbf{x} = \boldsymbol{\Phi}^{*T}\mathbf{M}\boldsymbol{\Phi}^*\boldsymbol{\eta} , \qquad (6.20)$$

i.e.,

$$\boldsymbol{\Phi}^{*T}\mathbf{M}\mathbf{x} = \overline{\mathbf{M}}\boldsymbol{\eta}. \qquad (6.21)$$

Premultiplying then both sides by the inverse of the matrix of the modal masses, it follows

$$\boldsymbol{\eta}_{m\times 1} = \overline{\mathbf{M}}^{-1}_{m\times m}\boldsymbol{\Phi}^*_{m\times n}{}^T\mathbf{M}_{n\times n}\mathbf{x}_{n\times 1} . \qquad (6.22)$$

Equation (6.22) is the required inverse modal transformation. As required, the transformation matrix

$$\overline{\mathbf{M}}^{-1}_{m\times m}\boldsymbol{\Phi}^*_{m\times n}{}^T\mathbf{M}_{n\times n}$$

has m rows and n columns.

When studying the response to an excitation due to the motion of the supporting points, it is usually sufficient to consider the few modes characterized by a high value of the corresponding modal participation factor. The order of vector $\mathbf{r}_j$ is in this case m and all the considerations discussed in Section 4.5 still hold, but Eq. (4.56) is only approximated. The precision that can be attained considering only a limited number of modes when computing the response to an excitation due to the motion of the constraints is measured by the approximation with which the sum in Eq. (4.56) approximates the total mass of the system.

The response computed by considering only the first m modes can be inaccurate even at low frequency if the static deformation has a shape which, once expressed in modal coordinates, has non-negligible contributions due to any mode of order higher than m. To account for the contribution of the modes which are neglected to the low-frequency response without having to use a high number of modes, it is possible to operate as follows.

Assuming that the eigenvectors are normalized in such a way that the modal masses have a unit value, the modal responses can be obtained from Eqs. (4.45) in which a harmonic forcing function is introduced

$$\eta_{0i} = \frac{\overline{f_{0i}}}{\omega_{ni}^2 - \omega^2} = \frac{1}{\omega_{ni}^2 - \omega^2}\,\mathbf{q}_i^T\mathbf{f}_0 \qquad \text{for } i = 1, 2, ..., n . \qquad (6.23)$$

By transforming the response from the modal coordinates back to the physical ones, it follows

$$\mathbf{x} = \sum_{i=1}^n \eta_{0i}\mathbf{q}_i = \left(\sum_{i=1}^n \frac{1}{\omega_{ni}^2 - \omega^2}\mathbf{q}_i\mathbf{q}_i^T\right)\mathbf{f}_0 . \qquad (6.24)$$

By subdividing the sum into two parts, related to the first m modes and to all other ones, it follows

$$\mathbf{x} = \left(\sum_{i=1}^{m} \frac{1}{\omega_{ni}^2 - \omega^2} \mathbf{q}_i \mathbf{q}_i^T \right) \mathbf{f}_0 + \left(\sum_{i=m+1}^{n} \frac{1}{\omega_{ni}^2 - \omega^2} \mathbf{q}_i \mathbf{q}_i^T \right) \mathbf{f}_0 . \qquad (6.25)$$

The first term is the approximated response computed using the first m modes, while the second term can be considered as the error due to considering a reduced number of modes. If the natural frequency of the $(m+1)$th mode is much higher than the excitation frequency ω (say at least 10 times), $\omega_{ni}^2 \gg \omega^2$ for all the modes included in the second term. The error can then be computed by neglecting ω^2 with respect to ω_{ni}^2

$$\mathbf{x} \approx \left(\sum_{i=1}^{m} \frac{1}{\omega_{ni}^2 - \omega^2} \mathbf{q}_i \mathbf{q}_i^T \right) \mathbf{f}_0 + \left(\sum_{i=m+1}^{n} \frac{1}{\omega_{ni}^2} \mathbf{q}_i \mathbf{q}_i^T \right) \mathbf{f}_0 . \qquad (6.26)$$

If for instance ω_{ni} for the $(m+1)$th mode is 10 times ω, the approximation in the computation of the error is less (much less) than 1%.

The response to a static force vector $\mathbf{f}_0$

$$\mathbf{x}_{st} = \mathbf{K}^{-1} \mathbf{f}_0 \qquad (6.27)$$

can be expressed in terms of modal coordinates

$$\mathbf{x}_{st} = \left(\sum_{i=1}^{n} \frac{1}{\omega_{ni}^2} \mathbf{q}_i \mathbf{q}_i^T \right) \mathbf{f}_0 = \left(\sum_{i=1}^{m} \frac{1}{\omega_{ni}^2} \mathbf{q}_i \mathbf{q}_i^T \right) \mathbf{f}_0 + \left(\sum_{i=m+1}^{n} \frac{1}{\omega_{ni}^2} \mathbf{q}_i \mathbf{q}_i^T \right) \mathbf{f}_0 . \qquad (6.28)$$

From Eqs. (6.27) and (6.28) it follows

$$\left(\sum_{i=m+1}^{n} \frac{1}{\omega_{ni}^2} \mathbf{q}_i \mathbf{q}_i^T \right) \mathbf{f}_0 = \mathbf{K}^{-1} \mathbf{f}_0 - \left(\sum_{i=1}^{m} \frac{1}{\omega_{ni}^2} \mathbf{q}_i \mathbf{q}_i^T \right) \mathbf{f}_0 . \qquad (6.29)$$

Finally, introducing Eq. (6.29) into Eq. (6.26) it follows

$$\mathbf{x} \approx \left(\sum_{i=1}^{m} \frac{1}{\omega_{ni}^2 - \omega^2} \mathbf{q}_i \mathbf{q}_i^T \right) \mathbf{f}_0 - \left(\sum_{i=1}^{m} \frac{1}{\omega_{ni}^2} \mathbf{q}_i \mathbf{q}_i^T \right) \mathbf{f}_0 + \mathbf{K}^{-1} \mathbf{f}_0 . \qquad (6.30)$$

Note that only the first m eigenvectors appear in Eq. (6.30).

The sum of the first two terms have an immediate physical meaning: it is the difference between the dynamic and the static response, computed using the first m modes. As a whole, the meaning of Eq. (6.30) is then clear: the response can be computed by accounting for the dynamic response of the lower order modes, plus the static response of the other ones.

The difference between the dynamic and static response can be immediately computed as

$$\left(\sum_{i=1}^{m} \frac{1}{\omega_{ni}^2 - \omega^2} \mathbf{q}_i \mathbf{q}_i^T\right) \mathbf{f}_0 - \left(\sum_{i=1}^{m} \frac{1}{\omega_{ni}^2} \mathbf{q}_i \mathbf{q}_i^T\right) \mathbf{f}_0 =$$

$$= \left[\sum_{i=1}^{m} \frac{\omega^2}{\omega_{ni}^2} \left(\frac{1}{\omega_{ni}^2 - \omega^2}\right) \mathbf{q}_i \mathbf{q}_i^T\right] \mathbf{f}_0 = \sum_{i=1}^{m} \frac{\omega^2}{\omega_{ni}^2} \eta_{0i} \mathbf{q}_i. \quad (6.31)$$

An approximated computation of the response can thus be performed by first obtaining the first m eigenvectors and modal responses η_{0i} and then recombining them using the relationship

$$\mathbf{x} \approx \mathbf{\Phi}^{**} \boldsymbol{\eta}^*, \quad (6.32)$$

where $\mathbf{\Phi}^{**}$ is a modified reduced matrix of the eigenvectors obtained by adding a further column to $\mathbf{\Phi}^*$ in which the first Ritz vector (see next section, Eq. (6.36))

$$\mathbf{r}_1 = \mathbf{K}^{-1} \mathbf{f}_0,$$

is included

$$\mathbf{\Phi}^{**} = [\mathbf{q}_1, \mathbf{q}_2, \ldots, \mathbf{q}_m, \mathbf{r}_1]. \quad (6.33)$$

$\boldsymbol{\eta}^*$ is the vector of the modified modal coordinates

$$\boldsymbol{\eta}^* = [\frac{\omega^2}{\omega_{n1}^2} \eta_{01}, \frac{\omega^2}{\omega_{n2}^2} \eta_{02}, \ldots, \frac{\omega^2}{\omega_{nm}^2} \eta_{0m}, 1]. \quad (6.34)$$

Example 6.2 *Consider the torsional system studied in Example 1.2, and compute the response of the first inertia (point 1) to a harmonic torque with unit amplitude applied in the same point using the modal approach.*
The modal analysis of the system was performed in Example 4.3. Let $\mathbf{\Phi}$ be the matrix of the eigenvectors normalized in such a way that the modal masses have a unit value. Its value is here repeated together with the modal mass matrix

$$\mathbf{\Phi} = \begin{bmatrix} 0.23728 & 0.15342 & 0.95925 \\ 0.45004 & 0.16837 & -0.13825 \\ 0.51681 & -1.31383 & 0.08228 \end{bmatrix},$$

$$\overline{\mathbf{K}}= \begin{bmatrix} 1.03353 & 0 & 0 \\ 0 & 9.02522 & 0 \\ 0 & 0 & 21.44125 \end{bmatrix}.$$

The modal contributions to the response in point 1 due to the ith mode can be readily computed as

$$x_{01i} = q_{1i}\eta_1 = q_{1i}\frac{1}{\omega_{ni}^2 - \omega^2}\mathbf{q}_i^T \mathbf{f}_0 .$$

The responses are reported in Fig. 6.6a. From the plot it is clear that the third mode is quite important in determining the response also at low frequency, and the third mode component of the static response is quite strong. The second mode is comparatively unimportant, except close to its own resonance. In these conditions, the modal computation of the response performed including a single mode (Fig. 6.6b) or two modes (Fig. 6.6c) is quite inaccurate even at low frequency. Only in the vicinity of the first resonance (the first and the second in Fig. 6.6c) such modal computations can be considered acceptable. The modal computation was repeated using the modified matrix and vector defined in Eqs. (6.33) and (6.34). In this case, as shown in Fig. 6.6d, very accurate results are obtained even with a single mode up to the antiresonance ($\omega \approx 1.5$). At higher frequency, the third mode starts to show strong dynamic effects and reducing its response to the static one becomes unsatisfactory. Taking into account also the second mode does not change essentially the picture.

This example is in a way not typical, since the three natural frequencies are quite close to each other and, in this condition, the high-order modes extend their influence at frequencies as low as the natural frequencies of the first modes. At any rate, it shows that the static component of the response of the modes resonating at high frequency can be too strong to be neglected.

6.4 Coordinate transformation based on Ritz vectors

The dynamic behavior of systems with many degrees of freedom was studied in the preceding sections using either physical coordinates or modal coordinates. It is, however, obvious that any other coordinate transformation can be applied and that any non-singular matrix of order n could be used to perform a coordinate transformation. This statement simply means that any set of linearly independent vectors can be assumed as a reference frame in the space of configurations. The modal transformation based on the eigenvectors has the drawbacks of requiring the solution of an eigenproblem, which sometimes is quite complex, and often requiring a large number of modes to compute the response to a generic forcing

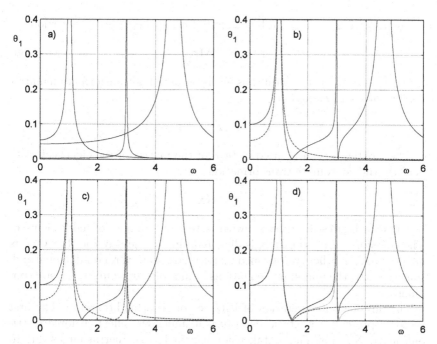

FIGURE 6.6. Response of the three degrees of freedom system of Example 1.2. (a) Contributions of the 3 modal systems to the response in point 1. (b) Response in mode 1, computed from the complete equations (*full line*) and using only the first modal response (*dashed line*). (c) As (b), but with the first two modal responses. (d) As (b), but taking into account the static response, plus one mode (*dashed line*) and plus two modes (*dotted line*).

function. When the response to an excitation due to the motion of the supports has to be computed, this means that many modes have a modal participation factor high enough to prevent from neglecting them. In this case, the use of Ritz vectors constitutes a different choice worthy of consideration.

Consider a multi-degree-of-freedom system excited in such a way that there is only one input $u(t)$. In the configurations space, the forcing function can be expressed as $\mathbf{f}(t) = \mathbf{f}_0 f(t)$. The first Ritz vector is defined by the equation

$$\mathbf{r}_1 = \mathbf{K}^{-1}\mathbf{f}_0 \,, \qquad (6.35)$$

and then it coincides with the static deflected shape under the effect of the constant force distribution $\mathbf{f}_0$. Ritz vectors, like eigenvectors, are normalized. The simplest way to normalize Ritz vectors is by making the products $\mathbf{r}^T\mathbf{M}\mathbf{r}$ equal to unity. This can easily be performed by dividing each Ritz vector $\mathbf{r}$ by the square root of $\mathbf{r}^T\mathbf{M}\mathbf{r}$.

The following vectors can be computed using the recursive equation

$$\mathbf{r}_i = \mathbf{K}^{-1}\mathbf{M}\mathbf{r}_{i-1}. \tag{6.36}$$

Once a set of m Ritz vectors has been computed, a transformation matrix with n rows and m columns

$$\mathbf{R} = [\mathbf{r}_1, \mathbf{r}_2, \ldots, \mathbf{r}_m]$$

can be written. It can be used instead of the matrix of the eigenvectors to perform the coordinate transformation

$$\mathbf{x} = \mathbf{R}\bar{\bar{\mathbf{x}}}.$$

The coordinates $\bar{\bar{x}}_i$ are m, as when a reduced number of modes is used. The transformed mass matrix is diagonal, due to the way the Ritz vectors are derived. All other transformed matrices are, however, non-diagonal and, consequently, the undamped equations of motions are coupled (elastic coupling).

The physical interpretation of Ritz vectors is straightforward. The first vector represents the static deformation under the effect of force distribution $\mathbf{f}_0$. No allowance is taken for inertia forces. Inertia forces due to harmonic motion with frequency ω and deformed shape $\mathbf{r}_1$ are $\omega^2\mathbf{M}\mathbf{r}_1$. The second Ritz vector is then proportional to the deformed shape due to inertia forces consequent to a harmonic oscillation with a deformed shape corresponding to the first vector. In the same way, all other vectors are computed.

The advantage of Ritz vectors with respect to the eigenvectors of the undamped system in the computation of the time history of the response is then clear: Fewer coordinates are generally required, and the amount of computational work needed to compute them is much smaller. The equation of motion obtained using Ritz vectors can then be subjected to modal analysis or reduced as it will be seen in Chapter 10.

There are, however, also disadvantages. First, it is common to perform an eigenanalysis before computing the time history of the response to obtain the natural frequencies and the mode shapes. In this case, the modal transformation involves very little additional computational work. In the undamped case, the equations of motion obtained through modal transformation are uncoupled, while, through Ritz vectors, a set of equations with elastic coupling is obtained. When damping is taken into consideration, and even more when nonlinearities are included in the model, the number of Ritz vectors needed can increase, and it is very difficult to assess how many must be considered, as happens with the true eigenvectors.

Although they are sometimes used in the computation of the response of structures to seismic excitation, Ritz vectors are not widely used in structural dynamics.

6.5   Response to periodic excitation

When a periodic excitation $F(t)$ with period T acts on the system, the steady-state response can easily be computed by decomposing the forcing function in a Fourier series

$$F(t) = a_0 + \sum_{i=1}^{n} a_i \cos\left(\frac{2\pi i}{T}\right) + \sum_{i=1}^{n} b_i \sin\left(\frac{2\pi i}{T}\right). \qquad (6.37)$$

The coefficients of the Fourier series can be obtained from function $F(t)$ using the formulae

$$
\begin{aligned}
a_0 &= \frac{1}{T} \int_0^T F(t)\,dt \\
a_i &= \frac{1}{T} \int_0^T F(t) \cos\left(\frac{2\pi i}{T}t\right) dt \\
b_i &= \frac{1}{T} \int_0^T F(t) \sin\left(\frac{2\pi i}{T}t\right) dt.
\end{aligned}
\qquad (6.38)
$$

Because the system is linear, the response to the poly-harmonic excitation (6.37) can be obtained by adding the responses to all terms of the forcing functions. Since in the case of undamped systems the frequency response is real, the following expression for the particular integral of the equation of motion can be obtained:

$$x(t) = \frac{1}{k}a_0 + \sum_{i=1}^{n} a_i H(\omega_i) \cos(\omega_i t) + \sum_{i=1}^{n} b_i H(\omega_i) \sin(\omega_i t), \qquad (6.39)$$

where the frequency of the ith harmonic of the forcing function is

$$\omega_i = \frac{2\pi i}{T}.$$

The complete solution can thus be obtained by adding the particular integral expressed by Eq. (6.39) to the complementary function

$$x = K^* e^{i\omega_n t}$$

and computing the complex constant K^* from the initial conditions.

6.6   Exercises

Exercise 6.1 *A machine whose mass is m contains a rotor with an eccentric mass m_e at a radius r_e running at speed Ω_1. Compute the stiffness of the supporting elements in such a way that the natural frequency of the machine is equal* to 1/8 of the forcing frequency due to the rotor; compute the maximum amplitude of the force transmitted to the supports at all speeds up to the operating speed.

Add a second unbalanced rotor identical to the first running in the same direction at a slightly different speed Ω_2. Write the equation of motion and show that a beat takes place. Compute the frequency of the beat and plot the time history of the response (assuming that the free part of the response has already decayed away) starting from the time at which the two unbalances are in phase. Data: $m = 500$ kg, $m_e = 0.050$ kg, $r_e = 1$ m, $\Omega_1 = 3,000$ rpm, $\Omega_2 = 3,010$ rpm.

Exercise 6.2 *Compute the time history of the system of Exercise 4.2 for an excitation due to the motion of the supports* $x_A = x_{A_0} \sin(\omega t)$. *Compute the time history of the response using physical and modal coordinates (neglect the free motion part of the response). Data:* $x_{A_0} = 5$ mm, $\omega = 30$ rad/s.

Exercise 6.3 *Consider the system of Example 1.4. The points at which the two pendulums are connected move together in the x-direction with a harmonic time history* $x_A = x_B = x_{A_0} \sin(\omega t)$, *with* $x_{A_0} = 30$ mm, $\omega = 1$ Hz.

Compute the response of the system, assuming that at time $t = 0$ the whole system is at a standstill. Compute the modal participation factors and assess whether an excitation of the type here assumed, with the initial condition described, gives way to a beat. Data: $m = 1$ kg, $l = 600$ mm, $k = 2$ N/m, $g = 9.81$ m/s^2.

Exercise 6.4 *An instrument, whose mass is 20 kg must be mounted on a space vehicle through a cantilever arm of annular cross-section made of light alloy, 600 mm long and with inner and outer diameters 100 and 110 mm respectively.*

Using a simple model with a single degree of freedom, check the ability of the arm to withstand a harmonic excitation due to the motion of the supporting point. Let the intensity of the excitation be defined by

- *Amplitude 10 mm in the frequency range 5–8.5 Hz*

- *Acceleration 3 g in the frequency range 8.5–35 Hz*

- *Acceleration 1 g in the frequency range 35–50 Hz.*

The stresses due the mentioned excitation must not exceed

- *The ultimate strength (328 MN/m^2) divided by a safety factor of 1.575*

- *The yield strength (216 MN/m^2) divided by a safety factor of 1.155*

- *The allowable fatigue strength for a duration of 10^7 cycles (115 MN/m^2)*

The relative displacement between the instrument at the end of the beam and the supporting structure must not exceed 2 mm. Material data: $E = 72 \times 10^9$ N/m^2; $\rho = 2,800$ kg/m^3.

7

Forced Response in the Frequency Domain: Damped Systems

When damping is considered, the response of a linear system to a harmonic excitation is still harmonic in time, but is not in phase with the excitation. If damping is small, there is still a well-defined resonance (or many of them, depending on number of degrees of freedom), but its amplitude remains finite. If damping is large, one or more resonance peaks may disappear altogether.

7.1 System with a single degree of freedom: steady-state response

The response of a damped system with a single degree of freedom can be computed following the same lines seen in Chapter 6 for the undamped system. The excitation and the response can be written in the form

- Force excitation

$$F(t) = f_0 e^{i\omega t} \ . \tag{7.1}$$

- Excitation due to motion of the constraint

$$x_A(t) = x_{A_0} e^{i\omega t} \ . \tag{7.2}$$

- Response

$$x(t) = x_0 e^{i\omega t} \ . \tag{7.3}$$

As it was stated for the case of undamped systems, force $F(t)$ is a real quantity and should be expressed as $F = \Re(f_0 e^{i\omega t})$; in the same way, the expression of the displacements should mention explicitly the real part, since the complex notation is used to express quantities that have a harmonic time history as projections on the real axis of vectors that rotate in the complex plane. The symbol $\Re$ is, however, usually omitted.

Phasing is much more important for damped systems than for conservative ones, since damping causes the response to be out of phase with respect to the excitation. The amplitudes f_0 and x_0 are then complex quantities, with different phasing as shown in Fig. 7.1.

Remark 7.1 *The response to a harmonic excitation is harmonic, with the same frequency of the forcing function but out of phase with respect to the latter.*

By introducing a harmonic time history for both excitation and response, the differential equation of motion can be transformed into an algebraic equation yielding the complex amplitude of the response

$$\left(-m\omega^2 + i\omega c + k\right)x_0 = \begin{cases} f_0, \\ (i\omega c + k)x_{A_0}, \\ -m\omega^2 x_{A_0}, \end{cases} \tag{7.4}$$

for excitation provided by a force, by the motion of the supporting point A using an inertial coordinate, and by the motion of the supporting point

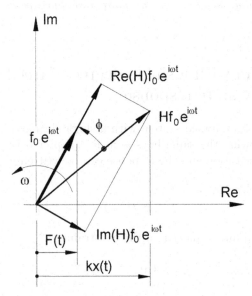

FIGURE 7.1. Response of a system with viscous damping as seen in the complex plane.

A using a relative coordinate, respectively. The coefficient of the unknown x_0 in Eq. (7.4) is the dynamic stiffness of the system already defined in Chapter 3

$$k_{dyn} = \left(-m\omega^2 + ic\omega + k \right) = k\left[1 - \left(\frac{\omega}{\omega_n} \right)^2 + 2i\zeta \left(\frac{\omega}{\omega_n} \right) \right]. \qquad (7.5)$$

Remark 7.2 *The dynamic stiffness, as well as its reciprocal, the dynamic compliance or receptance, is complex.*

The real part of the frequency response $H(\omega)$ gives the component of the response that is in phase with the excitation. The imaginary part gives the component in quadrature, which lags the excitation by a phase angle of 90°. The expressions for the real and imaginary parts of $H(\omega)$, its amplitude, and phase are

$$\Re(H) = k\frac{k - m\omega^2}{(k - m\omega^2)^2 + c^2\omega^2} = \frac{1 - \left(\frac{\omega}{\omega_n} \right)^2}{\left[1 - \left(\frac{\omega}{\omega_n} \right)^2 \right]^2 + \left(2\zeta\frac{\omega}{\omega_n} \right)^2},$$

$$\Im(H) = k\frac{-c\omega}{(k - m\omega^2)^2 + c^2\omega^2} = \frac{-2\zeta\frac{\omega}{\omega_n}}{\left[1 - \left(\frac{\omega}{\omega_n} \right)^2 \right]^2 + \left(2\zeta\frac{\omega}{\omega_n} \right)^2},$$

$$|H| = \frac{k}{\sqrt{(k - m\omega^2)^2 + c^2\omega^2}} = \frac{1}{\sqrt{\left[1 - \left(\frac{\omega}{\omega_n} \right)^2 \right]^2 + \left(2\zeta\frac{\omega}{\omega_n} \right)^2}},$$

$$(7.6)$$

$$\Phi = \arctan\left(\frac{-c\omega}{k - m\omega^2} \right) = \arctan\left[\frac{-2\zeta\left(\frac{\omega}{\omega_n} \right)}{1 - \left(\frac{\omega}{\omega_n} \right)^2} \right].$$

The situation in the complex plane at time t is described in Fig. 7.1.

The absolute value of the frequency response is the magnification factor. It is plotted together with the phase angle Φ as a function of the forcing frequency in Figs. 7.2a and b for different values of the damping ratio ζ. Logarithmic axes are often used, and the scale of the ordinates is expressed in decibels (Fig. 7.2c). This plot is referred to as the *Bode diagram*.

The resonance occurs when the excitation frequency ω is close to the natural frequency of the undamped system ω_n, but does not exactly coincide with it, and its amplitude is limited. In this zone the damping of the system, however small it may be, becomes the governing factor because, at resonance, the inertia force exactly balances the elastic force and, consequently, only the damping force can balance the excitation $F(t)$.

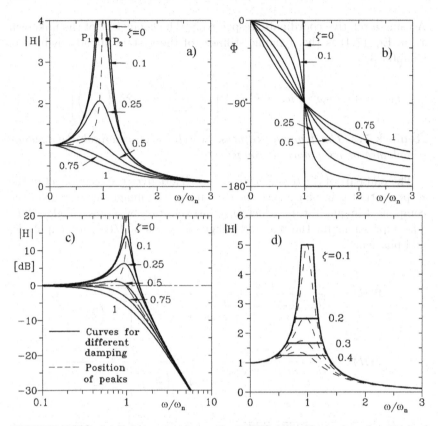

FIGURE 7.2. Bode diagram, i.e., magnification factor and phase as functions of the forcing frequency: (**a**) and (**b**) linear scales, (**c**) logarithmic scale for frequency and dB scale for amplitudes. (**d**) Frequency response of lightly damped systems approximated by using the response of the undamped system and shaving the peak at the value expressed by Eq. (7.7). Comparison with the exact solution.

Remark 7.3 *In a range close to the resonance, which is said to be controlled by damping, the presence of damping cannot be neglected; in the other frequency ranges, the behavior of the system can often be very well approximated using an undamped model.*

Remark 7.4 *At the natural frequency of the undamped system, the phase lag is exactly 90°, regardless of the value of the damping.*

If $\zeta < 1/2$, the frequency at which the peak amplitude occurs is

$$\omega_p = \omega_n \sqrt{1 - 2\zeta^2}\,.$$

It shifts toward the lower values of ω with increasing damping and does not coincide with the frequency of the free oscillations of the system. For greater values of ζ, the curve $H(\omega)$ does not show a peak and the maximum value occurs at $\omega = 0$. If the system is lightly damped and ζ^2 is negligible compared with unity, then the maximum values of the amplitude and magnification factor are, respectively,

$$|x_0|_{\max} \approx \frac{f_0}{c\omega_n}, \qquad |H|_{\max} \approx \frac{1}{2\zeta}. \qquad (7.7)$$

The term

$$Q = \frac{1}{2\zeta} \qquad (7.8)$$

is often called the *quality factor* and symbol Q is used to represent it.

On the curve obtained for $\zeta = 0.1$ in Fig. 7.2a, points P_1 and P_2, at which the amplitude is equal to the peak amplitude divided by $\sqrt{2}$, are reported. They are often defined as half-power points and correspond to an attenuation of about 3 dB with respect to the maximum amplitude. The frequency interval $\Delta\omega$ between points P1 and P2 is often called the *half-power bandwidth* and is used as a measure of the sharpness of the resonance peak. If damping is small enough to allow the usual simplifications (i.e., ζ^2 is negligible compared with unity), the frequencies corresponding to such points and the half-power bandwidth are

$$\omega_{P_1} \approx \omega_n \sqrt{1 - 2\zeta}, \qquad \omega_{P_2} \approx \omega_n \sqrt{1 + 2\zeta}, \qquad \Delta\omega \approx 2\zeta\omega_n. \qquad (7.9)$$

The frequency response of a lightly damped system can be approximated by the frequency response of the corresponding undamped system except for the frequency range spanning from point P_1 to point P_2, where the amplitude can be considered constant, its value being expressed by Eq. (7.7). As shown in Fig. 7.2d, this approximation still holds for values of damping as high as $\zeta = 0.25 - 0.30$.

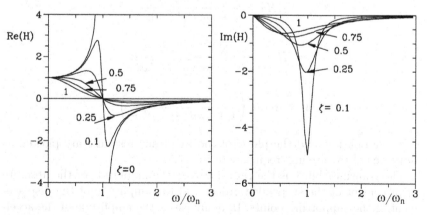

FIGURE 7.3. Real and imaginary parts of the frequency response as functions of the forcing frequency.

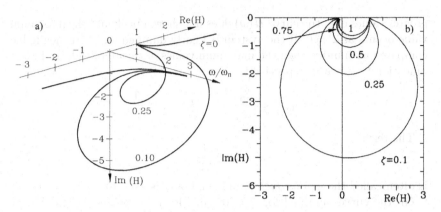

FIGURE 7.4. (a) Same as Fig. 7.3, but as a tridimensional plot. (b) Nyquist diagram for a system with a single degree of freedom.

Instead of plotting the amplitude and the phase of the frequency response, it is possible to separately plot its real and imaginary parts (Fig. 7.3). The two plots of Fig. 7.3 can be combined in the three-dimensional plot of Fig. 7.4a. The projection of the latter on the complex plane is the so-called Nyquist diagram (Fig. 7.4b).

If the excitation is provided by the harmonic motion of the supporting point A, a frequency response $H(\omega) = x_0/x_{A_0}$ can be defined. By separating the real part from the imaginary part, the following values of the magnification factor and phase lag are readily obtained:

$$\Re(H) = \frac{k(k - m\omega^2) + c^2\omega^2}{(k - m\omega^2)^2 + c^2\omega^2},$$

$$\Im(H) = \frac{-cm\omega^3}{(k - m\omega^2)^2 + c^2\omega^2},$$

$$|H| = \frac{\sqrt{k^2 + c^2\omega^2}}{\sqrt{(k - m\omega^2)^2 + c^2\omega^2}},$$

$$\Phi = \arctan\left(\frac{-cm\omega^3}{k(k - m\omega^2) + c^2\omega^2}\right).$$

(7.10)

The amplitude and the phase of the frequency response are plotted as functions of the forcing frequency in Fig. 7.5.

The transmissibility is the ratio between the amplitude of the absolute displacement of the suspended object and the amplitude of the displacement of the supporting points. In many cases, the amplitude of the acceleration is more important than the amplitude of the displacement (Fig. 7.6a).

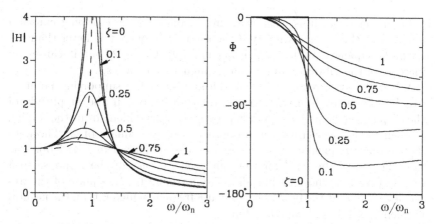

FIGURE 7.5. Same as Fig. 7.2, but with the excitation provided by the harmonic motion of the supporting point. *Full line* indicates curves for different values of damping; *dashed line* indicates line connecting the peaks.

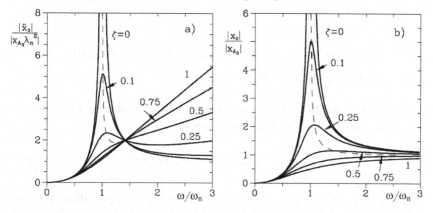

FIGURE 7.6. Non-dimensional response of a system excited by the motion of the supporting point: (a) amplitude of the absolute acceleration of point P as a function of the driving frequency, (b) displacement expressed in relative coordinates. The same response holds for the case of a system excited by a forcing function whose amplitude is proportional to the square of the frequency. *Full line* indicates curves for different values of damping; *dashed line* indicates line connecting the peaks.

Another problem related to the insulation of mechanical vibrations is that of reducing the excitation exerted on the supporting structure by a rigid body on which a force variable in time is acting. The ratio between the amplitude of the force exerted by the spring–damper system on the supporting point $kx + c\dot{x}$ and the amplitude of the excitation $F(t)$ is also referred to as transmissibility of the suspension. As in the case of undamped systems, the value of the transmissibility so defined is the same as obtained in Eq. (7.10) and the two ratios are referred to by the same name.

From Fig. 7.6a, it is clear that any increase of the damping causes a decrease of the transmissibility if the exciting frequency is lower than the natural frequency of the suspension multiplied by $\sqrt{2}$, but it causes an increase of the vibration amplitude at higher frequencies.

When the system is excited by the motion of the supporting point, it can be expedient to resort to relative coordinates. If the amplitude of the displacement of the supporting point is independent from the forcing frequency, the acceleration is proportional to the square of ω. The non-dimensional frequency response for this case is shown in Fig. 7.6b. The same figure can be used for the more general case of the response to a forcing function whose amplitude is proportional to the square of the frequency. Note that the peak is located at a frequency higher than the natural frequency of the undamped system.

All frequency responses, like the mobility, the mechanical impedance, the inertance, and its reciprocal, the dynamic mass, are complex in the case of damped systems.

7.2   System with a single degree of freedom: nonstationary response

The complete solution of the equation of motion is obtained by adding a particular integral to the complementary function

$$x = K^* e^{-\zeta \omega_n t} e^{i \omega_n \sqrt{1-\zeta^2}t} + H(\omega)\frac{f_0}{k}e^{i\omega t} \qquad (7.11)$$

and computing the complex constant K^* from the initial conditions. Owing to damping, the first term of Eq. (7.11) now tends to zero, often quite quickly, while the second one has a constant amplitude; as a consequence, when studying the response of a damped system to harmonic excitation, usually only the latter is considered. As already stated, there are, however, cases in which the initial transient cannot be neglected, particularly when dealing with lightly damped systems or when the forcing function is applied to a system that is at rest: in the latter case oscillations with growing amplitude usually result, until the steady-state conditions are reached.

To show this effect, the cases already seen in Fig. 6.3 but with a damping ratio $\zeta = 0.2$ are studied (Fig. 7.7). Like in the previous study, the system is at standstill in the origin ($x_0 = 0$, $v_0 = 0$) at time $t = 0$ and is excited by a force with only a sine components ($f_1 = 0$). Four different values of ratio ω/ω_n between the forcing frequency and the natural frequency of the undamped system are considered. Clearly the free oscillation damps out after a few oscillations and a solution coinciding with the forced response is obtained. The beat quickly disappears.

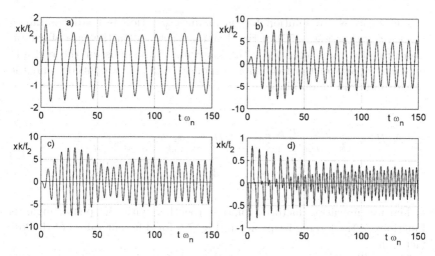

FIGURE 7.7. Same as Fig. 6.3 but for a system with a damping ratio $\zeta = 0.2$.

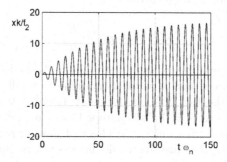

FIGURE 7.8. Resonant response of a system with a single degree of freedom, with $\zeta = 0.2$.

The same effect occurs when the frequency of the forcing function coincides with the natural frequency. The time history for the same case studied in Fig. 6.4, but for a damped system with $\zeta = 0.2$ is shown in Fig. 7.8.

The amplitude starts growing linearly, but then settles at the steady-state value.

7.3   System with structural damping

The hysteretic damping model can be used only in an approximated way in the case of free vibrations that for damped systems are necessarily decaying in time. On the contrary, it is perfectly adequate to studying the steady-state response to a harmonic forcing function, since its time history is exactly harmonic.

The steady-state solution for harmonic excitation is readily obtained from the expression (3.57) of the dynamic stiffness: In the case of excitation provided by a force $F(t)$, the following expressions for the real and imaginary parts of the frequency response $H(\omega)$, its amplitude, and the phase angle can be obtained:

$$\Re(H) = \frac{k(k - m\omega^2)}{(k - m\omega^2)^2 + k^2\eta^2}, \quad |H| = \frac{k}{\sqrt{(k - m\omega^2)^2 + k^2\eta^2}},$$

$$\Im(H) = \frac{-k^2\eta}{(k - m\omega^2)^2 + k^2\eta}, \quad \Phi = \arctan\left(\frac{-k\eta}{k - m\omega^2}\right). \tag{7.12}$$

The magnification factor is plotted together with the phase angle as functions of the forcing frequency in Fig. 7.9 for different values of the loss factor η.

The quality factor of a system with a single degree of freedom with structural damping is simply given by

$$Q = |H|_{max} = \frac{1}{\eta}. \tag{7.13}$$

If damping is small, as is usually the case, the shift of the resonance peak between viscous and structural damping is small and the behavior of systems with the two different types of damping is, at least close to the peak, similar. Since in lightly damped systems, i.e., when η^2 can be neglected compared with unity, the effect of damping is important only near the resonance, it is possible to define a constant equivalent damping as

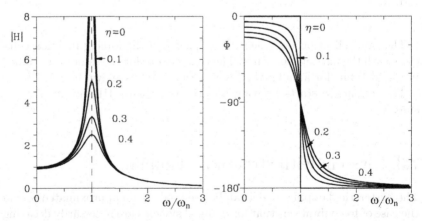

FIGURE 7.9. Same as Fig. 7.2, but for a system with structural damping. *Full line* indicates curves for different values of damping; *dashed line* indicates line connecting the peaks.

$$c_{eq} = \frac{\eta k}{\omega_n}, \qquad \zeta_{eq} = \frac{\eta}{2}. \tag{7.14}$$

In this way, the concept of hysteretic damping can be extended in a simple, although approximated, way also to time-domain equations.

The expression 2ζ is sometimes called "loss factor" in systems with viscous damping.

As an alternative, a Maxwell–Weichert damping tuned at the natural frequency of the system (see Chapter 3) can be used. If a single spring–damper series is put in parallel to the spring with stiffness k, the data of the spring–damper series are

$$k_1 = 2\eta k \ , \ c_1 = \frac{2\eta k}{\omega_n} . \tag{7.15}$$

In case more springs and dampers are used, the values of the relevant parameters found in Section 3.6.2 can be used.

7.4   System with many degrees of freedom

The response of a system with many degrees of freedom to a harmonic forcing function can be computed by writing the generic harmonic excitation in the form

$$\mathbf{f}(t) = \mathbf{f}_0 e^{i\omega t}$$

and the response in the form

$$\mathbf{x}(t) = \mathbf{x}_0 e^{i\omega t} .$$

The differential equation of motion (3.6) can thus be transformed into the algebraic equation

$$\mathbf{K}_{dyn}\mathbf{x}_0 = \mathbf{f}_0. \tag{7.16}$$

The dynamic stiffness matrix

$$\mathbf{K}_{dyn} = -\omega^2 \mathbf{M} + \mathbf{K} + i\omega \mathbf{C} \tag{7.17}$$

is complex when damping is present.

Both vectors $\mathbf{x}_0$ and $\mathbf{f}_0$ are in general complex, and Eq. (7.16) can be rewritten by separating its real and imaginary parts

$$\begin{bmatrix} -\omega^2 \mathbf{M} + \mathbf{K} & -\omega \mathbf{C} \\ \omega \mathbf{C} & -\omega^2 \mathbf{M} + \mathbf{K} \end{bmatrix} \begin{Bmatrix} \Re\left(\mathbf{x}_0\right) \\ \Im\left(\mathbf{x}_0\right) \end{Bmatrix} = \begin{Bmatrix} \Re\left(\mathbf{f}_0\right) \\ \Im\left(\mathbf{f}_0\right) \end{Bmatrix}. \tag{7.18}$$

As opposite to what happens with undamped systems, the equation dealing with the real parts of the forcing function and of the response is not uncoupled with that dealing with the imaginary parts. As a consequence, $\mathbf{x}_0$ is generally not real even if $\mathbf{f}_0$ is real

Remark 7.5 *Even if all the harmonic exciting forces are in phase, i.e., the excitation is coherently phased, the response of the system, although harmonic, is not coherently phased. Not even if the damping is proportional do the various parts of the system oscillate in phase when subjected to forces that are in phase with each other.*

To fully understand the meaning of a complex vector x_0, consider the case of a massless beam on which a number of masses are located (Fig. 7.10a). Consider a representation in which the plane of oscillation of the system (the vertical plane in Figs. 7.10b and c) is the real plane and the plane perpendicular to it is the imaginary plane. Any plane perpendicular to them can be considered a complex plane, in which the real and imaginary axes are defined by the intersections with the real and imaginary planes.

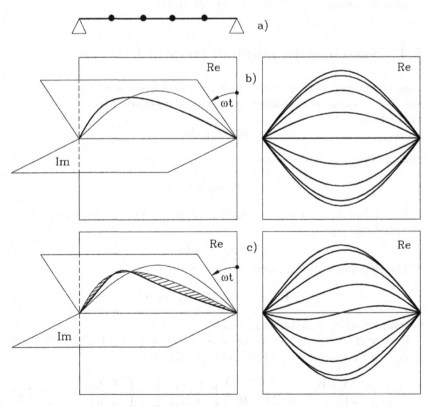

FIGURE 7.10. Meaning of the complex displacement vector x_0: (a) sketch of a beam modeled as a massless beam with concentrated masses, (b) undamped case: the deflected configuration is the projection on the real plane of a line lying on a plane that rotates at angular speed ω, (c) damped case: the *line* that generates the deflected shape is skew.

At the location of each mass, there is a complex plane in which the vector representing the displacement of the mass rotates.

If, as in the case of undamped systems with coherently phased excitation, vector $\mathbf{x}_0$ is real, the situation in a space in which the coordinate planes are the real and imaginary planes is that shown in Fig. 7.10b. The actual deformed shape is then the projection on the real plane of a planar line that rotates at the angular speed ω. The shapes it takes at various instants are consequently all similar; only their amplitudes vary in time.

If vector $\mathbf{x}_0$ is complex (Fig. 7.10c), the deformed shape is the projection on the real plane of a rotating skew line. Consequently, its shape varies in time and no stationary point of minimum deformation (node) or maximum deformation (loop or antinode) exists. As already seen, this last situation also characterizes the case of proportional damping and all the cases that can be reduced to it, at least in an approximate way.

The effect of damping on the frequency responses is that of reducing the resonance peaks and increasing the amplitude at the antiresonances. If the system is highly damped, some of the peaks may disappear completely. The Nyquist diagrams usually have as many loops as degrees of freedom, if the system is lightly damped. With increasing damping, some loops can disappear.

7.5   Modal computation of the response

In case of proportionally damped multi-degree-of-freedom systems, the equations of motion for forced vibrations can be uncoupled and the study reduces to the computation of the response of n uncoupled linear damped systems, like in the case of undamped system. The relevant uncoupled equations of motion are Eqs. (5.51) and Eqs. (5.52), depending on how the eigenvectors have been normalized.

If damping is small, the response of each mode can consequently be approximated as shown in Fig. 7.2d. However, while the amplitudes are approximated very well in this way, the error in the computation of the phases may be large.

If the forcing frequency is close to one of the natural frequencies, the shape of the response is usually very close to the relevant mode shape. Because the mode shapes used for the modal transformation are the real mode shapes of the undamped system and not the complex modes of the damped system, the modal response of the resonant mode is coherently phased even if the forces acting on the system are not. However, even if damping is proportional, the response is not a pure mode shape even exactly in resonance, because the amplitude of the resonant mode is not infinitely larger than the amplitude of the other modes, as would occur in undamped systems.

Remark 7.6 *At resonance, the phase lag between the modal force and the modal response of the resonant mode is exactly* $90°$.

Also in the case of damped systems it is possible to take into account only a limited number of modes, usually the lower order ones. The problem of taking into account the contribution of the neglected (higher order) modes to the static deformation is the same as already seen for the undamped system.

Equation (6.23) yielding the ith modal response can be written in the form

$$\eta_{0i} = \frac{1}{\omega_{ni}^2 - \omega^2 + 2i\zeta\omega\omega_{ni}} \, \mathbf{q}_i^T \mathbf{f}_0 \qquad \text{for } i = 1, 2, ..., n, \tag{7.19}$$

in which the eigenvectors have been assumed to be normalized in such a way that the modal masses have unit values. By operating in the same way seen for the undamped system, the approximated value of the response (6.30) becomes

$$\mathbf{x} \approx \left(\sum_{i=1}^{m} \frac{1}{\omega_{ni}^2 - \omega^2 + 2i\zeta\omega\omega_{ni}} \mathbf{q}_i \mathbf{q}_i^T \right) \mathbf{f}_0 - \left(\sum_{i=1}^{m} \frac{1}{\omega_{ni}^2} \mathbf{q}_i \mathbf{q}_i^T \right) \mathbf{f}_0 + \mathbf{K}^{-1} \mathbf{f}_0, \tag{7.20}$$

i.e.,

$$\mathbf{x} \approx \sum_{i=1}^{m} \frac{\omega^2 - 2i\zeta\omega\omega_{ni}}{\omega_{ni}^2} \eta_{0i} \mathbf{q}_i + \mathbf{K}^{-1} \mathbf{f}_0. \tag{7.21}$$

In the same way already seen for undamped systems, an approximated computation of the response can thus be performed by first obtaining the first m eigenvectors (of the undamped system) and computing the first m modal responses η_{0i} and then recombining them using the relationship

$$\mathbf{x} \approx \mathbf{\Phi}^{**} \boldsymbol{\eta}^*, \tag{7.22}$$

where

$$\mathbf{\Phi}^{**} = [\mathbf{q}_1, \mathbf{q}_2, \ldots, \mathbf{q}_m, \mathbf{r}_1] \tag{7.23}$$

is the same seen in Section 6.3 and $\boldsymbol{\eta}^*$ is the vector of the modified modal coordinates

$$\boldsymbol{\eta}^* = [\frac{\omega^2 - 2i\zeta\omega\omega_{n1}}{\omega_{n1}^2} \eta_{01}, \ldots, \frac{\omega^2 - 2i\zeta\omega\omega_{nm}}{\omega_{nm}^2} \eta_{0m}, 1]. \tag{7.24}$$

Although modal uncoupling can be applied exactly only in the case of proportional damping, such a procedure can often be applied in cases where the equations of motion could not be uncoupled theoretically. In particular, if the system is lightly damped, as is common in structural dynamics, the response of each mode to a harmonic excitation is close to that of the corresponding undamped system except in a narrow frequency range centered on the resonance of the mode itself, i.e., except in that frequency range in which the response is governed by damping.

Remark 7.7 *This statement amounts to state that only the elements on the main diagonal of the modal-damping matrix are important in determining the response of the system, and even them only close to the resonances. Neglecting the out-of-diagonal elements (even if they are of the same order of magnitude of those which are retained) is then acceptable.*

The eigenvectors of the undamped system can be used as a reference frame in the space of the configurations; consequently, the motion of the damped system can be expressed in terms of modal coordinates, whether or not the damping is proportional or small. If modal coupling is strong, all eigenvectors can be present in the response of the system at any frequency and it is not possible to understand a priori how much each of them affects the global response.

When damping is high enough to prevent from neglecting modal coupling, the iterative procedure based on Eq. (5.53) can be used. The relevant equation in the frequency domain is

$$\left(-\omega^2\overline{\mathbf{M}} + i\omega\overline{\mathbf{C}}_p + \overline{\mathbf{K}}\right)\boldsymbol{\eta}_0 = -i\omega\overline{\mathbf{C}}_{np}\boldsymbol{\eta}_0 + \overline{\mathbf{f}}_0 . \tag{7.25}$$

The equation obtained by neglecting matrix $\mathbf{C}_{np}$ is first solved. A solution $\boldsymbol{\eta}_0^{(0)}$, corresponding to a system with generalized proportional damping, is thus obtained. This solution is introduced on the right-hand side of Eq. (7.25), and a second-approximation solution $\boldsymbol{\eta}_0^{(1)}$ is obtained. The iterative procedure can continue until the difference between two subsequent solutions is smaller than any given small quantity. Either a Jacobi or a Gauss Siedel iterative scheme can be used; the second is generally faster (see Appendix A). The convergence of the iterative scheme is fast, even if the distribution of damping is far from being proportional. The simplification of the computations obtainable in this way is noticeable, particularly in the case of systems with many degrees of freedom.

Example 7.1 *Compute the element H_{33} of the frequency response of the system in Example 1.2, taking also damping into account.*

The frequency response H_{33} is reported in Fig. 7.11, comparing the results directly obtained with those earlier computed using the values of the modal damping, to obtain the modal responses and then transforming the results to physical coordinates.

The two curves are almost everywhere exactly superimposed, showing the very good approximation obtainable when using modal damping. The dashed line refers to the undamped system, for comparison.

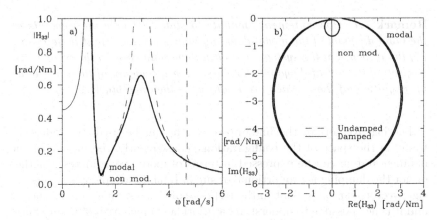

FIGURE 7.11. Frequency response H_{33}: (a) amplitude as a function of frequency; (b) Nyquist diagram.

The first mode is less damped than the other two, and the third one is so much damped that the resonance peak disappears completely. It must be noted that the third peak is, at any rate, very narrow in the response of the undamped system. In the Nyquist plot (Fig. 7.11b) the first peak generates a loop that is far larger than the one related to the second resonance.

7.6   Multi-degrees of freedom systems with hysteretic damping

As stated in Section 3.4.1, the stiffness matrix $\mathbf{K}^*$ is complex, with a real part $\mathbf{K}'$ defining the conservative properties of the system and an imaginary part $\mathbf{K}''$ defining the dissipative properties. The dynamic stiffness matrix for a system with structural damping is

$$\mathbf{K}_{dyn} = -\omega^2\mathbf{M} + \mathbf{K}' + i\mathbf{K}''.$$

By separating the real and imaginary parts of vectors $\mathbf{x}_0$ and $\mathbf{f}_0$, the response can be computed from the equation

$$\begin{bmatrix} -\omega^2\mathbf{M} + \mathbf{K}' & -\mathbf{K}'' \\ \mathbf{K}'' & -\omega^2\mathbf{M} + \mathbf{K}' \end{bmatrix} \left\{ \begin{array}{c} \Re(\mathbf{x}_0) \\ \Im(\mathbf{x}_0) \end{array} \right\} = \left\{ \begin{array}{c} \Re(\mathbf{f}_0) \\ \Im(\mathbf{f}_0) \end{array} \right\}. \qquad (7.26)$$

If the loss factor is constant throughout the system, matrices $\mathbf{K}''$ and $\mathbf{K}'$ are proportional and the complex stiffness matrix reduces to

$$\mathbf{K}^* = (1 + i\eta)\mathbf{K}. \qquad (7.27)$$

This is a form similar to that of proportional damping, and the equations of motion can be uncoupled exactly, yielding n equations of motions of the type seen for systems with a single degree of freedom with structural damping.

In the case of systems with many degrees of freedom, it is also possible to define an equivalent viscous damping matrix equal to $\mathbf{K}''$ divided by ω. Because systems with structural damping are very lightly damped, the effects of damping are restricted only in the fields of frequency that are close to the natural frequencies, and the modal uncoupling holds with good approximation. The behavior of the system can thus be studied by uncoupling the equations of motion using the eigenvectors of the undamped system and introducing a constant equivalent damping that does not depend on the frequency

$$\overline{C}_{ieq} = \frac{\overline{K_i}''}{\omega_{ni}}, \qquad \zeta_{ieq} = \frac{\eta_i}{2}. \tag{7.28}$$

The modal damping can easily be measured during a dynamic test by measuring the amplitude at resonance or the half-power bandwidth, or evaluated from data that can be found in the literature.

The equivalent viscous damping matrix in physical coordinates can thus be obtained by

- computing the eigenvectors of the undamped system;

- computing $\overline{\mathbf{K}}''$;

- computing the modal equivalent damping of the various modes $\overline{C}_{ieq}$;

- performing the back-transformation to the physical coordinates

$$\mathbf{C}_{eq} = \mathbf{\Phi}^{-T} \overline{C}_{ieq} \mathbf{\Phi}^{-1}. \tag{7.29}$$

Again, a Maxwell–Weichert damper can be associated to each modal system. The values of the stiffnesses and damping coefficients of the various springs and dampers can be computed in the usual way.

7.7   Response to periodic excitation

The response to a periodic excitation $F(t)$ (with period T) expressed by Eq. (6.37) can be computed in the same way seen for undamped system, with the difference that now the frequency response $H(\omega)$ is complex. By separating the real and imaginary parts of the frequency response, the following expression for the particular integral of the equation of motion is easily obtained:

$$x(t) = \frac{1}{k} \left\{ a_0 + \sum_{i=1}^{n} [a_i \Re(H(\omega_i)) + b_i \Im(H(\omega_i))] \cos(\omega_i t) + \right.$$
$$\left. + \sum_{i=1}^{n} [b_i \Re(H(\omega_i)) - a_i \Im(H(\omega_i))] \sin(\omega_i t) \right\} , \quad (7.30)$$

where the frequency of the ith harmonic of the forcing function is

$$\omega_i = \frac{2\pi i}{T} .$$

7.8   The dynamic vibration absorber

A dynamic vibration absorber is basically a spring–mass–damper system that is added to any vibrating system with the aim of reducing the amplitude of the vibrations of the latter. If the damper or the spring is missing, an undamped vibration absorber or a Lanchester damper (springless vibration absorber) is obtained.

Consider a system consisting of a mass m suspended on a spring with stiffness k on which a force varying harmonically in time with frequency ω and maximum amplitude f_0 is acting. The vibration absorber, consisting of a second mass m_s, a spring of stiffness k_s, and a damper with damping coefficient c, is connected to mass m (Fig. 7.12a). The equation yielding the amplitude of the harmonic response of the system is

$$\left[-\omega^2 \begin{bmatrix} m_s & 0 \\ 0 & m \end{bmatrix} + \begin{bmatrix} k_s & -k_s \\ -k_s & k_s + k \end{bmatrix} + i\omega \begin{bmatrix} c & -c \\ -c & c \end{bmatrix} \right] \begin{Bmatrix} x_{so} \\ x_0 \end{Bmatrix} = \begin{Bmatrix} 0 \\ f_0 \end{Bmatrix} .$$

By introducing the mass ratio μ, the stiffness ratio χ, the tuning ratio τ, and the nondimensional frequency ω^*

$$\mu = \frac{m_s}{m} , \quad \chi = \frac{k_s}{k} , \quad \tau = \frac{\chi}{\mu} , \quad \omega^* = \frac{\omega}{\omega_n} = \omega \sqrt{\frac{m}{k}} ,$$

the frequency response $H_{22}(\omega)$ can be easily computed:

$$|H_{22}| = \frac{1}{k} \sqrt{\frac{\left(\tau - \omega^{*2}\right)^2 + c^{*2}\omega^{*2}}{f^2(\omega^*) + c^{*2}\omega^{*2}g^2(\omega^*)}} , \quad (7.31)$$

where

$$f(\omega^*) = \omega^{*4} - \omega^{*2}(1 + \chi + \tau) , \quad g(\omega^*) = 1 - \omega^{*2}(1 + \mu),$$

and

$$c^* = \frac{c}{\mu\sqrt{km}} .$$

The tuning ratio is the square of the ratio between the natural frequency of the original system ω_n and that of the vibration absorber. If the vibration absorber is undamped, the amplitude of the motion of mass m is vanishingly small if

$$\tau = \omega^{*2} \, ,$$

i.e., the square of the nondimensional frequency of the excitation coincides with the tuning ratio. The frequency response computed for a mass ratio $\mu = 0.2$ and a tuning ratio

$$\chi = \frac{\mu}{(1 + \mu)^2}$$

is shown in Fig. 7.12b, curve labeled $c = 0$. The presence of an undamped vibration absorber is successful in completely damping the vibration at a given frequency but produces two new resonance peaks at the frequencies at which function $f(\omega^*)$ vanishes (Fig. 7.12c).

The working of the undamped vibration absorber can be easily understood by noting that at the frequency at which the vibration absorber is tuned, the motion of mass m_s is large enough to produce a force on mass m that balances force F. Consequently, the amplitude of the motion of mass m_s increases when the mass ratio μ decreases and tends to infinity when m_s tends to zero.

If the amplitude of motion of mass m is to be reduced also outside a narrow range near frequency ω_n, the use of a damper is mandatory.

All response curves, obtained with any value of c, pass through points A, B, and, C and lie in the shaded zone in Fig. 7.12b bounded by the two limiting cases of the undamped system and that with infinitely large damping. The latter coincides with a system with a single degree of freedom with mass $m + m_s$ and stiffness k. Such curves have a maximum in the zone included between points B and C in the case of high damping and two maxima outside the field BC in the case of small damping.

A reasonable way of optimizing the vibration absorber is to look for a value of the damping causing the maxima to coincide with points B and C and to tune the system (i.e., select the value of k_s) in a way so as to obtain the same value of the response in B and C.

The latter condition can be shown to be obtained[1] for the optimum value of the tuning ratio

$$\chi_{opt} = \frac{\mu}{(1 + \mu)^2} \, .$$

Strictly speaking, no value of the damping can cause the two peaks to be located simultaneously at points B and C. The value of the damping allowing one to meet this condition with good approximation and the approximated value of the maximum amplitude are

---

[1]D. Hartog, *Mechanical Vibrations*, McGraw-Hill, New York, 1956, pp. 93ff

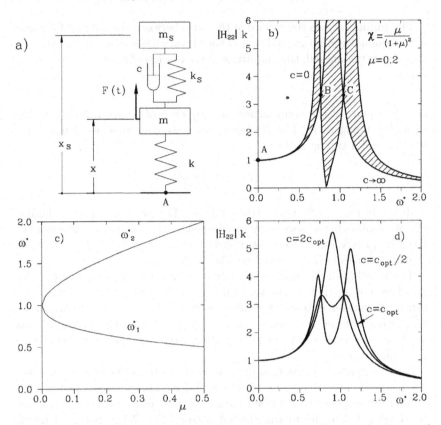

FIGURE 7.12. Vibration absorber applied to a system with a single degree of freedom: (a) sketch of the system; (b) limiting cases (damping tending to zero and infinity) for systems with optimum tuning; (c) natural frequencies of the undamped system as functions of the mass ratio μ; (d) amplitude of the response of the system in (b), but with three different values of damping.

$$c_{opt} = \sqrt{km}\sqrt{\frac{3\mu^3}{2(1+\mu)^3}}, \qquad |H|_{max} = \sqrt{1+\frac{2}{\mu}}.$$

The frequency response of a system with mass ratio $\mu = 0.2$, optimum tuning ratio, and three values of damping is plotted in Fig. 7.12d.

Another different type of dynamic vibration absorber is the so-called Lanchester damper, which consists of a mass connected to the system through a damper. Originally it had a dry-friction damper, but if the damper is of the viscous type, it is basically a damped vibration absorber without restoring spring. The frequency response can be easily computed, obtaining

$$|H_{22}| = \frac{1}{k} \sqrt{\frac{\omega^{*2} + c^{*2}}{\omega^{*2}(\omega^{*2} - 1)^2 + \omega^{*2}[1 - (\mu + 1)\omega^{*2}]^2}}.$$

Following the procedure used for the preceding case, the optimum value of the damping and the corresponding maximum value of the frequency response can be computed:

$$c_{opt} = \sqrt{km}\sqrt{\frac{2\mu^2}{(2 + \mu)(1 + \mu)}}, \qquad |H|_{max} = 1 + \frac{2}{\mu}.$$

A comparison between the frequency responses of a system with a single degree of freedom with a dynamic and a Lanchester vibration absorber is shown in Fig. 7.13. In both cases the mass ratio is $\mu = 0.2$.

Example 7.2 *Consider the system with two degrees of freedom shown in Fig. 7.14a. The data are the following: $m_1 = m_2 = 5$ kg, $k_1 = k_2 = k_3 = 5$ kN/m. The system is excited by a harmonic force f_1 applied on mass m_1.*

A dynamic vibration absorber is located on mass m_1. Assuming that the mass of the vibration absorber in $m_s = 1$ kg, compute the stiffness k_s and the damping c_s of the vibration absorber that minimizes the dynamic response on mass m_1. Plot the frequency response and compare it with that of the undamped system.

Repeat the computation for a Lanchester damper.

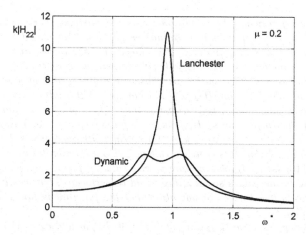

FIGURE 7.13. Comparison between the frequency responses of a system with a single degree of freedom with a dynamic and a Lanchester vibration absorber.

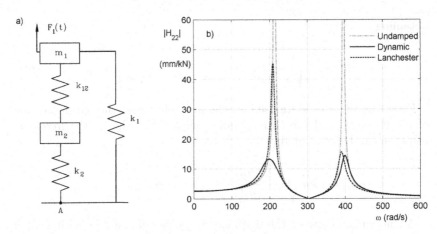

FIGURE 7.14. Dynamic vibration absorber. (a) System on which the vibration absorber is applied. (b) Frequency responses of the dynamic vibration absorber and of the Lanchester damper.

Once the dynamic vibration absorber is located on mass m_1, the dynamic stiffness of the system is

$$
\mathbf{K}_{dyn} = -\omega^2 \begin{bmatrix} m_s & 0 & 0 \\ 0 & m_1 & 0 \\ 0 & 0 & m_2 \end{bmatrix} + i\omega \begin{bmatrix} c_d & -c_d & 0 \\ -c_d & c_d & 0 \\ 0 & 0 & 0 \end{bmatrix} +
$$

$$
+ \begin{bmatrix} k_d & -k_d & 0 \\ -k_d & k_d + k_1 + k_{12} & -k_{12} \\ 0 & -k_{12} & k_2 + k_{12} \end{bmatrix}.
$$

The frequency response to be minimized is the modulus of element H_{22} of matrix $H(\omega)$. The optimization cannot be performed in closed form owing to the complexity of the analysis. It is, however, easy to use any numerical optimization method, and in the following the simplest approach is followed. A number of values of k_s and c_s are chosen in a given interval, and for each pair of values the function $|H_{22}(\omega)|$ is computed. A three-dimensional plot of its maximum value $\max(|H_{22}|)$ as a function of k_s and c_s is obtained. A map of the surface $\max(|H_{22}|)(k_s, c_s)$ for $50 < c_s < 300$ Ns/m (with increments 5 Ns/m) and $10 < k_s < 60$ kN/m (with increments 1 kN/m) is reported in Fig. 7.15a. It is clearly a valley, with a minimum for $k_s \approx 40$ kN/m and $c_s \approx 110$ Ns/m. To obtain more precise values a further computation in a range close to the values so identified could be performed, but was considered useless mostly due to the consideration that the function is fairly flat about the minimum.
The frequency response $|H_{22}(\omega)|$ for $k_s = 40$ kN/m and $c_s = 110$ Ns/m is shown in Fig. 7.14b. As it could be expected, the peaks have the same height, showing that the optimization was performed accurately.

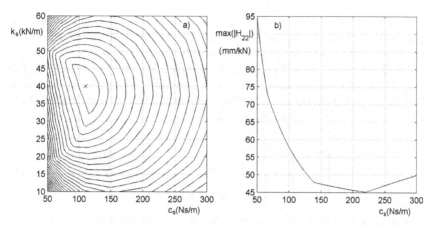

FIGURE 7.15. Dynamic vibration absorber (**a**) and Lanchester damper (**b**). Plots used to optimize the parameters.

The study was repeated for the Lanchester damper. Now $k_s = 0$ and there is just one parameter, c_s. The one-dimensional optimization is easy but, although a closed-form solution could be obtained without great computational difficulties, the function $\max\left(|H_{22}|\right)(c_s)$ for $50 < c_s < 300$ Ns/m (with increments 1 Ns/m) was plotted (Fig. 7.15b). The value $c_s = 210$ Ns/m was chosen.
The frequency response $|H_{22}(\omega)|$ for the selected value of c_s is shown in Fig. 7.14b. In this case it is impossible to reduce the height of the first peak as much as in the previous case, while the second peak has roughly the same height as in the case of the dynamic vibration absorber.

7.9 Parameter identification

In the preceding sections, attention was paid to the computation of the dynamic response of a system whose characteristics are known. Very often, however, the opposite problem must be solved: The behavior of the system has been investigated experimentally and a mathematical model has to be obtained from the experimental results. First consider the case of a system with a single degree of freedom and assume that the response $x(t)$ and the excitation $F(t)$ are known in a number m of different instants and that the corresponding velocities and accelerations are also known. By writing the equation of motion of the system m times, the following equation can be obtained

$$
\begin{bmatrix}
\ddot{x}_1 & \dot{x}_1 & x_1 \\
\ddot{x}_2 & \dot{x}_2 & x_2 \\
\cdots & \cdots & \cdots \\
\ddot{x}_m & \dot{x}_m & x_m
\end{bmatrix}
\left\{ \begin{array}{c} m \\ c \\ k \end{array} \right\}
= \left\{ \begin{array}{c} f_1 \\ f_2 \\ \cdots \\ f_m \end{array} \right\}. \tag{7.32}
$$

Equation (7.32) is a set of m linear equations with three unknowns, the parameters of the system to be determined. A subset of three equations is then required to solve the problem. Actually, the situation is more complex. All measurements are affected by some errors, and the results obtainable from a set of three measurements are unreliable. To obtain more reliable results, it is better to retain all rows of the matrix of the coefficients of Eq. (7.32) and to resort to its pseudo-inverse

$$
\left\{ \begin{array}{c} m \\ c \\ k \end{array} \right\} = \left[\begin{array}{ccc} \ddot{x}_1 & \dot{x}_1 & x_1 \\ \ddot{x}_2 & \dot{x}_2 & x_2 \\ \ldots & \ldots & \ldots \\ \ddot{x}_m & \dot{x}_m & x_m \end{array} \right]^{\dagger} \left\{ \begin{array}{c} f_1 \\ f_2 \\ \ldots \\ f_m \end{array} \right\}. \tag{7.33}
$$

The pseudo-inverse $\mathbf{A}^{\dagger}$ of matrix $\mathbf{A}$ can be computed as

$$
\mathbf{A}^{\dagger} = (\mathbf{A}^T \mathbf{A})^{-1} \mathbf{A}^T,
$$

but this simple approach based on matrix inversion is increasingly less efficient for large matrices. Algorithms based on singular value decomposition or QR factorization are both more accurate and more computationally efficient.

To avoid introducing the velocities and accelerations together with the displacements into Eq. (7.33), it is possible to work in the frequency domain. In this case, the complex amplitudes of the response $x_0(\omega)$ and the corresponding complex amplitudes of the excitation $f_0(\omega)$ at m values of the frequency are measured, and Eq. (7.32) can be transformed into a set of m complex equations or $2m$ real equations

$$
\left[\begin{array}{ccc} -\omega^2 x_{0_1} & i\omega x_{0_1} & x_{0_1} \\ -\omega^2 x_{0_2} & i\omega x_{0_2} & x_{0_2} \\ \ldots & \ldots & \ldots \\ -\omega^2 x_{0_m} & i\omega x_{0_m} & x_{0_m} \end{array} \right] \left\{ \begin{array}{c} m \\ c \\ k \end{array} \right\} = \left\{ \begin{array}{c} f_{0_1} \\ f_{0_2} \\ \ldots \\ f_{0_m} \end{array} \right\}. \tag{7.34}
$$

In the case of systems with many degrees of freedom, everything gets more complex as the number of parameters to be estimated becomes greater, but the computations can follow the same lines shown for systems with a single degree of freedom. Also, in this case both time-domain and frequency-domain methods are possible, and many procedures have been proposed and implemented.

The identification of the modal parameters of large mechanical systems is the main object of experimental modal analysis, which is, in itself, a specialized branch of mechanics of vibrations. It has been the subject of many books and papers in recent years. The algorithms used are often influenced by the hardware that is available for the acquisition of relevant data and subsequent computations. The recent advances in the field of computers and electronic instrumentation are causing steady advancements to take place in this field.

7.10   Exercises

Exercise 7.1 *Consider the same machine of Exercise 6.1 (with a single rotor). Use an elastomeric supporting system that, when tested at the frequencies of 50 and 200 Hz, is found to have a stiffness of 500 and 800 kN/m, respectively, and a damping of 980 and 450 Ns/m, respectively. Using the models of complex stiffness and complex damping to model the supporting system, compute the force exerted by the machine on the supporting structure at all speeds up to the maximum operating speed.*

Exercise 7.2 *Study the effect of an undamped vibration absorber with mass $m_a = 0.8$ kg applied to mass m_1 of Exercise 4.4. Tune it on the first natural frequency of the system. Plot the dynamic compliance H_{22} with and without the dynamic vibration absorber.*

Add a viscous damper between mass m_1 and the vibration absorber. Compute the response of the system with various values of the damping coefficient, trying to minimize the amplitude of the displacement of mass m_2 in a range of frequency between 5 and 30 rad/s.

Exercise 7.3 *Evaluate the elastic, inertial, and damping characteristics of a system with a single degree of freedom using a simple exciter provided of an eccentric mass. The mass of the exciter is 5 kg and the eccentric mass of 0.020 kg is located at a radius of 100 mm (Fig. 7.16)*

Two tests are run at speeds of 100 and 200 rpm, recording the following values of the amplitude and phase of the response:

Test 1: $\omega_1 = 100$ rpm $= 10.47$ rad/s $= 1.667$ Hz; $x_{01} = 0.011$ mm, $\phi_1 = -7°$;

Test 2: $\omega_2 = 200$ rpm $= 20.94$ rad/s $= 3.333$ Hz; $x_{02} = 0.165$ mm, $\phi_2 = -64°$.

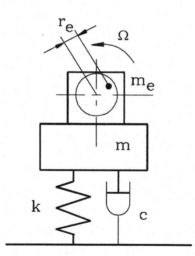

FIGURE 7.16. Sketch of a system with a single degree of freedom with the exciter.

Exercise 7.4 *Consider the damped system with 3 degrees of freedom studied in Examples 1.2 and 3.1. Assume that the excitation is due to the rotation of the supporting structure about the axis of the beams, with law*

$$\theta_A = \theta_0 \sin(\omega t).$$

Compute the response in terms of physical coordinates $(\theta_{0i}/\theta_{0A})$ and of modal coordinates (η_{0i}/θ_{0A}) in the range $0 \leq \omega \leq 6$.

Exercise 7.5 *Compute the forced response of the system of Exercise 4.4 with a damper added between masses m_1 and m_2 when excited by a motion of the supporting point*

$$x_A = x_{A_0} \sin(\omega t).$$

Use both a non-modal and an iterative approach. Add a further damper between mass m_2 and point A, equal to the one already existing between mass m_1 and m_2, and repeat the analysis. Data: $c_{12} = 40$ Ns/m, $x_{A_0} = 5$ mm, $\omega = 30$ rad/s.

8

Response to Nonperiodic Excitation

The response to a nonperiodic excitation can be computed in closed form only in a few selected cases. In general the solution can be obtained through Laplace transforms or by integrating numerically the equations of motion in the time domain, with the latter approach becoming increasingly popular.

8.1 Impulse excitation

When a large force acts on the system for a short time, as in the case of shock loads, the impulsive model that assumes that a force tending to infinity acts for a time tending to zero can be used. This model is based on the unit-impulse function $\delta(t)$ (or Dirac's δ), defined by the relationships

$$\begin{cases} \delta = 0 & \text{for } t \neq 0 \\ \delta = \infty & \text{for } t = 0 \end{cases} \qquad \int_{-\infty}^{\infty} \delta(t)dt = 1 \ . \qquad (8.1)$$

The impulse excitation can thus be expressed as

$$F = f_0 \delta(t) \ .$$

Remark 8.1 *The impulse function δ has the dimension of the reciprocal of a time $[s^{-1}]$ and f_0 has the dimensions of an impulse $[Ns]$. Because the impulse of the function $\delta(t)$ has a unit value, the value of f_0 is that of the total impulse of force $F(t)$.*

The response to an impulse excitation is easily computed: It is sufficient to observe that in the infinitely short period of time in which the impulsive force acts, all other forces are negligible compared to it. The momentum theorem can be applied to compute the conditions of the system just after the impulsive force has been applied from those related to the instant before its application.

The position x_0 after the impulse is equal to that before the impulse, while the velocity v_0 is equal to the one before the impulse plus an increment due to an increase of momentum equal to the impulse. Assuming that before the impulse the system with a single degree of freedom is at rest in the origin, it follows that

$$\begin{cases} x_0 = 0, \\ v_0 = \dfrac{f_0}{m} \end{cases}. \tag{8.2}$$

The time history can be computed from the equations governing the free behavior of the system, obtaining

$$x(t) = \frac{f_0}{m\omega_n} h(t), \tag{8.3}$$

where

$$\begin{cases} h(t) = \dfrac{1}{\sqrt{1-\zeta^2}} e^{-\zeta\omega_n t} \sin\left(\omega_n \sqrt{1-\zeta^2}\, t\right), \\[2mm] h(t) = \omega_n t e^{-\omega_n t}, \\[2mm] h(t) = \dfrac{1}{2\sqrt{1-\zeta^2}} \left\{ -e^{-\left[\zeta+\sqrt{1-\zeta^2}\right]\omega_n t} + e^{-\left[\zeta-\sqrt{1-\zeta^2}\right]\omega_n t} \right\} \end{cases}.$$

The three expressions of the (nondimensional) impulse response $h(t)$ hold for underdamped, critically damped, and overdamped systems, respectively. The impulse responses with different values of the damping ratio ζ are shown in nondimensional form in Fig. 8.1.

The impulse response $h(t)$ completely characterizes the system. Its Laplace transform can be immediately computed by multiplying the Laplace transform of the Dirac's δ by the transfer function of the system:

$$\tilde{h}(s) = G(s)\tilde{\delta}(s). \tag{8.4}$$

Because the Laplace transform of the Dirac's δ (see Appendix B) has a unit value, the Laplace transform of the impulse response coincides with the transfer function of the system.

In the case of multi-degrees-of-freedom systems it is possible to define a matrix of the impulse responses $\mathbf{H}(t)$: its Laplace transform coincides with the transfer matrix $\mathbf{G}(s)$ and its Fourier transform with the frequency response $\mathbf{H}(\omega)$.

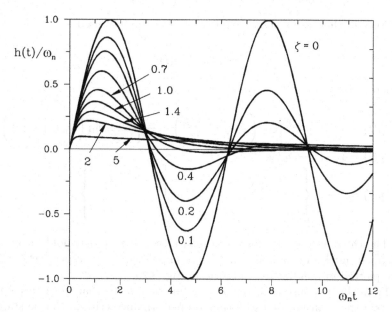

FIGURE 8.1. Response to an impulse excitation $h(t)$ for different values of the damping ratio ζ.

8.2 Step excitation

Another case for which a closed-form solution is available is that of the response to a step excitation. The unit step function $u(t)$ can be defined by the expression

$$\begin{cases} u = 0 & \text{for } t < 0, \\ u = 1 & \text{for } t \geq 0, \end{cases} \tag{8.5}$$

and is just the integral of the impulse function $\delta(t)$.

Its Laplace transform is then

$$\tilde{u}(s) = \frac{1}{s}\tilde{\delta}(s) = \frac{1}{s}. \tag{8.6}$$

The response of the system to the excitation

$$F = f_0 u(t)$$

can be computed by adding the solution obtained for free oscillations to the steady-state response to the constant force f_0.

Remark 8.2 *The step function u is nondimensional, while in the present case f_0 has the dimension of a force (N).*

Also, in this case a simple expression is commonly used

$$x(t) = \frac{f_0}{k} g(t) \,, \tag{8.7}$$

where

$$
\begin{cases}
g(t) = 1 - e^{-\zeta \omega_n t} \left[\cos \left(\omega_n \sqrt{1 - \zeta^2} t \right) + \frac{\zeta}{\sqrt{1 - \zeta^2}} \sin \left(\omega_n \sqrt{1 - \zeta^2} t \right) \right] , \\
g(t) = 1 - (1 - \omega_n t) e^{-\omega_n t} , \\
g(t) = 1 - \frac{1}{2} \left\{ -e^{-\left[\zeta + \sqrt{1 - \zeta^2} \right] \omega_n t} + e^{-\left[\zeta - \sqrt{1 - \zeta^2} \right] \omega_n t} \right\} .
\end{cases}
$$

The three expressions for the response to unit step $g(t)$ hold for under-damped, critically damped, and overdamped systems, respectively. They are plotted in nondimensional form in Fig. 8.2a.

From the response to a step forcing function, some characteristics of the system that can be used to formulate performance criteria can be stated. With reference to Fig. 8.2b, they are the peak time T_p (time required for the response to reach its peak value), the rise time T_r (time required for the response to rise from 10 to 90% of the steady-state value, sometimes from 5 to 95% or from 0 to 100%), the delay time T_d (time required for the response to reach 50% of the steady-state value), the setting time T_s (time required for the response to settle within a certain range, usually 5% but sometimes 2%, of the steady-state value), and the maximum overshot

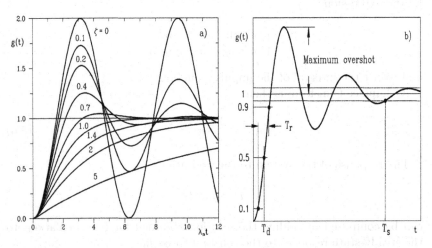

FIGURE 8.2. (a) Response to a step forcing function $g(t)$; (b) dynamic charac-teristics of a system with a single degree of freedom obtained from the response to a step forcing function.

(maximum deviation of the response with respect to the steady-state value). The last item is usually expressed as a percentage of the steady-state value.

The maxima and minima of the time history of an underdamped system are easily obtained by differentiating function $g(t)$ with respect to time and setting the derivative to zero:

$$\frac{dg}{dt} = -e^{\cdot -\zeta\omega_n t}\left[\frac{\sqrt{1+\zeta^2}}{\omega_n}\sin\left(\omega_n\sqrt{1-\zeta^2}t\right)\right] = 0 \ .$$

The solution of this equation is

$$\omega_n\sqrt{1-\zeta^2}t = i\pi \quad \text{for } i = 0, \ 1, \ 2, \ .... $$

Function $g(t)$ has a minimum when i is even and a maximum when i is odd. The value at the ith extremum is

$$g(t) = 1 - e^{-\frac{i\pi\zeta}{\sqrt{1-\zeta^2}}}(-1)^i \ . \tag{8.8}$$

The maximum overshot is thus

$$\frac{g(t)}{\lim_{t\to\infty}g(t)} = e^{-\frac{\pi\zeta}{\sqrt{1-\zeta^2}}} \ . \tag{8.9}$$

Its value is plotted as a function of ζ in Fig. 8.3.

Also in the case of the response to step excitation it is possible to extend the result here obtained to multi-degrees-of-freedom systems. A matrix of the step responses $\mathbf{G}(t)$ can thus be defined.

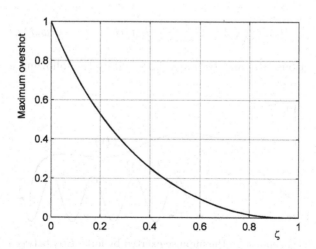

FIGURE 8.3. Maximum overshot as a function of ζ.

8.3   Duhamel's integral

A different approach, applicable to both periodic and nonperiodic forcing functions, is the use of Duhamel's or convolution integral. The impulse of force $F(t)$ acting on the system, computed between time τ and time $\tau + d\tau$ (Fig. 8.4), is simply $F(\tau)d\tau$.

The response of the system to such an impulse can be easily expressed in the form

$$d[x(t)] = \frac{F(\tau)d\tau}{m\omega_n}h(t - \tau), \qquad (8.10)$$

where function $h(t)$ is the response to a unit impulse defined earlier.

The response to the forcing function $F(t)$ can be computed by adding (or better, integrating, as there is an infinity of vanishingly small terms) the responses to all the impulses taking place at all times up to time t

$$x(t) = \int_0^t d[x(t)]d\tau = \frac{1}{m\omega_n} \int_0^t F(\tau)h(t - \tau)d\tau. \qquad (8.11)$$

The integral of Eq. (8.11), usually referred to as Duhamel's integral, allows the computation of the response of any linear system to a force $F(t)$ with a time history of any type. Only in a few selected cases the integration can be performed in closed form; however, the numerical integration of Eq. (8.11) is simpler than the direct numerical integration of the equation of motion.

By introducing the impulse response of an underdamped system, the particular integral of the equation of motion can be expressed in the more compact form

$$x(t) = A(t)\sin\left(\sqrt{1 - \zeta^2}\omega_n t\right) - B(t)\cos\left(\sqrt{1 - \zeta^2}\omega_n t\right), \qquad (8.12)$$

where functions $A(t)$ and $B(t)$ are expressed by the following integrals:

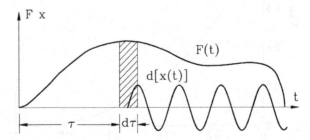

FIGURE 8.4. Response to the impulse exerted by force $F(t)$ between time τ and $\tau + d\tau$.

$$A(t) = \frac{1}{m\omega_n\sqrt{1-\zeta^2}}e^{\zeta\omega_n t}\int_0^t F(\tau)e^{\zeta\omega_n\tau}\cos\left(\sqrt{1-\zeta^2}\omega_n\tau\right)d\tau,$$

$$B(t) = \frac{1}{m\omega_n\sqrt{1-\zeta^2}}e^{\zeta\omega_n t}\int_0^t F(\tau)e^{\zeta\omega_n\tau}\sin\left(\sqrt{1-\zeta^2}\omega_n\tau\right)d\tau.$$

$$(8.13)$$

Example 8.1 *Check the ability of the system studied in Example 4.1 to withstand a shock corresponding to the prescriptions of MIL-STD 810 C, basic design, i.e., an acceleration of the supporting structure that increases linearly in time from $t = 0$ to $t_1 = 11$ ms up to a value of 20 g to drop subsequently to zero. The stresses due to the shock must not exceed the ultimate strength of the material, 328 MN/m^2.*
As the period of the free oscillations of the system is

$$T = \frac{2\pi}{\omega_n} = \frac{2\pi}{337} = 18.6 \ ms,$$

the duration of the shock is not much shorter than the period of the free oscillations, and good accuracy cannot be expected if the shock is studied as an impulse loading. The computation will, therefore, be performed using both an impulsive model and the Duhamel's integral.
The force acting on the beam is given by the mass multiplied by the acceleration. It increases in time from 0 to the value

$$ma_{max} = 3,924 \ N,$$

which is reached after 11 ms. The total impulse is equal to 21.6 Ns. If damping is neglected, the impulsive model (Eq. (8.3) with f_0=21.6 Ns) yields an amplitude of the harmonic motion that follows the impulse

$$x_0 = 3.2 \times 10^{-3} \ m \ .$$

The maximum value of the stress, which takes place at the clamped end, can be computed by dividing the maximum value of the bending moment klx_0 by the section modulus of the beam, obtaining

$$\sigma_{max} = 105.6 \times 10^6 \ N/m^2 \ .$$

To compute the displacement through the Duhamel's integral, function $F(t)$ must be explicitly computed:

$$F = \frac{ma_{max}}{t_1} = 357,000 \ t \quad for \quad 0 \leq t \leq 0.011,$$

$$F = 0 \quad for \quad t > 0.011.$$

By neglecting the presence of damping, which gives a conservative result, functions $A(t)$ and $B(t)$ are, for the first 11 ms,

$$A(t) = \frac{a_{max}}{t_1 \omega_n} \int_0^t \tau \cos(\omega_n \tau)\, d\tau = \frac{a_{max}}{t_1 \omega_n^3} \left[\omega_n t \sin(\omega_n t) + \cos(\omega_n t) - 1 \right],$$

$$B(t) = \frac{a_{max}}{t_1 \omega_n} \int_0^t \tau \sin(\omega_n \tau)\, d\tau = \frac{a_{max}}{t_1 \omega_n^3} \left[-\omega_n t \cos(\omega_n t) + \sin(\omega_n t) \right].$$

By introducing the values of $A(t)$ and $B(t)$ into Eq. (8.12), and remembering that the system is undamped, it follows that

$$x(t) = \frac{a_{max}}{t_1 \omega_n^3} \left[\omega_n t - \sin(\omega_n t) \right] = 0.466 \times 10^{-3} \left[\omega_n t - \sin(\omega_n t) \right].$$

When no more exciting force is present, i.e., after 11 ms, the values of $A(t)$ and $B(t)$ remain constant. By introducing a value of time $t=11$ ms in the expressions of $A(t)$ and $B(t)$, the following equation for the free motion of the system is obtained

$$x(t) = -0.0018 \sin(\omega_n t) - 0.0012 \cos(\omega_n t).$$

The response of the system is plotted in Fig. 8.5. The maximum value of the displacement and the corresponding value of the maximum stress are, respectively,

$$d_{max} = 2.2 \ mm, \qquad \sigma_{max} = 72.6 \ MN/m^2.$$

The value of the stress is far smaller than the allowable value and, consequently, it is not necessary to repeat the computation taking into account the presence of damping.

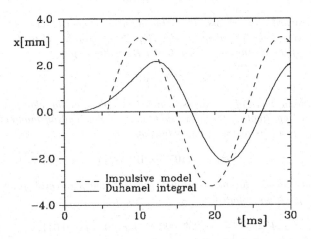

FIGURE 8.5. Time history of the response of the system. Impulsive model and results obtained through Duhamel's integral.

Remark 8.3 *As predicted, the impulsive model in this case does not allow a good approximation of the results, because the duration of the shock is not much shorter than the period of the free oscillations.*

8.4   Solution using the transition matrix

In the case of multi-degrees-of-freedom systems, the response to a generic input $u(t)$ can be expressed using the transition matrix $e^{\mathcal{A}t}$ in the form

$$\mathbf{z}(t) = e^{\mathcal{A}t}\mathbf{z}_0 + \int_0^t e^{\mathcal{A}(t-\tau)}\mathcal{B}\mathbf{u}(\tau)d\tau , \tag{8.14}$$

which can be regarded as a generalization of convolution integral. Some difficulties can be encountered in computing the transition matrix; they increase with increasing time t and with increasing absolute value of the highest eigenvalue of the dynamic matrix $\mathcal{A}$. The time interval t can be subdivided into subintervals and Eq. (8.14) can be applied in sequence, one subinterval after the other. If the input is constant at the value $\mathbf{u}_0$ in the subinterval from time t_0 to time t_1, the state $\mathbf{z}_1$ at the end can be computed from the one at the beginning $\mathbf{z}_0$ as

$$\mathbf{z}_1 = e^{\mathcal{A}(t_1-t_0)}\left\{\mathbf{z}_0 + \mathcal{A}^{-1}\left[\mathbf{I} - e^{-\mathcal{A}(t_1-t_0)}\right]\mathcal{B}\mathbf{u}_0\right\} . \tag{8.15}$$

If the input varies linearly from $\mathbf{u}_0$ to $\mathbf{u}_1$, the integral can be solved in closed form, yielding

$$\mathbf{z}_1 = e^{\mathcal{A}(t_1-t_0)}\left(\mathbf{z}_0 + \mathbf{R}\mathbf{u}_0 + \mathbf{S}\mathbf{u}_1\right) , \tag{8.16}$$

where

$$\mathbf{R} = \mathcal{A}^{-1}\left\{\mathbf{I} - \frac{1}{t_1 - t_0}\mathcal{A}^{-1}\left[\mathbf{I} - e^{-\mathcal{A}(t_1-t_0)}\right]\right\}\mathcal{B} ,$$

$$\mathbf{S} = \mathcal{A}^{-1}\left\{\frac{1}{t_1 - t_0}\mathcal{A}^{-1}\left[\mathbf{I} - e^{-\mathcal{A}(t_1-t_0)}\right] - e^{-\mathcal{A}(t_1-t_0)}\right\}\mathcal{B} .$$

By resorting to the left and right eigenvectors, the equations of motion can be easily uncoupled. Each time history of the modal state variables is of the type

$$\overline{z}_i(t) = \overline{z}_{i_0}e^{s_i t} + \int_0^t e^{s_i(t-\tau)}\mathbf{q}_{Li}^T\mathcal{B}\mathbf{u}(\tau)d\tau . \tag{8.17}$$

8.5   Solution using Laplace transforms

Once the Laplace transform of the time history of the excitation $f(s)$ is known, the Laplace transform of the response $x(s)$ can be easily computed:

$$x(s) = \frac{f(s)}{s^2 m + sc + k} - \frac{m\dot{x}(0) + (ms + c)x(0)}{s^2 m + sc + k}, \qquad (8.18)$$

or, in case of multi-degrees-of-freedom systems

$$\mathbf{x}(s) = \left[s^2 \mathbf{M} + s\mathbf{C} + k\mathbf{K} \right]^{-1} \left[\mathbf{F}(s) - \mathbf{M}\dot{\mathbf{x}}(0) - (\mathbf{M}s + \mathbf{C})\mathbf{x}(0) \right]. \qquad (8.19)$$

Since the Laplace transforms of the most common functions $f(t)$ are tabulated (see, for example, Appendix B), Eq. (8.18) can be used to compute the Laplace transform of the response of the system. The time history $x(t)$ can then be obtained through the inverse transformation or, more simply, by using Laplace transform tables.

Remark 8.4 *The main limitation of the Laplace transform approach is that of being restricted to the solution of linear differential equations with constant coefficients.*

Example 8.2 *Repeat the computation of the shock response of Example 8.1 through Laplace transforms.*
From time $t = 0$ to time $t_1 = 11$ ms, the forcing function is

$$F(t) = \frac{ma_{max}}{t_1} t.$$

Its Laplace transform can be found on the table in Appendix B:

$$\tilde{F}(s) = \frac{ma_{max}}{t_1} \frac{1}{s^2}.$$

By multiplying the Laplace transform of the input by the transfer function of the system it follows that

$$\tilde{x}(s) = G(s)\tilde{F}(s) = \frac{ma_{max}}{t_1} \frac{1}{s^2(ms^2 + k)},$$

i.e.,

$$\tilde{x}(s) = \frac{a_{max}}{t_1} \frac{1}{s^2(s^2 + \omega_n^2)}.$$

From the table a Laplace transform pair

$$\tilde{f}(s) = \frac{\omega^3}{s^2(s^2 + \omega^2)}, \quad f(t) = \omega t - \sin(\omega t)$$

can be found and the displacement can thus be computed

$$x(t) = \frac{a_{\max}}{\omega_n^3 t_1} \left[\omega_n t - \sin(\omega_n t)\right] .$$

This expression coincides with that obtained in Example 8.1 using the Duhamel's integral.
The corresponding velocity is

$$\dot{x}(t) = \frac{a_{\max}}{\omega_n^2 t_1} \left[1 - \cos(\omega_n t)\right] .$$

At time $t = t_1$ their values are

$$x(t) = 1.98 \ mm, \ \dot{x}(t) = 0.290 \ m/s .$$

After time t_1 the motion is free

$$x(t) = x_1 \cos(\omega_n t) + x_2 \sin(\omega_n t) .$$

Stating that at time t_1 the displacement and the velocity are those computed above

$$\begin{cases} x(t_1) = x_1 \cos(\omega_n t_1) + x_2 \sin(\omega_n t_1) = 0.00198, \\ \dot{x}(t_1) = -x_1 \omega_n \sin(\omega_n t_1) + x_2 \omega_n \cos(\omega_n t_1) = 0.290 , \end{cases}$$

the values of x_1 and x_2 are readily computed, obtaining

$$x_1 = -0.0012 \ m, \ x_2 = -0.0018 \ m,$$

that coincide with the values computed in Example 8.1.

8.6 Numerical integration of the equations of motion

An increasingly popular approach to the computation of the time history of the response from the time history of the excitation is the *numerical integration* of the equation of motion.

Remark 8.5 *While all other approaches seen (Laplace transform, Duhamel's integral, etc.) can be applied only to linear systems, the numerical integration of the equation of motion can also be performed for nonlinear systems (see Part II).*

Remark 8.6 *Any solution obtained through the numerical approach must be considered the result of a* numerical experiment *and usually gives little general insight to the relevant phenomena. The numerical approach does not substitute other analytical methods, but rather provides a very powerful tool to deal with cases that cannot be studied in other ways.*

Once the equations of motion of the system have been stated, the first choice is between numerical integration in the state or in the configuration space, i.e., between the numerical integration of a set of first- or second-order equations. Today the former approach is more popular, and most 'simulators', i.e., computer programs performing numerical simulations, operate in the state space.

There are many different methods that can be used to perform the integration of the equation of motion (see Appendix A). All of them operate following the same guidelines: The state of the system at time $t+\Delta t$ is computed from the known conditions that characterize the state of the system at time t. The finite time interval Δt must be small enough to allow the use of simplified expressions of the equation of motion without incurring errors that are too large. The mathematical simulation of the motion of the system is performed step by step, increasing the independent variable t with subsequent finite increments Δt.

Most simulation programs available commercially do not require the user to choose the time increment Δt, but adapt the time increment to the situation until the errors are kept within stated limits. At any rate, being immaterial who chooses the value of Δt, there are two criteria that must be satisfied:

- the time increment must be small enough for the integration algorithm to be stable, and

- the time increment must be small enough for the integration errors to be small enough.

These criteria may be independent of each other.

Since the size of the integration increment must be compared with the periods of the oscillations of the system, these criteria must be evaluated in each case. In particular, the oscillations that must be accounted for are not only the actual ones that are excited by the forcing functions and the initial conditions, but also those potentially occurring at all the natural frequencies of the system, even if they are not excited in the particular simulation considered. The latter may actually be excited by numerical errors, and then amplified by numerical instabilities, until they cause the whole integration procedure to fail.

Numerical integration methods may be either unconditionally or conditionally stable: the former are stable for any value of the time increment (although losing precision with increasing Δt), while the latter are stable only if the time increment is small enough.

When integrating the equations of motion of a single-degree-of-freedom system the stability of the algorithm is not usually a problem: If the time increment is small enough to yield the required precision, it is usually small enough to behave in a stable way as well.

When on the contrary there are many degrees of freedom, and then also many natural frequencies, some high-frequency modes may be little excited and their contribution to the overall response may be negligible. If the algorithm is unconditionally stable, the time increments may be small when compared with the periods of the modes of interest (to achieve the required precision), but large when compared with the higher, non-excited modes. The errors in the evaluation of the contribution of the latter have little importance, and they do not cause instability problems.

When using conditionally stable algorithms, on the contrary, the high-frequency modes can drive the integration to instability, and the time increments must be smaller when compared with the periods of all modes.

Systems containing both low- and high-frequency components at the same time, like when modeling a phenomenon that contains dynamics with widely different timescales, are said to be *stiff* and their numerical study requires either very short time increments or particularly stable integration algorithms.

A possible approach is to transform the equations into modal coordinates before attempting the integration. If the modes can be uncoupled in an exact or approximated way, different time steps can be used for the various modes, thus greatly reducing the integration time.

Even if the modes are coupled, a reduced set of modal equations can be used, at least in an approximated way. If the modes resonating at high frequency are left out, longer time steps can be used also when conditionally stable algorithms are employed. This is usually impossible for *stiff* systems, where both high- and low-frequency dynamics are usually important.

8.7   Exercises

Exercise 8.1 *A spring–mass system is excited by a force whose time history is given by the equation*

$$F = |f_0 \sin(\omega t)| .$$

At time $t = 0$ the system is at rest in the equilibrium position. Compute the time history of the response (a) by numerical integration and (b) by computing a Fourier series for the forcing function $F(t)$. Data: $m = 1$ kg, $k = 1,000$ N/m, $f_0 = 2$ N, $\omega = 10$ rad/s.

Exercise 8.2 *The same spring–mass system studied in Exercise 8.1 but with a damper added is excited by the same force with time history given by the equation*

$$F = |f_0 \sin(\omega t)| .$$

At time $t = 0$ the system is at rest in the equilibrium position. Compute the time history of the response (a) by numerical integration and (b) by computing a Fourier series for the forcing function $F(t)$, and compare the results with those *of the undamped system. Data: $m = 1$ kg, $k = 1,000$ N/m, $c = 6$ Ns/m, $f_0 = 2$ N, $\omega = 10$ rad/s.*

Exercise 8.3 *Add a hysteretic damping with loss factor $\eta = 0.05$ to all springs of the system of Exercise 4.4.*

Plot the four values of the dynamic compliance matrix as a function of the driving frequency.

Exercise 8.4 *Consider the system with hysteretic damping studied in Exercise 8.3. Repeat the computation of the dynamic compliance matrix by using*

- *an equivalent damping function of frequency*

- *a constant equivalent damping*

Compare the results. Repeat the computation with a value of the loss factor 10 times larger and compare again the results.

Exercise 8.5 *Consider the system studied in Example 8.1. Repeat the computations for a shock having a duration of 1 ms and a peak acceleration of 200 g. Compare the results obtained using the impulse model with those obtained through the Duhamel's integral.*

Exercise 8.6 *A spring–mass system is excited by a shock load with a duration of 11 ms and a linearly increasing intensity with peak acceleration of 20 g. Compute the response through the impulse model and Duhamel's integral and compare the results. Perform the numerical integration of the equation of motion, with time steps of $0.05, 0.5$, and 5 ms and compare the results. Data: $m = 20$ kg, $k = 2 \times 10^6$ N/m.*

Exercise 8.7 *The system with hysteretic damping of Exercise 8.3 is excited by a shock applied through the supporting point A. The acceleration of point A has the same time history described in Example 8.1.*

Compute the time histories of the displacement of both masses through numerical integration in time without resorting to modal coordinates. Repeat the computation using modal coordinates and Duhamel's integral.

Exercise 8.8 *Repeat the study of Exercise 8.7, with a value of the loss factor $\eta = 0.1$. Perform only the non-modal computation.*

9

Short Account of Random Vibrations

The study of the response to periodic and nonperiodic excitation is here complemented by a short account on random vibration, aimed more at defining some basic concepts than at dealing with the subject in any detail.

9.1 General considerations

There are many cases where the forcing function acting on a dynamic system has a very complex time history, that cannot be reduced to a simple periodic pattern and cannot be defined in a closely deterministic way. This implies mainly that records of the excitation and of the response of the system obtained in conditions that are alike differ from each other in a substantial way.

Typical examples are seismic excitation on buildings, excitation on the structures of ground vehicles due to road irregularities, and excitation of the structure of ships due to sea waves. For all these cases, and many other similar ones, the term random vibration is commonly used.

In all these cases the time history of the excitation can be measured, and if enough experimental data are taken, it is possible to perform a statistical analysis.

The study of the response of dynamic systems to an excitation of this kind is quite complex and two different approaches can be used.

A simple approach, at least from a conceptual point of view, is to operate in the time domain, computing the response of the system by numerically integrating the equations of motion using the experimental time history of the excitation as an input. In this way, however, only a limited insight of the phenomena involved is obtained, and computations becomes rapidly heavy if many experimental data are used in the attempt of understanding the behavior of the system with some generality.

A much better, and usually simpler in practice, way is to perform a statistic analysis of the input: the statistic parameters of the outputs of the system can be obtained from the statistic parameters of the input and this is usually sufficient to verify whether the system complies with the prescribed functionality and safety requirements.

This involves mostly frequency-domain computations and requires a good background in statistics. There are many excellent books devoted to the subject of random vibrations where the interested reader can find a more complete analysis. In this chapter only a brief outline, mainly on the qualitative aspects of the relevant phenomena, will be given.

9.2   Random forcing functions

Consider a generic function $y(t)$. Given a sample whose duration is T, the average value in time, and the *mean square* can be defined by the obvious relationships

$$
\begin{aligned}
\bar{y} &= \frac{1}{T} \int_0^T y(t)dt, \\
\overline{(y^2)} &= \frac{1}{T} \int_0^T y^2(t)dt \ .
\end{aligned}
\tag{9.1}
$$

The square root of the mean square is usually defined as the *root mean square* (in short r.m.s.) value

$$
y_{rms} = \sqrt{\overline{(y^2)}} = \sqrt{\frac{1}{T} \int_0^T y^2(t)dt} \ .
\tag{9.2}
$$

The *variance*, usually referred to by the symbol σ^2, is defined as

$$
\sigma^2 = \overline{(y - \bar{y})^2} = \frac{1}{T} \int_0^T \left[y(t) - \bar{y} \right]^2 dt \ .
\tag{9.3}
$$

In many cases the input is assumed to have a zero average value. This can always be obtained by subtracting to the sample its average value, which amounts to separate the static behavior of the system from its dynamic behavior. This approach is possible only in the case of linear systems.

The variance in this case coincides with the mean square.

The square root of the variance is the *standard deviation* σ. When the average is equal to zero, the standard deviation coincides with the r.m.s. value.

In case of an harmonic function with zero average (i.e. a sine or cosine wave), assuming that T is a multiple of the period, the mean square and the standard deviation are

$$\overline{(y^2)} = \frac{1}{T} \int_0^T y_{max}^2 \cos^2{(\omega t)} \, dt = \frac{1}{2} y_{max}^2 \, ,$$

$$\sigma = y_{rms} = \frac{1}{\sqrt{2}} y_{max} \, .$$

These statistical parameters in general depend on the sample used for the analysis.

When studying random excitation, a first assumption is that the phenomenon is stationary, i.e., its characteristics do not change when the study is performed starting at different times. Another assumption is that of *ergodicity*, a complex statistical property that in the present case can be summarized by stating that any sample can be considered typical of the whole set of available samples.

Under these assumptions, the average, the r.m.s. value, the variance, and all other statistical parameters can be considered independent of the particular sample used for their computation. These are oversimplifications of a more complex phenomenon, but in most cases they allow for results that are in close accordance with experimental evidence to be obtained.

If the phenomenon is stationary and ergodic, the values of the average and of the variance computed in time T are coincident with the same values obtained for T tending to infinity.

Another very important statistical parameter is the autocorrelation function

$$\Psi(\tau) = \lim_{T \to \infty} \frac{1}{T} \int_0^T y(t) y(t + \tau) dt \, , \tag{9.4}$$

which states how the value of function $y(t)$ at time t is linked with the value it takes at time $t + \tau$.

Under the above-mentioned assumptions, for a zero-mean phenomenon, it follows that

$$\Psi(0) = \sigma^2 = y_{rms}^2 \, . \tag{9.5}$$

The autocorrelation function of an harmonic function such as the sine or cosine waves considered above is

$$\Psi(\tau) = \frac{1}{2} y_{max}^2 \cos{(\tau)} \, .$$

In the case of an ideal random phenomenon, in which the value of function $y(t)$ in every instant is completely independent of the value it takes in any other instant, the autocorrelation is equal to zero for every value of τ except, as already stated, $\tau = 0$. As it will be seen below, the autocorrelation function of an actual random phenomenon, although tending to zero with increasing absolute value of τ, is not zero.

A spectral analysis of the signal can be performed using the Fourier transform of function $y(t)$:

$$Y(\omega) = \int_{-\infty}^{\infty} y(t) e^{-i\omega t} dt .$$ (9.6)

Instead of using the Fourier transform of function $y(t)$, the information regarding the frequency content of a random variable is the *power spectral density* $S(\omega)$, often indicated with the acronym PSD. The power spectral density is correctly defined as the Fourier transform of the autocorrelation function

$$S(\omega) = \int_{-\infty}^{\infty} \Psi(\tau) e^{-i\omega \tau} d\tau ,$$ (9.7)

but it can be also defined as the square of the modulus of the Fourier transform of function $y(t)$, multiplied by a suitable constant

$$S(\omega) = \frac{1}{2\pi} |Y(\omega)|^2 .$$ (9.8)

The term power is here used loosely: Because the power of an harmonic signal is proportional to the square of the amplitude, it just stands for amplitude squared.

If the random vibration is excited by a force, the dimension of its power spectral density $S(\omega)$ is that of the square of a force divided by a frequency. In S.I. units it is therefore measured in $N^2/(rad/s) = N^2s/rad$ or in N^2/Hz. If the system is excited by the motion of the supporting point, the forcing function is an acceleration and its power spectral density is measured in $(m/s^2)^2/(rad/s) = m^2/s^3 rad$ or in g^2/Hz.

The integral of function $S(\omega)$ is the variance of function $y(t)$, i.e., if the average value is equal to zero, the square of its r.m.s. value

$$y_{rms} = \sqrt{\int_{-\infty}^{\infty} S(\omega) d\omega} .$$ (9.9)

The power spectral density is here defined for both positive and negative values of the frequency ω (two-sided power spectral density). Often, on the contrary, a one-sided power spectral density is defined, limited to positive values of the frequency.

Remark 9.1 *A random forcing function is usually defined as a narrow-band or wide-band excitation, depending on the width of the frequency range involved.*

9.3   White noise

The simplest type of random excitation is a random forcing function with constant power spectral density. This type of forcing function, which contains all possible frequencies in the same measure, is often referred to as *white noise*. Its autocorrelation function has a zero value for all values of τ and goes to infinity for $\tau = 0$. It is, therefore, a Dirac's impulse function $\delta(\tau)$.

Remark 9.2 *A true white noise, with a spectrum extending for the whole frequency range from 0 to infinity, is just a mathematical model, since its r.m.s. value would be infinitely large, as also shown by the fact that Ψ is a Dirac's δ, i.e. $\Psi(0)$ is infinite.*

A more realistic random excitation is a band limited white noise, i.e., an excitation with a power spectral density (Fig. 9.1a) expressed as

$$
\begin{aligned}
S &= S_0 &&\text{for } |\omega| \le \omega_0 , \\
S &= 0 &&\text{for } |\omega| > \omega_0 .
\end{aligned}
\tag{9.10}
$$

Its r.m.s. value is thus

$$
y_{rms} = \sqrt{2 S_0 \omega_0},
\tag{9.11}
$$

and its autocorrelation function, shown in Fig. 9.1b, is

$$
\Psi(\tau) = 2 S_0 \frac{\sin (\omega_0 \tau)}{\tau} .
\tag{9.12}
$$

Often the power spectral density of the white noise is truncated both at a minimum and at a maximum frequency. It is assumed to have the shape of

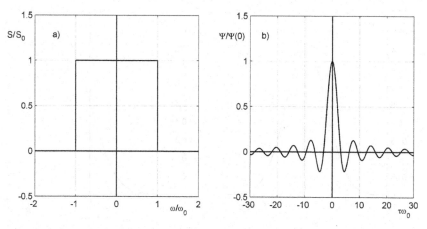

FIGURE 9.1. Nondimensional power spectral density (**a**) and autocorrelation function (**b**) of a band limited white noise.

a trapezium in a bi-logarithmic plane and then, while being constant in the central frequency field, it decays on both side with a constant slope (see, for instance, Fig. 9.6a), usually measured in dB/oct (decibel per octave) or dB/dec (decibel per decade). In other cases the power spectral density is considered as constant at different values in various frequency ranges, with straight ramps (in a bi-logarithmic plane) in between.

9.4   Probability distribution

The quantities defined earlier are not yet sufficient to completely characterize a random forcing function. It is also necessary to define a function expressing the probability density function $p(y)$ related to the amplitude.

The probability density function $p(y)$ is defined as the probability that a random variable $y(t)$ takes a value included between y and $y + dy$.

Usually, such a function is assumed to be a normal or Gaussian probability density

$$p(y) = \frac{1}{\sqrt{2\pi}\sigma} e^{-\frac{(y-\mu)^2}{2\sigma^2}}, \tag{9.13}$$

where μ is the mean and σ is the standard deviation (Fig. 9.2a).

The probability that $y(t)$ is included in the interval $[\mu - \sigma, \mu + \sigma]$ can be computed by integrating the probability density in that interval: it is 0.683. The probabilities it lies within a band of semi-amplitude 2σ and 3σ are, respectively, 0.954 and 0.997. There is, however, no interval in which the probability is 1: This means that there is a non-zero probability that function $y(t)$ reaches any value, however large.

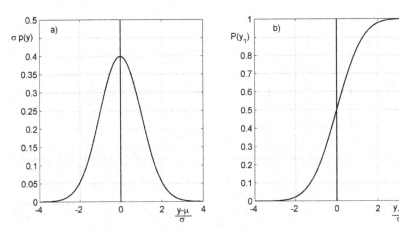

FIGURE 9.2. Nondimensional plot of the normal probability density function (**a**) and the normal probability distribution function (**b**).

Remark 9.3 *This is clearly just a theoretical statement, resulting from approximating the actual probability density with a normal distribution. All real world phenomena are limited and the 'tails' of the probability density function must be cut somewhere, but this can done only by reasoning on the physical significance and properties of function $y(t)$.*

Sometimes the suggestion is to cut the normal distribution at $y - \mu = \pm 3\sigma$, or at $y - \mu = \pm 5\sigma$, but these are anyway arbitrary statements.

The probability that

$$y(t) \leq y_1 \tag{9.14}$$

is the probability distribution function $P(y_1)$. In the case of a normal distribution it follows

$$P(y_1) = \int_{-\infty}^{y_1} p(y)dy = \frac{1}{\sqrt{2\pi}\sigma} \int_{-\infty}^{y_1} e^{-\frac{(y-\mu)^2}{2\sigma^2}} dy \ . \tag{9.15}$$

The integral cannot be performed in closed form, but series for both $(y - \mu)/\sigma$ small and large exist. A plot obtained numerically is shown in Fig. 9.2b.

To reach a probability exactly equal to 1 an infinitely large value of y_1 must be reached.

Remark 9.4 *When a random function is stationary, ergodic, and characterized by a normal probability distribution, the average (which is here assumed to be equal to zero), the variance, and the power spectral density characterize completely the function.*

9.5 Response of linear systems

Consider a linear system with a single degree of freedom on which a random forcing function (in terms of applied force or of displacement of the supporting point) is acting and assume that it is normal, stationary, and ergodic. The behavior of the system is completely characterized by its frequency response $H(\omega)$, which is complex if the system is damped.

The response can be measured in terms of displacement $x(t)$, velocity, or acceleration, but also of stresses in any point of interest of the system. It has a random nature as well, with the same characteristics of stationarity and ergodicity and the same normal probability distribution as the forcing function. Also, the mean value of the response is equal to zero.

The power spectral density of the response can be computed directly from the power spectral density of the excitation and the frequency response of the system:

$$S_x(\omega) = S_f(\omega)|H(\omega)|^2 \ . \tag{9.16}$$

200   9. Short Account of Random Vibrations

The r.m.s. value of the response is thus

$$x_{rms} = \sqrt{\int_{-\infty}^{\infty} S_x(\omega)d\omega}. \tag{9.17}$$

Assume that a force with a white noise power spectral density acts on a linear, single degree of freedom system with a damping ratio ζ. The frequency response is

$$|H(\omega)| = \frac{1}{k\sqrt{\left[1-\left(\frac{\omega}{\omega_n}\right)^2\right]^2 + \left(2\zeta\frac{\omega}{\omega_n}\right)^2}}, \tag{9.18}$$

and thus the power spectral density of the displacement is (Fig. 9.3a)

$$S_x(\omega) = \frac{S_0}{k^2}\frac{1}{\left[1-\left(\frac{\omega}{\omega_n}\right)^2\right]^2 + \left(2\zeta\frac{\omega}{\omega_n}\right)^2}. \tag{9.19}$$

The power spectral densities are now expressed, respectively, in N²s/rad and m²s/rad.

The r.m.s. value of the response is in this case

$$x_{rms} = \frac{\sqrt{S_0}}{k}\sqrt{\int_{-\infty}^{\infty} \frac{1}{\left[1-\left(\frac{\omega}{\omega_n}\right)^2\right]^2 + \left(2\zeta\frac{\omega}{\omega_n}\right)^2}d\omega}, \tag{9.20}$$

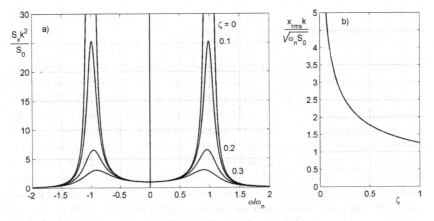

FIGURE 9.3. Response of a linear single degree of freedom system excited by a random force with white noise spectrum. (a) Nondimensional power spectral density of the response. (b) Nondimensional r.m.s. value of the response as a function of the damping ratio.

i.e.,

$$x_{rms} = \frac{1}{k}\sqrt{\frac{\pi S_0 \omega_n}{2\zeta}}. \tag{9.21}$$

If the excitation is provided by the motion of the supporting point, and then the power spectral density of the white noise excitation S_0 is referred to the acceleration (expressed in $(m/s)^2/rad/s$ or g^2/Hz), the r.m.s. value of the displacement is

$$x_{rms} = \sqrt{\frac{\pi S_0}{2\zeta \omega_n^3}}. \tag{9.22}$$

The r.m.s. value of the response is plotted as a function of the damping ratio in Fig. 9.3b. Provided that the system is damped, it remains limited, even if the excitation is a theoretical white noise having an infinitely large r.m.s. value.

From Fig. 9.3a it is clear that a lightly damped linear system with a single degree of freedom acts as a sort of filter, amplifying the input signal in a very narrow band about the resonant frequency and cutting off all other components. The lower is the damping of the system, the narrower is the passing band.

The response of a lightly damped system is thus a narrow-band random vibration. The time history of the response follows the pattern sketched in Fig. 9.4: an oscillation that is almost harmonic with randomly variable amplitude and slightly variable frequency. The frequency is very close to the natural frequency of the system.

Under these conditions it is possible to define an envelope of the time history and thus the probability density for a generic peak to be higher than the r.m.s. by a given factor. If the excitation (and then also the response) is normal, such a probability density follows the Rayleigh distribution

FIGURE 9.4. Pattern of the time history of the response of a lightly damped system to a random forcing function of the white-noise type.

$$p\left(\frac{x_p}{x_{rms}}\right) = \frac{x_p}{x_{rms}^2} e^{-\frac{x_p^2}{2x_{rms}^2}} \qquad (9.23)$$

plotted in Fig. 9.5a. From the figure it is clear that very low and very high values of the peaks are unlikely and that the maximum probability is that of having peaks roughly as high as the r.m.s. value.

Again, there is a nonzero probability that a peaks reaches any value, however high. This has no physical meaning and comes from having assumed a normal probability distribution. In practice the plot must be cut at a certain value of x_p/x_{rms}.

By integrating the probability density function it is possible to compute the probability that a peak is higher than the generic value x_{max}

$$P\left(\frac{x_{max}}{x_{rms}}\right) = e^{-\frac{x_{max}^2}{2x_{rms}^2}}, \qquad (9.24)$$

that is plotted in Fig. 9.5b. It is a Gaussian distribution, but in the figure it is shown in logarithmic scale to show better how it drops quickly for high values of the peak.

From Eq. (9.24) it is possible to directly compute the probability that the maximum amplitude of the response reaches any given value in a given working time. Because the response is a narrow-band random signal, its frequency is very close to the natural frequency of the system and the number of oscillations taking place in time t is $t\omega_n/2\pi$. The probability that in one of these periods the peak value is greater than x_{max} is thus

$$P\left(\frac{x_{max}}{x_{rms}}\right) = \frac{t\omega_n}{2\pi} e^{-\frac{x_{max}^2}{2x_{rms}^2}}. \qquad (9.25)$$

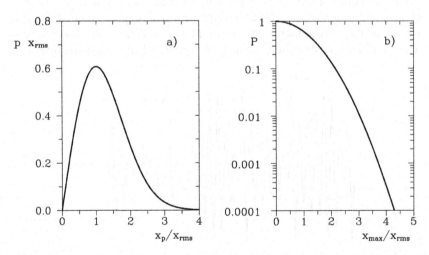

FIGURE 9.5. (a) Probability density of the peak values of a narrow-band random response and (b) probability that one of the peaks is higher than a given value x_{max}. Note the logarithmic scale in Fig. 9.5b.

Example 9.1 *Check the ability of the system studied in Example 4.1 to withstand for a time of 30s the random excitation provided by the motion of the supporting point defined in Fig. 9.6a. The (one-sided) power spectral density is*

- *constant in a frequency range between 100 and 250 Hz at a value of 0.03 g^2/Hz;*

- *increases at 9 dB/oct between 20 and 100 Hz; and*

- *decreases at −15 dB/oct between 250 and 2,000 Hz (Fig. 9.6a).*

The stresses must not exceed

- *the ultimate strength of 328 MN/m^2 divided by a safety factor of 1.575; or*

- *the yield strength of 216 MN/m^2 divided by a safety factor of 1.155; or*

- *the allowable fatigue strength for the prescribed duration (115 MN/m^2 for 10^7 cycles).*

The relative displacement between the instrument at the end of the beam and the supporting structure must not exceed 4 mm.

The power spectral density of the excitation includes the natural frequency of the system. As a consequence, the computation of the response cannot be performed when neglecting the presence of damping, which will be assumed to be of the hysteretic type. For safety, a low value of the loss factor $\eta = 0.01$ will be assumed, so that conservative results will be obtained.

Consider the first frequency field. The relationship between power spectral density and frequency is linear in a bi-logarithmic plane; its expression is then of the type

$$S(\omega) = a\omega^n \ .$$

As the power spectral density increases of 9 dB/oct and at a frequency of 100 Hz its value is 0.03 g^2/Hz = 2.88 $(m/s^2)^2/Hz$, its expression is

$$S(\omega) = 2.88 \times 10^{-6}\omega^3 \ ,$$

where frequencies are measured in Hz.

In the range between 100 and 250 Hz, the power spectral density is constant at a value 2.88 $(m/s^2)^2/Hz$, while between 250 and 2,000 Hz, where the power spectral density decreases at −15 dB/oct, the expression

$$S(\omega) = 2.81 \times 10^{12}\omega^{-5}$$

is readily obtained. The r.m.s. value of the acceleration is obtained by integrating the power spectral density between 20 and 2,000 Hz:

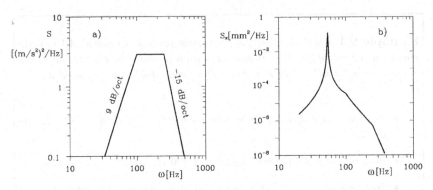

FIGURE 9.6. Power spectral density of the excitation (**a**) and the response (**b**) of the system of Example 9.1.

$$a_{rms} = 26.15 \ m/s^2 \ .$$

Because the response has to be computed in terms of relative displacement, the frequency response can be used, substituting the inertial force ma for the external force F. The power spectral density of the response can be immediately computed for the three frequency ranges by multiplying the power spectral density of the excitation by the square of the frequency response

$$S_x(\omega) = 2.23 \times 10^{-16} \frac{\omega^3}{\left[1-\left(\frac{\omega}{\omega_n}\right)^2\right]^2 + \eta^2},$$

$$S_x(\omega) = 2.23 \times 10^{-10} \frac{1}{\left[1-\left(\frac{\omega}{\omega_n}\right)^2\right]^2 + \eta^2},$$

$$S_x(\omega) = 2.17 \times 10^2 \frac{\omega^{-5}}{\left[1-\left(\frac{\omega}{\omega_n}\right)^2\right]^2 + \eta^2}.$$

where frequencies are expressed in Hz and power spectral densities in m^2/Hz. Note that in the above equations the damping has been assumed to be hysteretic, which is inconsistent with the fact that hysteretic damping loses any meaning when the motion is not harmonic. Here however the system is lightly damped and the response is a very narrow-band response centered on the resonant frequency: The use of hysteretic damping or of any sort of equivalent damping leads to very similar results.
The power spectral density of the response is plotted in Fig. 9.6b. The r.m.s. value of the response can be computed by integrating the power spectral density:

$$x_{rms} = 0.539 \ mm \ .$$

By comparing the contributions of the three integrals related to the various frequency ranges, it is clear that the only frequency field that contributes significantly to the response is the first one, because the natural frequency of the system falls in it and the response is of the narrow-band type.

The r.m.s. value of the stress, computed in the same way as in Example 3.1, is

$$\sigma_{rms} = 17.8 \times 10^6 \ N/m^2 \ .$$

The narrow-band response can be assimilated to a harmonic oscillation with random varying amplitude and frequency equal to the natural frequency 53.7 Hz. The total number of cycles occurring in the prescribed 1.930s is 1,611. As the r.m.s. value of the stress is 6.5 times smaller than the fatigue strength at 10^7 cycles, the third condition is surely satisfied. For the first condition, the allowable stress is 208 MN/m^2. The r.m.s. value is, therefore, 11.7 times smaller than the allowable value of the stress. The probability that the stress reaches the allowable value in the prescribed 30s can be computed using Eq. (9.25), obtaining

$$P = 3.6 \times 10^{-27} \ .$$

The situation is slightly more critical for the second condition, regarding the yield strength: The allowable strength is 187 MN/m^2, the ratio between the allowable stress and the r.m.s. value is 10.5, and the probability of reaching the critical condition in the prescribed time is 1.7×10^{-21}.

The probability that the structure fails under the effects of the random excitation prescribed is extremely low. For the critical condition on the displacement, the probability is also very low, namely, 1.8×10^{-9}.

Remark 9.5 *In the example, the resonant frequency lies outside the frequency range where the power spectral density of the excitation is constant. The results computed assuming that the excitation is a theoretical white noise would be quite far from the correct ones obtained here.*

If the excitation is a band-limited white noise (Fig. 9.1) with an upper limitation at frequency ω_0, the expression for the r.m.s. value of the response is

$$x_{rms} = \frac{\sqrt{S_0 \omega_n}}{k} \sqrt{\frac{f(\omega_0/\omega_n)}{2\zeta\sqrt{-1+\zeta^2}}} \ , \tag{9.26}$$

where

$$f\left(\frac{\omega_0}{\omega_n}\right) = -\left(\zeta - \sqrt{-1+\zeta^2}\right) \arctan\left(\frac{\omega_0/\omega_n}{\zeta+\sqrt{-1+\zeta^2}}\right) +$$

$$+ \left(\zeta + \sqrt{-1+\zeta^2}\right) \arctan\left(\frac{\omega_0/\omega_n}{\zeta-\sqrt{-1+\zeta^2}}\right) .$$

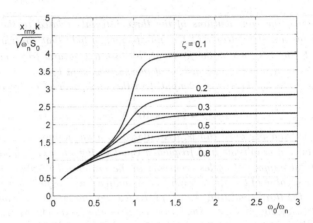

FIGURE 9.7. Nondimensional r.m.s. value of the response to a band-limited white noise as a function of the ratio ω_0/ω_n. The *dashed lines* are the asymptotes for $\omega_0/\omega_n \to \infty$, i.e., for an ideal white noise.

Remark 9.6 *For underdamped systems function $f(\omega_0/\omega_n)$ is imaginary, but x_{rms} is anyway real, as expected.*

The r.m.s. value of the response to a band-limited white noise is reported in Fig. 9.7 as a function of ratio ω_0/ω_n. From the plot it is clear that if the resonance peak is included in the band, i.e. $\omega_0/\omega_n > 1$, the ideal white-noise model yields fairly good results. If $\omega_0/\omega_n > 2$ the fact that the band is limited has practically no consequence on the results.

If the system has many degrees of freedom, the power spectral density of the response can still be computed using Eq. (9.16), with the only difference that now there are many responses and there may be many inputs.

If the system is lightly damped each frequency response $H_{ij}(\omega)$ has a number of peaks, and thus there are several bands in which the excitation may be amplified.

Example 9.2 *Quarter-car model.*
One of the simplest models used to study the dynamic behavior of motor vehicle suspensions is the so-called quarter car with two degrees of freedom (Fig. 9.9a). The upper mass simulates the part of the mass of the car body (the sprung mass) that can be considered supported by a given wheel, and the lower one simulates the wheel and all the parts that can be considered as rigidly connected with the unsprung mass.

The two masses are connected by a spring–damper system simulating the spring of the suspension and the shock absorber. The unsprung mass is connected to the ground with a second spring simulating the stiffness of the tire. The point at which the tire contacts the ground is assumed to move in a vertical direction with a given law $h(t)$, in order to simulate the motion on uneven ground. Assume the following values of the parameters: Sprung mass $m_s = 250$ kg; unsprung mass $m_u = 25$ kg; spring stiffness $k = 25$ kN/m; tire stiffness $k_t = 100$ kN/m; damping coefficient of the shock absorber $c = 2,150$ Ns/m. The following analyses will be performed:

- modal analysis;
- computation of the frequency response; and
- computation of the response of the system due to the road excitation when travelling at a sped of 30 m/s on a road whose surface can be considered normal following the ISO standards

Modal analysis

If coordinates x_s and x_u are defined with reference to an inertial frame, the equation yielding the response of the system to a harmonic excitation with frequency ω and amplitude h_0 is

$$\left[-\omega^2 \begin{bmatrix} m_s & 0 \\ 0 & m_n \end{bmatrix} + \begin{bmatrix} k & -k \\ -k & k+k_t \end{bmatrix} + i\omega \begin{bmatrix} c & -c \\ -c & c \end{bmatrix} \right] \left\{ \begin{matrix} x_{s_0} \\ x_{n_0} \end{matrix} \right\} =$$

$$= \left\{ \begin{matrix} 0 \\ k_t h_0 \end{matrix} \right\}$$

where

$$\mathbf{M} = 25 \begin{bmatrix} 10 & 0 \\ 0 & 1 \end{bmatrix}, \qquad \mathbf{K} = 25,000 \begin{bmatrix} 1 & -1 \\ -1 & 5 \end{bmatrix},$$

$$\mathbf{C} = 2,150 \begin{bmatrix} 1 & -1 \\ -1 & 1 \end{bmatrix},$$

$$\mathbf{f} = 100,000\, h_0 \left\{ \begin{matrix} 0 \\ 1 \end{matrix} \right\}, \qquad \mathbf{D} = \mathbf{M}^{-1}\mathbf{K} = \begin{bmatrix} 100 & -100 \\ -1,000 & 5,000 \end{bmatrix}.$$

The characteristic equation that allows the computation of the natural frequencies of the undamped system is

$$\begin{bmatrix} 100 - \omega^2 & -100 \\ -1,000 & 5,000 - \omega^2 \end{bmatrix} = 0,$$

i.e.,

$$\omega^4 - 5,100\omega^2 + 400,000 = 0.$$

Its solutions and the values of the natural frequencies are

$$\omega_1^2 = 79.676 \qquad (\omega_1 = 8.926\ rad/s = 1.421\ Hz),$$
$$\omega_2^2 = 5,020.32 \qquad (\omega_2 = 70.85\ rad/s = 11.28\ Hz).$$

The eigenvectors can be computed directly by introducing the eigenvalues into the dynamic matrix and stating equal to unity one of their values, e.g., the second

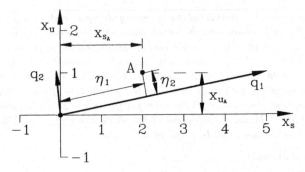

FIGURE 9.8. Eigenvectors in the space of the configurations. Point A, with coordinates $\mathbf{x} = [\ 2\quad 1\ ]^T$ and $\eta = \mathbf{\Phi}^{-1}\mathbf{x} = [\ 0.47.83\quad 0.1952\ ]^T$ is also shown as an example.

$$\begin{bmatrix} 100 - \omega^2 & -100 \\ -1,000 & 5,000 - \omega^2 \end{bmatrix} \begin{Bmatrix} x \\ 1 \end{Bmatrix} = 0.$$

The second equation yields

$$x = 5 - \frac{\omega^2}{1000}.$$

The two eigenvectors are then

$$\mathbf{q}_1 = \begin{Bmatrix} 4.9203 \\ 1 \end{Bmatrix}, \quad \mathbf{q}_2 = \begin{Bmatrix} -0.020324 \\ 1 \end{Bmatrix}.$$

The eigenvectors are represented in Fig. 9.8 in the space of the configurations. They are clearly not orthogonal with respect to each other.
The modal masses are thus easily computed, obtaining 6,052.4 and 25.103, respectively. These values can be used to normalize the eigenvectors, obtaining

$$\mathbf{\Phi} = \begin{bmatrix} 0.062346 & -0.0040564 \\ 0.012854 & 0.199588 \end{bmatrix}.$$

The modal masses have unit values, while the modal stiffnesses are equal to the squares of the natural frequencies. The modal damping matrix

$$\overline{\mathbf{C}} = \mathbf{\Phi}^T \mathbf{C} \mathbf{\Phi} = \begin{bmatrix} 5.4596 & -22.0634 \\ -22.0634 & 89.1627 \end{bmatrix}$$

is not diagonal and can be split into its proportional and nonproportional parts:

$$\overline{\mathbf{C}}_p = \begin{bmatrix} 5.4596 & 0 \\ 0 & 89.1627 \end{bmatrix}, \quad \overline{\mathbf{C}}_{np} = \begin{bmatrix} 0 & -22.0634 \\ -22.0634 & 0 \end{bmatrix}.$$

The inverse modal transformation yields the following expressions of the two parts of the original damping matrix

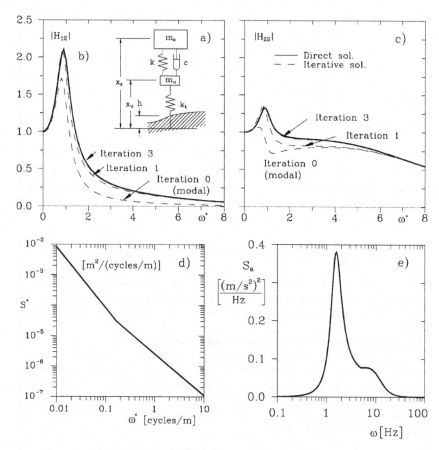

FIGURE 9.9. (a) Scheme of the quarter car model; (b), (c) frequency response; (d) power spectral density of a normal road following ISO standards; and (e) power spectral density of the acceleration at a speed of 30 m/s on the road profile whose power spectral density is given in (d).

$$\mathbf{C}_p = \begin{bmatrix} 1.445 & -424 \\ -424 & 2,221 \end{bmatrix}, \quad \mathbf{C}_{np} = \begin{bmatrix} 705 & -1,726 \\ -1,726 & -71 \end{bmatrix}.$$

The distribution of the damping is far from being proportional and the non-proportional damping matrix is nonpositive definite. Matrix C_p is, however, strictly proportional, with coefficients $\alpha = 4.084$ and $\beta = 0.01696$. The modal forces are then obtained:

$$\bar{\mathbf{f}} = \mathbf{\Phi}^T \mathbf{f} = h_0 \begin{Bmatrix} 1,285 \\ 19.959 \end{Bmatrix}.$$

Computation of the frequency response

The frequency response of the system can be computed directly from the equation of motion, obtaining

$$|H_{12}| = \frac{x_{s0}}{h_0} = k_p \sqrt{\frac{k^2 + c^2\omega^2}{f^2(\omega)c^2\omega^2 g^2(\omega)}} \,,$$

$$|H_{22}| = \frac{x_{n0}}{h_0} = k_p \sqrt{\frac{(k - m\omega^2)^2 + c^2\omega^2}{f^2(\omega)c^2\omega^2 g^2(\omega)}} \,,$$

where

$$f(\omega) = m_s m_n \omega^4 - [k_p m_s + k(m_s + m_n)]\omega^2 + kk_p \,,$$

$$g(\omega) = (m_s + m_n)\omega^2 - k_p \,.$$

The response is plotted in Fig. 9.9b and c. The same result can also be obtained through an iterative procedure starting from modal uncoupling. Using the Jacobi method, the modal coordinates at the ith iteration can be obtained from those at the (i − 1)-th through the formulae

$$\eta_1^{(i)} = \frac{\bar{f}_1 - i\omega\bar{C}_{12}\eta_2^{(i-1)}}{\bar{K}_1 - \omega^2\bar{M}_1 + i\omega\bar{C}_{11}} \,, \qquad \eta_2^{(i)} = \frac{\bar{f}_2 - i\omega\bar{C}_{21}\eta_1^{(i-1)}}{\bar{K}_2 - \omega^2\bar{M}_2 + i\omega\bar{C}_{22}} \,.$$

The response, computed after the first and third iterations, is plotted in Fig. 9.9b and c. Three iterations are enough to obtain a good approximation even in the current case, where damping is neither small nor close to being proportional.

Response to random excitation (road excitation)

The power spectral density of a normal road following ISO standards is shown in Fig. 9.9d. The power spectral density of a road profile S^* is usually measured in $m^2/(cycles/m)$ and is a function of the space frequency ω^*, in cycles/m. Once the speed V of the vehicle has been stated, it is possible to obtain the various quantities defined with reference to the time frequency from the space frequency through the formulas

$$\omega = V\omega^* \,, \quad S = S^*/V \,.$$

The curve of Fig. 9.9d can be expressed by the equation

$$S^* = a\omega^{*n} \,,$$

where constants a and n are

$$a = 8.33 \times 10^{-7} \ m \,,$$

$$n = -2 \ \ if \ \omega^* \leq 1/6 \ cycles/m \,,$$

i.e., for road undulations with wavelength greater than 6 m, and

$$a = 2.58 \times 10^{-6} \ m^{1,63} \,,$$

$$n = -1.37 \ \ if \ \omega^* > 1/6 \ cycles/m \,,$$

i.e., for short wavelength irregularities.
The (one-sided) power spectral density (with reference to the time frequency) is

$$S = a'\omega^n \,,$$

where

$$a' = \frac{a}{V^{n+1}} \ .$$

At a speed of 30 *m/s it follows*

$$a' = 2.5 \times 10^{-5} \ m^3 \ s^{-2} \quad for \ \ \omega \leq 5 \ Hz \ ,$$

$$a' = 9.07 \times 10^{-6} \ m^{2,37} \quad for \ \ \omega > 5 \ Hz \ .$$

The power spectral density of the acceleration of the sprung mass can be easily computed by multiplying the power spectral density of the road profile by the square of the inertance of the suspension. Because the inertance is equal to the dynamic compliance H_{12} (computed above) multiplied by ω^2, or, better, by $4\pi^2\omega^2$ if ω is measured in Hz, it follows that

$$S_a = S \, 16\pi^4\omega^4 H_{12}^2 = 16\pi^4 a' \omega^{n+4} H_{12}^2 \ .$$

To obtain the power spectral density of the acceleration in $(m/s^2)^2/Hz$, the frequency must be expressed in Hz, while in the equation yielding the dynamic compliance H_{12}, it is expressed in rad/s. The result is shown in Fig. 9.9e. The r.m.s. value of the acceleration can be obtained by integrating the power spectral density and extracting the square root

$$a_{rms} = 1.21 m/s^2 = 0.123g \ .$$

9.6   Exercises

Exercise 9.1 *Repeat the computations of the response of the system of Example* 9.1, *by substituting the hysteretic damping with an equivalent viscous damping, computed at the resonant frequency. Compare the results with those obtained in the example.*

Exercise 9.2 *A spring–mass–damper system is excited by a force whose time history is random. The power spectral density of the excitation is constant at a value of* 10,000 N^2/Hz *between the frequencies of* 50 *and* 200 *Hz. At lower frequencies it increases at* 12 *dB/oct, while a decay of* 12 *dB/oct takes place at frequencies higher than* 200 *Hz. Compute the power spectral density of the response and the r.m.s. values of both excitation and response. Data: $m = 10$ kg, $k = 5$ MN/m, $c = 3$ kNs/m.*

Exercise 9.3 *Repeat the computations of Exercises* 9.1 *and* 9.2 *assuming a theoretical unlimited white noise and compare the results with those already obtained. Discuss similarities and differences.*

Exercise 9.4 *Consider the torsional system with* 3 *degrees of freedom of Fig.* 1.8, *already studied in Exercises* 1.2, 2.1 *and* 2.3, *with damping added as in*

Exercise 3.1, excited by a motion of the supporting point like in Exercise 7.4. The excitation is random with a band-limited white noise with a power spectral density of the rotation $S_\theta = 10^{-6}$ rad^2/Hz up to a frequency of 8 Hz. Plot the power spectral density of the response of the third disc and compute numerically the r.m.s. value of the rotation at the same point.

10

Reduction of the Number of Degrees of Freedom

The mathematical models of many real-world systems are quite complicated and may include a large number of degrees of freedom, in particular when they are generated automatically starting from drawings or other geometric information. Techniques aimed at reducing the size of the model without loosing important information on the behavior of the system are then increasingly applied in various stages of the dynamic analysis.

10.1   General considerations

As already stated, when performing a dynamic analysis, there is a great advantage in reducing the size of the problem, particularly when using methods like the finite element method (see Chapter 15), which usually yield models with a large number of degrees of freedom. It is not uncommon to use models with thousands or even millions of degrees of freedom: When performing static analysis, this does not constitute a problem for modern computers, but the solution of an eigenproblem of that size can still be a formidable task.

Moreover, when using displacement methods, i.e., methods that first solve the displacements and then compute stresses and strains as derivatives of the displacements, displacements, and all other entities directly linked with them like mode shapes and natural frequencies, are obtained with much greater precision, for a given model, than stresses and strains. Conversely, this means that much more detailed models are needed when solving the stress field, which is typical of static problems, than when searching natural frequencies and mode shapes.

Remark 10.1 *Because it is often expedient to use the same model for the static and dynamic analysis, a reduction of the number of degrees of freedom for dynamic solution is useful, particularly when only a limited number of natural frequencies are required.*

In particular, reduced order models are particularly useful in the early stages of the analysis, when many details of the systems have not yet been exactly defined. While performing tests on the prototypes, simple models that can be solved in a short time on small computers may be of help too. As it will be seen in Chapter 11, simplified models may also be implemented in the control software.

Reduced models can be useful in both time-domain and frequency-domain computations. Their use is generally restricted to linear system but, as will be shown in Part II, they find applications also in the approximated solution of nonlinear models.

Two approaches may be used when computing the natural frequencies: reducing the size of the model or leaving the model as is and using algorithms, such as the subspace iteration method, that search only the lowest natural frequencies. Although the two are more or less equivalent, the first leaves the choice of which degrees of freedom to retain to the user, while the second operates automatically. As a consequence, a skilled operator can use advantageously reduction techniques, which allow good results to be obtained with a small number of degrees of freedom. A general-purpose code for routine computations, sometimes used by not much experienced analysts, on the contrary, can advantageously use the second approach. Only the first approach is dealt with here: The second is studied in Appendix A, together with other solution techniques.

Remark 10.2 *Before computers were available, remarkable results were obtained using models with very few (often a single) degrees of freedom, but this required great computational ability and physical insight.*

10.2   Static reduction of conservative models

Static reduction is based on the subdivision of the generalized coordinates x_i into two sets: master degrees of freedom contained in vector $\mathbf{x}_1$ and slave degrees of freedom contained in $\mathbf{x}_2$. Assuming that the master degrees of freedom are m (n is the total number of degrees of freedom), the stiffness matrix and the nodal force vector can be partitioned accordingly, and the equation expressing the static problem becomes

$$
\begin{bmatrix} \mathbf{K}_{11_{m \times m}} & \mathbf{K}_{12_{m \times (n-m)}} \\ \mathbf{K}_{21_{(n-m) \times m}} & \mathbf{K}_{22_{(n-m) \times (n-m)}} \end{bmatrix} \begin{Bmatrix} \mathbf{x}_1 \\ \mathbf{x}_2 \end{Bmatrix} = \begin{Bmatrix} \mathbf{f}_1 \\ \mathbf{f}_2 \end{Bmatrix}. \tag{10.1}
$$

Matrices $\mathbf{K}_{11}$ and $\mathbf{K}_{22}$ are symmetrical, while $\mathbf{K}_{12} = \mathbf{K}_{21}^T$ are neither symmetrical nor square. Solving the second set of Eqs. (10.1) in $\mathbf{x}_2$, the following relationship linking the slave to the master coordinates is obtained:

$$\mathbf{x}_2 = -\mathbf{K}_{22}^{-1}\mathbf{K}_{21}\mathbf{x}_1 + \mathbf{K}_{22}^{-1}\mathbf{f}_2 \,. \tag{10.2}$$

Introducing Eq. (10.2) into Eq. (10.1), the latter yields

$$\mathbf{K}_{cond}\mathbf{x}_1 = \mathbf{f}_{cond} \,, \tag{10.3}$$

where

$$\left\{ \begin{array}{l} \mathbf{K}_{cond} = \mathbf{K}_{11} - \mathbf{K}_{12}\mathbf{K}_{22}^{-1}\mathbf{K}_{12}^T \,, \\[2mm] \mathbf{f}_{cond} = \mathbf{f}_1 - \mathbf{K}_{12}\mathbf{K}_{22}^{-1}\mathbf{f}_2 \,. \end{array} \right.$$

Equation (10.3) yields the master generalized displacements $\mathbf{x}_1$. The slave displacements can be obtained directly from Eq. (10.2) simply by multiplying some matrices.

Remark 10.3 *When used to solve a static problem, static reduction yields exact results, i.e., the same results that would be obtained from the complete model.*

The subdivision of the degrees of freedom between vectors $\mathbf{x}_1$ and $\mathbf{x}_2$ can be based on different criteria. The master degrees of freedom can simply be those in which the user is directly interested. Another choice can be physically subdividing the structure in two parts. This practice can be generalized by subdividing the coordinates into many subsets and is generally known as *solution by substructures* or substructuring.

In particular, substructuring can be expedient when the structure can be divided into many parts that are all connected to a single frame. If the generalized displacements of the connecting structure or frame are listed in vector $\mathbf{x}_0$ and those of the various substructures are included in vectors $\mathbf{x}_i$, the equation for the static solution of the complete structure has the form

$$\begin{bmatrix} \mathbf{K}_{00} & \mathbf{K}_{01} & \mathbf{K}_{02} & \dots \\ & \mathbf{K}_{11} & \mathbf{0} & \dots \\ & & \mathbf{K}_{22} & \dots \\ & \text{symm.} & & \dots \end{bmatrix} \left\{ \begin{array}{c} \mathbf{x}_0 \\ \mathbf{x}_1 \\ \mathbf{x}_2 \\ \dots \end{array} \right\} = \left\{ \begin{array}{c} \mathbf{f}_0 \\ \mathbf{f}_1 \\ \mathbf{f}_2 \\ \dots \end{array} \right\}. \tag{10.4}$$

The equations related to the ith substructure can be solved as

$$\mathbf{x}_i = -\mathbf{K}_{ii}^{-1}\mathbf{K}_{i0}\mathbf{x}_0 + \mathbf{K}_{ii}^{-1}\mathbf{f}_i \,. \tag{10.5}$$

The generalized displacements of the frame can be obtained using an equation of the type of Eq. (10.3) where the condensed matrices are

$$\left\{ \begin{array}{l} \mathbf{K}_{cond} = \mathbf{K}_{00} - \displaystyle\sum_{\forall i} \mathbf{K}_{0i}\mathbf{K}_{ii}^{-1}\mathbf{K}_{0i}^T \,, \\[4mm] \mathbf{f}_{cond} = \mathbf{f}_0 - \displaystyle\sum_{\forall i} \mathbf{K}_{0i}\mathbf{K}_{ii}^{-1}\mathbf{f}_i \,. \end{array} \right. \tag{10.6}$$

As already stated, static reduction does not introduce any further approximation into the model. A similar reduction can be used in dynamic analysis without introducing approximations only if no generalized inertia is associated with the slave degrees of freedom. In this case, static reduction is advisable because the mass matrix of the original system is singular, and the condensation procedure allows removal of the singularity.

10.3   Guyan reduction

The so-called *Guyan reduction* is based on the assumption that the slave generalized displacements x_2 can be computed directly from master displacements x_1, neglecting inertia forces and external forces f_2. In this case, Eq. (10.2), without the last term, can also be used in dynamics.

By partitioning the mass matrix in the same way seen for the stiffness matrix, the kinetic energy of the structure can be expressed as

$$\mathcal{T} = \frac{1}{2} \left\{ \begin{array}{c} \dot{x}_1 \\ -K_{22}^{-1}K_{21}\dot{x}_1 \end{array} \right\}^T \left[\begin{array}{cc} M_{11} & M_{12} \\ M_{21} & M_{22} \end{array} \right] \left\{ \begin{array}{c} \dot{x}_1 \\ -K_{22}^{-1}K_{21}\dot{x}_1 \end{array} \right\} , \quad (10.7)$$

i.e.,

$$\mathcal{T} = \frac{1}{2}\dot{x}_1^T \left[\begin{array}{cc} I & -K_{22}^{-1}K_{21} \end{array} \right] \left[\begin{array}{cc} M_{11} & M_{12} \\ M_{21} & M_{22} \end{array} \right] \left\{ \begin{array}{c} I \\ -K_{22}^{-1}K_{21}\dot{q}_1 \end{array} \right\} \dot{x}_1 . \quad (10.8)$$

The kinetic energy is thus

$$\mathcal{T} = \frac{1}{2}\dot{x}_1^T M_{cond}\dot{x}_1 , \quad (10.9)$$

where the condensed mass matrix is

$$M_{cond} = M_{11} - M_{12}K_{22}^{-1}K_{12}^T - \left[M_{12}K_{22}^{-1}K_{12}^T\right]^T +$$
$$+ K_{12}K_{22}^{-1}M_{22}K_{22}^{-1}K_{12}^T . \quad (10.10)$$

Guyan reduction is not much more demanding from a computational viewpoint than static reduction because the only matrix inversion is that of K_{22}, which has already been performed for the computation of the condensed stiffness matrix. If matrix M is diagonal, two of the terms of Eq. (10.10) vanish.

Although approximate, it introduces errors that are usually small, at least if the choice of the slave degrees of freedom is appropriate. Inertia forces related to slave degrees of freedom are actually not neglected, but their contribution to the kinetic energy is computed from a deformed configuration obtained on the basis of the master degrees of freedom alone.

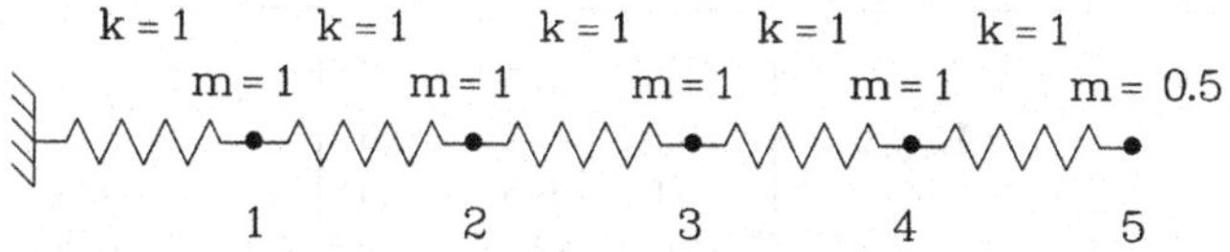

FIGURE 10.1. Sketch of a system with 5 degrees of freedom.

Remark 10.4 *If the relevant mode shapes are only slightly influenced by the presence of some of the generalized masses or if some parts of the structure are so stiff that their deflected shape can be determined by a few generalized coordinates, the results can be very accurate, even when few master degrees of freedom are used.*

Example 10.1 *Consider the conservative system with 5 degrees of freedom of Fig. 10.1. Compute the natural frequencies using a model with all degrees of freedom, and then repeat the computation using static and Guyan reductions taking the displacements of points 2 and 4 as slave degrees of freedom. The mass and stiffness matrices of the system are*

$$
\mathbf{K} = \begin{bmatrix} 2 & -1 & 0 & 0 & 0 \\ -1 & 2 & -1 & 0 & 0 \\ 0 & -1 & 2 & -1 & 0 \\ 0 & 0 & -1 & 2 & -1 \\ 0 & 0 & 0 & -1 & 1 \end{bmatrix}, \quad \mathbf{M} = \begin{bmatrix} 1 & 0 & 0 & 0 & 0 \\ 0 & 1 & 0 & 0 & 0 \\ 0 & 0 & 1 & 0 & 0 \\ 0 & 0 & 0 & 1 & 0 \\ 0 & 0 & 0 & 0 & 0.5 \end{bmatrix}.
$$

By directly solving the eigenproblem, the following matrix of the eigenvalues is obtained:

$$
[\omega^2] = diag \begin{bmatrix} 0.0979 & 0.8244 & 2.000 & 3.176 & 3.902 \end{bmatrix},
$$

i.e.,

$$
\omega_n = 0.3129, \quad 0.9080, \quad 1.4142, \quad 1.7820, \quad 1.9754 .
$$

Since the slave degrees of freedom are the second and the fourth, the partitioned matrices are

$$
\mathbf{K}_{11} = \begin{bmatrix} 2 & 0 & 0 \\ 0 & 2 & 0 \\ 0 & 0 & 1 \end{bmatrix}, \quad \mathbf{M}_{11} = \begin{bmatrix} 1 & 0 & 0 \\ 0 & 1 & 0 \\ 0 & 0 & 0.5 \end{bmatrix},
$$

$$
\mathbf{K}_{12} = \begin{bmatrix} -1 & 0 \\ -1 & -1 \\ 0 & -1 \end{bmatrix}, \quad \mathbf{K}_{22} = \begin{bmatrix} 2 & 0 \\ 0 & 2 \end{bmatrix}, \quad \mathbf{M}_{22} = \begin{bmatrix} 1 & 0 \\ 0 & 1 \end{bmatrix}.
$$

The reduced stiffness matrix is

$$\mathbf{K}_{cond} = \begin{bmatrix} 1.5 & -0.5 & 0 \\ -0.5 & 1 & -0.5 \\ 0 & -0.5 & 0.5 \end{bmatrix}.$$

Only three natural frequencies can be found using static reduction, since the system is reduced to a system with 3 degrees of freedom. The natural frequencies from the eigenproblem involving matrices M_{11} and K_{cond} are reported in the table below

	Exact	Static reduction	Guyan reduction
Mode 1	0.3129	0.4370	0.3167
Mode 2	0.9080	1.1441	0.9864
Mode 3	1.4142	1.4142	1.4142
Mode 4	1.7820	–	–
Mode 5	1.9754	–	–

The errors on the first two natural frequencies are 40% and 26%: as predictable static reduction yields very poor results. The fact that the third natural frequency is obtained correctly is an anomaly, due to the fact that the third eigenvector of the full model has zero amplitudes at the slave degrees of freedom. In this condition neglecting the corresponding masses does not lead to any error.
The reduced mass matrix obtained through the Guyan reduction is

$$\mathbf{M}_{cond} = \begin{bmatrix} 1.25 & 0.25 & 0 \\ 0.25 & 1.50 & 0.25 \\ 0 & 0.25 & 0.75 \end{bmatrix}.$$

Also using Guyan reduction only three natural frequencies can be found. The natural frequencies from the eigenproblem involving matrices M_{cond} and K_{cond} are reported in the table above.
The errors on the first two natural frequencies are 1.2% and 8.6%.
Considering that the masses associated to the slave degrees of freedom are equal to those associated to the master coordinates, the precision with which the first natural frequency is computed is an interesting result. The precision on the second one is smaller as expected. Again the third natural frequency is evaluated with no error, for the reason seen above.

10.4   Damped systems

In a way similar to that seen for the mass matrix, viscous, or structural damping matrices can be reduced using Eq. (10.10) in which $\mathbf{M}$ has been substituted with $\mathbf{C}$ and $\mathbf{K}''$, respectively. The reduced viscous damping matrix is thus

$$\mathbf{C}_{cond} = \mathbf{C}_{11} - \mathbf{C}_{12}\mathbf{K}_{22}^{-1}\mathbf{K}_{12}^T - \left[\mathbf{C}_{12}\mathbf{K}_{22}^{-1}\mathbf{K}_{12}^T\right]^T +$$
$$+\mathbf{K}_{12}\mathbf{K}_{22}^{-1}\mathbf{C}_{22}\mathbf{K}_{22}^{-1}\mathbf{K}_{12}^T. \tag{10.11}$$

The reduction of damping matrices introduces errors that depend on the choice of the slave degrees of freedom but are usually small when the degrees of freedom in which viscous dampers are applied or, in the case of hysteretic damping, where the loss factor of the material changes, are not eliminated.

Small errors are also introduced when the generalized coordinates of the slave degrees of freedom are determined with good accuracy by some master displacement, as in the case of very stiff parts of the structure.

10.5   Dynamic reduction

While the reduction techniques seen above can be used for both frequency-domain and time-domain computations, when only frequency-domain solutions are searched the reduction can be operated directly on the dynamic stiffness matrix (*dynamic reduction*). This procedure does not introduce approximations, but the frequency appears explicitly in the matrices that must be inverted and multiplied.

There is no difficulty in reducing also the complex dynamic stiffness matrix resulting from a damped system.

Because it is not possible to perform the inversion of matrix $\mathbf{K}_{dyn_{22}}$ leaving the frequency unspecified (except if the slave degrees of freedom are only 2 or 3), a numerical value of the frequency must be stated. While this does not give any problem when computing the frequency response of the system, it is impossible to compute the natural frequencies in this way, except if operating by trial and error.

Dynamic reduction has also an application in the computation of the approximated response of multi-degrees-of-freedom nonlinear systems: It will be further studied in Chapter 20.

10.6   Modal reduction

Performing the modal transformation and then neglecting a number of modes may be considered as a different approach to the reduction of the number of degrees of freedom of the system.

Using the reduced matrix of the eigenvectors $\mathbf{\Phi}^*$ instead of the full matrix $\mathbf{\Phi}$ introduces two types of errors:

- errors due to neglecting the contribution of the neglected modes and

- errors due to neglecting modal coupling.

The second cause is not present if the modes are exactly uncoupled, like in the case of undamped systems or damped systems with generalized proportional damping. Often high-frequency modes are determined more by how the model has been obtained than by the actual characteristics of the physical system, and then neglecting them no further approximations, beyond those already present in the model, are introduced.

Moreover, the use of the modified reduced matrix $\mathbf{\Phi}^{**}$ (see Section 7.5) can make the errors of the first type to be quite small, and a system with a large number of degrees of freedom can be studied with precision using only a small number of modes.

When damping (or the gyroscopic and circulatory matrices $\mathbf{G}$ and $\mathbf{H}$) couple the modal equations, it is difficult to assess how many modes are needed to obtain the required precision. This issue depends on many factors, like how large is damping (or the gyroscopic and circulatory effects), how many natural frequencies are included in the range of the exciting frequencies, how spaced are they are, etc.

10.7   Component-mode synthesis

When substructuring is used, the degrees of freedom of each substructure can be divided into two sets: internal degrees of freedom and boundary degrees of freedom. The latter are all degrees of freedom that the substructure has in common with other parts of the structure. They are often referred to as constraint degrees of freedom because they express how the substructure is constrained to the rest of the system. Internal degrees of freedom are those belonging only to the relevant substructure. The largest possible reduction scheme is that in which all internal degrees of freedom are considered slave coordinates and all boundary degrees of freedom are considered master coordinates. In this way, however, the approximation of all modes in which the motion of the internal points of the substructure with respect to the motion of its boundary is important, can be quite rough.

A simple way to avoid this drawback is to also consider as master coordinates, together with the boundary degrees of freedom, some of the modal coordinates of the substructure constrained at its boundary. This procedure would obviously lead to exact results if all modes were retained, but because the total number of modes is equal to the number of internal degrees of freedom, the model obtained has as many degrees of freedom as the original model. As usual with modal practices, the computational advantages grow, together with the number of modes that can be neglected.

The relevant matrices are partitioned as seen for reduction techniques, with subscript 1 referring to the boundary degrees of freedom and subscript 2 to the internal degrees of freedom. The displacement vector $\mathbf{x}_2$ can be assumed to be equal to the sum of the constrained modes $\mathbf{x}_2'$, i.e., the deformation pattern due to the displacements $\mathbf{x}_1$ when no force acts on the substructure, plus the constrained normal modes $\mathbf{x}_2''$, i.e., the natural modes of free vibration of the substructure when the boundary generalized displacements $\mathbf{x}_1$ are equal to zero.

The constrained modes $\mathbf{x}_2'$ can be expressed by Eq. (10.2) once the force vector $\mathbf{f}_2$ is set equal to zero:

$$\mathbf{x}_2' = -\mathbf{K}_{22}^{-1}\mathbf{K}_{21}\mathbf{x}_1\,.$$

The constrained normal modes can easily be computed by solving the eigenproblem

$$\left(-\omega^2\mathbf{M}_{22} + \mathbf{K}_{22}\right)\mathbf{x}_2'' = 0\,.$$

Once the eigenproblem has been solved, the matrix of the eigenvectors Φ can be used to perform the modal transformation

$$\mathbf{x}_2'' = \Phi\boldsymbol{\eta}_2\,.$$

The generalized coordinates of the substructure can thus be expressed as

$$\left\{\begin{array}{c} \mathbf{x}_1 \\ \mathbf{x}_2 \end{array}\right\} = \left\{\begin{array}{c} \mathbf{x}_1 \\ -\mathbf{K}_{22}^{-1}\mathbf{K}_{21}\mathbf{x}_1 + \Phi\boldsymbol{\eta}_2 \end{array}\right\} = \\ = \left[\begin{array}{cc} \mathbf{I} & 0 \\ -\mathbf{K}_{22}^{-1}\mathbf{K}_{21} & \Phi \end{array}\right]\left\{\begin{array}{c} \mathbf{x}_1 \\ \boldsymbol{\eta}_2 \end{array}\right\} = \Psi\left\{\begin{array}{c} \mathbf{x}_1 \\ \boldsymbol{\eta}_2 \end{array}\right\}\,. \tag{10.12}$$

Equation (10.12) represents a coordinate transformation, allowing the expression of the deformation of the internal part of the substructure in terms of constrained and normal modes. Matrix Ψ expressing this transformation can be used to compute the new mass, stiffness, and, where needed, damping matrices and the force vector

$$\mathbf{M}^* = \Psi^T\mathbf{M}\Psi\,, \quad \mathbf{K}^* = \Psi^T\mathbf{K}\Psi\,, \\ \mathbf{C}^* = \Psi^T\mathbf{C}\Psi\,, \quad \mathbf{f}^* = \Psi^T\mathbf{f}\,. \tag{10.13}$$

If there are m constrained coordinates and n internal coordinates and if only k constrained normal modes are considered ($k < n$), then the size of the original matrices $\mathbf{M}$, $\mathbf{K}$, ... is $m + n$, while that of matrices $\mathbf{M}^*$, $\mathbf{K}^*$, ... is $m + k$.

The main advantage of component-mode synthesis and substructuring is allowing the construction of the model and the analysis of the various parts of a large structure in an independent way. The results can then be assembled in a way similar to what will be seen in the context of the Finite Element Method (see Chapter 15) and the behavior of the structure can be assessed from that of its parts. If this is done, however, the connecting nodes must be defined in such a way that the same boundary degrees of freedom are considered in the analysis of the various parts. It is, however, possible to use algorithms allowing to connect otherwise incompatible meshes.

Remark 10.5 *All the methods discussed in this section, which are closely related to each other, can be found in the literature in a variety of versions.*

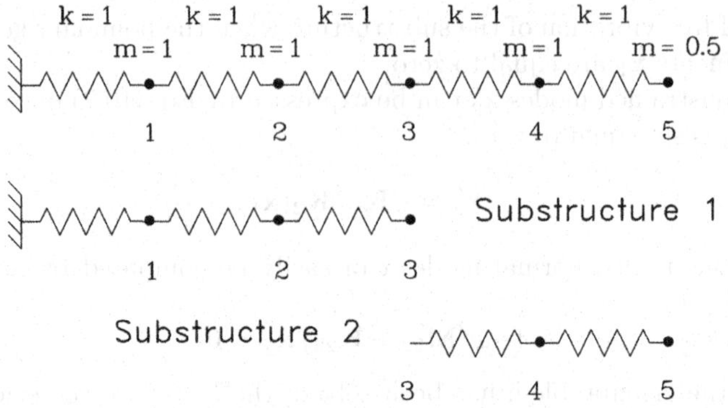

FIGURE 10.2. Sketch of the system and values of the relevant parameters.

Although they are general for discrete systems, they are mostly used in connection with the Finite Element Method, owing to the large number of degrees of freedom typical of the models based on it.

Example 10.2 *Consider again the discrete system already studied in Example 10.1 and sketched in Fig. 10.2. Study its dynamic behavior using component-mode synthesis retaining different numbers of modes.*

The total number of degrees of freedom of the system is five and the complete mass and stiffness matrices are those shown in the previous example.

The structure is then subdivided into two substructures and the analysis is accordingly performed.

Substructure 1 includes nodes 1, 2, and 3 with the masses located on them. The displacements at nodes 1 and 2 are internal coordinates, while the displacement at node 3 is a boundary coordinate. The mass and stiffness matrix of the substructure, partitioned with the boundary degree of freedom first and then the internal ones, are

$$
K = \left[\begin{array}{c|cc} 1 & -1 & 0 \\ \hline -1 & 2 & -1 \\ 0 & -1 & 2 \end{array} \right], \qquad M = \left[\begin{array}{c|cc} 1 & 0 & 0 \\ \hline 0 & 1 & 0 \\ 0 & 0 & 1 \end{array} \right].
$$

The matrix of the eigenvectors for the internal normal modes can be easily obtained by solving the eigenproblem related to matrices with subscript 22 and, by retaining all modes, matrices K^* and M^* of the first substructure can be computed as follows:

$$
\Phi = \left[\begin{array}{cc} \dfrac{\sqrt{2}}{2} & -\dfrac{\sqrt{2}}{2} \\ \dfrac{\sqrt{2}}{2} & \dfrac{\sqrt{2}}{2} \end{array} \right], \qquad K^* = \left[\begin{array}{c|cc} 0.3333 & 0 & 0 \\ \hline 0 & 1 & 0 \\ 0 & 0 & 3 \end{array} \right],
$$

$$\mathbf{M}^* = \begin{bmatrix} 1.556 & 0.7071 & -0.2357 \\ \hline 0.7071 & 1 & 0 \\ -0.2357 & 0 & 1 \end{bmatrix}.$$

Substructure 2 includes nodes 3, 4, and 5 with the masses located on nodes 4 and 5. The mass located on node 3 has already been taken into account in the first substructure and must not be considered again. The displacements at nodes 4 and 5 are internal coordinates, while the displacement at node 3 is a boundary coordinate. The mass and stiffness matrix of the substructure, partitioned with the boundary degree of freedom first and then the internal ones, are

$$K = \begin{bmatrix} 1 & -1 & 0 \\ \hline -1 & 2 & -1 \\ 0 & -1 & 1 \end{bmatrix} , \qquad M = \begin{bmatrix} 0 & 0 & 0 \\ \hline 0 & 1 & 0 \\ 0 & 0 & 0.5 \end{bmatrix} .$$

Operating as seen for the first substructure, it follows

$$\mathbf{\Phi} = \begin{bmatrix} \dfrac{\sqrt{2}}{2} & -\dfrac{\sqrt{2}}{2} \\ 1 & 1 \end{bmatrix} , \qquad \mathbf{K}^* = \begin{bmatrix} 0 & 0 & 0 \\ \hline 0 & 0.5858 & 0 \\ 0 & 0 & 3.4142 \end{bmatrix} ,$$

$$\mathbf{M}^* = \begin{bmatrix} 1.5 & 1.2071 & -0.2071 \\ \hline 1.2071 & 1 & 0 \\ -0.2071 & 0 & 1 \end{bmatrix} .$$

The substructures can be assembled in the same way as the elements (see Chapter 15). A map, that is a table in which the correspondence between the generalized coordinates of each substructure and those of the system as a whole, can be written

Subst. d.o.f.	1	2	3		
1 type	boundary	modal	modal		
Subst. d.o.f.	1			2	3
2 type	boundary			modal	modal
Global d.o.f.	1	2	3	4	5

The following global stiffness and mass matrices can thus be obtained

$$\mathbf{K}^* = \begin{bmatrix} 0.3333 & 0 & 0 & 0 & 0 \\ \hline 0 & 1 & 0 & 0 & 0 \\ 0 & 0 & 3 & 0 & 0 \\ 0 & 0 & 0 & 0.5858 & 0 \\ 0 & 0 & 0 & 0 & 3.4142 \end{bmatrix} ,$$

$$\mathbf{M}^* = \begin{bmatrix} 1.5 & 0.7071 & -0.2357 & 1.2071 & -0.2071 \\ \hline 0.7071 & 1 & 0 & 0 & 0 \\ -0.2357 & 0 & 1 & 0 & 0 \\ 1.2071 & 0 & 0 & 1 & 0 \\ -0.2071 & 0 & 0 & 0 & 1 \end{bmatrix} .$$

The matrices have been partitioned in such a way to separate the boundary displacement degree of freedom from the modal degrees of freedom. If no modal coordinate is considered, the component-mode synthesis coincides with Guyan reduction, with only one master degree of freedom.

If the third and fifth rows and columns are cancelled, only one internal normal mode is taken into account for each substructure. If the matrices are taken into account in complete form, all modes are considered and the result must coincide, except for computational approximations, with the exact ones. The results obtained in terms of the square of the natural frequency (those related to the complete model are taken from Example 10.1) are

Size of matrices	5 (exact)	1 (Guyan red.)	3 (1 mode)	5 (2 modes)
Mode 1	0.3129	0.3303	0.3129	0.3129
Mode 2	0.9080	–	0.9080	0.9080
Mode 3	1.4142	–	1.4883	1.4142
Mode 4	1.7820	–	–	1.7821
Mode 5	1.9754	–	–	1.9753

10.8   Exercises

Exercise 10.1 *Consider the system with two degrees of freedom of Exercise 4.4, made by masses m_1 and m_2, connected by a spring k_{12} between each other and springs k_1 and k_2 to point A. Write the reduced stiffness and mass matrix through Guyan reduction considering the displacement of mass m_2 as master degree of freedom. Compute the expression for the natural frequency and compare it with the lowest natural frequency obtained from the complete model. Evaluate their numerical value using the data below. Repeat the analysis using the displacement of mass m_1 as the master degree of freedom. Data: $m_1 = 5$ kg, $m_2 = 10$ kg, $k_1 = k_2 = 2$ kN/m, $k_{12} = 4$ kN/m.*

Exercise 10.2 *Add two dampers with a damping coefficient $c = 0.1$ to the system of Fig. 10.2 between nodes 2 and 3 and nodes 4 and 5. Compute the complex frequencies both using the complete model and through Guyan reduction, assuming that the displacement of node 1 is a slave degree of freedom. Repeat the computation assuming that also the displacement of node 4 is a slave degree of freedom.*

Exercise 10.3 *Repeat the study of Exercise 10.1 resorting to the component-mode synthesis. Compare the results with those of the previous exercise.*

Exercise 10.4 *Consider again the undamped system of Fig. 10.2. Compute the natural frequencies using the component-mode synthesis by considering the subsystem made of all masses and all spring except the first one as a substructure and then constraining it to the ground through the first spring. Consider a different number of modal coordinates.*

11
Controlled Linear Systems

Structures and machines are increasingly provided with control system, often active ones, that influence deeply their dynamic behavior. When this is the case, the dynamics of the system cannot be studied without a good knowledge of the performance of all components of the control loop, such as sensors, actuators, controllers, power amplifiers. The correct approach is thus to undertake the design and analysis of the system at a global level, including all components into models that take into account all of them in the required detail.

11.1 General considerations

Consider a structure[1] provided with a number of actuators and modeled as a discrete system (Fig. 11.1). Its equation of motion in the space of the configuration can be expressed in the form

$$\mathbf{M}\ddot{\mathbf{x}} + \mathbf{C}\dot{\mathbf{x}} + \mathbf{K}\mathbf{x} = \mathbf{f}_c + \mathbf{f}_e \, , \qquad (11.1)$$

where the control forces exerted by the actuators and the disturbances or external forces can be expressed, respectively, as

$$\mathbf{f}_c = \mathbf{T}_c \mathbf{u}_c \, , \qquad \mathbf{f}_e = \mathbf{T}_e \mathbf{u}_e \, ,$$

---

[1]In control terminology the controlled system is usually referred to as *plant*. In the following sections the more specific term *structure* will also be used: No attempt to deal with control theory in general is intended.

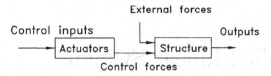

FIGURE 11.1. Block diagram of a structure on which both control and external forces are acting.

where $\mathbf{u}_c$ and $\mathbf{u}_e$ are the control and external inputs. Matrix $\mathbf{T}_c$ is sometimes referred to as the *control influence matrix*.

The corresponding state and output equations are

$$\begin{cases} \dot{\mathbf{z}} = \mathcal{A}\mathbf{z} + \mathcal{B}_c\mathbf{u}_c(t) + \mathcal{B}_e\mathbf{u}_e(t), \\ \mathbf{y}(t) = \mathcal{C}\mathbf{z}, \end{cases} \qquad (11.2)$$

where the inputs are assumed not to directly influence the outputs through the direct link matrix $\mathcal{D}$. The input gain matrices are simply

$$\mathcal{B}_c = \begin{bmatrix} \mathbf{M}^{-1}\mathbf{T}_c \\ \mathbf{0} \end{bmatrix}, \qquad \mathcal{B}_e = \begin{bmatrix} \mathbf{M}^{-1}\mathbf{T}_e \\ \mathbf{0} \end{bmatrix}.$$

The designer must deal with the whole system, made of the structure as well as the actuators, the control system that must provide the latter with the control inputs, the sensors and, in the case of active systems, the source of power for the actuators.

When designing an active system, the primary concern is often shifted from obtaining the required response to achieving stability. Or, better, stability becomes a prerequisite that must be fulfilled before thinking about performances.

This is new compared with the usual approach to structural dynamics: In most structures, stability is taken for granted, because the structure can only dissipate energy and free vibrations are bound to die out sooner or later. The designer has to provide sufficient damping, but at least he is sure that the motion is stable.

Notable exceptions are the cases in which the structure can absorb energy from its environment, as with aeroelastic structures or rotating machines. If even a small fraction of the energy from the aerodynamic field or kinetic energy linked with rotation seeps into the vibration, strong self-excited vibrations may take place. In these cases, stability has always been a primary concern.

For ideal and *co-located*[2] active systems, a theorem that ensures marginal stability exists, but most real-world active systems are prone to instability: When the structure is acted on by actuators of any type, the control system

---

[2]Sensors and actuators are said to be co-located when the generalized force due to the latter corresponds to the generalized coordinate measured by the former.

must avoid supplying energy to any vibration motion, otherwise an unstable behavior may occur.

The behavior of any active structure is the result of the integration of the behavior of the structural subsystem with that of the controller, the actuators, and the sensors and the only reasonable approach is to design the system as a whole. To add a controller to an already existing structure or one designed without taking into account the presence of the former may lead to performances far from the expected.

Structural dynamics, control engineering, transducer design, and electronics must merge from the beginning with that interdisciplinary approach often referred to as *mechatronics*.

11.2   Control systems

The science of control systems and the related technology saw enormous advances in recent decades, and it is impossible to summarize them satisfactorily in a text on structural dynamics. Only a few remarks on control systems will be reported here, limited to what can be useful in the context of structural control; the interested reader can find the relevant information in many textbooks on the subject.

Classical control theory deals with linear, or at least linearized, control systems. The basic tools are those typical of linear system dynamics, namely block diagrams, phase- or state-space equations, transfer functions, and eigenstructure analysis.

The control systems used for structural control can be based on transducers (sensors and/or actuators) of different types, such as mechanical, electrical, hydraulic, and pneumatic. Recently, however, electronic-based systems are becoming more common, both for all-electrical applications and in the form of electro-mechanical, electrohydraulic, and so on, applications. The electronic part can be based on analog or digital circuits; the former are preferred for simpler applications, where they are still cheaper than the latter. With the diffusion of microprocessors, digital techniques became more common, particularly for their flexibility and ability to perform very complex tasks.

Independently from their physical configuration, control systems can be divided into two categories: passive and active control systems. The first operate without any external energy supply, using the energy stored in the structure as potential or kinetic energy as a consequence of its dynamic response to supply the control forces.

Passive devices in many cases act as dampers. For example, piezoceramic materials can be used as both sensors and actuators: If a piezoelectric element is simply shunted by a resistor, a sort of electric damper is obtained. By also introducing an inductor into the circuit, the capacitance of the piezoelectric element forms, together with the other components of the circuit, a resonant inductor–resistor–capacitor system whose transfer function can be designed to obtain performance similar to that of a damped vibration absorber. Piezoelectric elements can also be used in active systems, as both sensors and actuators. Layers of piezoelectric materials can also perform as distributed sensory or actuating devices.

Active control systems are equipped with an external power supply to provide the control forces. There are cases in which the amount of control energy is minimal and in this case the term semi-active systems is sometimes used. Active systems can be either manual or automatic, but only the latter are important in structural control, particularly if true dynamic control is required.

Both active and passive control systems can operate as open-loop or closed-loop systems. Open-loop control systems, sometimes referred to as *predetermined control systems*, react to the variation of the input parameters of the controlled system without actually measuring its output parameters in order to check whether the response of the system conforms with the required values. A system that changes the stiffness of the supports of a rotor depending on its angular velocity, for instance, is an open-loop system.

In closed-loop or feedback systems the control system monitors the outputs of the controlled system, compares their actual values with predetermined reference values, and uses the information so obtained to perform the control action. Closed- and open-loop techniques can be used simultaneously, as in the case of feedback systems in which the rapidity of the response of the controller is incremented by also monitoring the excitation and using this information to help controlling the system (feedforward technique).

An example of an active closed-loop control system taken outside the field of structural control is the driver of a vehicle. The control action is performed by comparing the trajectory and the other parameters of the motion with those the driver aims to obtain and then acting on the controls to supply the required corrections. A blind driver who tries to drive home relying on his knowledge of the road would be an example of open-loop control.

An active magnetic bearing is another example of an active closed-loop control system.

Most passive systems are closed-loop systems. Closed-loop systems are usually preferred when the control is required to react to unknown external or internal disturbances, but they are usually more complex and costly than open-loop systems.

There are cases, like when the control is performed by mechanical means (for instance, a Watt's regulator), where it may be arbitrary to state where the control system ends and where the plant starts.

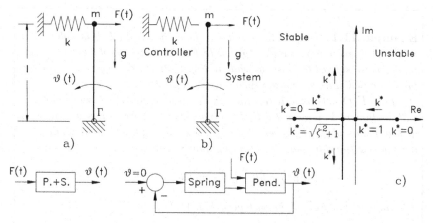

FIGURE 11.2. Inverted pendulum stabilized by a linear spring: (a) the whole assembly is considered as the system; (b) the pendulum is considered as the system and the spring is considered as the controller; (c) root locus in nondimensional form.

Another common distinction for control systems is that between regulators and servomechanisms or tracking systems. A regulator is used to maintain the system in a predetermined condition, which can be an equilibrium position, a velocity, or an acceleration, in spite of external disturbance. The spring of the inverted pendulum in Fig. 11.2 can be considered a regulator, even if it cannot achieve its goal with precision in the presence of a constant disturbance: If a constant force F acts on the pendulum (the controlled system), the spring cannot restore the position with $\theta = 0$.

The input to the system controlled by a regulator is the condition to be maintained, and it is usually referred to as the *set point*. It is a constant, but it can be changed in many cases from one constant value to another.

In the case of a servomechanism, the reference input changes in time and the control system tries to obtain an output of the controlled system that follows the reference. In this case the output can also be a position, a velocity, an acceleration, or any other relevant quantity.

A control system can be a single-input single-output (SISO) or multiple-input multiple-output (MIMO) or multivariable system. In the first case, the control system reacts to only one of the outputs of the controlled system, acting on just one input of the plant u_c: They are clearly not restricted to systems with a single degree of freedom.

MIMO control systems use a number of outputs of the plant to act on a number of control inputs. When each input separately controls a single output, the term *decentralized control* is used. Decentralized control is sometimes resorted to with the aim of weakening the coupling of a MIMO plant.

Example 11.1 *Consider the pendulum shown in Fig. 11.2. It is stabilized in its inverted position by a linear spring. In Fig. 11.2a the pendulum–spring assembly is considered the system, while in Fig. 11.2b the pendulum is considered the controlled system, and the spring is the controller.*

If a force $F(t)$ acts as a disturbance on the pendulum and a viscous damping with coefficient Γ is provided by the hinge, the linearized equation for small displacements about the inverted upright position, obtained as dynamic equilibrium equation for rotations about the hinge point, is

$$ml^2\ddot\theta + \Gamma\dot\theta + (kl^2 - mgl)\theta = lF(t).$$

The transfer function of the system of Fig. 11.2a is then

$$G(s) = \frac{1}{ml^2 s^2 + \Gamma s + kl^2 - mgl}.$$

If the scheme of Fig. 11.2b is considered, the equation of motion of the system is

$$ml\ddot\theta + \frac{\Gamma}{l}\dot\theta - mg\theta = F_c + F(t),$$

where F_c is the control force and the equation of the control system is

$$F_c = -kl\theta.$$

The characteristic equation of the controlled system, including the controller, is in nondimensional form

$$s^{*^2} + 2\zeta s^* + k^* - 1 = 0,$$

where

$$s^* = s\sqrt{\frac{l}{g}}, \qquad \zeta = \frac{\Gamma}{2m\sqrt{gl^3}}, \qquad k^* = k\frac{l}{mg}.$$

The root locus is plotted in Fig. 11.2c. The breakaway point occurs for

$$k^* = \sqrt{1 + \zeta^2},$$

and the system is stable if

$$k^* > 1, \quad i.e., \ if \ k > \frac{mg}{l}.$$

The last condition could be obtained directly from the equation of motion, stating that the total stiffness of the system must be positive to have stable behavior.

11.3   Controllability and observability

Consider the linear system whose behavior is described by Eqs. (11.2). The possibility of controlling it through the control inputs $\mathbf{u}_c$ and of knowing its state from the observation of its outputs $\mathbf{y}$ is defined as its controllability and observability.

If it is possible to determine a law for the inputs $\mathbf{u}_c(t)$ defined from time t_0 to time t_f, which allows the system to be driven from the initial state to a desired state and in particular to a state with all state variables equal to zero, the system is said to be *controllable*. If this can be done for any arbitrary initial time t_0 and initial state, then the system is completely controllable. To check whether the system is controllable, it is possible to write the controllability matrix

$$\mathbf{H} = \begin{bmatrix} \mathcal{B}_c & \mathcal{A}\mathcal{B}_c & \mathcal{A}^2\mathcal{B}_c & \cdots & \mathcal{A}^{n-r}\mathcal{B}_c \end{bmatrix}, \tag{11.3}$$

where r is the number of control inputs, i.e., the number of columns of matrix $\mathcal{B}_c$, and n is the number of states.

If matrix $\mathbf{H}$, which has a total of n rows and $r \times (n - r + 1)$ columns, has rank n, the system is completely controllable. The controllability matrix can also be written considering products $\mathcal{A}^i\mathcal{B}_c$ until $i = n - 1$ instead of $i = n - r$. In this way, the number of columns of the controllability matrix is $r \times n$ instead of $r \times (n - r + 1)$, but nothing is changed, because the added columns are linear combinations of the others.

In a similar way, a linear system is said to be *observable* if it is possible to determine its state at time t_0 from the laws of the inputs $\mathbf{u}(t)$ and the outputs $\mathbf{y}(t)$ defined from time t_0 to time t_f. If this can be done for any arbitrary initial time t_0 and initial state, then the system is completely observable. To check whether a linear system with fixed parameters is observable, it is possible to write the observability matrix

$$\mathbf{O} = \begin{bmatrix} \mathcal{C}^T & \mathcal{A}^T\mathcal{C}^T & (\mathcal{A}^T)^2\mathcal{C}^T & \cdots & (\mathcal{A}^T)^{n-m}\mathcal{C}^T \end{bmatrix}, \tag{11.4}$$

where m is the number of outputs, i.e., the number of rows of matrix $\mathcal{C}$.

If such a matrix, which has a total of n rows and $n \times m$ columns, has rank n, the system is completely observable.

It is easy to verify that in the case of a linear structural system with a single degree of freedom both conditions for observability and controllability are always verified. This is, however, not necessarily the case for systems with many degrees of freedom, where observability and controllability must be checked in each case.

The study of the controllability and observability matrices allows one to obtain a sort of on–off answer on the issue of whether a structure can be controlled by a given set of actuators or observed by a set of sensors but does not state how controllable or observable it is. Other criteria allowing one to obtain a measure of the controllability and observability of a plant, and then allowing to search for the best position of transducers, are described in the literature.[3]

It is possible to define the controllability and observability starting from the equations written in modal coordinates; in this way it is possible to state which modes can be controlled and observed by a given set of transducers. The obvious general rule is that a mode cannot be either observed or controlled by a transducer located near one of the nodes of the relevant mode shape.

11.4   Open-loop control

Consider the structure in Fig. 11.1. The actuators added to provide suitable control forces $\mathbf{F}_c$ are driven by a controller to which a number of reference inputs $\mathbf{r}(t)$ are provided (block diagram in Fig. 11.3a). If the controller is linear, the control input can thus be expressed as

$$\mathbf{u}_c = \mathbf{K}_r \mathbf{r}(t) \,. \tag{11.5}$$

A more complex example of an open-loop system is a system with input compensation, in which a device supplies a set of control inputs $\mathbf{u}_c$ that are functions not only of the reference inputs $\mathbf{r}(t)$ but also of the external forces $\mathbf{f}_e(t)$, or better, of the external inputs $\mathbf{u}_e(t)$, applied to the system (Figs. 11.3b, c)

$$\mathbf{u}_c = \mathbf{K}_r \mathbf{r}(t) + \mathbf{K}_e \mathbf{u}_e(t) \,. \tag{11.6}$$

The matrices of the gains of the control system $\mathbf{K}_r$ and $\mathbf{K}_e$ have as many rows as the control inputs and as many columns as the reference and external inputs, respectively. The total state equation of the controlled system can be obtained by introducing Eqs. (11.6) into Eq. (11.2):

$$\begin{cases} \dot{\mathbf{z}} = \mathcal{A}\mathbf{z} + (\mathcal{B}_c \mathbf{K}_e + \mathcal{B}_e)\, \mathbf{u}_e(t) + \mathcal{B}_c \mathbf{K}_r \mathbf{r}(t) \,, \\ \mathbf{y}(t) = \mathcal{C}\mathbf{z} \,. \end{cases} \tag{11.7}$$

In general, an open-loop system relies on the model of the plant to obtain a command input that, supplied to it, causes the output to follow a desired pattern. This strategy requires very good knowledge of the dynamics of the controlled system and is usually applied only as a feedforward component in conjunction with a feedback controller.

Remark 11.1 *The free response of the system is not affected by the presence of the control system, which plays a role only in determining the forced response. This feature is, however, strictly linked with the complete linearity of the system.*

---

[3]See for example J.L. Junkins, Y. Kim, *Introduction to Dynamics and Control of Flexible Structures*, AIAA, Washington DC, 1993.

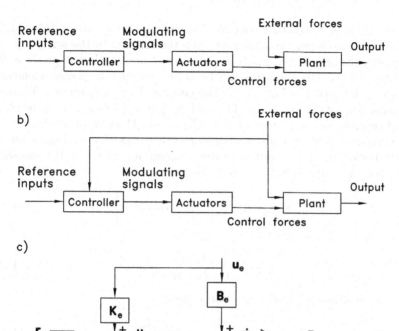

FIGURE 11.3. Open-loop control of the plant of Fig. 11.1: (a) tracking system in which the control input tries to follow an external reference; (b) open-loop control with external input compensation; (c) block diagram of the system shown in (b).

By resorting to Laplace transform and assuming that the gains of the control system are considered functions of the Laplace variable s and that at time $t = 0$ the value of all state variables is zero, the equation of motion becomes

$$\begin{cases} (s\mathbf{I} - \mathcal{A})\,\mathbf{z}\,(s) = (\mathcal{B}_c\mathbf{K}_e + \mathcal{B}_e)\,\mathbf{u}_e(s) + \mathcal{B}_c\mathbf{K}_r\mathbf{r}(s)\,, \\ \mathbf{y}(s) = \mathcal{C}\mathbf{z}\,(s)\,. \end{cases} \qquad (11.8)$$

If $\mathbf{r}(t) = 0$, the input–output relationship reduces to

$$\mathbf{y}(s) = \mathcal{C}\,(s\mathbf{I} - \mathcal{A})^{-1}\,(\mathcal{B}_c\mathbf{K}_e + \mathcal{B}_e)\,\mathbf{u}_e(s)\,. \qquad (11.9)$$

The transfer function linking the outputs of the system with the external inputs can be easily computed:

$$\mathbf{G}(s) = \mathcal{C}\,(s\mathbf{I} - \mathcal{A})^{-1}\,(\mathcal{B}_c\mathbf{K}_e(s) + \mathcal{B}_e)\,. \qquad (11.10)$$

11.5   Closed-loop control

Consider now a structure controlled by a closed-loop system (Fig. 11.4). The reference inputs $\mathbf{r}(t)$ interact with the outputs of the sensors $\mathbf{y}_s$ to supply suitable control inputs $\mathbf{u}_c$ to a set of actuators that produce the control forces. The actuators can be active systems, or passive elements, such as the spring in Fig. 11.2b. The general block diagram of a feedback system is shown in Fig. 11.2b. The open-loop transfer functions of the plant and control system are indicated as $\mathbf{G}_{ol}(s)$ and $\mathbf{H}(s)$, respectively.

Consider a SISO system. If no external disturbances are acting, i.e., $\mathbf{u}_e(t) = 0$, the output $y(t)$ is linked to the reference input $r(t)$ by the following relationship, written in the Laplace domain

$$y(s) = G_{ol}(s)\left[r(s) - H(s)y(s)\right], \tag{11.11}$$

i.e.,

$$y(s) = \frac{G(s)}{1 + G(s)H(s)}r(s). \tag{11.12}$$

The closed-loop transfer function is then

$$G_{cl}(s) = \frac{y(s)}{r(s)} = \frac{G(s)}{1 + G(s)H(s)}. \tag{11.13}$$

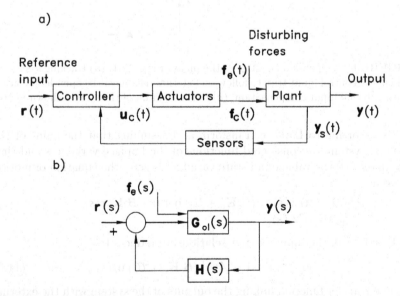

FIGURE 11.4. (a) Block diagram of the multi-degree-of-freedom structure in Fig. 11.1 controlled by a feedback system; (b) general block diagram of a feedback system in the Laplace domain.

In the case of MIMO systems, the input–output relationship is

$$\mathbf{y}(s) = \left[\mathbf{I} + \mathbf{G}_{ol}(s)\mathbf{H}(s)\right]^{-1}\mathbf{G}(s)\mathbf{r}(s). \tag{11.14}$$

The state equation of the controlled system is still Eq. (11.2), but now the control inputs $\mathbf{u}_c$ are determined by the outputs $\mathbf{y}(t)$ of the system and the reference inputs $\mathbf{r}(t)$

$$\mathbf{u}_c = \mathbf{K}_r\mathbf{r} - \mathbf{K}_y\mathbf{y}, \tag{11.15}$$

where the size of all matrices and vectors depends on the number of control and reference inputs and the number of outputs of the system. The state equation of the controlled system is thus

$$\dot{\mathbf{z}} = \left(\mathcal{A} - \mathcal{B}_c\mathbf{K}_y\mathcal{C}\right)\mathbf{z} + \mathcal{B}_c\mathbf{K}_r\mathbf{r}(t) + \mathcal{B}_e\mathbf{u}_e(t). \tag{11.16}$$

If the control system is a regulator, vector $\mathbf{r}$ contains just the constants that define the set point. If the aim of the control system is to maintain the structure in the static equilibrium position in spite of the presence of the perturbing inputs $\mathbf{u}_e(t)$, the reference inputs are equal to zero, and the equation of motion can be simplified.

Remark 11.2 *Note that the presence of the control loop affects the free behavior of the system as well as the forced response. The very stability of the system can be affected and, while the control system can be used to increase the stability of the structure or to give an artificial stability to an unstable plant, the behavior of the system must be carefully studied to avoid that unwanted instabilities are caused by the feedback loop.*

The block diagram corresponding to Eq. (11.16) is shown in Fig. 11.5. This type of feedback is usually referred to as *output feedback*, because the loop is closed using just the outputs of the system. The design of such a

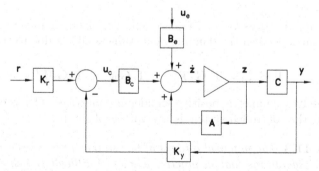

FIGURE 11.5. Block diagram of the multi-degrees-of-freedom structure in Fig. 11.4 controlled by an output feedback system.

control system consists of determining the gain matrices $\mathbf{K}_y$ and $\mathbf{K}_r$, which enable the system to behave in the desired way.

In the case of structural control, it is possible to write the equation of motion of the controlled system in the configuration space. Assume that the reference input is vanishingly small (i.e., the structure must be kept in the undeformed configuration); the output of the system can be written in the form

$$\mathbf{y}(t) = \begin{bmatrix} \mathcal{C}_1 & \mathbf{0} \\ \mathbf{0} & \mathcal{C}_2 \end{bmatrix} \begin{Bmatrix} \dot{\mathbf{x}} \\ \mathbf{x} \end{Bmatrix} \qquad (11.17)$$

and the gain matrix $\mathbf{K}_y$ has the form

$$\mathbf{K}_e = \begin{bmatrix} \mathbf{G}_1 & \mathbf{0} \\ \mathbf{0} & \mathbf{G}_2 \end{bmatrix}. \qquad (11.18)$$

These assumptions correspond to a separation between the control inputs linked to the position and the velocity outputs of the system. They lead to control forces that can be expressed as

$$\mathbf{f}_c = -\mathbf{T}_c \mathbf{G}_1 \mathcal{C}_1 \dot{\mathbf{x}} - \mathbf{T}_c \mathbf{G}_2 \mathcal{C}_2 \mathbf{x}. \qquad (11.19)$$

The equation of motion of the controlled system is then

$$\mathbf{M}\ddot{\mathbf{x}} + (\mathbf{C} + \mathbf{T}_c \mathbf{G}_1 \mathcal{C}_1)\dot{\mathbf{x}} + (\mathbf{K} + \mathbf{T}_c \mathbf{G}_2 \mathcal{C}_2)\mathbf{x} = \mathbf{f}_e. \qquad (11.20)$$

Gains $\mathbf{G}_1$ and $\mathbf{G}_2$ then cause an increase of the damping and stiffness characteristics of the structure, respectively.

If the actuators and sensors are co-located, i.e., the generalized forces exerted by the actuators correspond to the generalized displacements or velocities read by the sensors, the output gain matrices are equal to the transpose of the control influence matrix:

$$\mathcal{C}_1 = \mathcal{C}_2 = \mathbf{T}_c^T.$$

If the gain matrices $\mathbf{G}_1$ and $\mathbf{G}_2$ are fully populated, symmetric positive defined matrices, then the damping and stiffness effects due to the control system

$$\mathbf{T}_c \mathbf{G}_1 \mathcal{C}_1 = \mathbf{T}_c \mathbf{G}_1 \mathbf{T}_c^T, \quad \mathbf{T}_c \mathbf{G}_2 \mathcal{C}_2 = \mathbf{T}_c \mathbf{G}_2 \mathbf{T}_c^T$$

are themselves symmetric positive semidefinite matrices. This is enough to state that the controlled system is asymptotically stable.

Remark 11.3 *The practical interest of this statement is, however, reduced by the considerations that in practice it may be difficult to achieve a perfect sensor–actuator co-location and that in actual systems the behavior of sensors and actuators differs from the ideal behavior here assumed.*

By resorting to Laplace transforms and assuming that at time $t = 0$ the value of all state variables is zero, the closed-loop transfer function linking the outputs of the system $\mathbf{y}(s)$ to the external $\mathbf{u}_e(s)$ inputs when all reference inputs are equal to zero can be easily computed. For a system with output feedback, it follows that

$$\mathbf{G}(s) = \mathcal{C}\left(s\mathbf{I} - \mathcal{A} + \mathcal{B}_c\mathbf{K}_e(s)\mathcal{C}\right)^{-1}\mathcal{B}_e. \qquad (11.21)$$

Also in this case, the gains of the control system are considered functions of the Laplace variable s. The poles of the closed-loop system are then the roots of the characteristic equation

$$\det\left(s\mathbf{I} - \mathcal{A} + \mathcal{B}_c\mathbf{K}_e(s)\mathcal{C}\right) = 0. \qquad (11.22)$$

Similar equations hold for systems with state feedback and for the transfer functions linking the output with the reference input $\mathbf{r}(s)$.

Example 11.2 *Consider the inverted pendulum in Fig. 11.2. The system has a single degree of freedom and, hence, $n = 2$. The vectors and matrices included in the equation of motion of the controlled system are*

$$z = \left\{ \begin{array}{c} \dot{\theta} \\ \theta \end{array} \right\}, \quad \mathcal{A} = \left[\begin{array}{cc} -\dfrac{\Gamma}{ml^2} & \dfrac{g}{l} \\ 1 & 0 \end{array} \right],$$

$$\mathcal{B}_e = \mathcal{B}_c = \left\{ \begin{array}{c} \dfrac{1}{ml} \\ 0 \end{array} \right\}, \quad u_e = F.$$

The output and the control parameter of the system coincide with the angular displacement θ, and the gain matrix K_y states the dependence of the generalized force exerted by the spring as a function of the angular displacement:

$$y = \theta, \quad u = \theta, \quad \mathcal{C} = [0, 1], \quad K_y = kl^2.$$

Because the set point is characterized by $\theta = 0$, it follows that $r = 0$. Introducing the aforementioned values of all the relevant matrices and vectors into Eq. (11.16) the same equation of motion of the controlled system seen in Example 11.1 is obtained.

11.6   Basic control laws

In the preceding sections the controller was assumed to provide a set of control inputs $\mathbf{u}_c$ to the controlled system, which are determined just as linear combinations of the outputs of the system $\mathbf{y}$ and the control inputs

$\mathbf{r}(t)$. In the case of output control, if the number of reference inputs r is equal to the number m of outputs of the plant and the two gain matrices $\mathbf{K}_y$ and $\mathbf{K}_r$ are equal, the control law can be further simplified

$$\mathbf{u}_c = \mathbf{K}_p \mathbf{e}_y = \mathbf{K}_p(\mathbf{r} - \mathbf{y}), \qquad (11.23)$$

where the elements of vector $\mathbf{e}_y(t)$ are the errors of the output of the system and matrix $\mathbf{K}_p$ contains the gains of the control system. This type of control is said to be *proportional control* (hence, the gain in this case can be called *proportional gain*) and provides a large corrective action when the instantaneous errors are large. It has the main advantage of being very simple but has several disadvantages as well, such as a lack of precision in certain instances and the possibility of producing instability. In a certain sense, it stiffens the system.

A different choice, always within the frame of linear systems, is the so-called derivative control, which, in the case of SISO systems, can be summarized in the form

$$u_c = K_d \frac{de_y}{dt}. \qquad (11.24)$$

This type of control reacts more to the increase of error than to the error itself and provides large corrective actions when the errors increase at a high rate. It provides a sort of damping to the system and enhances stability, but it is insensitive to constant errors and not very sensitive to errors that accumulate slowly. It is prone to cause a drift in the output of the system, but this disadvantage is not critical in structural control, where the control system must prevent vibrations, and its reaction to static or quasi-static forces is often of little importance. In harmonic motion, its effectiveness increases with the frequency of the perturbation.

A third possibility is the so-called integral control, which can be summarized in the form

$$u_c = K_i \int_0^t e_y(t)dt. \qquad (11.25)$$

It reacts to the accumulation of errors and causes a slow-reacting control action. Its disadvantages are mainly that it is insensitive to high frequencies and prone to cause instability.

Remark 11.4 *The control laws here described are only theoretical abstractions because they assume that the sensors, actuators, controllers, and all other components of the control loop are ideal, having no delays and behaving in a perfectly linear way no matter how high their input is and how fast their action is required to be.*

Due to the different characteristics of the control laws, often a law that combines the aforementioned control strategies, and possibly others based on higher-order derivatives or integrals, is used. This type of control is usually referred to as *proportional–integral–derivative* (PID) control.

By combining the already reported control laws, the control input due to a SISO PID controller is

$$u_c = K_p e_y + K_d \frac{de_y}{dt} + K_i \int_0^t e_y(t)dt = K_p \left(e_y + T_d \frac{de_y}{dt} + \frac{1}{T_i} \int_0^t e_y(t)dt \right),$$

(11.26)

where the *prediction* or *derivative time* and the *reset time* are, respectively,

$$T_d = \frac{K_d}{K_p} \quad \text{and} \quad T_i = \frac{K_p}{K_i}.$$

In the Laplace domain, the relationship between the control input $u_c(s)$ and the error $e_y(s)$ for a SISO PID control system can be expressed in the form

$$u_c(s) = P(s)e_y(s) = K_p \left(1 + sT_d + \frac{1}{sT_i} \right) e_y(s).$$

(11.27)

Consider a plant governed by a set of second-order linear differential equation, and apply to it an ideal MIMO PID controller, whose control action is expressed as

$$u_c = \mathbf{K}_p \mathbf{e}_y + \mathbf{K}_d \frac{d\mathbf{e}_y}{dt} + \mathbf{K}_i \int_0^t \mathbf{e}_y(t)dt.$$

(11.28)

The equation of motion of the system in the configuration space (11.1) is still

$$\mathbf{M}\ddot{\mathbf{x}} + \mathbf{C}\dot{\mathbf{x}} + \mathbf{K}\mathbf{x} = \mathbf{f}_c + \mathbf{f}_e,$$

where

$$\mathbf{f}_c = \mathbf{T}_c \left[\mathbf{K}_p \mathbf{e}_y + \mathbf{K}_d \frac{d\mathbf{e}_y}{dt} + \mathbf{K}_i \int_0^t \mathbf{e}_y(t)dt \right], \quad \mathbf{f}_e = \mathbf{T}_e \mathbf{u}_e.$$

What is said above implicitly assumes that only the displacements $\mathbf{x}$ contribute to the output. If also the velocities $\dot{\mathbf{x}}$ contribute to the outputs, the roles of the proportional, derivative, and integrative control mix up: the proportional action has also a derivative effect, the integral action has a component depending on the displacements, and the derivative action reacts also to accelerations. To avoid this, in the following the output is assumed to depend only on the displacements:

$$\mathbf{y} = \mathbf{T}_s \mathbf{x},$$

where $\mathbf{T}_s$ is a matrix that states how the sensors are located with respect to the generalized coordinates. If the sensors and actuators are co-located,

$$\mathbf{T}_s = \mathbf{T}_c^T.$$

(11.29)

By introducing the control law into the equation of motion and assuming that the vector of the references is a generic known function of time $\mathbf{r}(t)$, the latter becomes

$$\mathbf{M}\ddot{\mathbf{x}} + [\mathbf{C} + \mathbf{T}_c\mathbf{K}_d\mathbf{T}_s]\dot{\mathbf{x}} + [\mathbf{K} + \mathbf{T}_c\mathbf{K}_p\mathbf{T}_s]\mathbf{x} + \mathbf{T}_c\mathbf{K}_i\mathbf{T}_s\int_0^t \mathbf{x}dt = \quad (11.30)$$

$$= \mathbf{T}_c\mathbf{K}_p\mathbf{r} + \mathbf{T}_c\mathbf{K}_d\dot{\mathbf{r}} + \mathbf{T}_c\mathbf{K}_i\int_0^t \mathbf{r}dt + \mathbf{T}_e\mathbf{u}_e \,.$$

The derivative action is thus similar to damping: if the gain matrix $\mathbf{K}_d$ is symmetrical and the sensors and actuators are co-located, the total damping effect is expressed by a symmetric matrix. If it is positive defined it is true damping.

The proportional action is a restoring action similar to stiffness: if the gain matrix $\mathbf{K}_p$ is symmetrical and the sensors and actuators are co-located, the total damping effect is expressed by a symmetric matrix and the system is non-circulatory. If it is positive defined it is a true restoring action.

The integral action is different from either stiffness or damping and introduces a behavior different from that obtainable from noncontrolled systems.

The integral of vector $\mathbf{x}$ appears explicitly in Eq. (11.30): A reworking of the equation is then required to reduce it to a standard differential equation. This can be done by adding a number of auxiliary states either by differentiating once more the equation and introducing the derivatives of the velocities as variables or by introducing the auxiliary variables

$$\mathbf{w} = \int_0^t \mathbf{x}dt$$

together with the velocities

$$\mathbf{v} = \dot{\mathbf{x}} \,.$$

The state equations are thus

$$\begin{cases} \dot{\mathbf{v}} = -\mathbf{M}^{-1}[\mathbf{C} + \mathbf{T}_c\mathbf{K}_d\mathbf{T}_s]\mathbf{v} - \mathbf{M}^{-1}[\mathbf{K} + \mathbf{T}_c\mathbf{K}_p\mathbf{T}_s]\mathbf{x} - \mathbf{M}^{-1}\mathbf{T}_c\mathbf{K}_i\mathbf{T}_s\mathbf{w} + \\ \qquad + \mathbf{M}^{-1}\mathbf{T}_c\mathbf{K}_p\mathbf{r} + \mathbf{M}^{-1}\mathbf{T}_c\mathbf{K}_d\dot{\mathbf{r}} + \mathbf{M}^{-1}\mathbf{T}_c\mathbf{K}_i\int_0^t \mathbf{r}dt + \mathbf{M}^{-1}\mathbf{T}_e\mathbf{u}_e \\ \dot{\mathbf{x}} = \mathbf{v} \\ \dot{\mathbf{w}} = \mathbf{x} \,, \end{cases}$$

$$(11.31)$$

or, to write them in standard form,

$$\dot{\mathbf{z}}^* = \mathcal{A}^*\mathbf{z}^* + \mathcal{B}_p^*\mathbf{r}(t) + \mathcal{B}_d^*\dot{\mathbf{r}}(t) + \mathcal{B}_i^*\int_0^t \mathbf{r}(t)dt + \mathcal{B}_e^*\mathbf{u}_e(t) \,, \qquad (11.32)$$

where the augmented state vector is

$$\mathbf{z}^* = \left\{ \begin{array}{c} \mathbf{v} \\ \mathbf{x} \\ \mathbf{w} \end{array} \right\}$$

and the augmented matrices are

$$
\mathcal{A}^* = \begin{bmatrix} -\mathbf{M}^{-1}\left[\mathbf{C} + \mathbf{T}_c\mathbf{K}_d\mathbf{T}_s\right] & -\mathbf{M}^{-1}\left[\mathbf{K} + \mathbf{T}_c\mathbf{K}_p\mathbf{T}_s\right] & -\mathbf{M}^{-1}\mathbf{T}_c\mathbf{K}_i\mathbf{T}_s \\ \mathbf{I} & 0 & 0 \\ 0 & \mathbf{I} & 0 \end{bmatrix},
$$

$$
\mathcal{B}_k^* = \begin{bmatrix} \mathbf{M}^{-1}\mathbf{T}_k\mathbf{K}_p \\ 0 \\ 0 \end{bmatrix} \quad \text{for } k = p, d, i, \quad \mathcal{B}_e^* = \begin{bmatrix} \mathbf{M}^{-1}\mathbf{T}_e \\ 0 \\ 0 \end{bmatrix}.
$$

Consider a single-degree-of-freedom system, controlled by a SISO PID co-located controller. The equation of motion is

$$
m\ddot{x} + c^*\dot{x} + k^*x + k_i \int_0^t x\,dt = k_p r + k_d \dot{r} + k_i \int_0^t r\,dt + f(t), \qquad (11.33)
$$

where c^* and k^* are the total damping and stiffness, including also the control actions. In the Laplace domain it becomes

$$
\left(ms^2 + c^*s + k^* + \frac{k_i}{s}\right)x(s) = \left(k_p + sk_d + \frac{k_i}{s}\right)r(s) + f(s), \qquad (11.34)
$$

Assuming as output the displacement and as input the reference $r(t)$, the transfer function is

$$
G(s) = \frac{x(s)}{r(s)} = \frac{s^2 k_d + sk_p + k_i}{ms^3 + c^*s^2 + k^*s + k_i}. \qquad (11.35)
$$

The system is a third-order system, and its characteristic equation is

$$
ms^3 + c^*s^2 + k^*s + k_i = 0. \qquad (11.36)
$$

It can be written in the following nondimensional form:

$$
s^{*3} + 2\zeta s^{*2} + s^* + k_i^* = 0, \qquad (11.37)
$$

where

$$
s^* = s\sqrt{\frac{m}{k^*}}, \quad \zeta = \frac{c^*}{2\sqrt{k^*m}}, \quad k_i^* = k_i\sqrt{\frac{m}{k^{*3}}}. \qquad (11.38)
$$

The characteristic equation has a real solution

$$
s = \frac{b}{6} + \frac{8\zeta^2}{3b} + \frac{2\zeta}{3}, \qquad (11.39)
$$

and two complex-conjugate solutions

$$
s = -\frac{b}{12} - \frac{4\zeta^2}{3b} + \frac{2\zeta}{3} \pm i\frac{\sqrt{3}}{2}\left(\frac{b}{6} - \frac{8\zeta^2}{3b}\right), \qquad (11.40)
$$

where

$$b = \sqrt[3]{-108 - 108k_i^* + 64\zeta^3 + 12\sqrt{81 + 162k_i^* - 96\zeta^3 + 81k_i^{*2} - 96\,k_i^*\zeta^3}}\,.$$
(11.41)

The roots locus for $0.1 < \zeta < 0.8$ and $0 < k_i^* < 1$ is shown in Fig. 11.6.

The dashed line refers to an equivalent PI system. The lines for the PID systems for various values of ζ branch from the dashed line toward the right: The stability is thus decreased by the presence of the integral action and in certain cases (low ζ and high k_i^*) the system becomes unstable. If a similar plot but for $\zeta > 1$ were shown, it could be seen that the presence of an integrative action can cause an oscillatory behavior.

If the PID control is used to keep the plant in its equilibrium position and thus $r = 0$, the transfer function between the displacement and the external input is

$$G(s) = \frac{x(s)}{f(s)} = \frac{s}{ms^3 + c^*s^2 + k^*s + k_i}\,.$$
(11.42)

or, in nondimensional form,

$$k^*G(s^*) = \frac{s^*}{s^{*3} + 2\zeta s^{*2} + s^* + k_i^*}\,.$$
(11.43)

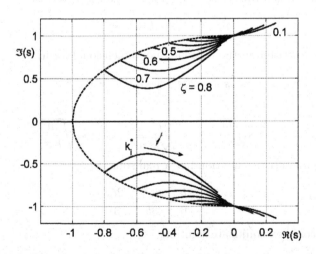

FIGURE 11.6. Roots locus for a single-degree-of-freedom system controlled by a PID SISO controller with $0.1 < \zeta < 0.8$ and $0 < k_i^* < 1$. The *dashed line* is the roots locus for a system controlled by a PD controller.

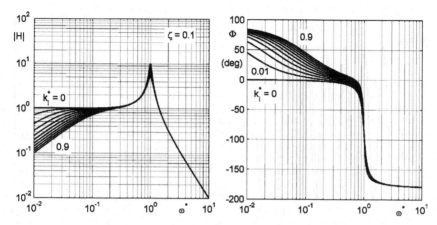

FIGURE 11.7. Bode plot of the nondimensional frequency response for a system with a PID controller with $\zeta = 0.1$ and $k_i^* = 0$, 0.01, 0.1, 0.2, 0.3, 0.4 , 0.5, 0.6, 0.7, 0.8, 0.9.

The bode plot of the nondimensional frequency response is shown in Fig. 11.7. The integrative action reduces slightly the height of the resonance peak, but above all introduces a zero in the origin and consequently the response is very little affected by low-frequency disturbances.

Example 11.3 *To understand the effect of the various control gains using a simple system with a single degree of freedom, consider the prismatic, homogeneous beam hinged at one end of Fig. 11.8a. Study the cases in which the hinge is controlled by an ideal proportional, PD, and a PID controller. The data are $l = 1$ m, $m = 5$ kg, $g = 9.81$ m/s^2.*

The system is so simple that the study can be performed in the configuration space. Since the beam is prismatic, the center of mass is at mid-length and the moment of inertia about the hinge is

$$J = \frac{ml^2}{3} \; .$$

The equation of motion is

$$J\ddot{\theta} + \frac{mgl}{2} \cos(\theta) = T \; ,$$

where T is the control torque.

Assume that the sensor detects the value of angle θ ($y = \theta$) and that the controller is required to bring the beam at angle θ_0 and to keep it there ($r = \theta_0 = $ constant).

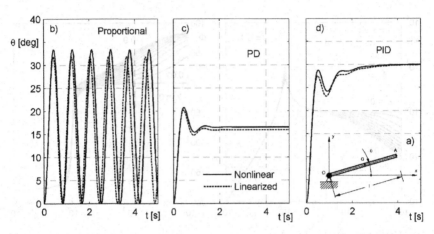

FIGURE 11.8. Prismatic, homogeneous beam hinged at one end. (a) Sketch of the system; (b) time history of angle θ for a proportional control; (c) time history of angle θ for a PD control; (d) time history of angle θ for a PID control.

Proportional controller

Using the expression of the error given by Eq. (11.23), the control torque is

$$T = -K_p \left(\theta - \theta_0 \right) .$$

The equation of motion of the controlled system is then

$$J\ddot{\theta} + K_p\theta + \frac{mgl}{2} \cos \left(\theta \right) = K_p\theta_0 .$$

The equation of motion is nonlinear due to presence of the cosine of angle θ. Only if θ is small can it be linearized. The proportional gain plays the same role of stiffness: The larger it is, the quicker the tendency toward the reference position is, but also the stronger is the oscillatory behavior of the system. The position at rest can be computed by assuming that $\ddot{\theta}$ and $\dot{\theta}$ vanish, obtaining

$$\theta + \frac{mgl}{2K_p} \cos \left(\theta \right) = \theta_0 .$$

A proportional controller is unable to reach the reference position if the system is subjected to external forces. The final position can be written as

$$\theta = \theta_0 + \Delta\theta ,$$

where $\Delta\theta$ is the error in the final position. The equation yielding the error can thus be written as

$$\Delta\theta + \frac{mgl}{2K_p} \cos \left(\theta_0 + \Delta\theta \right) = 0 .$$

If $\Delta\theta$ is small,

$$\cos \left(\theta_0 + \Delta\theta \right) \approx \cos \left(\theta_0 \right) - \Delta\theta \sin \left(\theta_0 \right) ,$$

the error can be computed easily:

$$\Delta\theta = -\frac{mgl\cos(\theta_0)}{2K_p - mgl\sin(\theta_0)} \ .$$

Since the equation of motion is nonlinear, a numerical value of the gain is assumed, $K_p = 100\ N\ m/rad$, and the equation of motion is integrated numerically.

Using a reference value $\theta_0 = 30°$, the final value of angle θ is equal to $16.52°$. The steady-state error is thus large: $\Delta\theta = 13.48°$. The linearized expression for the error yields $\Delta\theta = 13.87°$, which is fairly good, considering that the error is not really small.

Consider the arm at rest with $\theta = 0$ and apply the reference $\theta_0 = 30°$. The results of the numerical integration are shown in Fig. 11.8b, for both the nonlinear and the linearized cases. Strong undamped oscillation about the steady-state value occurs.

PD controller

By adding a derivative control action, and remembering that the reference θ_0 is constant, the equation of motion of the controlled system is

$$J\ddot{\theta} + K_d\dot{\theta} + K_p\theta + \frac{mgl}{2}\cos(\theta) = K_p\theta_0.$$

The derivative gain plays the same role of damping: its action is needed to avoid strong vibration, but it has no effect on the final position at rest and thus also a PD controller is unable to reach the reference position if the system is subjected to external forces.

Assuming that the derivative gain is $K_d = 10\ N\ m\ s/rad$ and that the proportional gain and the reference value are the same as above, the results of the numerical integration of the equation of motion are those shown in Fig. 11.8c. The steady-state value of $16.52°$ is now quickly reached.

PID controller

The control torque now is

$$T = -K_p(\theta - \theta_0) - K_d\dot{\theta} - K_i\int_0^t (\theta - \theta_0)\ du\ .$$

The equation of motion of the controlled system is thus

$$J\ddot{\theta} + K_d\dot{\theta} + K_p\theta + K_i\int_0^t \theta\ du + \frac{mgl}{2}\cos(\theta) = K_p\theta_0 + K_i\theta_0 t\ .$$

The equation of motion is then an integro-differential equation and must be written in the state space. Introducing two auxiliary variables

$$v = \dot{\theta}\ , \qquad w = \int_0^t \theta\ du\ ,$$

the equation becomes

$$J\dot{v} + K_d v + K_p\theta + K_i w + \frac{mgl}{2}\cos(\theta) = K_p\theta_0 + K_i\theta_0 t\ .$$

or, in matrix form,

$$\left\{\begin{array}{c} \dot{v} \\ \dot{\theta} \\ \dot{w} \end{array}\right\} = \left[\begin{array}{ccc} -\frac{K_d}{J} & -\frac{K_p}{J} & -\frac{K_i}{J} \\ 1 & 0 & 0 \\ 0 & 1 & 0 \end{array}\right] \left\{\begin{array}{c} v \\ \theta \\ w \end{array}\right\} + $$

$$+ \left\{\begin{array}{c} \frac{mgl}{2J}\cos{(\theta)} + \frac{K_p}{J}\theta_0 + \frac{K_i}{J}\theta_0 t \\ 0 \\ 0 \end{array}\right\}.$$

The results of the numerical integration obtained assuming an integrative gain $K_i = 100\ N\ m/rad$ and the same reference value $\theta_0 = 30°$ are shown in Fig. 11.8d.

The steady-state value now coincides with the reference value and is quickly reached, although with some damped oscillation.

The nondimensional parameters of the system are $\zeta = 0.39$ and $k_i^* = 0.13$ and the poles are $s_1 = -1.099\ 1/s$ and $s_{2,3} = -2.451 \pm 6.972\ i\ 1/s$.

11.7   Delayed control

The control action is usually quite different from the ideal control law defined above. A first effect, which is normally unwanted, is due to the impossibility, for all the control system–actuator combinations, to react in an infinitely fast way to the inputs provided by the sensors. Consider for instance a proportional controller, which reacts to the error

$$\mathbf{e}_y = \mathbf{r} - \mathbf{y}$$

with a delay τ. The control action is thus

$$\mathbf{u}_c(t) = \mathbf{K}_p \mathbf{e}_y(t - \tau) = \mathbf{K}_p \left[\mathbf{r}(t - \tau) - \mathbf{y}(t - \tau)\right]. \tag{11.44}$$

Assuming, as usual, a time history for the error of the kind

$$\mathbf{e}_y = \mathbf{e}_{y0} e^{st},$$

it follows that

$$\mathbf{e}_y(t - \tau) = \mathbf{e}_{y0} e^{s(t-\tau)} = \mathbf{e}_{y0} e^{st} e^{-s\tau} = \mathbf{e}_y(t) e^{-s\tau}. \tag{11.45}$$

The Laplace domain relationship between the control input $u_c(s)$ and the error $e_y(s)$ in a proportional controller with delay is thus

$$\mathbf{u}_c(s) = \mathbf{K}_p \mathbf{e}_y(s) e^{-s\tau} \tag{11.46}$$

and thus the transfer function of the control system is

$$\frac{u_c(s)}{e_y(s)} = k e^{-s\tau}. \tag{11.47}$$

If the delay is short enough and not too large values of s are considered, the exponential can be replaced by its Taylor series truncated at the first term, obtaining the following value of the transfer function

$$\frac{u_c(s)}{e_y(s)} = k\frac{1}{1+s\tau}.$$ (11.48)

This is equivalent to expressing the delay of the control system in the time domain by replacing $e_y(t - \tau)$ with its truncated Taylor series

$$\mathbf{u}_c(t) = \mathbf{K}_p\mathbf{e}_y(t - \tau) = \mathbf{K}_p[\mathbf{e}_y(t) - \tau\dot{\mathbf{e}}_y(t)].$$ (11.49)

Consider now a linear system provided of a proportional output control with delay τ and resort to the first-approximation expression (11.49). The equation of motion of the system in the configuration space is again Eq. (11.1), and the control forces are

$$\mathbf{f}_c = \mathbf{T}_c\mathbf{K}_p(\mathbf{e}_y - \tau\dot{\mathbf{e}}_y) = \mathbf{T}_c\mathbf{K}_p(\mathbf{r} - \tau\dot{\mathbf{r}} - \mathbf{y} + \tau\dot{\mathbf{y}}).$$

Assuming again that the output depends only on the displacements

$$\mathbf{y} = \mathbf{T}_s\mathbf{x},$$

by introducing the control and disturbance forces, the configuration space equation of the system is

$$\mathbf{M}\ddot{\mathbf{x}} + (\mathbf{C} - \tau\mathbf{T}_c\mathbf{K}_p\mathbf{T}_s)\dot{\mathbf{x}} + (\mathbf{K}+\mathbf{T}_c\mathbf{K}_p\mathbf{T}_s)\mathbf{x} = \mathbf{T}_c\mathbf{K}_p(\mathbf{r} - \tau\dot{\mathbf{r}}) + \mathbf{T}_e\mathbf{u}_e.$$

Remark 11.5 *The control delay has clearly the effect of reducing damping (at least if matrix $\mathbf{T}_c\mathbf{K}_p\mathbf{T}_s$ is positive defined, as it must be if the proportional control has to increase the stability of the system), i.e., it increases the tendency of the system to oscillate and if large enough (or if the system is little damped) it may cause instability. Although obtained from a first-approximation model, this consideration has a general value.*

Similar results can be obtained by introducing the delay also in PD or PID controllers.

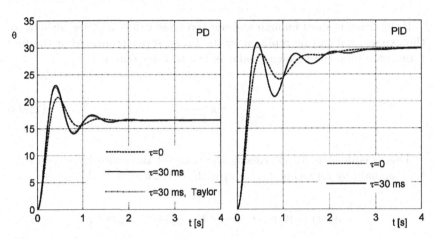

FIGURE 11.9. Prismatic, homogeneous beam hinged at one end. Time history of angle θ for a proportional and a PD control with a delay of 30 ms.

Example 11.4 *Study the same controlled beam of Example 11.3, but insert a delay in the control loop. Assume a value of 30 ms for the overall delay.*
After inserting the delay in the control loop, the system with a proportional control was studied first, by numerically integrating the equation of motion. The system was already at the verge of instability without delay; as soon as the delay was added it showed, as expected, a strongly unstable oscillatory behavior, and no attempt to study it in detail was done.
The response of the system with PD controller is compared with that of the system without delay in Fig. 11.9. The presence of the delay increases the oscillatory nature of the response, but the derivative control is successful in insuring stability. The dotted line was obtained by using a truncated Taylor series in τ (Eq. (11.49)). With this value of the delay the result is almost completely superimposed to that obtained by numerical integration of the equation of motion. With a higher delay, less accurate results would have been obtained. Also the response obtained using a PID control shows similar oscillations.

11.8 Control laws with frequency-dependent gains

Equation (11.27) defines the gains of the control system as functions of s in the Laplace domain.

In general, the gains may be functions of the Laplace variable s or, in the case of harmonic response, of the frequency ω, which depend on the actual physical configuration of the control system. As shown in the previous section, the delay of the control action can be modeled in this way.

The system described by Eq. (11.48) is a simple first-order linear system, whose parameters k and τ have an immediate physical meaning: the first is the gain of the control system; the second is the time constant for a step input.

The response $g(t)$ to a unit-step input, as defined by Eq. (8.5), is reported in nondimensional form in Fig. 11.10. It is simply

$$g(t) = k \left(1 - e^{-t/\tau} \right). \tag{11.50}$$

The frequency response of a first-order system can be easily obtained from the transfer function (11.48) by substituting $i\omega$ to s. Separating the real part from the imaginary part, its expression is

$$\Re[H(\omega)] = \frac{k}{1 + \tau^2\omega^2}, \qquad |H(\omega)| = \frac{k}{\sqrt{1 + \tau^2\omega^2}},$$

$$\tag{11.51}$$

$$\Im[H(\omega)] = -\frac{k\tau\omega}{1 + \tau^2\omega^2}, \qquad \Phi = \arctan(-\tau\omega).$$

The bode diagram of the frequency response is plotted in Fig. 11.11. The circular frequency

$$\omega_g = \frac{1}{\tau}$$

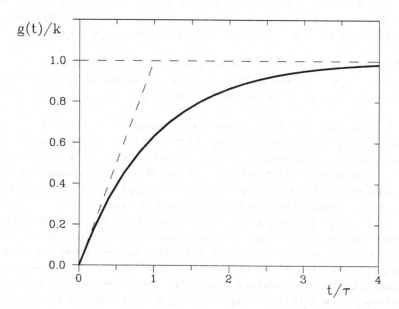

FIGURE 11.10. Nondimensional response of a first-order system to a unit step.

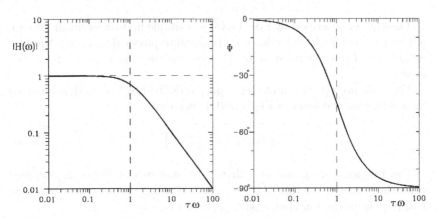

FIGURE 11.11. Bode diagram of the frequency response of a first-order system.

is the so-called cut-off frequency. The period corresponding to the cut-off frequency defined here does not directly coincide with the time constant but is

$$T_g = 2\pi\tau \ .$$

The first-order system acts as a low-pass filter with an attenuation, at frequencies higher than the cut-off frequency, of about 6 dB/oct or 20 dB/dec. It also introduces a phase lag of the response that goes from zero at very low frequency to 90° at a frequency tending to infinity. At the cut-off frequency the phase lag is 45°. The effectiveness of the control system is quickly reduced at frequencies higher than the cut-off frequency.

Remark 11.6 *The time constant of the control system–actuators combination is a most important parameter that can dictate the choice of a particular layout, especially when high-frequency operation is required.*

In most cases the control system can be modeled as a first-order system only as a very rough approximation. There are usually limits above which a linear model is no longer possible: All types of control systems can only supply an output that is limited in magnitude. When the maximum response is attained, the saturation phenomenon occurs and the gain decreases with increasing input, giving way to a nonlinear response.

Apart from the presence of a cut-off frequency, a dependence of the control law on the frequency can be purposely devised by adding a compensator with an appropriate law $K(s)$. For example, if the disturbance frequencies are well determined, an ideal control system would have a very small gain except in correspondence to the frequencies of the disturbances to be suppressed. On the contrary, a support that insulates a device from external disturbances should behave as a very stiff system at all frequencies except those that must be suppressed, where its stiffness should approach zero.

The use of a compensator allows the frequency response of the system to be tailored to fulfill the design goals of the particular application, to both achieve the required performance and ensure adequate robustness. This can be done easily with electronic control systems, which can be designed to obtain almost arbitrary transfer functions, provided the required output does not exceed their maximum limits.

Example 11.5 *Consider the linear mechanical system with a single degree of freedom shown in Fig. 11.12a. The control force is supplied by an actuator acting on an auxiliary spring of stiffness k_1 and governed by an active control system.*
With reference to the figure, the equation of motion of the system is

$$m\ddot{x} + c\dot{x} + kx = F(t) + F_c(t).$$

The control force can be expressed as $F_c = -k_1(x - u)$, and the equation of motion of the controlled system is

$$m\ddot{x} + c\dot{x} + (k + k_1)x = k_1 u + F(t).$$

Consider a closed-loop control system in which the displacement $u(t)$ is a linear function of the displacement x. The block diagram, in which the gain of the control system P is referred to as $G_c(s)$, is shown in Fig. 11.12b. This type of control is a regulator, in the sense described in Section 11.2 with a set point corresponding to $x = 0$. The transfer function of the total system is

$$G(s) = \frac{1}{ms^2 + cs + k + k_1 + k_1 G_c(s)}.$$

In the case of proportional control, in which the displacement of the actuator is proportional to the displacement of the system $u = -G_c x$, the control system actually adds a spring with stiffness equal to k_1 multiplied by the gain. If a derivative control is applied, the control system is equivalent to a damper, with damping coefficient equal to k_1 multiplied by the gain. Note that in this case the gain is not a nondimensional quantity.
Other cases of interest are shown in Figs. 11.12c and e, in which the actuator of the control system acts through a damper or on an auxiliary mass. The last layout is usually referred to as an active vibration absorber. The expressions of the control force are

$$F_c = -c_1(\dot{x} - \dot{u}) \quad and \quad F_c = -m_1(\ddot{x} - \ddot{u}).$$

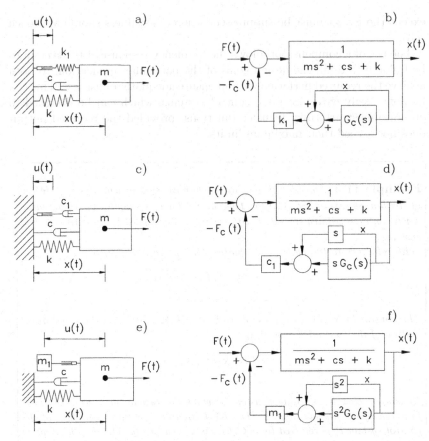

FIGURE 11.12. (**a**) Sketch and (**b**) block diagram of a system with a single degree of freedom controlled by an actuator through an auxiliary spring, (**c**) and (**d**) an auxiliary damper, or (**e**) and (**f**) an auxiliary mass.

If the control law is of the proportional type ($u = -G_c x$), the transfer functions of the two systems are, respectively,

$$G(s) = \frac{1}{ms^2 + (c + c_1)s + k + s c_1 G_c(s)},$$

$$G(s) = \frac{1}{(m + m_1)s^2 + cs + k + s^2 m_1 G_c(s)}.$$

In the first case, a proportional control introduces an active damping into the system, an integral control law in which $u(t)$ is proportional to the integral of the displacement or in which $\dot{u}(t)$ is proportional to $x(t)$ introduces an active stiffness. If a derivative control is used, the control force is proportional to the acceleration, and the effect is similar to that of an added mass.

11.9   Robustness of the controller

In many cases, some of the parameters of the system are not well known or are prone to change in time. There are many examples in the field of structural dynamics, such as the case of hysteretic damping, which is usually poorly known and is affected by many parameters in a way that is often impossible to control, or that of the wandering unbalance, shown by many rotors. One of the advantages of a feedback control system is that they can usually compensate for these unwanted effects, because they measure directly the outputs and act to keep them within stated limits. Feedback control systems can even compensate for the uncertainties and variations of their own parameters.

A system that is little affected by changes of operating conditions, by parameter variations, and by external disturbances is said to be robust and robustness is one of the basic requisites of control systems.

Generally speaking, the sensitivity of the quantity q to the variations of parameter α is measured by the derivative $\partial q / \partial \alpha$. By computing the sensitivities of the relevant characteristics of the system (eigenvalues, frequency responses, etc.) to the variation of the critical parameters, it is possible to assess its robustness.

The root sensitivity $S_{i,\alpha}$ of the ith root s_i of the transfer function to parameter α, for example, can be defined as

$$S_{i,\alpha} = \alpha \left(\frac{\partial s_i}{\partial \alpha} \right) = \frac{\partial s_i}{\partial \log(\alpha)} . \tag{11.52}$$

The derivative is, in many cases, computed numerically, by giving a small variation to the parameter under study and computing a new value of the relevant characteristic of the system. If the dynamic matrix $\mathcal{A}$ can be differentiated with respect to parameter α, the sensitivity of the ith eigenvalue s_i of the dynamic matrix $\partial s_i / \partial \alpha$ can be computed in closed form

$$\frac{\partial s_i}{\partial \alpha} = \mathbf{q}_{Li}^T \frac{\partial \mathcal{A}_i}{\partial \log(\alpha)} \mathbf{q}_{Ri} , \tag{11.53}$$

where $\mathbf{q}_{Li}$ and $\mathbf{q}_{Ri}$ are, respectively, the ith left and right eigenvectors. Similar but more complicated expressions can be found in the literature for the eigenvector sensitivity.

11.10   State feedback and state observers

A more complete type of feedback is the state feedback shown in Fig. 11.13, in which all state variables are used to close the loop. The control inputs are obtained in this case through the equation

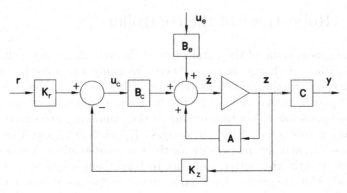

FIGURE 11.13. Block diagram of the multi-degrees-of-freedom structure in Fig. 11.4 controlled by a state feedback system.

$$\mathbf{u}_c = \mathbf{K}_r \mathbf{r} - \mathbf{K}_z \mathbf{z},$$

and the product $\mathbf{K}_y \mathcal{C}$ in Eq. (11.16) must be substituted by matrix $\mathbf{K}_z$.

Usually state feedback is regarded as an ideal situation, but for practical reasons, linked with the impossibility of measuring many state variables, output feedback is applied.

The alternative is to use a device that estimates the state variables of the plant, performing a sort of simulation of its behavior (Fig. 11.14; the figure refers to a regulation problem in which the reference inputs $\mathbf{r}$ are equal to zero). It is usually referred to as a *state observer*. This is possible if the system is observable, in the sense defined in Section 11.3.

The behavior of the observer is defined by matrices $\mathcal{A}_0$, $\mathcal{B}_0$, and $\mathcal{C}_0$. If they were equal to the matrices of the system, the observer would be an exact model of the plant. In general, this is impossible and the observer is only an approximated model, often a reduced order one. In structural control, if digital techniques are used, the observer can be a finite element model, perhaps reduced using Guyan or similar reduction procedures.

Usually the term *model-based control* is used when the control algorithm is based on a mathematical model of the plant, running in real time on the microprocessor on which the control system is based.

To ensure that the observer evolves in time like the actual plant, a feedback is introduced: The difference between the output of the observer $\hat{\mathbf{y}}$ and that of the system $\mathbf{y}$ is introduced, through a gain matrix $\mathbf{K}_0$, into the observer.

The states of the observer $\hat{\mathbf{z}}$ are readily available, and the states of the plant are not such (this is why the observer is used) and the control feedback is closed through the control gain matrix $\mathbf{K}_c$.

The equations allowing the study of the closed-loop behavior of the system of Fig. 11.14 are

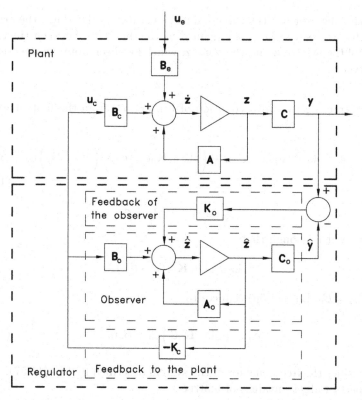

FIGURE 11.14. Block diagram of the multi-degrees-of-freedom structure in Fig. 11.4 controlled by a regulator with a state observer.

$$\begin{cases} \dot{\mathbf{z}} = \mathcal{A}\mathbf{z} + \mathcal{B}_c\mathbf{u}_c + \mathcal{B}_e\mathbf{u}_e\,, \\ \mathbf{y} = \mathcal{C}\mathbf{z}\,, \\ \dot{\hat{\mathbf{z}}} = \mathcal{A}_0\hat{\mathbf{z}} + \mathcal{B}_0\mathbf{u}_c + \mathbf{K}_0(\mathbf{y} - \hat{\mathbf{y}})\,, \\ \hat{\mathbf{y}} = \mathcal{C}_0\hat{\mathbf{z}}\,, \\ \mathbf{u}_c = -\mathbf{K}_c\hat{\mathbf{z}}\,. \end{cases} \qquad (11.54)$$

The observer with its feedback branches constitutes the regulator of the plant, which can be considered a system that has the output of the plant $\mathbf{y}$ as input and outputs the control inputs $\mathbf{u}_c$, which are fed back to the plant. The last three equations of Eq. (11.54), which define the behavior of the regulator, can be written in the form

$$\begin{cases} \dot{\hat{\mathbf{z}}} = \mathcal{A}_{reg}\hat{\mathbf{z}} + \mathcal{B}_{reg}\mathbf{y}, \\ \mathbf{u}_c = \mathcal{C}_{reg}\hat{\mathbf{z}}. \end{cases} \qquad (11.55)$$

where

$$\mathcal{A}_{reg} = \mathcal{A}_0 + \mathbf{K}_0\mathcal{C}_0 - \mathcal{B}_0\mathbf{K}_c\,, \qquad \mathcal{B}_{reg} = \mathbf{K}_0\,, \qquad \mathcal{C}_{reg} = -\mathbf{K}_c\,.$$

If the observer has the same number of states as the plant, the error of the observer $\mathbf{e}_0$ can be easily defined as the difference between the actual state of the system $\mathbf{z}$ and the state $\hat{\mathbf{z}}$, which has been approximated by the observer

$$\mathbf{e}_0 = \mathbf{z} - \hat{\mathbf{z}} \ .$$

By subtracting the third equation of Eq. (11.54) from the first, it follows that

$$\dot{\mathbf{e}}_0 = \left[\mathcal{A} - \mathcal{A}_0 - \mathbf{K}_0 \mathcal{C}_0 - \mathcal{C} \right] \mathbf{z} + \left(\mathcal{A}_0 - \mathbf{K}_0 \mathcal{C}_0 \right) \mathbf{e}_0 + \left(\mathcal{B}_c - \mathcal{B}_0 \right) \mathbf{u}_c + \mathcal{B}_e \mathbf{u}_e \ . \tag{11.56}$$

If

$$\mathcal{A}_0 = \mathcal{A} \ , \qquad \mathcal{C}_0 = \mathcal{C}$$

(actually it is enough that

$$\mathcal{A}_0 = \mathcal{A} + \mathbf{K}_0 (\mathcal{C}_0 - \mathcal{C}) \ ),$$

and $\mathcal{B}_0 = \mathcal{B}_c$, Eq. (11.56) reduces to

$$\dot{\mathbf{e}}_0 = \left(\mathcal{A} - \mathbf{K}_0 \mathcal{C} \right) \mathbf{e}_0 + \mathcal{B}_e \mathbf{u}_e \ . \tag{11.57}$$

Consider the free behavior of the system described by Eq. (11.57). If the real part of all eigenvalues of matrix

$$\mathcal{A}_0 = \mathcal{A} - \mathbf{K}_0 \mathcal{C}$$

is negative, the error of the observer tends to zero for time tending to infinity and the observer is asymptotically stable. The observer can thus be designed by stating a set of eigenvalues of matrix $\mathcal{A}_0$, whose real parts are negative and that minimize the time delay between the estimated and actual state vectors. It is a common suggestion[4] that the poles of the observer (i.e., the eigenvalues of matrix $\mathcal{A}_0$) be placed on the complex plane on the left of the poles of the controlled structure. However, a fast observer obtained in this way can be prone to amplify the disturbances that are always present in the signals from the sensors. The choice of the poles must be a trade-off between the requirements of quick response and disturbance rejection.

The computation of matrix $\mathbf{K}_0$, which causes matrix $\mathcal{A}_0$ to yield the required eigenvalues, can be performed using the pole-placement procedure, provided the rank of the observability matrix $\mathbf{O}$ is n, i.e., the system is observable. This procedure is straightforward if the system has a single

---

[4]See, for example, H.H.E. Leipholz, M. Abdel Rohman, *Control of Structures*, Martinus Nijhoff, Dordrecht, 1986, p. 106.

output, while for multivariable systems some arbitrary choices must be made.

The state equation of the complete system shown in Fig. 11.14 is

$$\left\{ \begin{array}{c} \dot{\mathbf{z}} \\ \dot{\mathbf{e}}_0 \end{array} \right\} = \left[\begin{array}{cc} \mathcal{A} - \mathcal{B}_c \mathbf{K}_c & \mathcal{B}_c \mathbf{K}_c \\ \mathbf{0} & \mathcal{A} - \mathbf{K}_0 \mathcal{C} \end{array} \right] \left\{ \begin{array}{c} \mathbf{z} \\ \mathbf{e}_0 \end{array} \right\} + \left\{ \begin{array}{c} \mathcal{B}_e \mathbf{u}_e \\ \mathcal{B}_e \mathbf{u}_e \end{array} \right\}. \tag{11.58}$$

The characteristic equation of the closed-loop system is

$$\det \left[\begin{array}{cc} s\mathbf{I} - \mathcal{A} + \mathcal{B}_c \mathbf{K}_c & \mathcal{B}_c \mathbf{K}_c \\ \mathbf{0} & s\mathbf{I} - \mathcal{A} + \mathbf{K}_0 \mathcal{C} \end{array} \right] = 0, \tag{11.59}$$

i.e.,

$$\det \left(s\mathbf{I} - \mathcal{A} + \mathcal{B}_c \mathbf{K}_c \right) \det \left(s\mathbf{I} - \mathcal{A} + \mathbf{K}_0 \mathcal{C} \right) = 0. \tag{11.60}$$

The roots of Eq. (11.60) are the eigenvalues of the controlled system plus the eigenvalues of the observer and the relevant parts of the characteristic equation can be solved separately.

11.11   Control design

To prescribe the free behavior of the controlled system and of the observer in terms of natural frequencies and decay rates, the n eigenvalues $s_1, s_2, \ldots,$ s_n of the dynamic matrix $\mathcal{A} - \mathcal{B}_c \mathbf{K}_c$ (for the former) and of the dynamic matrix $\mathcal{A} - \mathbf{K}_0 \mathcal{C}$ (for the second) can be stated. Note that if they are complex, there must be a number of conjugate pairs in order to obtain real values of the gains. The characteristic equation of the controlled system (corresponding to the first part of Eq. (11.60)) is

$$\det \left(s\mathbf{I} - \mathcal{A} + \mathcal{B}_c \mathbf{K}_c \right) = (s - s_1)(s - s_2) \ldots (s - s_n) = 0. \tag{11.61}$$

The n eigenvalues can be used to compute the n coefficients of the characteristic polynomial. By equating the expressions of the coefficients on the left-hand side of Eq. (11.61) with the coefficients so computed, n equations, which can be used to compute the n unknown elements of matrix $\mathbf{K}_c$, are obtained. The computation of the unknowns can be performed easily using Ackermann's formula

$$\mathbf{K}_c = \left[\begin{array}{cccccc} 0 & 0 & 0 & \ldots & 0 & 1 \end{array} \right] \mathbf{H}^{-1} \mathbf{N}, \tag{11.62}$$

where $\mathbf{H}$ is the controllability matrix defined by Eq. (11.3), and matrix $\mathbf{N}$ is obtained from the coefficients $a_0, a_1, \ldots, a_{n-1}$ of the characteristic polynomial (subscripts refer to the power of the unknown) through the formula

$$\mathbf{N} = \mathcal{A}^n + a_{n-1} \mathcal{A}^{n-1} + a_{n-2} \mathcal{A}^{n-2} + \cdots + a_0 \mathbf{I}. \tag{11.63}$$

If the controllability matrix, which in the case of a single-input system is a square matrix, is not singular, there is no difficulty in computing the gains of the control system $\mathbf{K}_c$. However, this is essentially a theoretical statement because in practice the numerical evaluation of matrix $\mathbf{K}_c$ for systems whose order of the dynamic matrix is higher than a few units can be affected by large numerical errors.

Apart from the computational difficulties encountered when applied to MIMO systems, the pole assignment method allows one to state the poles of the controlled system but does not allow one to control its eigenvectors. This can result in badly conditioned eigenvector matrices, which is highly undesirable. It is much more convenient to resort to methods that are generally referred to as *eigenstructure assignment* procedures, which allow to state, within ample limits, both the eigenvalues and the eigenvectors of the controlled system. For a detailed presentation of such methods, the reader is again advised to refer to specialized textbooks on structural control.

Optimal control techniques are mostly based on the minimization of performance indices, which include the error of the control system but can take into account the costs or energy needed to perform the control action. A common definition of the performance indices is

$$ J = \int_0^\infty \left[\mathbf{z}^T(t)\mathbf{Q}\mathbf{z}(t) + \mathbf{u}_c^T(t)\mathbf{R}\mathbf{u}_c(t) \right] dt . \qquad (11.64) $$

This formulation, to be applied to systems in which the desired condition is that with all state variables equal to zero, minimizes a linear combination of the integrals in time of the deviation from this condition (i.e., of the control errors) and of the control inputs, which are in some way linked with the energy needed for the control function. The various state variables and control inputs can have different importance, as shown by the weight matrices $\mathbf{Q}$ and $\mathbf{R}$ introduced into the definition of the performance index.

Under wide assumptions, it is possible to demonstrate that the feedback gain matrix $\mathbf{K}_c$ for an optimal state feedback can be written in the form

$$ \mathbf{K}_c = \mathbf{R}^{-1}\mathcal{B}_c^T\mathbf{P} , \qquad (11.65) $$

where $\mathbf{P}$ is the solution of the algebraic Riccati equation

$$ \mathcal{A}^T\mathbf{P} + \mathbf{P}\mathcal{A} - \mathbf{P}\mathcal{B}_c\mathbf{R}^{-1}\mathcal{B}_c^T\mathbf{P} + \mathbf{Q} = \mathbf{0} . \qquad (11.66) $$

The regulator so obtained is referred to as an optimal linear quadratic regulator, where *linear* refers to the linearity of the controlled system (plant + controller) and *quadratic* refers to the quadratic performance index. Similar considerations can be made for the observer, if $\mathbf{y}$ and $\hat{\mathbf{z}}$ are substituted for $\mathbf{z}$ and $\mathbf{u}_c$ into the performance index (11.64).

The detailed design of the control system and observer is, however, well beyond the scope of this text, and the reader is advised to refer to the many texts in which the optimal control theory and the solution of algebraic Riccati equations are dealt with.[5]

11.12   Modal approach to structural control

Structural models usually have many degrees of freedom, particularly if they have been obtained through discretization techniques, like the finite element method (see Chapter 15). They are thus impractical for the design of the control system, which usually requires to deal with low-order models of the plant. Moreover, these models have a large number of high-frequency modes with little physical meaning, because they are due more to the discretization procedure than to the structure itself.

A very effective way to obtain a reduced order model for the design of the control system is that of using the Guyan reduction: The model of the plant can be condensed until a small number of master degrees of freedom are retained. The comparison of the open-loop natural frequencies and vibration modes of the reduced order model with the corresponding ones obtained from the full model allows verification of the validity of this approach in any relevant case.

The approach most frequently resorted to is, however, that of using modal coordinates and retaining only a reduced number of modes.

The equation of motion of the controlled structure, i.e., of the structure on which the control forces $\mathbf{f}_c$ are acting, can be reduced in modal form by resorting to the eigenvectors of the undamped system

$$\overline{\mathbf{M}}\ddot{\boldsymbol{\eta}} + \overline{\mathbf{C}}\dot{\boldsymbol{\eta}} + \overline{\mathbf{K}}\boldsymbol{\eta} = \boldsymbol{\Phi}^T\mathbf{f}_c + \boldsymbol{\Phi}^T\mathbf{f}_e . \qquad (11.67)$$

Remark 11.7 *As it will be seen later, Eq. (11.67) holds not only for discrete systems but also for continuous ones. The only difference is that in the latter case there is, theoretically, an infinity of modes.*

Remark 11.8 *The modal-damping matrix* $\overline{\mathbf{C}}$ *is often obtained by directly assessing the modal damping of the various modes and hence is usually diagonal.*

The modal mass matrix may reduce to an identity matrix if the eigenvectors are normalized in such a way that the modal masses have a unit value.

When computed from the characteristics of the system and not directly by assessing the damping of the various modes, the modal-damping matrix is not, in general, a diagonal matrix, and the modes are not uncoupled.

---

[5]See, for example, the already-mentioned book J.L. Junkins, Y. Kim, *Introduction to Dynamics and Control of Flexible Structures*, AIAA, Washington DC, 1993.

However, as structures are in most cases very little damped, the coupling of the modes is fairly weak and the modal-damping matrix can be approximated by a diagonal matrix simply by neglecting its out-of-diagonal elements.

The modal equation in the phase space is then

$$\dot{\mathbf{z}} = \mathcal{A}\mathbf{z} + \mathcal{B}_c\mathbf{u}_c(t) + \mathcal{B}_e\mathbf{u}_e(t)\,,$$
$$\mathbf{y} = \mathcal{C}\mathbf{z}\,, \qquad (11.68)$$

where

$$\mathbf{z} = \left\{ \begin{array}{c} \dot{\eta} \\ \eta \end{array} \right\}\,, \qquad \mathcal{A} = \left[\begin{array}{cc} -\overline{\mathbf{C}} & -\overline{\mathbf{K}} \\ \mathbf{I} & \mathbf{0} \end{array} \right]\,,$$

$$\qquad (11.69)$$

$$\mathcal{B}_c = \left[\begin{array}{c} \Phi^T\mathbf{T}_c \\ \mathbf{0} \end{array} \right]\,, \qquad \mathcal{B}_e = \left[\begin{array}{c} \Phi^T\mathbf{T}_e \\ \mathbf{0} \end{array} \right]\,,$$

and the output gain matrix links the outputs $\mathbf{y}$ of the system with the modal coordinates.

Remark 11.9 *If the damping matrix is diagonal, the first Eq. (11.68) uncouples in a number of sets of two equations, each dealing with a single modal coordinate. This uncoupling, which is usually only approximate, does not hold any more once a closed-loop control system is included in the equation.*

The coupling due to the control system can be very strong, much stronger than that linked with damping which is often neglected. In a way, the effect of the action of the control system can often be assimilated to that of very high damping, and some of the modes of the controlled system can even be overdamped: No small-damping assumption holds anymore. Apart from the strong coupling due to the added damping, the presence of the control system changes the mode shapes; it is true that the new mode shapes can be expressed as combinations of the modes of the uncontrolled system, but this further increases modal coupling. Due to this coupling, the results obtained by considering only a limited number of modes in the design of the control system are only an approximation of the behavior of the complete model.

In particular, the errors introduced by the fact that the outputs $\mathbf{y}$ of the system are also influenced by the modes that have been neglected are referred to as *observation spillover*. Those introduced by the fact that the control forces due to the control system do influence not only the modes considered but also the neglected ones are referred to as *control spillover*. Spillover can cause the system to behave in ways different from the predicted one and in some cases can even cause instability.

Remark 11.10 *Active systems are prone to instability, because energy is supplied to the structure, and if this occurs in a wrong way, it can cause the excitation of some modes instead of damping them.*

If an observer is used to approximate only a reduced number of modes, it is better to use an equation that considers the vectors $\mathbf{z}$ (of order n, now containing the modal coordinates and their derivatives) and $\hat{\mathbf{z}}$ (of order n_r, with $n_r < n$, containing the estimated values of the modal coordinates on which control is performed and their derivatives), which are state variables of the system, instead of Eq. (11.58), considering $\mathbf{z}$ and the errors $\mathbf{e}_0$

$$\left\{ \begin{array}{c} \dot{\mathbf{z}} \\ \dot{\hat{\mathbf{z}}} \end{array} \right\} = \left[\begin{array}{cc} \mathcal{A} & -\mathcal{B}_c\mathbf{K}_c \\ \mathbf{K}_0\mathcal{C} & \mathcal{A} - \mathcal{B}_0\mathbf{K}_c - \mathbf{K}_0\mathcal{C}_0 \end{array} \right] \left\{ \begin{array}{c} \mathbf{z} \\ \hat{\mathbf{z}} \end{array} \right\} + \left\{ \begin{array}{c} \mathcal{B}_e\mathbf{u}_e \\ 0 \end{array} \right\} . \tag{11.70}$$

The various matrices have different sizes, and they combine in such a way as to yield a closed-loop dynamic matrix that has $n + n_r$ rows and columns. Its eigenvalues are the closed-loop roots of the system and their study allows the effects of spillover to be predicted.

Although spillover is an actual danger plaguing the modal approach to the design of the control system for a structure, modal control has been used successfully in many cases; it allows the solution of the problem of designing the control system in a simple and straightforward way, particularly when the number of modes to be considered is low. Even if modal control is widely accepted, the possible dangers due to spillover must always be kept in mind, and an evaluation of the effects of higher-order modes is in many cases advisable.

Remark 11.11 *Equations (11.68) and (11.70) also hold in the case of continuous systems, once their modal characteristics have been obtained.*

Remark 11.12 *Equations (11.68) and (11.70) contain only a form of proportional control, but since in the latter the control forces are also linked to the modal velocities included into the modal states, the control system provides a form of damping as in the case of derivative control.*

11.13 Exercises

Exercise 11.1 *Consider the damped inverted pendulum in Fig. 11.2. Plot the root locus with the following data. Compute the value of the stiffness of the spring in such a way that the natural frequency of the system is 50 Hz, and the damping ratio of the system. Compute also the time history following a shock due to the impact of a ball of mass of 1 kg traveling at 30 m/s on the bob of the pendulum. Assume a perfectly elastic shock. Data: length $l = 1$ m, $m = 2$ kg, $g = 9.806$ m/s^2, $\Gamma = 100$ N m/rad.*

Exercise 11.2 *Substitute the spring in Example 11.1 with an actuator driven by a proportional–derivative control system and a transducer reading angle θ. Compute the gains in such a way that the frequency of the damped oscillations of the system is 50 Hz and the damping ratio is 0.2. Compute the response to the same impulse studied in Exercise 11.1.*

Exercise 11.3 *Consider the torsional system with 3 degrees of freedom of Fig. 1.8, already studied in Exercises 1.2 , 2.1, and 2.3, and add a control system that exerts a moment on disc 2. Consider as input to the system the rotation θ_A of the supporting point and as output the rotation of disc 2. Compute the matrices of the system and check whether it is controllable and observable. Give the system a step input with a rotation $\theta_A = 10°$, and assume as new reference position the static equilibrium configuration rotated of angle θ_A. Compute the gains in such a way that the overshot in the first mode is not greater than 50%, assuming an output feedback. Compute the time histories of the response at all discs.*

Exercise 11.4 *Consider the quarter-car model of Example 9.2. Substitute the shock absorber with a 'skyhook damper', i.e., an active device that exerts on the sprung mass a force proportional to its absolute vertical velocity. Assume that the feedback loop is based on the vertical acceleration of the sprung mass. Compute the frequency response of the system; plot the maximum value of the acceleration of the sprung mass as a function of the gain of the control system and choose the parameters that allow the minimum value of the peak acceleration to be obtained. Compute the variable component of the vertical force exerted by the tire on the road and the value of the gain that minimizes its peak value.*

Exercise 11.5 *Compute the power spectral density of the response of the active suspension obtained in Exercise 11.4 (with the value of the gain minimizing the vertical acceleration) when traveling on a normal road following ISO standards (Fig. 9.9d). Compare the results with those obtained in Example 9.2 for a conventional suspension.*

12
Vibration of Beams

At a macroscopic scale most real life objects can be accurately modeled as elastic continuums. The study of the vibration dynamics of a continuous system can however be performed only when its geometrical configuration is simple, and the simplest elastic continuum is a beam. In spite of its simplicity, the study of vibrating beams yields interesting insights on the behavior of a much wider class of systems.

12.1 Beams and bars

The main feature of the discrete systems studied in the preceding chapters is that a finite number of degrees of freedom is sufficient to describe their configuration. The ordinary differential equations (ODEs) of motion can be easily substituted by a set of linear algebraic equations: The natural mathematical tool for the study of linear discrete systems is matrix algebra.

The situation is different when a deformable elastic body is studied as a continuous system: Because they can be thought as being constituted by an infinity of points, they have an infinity of degrees of freedom and the resulting mathematical models are made by partial (derivatives) differential equations (PDEs).

Because no general approach to the dynamics of an elastic body is feasible, many different models for the study of particular classes of structural elements (beams, plates, shells, etc.) have been developed. In this chapter, only beams will be studied: This choice is only in part due to the fact that many machine elements can be studied as beams (shafts, blades, springs, etc.); it comes from the need of showing some general properties of continuous systems in the simplest case to gain a good insight on the properties of deformable bodies.

The study of the elastic behavior of beams dates back to Galileo, with important contributions by Daniel Bernoulli, Euler, De Saint Venant, and many others. A *beam* is essentially an elastic solid in which one dimension is prevalent over the others. Often the beam is prismatic (i.e., the cross-sections are all equal), homogeneous (i.e., with constant material characteristics), straight (i.e., its axis is a part of a straight line), and untwisted (i.e., the principal axes of elasticity of all sections are equally directed in space). The unidimensional nature of beams allows simplification of the study: Each cross-section is considered as a rigid body whose thickness in the axial direction is vanishingly small; it has 6 degrees of freedom, three translational and three rotational. The problem is then reduced to a unidimensional problem, in the sense that a single coordinate, namely the axial coordinate, is required.

Setting the z-axis of the reference frame along the axis of the beam (Fig. 12.1), the six generalized coordinates of each cross-section are the axial displacement u_z, the lateral displacements u_x and u_y, the torsional rotation ϕ_z about the z-axis, and the flexural rotations ϕ_x and ϕ_y about axes x and y. Displacements and rotations are assumed to be small, so rotations can be regarded as vector quantities, which simplify all rotation matrices by linearizing trigonometric functions. The three rotations will then be considered components of a vector in the same way as the three displacements are components of vector **u**. The generalized forces acting on each cross-section and corresponding to the 6 degrees of freedom defined earlier are the axial force F_z, shear forces F_x and F_y, the torsional moment M_z about the z-axis, and the bending moments M_x and M_y about the x- and y-axes.

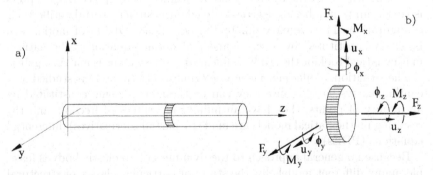

FIGURE 12.1. Straight beam. (a) Sketch and reference frame; (b) generalized displacements and forces on a generic cross-section.

TABLE 12.1. Degrees of freedom and generalized forces in beams.

Type of behavior	Degrees of freedom	Generalized forces
Axial	Displacement u_z	Axial force F_z
Torsional	Rotation ϕ_z	Torsional moment M_z
Flexural	Displacement u_x	Shearing force F_x
(xz-plane)	Rotation ϕ_y	Bending moment M_y
Flexural	Displacement u_y	Shearing force F_y
(yz-plane)	Rotation ϕ_x	Bending moment M_x

From the aforementioned assumptions it follows that all normal stresses in directions other than z (σ_x and σ_y) are assumed to be small enough to be neglected. When geometric and material parameters are not constant along the axis, they must change at a sufficiently slow rate not to induce stresses σ_x and σ_y, which could not be considered in this model. If the axis of the beam is assumed to be straight, the axial translation is uncoupled from the other degrees of freedom, at least in first approximation. A beam that is loaded only in the axial direction and whose axial behavior is the only one studied is usually referred to as a *bar*. The torsional–rotational degree of freedom is uncoupled from the others only if the area center of all cross-sections coincides with their shear center, which happens if all cross-sections have two perpendicular planes of symmetry. If the planes of symmetry of all sections are equally oriented (the beam is not twisted) and the x- and y-axes are perpendicular to such planes, the flexural behavior in the xz-plane is uncoupled from that in the yz-plane. The coupling of the degrees of freedom in straight, untwisted beams with cross-sections having two planes of symmetry is summarized in Table 12.1.

12.2   Axial behavior of straight bars

Consider a bar as defined in the preceding section made of an isotropic linear elastic material. An axial distributed force $f_z(z,t)$ (being a force per unit length in the SI it is measured in N/m) acts on each point of the bar. The assumption that the cross-section behaves as a rigid body is clearly a simplification of the actual behavior, since a longitudinal stretching or compression is always accompanied by deformations in the plane of the cross-section, but the assumption of small deformations makes these side effects negligible. The position of each point in the original configuration is defined by its coordinate z, while its displacement has only one component u_z.

12.2.1 *Equation of motion*

Consider a length dz of a beam centered on the point with coordinate z (Fig. 12.2). The inertia force due to motion in z-direction is

$$\ddot{u}_z dm = \rho A \ddot{u}_z dz\,.$$

The resultant of the axial force F_z exerted by the other parts of the bar is

$$F_z + \frac{1}{2}\frac{\partial F_z}{\partial z}dz - \left(F_z - \frac{1}{2}\frac{\partial F_z}{\partial z}dz\right) = \frac{\partial F_z}{\partial z}dz.$$

The external force acting on the same length of the beam is

$$f_z(z,t)dz.$$

The dynamic equilibrium equation can thus be written in the form

$$\rho A \ddot{u}_z = \frac{\partial F_z}{\partial z} + f_z(z,t)\,. \tag{12.1}$$

The axial force F_z is easily linked with the displacement by the usual formula from the theory of elasticity

$$F_z = A\sigma_z = EA\epsilon_z = EA\frac{\partial u_z}{\partial z}\,. \tag{12.2}$$

By introducing Eq. (12.2) into Eq. (12.1), the dynamic equilibrium equation is obtained:

$$m(z)\frac{d^2 u_z}{dt^2} - \frac{\partial}{\partial z}\left[k(z)\frac{\partial u_z}{\partial z}\right] = f_z(z,t)\,, \tag{12.3}$$

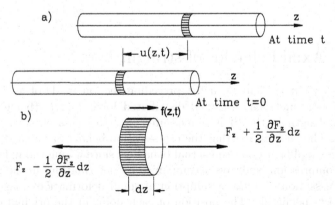

FIGURE 12.2. Axial behavior of a straight bar: **(a)** system of reference and displacement; **(b)** forces acting on the length dz of the bar.

where the mass per unit length of the bar and the axial stiffness are, respectively,

$$m(z) = \rho(z)A(z), \qquad k(z) = E(z)A(z). \qquad (12.4)$$

The boundary conditions must be stated following the actual conditions at the ends of the bar: if, for instance, they are clamped, the displacement u_z must be equated to zero for $z = 0$ and $z = l$ (where l is the length of the beam and the origin is set at the left end).

If the bar is prismatic and homogeneous, i.e., all characteristics are constant, Eq. (12.3) reduces to

$$\frac{d^2 u_z}{dt^2} - \frac{E}{\rho}\frac{\partial^2 u_z}{\partial z^2} = f_z(z,t). \qquad (12.5)$$

12.2.2 Free behavior

The homogeneous equation associated with Eq. (12.5) describing the free behavior of the bar

$$\frac{d^2 u_z}{dt^2} - \frac{E}{\rho}\frac{\partial^2 u_z}{\partial z^2} = 0 \qquad (12.6)$$

is a wave propagation equation in a one-dimensional medium and constant E/ρ is the square of the speed of propagation of the waves, i.e., the speed of sound in the material

$$v_s = \sqrt{\frac{E}{\rho}}. \qquad (12.7)$$

It can be used to study the propagation of elastic waves in the bar, such as those involved in shocks and other phenomena of interest in the field of structural dynamics. For the study of most problems related to vibration dynamics, stationary solutions are, however, more useful.

Remark 12.1 *The two approaches lead to the same results, because it is possible to describe the propagation of elastic waves in terms of superimposition of stationary motions and vice versa. It is not, however, immaterial which way is followed in the actual study of any particular phenomena, because the complexity of the analysis can be quite different.*

As an example, in the study of the propagation of the elastic waves caused by a shock, the number of modes, which must be considered when the relevant phenomena are studied as the sum of stationary solutions, can be exceedingly high, and then it is advisable to consider the equation of motion as wave equation. In most cases, however, the dynamic behavior of continuous systems can be described accurately by adding only a limited number of modes.

The solution of the homogeneous equation (12.6) can be expressed as the product of a function of time by a function of the space coordinate

$$u_z(z,t) = q(z)\eta(t). \qquad (12.8)$$

Introducing Eq. (12.6) into the homogeneous equation (12.3) and separating the variables, it follows that

$$\frac{1}{\eta(t)} \frac{d^2\eta(t)}{dt^2} = \frac{1}{m(z)q(z)} \frac{d}{dz}\left[k(z)\frac{dq(z)}{dz}\right].$$ (12.9)

The function on the left-hand side depends on time but not on the space coordinate z. Conversely, the function on the right-hand side is a function of z but not of t. The only possible way of satisfying Eq. (12.9) for all values of time and of coordinate z is to state that both sides are constant and that the two constants are equal. This constant can be indicated as $-\omega^2$.

The condition on the function of time on the left-hand side is thus

$$\frac{1}{\eta(t)} \frac{d^2\eta(t)}{dt^2} = \text{constant} = -\omega^2,$$ (12.10)

i.e.,

$$\frac{d^2\eta}{dt^2} + \omega^2\eta = 0.$$ (12.11)

Neglecting a proportionality constant that will be introduced into function $q(z)$ later, it yields a harmonic oscillation with frequency ω

$$\eta(t) = \sin(\omega t + \phi).$$ (12.12)

The solution of the equation of motion for free axial oscillations of the bar is

$$u(z,t) = q(z)\sin(\omega t + \phi).$$ (12.13)

Function $q(z)$ is said to be the *principal function*. Each point of the bar performs a harmonic motion with frequency ω, while the amplitude is given by function $q(z)$.

Remark 12.2 *The resultant motion is then a standing wave, with all points of the bar vibrating in phase.*

By introducing Eq. (12.13) into Eq. (12.9), it follows that

$$\frac{d}{dz}\left[k(z)\frac{dq(z)}{dz}\right] + \omega^2 m(z)q(z) = 0,$$ (12.14)

or, in the case of constant parameters,

$$\frac{d^2q(z)}{dz^2} + \frac{\omega^2}{v_s^2}q(z).$$ (12.15)

Equations (12.14) and (12.15) are eigenproblems. The second, for example, states that the second derivative of function $q(z)$ (with respect to the space coordinate z) is proportional to the function itself, the constant of proportionality being

$$\frac{\omega^2}{v_s^2} = \omega^2 \frac{\rho}{E}.$$

(12.16)

The values of such a constant that allow the equation to be satisfied by a solution other than the trivial solution $q(z) = 0$ are the eigenvalues, and the corresponding functions $q(z)$ are the eigenfunctions. The first equation (12.14), although more complex, has a similar meaning.

Remark 12.3 *The eigenvalues are infinite in number, and the general solution of the equation of motion (12.14) can be obtained as the sum of an infinity of terms of the type of Eq. (12.13).*

Remark 12.4 *The eigenfunctions $q_i(z)$ are defined only as far as their shape is concerned, exactly as was the case for eigenvectors. The amplitude of the various modes can be computed only after the initial conditions have been stated.*

Remark 12.5 *Although the number of eigenfunctions, and hence of modes, is infinite, a small number of principal functions is often sufficient to describe the behavior of an elastic body with the required precision, in a way that is similar to what has already been said for eigenvectors.*

Remark 12.6 *Eigenfunctions have some of the properties seen for eigenvectors, particularly that of orthogonality with respect to the mass $m(z)$ and to the stiffness $k(z)$.*

In the case of a bar, the property of orthogonality with respect to mass and stiffness means that if $q_i(z)$ and $q_j(z)$ are two distinct eigenfunctions, it follows that if $(i \neq j)$,

$$\int_0^l m(z)q_i(z)q_j(z)dz = 0, \qquad \int_0^l k(z)\frac{dq_i(z)}{dz}\frac{dq_j(z)}{dz}dz = 0.$$

(12.17)

As was the case for eigenvectors, eigenfunctions are not directly orthogonal:

$$\int_0^l q_i(z)q_j(z)dz \neq 0,$$

except when the mass $m(z)$ is constant along the beam. If $i = j$, the integrals of Eq. (12.17) do not vanish:

$$\int_0^l m(z)[q_i(z)]^2 dz = \overline{M}_i \neq 0, \qquad \int_0^l k(z)\left[\frac{dq_i(z)}{dz}\right]^2 dz = \overline{K}_i \neq 0.$$

(12.18)

Equation (12.17) can be proved by writing Eq. (12.14) for two different eigenfunctions $q_i(z)$ and $q_j(z)$ (with the corresponding eigenvalues ω_i

and ω_j) and then multiplying the first one by $q_j(z)$ and the second one by $q_i(z)$

$$-\omega_i^2 m(z)q_i(z)q_j(z) = q_j(z)\frac{d}{dz}\left[k(z)\frac{dq_i(z)}{dz}\right], \tag{12.19}$$

$$-\omega_j^2 m(z)q_j(z)q_i(z) = q_i(z)\frac{d}{dz}\left[k(z)\frac{dq_j(z)}{dz}\right]. \tag{12.20}$$

By subtracting Eq. (12.20) from Eq. (12.19) and integrating over the whole length of the beam, it follows

$$-(\omega_i^2 - \omega_j^2)\int_0^l m(z)q_i(z)q_j(z)dz = \int_0^l q_j(z)\frac{d}{dz}\left[k(z)\frac{dq_i(z)}{dz}\right]dz+$$
$$-\int_0^l q_j(z)\frac{d}{dz}\left[k(z)\frac{dq_i(z)}{dz}\right]dz.$$
$$\tag{12.21}$$

The two expressions at the right-hand side can be integrated by parts. The first one yields

$$\int_0^l q_j(z)\frac{d}{dz}\left[k(z)\frac{dq_i(z)}{dz}\right]dz = \left[q_j(z)k(z)\frac{dq_i(z)}{dz}\right]_0^l +$$
$$-\int_0^l k\frac{dq_i(z)}{dz}\frac{dq_j(z)}{dz}dz. \tag{12.22}$$

The ends of the bar may be either free or clamped: in the first case the axial force, and hence the axial strain, vanishes, while in the latter the displacement is equal to zero:

$$\text{either } q(z) = 0 \text{ or } \frac{dq(z)}{dz} = 0 \text{ for } z = 0 \text{ and } z = l.$$

It then follows

$$\int_0^l q_j(z)\frac{d}{dz}\left[k(z)\frac{dq_i(z)}{dz}\right]dz = -\int_0^l k\frac{dq_i(z)}{dz}\frac{dq_j(z)}{dz}dz,$$
$$\int_0^l q_i(z)\frac{d}{dz}\left[k(z)\frac{dq_j(z)}{dz}\right]dz = -\int_0^l k\frac{dq_i(z)}{dz}\frac{dq_j(z)}{dz}dz. \tag{12.23}$$

By introducing Eq. (12.20) into Eq. (12.21), it follows

$$-(\omega_i^2 - \omega_j^2)\int_0^l m(z)q_i(z)q_j(z)dz = 0, \tag{12.24}$$

and then, if $\omega_i \neq \omega_j$,

$$\int_0^l m(z)q_i(z)q_j(z)dz = 0, \tag{12.25}$$

which proves the m-orthogonality property.

The k-orthogonality property can be proved in a similar way, by dividing Eqs. (12.20) and (12.19) by ω_i^2 and ω_j^2, respectively, subtracting and integrating over the whole length of the beam.

Constants $\overline{M}_i$ and $\overline{K}_i$ are the modal mass and the modal stiffness, respectively, for the ith mode.

Remark 12.7 *The meaning of the modal mass and stiffness of continuous systems is exactly the same as for discrete systems; the only difference is that in the current case the number of modes, and then of modal masses and stiffiness, is infinite.*

Any deformed configuration of the system $u(z,t)$ can be expressed as a linear combination of the eigenfunctions. The coefficients of this linear combination, which are functions of time, are the modal coordinates $\eta_i(t)$:

$$u(z,t) = \sum_{i=0}^{\infty} \eta_i(t) q_i(z) \,. \tag{12.26}$$

Equation (12.26) expresses the modal transformation for continuous systems and is equivalent to the first equation (4.37). The inverse transformation, needed to compute the modal coordinates $\eta_i(t_k)$ corresponding to any given deformed configuration $u(z,t_k)$ occurring at time t_k, can be obtained through a procedure closely following that used to obtain Eq. (6.22). Multiplying Eq. (12.26) by the jth eigenfunction and by the mass distribution $m(z)$ and integrating on the whole length of the bar, it follows that

$$\int_0^l \left[m(z) q_j(z) u(z,t_0) \right] dz = \sum_{i=0}^{\infty} \eta_i(t_0) \int_0^l \left[m(z) q_j(z) q_i(z) \right] dz \,. \tag{12.27}$$

Of the infinity of terms on the right-hand side, only the term with $i = j$ does not vanish and the integral yields the jth modal mass.

Remembering the definition of the modal masses, it then follows that

$$\eta_i(t_0) = \frac{1}{\overline{M}_j} \int_0^l \left[m(z) q_j(z) u(z,t_0) \right] dz \,, \tag{12.28}$$

which can be used to perform the inverse modal transformation, i.e., to compute the modal coordinates corresponding to any deformed shape of the system.

Eigenfunctions can be normalized in several ways, one being that leading to unit values of the modal masses. This is achieved simply by dividing each eigenfunction by the square root of the corresponding modal mass.

12.2.3 Forced response

The vibration of the bar under the effect of the forcing function $f_z(z,t)$ can be obtained by solving the complete equation (12.3), whose general solution can be expressed as the sum of the complementary function obtained earlier and a particular integral of the complete equation. Owing to the orthogonality properties of the normal modes $q_i(z)$, the latter can be expressed as a linear combination of the eigenfunctions. Equation (12.26) then also holds in the case of forced motion of the system.

By introducing the modal transformation (12.26) into the equation of motion (12.3), the latter can be transformed into a set of an infinite number of equations in the modal coordinates η_i

$$\overline{M}_i\ddot{\eta}_i(t) + \overline{K}_i\eta_i(t) = \overline{f}_i(t),\tag{12.29}$$

where the ith modal force is defined by the following formulas

$$\overline{f}_i(t) = \int_0^l q_i(z)f(z,t)dz, \qquad \overline{f}_i(t) = \sum_{k=1}^m q_i(z_k)f_k(t),\tag{12.30}$$

holding in the cases of a continuous force distribution and m concentrated axial forces $f_k(t)$ acting on points of coordinates z_k, respectively.

Remark 12.8 *Equation* (12.30) *corresponds exactly to the definition of the modal forces for discrete systems* (4.44).

Equation (12.29) can be used to study the forced response of a continuous system to an external excitation of any type by reducing it to a number of systems with a single degree of freedom. In the case of continuous systems, their number is infinite, but usually few of them are enough to obtain the results with the required precision.

If the excitation is provided by the motion of the structure to which the bar is connected, it is expedient to resort to a coordinate system that moves with the supporting points. In the case of the axial behavior of a bar, only the motion of the supporting structure in the axial direction is coupled with its dynamic behavior.

If the origin of the coordinates is displaced by the quantity z_A, the absolute displacement in z-direction $u_{z_{iner}}(z,t)$ is linked to the relative displacement $u_z(z,t)$ by the obvious relationship

$$u_{z_{iner}}(z,t) = u_z(z,t) + z_A(t).$$

By writing Eq. (12.3) using the relative displacement, it follows

$$m(z)\frac{d^2u_z}{dt^2} - \frac{\partial}{\partial z}\left[k(z)\frac{\partial u_z}{\partial z}\right] = -m(z)\ddot{z}_A.\tag{12.31}$$

This result is similar to that obtained for discrete systems: The excitation due to the motion of the constraints can be dealt with by using relative coordinates and applying an external force distribution equal to $-m(z)\ddot{z}_A$.

The modal forces can be readily computed through Eq. (12.30)

$$\overline{f}_i(t) = -r_i\ddot{z}_A \,, \tag{12.32}$$

where

$$r_i = \int_0^l q_i(z)m(z)dz$$

are the modal participation factors related to the axial motion of the bar.

Remark 12.9 *The physical meaning of the modal participation factors is the same as that already seen for discrete systems.*

12.2.4 Prismatic homogeneous bars

If the bar is prismatic and homogeneous, i.e., all characteristics are constant along the axial coordinate, the general solution of the second equation (12.14) is

$$q(z) = C_1 \sin\left(\frac{\omega}{v_s}z\right) + C_2 \cos\left(\frac{\omega}{v_s}z\right). \tag{12.33}$$

Equation (12.33) expresses the mode shapes of longitudinal vibration of the bar, after constants C_1 and C_2 have been computed from the boundary conditions.

First consider the case in which both ends of the bar are free: The stress σ_z and, hence, the strain

$$\epsilon_z = \frac{\partial u_z}{\partial z}$$

vanishes at both ends

$$\left[\frac{dq(z)}{dz}\right]_{z=0} = \left[\frac{dq(z)}{dz}\right]_{z=l} = 0 \,, \tag{12.34}$$

if the origin is set at the left end of the bar. With simple computations, Eq. (12.34) yields

$$\frac{\omega}{v_s}\left[C_1 \cos\left(\frac{\omega}{v_s}z\right) - C_2 \sin\left(\frac{\omega}{v_s}z\right)\right] = 0 \text{ for } z = 0, \; z = l, \tag{12.35}$$

i.e.,

$$\begin{cases} C_1 = 0 \,, \\[2mm] C_1 - C_2 \sin\left(\frac{\omega}{v_s}l\right) = 0 \,. \end{cases} \tag{12.36}$$

The second equation (12.36) leads to a solution different from the trivial solution $C_2 = 0$ only if

$$\frac{\omega_i l}{v_s} = i\pi \, , \tag{12.37}$$

i.e., if

$$\omega_i = i\pi \frac{v_s}{l} = \frac{i\pi}{l} \sqrt{\frac{E}{\rho}} \, , \qquad i = 0, 1, 2, 3, \dots \, . \tag{12.38}$$

Equation (12.38) yields the natural frequencies of the longitudinal vibrations of the bar. The first natural frequency is 0, which corresponds to a rigid-body longitudinal motion. This was to be expected, because no axial constraint was stated. Constant C_2 is not determined, which is obvious because it is the amplitude of the mode of free vibration. Some of the mode shapes

$$q_i(z) = q_{i_0} \cos(i\pi\zeta) \, , \tag{12.39}$$

where

$$\zeta = \frac{z}{l} \, ,$$

are plotted in Fig. 12.3. The number of *nodes* (points in which the amplitude of motion vanishes) of each mode is equal to the order of the mode. The cases regarding other boundary conditions can be dealt with in a similar way. A general expression for the natural frequencies is

$$\omega_i = \frac{\beta_{a_i}}{l} \sqrt{\frac{E}{\rho}} \, , \tag{12.40}$$

where constants β_{a_i} depend on the boundary conditions

$$
\begin{aligned}
&\beta_{a_i} = i\pi, &&(i = 0, 1, 2, \dots), &&\text{free--free;} \\
&\beta_{a_i} = i\pi, &&(i = 1, 2, 3, \dots), &&\text{constrained--constrained;} \\
&\beta_{a_i} = (i + \tfrac{1}{2})\pi, &&(i = 0, 1, 2, \dots), &&\text{constrained--free.}
\end{aligned}
\tag{12.41}
$$

The mode shapes for the last two cases are, respectively,

$$q_i(z) = q_{i_0} \sin(i\pi\zeta) \, , \qquad q_i(z) = q_{i_0} \cos\left[\pi \left(i + \frac{1}{2} \right) \zeta \right] \, . \tag{12.42}$$

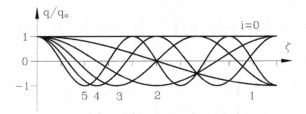

FIGURE 12.3. The first five normal modes of a prismatic homogeneous bar free at both ends. The plot must not confuse the reader: The displacements are in the axial direction.

Example 12.1 *Consider a prismatic homogeneous bar free at both ends and at rest. At time $t = 0$ a concentrated force $F_0(t)$ is applied at the right end. Compute the equations of motion in modal form.*
If the eigenfunctions (12.39) are normalized in such a way that their maximum amplitude is equal to one, all constants q_i have a unit value. The modal masses and stiffnesses are then

$$\overline{M}_i = \rho A \int_0^l [\cos(i\pi\zeta)]^2 \, dz = \begin{cases} \rho A l & \text{for } i = 0, \\ \tfrac{1}{2}\rho A l & \text{for } i = 1, 2, \dots, \end{cases}$$

$$\overline{K}_i = EA \int_0^l \left\{ \frac{d}{dz} [\cos(i\pi\zeta)] \right\}^2 \, dz = \begin{cases} 0 & \text{for } i = 0, \\ \frac{i^2\pi^2}{2l} EA & \text{for } i = 1, 2, \dots. \end{cases}$$

Note that the mass for the mode with $i = 0$, i.e., for the rigid-body motion in z-direction, is equal to the actual mass of the bar while the stiffness related to the same mode is zero because the bar is unrestrained.
The modal forces are

$$\overline{f}_i(t) = F_0(t)\cos(i\pi) = (-1)^i F_0(t).$$

All modes are then excited; the equations of motion are

$$\rho A l \ddot{\eta}_i = F_0(t)$$

for the first mode ($i = 0$) and

$$\frac{\rho A l}{2} \ddot{\eta}_i + i^2 \pi^2 \frac{EA}{2l} = (-1)^i F_0(t)$$

for all other modes.
Once force $F_0(t)$ has been stated, it is easy to obtain the modal coordinates as functions of time and superimpose the solutions to obtain the solution $u(z, t)$. The number of modes that need to be considered depends on function $F_0(t)$. If instead of a concentrated force at one end of the bar, a force $f(z, t)$ is applied that is constant along the bar, the obvious result is that the bar moves along the z-axis without any axial vibration because all points are subject to the same acceleration. In fact, if

$$f(z, t) = \frac{F_0(t)}{l} = \text{constant}$$

along the beam, the modal forces are

$$\overline{f}_i(t) = \int_0^l f(z, t)\cos(i\pi\zeta)dz = F_0(t)$$

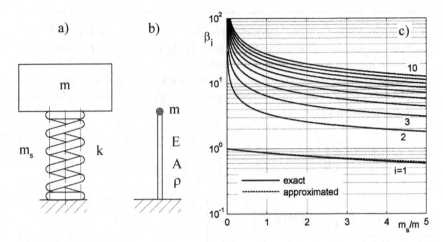

FIGURE 12.4. Sketch of a mass–spring system in which the mass of the spring has not been neglected (a) and of an equivalent system based on a bar (b). The ratio β is reported as a function of m_s/m in (c); the curve labeled as 'approximated' follows the rule of adding one-third of the mass of the spring to mass m.

---

for the first mode (i = 0) and

$$\overline{f}_i(t) = 0$$

for all other modes. Only the first mode, i.e., the rigid-body mode, is excited.

Example 12.2 *Consider a helical spring, with stiffness k and mass m_s, with a mass m at its end (Fig. 12.4a).*
Compute the natural frequencies of the system for different values of ratio m_s/m. If the mass of the spring is negligible (i.e., if $m_s/m \to 0$), the system reduces to a spring–mass system and it has only one natural frequency:

$$\omega_n = \sqrt{\frac{k}{m}} \ .$$

To account for the mass of the spring, the latter can be considered as a prismatic bar (Figure 12.4b), with mass

$$\rho A l = m_s$$

and axial stiffness

$$\frac{EA}{l} = k.$$

The two relationships above can be used to compute the density ρ, the area A, and the Young's modulus E of the bar equivalent to the spring. Note that one of the above characteristics can be stated arbitrarily. The axial displacement of the bar is

$$u_z = \eta(t)q(z) = \eta(t)\left[C_1 \sin\left(\frac{\omega}{v_s}z\right) + C_2 \cos\left(\frac{\omega}{v_s}z\right)\right].$$

The boundary condition for $z = 0$ is just

$$q(0) = 0 ,$$

yielding

$$C_2 = 0 .$$

At the other end $(z = l)$ it is possible to write a dynamic equilibrium equation stating that the force the beam exerts on the mass

$$EA(\epsilon)_{z=l} = EA\left(\frac{du_z}{dz}\right)_{z=l} = C_1 EA\frac{\omega}{v_s}\eta(t)\cos\left(\frac{\omega}{v_s}l\right)$$

is equal to the inertia force exerted by the mass

$$m\left(\ddot{u}_z\right)_{z=l} = m\ddot{\eta}(t)q(l) = C_1 m\omega^2\eta(t)\sin\left(\frac{\omega}{v_s}z\right).$$

This relationship yields

$$C_1 m\omega^2\eta(t)\sin\left(\frac{\omega}{v_s}l\right) = C_1 EA\frac{\omega}{v_s}\eta(t)\cos\left(\frac{\omega}{v_s}l\right),$$

which is satisfied for any value of C_1 and can be reduced to an equation

$$\frac{\omega l}{v_s}\tan\left(\frac{\omega l}{v_s}\right) = \frac{EAl}{mv_s^2} = \frac{m_s}{m},$$

allowing to compute the natural frequency ω of the system.
If β is the solution of the equation

$$x\tan(x) = \frac{m_s}{m},$$

the natural frequency is

$$\omega = \beta\frac{v_s}{l} = \frac{\beta}{l}\sqrt{\frac{E}{\rho}} ,$$

i.e.,

$$\omega = \beta\sqrt{\frac{k}{m}} .$$

The solution β of the equation is thus a factor for which the natural frequency of the mass–(massless) spring system must be multiplied to account for the mass of the spring.
If $m_s/m \to 0$, i.e., the mass of the spring $m_s = \rho Al$ is much smaller than mass m, also ratio

$$x = \frac{\omega}{v_s} l = \omega l \sqrt{\frac{\rho}{E}} = \omega l \sqrt{\frac{\rho A}{EA}}$$

is small and $\tan(x) \approx x$. The equation yielding the natural frequency is then

$$x^2 = \frac{m_s}{m} ,$$

i.e.,

$$\omega^2 l^2 \frac{\rho A}{EA} = \frac{\rho A l}{m} ,$$

which yields

$$\omega = \sqrt{\frac{EA}{ml}} = \sqrt{\frac{k}{m}} .$$

The other natural frequencies tend to the values seen for a clamped–clamped bar.

The values of β_i for the first 10 natural frequencies are plotted as functions of ratio m_s/m in Fig. 12.4c. If ratio m_s/m is small, the results are similar to that of the mass–spring system: The presence of the mass of the spring slightly lowers the natural frequency and the other natural frequencies, due to the vibrations of the spring, are much higher and are usually neglected.

For $m_s/m = \infty$, i.e., when the mass m is negligible if compared with the mass of the spring, the same values already obtained for a clamped-free bar are found.

For low values of m_s/m a first-approximation solution for the first natural frequency is easily obtained. Assuming that the displacement x of the mass $\rho A dl$ of the generic length dl of the spring located at a distance z from the fixed end is linked to the displacement x_m of mass m by the linear relationship

$$x = x_m \frac{z}{l} ,$$

the kinetic energy of the spring is

$$T_s = \frac{1}{2} \int_0^l \rho A \dot{x}_m^2 \left(\frac{z}{l}\right)^2 dl = \frac{1}{6} m_s \dot{x}_m^2 .$$

The total kinetic energy of the spring–mass system, that neglecting the mass of the spring

$$T_m = \frac{1}{2} m_m \dot{x}_m^2 ,$$

becomes

$$T = T_s + T_m = \frac{1}{2}\left(m_m + \frac{m_s}{3}\right)\dot{x}_m^2 .$$

The natural frequency of the spring–mass system is thus

$$\omega_n = \sqrt{\frac{k}{m_m + \frac{m_s}{3}}} .$$

This result yields the empirical rule of adding one-third of the mass of the spring to the mass m in the computation of the natural frequency. As shown in Fig. 12.4c it yields fairly accurate results even when the mass of the spring is several times the mass supported by it.

12.3   Torsional vibrations of straight beams

Consider a beam of the same type studied in the preceding section. Under the assumptions seen in Section 4.2.1, the torsional behavior of the beam is uncoupled from the axial and flexural behavior and can be studied separately. Using the elementary theory of torsion, the problem is one dimensional, because all parameters are functions only of the axial coordinate z. The dynamic equilibrium equation of a length dz of the beam is

$$\rho I_z \ddot{\phi}_z dz = \frac{\partial M_z}{\partial z} dz + m_z(z,t)dz \,, \tag{12.43}$$

where $m_z(z,t)$ is an arbitrary distribution of torsional moment acting on the beam.

Since the relationship linking the torsional generalized displacement (i.e., the rotation) with the torsional moment is

$$M_z = GI'_p \frac{\partial \phi_z}{\partial z} \,, \tag{12.44}$$

Equation (12.43) becomes

$$\rho I_z \frac{d^2 \phi_z}{dt} = \frac{\partial}{\partial z} \left[GI'_p \frac{\partial \phi_z}{\partial z} \right] + m_z(z,t) \,. \tag{12.45}$$

If the cross-section of the beam is circular or annular, the torsional moment of inertia I'_p coincides with the polar moment of inertia I_z while in other cases the two quantities are different; the values for I'_p for the cases of greater practical interest can be found in the literature. If the cross-section is elliptical with axes a and b, for instance, the following formula can be used:

$$I'_p = \frac{a^3 b^3}{5.1 \left(a^2 + b^2 \right)} \,, \tag{12.46}$$

while in the case of a rectangular cross-section with sides a and b ($a \leq b$)

$$I'_p = \frac{a^3 b}{\Psi} \,, \tag{12.47}$$

where Ψ is reported as a function of ratio a/b in Table 12.2.

Equation (12.45) is identical to Eq. (12.3), provided that

u_z	is substituted by	ϕ_z
ρA	is substituted by	ρI_z
EA	is substituted by	GI'_p
$f_z(z,t)$	is substituted by	$m_z(z,t)$,

the only difference being that the equations reported for prismatic homogeneous bars can be used only if the beam has a circular or annular

TABLE 12.2. Values of coefficient Ψ for the computation of I'_p of rectangular cross-sections.

b/a	1	1.5	2	3	4	6	10	∞
Ψ	7.14	5.10	4.37	3.80	3.56	3.34	3.19	3.00

cross-section, because the area (here the moments of inertia I_z and I'_p) was canceled from both the mass and the stiffness.

Since the equation of motion for the torsional behavior of beams is identical to that already seen for their axial behavior, all the considerations seen above for the latter hold also for the present case. The torsional behavior of beams will not be dealt with any further.

12.4   Flexural vibrations of straight beams: The Euler–Bernoulli beam

With the assumptions in Section 12.1, the flexural behavior in each lateral plane can be studied separately from the other degrees of freedom. The study of the flexural behavior is more complicated than that of the axial or torsional behavior, because in bending two of the degrees of freedom of each cross-section are involved. If bending occurs in the xz-plane, the relevant degrees of freedom are displacement u_x and rotation ϕ_y.

The simplest approach is that often defined as Euler–Bernoulli beam, based on the added assumptions that both shear deformation and rotational inertia of the cross-sections are negligible compared with bending deformation and translational inertia, respectively. These assumptions lead to a good approximation if the beam is very slender, i.e., if the thickness in the x-direction is much smaller than length l.

Remark 12.10 *The thickness in the x-direction must be, at any rate, small enough to use beam theory.*

12.4.1   Equations of motion

The equilibrium equation for translations in the x-direction of the length dz of the beam (Fig. 12.5) is readily obtained by equating the inertia force

$$\rho A \frac{d^2 u_x}{dt^2} dz$$

to the sum of the forces exerted by the other parts of the beam

$$\frac{\partial F_x}{\partial z} dz$$

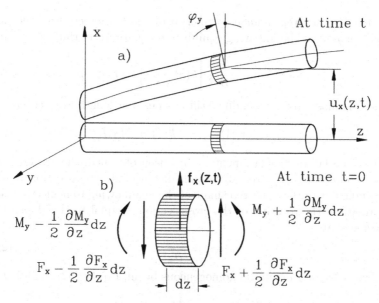

FIGURE 12.5. Flexural behavior of a straight beam in the xz-plane: (a) sketch of the system; (b) forces and moments acting on the length dz of the beam.

and the external force $f_x(z,t)dz$:

$$\rho A \frac{d^2 u_x}{dt^2} = \frac{\partial F_x}{\partial z} + f_x(z,t) \,. \tag{12.48}$$

If the rotational inertia of the length dz of the beam is neglected, and no distributed bending moment acts on the beam, the equilibrium equation for rotations about the y-axis of the length dz of the beam is obtained by equating the moment due to the shear forces acting at a distance dz to the sum of the moments exerted by the other parts of the beam

$$F_x dz + \frac{\partial M_y}{\partial z} dz = 0 \,. \tag{12.49}$$

By introducing Eq. (12.49) into Eq. (12.48), it follows

$$\rho A \frac{d^2 u_x}{dt^2} = -\frac{\partial^2 M_y}{\partial z^2} + f_x(z,t) \,. \tag{12.50}$$

The bending moment is proportional to the curvature of the inflected shape of the beam; neglecting shear deformation and using elementary beam theory, the latter coincides with the second derivative of the displacement u_x

$$M_y = EI_y \frac{\partial^2 u_x}{\partial z^2} \,, \tag{12.51}$$

where I_y is the area moment of inertia of the cross-section of the beam about its y-axis. The following equilibrium equation can thus be obtained:

$$m(z)\frac{d^2 u_x}{dt^2} + \frac{\partial^2}{\partial z^2}\left[k(z)\frac{\partial^2 u_z}{\partial z^2}\right] = f_x(z,t), \qquad (12.52)$$

where the mass and the bending stiffness per unit length are, respectively,

$$m(z) = \rho(z)A(z), \qquad k(z) = E(z)I_y(z). \qquad (12.53)$$

Once the lateral displacement u_x has been obtained, the second generalized coordinate ϕ_y is readily obtained: Since the cross-section remains perpendicular to the deflected shape of the beam owing to neglecting shear deformation, the rotation of the cross-section is equal to the slope of the inflected shape

$$\phi_y = \frac{\partial u_z}{\partial z}. \qquad (12.54)$$

In the case of a prismatic homogeneous beam, Eq. (12.52) reduces to

$$\rho A\frac{d^2 u_x}{dt^2} + EI_y\frac{\partial^4 u_x}{\partial z^4} = f_x(z,t). \qquad (12.55)$$

12.4.2   Free vibrations

Operating in the same way as seen for axial vibration, a steady-state solution of the homogeneous equation can be obtained as the product of a function of time by a function of the space coordinate

$$u_x(z,t) = q(z)\eta(t). \qquad (12.56)$$

Again function $\eta(t)$ can be shown to be harmonic

$$\eta(t) = \sin(\omega t + \phi). \qquad (12.57)$$

The principal function $q(z)$ can be obtained by introducing Eqs. (12.56) and (12.57) into Eq. (12.55)

$$-\omega^2 m(z)q(z) = \frac{d^2}{dz}\left[k(z)\frac{d^2 q(z)}{dz^2}\right], \qquad (12.58)$$

or, in the case of a prismatic homogeneous beam,

$$-\omega^2 q(z) = \frac{EI_y}{\rho A}\frac{d^4 q(z)}{dz^4}. \qquad (12.59)$$

The general solution of Eq. (12.59) is

$$q(z) = C_1 \sin(az) + C_2 \cos(az) + C_3 \sinh(az) + C_4 \cosh(az), \qquad (12.60)$$

where

$$a = \sqrt{\omega} \sqrt[4]{\frac{\rho A}{EI_y}}.$$ (12.61)

Constants C_i can be computed from the boundary conditions. In the present case four boundary conditions must be stated, which is consistent both with the order of the differential equation and with the number of degrees of freedom involved. Each end of the beam may be free, clamped, simply supported, or, a condition seldom accounted for, constrained in such a way to restrain rotations but not displacements.

At a free end displacement and rotations are free, but both the bending moment and the shear force must vanish. This can be expressed by the relationships

$$\frac{d^2q}{dz^2} = 0, \qquad \frac{d^3q}{dz^3} = 0.$$ (12.62)

If on the contrary an end is clamped, both the displacement and the rotation vanish

$$q = 0, \qquad \frac{dq}{dz} = 0.$$ (12.63)

A supported end is free to rotate, and hence the bending moment must vanish, but the displacement is constrained

$$q = 0, \qquad \frac{d^2q}{dz^2} = 0.$$ (12.64)

A further condition is the case where the end is free to move, and hence the shear force vanishes, but the rotation is constrained

$$\frac{dq}{dz} = 0, \qquad \frac{d^3q}{dz^3} = 0.$$ (12.65)

The boundary conditions can be written in the form

$$\mathbf{AC} = \mathbf{0},$$ (12.66)

where $\mathbf{C}$ is a vector containing the four unknown coefficients C_i of Eq. (12.60) and $\mathbf{A}$ is a square matrix with size 4×4, containing functions $\sin(az)$, $\cos(az)$, etc., computed at the two ends. Equation (12.66) is an homogeneous algebraic equation; for solutions different from the trivial solution $\mathbf{C} = \mathbf{0}$ to exist, the determinant of matrix $\mathbf{A}$, that contains constant a and hence the frequency ω, must vanish. The characteristic equation allowing to compute the natural frequencies is thus

$$\det(\mathbf{A}) = 0.$$ (12.67)

Example 12.3 *Compute the natural frequencies and the mode shapes of a beam supported at both ends.*
The conditions stating that the displacements vanish at both ends yield

$$q(0) = C_2 + C_4 = 0,$$

$$q(l) = C_1 \sin(al) + C_2 \cos(al) + C_3 \sinh(al) + C_4 \cosh(al) = 0.$$

The second derivative of function $q(z)$ is

$$\frac{d^2 q}{dz^2} = -a^2 C_1 \sin(az) - a^2 C_2 \cos(az) + a^2 C_3 \sinh(az) + a^2 C_4 \cosh(az),$$

and then the conditions stating that the bending moment vanishes at both ends yield

$$\frac{1}{a^2}\left(\frac{d^2 q}{dz^2}\right)_{z=0} = -C_2 + C_4 = 0,$$

$$\frac{1}{a^2}\left(\frac{d^2 q}{dz^2}\right)_{z=l} = -C_1 \sin(al) - C_2 \cos(al) + C_3 \sinh(al) + C_4 \cosh(al) = 0.$$

The conditions at the left end ($z = 0$) yield the set of equations

$$\begin{cases} C_2 + C_4 = 0 \\ -C_2 + C_4 = 0, \end{cases}$$

i.e.,

$$C_2 = 0, \qquad C_4 = 0.$$

The conditions at the other end ($z = l$) yield

$$\begin{cases} C_1 \sin(al) + C_3 \sinh(al) = 0 \\ -C_1 \sin(al) + C_3 \sinh(al) = 0, \end{cases}$$

i.e.,

$$\begin{bmatrix} \sin(al) & \sinh(al) \\ -\sin(al) & \sinh(al) \end{bmatrix} \begin{Bmatrix} C_1 \\ C_3 \end{Bmatrix} = \begin{Bmatrix} 0 \\ 0 \end{Bmatrix}.$$

To obtain a solution other than the trivial solution $C_1 = 0$, $C_3 = 0$, the determinant of the matrix of the coefficients of the set of linear equations in C_1 and C_3 must vanish

$$2 \sin(al) \sinh(al) = 0.$$

This equation can be easily solved in al

$$al = i\pi \qquad (i = 1, 2, 3, ...),$$

obtaining an infinity of natural frequencies

$$\omega = \frac{i^2 \pi^2}{l^2} \sqrt{\frac{E I_y}{\rho A}} \,.$$

Constant C_3 can be shown to be equal to 0, while constant C_1 is arbitrary and depends on the initial conditions. The eigenfunctions, normalized to yield a unit maximum amplitude, are

$$q(z) = \sin\left(i\pi \frac{z}{l}\right) \,.$$

Example 12.4 *Repeat the computation for a beam clamped at both ends. The conditions stating that the displacements vanish at both ends are the same as seen for the previous example:*

$$q(0) = C_2 + C_4 = 0 \,,$$

$$q(l) = C_1 \sin(al) + C_2 \cos(al) + C_3 \sinh(al) + C_4 \cosh(al) = 0 \,.$$

The second condition states that the rotation, i.e., the derivative of the displacement

$$\frac{dq}{dz} = aC_1 \cos(az) - aC_2 \sin(az) + aC_3 \cosh(az) + aC_4 \sinh(az) \,,$$

vanish at both ends, yielding

$$\frac{1}{a}\left(\frac{dq}{dz}\right)_{z=0} = C_1 + C_3 = 0 \,,$$

$$\frac{1}{a}\left(\frac{dq}{dz}\right)_{z=l} = C_1 \cos(al) - C_2 \sin(al) + C_3 \cosh(al) + C_4 \sinh(al) = 0 \,.$$

Matrix $\mathbf{A}$ is thus

$$\mathbf{A} = \begin{bmatrix} 0 & 1 & 0 & 1 \\ \sin(al) & \cos(al) & \sinh(al) & \cosh(al) \\ 1 & 0 & 1 & 0 \\ \cos(al) & -\sin(al) & \cosh(al) & \sinh(al) \end{bmatrix} \,,$$

and its determinant vanishes when

$$1 - \cos(al)\cosh(al) = 0 \,.$$

This equation cannot be solved in closed form, but its numerical solution is straightforward. The first four solutions, accurate to four decimal places, are

$$\beta_{f_i} = a_i l = 4.7301, \quad 7.8532, \quad 10.9956, \quad 14.1372 \, .$$

The natural frequencies are thus

$$\omega_i = \frac{\beta_{f_i}^2}{l^2} \sqrt{\frac{EI_y}{\rho A}} \, .$$

Since the hyperbolic cosine of a number greater than 1 is a large number (for instance, $\cosh(10) = 1.101 \times 10^4$), the characteristic equation is solved when $\cos(al) \approx 0$, i.e., al is an odd multiple of $\pi/2$. Discarding the value $\pi/2$, too small for being accurate enough, it follows

$$\beta_{f_i} = a_i l = \pi \left(i + \frac{1}{2} \right) ,$$

that yields

$$\beta_{f_i} = a_i l = 4.7124, \quad 7.8540, \quad 10.9956, \quad 14.1372 \, .$$

Starting from $i = 3$, the values so obtained are exact to four decimal places.

The procedure seen in the examples can be applied also for other boundary conditions. Usually the characteristic equation, which involves trigonometric and hyperbolic functions, cannot be solved in closed form, but there is no difficulty in obtaining numerical solutions.

A general expression for the natural frequencies with any boundary condition is

$$\omega_i = \frac{\beta_{f_i}^2}{l^2} \sqrt{\frac{EI_y}{\rho A}} \, , \tag{12.68}$$

where the values of constants $\beta_{f_i} = a_i l$ depend on the boundary conditions (Table 12.3).

The eigenfunctions, expressed with reference to the nondimensional coordinate

$$\zeta = \frac{z}{l} \tag{12.69}$$

TABLE 12.3. Values of constants $\beta_{f_i} = a_i l$ for the various modes with different boundary conditions.

Boundary condition	$i = 0$	$i = 1$	$i = 2$	$i = 3$	$i = 4$	$i > 4$
Free–free	0	4.730	7.853	10.996	14.137	$\approx (i + 1/2)\pi$
Supported–free	0	1.25π	2.25π	3.25π	4.25π	$(i + 1/4)\pi$
Clamped–free	–	1.875	4.694	7.855	10.996	$\approx (i - 1/2)\pi$
Supported–supported	–	π	2π	3π	4π	$i\pi$
Supported–clamped	–	3.926	7.069	10.210	13.352	$\approx (i + 1/4)\pi$
Clamped–clamped	–	4.730	7.853	10.996	14.137	$\approx (i + 1/2)\pi$

and normalized in such a way that the maximum value of the displacement is equal to unity, are, for different boundary conditions, as follows:

1. *Free–free.* Rigid-body modes:

$$q_0^I(\zeta) = 1 , \qquad q_0^{II}(\zeta) = 1 - 2\zeta .$$

Other modes:

$$q_i(\zeta) = \frac{1}{2N} \{\sin(\beta_i\zeta) + \sinh(\beta_i\zeta) + N [\cos(\beta_i\zeta) + \cosh(\beta_i\zeta)]\} ,$$

where

$$N = \frac{\sin(\beta_i) - \sinh(\beta_i)}{-\cos(\beta_i) + \cosh(\beta_i)} .$$

2. *Supported–free.* Rigid-body mode:

$$q_0(\zeta) = \zeta .$$

Other modes:

$$q_i(\zeta) = \frac{1}{2\sin(\beta_i)} \left[\sin(\beta_i\zeta) + \frac{\sin(\beta_i)}{\sinh(\beta_i)} \sinh(\beta_i\zeta) \right] .$$

3. *Clamped–free.*

$$q_i(\zeta) = \frac{1}{N_2} \{\sin(\beta_i\zeta) - \sinh(\beta_i\zeta) - N_1 [\cos(\beta_i\zeta) - \cosh(\beta_i\zeta)]\} ,$$

where

$$N_1 = \frac{\sin(\beta_i) + \sinh(\beta_i)}{\cos(\beta_i) + \cosh(\beta_i)} ,$$

$$N_2 = \sin(\beta_i) - \sinh(\beta_i) - N_1 [\cos(\beta_i) - \cosh(\beta_i)] .$$

4. *Supported–supported.*

$$q_i(\zeta) = \sin(i\pi\zeta) .$$

5. *Supported–clamped.*

$$q_i(\zeta) = \frac{1}{N} \left[\sin(\beta_i\zeta) - \frac{\sin(\beta_i)}{\sinh(\beta_i)} \sinh(\beta_i\zeta) \right] ,$$

where N is the maximum value of the expression within brackets and must be computed numerically.

6. *Clamped–clamped.*

$$q_i(\zeta) = \frac{1}{N_2} \left\{ \sin(\beta_i\zeta) - \sinh(\beta_i\zeta) - N_1 \left[\cos(\beta_i\zeta) - \cosh(\beta_i\zeta)\right] \right\},$$

where

$$N_1 = \frac{\sin(\beta_i) - \sinh(\beta_i)}{\cos(\beta_i) - \cosh(\beta_i)},$$

and N_2 is the maximum value of the expression between braces and must be computed numerically. The first four mode shapes (plus the rigid-body modes where they do exist) for each boundary condition are plotted in Fig. 12.6.

Other boundary conditions can be used also in the case of beams, as shown in the following examples.

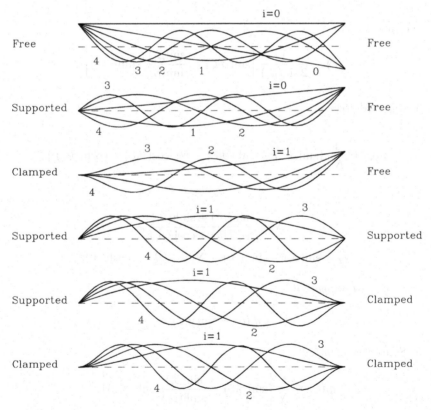

FIGURE 12.6. Normal modes of a straight beam with different end conditions. The first four modes plus the rigid-body modes, where they exist, are shown.

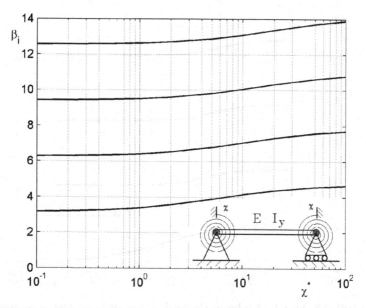

FIGURE 12.7. Sketch of a beam supported at both ends with two spring restraining rotations and values of β_1 to be inserted into Eq. (12.68) to compute the natural frequencies as functions of the ratio between the stiffness of the springs and that of the beam.

Example 12.5 *Consider a clamped–clamped beam, but assume that the constraints at the ends are not perfect and in particular are unable to constrain completely the rotational degrees of freedom. This situation can be modeled as a supported beam with a spring at each end reacting to rotation with a stiffness χ (Fig. 12.7).*
The boundary conditions state that the displacements at the end vanish, while the bending moment is proportional to the rotation through the stiffness

$$q = 0, \qquad EI_y \frac{d^2 q}{dz^2} = \pm\chi \frac{dq}{dz}.$$

The double sign accounts for the different situation at the two ends: the upper sign holds at the left end while the lower one at the right end.
The first condition yields again

$$C_2 + C_4 = 0,$$

$$C_1 \sin(al) + C_2 \cos(al) + C_3 \sinh(al) + C_4 \cosh(al) = 0,$$

while the second one yields

$$a(-C_2 + C_4) = \frac{\chi}{EI_y}(C_1 + C_3),$$

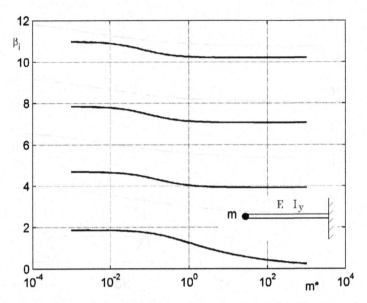

FIGURE 12.8. Sketch of a beam clamped at one end and with a concentrated mass m at the other. Values of β_i to be inserted into Eq. (12.68) to compute the natural frequencies as functions of the ratio between the added mass and the mass of the beam.

$$a\left[-C_1 \sin\left(al\right) - C_2 \cos\left(al\right) + C_3 \sinh\left(al\right) + C_4 \cosh\left(al\right)\right] =$$

$$= -\frac{\chi}{EI_y}\left[C_1 \cos\left(al\right) - C_2 \sin\left(al\right) + C_3 \cosh\left(al\right) + C_4 \sinh\left(al\right)\right].$$

The four boundary conditions can be written as a set of linear equations in C_i

$$\begin{bmatrix} 0 & 1 & 0 & 1 \\ s & c & S & C \\ \chi^* & al & \chi^* & -al \\ -als + \chi^* c & -alc - \chi^* s & alS + \chi^* C & alC + \chi^* S \end{bmatrix} \begin{Bmatrix} C_1 \\ C_2 \\ C_3 \\ C_4 \end{Bmatrix} = 0,$$

where $s = \sin\left(al\right)$, $c = \cos\left(al\right)$, $S = \sinh\left(al\right)$, $C = \cosh\left(al\right)$, and the nondimensional stiffness

$$\chi^* = \frac{\chi l}{EI_y}$$

is the ratio between the stiffness of the spring and that of the beam.
The characteristic equation expressing the condition that the determinant of the matrix of the coefficients of the set of linear equations in C_i vanishes can be solved numerically in al, obtaining the results shown in Fig. 12.7.
When $\chi^* \to 0$ the solution for a beam supported at both ends is obtained, while for $\chi^* \to \infty$ the solution is that for a beam clamped at both ends.

Example 12.6 *Consider a beam clamped at the right end and with a mass m at the left end (Fig. 12.8).*

The boundary conditions state that the displacement and rotation at the right end vanishes, like in the case of the clamped–clamped beam:

$$q(l) = C_1 \sin(al) + C_2 \cos(al) + C_3 \sinh(al) + C_4 \cosh(al) = 0,$$

$$\left(\frac{dq}{dz}\right)_{z=l} = a\left[C_1 \cos(al) - C_2 \sin(al) + C_3 \cosh(al) + C_4 \sinh(al)\right] = 0.$$

At the left end the bending moment vanishes:

$$EI_y \left(\frac{d^2q}{dz^2}\right)_{z=0} = 0,$$

i.e.,

$$-C_2 + C_4 = 0.$$

The shear force

$$EI_y \left(\frac{d^3q}{dz^3}\right)_{z=0} = EI_y a^3 \left(-C_1 + C_3\right)$$

must balance the inertia force

$$-m\omega^2 (q)_{z=0} - m\omega^2 (C_2 + C_4) = 0.$$

The relevant equation is thus

$$EI_y a^3 C_1 + m\omega^2 C_2 - EI_y a^3 C_3 + m\omega^2 C_4 = 0.$$

By introducing the nondimensional mass, i.e., the ratio between the mass m and the mass of the beam

$$m^* = \frac{m}{\rho A l},$$

and remembering that

$$a = \sqrt{\omega} \sqrt[4]{\frac{\rho A}{EI_y}},$$

this equation can be written as

$$-C_1 + m^* a l C_2 + C_3 + m^* a l C_4 = 0.$$

The matrix of the coefficients A is thus

$$\mathbf{A} = \begin{bmatrix} \sin(al) & \cos(al) & \sinh(al) & \cosh(al) \\ \cos(al) & -\sin(al) & \cosh(al) & \sinh(al) \\ 0 & -1 & 0 & 1 \\ 1 & m^* a l & -1 & m^* a l \end{bmatrix}.$$

The characteristic equation is thus

$$\det(\mathbf{A}) = 1 + \cos(al)\cosh(al) +$$

$$-m^* a l \left[\sin(al)\cosh(al) - \cos(al)\sinh(al)\right] = 0.$$

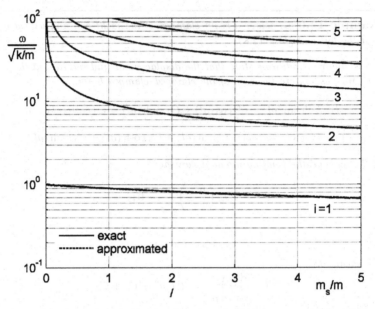

FIGURE 12.9. Natural frequencies of a mass–leaf spring system in which the mass of the spring has not been neglected, as a function of m_s/m. The *curve* labeled as 'approximated' (fully superimposed to the one labeled 'exact') follows the rule of adding 23% of the mass of the spring to mass m.

The characteristic equation expressing the condition that the determinant of matrix A vanishes can be solved numerically in al, obtaining the results shown in Fig. 12.8.
When $m^* \to 0$ the solution for a beam supported at one end is obtained, while for $m^* \to \infty$ the solution is that for a beam clamped at one end and supported at the other one.

Example 12.7 *Consider a leaf spring made of a prismatic and homogeneous beam clamped at one end and supporting the mass m at the other end. The system is very similar to the one studied in Example 12.6 (Fig. 12.8), but now the stress is laid on the action of the beam as a spring. Compute the natural frequencies of the system at changing values of the ratio between the mass of the spring m_s and the mass m.*
The stiffness of the spring is

$$k = \frac{3EI_y}{l^3}.$$

The mass ratio is

$$\frac{m_s}{m} = \frac{\rho Al}{m}.$$

> *The natural frequencies of the system are*
>
> $$\omega_i = \frac{\beta_i^2}{l^2}\sqrt{\frac{EI_y}{\rho A}} = \frac{\beta_i^2}{3}\sqrt{\frac{m}{m_s}}\sqrt{\frac{k}{m}}\,,$$
>
> *where coefficients β_i were computed in Eq. (12.6).*
> *The factors*
>
> $$\frac{\beta_i^2}{3}\sqrt{\frac{m}{m_s}}$$
>
> *are plotted as functions of the mass ratio m_s/m in Fig. 12.9, obtaining a plot not very different from that obtained for helical springs in Fig. 12.4c.*
> *Also here there is a simple rule for computing the first natural frequency: adding the 23% of the mass of the spring to mass m. As shown in the figure, this rule allows to compute the first natural frequency with good precision.*

Eigenfunctions also have orthogonality properties in the case of bending. While the property of orthogonality with respect to the mass has the same expression as in the case of axial vibration, the expression of orthogonality with respect to the stiffness is different:

$$\int_0^l m(z)q_i(z)q_j(z)dz = 0\,, \qquad \int_0^l k(z)\frac{d^2q_i(z)}{dz^2}\frac{d^2q_j(z)}{dz^2}dz = 0\,, \quad (12.70)$$

for $i \neq j$.

As a consequence, the definitions of modal masses and stiffnesses are

$$\overline{M}_i = \int_0^l m(z)\left[q_i(z)\right]^2 dz\,, \qquad \overline{K}_i = \int_0^l k(z)\left[\frac{d^2q_i(z)}{dz^2}\right]^2 dz\,. \quad (12.71)$$

If function $m(z)$ is constant, the eigenfunctions are orthogonal.

12.4.3 Forced response

The expressions of the modal forces $\overline{f}_i(t)$ are similar to those seen for the axial behavior

$$\overline{f}_i = \int_0^l q_i(z)f_x(z,t)dz\,. \quad (12.72)$$

In the case of excitation due to motion $x_A(t)$ of the supporting structure, the response of the beam can be computed by resorting to a reference frame fixed to the constraints and applying the modal forces

$$\overline{f}_i = -r_i\ddot{x}_A\,, \quad (12.73)$$

where

$$r_i = \int_0^l q_i(z)m(z)dz\,.$$

Example 12.8 *Consider the prismatic homogeneous beam shown in Fig. 12.10a. It is supported at one end; at the other end it is supported by an actuator that can supply both the static force to balance the self-weight and dynamic forces to control vibration. The control system will be studied using modal analysis, taking into account the first three modes, one of which is a rigid-body motion.*

Using the equations in Section 12.4 and the values reported in Table 12.3, the eigenvectors, normalized to yield unit values for the modal masses, the modal masses, and stiffnesses are as follows:
Mode 0 (rigid-body mode):

$$q_0(\zeta) = \zeta\sqrt{\frac{3}{m}}\,, \qquad \overline{M} = 1\,, \qquad \overline{K} = 0.$$

ith mode (any i):

$$q_i(\zeta) = \sqrt{\frac{2}{m}}\left[\sin(\beta_i\zeta) + \frac{\sin(\beta_i)}{\sinh(\beta_i)}\sinh(\beta_i\zeta)\right]\,,$$

$$\overline{M} = 1\,, \qquad \overline{K} = \beta_i^4\frac{EI_y}{ml^3}\,,$$

where

$$m = \rho Al$$

is the total mass of the beam, $\zeta = z/l$ is a nondimensional coordinate, and coefficients β_i are

$$\beta_i = (i + 1/4)\pi\,, \quad i.e., \quad \sin(\beta_i) = -1^i/\sqrt{2}\,.$$

The modal forces due to self-weight are

$$\overline{f}_0 = -g\frac{\sqrt{3m}}{2}\,, \qquad \overline{f}_i = 0 \ \forall i \neq 0\,.$$

The modal control forces are

$$\overline{f}_{c_0} = f_c\sqrt{\frac{3}{m}}\,, \qquad \overline{f}_{c_i} = 2F_c\frac{(-1^i)}{\sqrt{m}} \ \forall i \neq 0\,.$$

Comparing the expressions of the modal forces, it follows immediately that the static component of the control force is

$$\overline{f}_{c_{st}} = \frac{mg}{2}\,.$$

Note that all modal coordinates have a static value even if the modal forces due to self-weight vanish: In the equilibrium condition the static control force makes the modal coordinate of mode zero vanish while the others yield the static deflected shape

$$\eta_{i_{st}} = (-1)^i \frac{gl^3 \sqrt{m^3}}{(i+1/4)^4 \pi^4 EI_y} .$$

In the following an ideal PD feedback controller will be used, in conjunction with a feedforward compensation of the weight of the beam. Due to the linearity of the problem, the dynamic study will be performed neglecting the static forces and deflections; this is possible only within the linear range, particularly if no saturation occurs in the actuator.

By introducing the constant

$$\omega_1 = \pi^2 \sqrt{\frac{EI_y}{ml^3}} ,$$

which is nothing other than the first natural frequency of the same beam simply supported at both ends, matrices $\mathcal{A}$ and $\mathcal{B}_c$ reduce to

$$\mathcal{A} = \begin{bmatrix} \mathbf{0} & -\omega_1^2 \begin{bmatrix} 0 & 0 & 0 & \dots \\ 0 & 1,25^4 & 0 & \dots \\ 0 & 0 & 2,25^4 & \dots \\ \dots & \dots & \dots & \dots \end{bmatrix} \\ \mathbf{I} & \mathbf{0} \end{bmatrix} ,$$

$$\mathcal{B}_c = \frac{1}{\sqrt{m}} \left\{ \left\{ \begin{array}{c} \sqrt{3} \\ -2 \\ 2 \\ \dots \end{array} \right\} \\ \mathbf{0} \end{array} \right\} .$$

Consider as a first case an output-feedback control system with a single sensor at the free end of the beam yielding both displacement and velocity. The output vector y and matrix C are then

$$y = \left\{ \begin{array}{c} \dot{x}(l) \\ x(l) \end{array} \right\} ,$$

$$C = \frac{1}{\sqrt{m}} \begin{bmatrix} \sqrt{3} & -2 & 2 & \dots & 0 & 0 & 0 & \dots \\ 0 & 0 & 0 & \dots & \sqrt{3} & -2 & 2 & \dots \end{bmatrix} .$$

The control gain matrix K_y has one row, because it must supply the control input to a single actuator, and two columns, because there are two outputs of the system:

$$K_y = \begin{bmatrix} k_v & k_d \end{bmatrix} ,$$

where k_d is the proportionality constant between the force of the actuator and the displacement (dimensionally it is a stiffness) and k_v is the proportionality constant between the force and the velocity (dimensionally it is a damping coefficient). By introducing the nondimensional gains

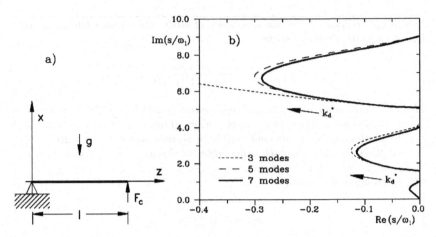

FIGURE 12.10. *Beam supported at one end and actively controlled at the other:* (a) *sketch of the system;* (b) *root loci.*

$$k_v^* = \frac{k_v}{\omega_1 m}, \qquad k_d^* = \frac{k_d}{\omega_1^2 m} = k_d \frac{l^3}{EI_y \pi^4},$$

the closed-loop dynamic matrix is

$$\mathcal{A}_{cl} = \begin{bmatrix} \omega_1 k_v^* \mathbf{R} & -\omega_1^2 \mathbf{S} \\ \mathbf{I} & \mathbf{0} \end{bmatrix},$$

where

$$\mathbf{R} = \begin{bmatrix} 3 & -2\sqrt{3} & 2\sqrt{3} & \cdots \\ -2\sqrt{3} & 4 & -4 & \cdots \\ 2\sqrt{3} & -4 & 4 & \cdots \\ \cdots & \cdots & \cdots & \cdots \end{bmatrix},$$

$$\mathbf{S} = \begin{bmatrix} 3k_d^* & -2\sqrt{3}k_d^* & 2\sqrt{3}k_d^* & \cdots \\ -2\sqrt{3}k_d^* & 1,25^4 + 4k_d^* & -4k_d^* & \cdots \\ 2\sqrt{3}k_d^* & -4k_d^* & 2,25^4 + 4k_d^* & \cdots \\ \cdots & \cdots & \cdots & \cdots \end{bmatrix}.$$

The nondimensional root loci for the first three modes with various values of the gain k_d^* are plotted in Fig. 12.10b. To state a relationship between the two elements of the gain matrix, the gain k_v^* has been assumed to comply with the relationship

$$k_v^* = 0.1\sqrt{k_d^*/3},$$

which corresponds to assuming a damping ratio $\zeta = 0.1$ for the controlled first mode if mode coupling is neglected.

Note that when the gain tends to zero the natural frequencies of the supported–free beam are obtained. If the gain tends to infinity, the actuator reduces to a rigid support and the natural frequencies of the supported–supported beam are obtained. In these two limiting cases the system is undamped, because no structural damping was taken into account.

Intermediate values of the frequency are obtained for finite values of the gain. The roots are complex and the negative real part evidences the damping role of the control system.

Another consideration regards spillover. The figure has been drawn by taking into account three, five, and seven modes. The response at the first mode is correct in all three cases; that at the second mode requires at least five modes to be computed correctly, but the errors due to spillover with three modes are not large. For the response at the third mode seven modes are required. Using five modes the results are still acceptable, but with three modes the errors due to spillover lead to unacceptable results.

Consider now a state feedback, obtained by resorting to a single displacement sensor at the free end and adding an observer to the system. Take into account only the first three modes.

Matrices A, $\mathcal{B}_c$, *and* $\mathcal{C}$ *are*

$$
A = \left[\begin{array}{cc} \mathbf{0} & -\omega_1^2 \left[\begin{array}{ccc} 0 & 0 & 0 \\ 0 & 1.25^4 & 0 \\ 0 & 0 & 2.25^4 \end{array} \right] \\ \mathbf{I} & \mathbf{0} \end{array} \right] ,
$$

$$
\mathcal{B}_c = \frac{1}{\sqrt{m}} \left\{ \left\{ \begin{array}{c} \sqrt{3} \\ -2 \\ 2 \\ 0 \end{array} \right\} \right\} ,
$$

$$
\mathcal{C} = \frac{1}{\sqrt{m}} \left[\begin{array}{cccccc} 0 & 0 & 0 & \sqrt{3} & -2 & 2 \end{array} \right] .
$$

By assuming a unit value for ω_1, *the controllability and observability matrices and their determinants are*

$$
\mathbf{H} = \frac{1}{\sqrt{m}} \left[\begin{array}{cccccc} \sqrt{3} & 0 & 0 & 0 & 0 & 0 \\ -2 & 0 & 4.883 & 0 & -11.92 & 0 \\ 2 & 0 & -51.26 & 0 & 1,314 & 0 \\ 0 & \sqrt{3} & 0 & 0 & 0 & 0 \\ 0 & -2 & 0 & 4.883 & 0 & -11.92 \\ 0 & 2 & 0 & -51.26 & 0 & 1,314 \end{array} \right] ,
$$

$$
\mathbf{O} = \frac{1}{\sqrt{m}} \left[\begin{array}{cccccc} 0 & \sqrt{3} & 0 & 0 & 0 & 0 \\ 0 & -2 & 0 & 4.883 & 0 & -11.92 \\ 0 & 2 & 0 & -51.26 & 0 & 1,314 \\ \sqrt{3} & 0 & 0 & 0 & 0 & 0 \\ -2 & 0 & 4.883 & 0 & -11.92 & 0 \\ 2 & 0 & -51.26 & 0 & 1314 & 0 \end{array} \right] ,
$$

$$
det(\mathbf{H}) = -1.01 \times 10^8 , \qquad det(\mathbf{O}) = -1.01 \times 10^8 .
$$

Assume that the six nondimensional eigenvalues of the controlled system are

$$
-0.2 \pm 0.6i , \quad -0.2 \pm 2i , \quad -0.5 \pm 5i ,
$$

and those of the observer are

$$-0.3 \pm 0.6i , \quad -0.5 \pm 2i , \quad -0.9 \pm 5i .$$

By using the pole-placement procedure, the following gain matrices for the controller and observer are obtained

$$\mathbf{K}_c = \omega_1 \sqrt{m} \left[\begin{array}{cccccc} 0.452 & -0.149 & 0.460 & 0.381\omega_1 & -0.784\omega_1 & 0.341\omega_1 \end{array} \right] ,$$

$$\mathbf{K}_0 = \omega_1 \sqrt{m} \left[\begin{array}{cccccc} 0.456\omega_1 & -1.011\omega_1 & 1.555\omega_1 & 0.746 & -0.268 & 0.786 \end{array} \right]^T .$$

A simple check shows that the required roots of the system are obtained. To check the errors due to spillover, the eigenvalues of the systems were again computed, taking into account 10 modes, only the first three of which are controlled using the observer and the controller defined earlier. The following 26 nondimensional eigenvalues have been obtained:

$$\begin{array}{lll}
-0.200 \pm 0.584i , & -0.303 \pm 0.621i , & -0.266 \pm 1.949i , \\
-0.598 \pm 2.071i , & -0.893 \pm 4.765i , & -0.482 \pm 5.108i , \\
0.0213 \pm 10.599i , & 0.0103 \pm 18.069i , & 0.0047 \pm 27.564i , \\
0.0024 \pm 39.063i , & 0.0013 \pm 52.563i , & 0.0008 \pm 68.063i , \\
0.0005 \pm 85.563i .
\end{array}$$

The first six are similar to the ones stated for the controller and the observer. The remaining ones are not very different from the corresponding natural frequencies of the uncontrolled system:

$$10.562 , \quad 18.063 , \quad 27.563 , \quad 39.063 , \quad 52.563 , \quad 68.063 , \quad 85.563 ,$$

but have a positive real part. Spillover causes a slight instability of the system, which, however, in the current case is due to the fact that no damping of the beam has been considered. The very small damping introduced by the beam itself should be sufficient to achieve stable behavior in all modes.

Example 12.9 *Consider the same system studied in Example 12.8. Study its response to a shock load of the type prescribed by MIL-STD 810 C, basic design (see Example 8.1) and compare the results with those obtained for a simply supported beam with the same dimensions.*
The total mass of the beam is $m = \rho Al = 3.124$ kg and the value of parameter ω_1 in the previous example is 147.74 rad/s. Assuming a unit value for the modal masses, the modal stiffnesses for the first eight modes (including the rigid-body mode) are

$$\overline{K}_0 = 0 , \qquad \overline{K}_3 = 2.435 \times 10^6 , \qquad \overline{K}_6 = 3.331 \times 10^7 ,$$
$$\overline{K}_1 = 53,290 , \qquad \overline{K}_4 = 7.121 \times 10^6 , \qquad \overline{K}_7 = 6.030 \times 10^7 .$$
$$\overline{K}_2 = 559,400 , \qquad \overline{K}_5 = 1.658 \times 10^7 ,$$

Because the uncontrolled system is very lightly damped, a modal damping approximation is accepted. The modal damping for the various modes can be computed as $C_i = \eta \omega_i$:

$$\overline{C}_0 = 0 , \qquad \overline{C}_3 = 15.605 , \qquad \overline{C}_6 = 57.710 ,$$
$$\overline{C}_1 = 2.308 , \qquad \overline{C}_4 = 26.683 , \qquad \overline{C}_7 = 77.655 .$$
$$\overline{C}_2 = 7.479 , \qquad \overline{C}_5 = 40.720 ,$$

By assuming the same eigenfrequencies of the three controlled modes as in Example 11.4, the following gain matrices for the controller and the observer are obtained:

$$\mathbf{K}_c = \left[\begin{array}{cccccc} 118.1 & -38.85 & 120.0 & 14,705 & -30,239 & 13,152 \end{array} \right] ,$$

$$\mathbf{K}_0 = \left[\begin{array}{cccccc} 17,571 & -38,985 & 59,983 & 195.86 & -69.88 & 205.28 \end{array} \right]^T .$$

As seen in Example 12.8, to check the errors due to spillover the eigenvalues of the systems were computed again, taking into account 10 modes. This computation is now performed taking into account damping in order to verify whether the damping of the system can counteract the destabilizing effects of spillover. The following 26 eigenvalues, expressed in rad/s, have been obtained:

$$-29.73 \pm 86.09i , \qquad -44.60 \pm 92.20i , \qquad -43.95 \pm 290.45i ,$$
$$-84.68 \pm 302.97i , \qquad -129.03 \pm 700.17i , \qquad -77.97 \pm 758.20i ,$$
$$-4.59 \pm 1,565.7i , \qquad -11.81 \pm 2,669.2i , \qquad -19.67 \pm 4,072.2i ,$$
$$-28.51 \pm 5,771.0i , \qquad -38.63 \pm 7,765.4i , \qquad -50.13 \pm 10,055i ,$$
$$-63.13 \pm 12,641i .$$

The first six are very similar to the ones for the controller and the observer and are almost not changed (to working accuracy) by the presence of damping. The remaining ones, which are not very different from the corresponding natural frequencies of the uncontrolled system, now have a negative real part. The damping of the system counteracts the instability due to spillover, even if the very low absolute value of some decay rates (particularly for the seventh mode) suggests a behavior with a small stability margin. The very low value assumed for structural damping accounts for this type of behavior.

The shock, applied to the frame that supports the beam, is prescribed as a linearly increasing acceleration, reaching 20 g in 11 ms and then decreasing abruptly to zero. The shock load is seen from the beam as a constant distributed load whose absolute value is given by the mass multiplied by the time-varying acceleration. Only the modal force related to the first mode (mode 0, rigid-body mode) is different from zero:

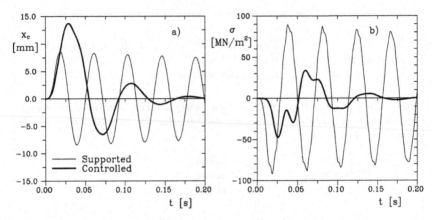

FIGURE 12.11. Time history of the displacement (**a**) and of the stress (**b**) at the center of the beam in Fig. 12.10 when a shock load of the type prescribed by MIL-STD 810 C, basic design, is applied to it; comparison with a simply supported beam.

$$\overline{F}_0 = -\frac{a\sqrt{3m}}{2} .$$

The response has been computed using Eq. (11.67), taking into account 20 state variables for the beam (10 modes) and 6 state variables for the observer, by integrating the equation in time using the Euler method. The external input u_e is the time-varying acceleration $-a$, while vector $\mathcal{B}_e$ has all elements vanishingly small except the first, which is

$$\mathcal{B}_{e_1} = \frac{\sqrt{3m}}{2} .$$

The response at the center of the beam has been reconstructed from the modal response and is plotted in Fig. 12.11a. In Fig. 12.11b the stress at the center of the beam, computed from the mode shapes and modal coordinates, has been plotted. In the same figures the values obtained applying the same shock to a similar beam simply supported at both ends are plotted for comparison.

From the figure it is clear that the active control system is very effective in damping out the free oscillations caused by the shock, while the simply supported beam gives way to a lower value of the maximum displacement. This is, however, to be expected, because the control system is assumed to be quite soft and then it allows a sort of large rigid-body motion under the effect of the shock.

The reduction of the stress peak is large, since the rigid-body motion, which absorbs most of the energy of the shock, does not contribute to the stresses. The time history of the force applied by the actuator at the free end of the beam is shown in Fig. 12.12.

This example has been reported only to show some features of structural control.

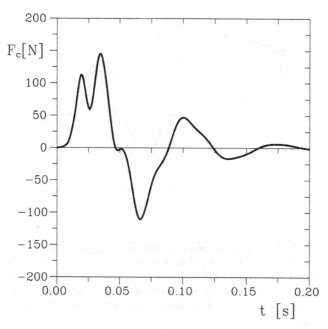

FIGURE 12.12. Time history of the control force of the beam studied in Fig. 12.11.

No attempt to optimize the control system or verify its feasibility from an energy viewpoint has been performed. A purely proportional control law with no cut-off frequency has been used: No actual control systems can follow this law; the highest controlled frequency is, however, about 120 Hz and any system with a cut-off frequency higher than this value would do the job.

Another source of nonlinearity here not accounted for is the saturation of the actuator. However, it is not difficult to include it into the model when the computation is based on the numerical integration of the equations of motion, like in this case.

Example 12.10 *Consider the same beam studied in Example 12.9. The beam is now a structural member moved by an actuator that exerts a torque at the hinged end (Fig. 12.13a). The beam is controlled by an open-loop system that governs the actuator torque following a predetermined pattern.*

Assume that the torque is controlled following either a square-wave pattern (bang–bang control) (Fig. 12.13b) or a more elaborate double-versine time history (Fig. 12.13c). Neglecting weight, compute the maximum torque needed to achieve a rotation of 45° (π/4 rad) in 0.5 s and the time history of the tip of the beam during and after the maneuver.

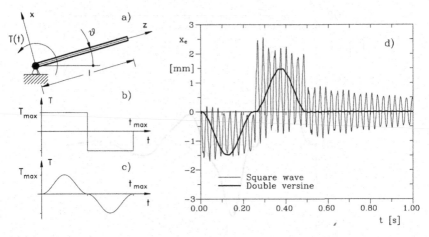

FIGURE 12.13. *Beam rotating about one of its ends under the effect of the driving torque $T(t)$:* **(a)** *sketch of the system at time t;* **(b)** *and* **(c)** *time histories of the control torque during the maneuver: square-wave and double-versine patterns;* **(d)** *time history of the displacement of the free end: displacement from the rigid-body position, computed by numerically integrating the equations of the first five modes.*

Rigid-body dynamics

The equation of motion of the beam as a rigid body is simply

$$\ddot{\theta} = \frac{T(t)}{J} ,$$

where

$$J = \frac{1}{3}ml^2$$

is the moment of inertia of the beam about the axis of rotation.
In the case of the bang–bang control the time history of the torque is

$$T = T_{max} \qquad\qquad for \ 0 \leq t \leq \frac{t_{max}}{2} ,$$

$$T = -T_{max} \qquad\qquad for \ \frac{t_{max}}{2} < t \leq t_{max} .$$

The square-wave law allows for the minimum travel time for a given value of the maximum driving torque. The control law is symmetrical in time and the speed of the beam at time t_{max} reduces to zero. By integrating the equation of motion, the time history of the displacement is

$$\theta = \frac{3T_{max}}{2ml^2}t^2 \qquad\qquad for \ 0 \leq t \leq \frac{t_{max}}{2},$$

$$\theta = \frac{3T_{max}}{4ml^2}\left(-t_{max}^2 + 4t_{max}t - 2t^2\right) \qquad for \ \frac{t_{max}}{2} \leq t \leq t_{max}.$$

The relationship linking the torque with the displacement θ_{max} and the time needed for the rotation is

$$T_{max} = \frac{4ml^2\theta_{max}}{3t_{max}^2} .$$

In the case of the double-versine control, the time history of the torque is

$$T = \frac{T_{max}}{2}\left[1 - \cos\left(\frac{4\pi}{t_{max}}t\right)\right] \quad \text{for } 0 \le t \le \frac{t_{max}}{2},$$

$$T = -\frac{T_{max}}{2}\left[1 - \cos\left(\frac{4\pi}{t_{max}}t\right)\right] \quad \text{for } \frac{t_{max}}{2} \le t \le t_{max}.$$

Also in this case the control law is symmetrical in time and the speed of the beam reduces to zero at time t_{max}. By integrating the equation of motion, the time history of the displacement is

$$\theta = \frac{3T_{max}}{4ml^2}\left\{t^2 + \frac{t_{max}^2}{8\pi^2}\left[\cos\left(\frac{4\pi}{t_{max}}t\right) - 1\right]\right\},$$

for $0 \le t \le t_{max}/2$ and

$$\theta = \frac{3T_{max}}{8ml^2}\left\{-t_{max}^2 + 4t_{max}t - 2t^2 - \frac{t_{max}^2}{4\pi^2}\left[\cos\left(\frac{4\pi}{t_{max}}t\right) - 1\right]\right\},$$

for $t_{max}/2 < t \le t_{max}$, respectively.
The relationship linking the torque with the displacement θ_{max} and the time needed for the rotation is

$$T_{max} = \frac{8ml^2\theta_{max}}{3t_{max}^2} .$$

The values of the maximum torque are then $T_{max} = 13.09$ Nm for the square wave and $T_{max} = 26.17$ Nm for the double versine.
Dynamic behavior of the beam
The dynamic behavior of the beam will be studied in the xz-reference frame of Fig. 12.13a. Such a frame is not inertial and, consequently, inertia forces due to its motion must be introduced into the equations.
An approach similar to that seen in Chapter 17 for discrete systems, but much more simplified, will be followed here. The position in an inertial (nonrotating) reference frame $x'z'$ of a general point of the beam, located at coordinate z, is

$$\left\{ \begin{array}{c} x' \\ z' \end{array} \right\} = \left\{ \begin{array}{c} z\sin(\theta) + u_x\cos(\theta) \\ z\cos(\theta) - u_x\sin(\theta) \end{array} \right\},$$

where $u_x(t)$ is the displacement due to the bending of the beam. By differentiating the position, the kinetic energy of a length dz of beam is readily obtained:

$$dT = \frac{1}{2}\rho A dz(\dot{\theta}^2 z^2 + \dot{u}_x^2 + u_x^2\dot{\theta}^2 + 2z\dot{\theta}\dot{u}_x).$$

By assuming that the law $\theta(t)$ is stated, performing the relevant derivatives with respect to u_z and $\dot{u}_z$, the inertial terms entering into the equation of motion for the bending behavior of the beam are

$$\rho A dz(\ddot{u}_x + z\ddot{\theta}^2 + u_x\dot{\theta}^2).$$

The first term is the usual term that appears in the equation of motion of the stationary beam. The second term is due to the acceleration of the beam and can easily be accounted for as an external excitation, because it does not contain the deformation of the beam. The third term is a 'centrifugal stiffening' effect, because of the centrifugal force due to rotation.

In the small movement studied, the acceleration is high but the angular velocity maintains a value low enough to neglect the last effect.

Moreover, the aim of the study is mainly to predict the free behavior of the beam after the required position is reached and the two neglected effects stop acting. Clearly they are not ininfluent, because the behavior after the beam has stopped depends on what happens during the motion, but the assumption that their effect is small can be accepted, at least in a first-approximation study.

The beam can be studied as a beam at standstill, under the effect of the moment $T(t)$ and the inertia force $-\rho A z\ddot{\theta}(t)$. The study will be performed using modal analysis. The mode shapes, modal masses, and stiffnesses have been computed in Example 12.8. The modal forces due to the inertia forces are

$$\overline{f}_{i_0} = -l\ddot{\theta}\sqrt{m/3} \; ; \quad \overline{f}_{i_i} = 0 \quad \forall i \neq 0.$$

The modal forces due to the driving torque $T(t)$ are

$$\overline{f}_{T_i} = T(t)\frac{dq_i(z)}{dz} \, ,$$

i.e.,

$$\overline{f}_{T_0} = \frac{T(t)}{l}\sqrt{\frac{3}{m}} \, , \qquad \overline{f}_{T_i} = \frac{T(t)}{l}\sqrt{\frac{2}{m}}\beta_i\left[1 + \frac{\sin(\beta_i)}{\sinh(\beta_i)}\right] \, .$$

Remembering that

$$\ddot{\theta} = \frac{3T}{ml^2} \, ,$$

the equation of motion for the rigid-body mode is simply

$$\ddot{\eta}_0 = \frac{T(t)}{l}\sqrt{\frac{3}{m}} - l\sqrt{\frac{3}{m}}\ddot{\theta} = 0 \, ,$$

i.e., because at time $t = 0$ both $\eta_0 = 0$ and $\dot{\eta}_0 = 0$, the first modal coordinate is always equal to zero. This result is obvious, because the driving torque has been computed by imposing the rigid-body motion of the beam. The equations of motion of the other modes are

$$\ddot{\eta}_i + \overline{C}_i\dot{\eta}_i + \overline{K}_i\eta_i = \frac{T(t)}{l}\sqrt{\frac{2}{m}}\beta_i\left[1 + \frac{\sin(\beta_i)}{\sinh(\beta_i)}\right] \, .$$

The displacement at the free end of the beam, with respect to the position of the rigid-body motion is simply

$$x_e = \sum_{i=1}^{\infty} \eta_i q_i(t) = \frac{2}{\sqrt{m}} \sum_{i=1}^{\infty} (-1)^i \eta_i \,.$$

The modal equations of motion can be solved by numerical integration or using Duhamel's integral. A solution based on the numerical integration of the first five modes is shown in Fig. 12.13d. From the figure, it is clear that the double-versine control succeeds in positioning the beam without causing long-lasting vibrations as in the case of the bang–bang control pattern. The latter strongly excites the first mode, which is little damped.

These results are linked with the particular application, because a control input of the versine type can also excite some modes; however, the fact that the square-wave control input is more prone to exciting vibrations than more smooth control laws is a general feature. In many cases, particular control laws that are much better than the square-wave and the versine laws are used, and much theoretical and experimental work has been devoted to identifying optimal control laws.

In a practical case, if positioning accuracy is required, a closed-loop control must be associated with the open-loop control shown here. If the response of the control device is fast enough, deformation modes can also be controlled.

12.5 Bending in the yz-plane

The situation in the yz-plane is clearly very similar to that already studied with reference to the xz-plane. The physical problem is the same; however, there are some differences in the equations due to the fact that in the xz-plane the positive direction for rotations is from z-axis to x-axis while in the yz-plane a rotation is positive if it goes from y-axis to z-axis (Fig. 12.14).

While in the xz-plane the rotation and the displacement in any point of an Euler–Bernoulli beam are linked by the relationship

$$\phi_y = \frac{du_x}{dz} \,,$$

the corresponding relationship for yz-plane is

$$\phi_x = -\frac{du_y}{dz} \,.$$

Remark 12.11 *The difficulties of notation arising from the differences existing in the two planes could be easily circumvented by assuming $-\phi_x$ instead of ϕ_x as the generalized coordinate for rotation. If the corresponding generalized force is taken as $-M_x$ instead of M_x, the situation in the two planes is exactly the same. This is, however, not free from drawbacks, as it will be shown, and is generally not done.*

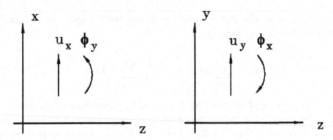

FIGURE 12.14. Positive directions for displacements and rotations in xz- and yz-planes.

12.6   Coupling between flexural and torsional vibrations of straight beams

The uncoupling between flexural and torsional behavior in straight beams was possible as a consequence of a number of assumptions, introduced to simplify the dynamic study of beams. If the cross-section does not possess two planes of symmetry perpendicular to each other, the above-mentioned uncoupling does not occur any more.

Consider, as an example, a straight, prismatic, and homogeneous beam, whose cross-section has only one plane of symmetry. Locate the reference frame in such a way that axes y and z are contained in the symmetry plane and study the flexural behavior in xz-plane (Fig. 12.15). Let z-axis pass through the shear centers of the cross-sections, i.e., the points having the property that when a bending force is applied to it, no torsional deformation results.

By definition, the elastic reaction F_y of the beam loaded in bending passes through the shear center C.

Consider the length dz of beam as a rigid body and write the dynamic equilibrium equations for motions involving the torsional degree of freedom and those linked to bending in xz-plane. Due to the very small length dz, the mass moments of inertia of such rigid body about its center of gravity can be approximated as

$$J_y = \rho I_y dz, \qquad J_z = \rho I_z dz, \qquad J_{xz} = 0 \;, \tag{12.74}$$

the latter relationship being due to symmetry.

The velocity of the center of mass and the angular velocity are

$$\mathbf{V} = \left\{ \begin{array}{c} \dot{u}_x - d\dot{\phi}_z \\ 0 \\ 0 \end{array} \right\} , \quad \mathbf{\Omega} = \left\{ \begin{array}{c} 0 \\ \dot{\phi}_y \\ \dot{\phi}_z \end{array} \right\} . \tag{12.75}$$

The kinetic energy of the rigid body is thus

$$\mathcal{T} = \frac{1}{2}\rho dz \left[A \left(\dot{u}_x^2 + d^2\dot{\phi}_z^2 - 2d\dot{u}_x\dot{\phi}_z \right) + I_y\dot{\phi}_y^2 + I_z\dot{\phi}_z^2 \right] . \tag{12.76}$$

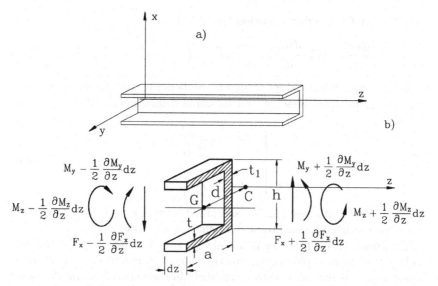

FIGURE 12.15. Coupled flexural–torsional behavior of a beam. (**a**) Sketch of a channel beam. (**b**) Forces and moments acting on the length dz of the beam with reference to the shear center C.

The generalized forces acting on the body with references to the generalized coordinates u_x, ϕ_y, and ϕ_z are

$$\begin{cases} Q_x = \dfrac{\partial F_x}{\partial z}dz, \\[2ex] Q_{\phi_y} = \dfrac{\partial M_y}{\partial z}dz + F_x dz, \\[2ex] Q_{\phi_z} = \dfrac{\partial M_z}{\partial z}dz. \end{cases} \qquad (12.77)$$

By introducing the kinetic energy and the generalized forces into the Lagrange equations, the following equations of motion are obtained

$$\begin{cases} \rho A \left(\dfrac{d^2 u_x}{dt^2} - d\dfrac{d^2 \phi_z}{dt^2} \right) = \dfrac{\partial F_x}{\partial z}, \\[2ex] \rho I_y \dfrac{d^2 \phi_y}{dt^2} = \dfrac{\partial M_y}{\partial z} + F_x, \\[2ex] \rho \left(Ad^2 + I_z \right) \dfrac{d^2 \phi_z}{dt^2} - \rho Ad\dfrac{d^2 u_x}{dt^2} = \dfrac{\partial M_z}{\partial z}. \end{cases} \qquad (12.78)$$

If shear deformation is neglected, the rotation ϕ_y of the cross-section is equal to the derivative of the displacement du_x/dz. By solving the second equation in F_x, the first equation transforms into

$$\rho A \frac{d^2 u_x}{dt^2} - \rho A d \frac{d^2 \phi_z}{dt^2} - \frac{\partial}{\partial z}\left[\rho I_y \frac{\partial}{\partial z}\left(\frac{d^2 u_x}{dt^2}\right)\right] = \frac{\partial^2 M_y}{\partial z^2}. \tag{12.79}$$

If the usual relationships between the bending and torsional moments and the generalized bending and torsional displacements (Eqs. 12.51 and 12.44)

$$\begin{cases} M_y = EI_y \dfrac{\partial^2 u_x}{\partial z^2} \\[4mm] M_z = GI'_p \dfrac{\partial \phi_z}{\partial z} \end{cases} \tag{12.80}$$

are introduced into Eqs. (12.78) and (12.79), two coupled differential equations for the bending and torsional behavior are obtained.

Equation (12.44) however is not always suited to describe the torsional behavior of beams, particularly in the case of thin-walled sections. If the cross-section is not allowed to warp freely in each station along the beam, some bending is always present together with torsion. In the case of Fig. 12.15, torsion is accompanied by bending of the flanges in the yz-plane. Timoshenko[1] showed that this behavior can be taken into account by substituting Eq. (12.44) by the following relationship

$$M_z = GI'_p \frac{\partial \phi_z}{\partial z} - R \frac{\partial^3 \phi_z}{\partial z^3}, \tag{12.81}$$

where R is the warping rigidity (as opposed to the torsional rigidity GI'_p) of the beam. In the case of Fig. 12.15 its value is

$$R = \frac{Dh^2}{2} = \frac{Eta^3 h^2}{24}, \tag{12.82}$$

or, taking into account also the effect of the web,

$$R = \frac{Dh^2}{2}\left(1 + \frac{t_1 h^3}{4 I_z}\right) = \frac{Eta^3 h^2}{24}\left(1 + \frac{t_1 h^3}{4 I_z}\right), \tag{12.83}$$

where D is the flexural rigidity of each flange.

By using Eq. (12.81) instead of Eq. (12.44), assuming that the beam is prismatic and homogeneous and neglecting the rotational inertia of the cross-section for rotations about y-axis, the equations of motion become

$$\begin{cases} \rho A \dfrac{d^2 u_x}{dt^2} - \rho A d \dfrac{d^2 \phi_z}{dt^2} = -EI_y \dfrac{\partial^4 u_x}{\partial z^4}, \\[4mm] \rho\left(A d^2 + I_z\right) \dfrac{d^2 \phi_z}{dt^2} - \rho A d \dfrac{d^2 u_x}{dt^2} = GI'_p \dfrac{\partial^2 \phi_z}{\partial^2 z} - R \dfrac{\partial^4 \phi_z}{\partial z^4}. \end{cases} \tag{12.84}$$

---

[1]S.H. Timoshenko, *Strength of Materials*, Van Nostrand, New York, 1978.

If a solution of the type

$$u_x = q_x(z)e^{i\omega t}, \qquad \phi_z = q_z(z)e^{i\omega t} \tag{12.85}$$

is assumed, the following equations for the coupled free vibrations of the system are obtained:

$$\begin{cases} -EI_y \dfrac{d^4 q_x}{dz^4} = \omega^2 \rho A \left(q_x - d q_z\right), \\[4mm] -GI_p' \dfrac{d^2 q_z}{d^2 z} - R\dfrac{d^4 q_z}{dz^4} = \omega^2 \rho \left[\left(Ad^2 + I_z\right) q_z - Adq_x \right]. \end{cases} \tag{12.86}$$

Only the case of a beam with simply supported ends will be studied in detail. Using as boundary conditions

$$\begin{cases} q_x = 0 \quad \dfrac{d^2 q_x}{dz^2} = 0 \quad \text{for } z = 0 \text{ and } z = l, \\[4mm] q_z = 0 \quad \dfrac{d^2 q_z}{dz^2} = 0 \quad \text{for } z = 0 \text{ and } z = l, \end{cases} \tag{12.87}$$

the eigenfunctions are

$$q_x = q_{x_{i0}} \sin\left(\frac{i\pi z}{l}\right), \qquad q_z = q_{z_{i0}} \sin\left(\frac{i\pi z}{l}\right). \tag{12.88}$$

Remark 12.12 *While the boundary conditions on displacement u_x are quite obvious, the second condition on angle ϕ_y comes from assuming that each flange behaves as a beam simply supported in yz-plane.*

Introducing the solution (12.88) into Eq. (12.86) they yield

$$\begin{bmatrix} \omega_{fi}^2 - \omega_i^2 & d\omega_i^2 \\[3mm] \dfrac{dA}{d^2 A + I_z}\omega_i^2 & \omega_{ti}^2 - \omega_i^2 \end{bmatrix} \left\{ \begin{array}{c} q_{x_{i0}} \\[2mm] q_{z_{i0}} \end{array} \right\} = \mathbf{0}, \tag{12.89}$$

where ω_{fi} and ω_{ti} are, respectively, the ith flexural and torsional uncoupled natural frequencies, i.e., the solutions of Eq. (12.89) when $d = 0$ (the shear center coincides with the center of mass)

$$\begin{cases} \omega_{fi} = \dfrac{i^2 \pi^2}{l^2} \sqrt{\dfrac{EI_y}{\rho A}}, \\[6mm] \omega_{ti} = \dfrac{i\pi}{l^2} \sqrt{\dfrac{i^2 \pi^2 R + GI_p' l^2}{\rho I_z}}. \end{cases} \tag{12.90}$$

The former coincides with the bending natural frequency of an Euler–Bernoulli beam supported at both ends, while the second one is different from that obtained for the torsional vibration of bars owing to the warping rigidity. If the warping rigidity is neglected, it coincides with that previously computed.

If no uncoupling is possible $(d \neq 0)$, the characteristic equation of the eigenproblem is

$$\left(\omega_{fi}^2 - \omega_i^2\right)\left(\omega_{ti}^2 - \omega_i^2\right) - \alpha\omega_i^4 = 0, \tag{12.91}$$

where

$$\alpha = \frac{d^2 A}{d^2 A + I_z} . \tag{12.92}$$

The coupled natural frequencies can thus be easily computed

$$\omega_i = \frac{\left(\omega_{fi}^2 + \omega_{ti}^2\right) \pm \sqrt{\left(\omega_{fi}^2 + \omega_{ti}^2\right)^2 - 4\left(1 - \alpha\right)\omega_{fi}^2\omega_{ti}^2}}{2\left(1 - \alpha\right)} . \tag{12.93}$$

If the cross-section has no plane of symmetry, a more complicated problem arises, since vibrations in xz- and yz-planes are both coupled with torsional vibration. All degrees of freedom except axial translation u_z must thus be studied together. Even more complex cases are found in engineering practice, particularly when dealing with twisted non-prismatic beams, as turbine, fan, or propeller blades.

12.7   The prismatic homogeneous Timoshenko beam

Both the rotational inertia of the cross-section and shear deformation were not taken into account in the Euler–Bernoulli beam model. In this section this assumption will be dropped, with reference only to a prismatic homogeneous beam.

A beam in which these effects are not neglected is usually referred to as a *Timoshenko beam*.

Shear deformation can be accounted for as a deviation of the direction of the deflected shape of the beam not accompanied by a rotation of the cross-section (Fig. 12.16). The latter is thus no more perpendicular to the deformed shape of the beam, and the rotation of the cross-section can be expressed as

$$\phi_y = \frac{\partial u_x}{\partial z} - \gamma_x . \tag{12.94}$$

The shear strain γ_x is linked to the shear force by the relationship

$$\gamma_x = \frac{\chi F_x}{GA} , \tag{12.95}$$

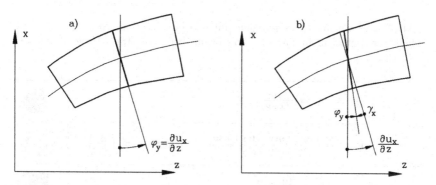

FIGURE 12.16. Effect of shear deformation on beam bending. (a) Euler–Bernoulli beam; (b) Timoshenko beam.

where the shear factor χ depends on the shape of the cross-section, even if there is not complete accordance on its value. For a circular beam, a value of 10/9 is usually assumed; for other shapes the expressions reported in Table 12.4 can be used.

Equation (12.94) can thus be written in the form

$$\phi_y = \frac{\partial u_x}{\partial z} - \frac{\chi F_x}{GA}. \tag{12.96}$$

The bending moment has no effect on shear deformation: If the latter is accounted for, the relationship linking the bending moment with the inflected shape of the beam becomes

$$M_y = EI_y \frac{\partial \phi_y}{\partial z}. \tag{12.97}$$

The rotational inertia of the cross-section is no more neglected, and the dynamic equilibrium equations for displacement and rotation of the length dz of the beam are

$$\begin{cases} \rho A \dfrac{d^2 u_x}{dt^2} = \dfrac{\partial F_x}{\partial z} + f_x(z,t), \\[2mm] \rho I_y \dfrac{d^2 \phi_y}{dt^2} = F_x + \dfrac{\partial M_y}{\partial z}. \end{cases} \tag{12.98}$$

By solving Eq. (12.96) in F_x and introducing it into Eq. (12.98), it follows

$$\begin{cases} \rho A \dfrac{d^2 u_x}{dt^2} = \dfrac{GA}{\chi} \left(\dfrac{\partial^2 u_x}{\partial z^2} - \dfrac{\partial \phi_y}{\partial z} \right), \\[3mm] \rho I_y \dfrac{d^2 \phi_y}{dt^2} = \dfrac{GA}{\chi} \left(\dfrac{\partial u_x}{\partial z} - \phi_y \right) + EI_y \dfrac{\partial^2 \phi_y}{\partial z^2}. \end{cases} \tag{12.99}$$

TABLE 12.4. Shear factors for some different cross-sections (from G.R. Cowper, The shear coefficient in Timoshenko's beam theory, *J. Appl. Mech.*, June, 1966, 335–340).

$\chi = \frac{7+6\nu}{6(1+\nu)}$	
$\chi = \frac{(7+6\nu)(1+m)^2+4m(5+3\nu)}{6(1+\nu)(1+m)^2}$ where $m = \left(\frac{d_i}{d_o}\right)^2$	
$\chi = \frac{12+11\nu}{10(1+\nu)}$	
$\chi = \frac{40+37\nu+m(16+10\nu)+\nu m^2}{12(1+\nu)(3+m)}$ where $m = \left(\frac{b}{a}\right)^2$	
$\chi = \frac{1.305+1.273\nu}{1+\nu}$	
$\chi = \frac{4+3\nu}{2(1+\nu)}$	
$\chi = \frac{48+39\nu}{20(1+\nu)}$	
$\chi = \frac{p+q\nu+30n^2m(1+m)+5\nu n^2 m(8+9m)}{10(1+\nu)(1+3m)^2}$ where $m = \frac{bt_1}{at_a}$, $n = \frac{b}{h}$	
$\chi = \frac{p+q\nu+10n^2\left[m(3+\nu)+3m^2\right]}{10(1+\nu)(1+3m)^2}$ where $m = \frac{bt_1}{at_a}$, $n = \frac{b}{h}$	
$\chi = \frac{p+q\nu}{10(1+\nu)(1+3m)^2}$ where $m = \frac{2A}{ht}$, $A =$ flange area	
$\chi = \frac{p'+q'\nu+30n^2m(1+m)+10\nu n^2 m\left(4+5m+m^2\right)}{10(1+\nu)(1+4m)^2}$ $m = \frac{bt_1}{ht_a}$, $n = \frac{b}{h}$	

$p = 12 + 72m + 150m^2 + 90m^3;\quad q = 11 + 66m + 135m^2 + 90m^3$
$p' = 12 + 96m + 276m^2 + 192m^3;\quad q' = 11 + 88m + 248m^2 + 216m^3$

By differentiating the second equation (12.99) with respect to z and eliminating ϕ_y, the following equation can be obtained:

$$EI_y\frac{\partial^4 u_x}{\partial z^4} - \rho I_y\left(1+\frac{E\chi}{G}\right)\frac{\partial^2}{\partial z^2}\left(\frac{\partial^2 u_x}{\partial t^2}\right) + \frac{\rho^2 I_y\chi}{G}\frac{d^4 u_x}{dt^4} + \rho A\frac{d^2 u_x}{dt^2} = 0,$$

$$(12.100)$$

and then, in the case of free oscillations with harmonic time history,

$$EI_y \frac{d^4 q(z)}{dz^4} + \rho\omega^2 I_y \left(1 + \frac{E\chi}{G}\right) \frac{d^2 q(z)}{dz^2} - \rho\omega^2 \left(A - \omega^2 \frac{\rho I_y \chi}{G}\right) q(z) = 0.$$
(12.101)

The same considerations regarding the form of the eigenfunctions (seen in Section 12.6) also hold in this case. If the beam is simply supported at both ends, the same eigenfunctions seen in the case of the Euler–Bernoulli beam still hold, and Eq. (12.101) can be expressed in nondimensional form as

$$\left(\frac{\omega}{\omega^*}\right)^4 - \left(\frac{\omega}{\omega^*}\right)^2 \frac{\alpha^2}{i^2 \pi^2 \chi^*} \left(1 + \chi^* + \frac{\alpha^2}{i^2 \pi^2}\right) + \frac{\alpha^4}{i^4 \pi^4 \chi^*} = 0, \qquad (12.102)$$

where ω_i^* is the ith natural frequency computed using the Euler–Bernoulli assumptions, the slenderness α of the beam is defined as

$$\alpha = l \sqrt{\frac{A}{I_y}} = \frac{l}{r} \qquad (12.103)$$

(r is the radius of inertia of the cross-section) and

$$\chi^* = \chi \frac{E}{G} . \qquad (12.104)$$

The results obtained from Eq. (12.102) for a beam with circular cross-section and material with $\nu = 0.3$ are reported as functions of the slenderness in Fig. 12.17.

Remark 12.13 *Both shear deformation and rotational inertia lower the value of the natural frequencies, the first being stronger than the second by a factor of about 3, as shown by Timoshenko.*[2]

Remark 12.14 *The effects of shear deformation and rotational inertia are increasingly strong with increasing order of the modes.*

Remark 12.15 *It must be remembered, at any rate, that even the so-called Timoshenko beam model is an approximation, because it is based on the usual approximations of the beam theory, and the very model of a one-dimensional solid is no more satisfactory when the slenderness is low. Because the beam behaves as if it were less and less slender with increasing order of the modes, high-order modes can be computed using the beam theory only if the slenderness is high enough.*

---

[2]S.P. Timoshenko et al., *Vibration Problems in Engineering*, Wiley, New York, 1974.

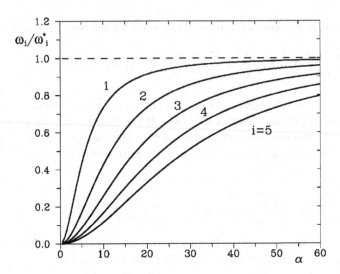

FIGURE 12.17. Effect of shear deformation on the first five natural frequencies of a simply supported beam. Ratio between the natural frequency computed taking into account rotational inertia and shear deformation and that computed using the Euler–Bernoulli model.

12.8 Interaction between axial forces and flexural vibrations of straight beams

The uncoupling mentioned in the preceding sections is based on the assumption of geometrical linearity, among others. When studying a linear problem, the equilibrium equations are written applying the loads to the body in its undeformed configuration. Consider a straight beam to which an axial force F_z, constant in time, is applied, and assume that the cross-section of the beam is such that torsional and lateral behavior and bending in the xz- and yz-planes are uncoupled. If the equilibrium equations in the xz-plane are written in the deflected position, the bending moment due to the axial force $F_z u_x(z,t)$ acting in the xz-plane must be taken into account. Similarly, in studying the behavior in the yz-plane, the moment $-F_z u_y(z,t)$ must be considered. If shear deformation and rotational inertia of the cross-sections are neglected, the dynamic equilibrium equation of a length dz of the beam for rotation about the y-axis (Eq. 12.49) becomes (Fig. 12.18)

$$F_x dz + \frac{\partial M_y}{\partial z} dz - F_z \frac{\partial u_x}{\partial z} dz = 0. \tag{12.105}$$

The equation of motion can thus be written in the form

$$\rho A \frac{d^2 u_x}{dt^2} + \frac{\partial^2}{\partial z^2} \left[E I_y \frac{\partial^2 u_x}{\partial z^2} \right] - F_z \frac{\partial^2 u_x}{\partial z^2} = f_x(z,t), \tag{12.106}$$

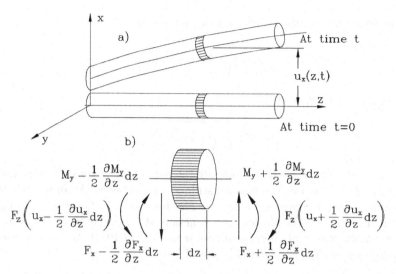

FIGURE 12.18. Effect of a constant axial force on the flexural behavior of a straight beam in the xz-plane; (a) sketch of the system; (b) forces and moments acting on the length dz of the beam.

or, for a prismatic homogeneous beam,

$$\rho A \frac{d^2 u_x}{dt^2} + EI_y \frac{\partial^4 u_x}{\partial z^4} - F_z \frac{\partial^2 u_x}{\partial z^2} = f_x(z,t). \qquad (12.107)$$

The solution of the homogeneous equation governing free oscillations is again of the type

$$u_x = \eta(t)q(z), \qquad (12.108)$$

where the time history $\eta(t)$ is harmonic. In the case of a prismatic homogeneous beam performing harmonic free oscillations with frequency ω, the equation of motion in the frequency domain is

$$-\omega^2 \rho A q(z) + EI_y \frac{d^4 q(z)}{dz^4} - F_z \frac{d^2 q(z)}{dz^2} = 0. \qquad (12.109)$$

If the beam is simply supported at both ends, the eigenfunctions are still those seen for the Euler–Bernoulli beam, and the natural frequencies are

$$\omega_i = \frac{i\pi}{l} \sqrt{\frac{EI_y}{\rho A} \left(\frac{i\pi}{l}\right)^2 + \frac{F_z}{\rho A}}. \qquad (12.110)$$

By introducing the natural frequency ω_i^*, computed without taking into account the axial force F_z,

$$\omega_i^* = \left(\frac{i\pi}{l}\right)^2 \sqrt{\frac{EI_y}{\rho A}},$$

the slenderness of the beam, and the axial stress

$$\alpha = l\sqrt{\frac{A}{I_y}} \quad , \quad \sigma_z = \frac{F_z}{A} \quad ,$$

Equation (12.106) yields

$$\frac{\omega_i}{\omega_i^*} = \sqrt{1 + \frac{\sigma_z}{E} \frac{\alpha^2}{i^2 \pi^2}} . \tag{12.111}$$

A tensile axial force then causes an increase of the natural frequency, i.e., acts as if it increases the flexural stiffness of the beam. An opposite effect is due to axial compressive forces. The first natural frequency obtained from Eq. (12.111) is plotted as a function of the axial strain $\epsilon_z = \sigma_z/E$ for various values of the slenderness α in Fig. 12.19. If the axial stress is compressive, Eq. (12.111) holds only if

$$|\sigma_z| < \frac{Ei^2\pi^2}{\alpha^2} . \tag{12.112}$$

When the axial stress reaches one of the mentioned values, the natural frequency reduces to zero; this means that no stable equilibrium condition is possible.

Remark 12.16 *When the stress is equal to the lowest of these values, buckling takes place and the expression obtained with $i = 1$ is the buckling stress of the beam.*

Remark 12.17 *An effect similar to that due to an axial force applied to the end of the beam is because of axial force fields, like the centrifugal field in propeller or turbine blades. Centrifugal field also causes a stiffening effect in discs and other parts of rotors.*

The limiting case for an infinitely slender beam under tensile axial force, i.e., a beam whose bending stiffness is negligible when compared with the stiffening effect of the axial force, is that of a vibrating taut string. If both the area of the cross-section of the string and the tensile force F_z are constant, the equation of motion (12.106) reduces to

$$\rho A \frac{d^2 u_x}{dt^2} = F_z \frac{\partial^2 u_x}{\partial z^2} . \tag{12.113}$$

The equation of motion is then identical to Eq. (12.9), which was obtained for the axial vibration of bars. The results obtained in Section 12.2.1 can be directly used by simply substituting the axial force F_z for the stiffness EA. Equation (12.113) has been obtained from the equation of motion of a beam subject to axial forces and, therefore, holds under the assumptions seen here. They are, however, not needed in the case of a vibrating

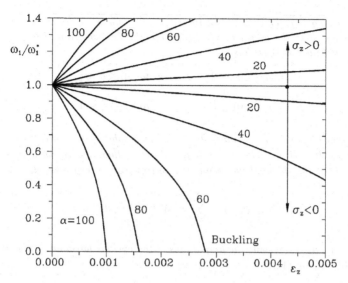

FIGURE 12.19. Effect of an axial force on the first natural frequency of a simply supported beam. Ratio between the natural frequency computed taking into account the axial force and that computed using Eq. (12.68) as a function of the axial strain for beams with different values of the slenderness α.

taut string, and the same equation holds if the lateral displacements are not small and the tensile force along the string is not constant, provided the material has a linear stress–strain characteristic.[3]

Example 12.11 *A string whose length is 500 mm has a cross-section of 3 mm^2 and is made of a material whose density is $\rho = 8,000$ kg/m³. Compute the tensile force in the string in such a way that the first natural frequency corresponds to the central A (nominal frequency 440 Hz) and the relative amplitudes of the harmonics of the vibration obtained by plucking the string at 1/3 of its length.*
The first natural frequency is

$$\omega_1 = \frac{\pi}{l}\sqrt{\frac{F_z}{\rho A}}\,.$$

By stating that it is equal to 440 Hz = 880π rad/s, it follows that $F_z = 4,646$ N. The other harmonics are simply multiples of the fundamental frequency.

---

[3]R. Buckley, *Oscillations and Waves*, Adam Hilger Ltd., Bristol, UK, 1985.

If the string is plucked at 1/3 of its length, the configuration at time $t = 0$ is a triangle, with the vertex at 1/3 of the length. The equation of the deformed shape at time $t = 0$ is thus

$$u_x(\zeta, 0) = 3\zeta \qquad \text{for } 0 \le \zeta \le \tfrac{1}{3},$$

$$u_x(\zeta, 0) = \tfrac{3}{2}(1 - \zeta) \qquad \text{for } \tfrac{1}{3} < \zeta \le 1.$$

This shape can easily be expressed as a linear combination of the mode shapes. Because all modes oscillate in phase and are at their maximum amplitude at time $t = 0$, the shape of the string at any time can be expressed as

$$u_x(z, t) = \sum_{i=1}^{\infty} a_i \sin(i\pi\zeta) \cos(\omega_i).$$

Constants a_i, expressing the amplitudes of the various harmonics, can be computed from Eq. (12.28)

$$a_i = \frac{m_t}{M_i} \int_0^1 u_x(\zeta, 0) \sin(i\pi\zeta) d\zeta =$$

$$= 6 \int_0^{1/3} \zeta \sin(i\pi\zeta) d\zeta + 3 \int_{1/3}^1 (1 - \zeta) \sin(i\pi\zeta) d\zeta.$$

The results for the first 10 natural frequencies are as follows:

Mode	ω_i [Hz]	a_i/a_1	Mode	ω_i [Hz]	a_i/a_1
1	440	1	6	2,640	0
2	880	0.250	7	3,080	0.0206
3	1,320	0	8	3,520	0.0158
4	1,760	−0.0628	9	3,960	0
5	2,200	−0.0402	10	4,400	−0.0102

Note that the third, sixth, ninth, ... harmonics are not excited. This could be predicted because the string was plucked in a spot where these harmonics have a node.

12.9   Exercises

Exercise 12.1 A transmission shaft is made of a circular steel tube and has two joints at the ends. Without considering the mass of the joints, compute the first five flexural natural frequencies of the shaft neglecting shear deformations and rotations of the cross-sections. Repeat the computations, taking into account shear deformations, and compare the results. Data: inner diameter = 80 mm, outer diameter $= 100$ mm, length $= 1.5$ m, $E = 2.1 \times 10^{11}$ N/m^2, $\nu = 0.3$, $\rho = 7,810$ kg/m^3.

Exercise 12.2 *A cylindrical chimney is* 40 m *tall, has an inner diameter of* 500 *mm, and has a wall thickness of* 15 mm. *Compute the flexural natural frequencies and mode shapes (neglecting the influence of compressive stresses due to self-weight). To perform a dynamic test, a small rocket attached to the top of the chimney and directed horizontally is fired. The thrust of the rocket instantaneously reaches the value of* 100 N, *lasts for* 5 s, *and then instantaneously drops to zero. Compute the time history of the various modes and the maximum stresses at the base of the stack. If necessary, assume that the damping is of structural type, with a loss factor* $\eta = 0.02$. *Use Duhamel's integral or numerical integration; if possible, also use an impulsive approach. Data:* $E = 2.1 \times 10^{11}$ N/m^2, $\nu = 0.3$, $\rho = 7,810$ kg/m^3.

Exercise 12.3 *Consider the chimney of Exercise 12.2. Compute the modal participation factors of the first five modes for an excitation provided by the horizontal motion of the supporting ground. Assume that the chimney is excited by a horizontal random motion of the ground, whose power spectral density is constant at* 0.003 g^2/Hz *between* 3 *and* 30 Hz, *increases at* 9 dB/oct *between* 0.1 *and* 3 Hz, *and drops at* -12 dB/oct *between* 30 *and* 100 Hz.[4] *Compute the spectrum of the response of the first five modes and the r.m.s. values of the displacement at the top and of the stresses at the base.*

Exercise 12.4 *Compute the natural frequencies of the beam sketched in Fig. 12.20. The beam, having a box structure, is supported at both ends by two cylindrical hinges whose axes lie in the horizontal plane. Data:* $l = 10$ m, $t = 10$ mm, $\rho = 7,810$ kg/m^3, $a = 500$ mm, $E = 2.1 \times 10^{11}$ N/m^2, $\nu = 0.3$.

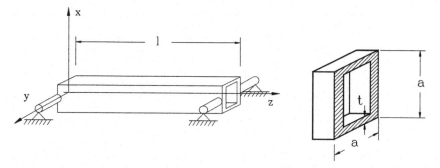

FIGURE 12.20. Exercise 4.4. Sketch of the beam.

---

[4]This spectrum is not meant to give any realistic indications on seismic design, but only as an example of computation of a random response.

Exercise 12.5 *Consider the beam studied as Exercise 12.4. A trolley carrying an eccentric rotor (of mass m_R, eccentricity ϵ) and rotating at speed Ω travels along the axis of the beam with speed V. The axis of rotation is horizontal and perpendicular to the axis of the beam and parallel to y-axis.*

Assuming that the center of mass of the rotor (neglecting the eccentricity) is on the axis of the beam, study the forced vibrations of the beam and evaluate the dynamic stresses. Neglect the inertia of the trolley and of the rotor, the rotational inertia of the beam, and shear deformations. Data: $m_R = 100$ kg, $\epsilon = 10$ mm, $V = 10$ m/s, $\Omega = 3,000$ rpm $= 314$ rad/s.

Exercise 12.6 *Repeat the computations related to the beam studied as Exercise 12.5, taking into account the structural damping of the material. Assume a value of the loss factor $\eta = 0.02$.*

Exercise 12.7 *A connecting rod has an annular cross-section and is hinged at its ends through two parallel cylindrical hinges. Compute the natural frequencies of the flexural vibrations in the plane containing the axes of the hinges and in a plane perpendicular to it. Repeat the computation when a compressive axial force of 10,000 N acts on the rod. Data: inner diameter $= 30$ mm, outer diameter $= 40$ mm, length $= 400$ mm, $E = 2.1 \times 10^{11}$ N/m^2, $\nu = 0.3$, $\rho = 7,810$ kg/m^3.*

Exercise 12.8 *Consider the taut string of Example 12.11. Compute the shift of the fundamental frequency due to a change of the tension equal to 10%. Compute the response of the string when it is plucked at 1/4 of its length (assume the original value of the tension). Assume that the point at 1/4 of the length of the string is displaced of 10 mm and released, and compute the actual amplitude of the various modes and the maximum value of the kinetic energy related to them. How is the energy distributed among the modes? Data: $l = 500$ mm, $A = 3$ mm^2, $\rho = 8,000$ kg/m^3.*

13

General Continuous Linear Systems

Apart from vibrating beams, other continuous models can be used to obtain closed-form solutions. Although in practical applications nowadays discretization techniques are more used than continuous models, a quick survey of the latter may be useful. Also the problems related to wave propagation in solids and fluids are touched.

13.1 Elastic continuums

Usually a deformable body is modeled as an elastic continuum or, in the case its behavior can be assumed to be linear, as a linear elastic continuum. It is clear that the elastic continuum is only a model, since no actual body is such at an atomic scale, but for most objects studied by structural dynamics the continuum model is more than adequate. Besides, it is to exclude (not only from a practical point of view but also for theoretical reasons) the possibility of using the discontinuous structure of matter to build a model with a very large, but finite, number of degrees of freedom.

An elastic body can be thought of as consisting of an infinity of points. To describe the undeformed (or initial) configuration of the body, a reference frame is set in space. Many problems can be studied with reference to a two-dimensional frame or, as seen in the previous chapter for beams, even a single coordinate, but there are cases in which a full tridimensional approach is required. The characteristics of the material are stated as functions of the position defined in all the portions of space (or plane or line)

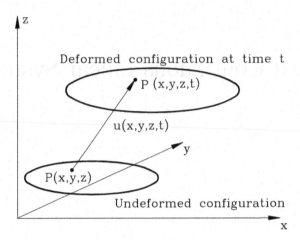

FIGURE 13.1. Deformation of an elastic continuum; reference frame and displacement vector.

that are occupied by the continuum. These functions need not, in general, be continuous.

The configuration at any time t can be obtained from the initial configuration once a vector function $\vec{u}(x, y, z, t)$ expressing the displacements of all points is known (Fig. 13.1). The displacement of a point is a vector, with a number of components equal to the number of dimensions of the reference frame. Usually the components of this vector are taken as the degrees of freedom of each point, even if in some cases a different choice is considered; the number of degrees of freedom of an elastic (or, more generally, deformable) body is thus infinite. The components of the displacement can be considered as continuous functions of space coordinates and time, and the theory of continuous functions is the natural tool for dealing with deformable continua.

Remark 13.1 *The function $\vec{u}(x, y, z, t)$ describing the displacement of the points of the body is differentiable with respect to time at least twice; the first derivative gives the displacement velocity and the second the acceleration. Usually, however, higher-order derivatives also exist.*

If displacements and rotations can be considered as small, to avoid introducing geometrical nonlinearities, the definition of stresses and strains used in elementary theory of elasticity can be stated. When the dynamics of an elastic body can be dealt with as a linear problem, as when the behavior of the material is linear and no geometrical nonlinearity is considered, an infinity of natural frequencies exist as a consequence of the infinity of degrees of freedom of the model.

Assuming that the forces acting on the body are expressed by the function $\mathbf{f}(x, y, z, t)$, the equation of motion can generally be written as

$$D[\mathbf{u}(x, y, z, t)] = \mathbf{f}(x, y, z, t), \qquad U[\mathbf{u}(x, y, z, t)]_B = 0, \qquad (13.1)$$

where the differential operator D completely describes the behavior of the body and operator U, defined on the boundary, states the boundary conditions (only homogeneous boundary conditions are described by Eq. (13.1)). The actual form of the differential operator can be obtained by resorting directly to the dynamic equilibrium equations or by writing the kinetic and potential energies and using Lagrange equations. The boundary conditions usually follow from geometrical considerations.

In the case of the axial behavior of beams studied in the previous chapter the differential operator D is defined by Eq. (12.3):

$$D(u_z) = m(z)\frac{d^2 u_z}{dt^2} - \frac{\partial}{\partial z}\left[k(z)\frac{\partial u_z}{\partial z}\right]. \qquad (13.2)$$

Since the problem is one dimensional this is one of the simplest cases and a closed-form solution could be found.

In general, the solution of Eq. (13.1) exists if an inverse operator D^{-1} can be defined,

$$\mathbf{u}(x, y, z, t) = D^{-1}[\mathbf{f}(x, y, z, t)], \qquad (13.3)$$

so that the time history of the displacements of all points of the body can be obtained from the time history of the forces applied to it.

Remark 13.2 *Equation (13.3) is just a formal statement; in most cases the relevant operator cannot be written in explicit form, particularly when the boundary conditions are not the simplest ones.*

The computational difficulties arise both from the differential equations themselves and, even more, from the boundary conditions. They can be overcome only in a few simple cases: Most problems encountered in engineering practice require dealing with complex structures and the use of continuous models describing them in detail is, consequently, ruled out.

The traditional approach was to resort to much simplified models, like the beam theory seen in the previous chapter or the plate theory seen here or to even more simplified discrete models.

For complex shapes the only feasible approach is the discretization of the continuum and then the application of the methods seen for discrete systems. The substitution of a continuous system, characterized by an infinity of degrees of freedom, with a discrete system, sometimes with a very large but finite number of degrees of freedom, is usually referred to as *discretization*. This step is of primary importance in the solution of practical problems, because the accuracy of the results depends largely on the adequacy of the discrete model to represent the actual system.

The fact that the solutions of continuous models have the same properties as the solutions of discrete models is a prerequisite for the success of discretization techniques. The eigenvectors obtained by discretization are m- and k-orthogonal exactly as the eigenfunctions of the original continuous system; any general solution of the discretized model can be expressed as a linear combination of the eigenvectors exactly like the general solution of the original system can be expressed in terms of eigenfunctions. Moreover, the modal decomposition of a continuous system is a sort of discretization that holds exactly, at least if an infinity of eigenfunctions are used.

The above-mentioned properties were demonstrated in the previous chapter for beams, but can be demonstrated, although with much greater difficulty, in general.

13.2   Flexural vibration of rectangular plates

Consider a plate that is thin enough to be modeled as a two-dimensional system: It can be considered as a two-dimensional equivalent of the beam studied in Chapter 12. In the case of plates, two basic formulations exist: In the simplest one, the so-called *Kirchoff plate*, corresponding to the Euler–Bernoulli beam, shear deformation is neglected and any line perpendicular to the plate is assumed to remain perpendicular to the deflected mid-surface during deformation. The second formulation, in which shear deformation is not neglected, is referred to as *Mindlin plate*, and is similar to the Timoshenko beam.

Even if the analysis is limited to the first formulation and the displacements are assumed to be small enough to allow a complete linearization of the problem, the analysis is far more complex than what was seen for beams. As a consequence, only a short account will be given here.[1]

A reference frame with axes x and y contained in the mid-plane of the plate is stated (Fig. 13.2a). Consider a small portion of plate, with length and width equal to dx and dy, and study the forces and moments exerted on it by the other parts of the plate though its lateral surfaces (Fig. 13.2b).

A shearing force (per unit length) F_y, a bending moment (per unit length) M_y, and a twisting moment (per unit length) M_{yx} act on the sides parallel to y-axis. In a similar way, the forces and moments acting on the sides parallel to x-axis are F_x, M_x, and M_{xy}.

While a sort of uncoupling between the 'in-plane' and the flexural behaviors can be stated for the plates as was seen for beams, and thus no in-plane forces are considered in the study of plate bending, flexural and

---

[1]For a complete description of the static behavior of plates, see S. Timoshenko, S. Woinowsky-Krieger, *Theory of Plates and Shells*, McGraw-Hill, New York, 1959.

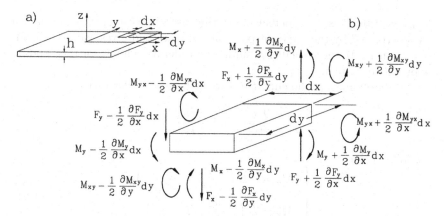

FIGURE 13.2. Plate: (a) geometrical definitions and system of reference; (b) forces and moments acting on a small portion of the plate.

twisting behavior cannot be uncoupled, as it was the case for torsional and flexural behavior of beams.

If a surface force distribution $f_z(x, y, t)$ acts on the plate, the dynamic equilibrium equation for translations in the z direction is

$$\rho h \frac{d^2 u_z}{dt^2} dxdy = \frac{\partial F_x}{\partial x} dxdy + \frac{\partial F_y}{\partial y} dxdy + f_z dxdy. \qquad (13.4)$$

Equation (13.4) is the two-dimensional equivalent of Eq. (12.48). Neglecting the rotational inertia of the cross-section and canceling all terms containing negligible quantities, the equilibrium equation to rotation about x- and y-axes are, respectively,

$$\begin{cases} \dfrac{\partial M_{xy}}{\partial x} + \dfrac{\partial M_x}{\partial y} + \partial F_x = 0 \\[3mm] \dfrac{\partial M_y}{\partial x} + \dfrac{\partial M_{xy}}{\partial y} - \partial F_y = 0. \end{cases} \qquad (13.5)$$

By introducing Eq. (13.5) into Eq. (13.4) and remembering that because $\tau_{xy} = \tau_{yx}$, $M_{xy} = -M_{yx}$, it follows that

$$\rho h \frac{d^2 u_z}{dt^2} = \frac{\partial^2 M_y}{\partial x^2} + 2 \frac{\partial^2 M_{xy}}{\partial x \partial y} - \frac{\partial^2 M_x}{\partial y^2} + f_z. \qquad (13.6)$$

The bending and twisting moments can be easily linked to the deformation of the plate. The bending moment M_y about y-axis is directly linked with the curvature of the plate in xz-plane, as it was the case of beams, but also to the curvature in yz-plane. The cross-section of a beam deflecting in xz-plane is subjected to a lateral deformation, due to lateral contraction, expressed by Poisson's coefficient. In the case of a plate such deformation is constrained by the adjacent parts of the plate and gives way to a bending moment M_x in yz-plane. A similar consideration holds for the bending moment in xz-plane.

By approximating the curvatures with the second derivatives

$$\frac{\partial^2 u_z}{\partial x^2} \quad \text{and} \quad \frac{\partial^2 u_z}{\partial y^2} \,,$$

it follows that

$$\begin{cases} M_y = -B \left(\dfrac{\partial^2 u_z}{\partial x^2} + \nu \dfrac{\partial^2 u_z}{\partial y^2} \right) \\ M_x = B \left(\dfrac{\partial^2 u_z}{\partial y^2} + \nu \dfrac{\partial^2 u_z}{\partial x^2} \right), \end{cases} \tag{13.7}$$

where

$$B = \frac{Eh^3}{12 \left(1 - \nu^2 \right)} \tag{13.8}$$

is the bending stiffness of the plate.

In a similar way, the twisting moment can be expressed as

$$M_{xy} = -B \left(1 - \nu \right) \frac{\partial^2 u_z}{\partial x \partial y}. \tag{13.9}$$

By introducing the expressions for the bending and twisting moments into the equation of motion (13.6), the latter yields

$$\frac{\rho h}{B} \frac{\partial^2 u_z}{\partial t^2} + \frac{\partial^4 u_z}{\partial x^4} + 2 \frac{\partial^4 u_z}{\partial x^2 \partial y^2} + \frac{\partial^4 u_z}{\partial y^4} = \frac{f_z}{B}. \tag{13.10}$$

The differential operator of Eq. (13.1) is thus

$$D = \left[\rho h \frac{\partial^2}{\partial t^2} + B \frac{\partial^4}{\partial x^4} + 2B \frac{\partial^4}{\partial x^2 \partial y^2} + B \frac{\partial^4}{\partial y^4} \right] u_z. \tag{13.11}$$

The force $f_z(x, y, t)$ must be set to zero in the study of free vibration and Eq. (13.10) reduces to a homogeneous equation. Its solution can be expressed as the product of a function of the space coordinates by a function of time:

$$u_z(x, , y, t) = q(x, y) \eta(t), \tag{13.12}$$

which is the two-dimensional equivalent of Eq. (12.8).

The function of time $\eta(t)$ is harmonic and the form of Eq. (12.12) can be assumed. Also in this case function $q(x, y)$ can be regarded as a *principal* or *normal* function and can be obtained from the equation

$$\omega^2 \frac{\rho h}{B} q(x, y) + \frac{\partial^4 q(x, y)}{\partial x^4} + 2 \frac{\partial^4 q(x, y)}{\partial x^2 \partial y^2} + \frac{\partial^4 q(x, y)}{\partial y^4} = 0. \tag{13.13}$$

An eigenproblem is thus obtained, although a far more complex one than those encountered in the study of one-dimensional systems. Remarkable difficulties can arise in the statement of boundary conditions, particularly when the geometrical shape of the boundary is not of the simplest type.

Consider the case of a rectangular plate of length a (in the x-direction) and width b (in the y-direction). If the lower-left corner is in the origin of the xy reference frame, the boundaries of the plate are the straight lines of equations $x = 0$, $x = a$, $y = 0$, and $y = b$. Assume that the plate is simply supported at its edges, i.e., that the displacement and curvature in the plane containing the edge are equal to zero. The bending moment along the edge must vanish as well. These three conditions can be shown to be equivalent to the two conditions[2]:

$$q = 0, \qquad \frac{\partial^2 q}{\partial x^2} = 0 \qquad \text{along the lines } x = 0 \text{ and } x = a$$

$$q = 0, \qquad \frac{\partial^2 q}{\partial y^2} = 0 \qquad \text{along the lines } y = 0 \text{ and } y = b .$$

$$(13.14)$$

The solution of the eigenproblem, even with the present very simple boundary conditions, is rather complex. It can be shown that the eigenvalues are[3]

$$\omega_{ij} = \pi^2 \left[\left(\frac{i}{a} \right)^2 + \left(\frac{j}{b} \right)^2 \right] \sqrt{\frac{B}{\rho h}} = \pi^2 \left[\left(\frac{i}{a} \right)^2 + \left(\frac{j}{b} \right)^2 \right] \sqrt{\frac{Eh^2}{12\rho(1 - \nu^2)}},$$

$$(13.15)$$

and the corresponding eigenfunctions are

$$q_{ij} = q_0 \sin \left(\frac{i\pi x}{a} \right)^2 \sin \left(\frac{j\pi y}{b} \right)^2 , \qquad i, j = 1, 2, \dots$$

Constants i and j are the number of half-waves of the deflected shape in the x- and y-directions, respectively. The eigenfunctions can be shown to possess the usual properties of orthogonality with respect to both stiffness and mass. Because the thickness and the material properties are constant, the eigenfunctions are strictly orthogonal. Some eigenfunctions are shown in Fig. 13.3. In the figure, the nodal lines, i.e., the lines along which the amplitude of the vibration vanishes, are clearly visible. They are straight lines; those parallel to the x-axis are $j - 1$, and those parallel to the y-axis are $i - 1$, excluding the boundaries of the plate, which are themselves nodal lines.

---

[2]For a detailed description of the boundary conditions, see S. Timoshenko, S. Woinowsky-Krieger, *Theory of Plates and Shells*, Mc Graw Hill, New York, 1959.

[3]L. Meirovitch, *Analytical Methods in Vibrations*, Macmillan, New York, 1967, p. 184.

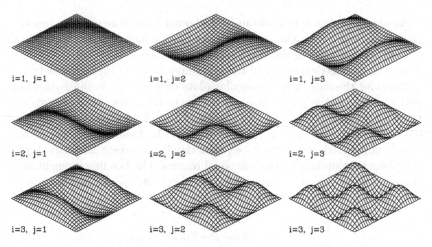

FIGURE 13.3. Nine eigenfunctions of a rectangular plate.

Remark 13.3 *If the plate is square (a = b), the eigenfrequency of order i, j is coincident with that of order j, i. In all cases in which coincident eigenfrequencies are found, any linear combinations of the eigenfunctions is itself an eigenfunction and often happens that those obtained from the computations are not the more significant or obvious ones.*

No other case will be studied here because, even in the case of circular plates, the analysis is fairly complex and it is necessary to resort to Bessel's functions. Actually, while in the study of the static behavior of circular plates symmetry can simplify the problem, this cannot be done in dynamics, as even if the system possesses some properties of symmetry and is loaded by a symmetric forcing function, free vibrations whose modes are not symmetrical can exist and can be excited by any small deviation from symmetry.

Example 13.1 *Consider a square plate simply supported at all sides; compute the first frequencies corresponding to modes with up to three half-waves in both directions. Because the second and third modes correspond to identical eigenfrequencies (i.e., to a multiple eigenvalue), find the linear combinations of the eigenfunctions yielding nodal lines coinciding with the diagonals of the square plate. Data: $a = b = 1$ m, $h = 10$ mm, $E = 2.1 \times 10^{11}$ N/m^2, $\rho = 7,810$ kg/m^3, $\nu = 0.3$. The natural frequencies can be computed using Eq. (13.15): $\omega_{11} = 98.6$ rad/s $= 15.7$ Hz, $\omega_{12} = \omega_{21} = 246.4$ rad/s $= 39.2$ Hz, $\omega_{13} = \omega_{31} = 492.8$ rad/s $= 78.4$ Hz, $\omega_{22} = 394.3$ rad/s $= 62.8$ Hz, $\omega_{23} = \omega_{32} = 640.7$ rad/s $= 102.0$ Hz, $\omega_{33} = 887.1$ rad/s $= 141.2$ Hz.*
The eigenfunctions related to the modes with $i = 1$, $j = 2$ and $i = 2$, $j = 1$ are

$$q_{12}(x, y) = q_0 \sin(\pi x) \sin(2\pi y)$$

and

$$q_{21}(x, y) = q_0 \sin(2\pi x) \sin(\pi y).$$

Any linear combination of these modes

$$K_1 \sin(\pi x) \sin(2\pi y) + K_2 \sin(2\pi x) \sin(\pi y)$$

is itself a mode. With simple trigonometric manipulations the expression of the generic linear combination becomes

$$2 \sin(\pi x) \sin(\pi y)[K_1 \cos(\pi y) + K_2 \cos(\pi x)].$$

The values of the coefficients yielding a mode shape whose nodal line coincides with the diagonal of the square expressed by the equation $x = y$ are readily obtained. The equation of the nodal line is simply obtained by equating to zero the expression of the mode shape given earlier.

By noting that the product outside brackets coincides with the first mode shape and never vanishes inside the plate, the equation of the nodal lines can be reduced to

$$K_1 \cos(\pi y) + K_2 \cos(\pi x) = 0.$$

Introducing the equation of the first diagonal of the plate $x = y$ in the equation so obtained, it follows that $K_1 + K_2 = 0$, i.e., $K_1 = -K_2$.

The equation of the other diagonal is $x = 1 - y$. By introducing this expression into the equation of the nodal lines it follows that $K_1 = K_2$. The mode shapes so obtained are shown in Fig. 13.4.

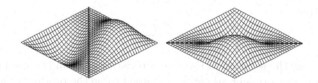

FIGURE 13.4. Mode shapes corresponding to the multiple eigenvalues chosen in such a way that the nodal lines coincide with the diagonals of the plate.

13.3 Vibration of membranes

A membrane is like a very thin plate, with a vanishingly small bending stiffness, in which the restoring force is supplied by *in-plane* tensile forces. Like the plate was the two-dimensional equivalent of the beam, the membrane is the two-dimensional equivalent of the taut string. Also in this case the equation of motion could be obtained by adding the in-plane forces to the plate, taking into account the nonlinear coupling between flexural and 'in-plane' behavior, linearizing and then reducing to zero the bending stiffness.

This procedure would however be exceedingly, and actually nonnecessarily, complex.

The dynamic equilibrium equation of a portion of membrane of length dx and width dy is the same already seen for plates (Eq. (13.4))

$$\rho h \frac{d^2 u_z}{dt^2} = \frac{\partial F_x}{\partial x} + \frac{\partial F_y}{\partial y} + f_z .$$

The shear forces cannot be obtained from the bending moments, which do not exist in a membrane. They can however be directly derived from the in-plane tensile forces, which are known. If T is the tensile force per unit length acting in the plane of the membrane, when the mid-plane is deflected to the deformed shape $u_z(x, y, z)$, the forces per unit length acting on the sides parallel to x- and y-axes are, respectively,

$$F_x = T \frac{\partial u_z}{\partial x} \quad , \quad F_y = T \frac{\partial u_z}{\partial y} . \tag{13.16}$$

By introducing Eq. (13.16) into Eq. (13.4), it follows that

$$\rho h \frac{d^2 u_z}{dt^2} = T \left(\frac{\partial^2 u_z}{\partial x^2} + \frac{\partial^2 u_z}{\partial y^2} \right) + f_z . \tag{13.17}$$

The study of the free oscillations can follow the same guidelines seen for the plates, with the only obvious difference of dealing with a simpler equation. The solution for free vibrations is again Eq. (13.12) and function $\eta(t)$ is harmonic. The normal functions $q(x, y)$ can thus be obtained from the equation

$$-\omega^2 \frac{\rho h}{T} q(x, y) = \frac{\partial^2 q(x, y)}{\partial x^2} + \frac{\partial^2 q(x, y)}{\partial y^2} . \tag{13.18}$$

In this case the solution of the characteristic equation is straightforward. The normal function $q(x, y)$ can be assumed to be the product of a function of x by a function of y

$$q(x, y) = q_1(x) q_2(y) \tag{13.19}$$

obtaining

$$\omega^2 \frac{\rho h}{T} + \frac{1}{q_1(x)} \frac{d^2 q_1(x)}{dx^2} + \frac{1}{q_2(y)} \frac{d^2 q_2(y)}{dy^2} . \tag{13.20}$$

Since this equation must hold for any values of x and y, it can be split into the two equations

$$\begin{cases} \dfrac{d^2 q_1(x)}{dx^2} = \alpha^2 q_1(x) \\[4mm] \dfrac{d^2 q_2(y)}{dy^2} = \beta^2 q_2(y), \end{cases} \tag{13.21}$$

where

$$\alpha^2 + \beta^2 = \omega^2 \frac{\rho h}{T}.$$

The general solution of Eq. (13.21) is

$$\begin{cases} q_1(x) = C_1 \sin(\alpha x) + C_2 \cos(\alpha x) \\ q_2(y) = C_3 \sin(\beta y) + C_2 \cos(\beta y) \end{cases} \quad (13.22)$$

Also in this case some difficulties can be encountered in stating the boundary conditions in all cases but that of the simplest geometrical shapes. If the membrane is rectangular, of length a and width b, the boundary conditions are simply stated assuming that

$$\begin{aligned} q_1(x) &= 0 \quad \text{for } x = 0 \text{ and } x = a \\ q_2(y) &= 0 \quad \text{for } y = 0 \text{ and } y = b. \end{aligned} \quad (13.23)$$

From the conditions at $x = 0$ and $y = 0$, it follows immediately that constants C_2 and C_4 are equal to 0. The conditions at $x = a$ and $y = b$ allow to find a solution different from the trivial solution only if

$$\alpha = \frac{i\pi}{a} \quad \text{and} \quad \beta = \frac{j\pi}{b} \quad \text{for } i, j = 1, 2, 3, \dots \quad (13.24)$$

The relationship linking constants α and β to the frequency ω allows to compute the eigenfrequencies of the membrane

$$\omega_{ij} = \pi \sqrt{\frac{T}{\rho h}} \sqrt{\frac{i^2}{a^2} + \frac{j^2}{b^2}} \quad \text{for } i, \ j = 1, 2, 3, \dots \quad (13.25)$$

The eigenfunctions are

$$q_{ij} = q_0 \sin\left(\frac{i\pi x}{a}\right)^2 \sin\left(\frac{j\pi y}{b}\right)^2, \quad i, \ j = 1, 2, \dots,$$

and coincide with those seen for the supported rectangular plate.

Remark 13.4 *While in the case of a vibrating taut string the natural frequencies are in harmonic proportion, i.e., are proportional to the sequence of the natural numbers, the natural frequencies of a membrane are spaced in a more complicated and less regular way. This partially explains the peculiar characteristics of the sound of musical instruments based on a vibrating membrane, like the drum, as opposite to those based on vibrating strings.*

Also in the case of membranes no other boundary conditions will be dealt with here, as even in the simple case of a circular membrane it is necessary to resort to Bessel's functions.

13.4   Propagation of elastic waves in taut strings

Consider the taut string studied in the last part of Section 12.8. The equation of motion of the string (12.113)

$$\frac{d^2 u_x}{dt^2} = \frac{F_z}{\rho A} \frac{\partial^2 u_x}{\partial z^2} \tag{13.26}$$

can be considered as a wave equation and constant

$$v = \sqrt{\frac{F_z}{\rho A}}$$

is the speed of propagation of the waves along the string. This can be readily shown by taking an arbitrary function

$$f(\zeta), \quad \text{where} \quad \zeta = z \pm vt,$$

and noting that its derivatives

$$\frac{d^2 f}{dt^2} = v^2 \frac{\partial^2 f}{\partial \zeta^2}; \qquad \frac{\partial^2 f}{\partial z^2} = \frac{\partial^2 f}{\partial \zeta^2} \tag{13.27}$$

satisfy Eq. (12.113). Any function of $z \pm vt$ is a solution of the equation of motion.

Since function $f(\zeta) = f(z)$ for time $t = 0$, it represents the initial deformed configuration of the string. As time goes on, the deformed configuration travels along the string toward the positive z-axis if the solution with sign $(-)$ is taken, or in the opposite direction if the solution with sign $(+)$ is considered. The motion of the deformed configuration, which does not involve any actual movement of matter in the z-direction, takes place with speed v. No change in shape takes place, apart from this displacement.

Remark 13.5 *The actual motion of the string takes place in the x-direction while the waves propagate in the z-direction. When this occurs, the wave is said to be a* transversal wave. *In the case of the longitudinal behavior of bars, which is governed by Eq. (12.9), however, the direction of the motion and that of propagation coincide and the wave is said to be a* longitudinal *wave.*

The general solution of Eq. (12.113) can be expressed as

$$u_x(z,t) = f_1(z + vt) + f_2(z - vt), \tag{13.28}$$

where f_1 and f_2 are two arbitrary functions that represent the shape of two waves traveling to the left and right, respectively. Owing to the complete linearity of the equation of motion, the solutions can simply be added to

each other, and when the two waves meet along the string, the relevant
amplitudes can simply be added for any value of time and z-coordinate.

In the analysis performed up to now, the string has been considered
infinitely long and no boundary condition has been assumed. Consider an
indefinite string on the positive half of the z-axis. Assume that the origin
($z = 0$) is a fixed point and that a general disturbance $u_x = f_1(z + vt)$
travels toward the origin (Fig. 13.5a). At a certain time the disturbance
reaches the end of the string, which is fixed (Fig. 13.5b).

The boundary condition can be introduced simply by adding a second
function $f_2(z - vt)$, i.e., by adding a second wave traveling to the right and
computing the unknown function f_2 in such a way that the displacement
in the origin is always equal to zero:

$$u_x(0, t) = f_1(z + vt) + f_2(z - vt) = 0 \text{ , for } z = 0, \text{ at any time,} \quad (13.29)$$

i.e.,

$$f_2(\zeta) = -f_1(-\zeta). \quad (13.30)$$

The complete solution of the equation of motion is

$$u_x(z, t) = f_1(z + vt) - f_1(-z + vt). \quad (13.31)$$

The first term is the original wave traveling toward the left and the
second term is a reflected wave, reversed in space and with opposite sign,
traveling toward the right. To show this reversal of shape, an unsymmetrical
disturbance is shown in Fig. 13.5. An easy way to visualize the reflection
of a wave at a fixed point is to consider the string as if it were infinite

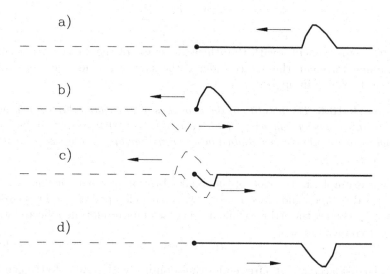

FIGURE 13.5. Reflection of a wave at a fixed end of a semi-indefinite string.

by adding a ghost string to the left of the origin. The original wave, after reaching the origin, passes to the ghost string, while the reflected wave, coming from the latter, passes on the actual string.

The solution of the equation of motion was first assumed to be of the type of Eq. (12.8) and then of Eq. (13.28), i.e., as a stationary oscillation and as a wave propagating along the system. Both have been claimed to be the general solution of the problem, so they must in some way be equivalent.

At time $t = 0$, give it a disturbance of the type

$$u_x(z, 0) = q_0 \sin\left(\frac{i\pi z}{l}\right) \tag{13.32}$$

to a string extending from $z = 0$ to $z = l$. The solution can be expressed as a wave traveling to the left and reflecting in the origin, a fixed point,

$$u_x(z, t) = q_0 \sin\left[\frac{i\pi}{l}(z + vt)\right] - q_0 \sin\left[\frac{i\pi}{l}(-z + vt)\right]. \tag{13.33}$$

Note that the boundary condition at the other end is satisfied because Eq. (13.33) yields $u_x(l, t) = 0$ for any value of t. By remembering some simple trigonometric identities, Eq. (13.33) can be transformed to

$$u_x(z, t) = 2q_0 \sin\left(\frac{i\pi z}{l}\right) \cos\left(\frac{i\pi v}{l}t\right). \tag{13.34}$$

The solution expressed by Eq. (13.34) is a standing wave because it does not move along the string and it coincides with the vibration at the ith natural frequency

$$\omega_i = \frac{i\pi v}{l} = \frac{i\pi}{l}\sqrt{\frac{F_z}{\rho A}}. \tag{13.35}$$

Its shape coincides with the ith mode shape of the string. A harmonic standing wave can thus be considered the sum of two identical harmonic waves traveling in opposite directions.

Remark 13.6 *The identity of the two ways of describing the free vibration of a continuous system seen for the case of a taut string is general and also applies to all other cases studied here, even if greater analytical difficulties can arise.*

For example, in the case of the axial vibration of a bar the ends can be clamped or free. The reflection of a wave at a clamped end follows closely what is seen for the strings while in the case of a free end the reflected wave is not changed in sign:

$$f_2(\zeta) = f_1(-\zeta). \tag{13.36}$$

The displacement at a free end is thus doubled with respect to the general displacement in the bar.

13.5 Propagation of sound waves in pipes

Although the detailed study of wave-propagation phenomena is well beyond the scope of this book, a simple model that allows the study of the
one-dimensional propagation of sound waves will be summarized. Consider
a pipe of cross-section A, filled with a gas with pressure p_0 and density
ρ_0. Assume that all parameters related to the local properties of the gas
(pressure p, density ρ, velocity in axial direction V) are functions of the
axial z-coordinate alone (Fig. 13.6).

This assumption is equivalent to those at the base of beam theory and
allows a description of the phenomenon of wave propagation using only one
space coordinate. A first equation can be stated by observing that the net
flow into the control volume of length dz must be equal to the increase of
the mass contained in the same volume:

$$\frac{d\rho}{dt} A\,dz\,dt = \left[(\rho V)_z - (\rho V)_{z+dz}\right] A\,dt\,. \tag{13.37}$$

This equation, generally referred to as the conservation of mass equation,
or simply the continuity equation, can be written as

$$\frac{d\rho}{dt} = -\frac{\partial(\rho V)}{\partial z}\,. \tag{13.38}$$

The momentum of the gas contained in the control volume $\rho V A\,dz$ can
change during time dt owing to the momentum of the gas entering the same
volume and the total impulse of pressure forces:

$$\frac{d(\rho V)}{dt} A\,dz\,dt = \left[(\rho V^2)_z - (\rho V^2)_{z+dz}\right] A\,dt + (p_z - p_{z+dz})\,A\,dt\,. \tag{13.39}$$

The equation expressing the conservation of momentum is thus

$$\frac{d(\rho V)}{dt} = -\frac{\partial(\rho V^2)}{\partial z} - \frac{\partial p}{\partial z}\,. \tag{13.40}$$

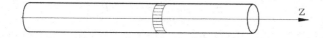

$$p - \frac{1}{2}\frac{\partial p}{\partial z}dz \qquad\qquad p + \frac{1}{2}\frac{\partial p}{\partial z}dz$$

$$\rho - \frac{1}{2}\frac{\partial \rho}{\partial z}dz \qquad\qquad \rho + \frac{1}{2}\frac{\partial \rho}{\partial z}dz$$

$$V - \frac{1}{2}\frac{\partial V}{\partial z}dz \qquad\qquad V + \frac{1}{2}\frac{\partial V}{\partial z}dz$$

FIGURE 13.6. Wave propagation in a pipe filled with gas.

A third equation expressing the usual relationship between pressure and density during adiabatic changes

$$\frac{p}{p_0} = \left(\frac{\rho}{\rho_0}\right)^{\gamma},$$

(13.41)

where γ is the ratio between the specific heat at constant pressure and volume, respectively, can be added. Without entering into details about the assumption of adiabatic wave propagation, Eq. (13.41) will be assumed to be an expression of the conservation of energy. Equations (13.38), (13.39), (13.40) and (13.41) are a set of three nonlinear equations describing sound-wave propagation in the pipe.

A linearization of these equations can, however, be performed. Pressure, density, and velocity can be expressed as small variations about the static values p_0, ρ_0, and 0 (the gas is at a standstill as an average):

$$p = p_0 + p_1 \ , \ \rho = \rho_0 + \rho_1, \quad \text{and} \quad V = V_1.$$

(13.42)

Functions $p_1(z,t)$, $\rho_1(z,t)$, and $V_1(z,t)$ are all assumed to be small quantities. The relevant equations can be simplified by neglecting the terms containing products of small quantities, obtaining

$$\begin{cases} \dfrac{d\rho_1}{dt} = -\rho_0 \dfrac{\partial V_1}{\partial z} \\[2mm] \dfrac{dV_1}{dt} = -\dfrac{1}{\rho_0}\dfrac{\partial p_1}{\partial z} \end{cases} \quad , \quad \frac{p_1}{p_0} = \gamma\frac{\rho_1}{\rho_0} \ .$$

(13.43)

By eliminating p_1 and ρ_1 these equations yield the following wave equation

$$\frac{d^2 V_1}{dt^2} = \gamma\frac{p_0}{\rho_0}\frac{\partial^2 V_1}{\partial z^2} \ .$$

(13.44)

The speed of propagation of the waves, i.e., the speed of sound in the gas, is

$$v_s = \sqrt{\frac{\gamma p_0}{\rho_0}} = \sqrt{\gamma R T},$$

(13.45)

where R is a constant depending on the nature of the gas and T is the absolute temperature. For air at standard temperature, pressure, and density, the value of γ is roughly 7/5, the pressure is about $p_0 = 1 \times 10^5$ Pa, and the density is $\rho_0 = 1.29$ kg/m^3; the speed of sound is thus $v_s = 330$ m/s.

In order to evaluate the approximations linked with linearization, consider a sound with an intensity of 140 dB, corresponding to the threshold of pain. The measure of sound pressure in decibels is defined as

$$20\log_{10}\left(\frac{p_1}{p_{ref}}\right) ,$$

where the reference pressure $p_{ref} = 2 \times 10^{-5}$ Pa corresponds roughly to the threshold of human hearing at a frequency of 1 kHz. With easy computations it is possible to show that the pressure at the threshold of pain is of about 200 Pa. The ratio p_1/p_0 for such a very loud sound is then about 2×10^{-3} . This shows that the linearized theory is quite acceptable for almost all practical applications.

This model can be applied to the propagation of sound in an infinitely long pipe. If the length of the pipe is limited, appropriate boundary conditions must be stated, and standing waves can be found. Note that the speed of propagation of sound does not depend on the frequency, at least in the case of the linearized theory. This is, however, not a general rule for all wave-propagation phenomena: An example of waves that are dispersive, i.e., their speed of propagation depends on the frequency, is that of water waves in deep waters. The waveform of any wave can be considered the superimposition of harmonic waves, each having its own frequency. If they propagate at different speeds, the waveform distorts continuously during propagation.

For a more detailed introduction to wave-propagation phenomena, the already-mentioned book by R. Buckley[4] can be referred to.

13.6 Linear continuous systems with structural damping

Within the validity of the model of the *complex modulus of elasticity*, structural damping can be simply accounted for by substituting the Young's modulus E or the shear modulus G with the complex moduli $E(1 + i\eta)$ and $G(1 + i\eta)$, respectively.

In the present section only homogeneous systems will be studied. In this case a sort of proportional structural damping is present and the normal modes found for the undamped system can be used to obtain a set of uncoupled equations of motion, whose number is theoretically infinite. In the more general case where the loss factor of the material is not constant, this approach is no more possible, and the complex differential equation of motion must be dealt with.

Consider as an example the bending behavior in xz-plane of a prismatic homogeneous beam. Taking into account the loss factor of the material, Eq. (12.59) transforms into

$$-\omega^2 q(z) = E(1 + i\eta)\frac{I_y}{\rho A}\frac{d^4 q(z)}{dz^4} . \tag{13.46}$$

---

[4]R. Buckley, *Oscillations and Waves*, Adam Hilger Ltd., Bristol, 1985.

Remark 13.7 *Only a frequency-domain equation has been written, since the hysteretic damping model can be used only in case of harmonic motion.*

This equation is identical to Eq. (12.59), except for the term $(1 + i\eta)$. If ω^* is a natural frequency of the undamped system, the complex frequency of the damped system is

$$\nu = is = \omega^* \sqrt{1 + i\eta}. \tag{13.47}$$

By separating the real and imaginary parts of the complex frequency and remembering that η is usually very small if compared with unity, the frequency and the decay rate are

$$
\begin{cases}
\omega = \Im(s) = \omega^* \sqrt{\dfrac{1 + \sqrt{1 + \eta^2}}{2}} \approx \omega^* \\[4mm]
\sigma = \Re(s) = -\omega^* \sqrt{\dfrac{-1 + \sqrt{1 + \eta^2}}{2}} \approx -\dfrac{\eta}{2}\omega^*.
\end{cases}
\tag{13.48}
$$

The frequencies of the damped free vibrations are then very close to those of the corresponding undamped system, while the decay rate is equal to the corresponding natural frequency multiplied by half of the loss factor of the material. The eigenfunctions of the undamped system can be used to obtain the modal masses, stiffness, and forces (the latter in case of forced vibrations).

13.7   Exercises

Exercise 13.1 *A chisel has a square cross-section with side $a = 10$ mm and length $l = 500$ mm. Compute the time needed for the pressure wave caused by a hammer striking the end of the chisel to reach the bit. Assume that the bit of the chisel is pressed against a rigid surface and can be considered perfectly clamped. If the blow of the hammer causes a stress wave whose profile can be assumed to be a trapezium, plot the profile of the incident wave and of the wave reflected back at different times.*

Exercise 13.2 *Compute the length of a pipe with closed ends in such a way that the air contained into it can vibrate with a fundamental frequency corresponding to the central A (nominal frequency 440 Hz). Compute the other natural frequencies and state whether they are multiples of the fundamental frequency. Data: $\gamma \approx 7/5$, $p_0 = 1.007 \times 10^5$ Pa, $\rho_0 = 1.29$ kg/m^3.*

Exercise 13.3 *Compute the natural frequencies of a rectangular steel membrane with ratio between the sides of 1:2, clamped at all edges. Data: $a = 0.5$ m, $b = 1$ m, $\rho = 7810$ kg/m^3, $t = 0.1$ mm, in-plane stress $\sigma = 1 \times 10^6$ Pa.*

Exercise 13.4 *Consider the membrane of the previous exercise. The supporting frame receives a shock defined by a duration of 11 ms and an acceleration increasing linearly from 0 to 20 g. State which modes are excited and compute through numerical integration in time the time history of the modal oscillations. Compute an upper bound for the maximum displacement at the center of the membrane.*

14
Discretization of Continuous Systems

A number of discretization techniques were developed with the aim of avoiding the overwhelming difficulties found in the solution of the partial derivatives equations related to continuum models. They are particularly useful when the shape of the continuous system, and hence the boundary conditions, is complex. Their aim is to replace a discrete model for the continuous one, i.e., to replace a model based on a set of ODEs for the continuous model made of PDEs.

14.1 Overview of discretization techniques

Over the last two centuries, many discretization techniques have been developed with the aim of replacing the equation of motion consisting of partial derivative differential equations (with derivatives with respect to time and space coordinates) with a set of linear ordinary differential equations containing only derivatives with respect to time. The resulting set of equations, generally of the second order, is of the same type as seen for discrete systems (hence the term *discretization*).

Another problem, which is now far less severe due to the growing power of computers, is linked with the size of the eigenproblem obtained through discretization techniques. Before the widespread use of electronic computers, the solution of an eigenproblem containing matrices whose order was greater than a few units was very difficult, if not insoluble. Many techniques were aimed at reducing the size of the eigenproblem or transforming it into a form that could be solved using particular algorithms. The possibility of solving very large eigenproblems without long or costly computations did actually change the basic approach to dynamic problems, making many popular methods obsolete.

At a first sight it would seem straightforward to discretize any continuous system through modal decomposition: The eigenfunctions allow to write a set of ordinary differential equations of motion in the modal coordinates. The fact that their number is theoretically infinite is not a problem: the approximation obtained by using an adequate (usually quite low) number of them is more than acceptable.

Unfortunately this strategy does not work: Searching for the eigenfunctions is usually the most difficult step in solving a continuous system, and the system must be discretized before and not after obtaining them.

It is difficult to make a satisfactory classification of discretization techniques, and perhaps impossible to make a complete one. It is, however, possible to attempt to group them in three classes.

The first class is that of the methods in which the deformed shape of the system is assumed to be a linear combination of n known functions of the space coordinates, defined in the whole space occupied by the body. These methods could be labeled as *assumed-modes methods*, owing to the similarity of these functions, which are arbitrarily assumed, with the eigenfunctions (i.e., the mode shapes) of the system.

The second class is that of the so-called *lumped-parameters methods*. The mass of the body is lumped in a certain number of rigid bodies (sometimes simply point masses) located at given stations in the deformable body. These lumped masses are then connected by massless fields that possess elastic, and sometimes damping, properties. Usually the properties of the fields are assumed to be uniform in space. Because the degrees of freedom of the lumped masses are used to describe the motion of the system, the model leads intuitively to a discrete system. While the mass matrix of such systems is easily obtained, it is often quite difficult to write the stiffness matrix, or, alternatively, its compliance matrix.

To avoid such difficulty, together with that linked to the solution of large eigenproblems, an alternative approach can be followed. Instead of dealing with the system as a whole, the study can start at a certain station and proceed station by station using the so-called *transfer matrices*. Methods based on transfer matrices were very common in the recent past, because they could be worked out with tabular manual computations or implemented on very small computers. Their limitations are now making them yield to the finite element method. This change between the two approaches is not yet complete, and many computer codes based on transfer matrices are still in use.

It is also possible to model a continuous system by lumping the elastic properties at given stations leaving the inertia properties as distributed along the fields. Such methods were proposed but seldom used.

A separate class can be assigned to the finite element method (FEM). In the FEM, the body is divided into a number of regions, called *finite elements*, as opposed to the vanishingly small regions used in writing the differential equations for continuous systems. The deformed shape of each finite element is assumed to be a linear combination of a set of functions of space coordinates through a certain number of parameters, considered as the degrees of freedom of the element.

Usually such functions of space coordinates (called *shape functions*) are quite simple and the degrees of freedom have a direct physical meaning: They are the generalized displacements at selected points of the element, usually referred to as *nodes*. The analysis then proceeds by writing a set of differential equations of the same type as those obtained for discrete systems.

The use of the FEM can be limited to writing the stiffness matrix to be introduced into a lumped-parameters approach; alternatively, it can be used to write the mass matrix too. In this case, the mass matrix is said to be *consistent*, because it is obtained in the same way as the stiffness matrix.

Although often considered a separate approach, the dynamic stiffness method will be regarded here as a particular form of the FEM in which the shape functions are obtained from the actual deflected shape in free vibration. The facts that it can be used only for free vibration or harmonic excitation and that the dynamic stiffness matrix it yields is a function of the frequency have prevented it from becoming very popular.

The FEM is gaining popularity for the unquestionable advantage of being implemented in the form of general-purpose codes, which can be used for static and dynamic analyses, and interfaced with CAD and CAM codes.

The finite difference method will not be dealt with, because it is considered more as a method for the solution of the differential equations obtained from continuous models than as a discretization technique yielding a model with many degrees of freedom from a continuous one. Actually, finite difference techniques are at present seldom used in structural dynamics.

The assumed-modes and lumped-parameters methods are dealt with in the present chapter, while a whole chapter will be dedicated to the FEM.

14.2   The assumed-modes methods

The displacement field $\vec{u}(x, y, z, t)$ of a general undamped continuous elastic body can be approximated by a linear combination of n arbitrarily assumed functions $\vec{q}_i(x, y, z)$, often referred to as *assumed modes*:

$$\vec{u}(x, y, z, t) = \sum_{i=1}^{n} a_i(t)\vec{q}_i(x, y, z) \,. \qquad (14.1)$$

Expression (14.1) yields exact results if the assumed functions $\bar{q}_i(x, y, z)$ are the eigenfunctions of the system and their number is infinite. In this case, functions $a_i(t)$ that can be considered the generalized coordinates expressing the deformation of the system are the modal coordinates.

If a finite number of coordinates are used, the results are clearly approximate. The expressions of the kinetic and potential energies of the system can be easily obtained from Eq. (14.1):

$$\mathcal{T} = \frac{1}{2}\dot{\mathbf{a}}^T \mathbf{M} \dot{\mathbf{a}}, \qquad \mathcal{U} = \frac{1}{2}\mathbf{a}^T \mathbf{K} \mathbf{a}, \qquad (14.2)$$

where $\mathbf{a}$ is a vector containing the n generalized coordinates $a_i(t)$ and $\mathbf{M}$ and $\mathbf{K}$ are square matrices of order n that depend on the inertial and elastic properties of the system, respectively, and on the assumed functions q_i.

In the same way, also the virtual work performed by a given force distribution $\vec{f}(x, y, z, t)$ for the virtual displacement

$$\delta \vec{u}(x, y, z) = \sum_{i=1}^{n} \delta a_i \vec{q}_i(x, y, z) \qquad (14.3)$$

corresponding to a virtual change of the generalized coordinates δa_i is

$$\delta \mathcal{L} = \int_V \sum_{i=1}^{n} \delta a_i \vec{f}(x, y, z, t) \times \vec{q}_i(x, y, z) dV. \qquad (14.4)$$

The equations of motion of the system can thus be computed through Lagrange equations

$$\mathbf{M}\ddot{\mathbf{a}} + \mathbf{K}\mathbf{a} = \mathbf{f}(t), \qquad (14.5)$$

where the generalized forces f_i are

$$f_i = \int_V \vec{f}(x, y, z, t) \times \vec{q}_i(x, y, z) dV.$$

Equation (14.5) is formally identical to the equation of motion for discrete systems. If the assumed functions $q_i(z)$ are the mode shapes, the mass and stiffness matrices are diagonal and the non-zero elements coincide with the modal masses and stiffnesses. Forces f_i are the modal forces.

Many methods were developed following the lines summarized here; the best known are the Ritz, Rayleigh–Ritz, and Galerkin methods. The aim of the Raleigh method, for example, is to compute an approximate value of the first natural frequency by transforming the original system, which can be a continuous system (if used as a true discretization method) or a system with many degrees of freedom (and then the method allows a drastic reduction in the number of degrees of freedom), into a system with a single degree of freedom.

Only one arbitrary mode shape is thus assumed. It must be chosen in such a way that it approximates the actual mode shape corresponding to the natural frequency that is looked for, usually the lowest natural frequency, and must be compatible with the constraints and the boundary conditions.

To assume a deformed shape different from the actual shape the system takes in free vibration amounts to implicitly adding other constraints, which have no physical meaning, and then to make the system stiffer than it actually is.

Remark 14.1 *Increasing the stiffness causes the relevant natural frequency to grow: the Rayleigh method yields a value for the lowest natural frequency which is always higher than the correct one. The closer the assumed mode is to the actual mode shape, the lower the value of the natural frequency computed through the Raleigh method is.*

Among the infinite number of possible shapes that can be assumed, that corresponding to the actual mode shape of the first natural frequency (the first eigenfunction or eigenvector, if the system is discrete) yields the lowest value of the natural frequency.

The method is very easily applied: A deflected shape is assumed, and the corresponding kinetic and potential energies are computed. The mass and stiffness of the system with a single degree of freedom are then obtained and the natural frequency is computed. The Raleigh method allows very quick computation of an approximate value of the first natural frequency that often is quite close to the correct one.

Its ease of application made it to survive the introduction of numerical methods that allow detailed analysis of very complex systems but involve computation times and costs that are largely greater.

Example 14.1 *Compute the first natural frequency of a prismatic, homogeneous simply supported beam using the Rayleigh method.*
Compare the result with that obtained using the beam theory. Neglect the rotational inertia of the cross-sections and shear deformation and assume a parabolic assumed-mode shape. Using the reference frame shown in Fig. 14.1, the parabolic assumed mode is expressed by the formula

$$q(\zeta) = \zeta(\zeta - 1) \, ,$$

where $\zeta = z/l$. The kinetic and potential energies corresponding to the deflected shape $aq(z)$ can be easily computed:

$$T = \frac{1}{2} \int_0^l \rho A \left[\dot{a}q(z)\right]^2 dz \, , \qquad U = \frac{1}{2} \int_0^l EI_y \left[a\frac{\partial^2 q(z)}{\partial dz^2}\right]^2 dz \, .$$

The mass and stiffness of the system with a single degree of freedom and its natural frequency are

$$m = \int_0^l \rho A \, [q(z)]^2 \, dz = \frac{\rho A l}{30}, \qquad k = \int_0^l EI_y \left[\frac{\partial^2 q(z)}{\partial dz^2} \right]^2 dz = \frac{4EI_y}{l^3},$$

$$\omega = \sqrt{\frac{k}{m}} = \frac{10.95}{l^2} \sqrt{\frac{EI_y}{\rho A}}.$$

The value computed through the beam theory is

$$\omega = \frac{\pi^2}{l^2} \sqrt{\frac{EI_y}{\rho A}} = \frac{9.87}{l^2} \sqrt{\frac{EI_y}{\rho A}}.$$

The Rayleigh method leads to a value 9.9% in excess of the correct one. Different results would have been obtained if a different assumed-mode shape had been used.

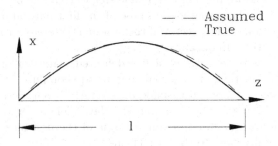

FIGURE 14.1. A simply supported beam; first mode shape and parabolic assumed shape. The two shapes have been normalized in order to have the same amplitude.

Example 14.2 Compute the first natural frequency of the simply supported shaft shown in Fig. 14.2a using a simplified approach based on the consideration that the central part is much stiffer than the two equal ends.
Data: $a = 200$ mm, $b = 100$ mm, $r_1 = 5$ mm, $r_2 = 20$ mm, $\rho = 7,810$ kg/m^3, and $E = 2.1 \times 10^{11}$ Pa.
For symmetry reasons, in the first mode the central part of the shaft remains parallel to the rotation axis.
The simplest way to model the system as a single degree of freedom system is to assume that the central part is a rigid body of mass

$$m = \rho A_2 b = \frac{\pi}{4} \rho d_2^2 b \, .$$

The two massless outer parts of the shaft can be assumed to be two beams with length a clamped to the central part of the shaft and supported at the end of the shaft. Each one has a stiffness

$$k = \frac{3EJ_1}{a^3} = \frac{3\pi E d_1^4}{64a^3}.$$

The natural frequency of the spring–mass system is thus

$$\omega = \sqrt{\frac{2k}{m}} = \sqrt{\frac{6EJ_1}{\rho A_2 a^3 b}} = 280.7 \ rad/s = 44.7 \ Hz \ .$$

Using the Rayleigh method, the assumed-mode shape can be expressed by a parabolic shape at the sides and as a straight line in the center (Fig. 14.2b, dotted line). The expression of function $q(z)$, normalized so that its maximum has a unit value, is

$$q(z) = \frac{-z^2 + 2az}{a^2} \ , \qquad\qquad for \ 0 \leq z \leq a$$

$$q(z) = 1 \ , \qquad\qquad for \ a \leq z \leq a+b$$

$$q(z) = \frac{-z^2 + 2z\,(a+b) - \left(b^2 + 2ab\right)}{a^2} \ , \qquad for \ a+b \leq z \leq 2a+b \ .$$

The mass of the single-degree-of-freedom model is thus

$$m = \int_0^{2a+b} \rho A \left[q(z)\right]^2 dz = \rho A_1 \int_0^a \left(\frac{-z^2 + 2az}{a^2}\right)^2 dz +$$

$$+\rho A_2 \int_a^{a+b} dz + \rho A_1 \int_{a+b}^{2a+b} \left[\frac{-z^2 + 2z\,(a+b) - \left(b^2 + 2ab\right)}{a^2}\right]^2 dz.$$

Performing all computations, it follows that

$$m = \frac{16}{15}\rho A_1 a + \rho A_2 b \ .$$

In a similar way, it follows

$$k = \int_0^l EI_y \left[\frac{\partial^2 q(z)}{\partial z^2}\right]^2 dz = \frac{8EI_1}{a^3} \ .$$

Note that the characteristics of the central part do not affect the stiffness. The natural frequency is thus

$$\omega = \sqrt{\frac{k}{m}} = \sqrt{\frac{8EJ_1}{\rho a^3 \left(\frac{16}{15}A_1 a + A_2 b\right)}} = 304.4 \ rad/s = 48.5 \ Hz \ .$$

A much better assumption is to use a sine function for the end parts of the shaft (full line in Fig. 14.2b):

$$q(z) = \sin\left(\frac{\pi z}{2a}\right) \ , \qquad for 0 \leq z \leq a$$

$$q(z) = 1 , \qquad\qquad for\ a \leq z \leq a + b$$

$$q(z) = \sin\left(\pi \frac{z^2 + b + 2a}{a} \right) , \quad for\ a + b \leq z \leq 2a + b .$$

Operating as seen before, it follows that

$$m = \int_0^{2a+b} \rho A \left[q(z) \right]^2 dz = \rho \left(A_1 a + A_2 b \right) .$$

The stiffness is also easily computed:

$$k = \int_0^l EI_y \left[\frac{\partial^2 q(z)}{\partial dz^2} \right]^2 dz = EI_1 a \left(\frac{\pi}{2a} \right)^4 .$$

The natural frequency is thus

$$\omega = \sqrt{\frac{k}{m}} = \frac{\pi^2}{4} \sqrt{\frac{EJ_1}{\rho a^3 \left(A_1 a + A_2 b \right)}} = 266.6\ rad/s = 42.4\ Hz .$$

The value computed through the FEM model based on 10 beam elements and on the consistent mass matrix (see Chapter 15) is

$$\omega = 264.6\ rad/s = 42.1\ Hz .$$

The results are compared in the following table, in which also the error with respect to the FEM solution with consistent mass matrix is reported as well. A first point is that all solutions yield results in excess to the correct one, except for the lumped-parameters FEM approach. A good guess is that correct solution is between 264.4 and 264.6 rad/s. The most simplified approach yields already quite good results.

	Spring–mass	Assumed modes		FEM (10 elements)	
		Parabolic.	Sine	Consistent	Lumped
ω_1 (rad/s)	280.7	304.4	266.6	264.6	264.4
ω_1 (Hz)	44.7	48.5	42.4	42.1	42.1
Error (%)	6.1	15.1	0.7	–	–0.07

The assumed-modes method yields results that depend much on the assumed function used. With parabolic deflected shape a 15% error is obtained, while with a sine function $q(z)$ the error reduces to less than 1%. This quite large difference is obtained with functions $q(z)$ which are apparently very close.

It is possible to increase its precision by introducing iterative procedures aimed at obtaining assumed modes that are closer to the actual first mode, by loading the structure with a distribution of inertia forces corresponding

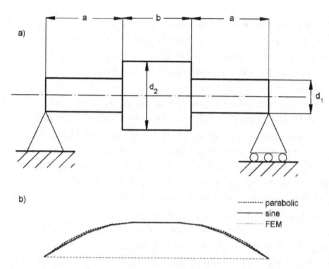

FIGURE 14.2. Shaft with a central boss: (**a**) sketch of the system; (**b**) inflected shape used for the computations of the first natural frequency (parabolic and sine) and first mode shape obtained through a FEM model.

to the deformation $q(x, y, z)$. The precision increases in this way, but the main advantage of the method, namely its simplicity, is lost.

Remark 14.2 *The greatest disadvantage of these approaches lies in their very principle: Because they are based on the approximation of the displacement field with functions defined in the whole field, they are not very well suited to model systems containing discontinuities or, at least, abrupt changes in characteristics, as often occurs in actual systems.*

The assumed-modes methods will not be dealt with in further detail, because their use in structural dynamics is now limited.

14.3   Lumped-parameters methods

Lumped-parameters methods are based on the discretization of continuous systems by discretizing their physical structure. The inertial properties are concentrated into a number of rigid bodies or point masses, located at chosen points, often called *stations* or *nodes*. They are connected by the fields to which the elastic properties of the structure are ascribed. There is no conceptual difficulty in also considering its damping properties. External force distributions are substituted by concentrated forces acting at the stations. The generalized displacements at the nodes are assumed to be generalized coordinates.

The determination of the mass matrix is more or less an arbitrary process, since it follows directly from the lumping of the system and is more a matter of common sense and engineering judgment than of structural analysis. The same can be said for the computation of the force vector.

It is obvious that some systems are better suited than others for this type of modeling: In some cases the mass is already lumped in the actual system and thus the model follows directly from the actual configuration. In other cases, however, the inertial properties of the actual system are distributed and many discretization schemes are possible. In this case it may be advisable to try different models to understand how the results are influenced by their complexity. The aim is clearly to reach the required precision with the simplest, and consequently the least costly, model.

The construction of the stiffness matrix is usually a difficult part of the computation. Traditionally, the compliance matrix **B** is obtained by computing the influence coefficients instead of the stiffness matrix **K**. It must be noted that the compliance matrix can be obtained only if the stiffness matrix is non-singular, i.e., if the system is constrained in such a way that no rigid-body motion is possible. If this does not occur, the compliance approach is not feasible, at least in a direct way.

Moreover, while the stiffness matrix of most structural systems has a band structure with a more or less narrow band, the compliance matrix does not have a similar structure: Its only regularity is usually that of being symmetrical.

For the aforementioned reasons, nowadays it is customary to resort to the stiffness approach and to use the FEM to compute the stiffness matrix. Many general-purpose FEM codes operate by default or as an option using the lumped-parameters approach instead of the consistent approach.

14.4   Transfer-matrices methods

The lumped-parameters methods, as described in the preceding section, involve long computations, mainly in two phases: the construction of the stiffness matrix (or of the compliance matrix) and the solution of the eigenproblem. To avoid such difficulties, a different approach to the computation of the natural frequencies was evolved: the transfer-matrices method. Two applications of this approach are well known and were very widely used, particularly when no automatic computing machines were in use: Holzer's *method* for torsional vibrations of shafts and *Myklestadt's method* for flexural vibrations of beam-like structures. Both were suited for tabular computations, even if they required much computational work, and when computers became available, computer codes based on them were written and widely used. They are still used, at least in particular cases. Their main limitation lies in the very principle of the transfer-matrices method, which is suited only for *in-line* systems, i.e., systems in which every station is linked by two fields to only two other stations, a leading station and a following station.

If the mass of the system is lumped in n nodes, the fields are $n - 1$. The computation starts at the first station (often called *station at the left*, as the structure is thought to go from left to right as a printed line) and ends at the last (right) station. It is clear that no branched systems or systems with multiple connections can be studied (unless by approximate approaches in which a branch is concentrated in the station where it stems from the main structure) and this is a serious drawback leading to the difficulty of writing general-purpose codes to deal with systems with different geometrical configurations.

The aforementioned limitations explain why the transfer-matrices method is now yielding to methods based on the study of the system as a whole, without having to go through it step by step from one station to the next.

The method is based on the definition of state vectors and transfer matrices. A *state vector* is a vector that contains the generalized displacements and forces related to the degrees of freedom that characterize the ends of each field, considered as insulated from the rest of the structure. Consequently, each field has two state vectors, one at its left end and one at its right end. The state vectors at the ends of a field are linked by the transfer matrix of the field

$$\mathbf{s}_{R_i} = \mathbf{T}_{f_i} \mathbf{s}_{Li}, \qquad (14.6)$$

where subscript i refers to the ith field and R and L designate the right and left ends, respectively. $\mathbf{T}_{f_i}$ is the transfer matrix of the ith field.

The left end of the ith field and the right end of the $(i - 1)$th field are located at the ith station, and between them there is the ith lumped mass. The corresponding state vectors are not coincident because the mass exerts generalized inertia forces on the node. They are linked by the transfer matrix of the ith station

$$\mathbf{s}_{Li} = \mathbf{T}_{n_i} \mathbf{s}_{R_{i-1}}. \qquad (14.7)$$

The station transfer matrix contains inertia forces due to the lumped mass that, in harmonic free vibrations, are functions of the square of the frequency of vibration. On the contrary, the field is massless and the field transfer matrix is independent of the frequency.

If there is a linear elastic constraint at the ith node, its stiffness can be introduced into the expression of the station transfer matrix. In this way it is possible to use the transfer-matrix method for systems in which the constraints are applied to nodes other than the first and last ones. The case of rigid constraints can be dealt with by introducing elastic constraints with very high stiffness.

The state vector at the left of the first station s_0 and that at the right of the last station s_n can be linked together by the equation

$$s_n = T_{n_n} T_{f_{n-1}} T_{n_{n-1}} T_{f_{n-2}} \cdots T_{n_2} T_{f_1} T_{n_1} s_0 = T_G s_0. \tag{14.8}$$

The overall transfer matrix

$$T_G = \prod_{i=n}^{1} T_i \tag{14.9}$$

is the product of all transfer matrices of all nodes and fields from the last to the first, in the correct order

Remark 14.3 *Matrix products depend on the order in which they are performed, so the overall transfer matrix must be computed by strictly following the aforementioned rule.*

The overall transfer matrix is a function of the frequency ω of the oscillations of the system or, better, of ω^2.

Some elements of the state vectors s_0 or s_n are known: If a degree of freedom is constrained, the corresponding generalized displacement is zero, while if it is free the corresponding generalized force vanishes. This leads to some simplifications, because some of the rows and columns of the overall transfer matrix can be canceled.

Equation (14.8) yields a solution different from the trivial one, with all state vectors equal to zero only if the overall transfer matrix is singular. Because such a matrix is a function of ω^2, this condition leads to an equation in ω^2, which coincides with the characteristic equation of the eigenproblem yielding the natural frequencies of the system.

The transfer-matrix approach is used to avoid solving an intricate eigenproblem and an alternative approach is usually followed. A value of the frequency ω is assumed, and the transfer matrices are computed and multiplied to obtain the overall transfer matrix. After canceling the rows and columns following the constraint conditions, its determinant is computed: If it vanishes, the frequency assumed is one of the natural frequencies. If this does not occur, a new frequency is assumed and the computation is repeated. By plotting the value of the determinant as a function of the frequency, it is easy to obtain as many natural frequencies as needed. Operating along these lines, no matrix of an order greater than that of the state vector (usually not higher than 4) must be dealt with. When the appropriate rows and columns of the overall transfer matrix have been canceled, the size of the determinant to be computed is usually not greater than 2. This explains why the method could be used without resorting to computers, even if long computations are usually involved.

Remark 14.4 *Cases in which the system is constrained in such a way that rigid-body motions are possible can be dealt with using the transfer-matrices method, obtaining a first natural frequency equal to zero.*

14.5 Holtzer's method for torsional vibrations of shafts

Consider a lumped torsional system, i.e., a system consisting of rigid discs connected by straight shafts possessing all the properties needed for the uncoupling of the torsional modes from flexural and axial ones (Fig. 14.3a). The system can result from the lumping of a more or less uniform continuous system or can model a system that is actually made of concentrated rotors and lightweight shafts. Where the torsional behavior is concerned, each station has only 1 degree of freedom, rotation ϕ_z. The order of the state vectors is then 2: It contains rotation ϕ_z and moment M_z.

The field transfer matrices are easily computed by considering that the moment at the left is equal to the moment at the right because no moment acts along the field, while the rotation at the right end is equal to the rotation at the left end increased by the twisting of the field

$$\begin{cases} \phi_{zR_i} = \phi_{zR_i} + \Delta\phi_{z_i} = \phi_{zR_i} + \dfrac{l_i}{G_i I'_{P_i}} M_{zL_i} \\ M_{zL_i} = M_{zR_i} . \end{cases} \tag{14.10}$$

The field transfer matrix is thus

$$\left\{ \begin{array}{c} \phi_z \\ M_z \end{array} \right\}_{R_i} = \begin{bmatrix} 1 & \dfrac{l_i}{G_i I'_{P_i}} \\ 0 & 1 \end{bmatrix} \left\{ \begin{array}{c} \phi_z \\ M_z \end{array} \right\}_{L_i} . \tag{14.11}$$

The station transfer matrix is obtained by considering that the rotation at the left of the ith station (i.e., the rotation at the right of the $(i-1)$th field) is equal to the rotation at the right, while the moment at the right of the station is equal to that at the left of the station, increased by the concentrated moment acting at the station. If no external moment acts on

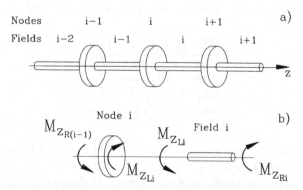

FIGURE 14.3. Holtzer method: (a) sketch of the system, (b) moments acting on the ith node and on the ith field.

the station, the latter is due only to the inertia reaction which, in harmonic motion, is proportional to the square of the frequency and to the rotation $\phi_{z_{L_i}}$. If the station is constrained by a torsional spring of stiffness χ_c, the moment $\chi_c \phi_{z_{L_i}}$ also acts on the node:

$$\begin{cases} \phi_{z_{R_i}} = \phi_{z_{R_i}} \\ M_{z_{L_i}} = M_{z_{R_i}} + \chi_c \phi_{z_{L_i}} - \omega^2 J_{z_i} \phi_{z_{L_i}}. \end{cases} \tag{14.12}$$

Since the state vector at the right of the ith station is the state vector at the left of the ith field and the state vector at the left of the ith station is the state vector at the right of the $(i-1)$th field, the equation defining the station transfer matrices is

$$\left\{ \begin{array}{c} \phi_z \\ M_z \end{array} \right\}_{L_i} = \left[\begin{array}{cc} 1 & 0 \\ -\omega^2 J_{z_i} + \chi_c & 1 \end{array} \right] \left\{ \begin{array}{c} \phi_z \\ M_z \end{array} \right\}_{R_{i-1}}. \tag{14.13}$$

Once the transfer matrices have been obtained, it is easy to multiply them to obtain the overall transfer matrix. Equation (14.8) takes the simple form

$$\left\{ \begin{array}{c} \phi_z \\ M_z \end{array} \right\}_n = \left[\begin{array}{cc} T_{11} & T_{12} \\ T_{21} & T_{22} \end{array} \right] \left\{ \begin{array}{c} \phi_z \\ M_z \end{array} \right\}_0. \tag{14.14}$$

The boundary conditions are easily assessed. If the end at the left is free, moment M_{z_0} must vanish. The second column is of no interest, because it multiplies a vanishing moment. If the end at the right is also free, as often happens with torsional systems, the second equation (14.14) reduces to

$$M_{z_n} = T_{21} \phi_{z_0} = 0, \tag{14.15}$$

yielding a solution different from the trivial one $\phi_{z_0} = 0$ only if $T_{21} = 0$.

Working in the same way for the other end conditions, the following equations yielding the natural frequencies can be obtained:

Right end	Left end	Equation
Free	Free	$T_{21} = 0$
Free	Clamped	$T_{22} = 0$
Clamped	Free	$T_{11} = 0$
Clamped	Clamped	$T_{12} = 0$

As already said, the solution is usually obtained numerically, plotting the appropriate element T_{ij} as a function of ω, and looking for the values of the frequency at which the curve crosses the frequency axis. The mode shapes can be obtained by setting an arbitrary value (usually a unit value) for M_{z_0}, if the left end is free, or for ϕ_{z_0} if it is clamped, and computing the state vectors after having introduced the value of the relevant natural frequency. Because only the shape of the modes is defined, the arbitrariness is not inconvenient, and it can be removed by normalizing the eigenvectors.

Example 14.3 *Compute the natural frequencies of the system shown in Examples 1.2 and 4.3 using Holtzer's method. Show that if the global transfer matrix is computed, it yields the same characteristic equation already obtained. Starting the numeration of the nodes from the constrained end, the following transfer matrices are obtained:*

$$\mathbf{T}_{f1} = \begin{bmatrix} 1 & 0.1 \\ 0 & 1 \end{bmatrix}, \qquad \mathbf{T}_{n2} = \begin{bmatrix} 1 & 0 \\ -\omega^2 & 1 \end{bmatrix},$$

$$\mathbf{T}_{f2} = \begin{bmatrix} 1 & 0.1 \\ 0 & 1 \end{bmatrix}, \qquad \mathbf{T}_{n3} = \begin{bmatrix} 1 & 0 \\ -4\omega^2 & 1 \end{bmatrix},$$

$$T_{f3} = \begin{bmatrix} 1 & 0.25 \\ 0 & 1 \end{bmatrix}, \qquad T_{n4} = \begin{bmatrix} 1 & 0 \\ -0.5\omega^2 & 1 \end{bmatrix}.$$

The global transfer matrix obtained by performing the relevant matrix multiplications is

$$T_G = \begin{bmatrix} 0.1\omega^4 - 1.35\omega^2 + 1 & 0.01\omega^4 - 0.235\omega^2 + 0.45 \\ -0.05\omega^6 + 1.075\omega^4 - 5.5\omega^2 & -0.005\omega^6 + 0.1575\omega^4 - 1.125\omega^2 + 1 \end{bmatrix}.$$

The right end is free while the left end is clamped and the condition allowing computation of the natural frequencies is $T_{22} = 0$, i.e.,

$$\omega^6 - 31.5\omega^4 + 225\omega^2 - 200 = 0,$$

which coincides with that obtained in Example 4.3.
The plot of element T_{22} of the global transfer matrix as a function of ω is reported in Fig. 14.4. The curve can be used to evaluate the natural frequencies without having to solve the characteristic equation.

In case of free–free boundary conditions a more usual formulation is based on the elimination of moments M_{zi} from Eqs. (14.7) and (14.14). Remembering the boundary condition at the left end, the following n equations are readily obtained:

$$\phi_{zi} = \left(1 - \frac{\omega^2 J_{zi}}{k_i} + \frac{k_{i-1}}{k_i}\right)\phi_{z(i-1)} - \frac{k_{i-1}}{k_i}\phi_{z(i-2)} \quad \text{for } i = 0,\ 1,\ ...,\ i-1,$$

$$(14.16)$$

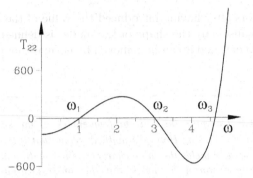

FIGURE 14.4. Plot used for the graphical computation of the natural frequencies of Example 4.3 using Holzer's method.

where ϕ_{zi} is the rotation at the right of the ith field, i.e., the rotation at the $(i+1)$th station and $k_i = G_i I'_{P_i}/l_i$ is the torsional stiffness of the ith field.

As usual, the computation starts by assuming a value of the frequency ω. The rotation ϕ_{z0} at the first station is arbitrarily assumed, usually by setting it equal to unity.

The first of Eq. (14.16)

$$\phi_{z1} = \left(1 - \frac{\omega^2 J_{zi}}{k_i}\right)\phi_{z0} \qquad (14.17)$$

yields the rotation at the second station. All rotations are computed step by step, until the rotation at the last station $\phi_{z(n-1)}$ is obtained.

From the boundary condition at the last station it follows that

$$\sum_{i=1}^{n} \omega^2 J_{z_i}\phi_{z(i-1)} = 0 , \qquad (14.18)$$

which is equivalent to Eq. (14.15) and is used to check whether the frequency is a natural frequency.

Remark 14.5 *The physical meaning of Eq. (14.18) is clear: Since no constraint acts on the system, inertia torques $\omega^2 J_{z_i}\phi_{z(i-1)}$ must be balanced.*

The value of

$$\sum_{i=1}^{n} \omega^2 J_{z_i}\phi_{z(i-1)}$$

is plotted against ω and the natural frequencies are obtained, as seen above.

14.6   Myklestadt's method for flexural vibrations of beams

Another well-known method based on the transfer-matrices approach is that introduced by Myklestadt for the flexural behavior of beams, originally developed for the dynamic study of aircraft wings. It will be presented here in a form different from that of the original paper by Myklestadt with the aim of organizing it in the way generally seen for transfer-matrices methods.

Consider the bending behavior of a beam, modeled as a lumped-parameters system, in the xz-plane. Because a moment of inertia J_y can be associated with each station, they are considered rigid bodies rather than point masses. Each end of a field is characterized by two degrees of freedom: displacement u_x and rotation ϕ_y; the order of the state vectors is four, and they contain the shearing force F_x and bending moment M_y as well. The computation of the transfer matrices follows the same guidelines seen for Holtzer's method. Considering the ith field, it follows that

- the shear force at the right end is equal to the shear force at the left end, because no force acts on the field

- the bending moment at the right end is equal to the bending moment at the left end, increased by the moment due to the shear force at the left end

- the displacement at the right end is equal to the displacement at the left end, increased by the displacements due to the rigid rotation and to the deformation of the field

- the rotation at the right end is equal to the rotation at the left end increased by the rotation due to bending deformation of the field

$$\begin{cases} u_{x_{R_i}} = u_{x_{L_i}} + l_i \phi_{y_{L_i}} + \Delta u_{xi} \\ \phi_{y_{R_i}} = \phi_{y_{L_i}} + \Delta \phi_{yi} \\ F_{x_{R_i}} = F_{x_{L_i}} \\ M_{y_{R_i}} = M_{y_{L_i}} + l_i F_{x_{L_i}} \end{cases} \qquad (14.19)$$

The increments of displacement and rotation Δu_{xi} and $\Delta \phi_{yi}$ due to the deformation of the field can be computed through the beam theory because the ith field can be considered as a prismatic beam clamped at the left end and loaded at the right end by a force $F_{x_{R_i}}$

$$\begin{cases} (\Delta u_{xi})_F = F_{x_{R_i}} \left(\dfrac{l_i^3}{3EI_y} + \dfrac{l_i \chi}{GA} \right) \\ (\Delta \phi_{yi})_F = F_{x_{R_i}} \dfrac{l_i^2}{2EI_y} \end{cases} \qquad (14.20)$$

and a moment $M_{y R_i}$

$$
\begin{cases}
(\Delta u_{xi})_M = M_{y R_i} \dfrac{l_i^2}{2 E I_y} \\[2mm]
(\Delta \phi_{yi})_M = M_{y R_i} \dfrac{l_i}{E I_y}
\end{cases}
. \qquad (14.21)
$$

The relationship linking the state vector at the left and at the right of a field can thus be written as

$$
\left\{
\begin{array}{c}
u_x \\ \phi_y \\ F_x \\ M_y
\end{array}
\right\}_{R_i}
=
\begin{bmatrix}
1 & l_i & -\dfrac{l_i^3}{6 E I_y} + \dfrac{l_i \chi}{G A} & \dfrac{l_i^2}{2 E I_y} \\[3mm]
0 & 1 & -\dfrac{l_i^2}{2 E I_y} & \dfrac{l_i}{E I_y} \\[3mm]
0 & 0 & 1 & 0 \\[2mm]
0 & 0 & -l_i & 1
\end{bmatrix}
\left\{
\begin{array}{c}
u_x \\ \phi_y \\ F_x \\ M_y
\end{array}
\right\}_{L_i}
. \qquad (14.22)
$$

Now consider the ith node. The displacement and the rotation at the left of the node are equal to the displacement and the rotation at the right. The force at the right is equal to the force at the left increased by the inertia force due to mass m_i which, in harmonic motion, is $-\omega^2 m_i u_x$; the moment at the right is equal to the moment at the left increased by the inertia torque $-\omega^2 J_{i_y} \phi_y$. If a constraint with stiffness k_c and angular stiffness χ_c is located at the node, the force at the left of the node must be increased by $k_c u_x$ and the moment must be increased by $\chi_c \phi_y$. Using matrix notation, the following expression for the transfer matrix of the ith node is obtained:

$$
\left\{
\begin{array}{c}
u_x \\ \phi_y \\ F_x \\ M_y
\end{array}
\right\}_{L_i}
=
\begin{bmatrix}
1 & 0 & 0 & 0 \\
0 & 1 & 0 & 0 \\
-\omega_i^2 m_i + k_c & 0 & 1 & 0 \\
0 & -\omega^2 J_{i_y} + \chi_c & 0 & 1
\end{bmatrix}
\left\{
\begin{array}{c}
u_x \\ \phi_y \\ F_x \\ M_y
\end{array}
\right\}_{R_{i-1}}
.
$$
$$(14.23)$$

Once the transfer matrices of all nodes and all fields have been obtained, there is no difficulty in computing the global transfer matrix and introducing the boundary conditions in a way that is similar to that seen for the Holtzer method.

Consider, for example, that both ends are supported: The displacement and the bending moment must vanish in both s_n and s_0. Equation (14.8) then becomes

$$
\left\{
\begin{array}{c}
0 \\ \phi_y \\ F_x \\ 0
\end{array}
\right\}_n
=
\begin{bmatrix}
T_{11} & T_{12} & T_{13} & T_{14} \\
T_{21} & T_{22} & T_{23} & T_{24} \\
T_{31} & T_{32} & T_{33} & T_{34} \\
T_{41} & T_{42} & T_{43} & T_{44}
\end{bmatrix}
\left\{
\begin{array}{c}
0 \\ \phi_y \\ F_x \\ 0
\end{array}
\right\}_0
. \qquad (14.24)
$$

The first and fourth columns of the global transfer matrix can be canceled, because they multiply elements equal to zero in the state vector. The first and last equations (14.8) reduce to the homogeneous equation

$$\begin{bmatrix} T_{12} & T_{13} \\ T_{42} & T_{43} \end{bmatrix} \begin{Bmatrix} \phi_y \\ F_x \end{Bmatrix}_0 = \begin{Bmatrix} 0 \\ 0 \end{Bmatrix}, \tag{14.25}$$

which yields a solution different from the trivial one only if

$$\det \begin{bmatrix} T_{12} & T_{13} \\ T_{42} & T_{43} \end{bmatrix} = T_{12}T_{43} - T_{42}T_{13} = 0. \tag{14.26}$$

In this case the possible boundary conditions are many; only a few of them are reported here:

Right end	Left end	Equation
Free	Free	$T_{31}T_{42} - T_{41}T_{32} = 0$
Free	Supported	$T_{32}T_{43} - T_{42}T_{33} = 0$
Free	Clamped	$T_{33}T_{44} - T_{43}T_{34} = 0$
Supported	Supported	$T_{12}T_{43} - T_{42}T_{13} = 0$
Clamped	Free	$T_{11}T_{22} - T_{21}T_{12} = 0$
Clamped	Clamped	$T_{13}T_{24} - T_{23}T_{14} = 0$

The value of the determinant of the matrix of the coefficients is plotted as a function of the frequency ω, and the values of the frequency for which it vanishes are obtained in a way similar to that seen for the Holtzer's method.

Remark 14.6 *The order in which the state variables are listed in the state vector can be different from the one shown here. Before using any formula from the literature the meaning of the subscripts of the elements of the transfer matrix must be verified.*

Example 14.4 *Compute the natural frequencies of the beam shown in Fig. 14.5a using Myklestadt's method. The beam has an annular cross-section with inner and outer diameters of 80 mm and 100 mm, respectively, and the dimensions shown in the figure are $l_1 = l_3 = 200$ mm and $l_2 = 800$ mm. The main characteristics of the material are $E = 2.1 \times 10^{11}$ N/m^2, $\rho = 7,810$ kg/m^3, and $\nu = 0.3$.*

The mass of the beam is lumped in seven stations (4.417 kg in all nodes except the first and last ones where there is a mass of 2.209 kg). In the second and sixth stations there are two supports. The rigid supports are modeled using elastic constraints with a stiffness of 1×10^{11} N/m. The computation of the global transfer matrix involves the multiplication of six field transfer matrices and seven node transfer matrices.

Because both ends are free, the plot of the determinant $T_{31}T_{42} - T_{41}T_{32}$ of the reduced global transfer matrix as a function of ω is reported in Fig. 14.5b, for a range up to the third natural frequency.

The natural frequencies, up to the fifth, computed by searching the points at which the plot of Fig. 14.5b (extended to higher frequencies) crosses the frequency axis, are

$\omega_1 = 2,158$ rad/s $= 343.4$ Hz, $\omega_2 = 5,091$ rad/s $= 810.2$ Hz,
$\omega_3 = 6,780$ rad/s $= 1,079.1$ Hz, $\omega_4 = 10,469$ rad/s $= 1,666.2$ Hz, and
$\omega_5 = 15,858$ rad/s $= 2,523.9$ Hz .

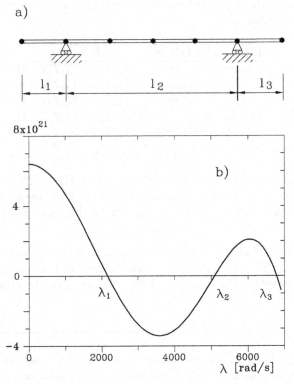

FIGURE 14.5. (a) Sketch of the beam of Example 14.4. (b) Plot used for the graphical computation of the natural frequencies using Myklestadt's method.

14.7   Exercises

Exercise 14.1 *Repeat the study of Example 4.3, computing also the three mode shapes of the system.*

Exercise 14.2 *Consider the transmission shaft of Exercise 12.1. Compute the first four flexural natural frequencies using the Myklestadt method. Lump the mass of the shaft in two, four, and eight stations.*

Exercise 14.3 *Consider the chimney studied in Exercise 12.2. Compute the first three flexural natural frequencies using the Myklestadt method. Lump the mass of the shaft in eight stations.*

Exercise 14.4 *Consider the beam of Example 4.1, but now take into account also the mass of the beam. Compute the first three flexural natural frequencies using the Myklestadt method, neglecting shear deformation. Lump the mass of the shaft in eight stations. Repeat the study taking into account shear deformation.*

15
The Finite Element Method

The finite element method is at present the most common discretization method. Its success is due to the possibility of using it for a wide variety of problems, but above all to the availability of computing machines of ever-increasing power. The method yields usually models with a large number (hundred thousands or millions) of degrees of freedom, but the ODEs so obtained are easily implemented in general-purpose codes for digital computers. It can be used for both time-domain and frequency-domain computations.

15.1   Element characterization

The finite element method is a general discretization method for the solution of partial derivative differential equations and, consequently, it finds its application in many other fields beyond structural static and dynamic analysis. The aim of this chapter is not to provide a complete survey of the method, which can be dealt with only in a specialized text, but simply to describe its main features, comparing it with the other discretization techniques.

The FEM is based on the subdivision of the structure into finite elements, i.e., into parts whose dimensions are not vanishingly small. Many different element formulations have been developed, depending on their shape and characteristics: beam elements, shell elements, plate elements, solid elements, and many others. A structure can be built by assembling elements of the same or different types, as dictated by the nature of the problem and by the capabilities of the computer code used.

The FEM is usually developed using matrix notation to obtain formulas easily translated into computer codes. The displacement is thus written as a vector of order 3 in the tridimensional space (sometimes of higher order, if rotations are also considered), and the equation expressing the displacement of the points inside each element is

$$\mathbf{u}(x, y, z, t) = \mathbf{N}(x, y, z)\mathbf{q}(t) \,, \tag{15.1}$$

where $\mathbf{q}$ is a vector where the n generalized coordinates of the element are listed and $\mathbf{N}$ is the matrix containing the shape functions. There are as many rows in $\mathbf{N}$ as in $\mathbf{u}$ and as many columns as the number n of degrees of freedom.

Usually the degrees of freedom of the elements are the displacements at given points, referred to as *nodes*. In this case, Eq. (15.1) reduces to the simpler form,

$$\left\{ \begin{array}{l} u_x(x, y, z, t) \\ u_y(x, y, z, t) \\ u_z(x, y, z, t) \end{array} \right\} = \left[\begin{array}{ccc} \mathbf{N}(x, y, z) & \mathbf{0} & \mathbf{0} \\ \mathbf{0} & \mathbf{N}(x, y, z) & \mathbf{0} \\ \mathbf{0} & \mathbf{0} & \mathbf{N}(x, y, z) \end{array} \right] \left\{ \begin{array}{l} \mathbf{q}_x(t) \\ \mathbf{q}_y(t) \\ \mathbf{q}_z(t) \end{array} \right\} , \tag{15.2}$$

where the displacements in each direction are functions of the nodal displacements in the same direction only. In this case matrix $\mathbf{N}$ has only one row and as many columns as the number of nodes of the element. Equation (15.2) has been written for a three-dimensional element; a similar formulation can also be easily obtained for one- or two-dimensional elements.

Each element is essentially the model of a small deformable solid. The behavior of the element is studied using an assumed-modes approach, as seen in Section 14.2. A limited number, usually small, of degrees of freedom is then substituted to the infinity of degrees of freedom of each element. Inside each element, the displacement $\bar{u}(x, y, z)$ of the point of coordinates x, y, z is approximated by the linear combination of a number n of arbitrarily assumed functions, the shape functions.

The shape functions are, as already stated, arbitrary. The freedom in the choice of such functions is, however, limited, because they must satisfy several conditions. A first requirement is a simple mathematical formulation: A set of polynomials in the space coordinates is thus usually assumed.

The results of the analysis converge toward the exact solution of the differential equations constituting the continuous model discretized by the FEM, with decreasing element size (i.e., with increasing number of elements) if the shape functions

- are continuous and differentiable up to the required order, which depends on the type of element;

- are able to describe rigid-body motions of the element leading to vanishing elastic potential energy;

- lead to a constant strain field when the overall deformation of the element dictates so; and

- lead to a deflected shape of each element that matches the shape of the neighboring elements.

The last condition means that when the nodes of two neighboring elements displace in a compatible way, all the interfaces between the elements must displace in a compatible way.

Another condition, not always satisfied, is that the shape functions are isotropic, i.e., do not show geometrical properties that depend on the orientation of the reference frame.

Sometimes not all these conditions are completely met; in particular, there are elements that fail to completely satisfy the matching of the deflected shapes of neighboring elements (see Section 15.4).

The nodes are usually located at the vertices or on the sides of the elements and are common to two or more of them, but points that are internal to an element may also be used.

The equation of motion of each element can be written following what was said about the assumed-modes methods. The strains can be expressed as functions of the derivatives of the displacements with respect to space coordinates. In general, it is possible to write a relationship of the type

$$\epsilon(x, y, z, t) = \mathbf{B}(x, y, z)\mathbf{q}(t), \qquad (15.3)$$

where ϵ is a column matrix in which the various elements of the strain tensor are listed (it is commonly referred to as a *strain vector* but it is such only in the sense that it is a column matrix) and $\mathbf{B}$ is a matrix[1] containing appropriate derivatives of the shape functions. $\mathbf{B}$ has as many rows as the number of components of the strain vector and as many columns as the number of degrees of freedom of the element.

If the element is free from initial stresses and strains and the behavior of the material is linear, the stresses are obtained from the strains as

$$\boldsymbol{\sigma}(x, y, z, t) = \mathbf{E}\epsilon = \mathbf{E}(x, y, z)\mathbf{B}(x, y, z)\mathbf{q}(t), \qquad (15.4)$$

where $\mathbf{E}$ is the stiffness matrix of the material. It is a symmetric square matrix whose elements can theoretically be functions of the space coordinates

---

[1]Matrix $\mathbf{B}$ (capital b) must not be confused with the influence coefficients matrix $\mathbf{B}$ (capital $\boldsymbol{\beta}$).

but are usually constant within the element. The potential energy of the element can be expressed as

$$\mathcal{U} = \frac{1}{2} \int_V \boldsymbol{\epsilon}^T \boldsymbol{\sigma} dV = \frac{1}{2} \mathbf{q}^T \left(\int_V \mathbf{B}^T \mathbf{E} \mathbf{B} dV \right) \mathbf{q}. \tag{15.5}$$

The integral in Eq. (15.5) is the stiffness matrix of the element

$$\mathbf{K} = \int_V \mathbf{B}^T \mathbf{E} \mathbf{B} dV. \tag{15.6}$$

Because the shape functions do not depend on time, the generalized velocities can be expressed as

$$\dot{\mathbf{u}}(x, y, z, t) = \mathbf{N}(x, y, z) \dot{\mathbf{q}}(t).$$

In the case where all generalized coordinates are related to displacements, the kinetic energy and the mass matrix of the element can be expressed as

$$\mathcal{T} = \frac{1}{2} \int_V \rho \dot{\mathbf{u}}^T \dot{\mathbf{u}} dV = \frac{1}{2} \dot{\mathbf{q}}^T \left(\int_V \rho \mathbf{N}^T \mathbf{N} dV \right) \dot{\mathbf{q}},$$

$$\mathbf{M} = \int_V \rho \mathbf{N}^T \mathbf{N} dV. \tag{15.7}$$

If some generalized displacements are rotations, Eq. (15.7) must be modified to introduce the moments of inertia, but its basic structure remains the same.

As already stated, the FEM is often used just to compute the stiffness matrix to be used in the context of the lumped-parameters approach. In this case, the consistent mass matrix (15.7) is not computed and a diagonal matrix obtained by lumping the mass at the nodes is used. The advantage is that of dealing with a diagonal mass matrix, whose inversion to compute the dynamic matrix is far simpler than that of the consistent mass matrix. The accuracy is, however, reduced or, better, a greater number of elements is needed to reach the same accuracy, and then the convenience between the two formulations must be assessed in each case.

Remark 15.1 *As a general rule, the consistent approach leads to values of the natural frequencies in excess with respect to those computed using the lumped-parameters approach. If enough elements are used, the frequencies obtained from the elastic continuum model lie in between those computed through the lumped-parameters and the consistent approach.*

If a force distribution $\mathbf{f}(x, y, z, t)$ acts on the body, the virtual work linked with the virtual displacement

$$\delta \mathbf{u} = \mathbf{N} \delta \mathbf{q}$$

and the nodal force vector can be expressed in the form

$$\delta\mathcal{L} = \int_V \delta\mathbf{u}^T\mathbf{f}(x,y,z,t)dV = \int_V \delta\mathbf{q}^T\mathbf{N}^T\mathbf{f}(x,y,z,t)dV,$$

$$\mathbf{f}(t) = \int_V \mathbf{N}^T\mathbf{f}(x,y,z,t)dV.$$

(15.8)

In a similar way, it is possible to obtain the nodal force vectors corresponding to surface force distributions or to concentrated forces acting on any point of the element.

The equation of motion of the element is then the usual one for discrete undamped systems

$$\mathbf{M\ddot{q}} + \mathbf{Kq} = \mathbf{f}(t).$$

(15.9)

Remark 15.2 *The equations of motion and the relevant matrices have been obtained here by closely following the assumed-modes approach, basically using Lagrange equations; this approach is neither the only one nor the most common.*

15.2   Timoshenko beam element

The beam element is one of the most common elements and is generally available in all computer codes. Several beam formulations have been developed that differ owing to both the theoretical formulation and the number of nodes and degrees of freedom per node. Some of them are Euler–Bernoulli element, i.e., do not take into account shear deformation, while others are Timoshenko elements.

The element that will be studied here is often referred to as the *simple Timoshenko beam*. It is a prismatic homogeneous beam, with a node at each end and six degrees of freedom per node, to which all considerations seen in Section 12.1 apply. The relevant geometrical definition and the reference frame used for the study are shown in Fig. 15.1.

Like in the beam studied in Chapter 12, each cross section has six degrees of freedom, three displacements, and three rotations. The total number of degrees of freedom of the element is thus 12. The vector of the nodal displacements, i.e., of the generalized coordinates of the element, is

$$\mathbf{q} = [u_{x_1}, u_{y_1}, u_{z_1}, \phi_{x_1}, \phi_{y_1}, \phi_{z_1}, u_{x_2}, u_{y_2}, u_{z_2}, \phi_{x_2}, \phi_{y_2}, \phi_{z_2}]^T.$$

(15.10)

Since the beam has the properties needed to perform a complete uncoupling between axial, torsional, and flexural behaviors in each of the coordinate planes, it is expedient to subdivide vector $\mathbf{q}$ into four smaller vectors:

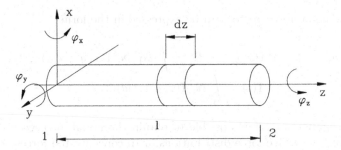

FIGURE 15.1. Beam element: geometrical definitions and reference frame.

$$\mathbf{q}_A = \left\{ \begin{array}{c} u_{z1} \\ u_{z2} \end{array} \right\}, \quad \mathbf{q}_T = \left\{ \begin{array}{c} \phi_{z1} \\ \phi_{z2} \end{array} \right\}, \quad \mathbf{q}_{F1} = \left\{ \begin{array}{c} u_{x1} \\ \phi_{y1} \\ u_{x2} \\ \phi_{y2} \end{array} \right\}, \quad \mathbf{q}_{F2} = \left\{ \begin{array}{c} u_{y1} \\ \phi_{x1} \\ u_{y2} \\ \phi_{x2} \end{array} \right\}.$$
(15.11)

Remark 15.3 *If the rotational degree of freedom* $-\phi_x$ *were used instead of* ϕ_x *the same equations could be used to describe the flexural behavior in both planes. This approach is, however, uncommon because it would make it more difficult to pass from the system of reference of the elements to that of the whole structure (See Section 15.9).*

By reordering the various generalized coordinates with the aim of clearly showing such uncoupling, vector $\mathbf{q}$ can be written as

$$\mathbf{q} = \left[\begin{array}{cccc} \mathbf{q}_A^T & \mathbf{q}_T^T & \mathbf{q}_{F1}^T & \mathbf{q}_{F2}^T \end{array} \right]^T.$$
(15.12)

The uncoupling between the various degrees of freedom makes it possible to split the matrix of the shape functions into a number of submatrices, most of which are equal to zero. The generalized displacement of the internal points of the element can be expressed in the form of Eq. (15.1) as

$$\mathbf{u}(z,t) = \left\{ \begin{array}{c} u_z \\ \phi_z \\ u_x \\ \phi_y \\ u_y \\ \phi_x \end{array} \right\} = \left[\begin{array}{cccc} \mathbf{N}_A & \mathbf{0} & \mathbf{0} & \mathbf{0} \\ \mathbf{0} & \mathbf{N}_T & \mathbf{0} & \mathbf{0} \\ \mathbf{0} & \mathbf{0} & \mathbf{N}_{F1} & \mathbf{0} \\ \mathbf{0} & \mathbf{0} & \mathbf{0} & \mathbf{N}_{F2} \end{array} \right] \left\{ \begin{array}{c} \mathbf{q}_A \\ \mathbf{q}_T \\ \mathbf{q}_{F1} \\ \mathbf{q}_{F2} \end{array} \right\}.$$
(15.13)

15.2.1   Axial behavior

The beam reduces to a bar for what the axial behavior is concerned.

Because each point of the element has a single degree of freedom, vector $\mathbf{u}$ has a single component u_z and matrix $\mathbf{N}_A$ has one row and two columns (the element has two degrees of freedom). Displacement u_z can be expressed as a polynomial in z or, better, in the nondimensional axial coordinate

$$\zeta = \frac{z}{l} \; :$$

$$u_z = a_0 + a_1\zeta + a_2\zeta^2 + a_3\zeta^3 + \dots . \tag{15.14}$$

The polynomial must yield the values of the displacements u_{z_1} and u_{z_2} at the left end (node 1, $\zeta=0$) and at the right end (node 2, $\zeta=1$), respectively. These two conditions allow computation of only two coefficients a_i and thus the polynomial expansion of the displacement must include only two terms, i.e., the constant and the linear terms:

$$u_z = a_0 + a_1\zeta . \tag{15.15}$$

By stating that the displacements at the two nodes are equal to u_{z_1} and u_{z_2}, it follows

$$\begin{cases} u_{z_1} = a_0 \\ u_{z_2} = a_0 + a_1 \; , \end{cases} \tag{15.16}$$

i.e.,

$$\begin{cases} a_0 = u_{z_1} \\ a_1 = -u_{z_1} + u_{z_2} \; . \end{cases} \tag{15.17}$$

The polynomial is thus

$$u_z = u_{z_1} - \zeta\left(u_{z_1} - u_{z_2}\right) \; , \tag{15.18}$$

yielding the following matrix of the shape functions:

$$\mathbf{N}_A = [1 - \zeta, \; \zeta] . \tag{15.19}$$

The axial strain ϵ_z can be expressed as

$$\epsilon_z = \frac{du_z}{dz} , \tag{15.20}$$

or, using vector ϵ, which in this case has only one element

$$\epsilon_z = \left[\frac{d}{dz}(1-\zeta), \; \frac{d}{dz}\zeta\right] \begin{Bmatrix} u_{z_1} \\ u_{z_2} \end{Bmatrix} . \tag{15.21}$$

Matrix

$$\mathbf{B} = \left[\frac{d}{dz}(1-\zeta), \; \frac{d}{dz}\zeta\right] = \frac{1}{l}[-1, \; 1] \tag{15.22}$$

has one row and two columns.

Also, vector σ and matrix $\mathbf{E}$ have only a single element: the axial stress σ_z and Young's modulus E, respectively. The stiffness and mass matrices can be obtained directly from Eqs. (15.6) and (15.7).

Remembering that

$$dV = A dz \ ,$$

they reduce to

$$\mathbf{K}_A = \int_0^l \mathbf{A}\mathbf{B}^T E \mathbf{B} dz = \frac{EA}{l} \int_0^1 \begin{bmatrix} 1 & -1 \\ -1 & 1 \end{bmatrix} d\zeta = \frac{EA}{l} \begin{bmatrix} 1 & -1 \\ -1 & 1 \end{bmatrix} ,$$
(15.23)

$$\mathbf{M}_A = \int_0^l \rho \mathbf{A}\mathbf{N}^T \mathbf{N} dz = \rho A l \int_0^1 \begin{bmatrix} (1-\zeta)^2 & \zeta(1-\zeta) \\ \zeta(1-\zeta) & \zeta^2 \end{bmatrix} d\zeta =$$

$$= \frac{\rho A l}{6} \begin{bmatrix} 2 & 1 \\ 1 & 2 \end{bmatrix} .$$
(15.24)

If the mass is lumped half in each node, the mass matrix is simply

$$\mathbf{M}_{A_l} = \frac{\rho A l}{2} \begin{bmatrix} 1 & 0 \\ 0 & 1 \end{bmatrix} .$$
(15.25)

If an axial force distribution $f_z(t)$ that is constant along the space coordinate z or a concentrated axial force $F_{z_k}(t)$ located in the point of coordinate z_k acts on the bar, the nodal force vector is, respectively,

$$\mathbf{f}(t) = l \left[\int_0^l \left\{ \begin{array}{c} (1-\zeta) \\ \zeta \end{array} \right\} d\zeta \right] f_z(t) = f_z(t) \frac{l}{2} \left\{ \begin{array}{c} 1 \\ 1 \end{array} \right\} ,$$
(15.26)

or

$$\mathbf{f}(t) = F_{z_k}(t) \left\{ \begin{array}{c} 1 - \dfrac{z_k}{l} \\ \dfrac{z_k}{l} \end{array} \right\} .$$
(15.27)

In the first case the distributed load has been reduced to two identical forces; each is equal to half of the total load acting on the bar.

Remark 15.4 *While the consistent mass matrix is different from the lumped mass matrix, in this case an identical force vector would have been obtained by simply lumping the load at the nodes. This is not, however, a general rule; in other cases the consistent approach leads to a load vector that is different from that obtained using the lumped-parameters approach.*

15.2.2   Torsional behavior

As stated in Section 2.2.3, the equations of motion governing the torsional behavior of beams are formally identical to those governing the axial behavior. Using this identity, the characterization of the beam element in torsion can be obtained from what has been seen for the axial behavior. Matrix $\mathbf{N}_T$ is identical to matrix $\mathbf{N}_A$

$$\mathbf{N}_T = [1 - \zeta, \, \zeta]$$

and the expressions of the relevant matrices and vectors are

$$\mathbf{M}_T = \frac{\rho I_p l}{6} \begin{bmatrix} 2 & 1 \\ 1 & 2 \end{bmatrix}, \qquad \mathbf{K}_T = \frac{G I'_p}{l} \begin{bmatrix} 1 & -1 \\ -1 & 1 \end{bmatrix},$$

$$\mathbf{f}(t)_T = \frac{1}{2} l m_z(t) \begin{Bmatrix} 1 \\ 1 \end{Bmatrix}. \tag{15.28}$$

15.2.3   Flexural behavior in the xz-plane

The expressions of the shape function are in this case more complex. Matrix $\mathbf{N}_{F1}$ has two rows and four columns because it must yield the displacement in the x-direction and the rotation about the y-axis of the generic cross section of the beam when it is multiplied by vector $\mathbf{q}_{F1}$, which contains four elements, namely, the displacements in the x-direction and the rotation about the y-axis of the two nodes. The simplest approach would be that of assuming polynomial expressions for the generalized displacements

$$\begin{cases} u_x = a_0 + a_1\zeta + a_2\zeta^2 + a_3\zeta^3 + \dots \\ \phi_y = b_0 + b_1\zeta + b_2\zeta^2 + b_3\zeta^3 + \dots. \end{cases} \tag{15.29}$$

These polynomial expressions must yield the values of the displacements u_{x1} and u_{x2} and of the rotations ϕ_{y1} and ϕ_{y2} at the left (node 1, $\zeta=0$) and at the right ends (node 2, $\zeta=1$). These four conditions allow computation of four coefficients a_i and b_i only. Each polynomial must thus include only two terms, and both rotations and displacements must vary linearly along the z-coordinate. This element formulation, although sometimes used, leads to the very severe problem of locking, i.e., to the possibility of grossly overestimating the stiffness of the element.

Although locking will not be dealt with in detail here (the reader can find a detailed discussion on this matter in any good textbook on FEM), an intuitive explanation can be seen immediately: If the beam is slender the rotation of each cross section is very close to the derivative of the displacement, as stated by the Euler–Bernoulli approach for slender beams. The polynomial shape functions, when truncated at the second term, do not allow the rotation to be equal to the derivative of the displacement and forces the curvature (approximated as the second derivative of the displacement) to vanish. It can be shown that in this way the displacements, i.e., the flexibility of the beam, are severely underestimated.

A simple cure for the problem is that of resorting to the Euler–Bernoulli formulation, i.e., neglecting shear deformation and then assuming that the rotation is coincident with the derivative of the displacement. In this case, only the polynomial for u_x needs to be stated, and the aforementioned four conditions at the nodes can be used to compute the four coefficients of a cubic expression of the displacement.

To avoid locking, a Timoshenko beam element can be formulated using the deformed shape computed using the continuous model assuming that only end forces are applied to the beam as shape functions. This Timoshenko beam element reduces to the Euler–Bernoulli element when the slenderness of the beam increases and no locking occurs. The relevant functions are

$$N_{11} = \frac{1 + \Phi(1 - \zeta) - 3\zeta^2 + 2\zeta^3}{1 + \Phi}, \quad N_{12} = l\zeta \frac{1 + \frac{1}{2}\Phi(1 - \zeta) - 2\zeta + \zeta^2}{1 + \Phi},$$

$$N_{13} = \zeta \frac{\Phi + 3\zeta - 2\zeta^2}{1 + \Phi}, \quad N_{14} = l\zeta \frac{-\frac{1}{2}\Phi(1 - \zeta) - \zeta + \zeta^2}{1 + \Phi},$$

$$N_{21} = 6\zeta \frac{\zeta - 1}{l(1 + \Phi)}, \quad N_{22} = \frac{1 + \Phi(1 - \zeta) - 4\zeta + 3\zeta^2}{1 + \Phi},$$

$$N_{23} = -6\zeta \frac{\zeta - 1}{l(1 + \Phi)}, \quad N_{24} = \frac{\Phi\zeta - 2\zeta + 3\zeta^2}{1 + \Phi},$$

$$(15.30)$$

where

$$\Phi = \frac{12EI_y\chi}{GAl^2}$$

is the ratio between the shear and the flexural flexibility of the beam.

When the slenderness of the beam increases, the value of Φ decreases, tending to zero for an Euler–Bernoulli beam.

In beam elements some of the generalized coordinates are related to rotations; as a consequence, Eqs. (15.6) and (15.7) cannot be used directly to express the stiffness and mass matrices.

The potential energy can be computed by adding the contributions due to bending and shear deformations. By using the symbols $\mathbf{N}_1$ and $\mathbf{N}_2$ to express the first and second rows of matrix $\mathbf{N}_{F1}$, respectively, the two contributions to the potential energy of the length dz of the beam are

$$d\mathcal{U}_f = \frac{1}{2}EI_y \left(\frac{d\phi_y}{dz}\right)^2 dz = \frac{1}{2}EI_y\{q\}^T \left[\frac{d}{dz}\mathbf{N}_2\right]^T \left[\frac{d}{dz}\mathbf{N}_2\right] \{q\}dz,$$

$$d\mathcal{U}_s = \frac{1}{2}\frac{GA}{\chi} \left(\phi_y - \frac{du_x}{dz}\right)^2 dz = \qquad (15.31)$$

$$= \frac{12EI_y}{2\Phi l^2}\{q\}^T \left[\mathbf{N}_2 - \frac{d}{dz}\mathbf{N}_1\right]^T \left[\mathbf{N}_2 - \frac{d}{dz}\mathbf{N}_1\right]\{q\}dz\,.$$

By introducing the shape functions into the expression of the potential energy and integrating, the bending stiffness matrix is obtained

$$\mathbf{K}_{F_1} = \frac{EI_y}{l^3(1+\Phi)}\begin{bmatrix} 12 & 6l & -12 & 6l \\ & (4+\Phi)l^2 & -6l & (2-\Phi)l^2 \\ & & 12 & -6l \\ \text{symm.} & & & (4+\Phi)l^2 \end{bmatrix}. \tag{15.32}$$

The kinetic energy of the length dz of the beam is

$$dT = \frac{1}{2}\rho A\dot{u}^2 dz + \frac{1}{2}\rho I_y \dot{\phi}_y^{\,2} dz =$$

$$= \frac{1}{2}\rho A\dot{\mathbf{q}}^T\mathbf{N}_1^T\mathbf{N}_1\dot{\mathbf{q}}dz + \frac{1}{2}\rho I_y\dot{\mathbf{q}}^T\mathbf{N}_2^T\mathbf{N}_2\dot{\mathbf{q}}dz\,. \tag{15.33}$$

The first term on the right-hand side is the translational kinetic energy, and the second expresses the rotational kinetic energy, which is often neglected in the case of slender beams. By introducing the expressions of the shape functions and integrating, the consistent mass matrix is obtained

$$\mathbf{M}_{F1} = \frac{\rho Al}{420(1+\Phi)^2}\begin{bmatrix} m_1 & lm_2 & m_3 & -lm_4 \\ & l^2m_5 & lm_4 & -l^2m_6 \\ & & m_1 & -lm_2 \\ \text{symm.} & & & l^2m_5 \end{bmatrix} +$$

$$+\frac{\rho I_y}{30l(1+\Phi)^2}\begin{bmatrix} m_7 & lm_8 & -m_7 & lm_8 \\ & l^2m_9 & -lm_8 & -l^2m_{10} \\ & & m_7 & -lm_8 \\ \text{symm.} & & & l^2m_9 \end{bmatrix}, \tag{15.34}$$

where

$$\begin{array}{ll} m_1 = 156 + 294\Phi + 140\Phi^2\,, & m_2 = 22 + 38.5\Phi + 17.5\Phi^2\,, \\ m_3 = 54 + 126\Phi + 70\Phi^2\,, & m_4 = 13 + 31.5\Phi + 17.5\Phi^2\,, \\ m_5 = 4 + 7\Phi + 3.5\Phi^2\,, & m_6 = 3 + 7\Phi + 3.5\Phi^2\,, \\ m_7 = 36\,, & m_8 = 3 - 15\Phi\,, \\ m_9 = 4 + 5\Phi + 10\Phi^2\,, & m_{10} = 1 + 5\Phi - 5\Phi^2\,. \end{array}$$

The consistent load vector due to a uniform distribution of shear force $f_x(t)$ or of bending moment $m_y(t)$ can be obtained directly from Eq. (15.8)

$$\mathbf{f}(t)_{F1} = l\left[\int_0^1 \mathbf{N}_{F1}^T d\zeta\right]\left\{\begin{array}{c} f_x(t) \\ m_y(t) \end{array}\right\} = \frac{lf_x(t)}{12}\left\{\begin{array}{c} 6 \\ l \\ 6 \\ -l \end{array}\right\} + \frac{m_y(t)}{1+\Phi}\left\{\begin{array}{c} -l \\ \dfrac{\Phi l}{2} \\ l \\ \dfrac{\Phi l}{2} \end{array}\right\}. \tag{15.35}$$

Remark 15.5 *In the present case the lumped load vector is different from the consistent one.*

15.2.4 Flexural behavior in the yz-plane

As already pointed out, the flexural behavior in the yz-plane must be studied using equations different from those used for the xz-plane, owing to the different signs of rotation. Matrix $\mathbf{N}_{F2}$ can, however, be obtained from matrix $\mathbf{N}_{F1}$ simply by changing the signs of elements with subscripts 12, 14, 21, and 23, and the mass and stiffness matrices related to plane yz are the same as those computed for plane xz, except for the signs of elements with subscripts 12, 14, 23, and 34 and their symmetrical ones. In the force vectors related to external forces (distributed or concentrated) or external moments, the signs of elements 2 and 4 or 1 and 3, respectively, must be changed. If the beam is not axially symmetrical and the elastic and inertial properties in the two planes are not coincident, different values of the moments of inertia and the shear factors must be introduced.

15.2.5 Global behavior of the beam

The complete expression of the mass and stiffness matrices and of the nodal force vector is respectively

$$
\mathbf{M} = \begin{bmatrix} \mathbf{M}_A & \mathbf{0} & \mathbf{0} & \mathbf{0} \\ \mathbf{0} & \mathbf{M}_T & \mathbf{0} & \mathbf{0} \\ \mathbf{0} & \mathbf{0} & \mathbf{M}_{F1} & \mathbf{0} \\ \mathbf{0} & \mathbf{0} & \mathbf{0} & \mathbf{M}_{F2} \end{bmatrix} , \tag{15.36}
$$

$$
\mathbf{K} = \begin{bmatrix} \mathbf{K}_A & \mathbf{0} & \mathbf{0} & \mathbf{0} \\ \mathbf{0} & \mathbf{K}_T & \mathbf{0} & \mathbf{0} \\ \mathbf{0} & \mathbf{0} & \mathbf{K}_{F1} & \mathbf{0} \\ \mathbf{0} & \mathbf{0} & \mathbf{0} & \mathbf{K}_{F2} \end{bmatrix} , \quad \mathbf{f} = \begin{Bmatrix} \mathbf{f}_A \\ \mathbf{f}_T \\ \mathbf{f}_{F1} \\ \mathbf{f}_{F2} \end{Bmatrix} .
$$

15.2.6 Effect of axial forces

One of the simplest geometric nonlinearity effects is the interaction between an axial load and the flexural behavior of beams. In Section 12.8 it was shown that if the axial load can be considered as a constant, it is possible to study this interaction using a linearized model. The same also holds when the FEM is used.

Consider the flexural vibration in the xz-plane of a Timoshenko beam element, subjected to a constant and known axial force F_a. Such a force must not be confused with the axial force $F_z(t)$ acting on the element as a consequence of bending, which is usually one of the unknowns in the problem. If force F_a is not known, an iterative procedure must be followed.

The axial force F_z due to the loads must be computed first, and then the same force can be introduced in further dynamic computations.

Taking into account the lateral inflection of the beam, the axial strain in correspondence to the neutral axis of the beam, which is usually expressed as

$$\epsilon_z = \partial u_z / \partial z,$$

becomes

$$\epsilon_z = \frac{\partial u_z}{\partial z} + \frac{1}{2}\left(\frac{\partial u_y}{\partial z}\right)^2. \tag{15.37}$$

If the small term linked with lateral deflection is also considered, the elastic potential energy due to the axial strain becomes

$$\mathcal{U} = \frac{1}{2}\int_0^l EA\epsilon_z^2 dz = \frac{1}{2}\int_0^l EA\left[\frac{\partial u_z}{\partial z} + \frac{1}{2}\left(\frac{\partial u_y}{\partial z}\right)^2\right]^2 dz \approx$$

$$\approx \frac{1}{2}\int_0^l EA\left(\frac{\partial u_z}{\partial z}\right)^2 dz + \frac{1}{2}\int_0^l EA\frac{\partial u_z}{\partial z}\left(\frac{\partial u_y}{\partial z}\right)^2 dz. \tag{15.38}$$

The last expression was obtained by neglecting the term in $(\partial u_y/\partial z)^4$. The first term of the potential energy has already been taken into account in the computation of the stiffness matrix of the element. The second one causes an increase of the potential energy, which can be considered by adding a suitable matrix, the so-called geometric stiffness matrix, or simply geometric matrix, to the stiffness matrix of the element. Assuming that the axial force $F_a = \epsilon_z EA$ is constant and using the shape functions (15.30), the increment of potential energy can be written in terms of matrix $\mathbf{K}_{g_1}$ for flexural behavior in the xz-plane in the form

$$\Delta\mathcal{U} = \frac{1}{2}\epsilon_z\int_0^l EA\left(\frac{\partial u_y}{\partial z}\right)^2 dz = \frac{1}{2}F_a\mathbf{q}_{F1}^T\mathbf{K}_{g_1}\mathbf{q}_{F1}, \tag{15.39}$$

where

$$\mathbf{K}_{g_1} = \frac{F_a}{30l(1+\Phi)^2}\begin{bmatrix} k_1 & lk_2 & -k_1 & lk_2 \\ & l^2k_3 & -lk_2 & -l^2k_4 \\ & & k_1 & -lk_2 \\ \text{symm.} & & & l^2k_3 \end{bmatrix},$$

$$k_1 = 36 + 60\Phi + 3\Phi^2, \qquad k_2 = 3,$$
$$k_3 = 4 + 5\Phi + 2.5\Phi^2, \qquad k_4 = 1 + 5\Phi + 2.5\Phi^2.$$

As for all other matrices describing flexural behavior, the geometric matrix related to the yz-plane differs from that computed for the xz-plane for the signs of elements 12, 14, 23, and 34 and their symmetricals, apart from being computed using the appropriate value of Φ.

Apart from being used to take into account the effects of axial forces on the flexural behavior, the geometric matrix allows the study of the elastic stability of a structure. Because an axial tensile force causes an increase of stiffness while a compressive force has the opposite effect, the value of the compressive axial force causing the stiffness matrix to be singular is the buckling load of the element or of the structure. Geometric matrices can also be defined for other types of elements.

15.3   Mass and spring elements

Consider a concentrated mass, or better, a rigid body located at the ith node. Let $\mathbf{q}$ be the vector of the generalized displacements of the relevant node, which may also contain rotations

$$\mathbf{q} = [u_x, u_y, u_z, \phi_x, \phi_y, \phi_z] \, ,$$

if the node is of the type of those seen in the case of beam elements.

In the latter case, if the mass also has moments of inertia, let the axes of the reference frame coincide with the principal axes of inertia of the rigid body. The mass matrix is thus diagonal

$$\mathbf{M} = \mathrm{diag}[m, m, m, J_x, J_y, J_z] \, . \tag{15.40}$$

A simpler expression is obtained when only the translational degrees of freedom of the node are considered.

Remark 15.6 *In many computer programs, different values of the mass can be associated with the various degrees of freedom. This can account for particular physical layouts, such as a mass constrained to move with the structure in one direction and not in others.*

Consider a spring element, i.e., an element that introduces a concentrated stiffness between two nodes, say node 1 and node 2, of the structure. When the nodes have a single degree of freedom, the generalized coordinates of the element are

$$q = [u_1, u_2]^T \, ,$$

and the stiffness matrix is

$$\mathbf{K} = \begin{bmatrix} k & -k \\ -k & k \end{bmatrix} . \tag{15.41}$$

If the nodes, like those used with beam elements, have three translational and three rotational degrees of freedom, three values of translational stiffness k_x, k_y, and k_z, and three values of rotational stiffness χ_x, χ_y, and χ_z can be stated and matrices similar to that in Eq. (15.41) can be written for each degree of freedom.

15.4   Plate element: Kirchoff formulation

Consider a rectangular flat plate element with constant thickness (Fig. 15.2a). The simplest formulation has four nodes at the corners; other more complex formulations have additional nodes at the center of the sides (8-node rectangular plate) or even at center of the element (nine nodes). Elements with two or more nodes on each side (12, 16, etc., nodes) to which 4, 9, etc., nodes inside the element may be added, are also used.

Each node has 3 degrees of freedom, the translation u_z in the z-direction and rotations ϕ_x and ϕ_y. The total number of degrees of freedom of the element in the figure is 12.

Consistently with the Kirchoff formulation (i.e., neglecting shear deformation; Section 13.2), rotations are strictly linked with translation in the z-direction

$$\phi_y = -\frac{\partial u_z}{\partial x} , \qquad \phi_x = \frac{\partial u_z}{\partial y} . \tag{15.42}$$

Using the nondimensional coordinates

$$\xi = \frac{x}{a} , \qquad \eta = \frac{y}{b} ,$$

the following polynomial expression for the displacement u_z can be assumed:

$$
\begin{aligned}
u_z = {} & a_1 + a_2\xi + a_3\eta + a_4\xi\eta + a_5\xi^2 + a_6\eta^2 + a_7\xi^2\eta + \\
& + a_8\xi\eta^2 + a_9\xi^3 + a_{10}\eta^3 + a_{11}\xi^3\eta + a_{12}\xi\eta^3 .
\end{aligned}
\tag{15.43}
$$

The 12 coefficients a_i can be easily computed as functions of the nodal displacements, obtaining the first line $\mathbf{N}_1$ of the shape functions matrix

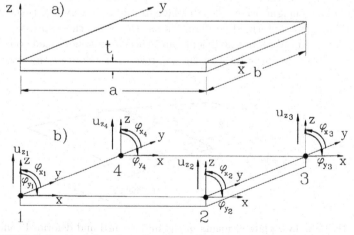

FIGURE 15.2. Rectangular plate element with four nodes: geometrical definitions and nodal displacements.

N, which has 3 lines and 12 columns. The polynomials expressing the rotations can be obtained by differentiating Eq. (15.43) with respect to x and y

$$\phi_y = -\frac{1}{a}\left(a_2 + a_4\eta + 2a_5\xi + 2a_7\xi\eta + a_8\eta^2 + 3a_9\xi^2 + 3a_{11}\xi^2\eta + a_{12}\eta^3\right),$$

$$\phi_x = -\frac{1}{b}\left(a_3 + a_4\xi + 2a_6\eta + a_7\xi^2 + 2a_8\xi\eta + 3a_{10}\eta^2 + a_{11}\xi^3 + 3a_{12}\xi\eta^2\right).$$

$$(15.44)$$

They allow to compute the remaining two lines **N**$_2$ and **N**$_3$ of the matrix of the shape functions.

The shape functions so obtained are *nonconforming*, i.e., they do not guarantee that the sides of two contiguous elements deform in a consistent way. This can be shown by simply writing the deformation of one of the sides parallel to the y-axis, for instance, the side between nodes 1 and 2 of element 1 in Fig. 15.3

$$u_z = a_1 + a_3\eta + a_6\eta^2 + a_{10}\eta^3,$$

$$\phi_y = -\frac{1}{a}\left(a_2 + a_4\eta + a_8\eta^2 + a_{12}\eta^3\right), \qquad (15.45)$$

$$\phi_x = -\frac{1}{b}\left(a_3 + 2a_6\eta + 3a_{10}\eta^2\right).$$

The deflected shape of the side depends on eight parameters and consequently cannot be univocally determined by the six generalized displacements of the two nodes at its ends. This means that although the nodes at the end of two neighboring sides displace in the same way, the two sides can assume different deformed configurations. However, the expressions of u_z and ϕ_x contain only four coefficients and these two generalized coordinates are defined at the ends by 4 of the degrees of freedom of the two nodes.

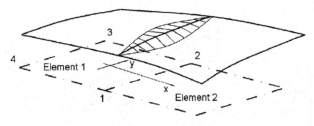

FIGURE 15.3. Two plate elements in the undeformed and deformed conditions. Since the shape functions are not conforming, rotations ϕ_y at the border between the elements are not consistent.

This means that the mentioned formulation guarantees that generalized displacements u_z and ϕ_x are consistent on the side between nodes 1 and 2. On the contrary rotation ϕ_y is not consistent.

The same considerations can be applied to the other sides of the element: While compatibility is insured for what displacements in z-directions and rotations about an axis perpendicular to the sides, no compatibility can be assessed for rotation about axes parallel to the sides. This means that the overall deformation can show a 'kink' on the sides of the elements.

Remark 15.7 *Such a 'kink' however cannot extend up to the nodes, since at the nodes all generalized displacements are compatible. By increasing the number of elements, this kink tends thus to disappear, and the errors introduced by the nonconformity of the shape functions decrease.*

Since the generalized displacements at the ith node can be expressed as

$$\mathbf{q}_i\left(t\right) = \left[\begin{array}{ccc} u_{z_i} & \phi_{x_i} & \phi_{y_i} \end{array}\right]^T, \tag{15.46}$$

the generalized displacements of the internal points of the elements are

$$\left\{\begin{array}{c} u_z\left(x,y,t\right) \\ \phi_x\left(x,y,t\right) \\ \phi_y\left(x,y,t\right) \end{array}\right\} = \left[\begin{array}{c} \mathbf{N}_1\left(x,y\right) \\ \mathbf{N}_2\left(x,y\right) \\ \mathbf{N}_3\left(x,y\right) \end{array}\right] \left\{\begin{array}{c} \mathbf{q}_1\left(t\right) \\ \mathbf{q}_2\left(t\right) \\ \mathbf{q}_3\left(t\right) \\ \mathbf{q}_4\left(t\right) \end{array}\right\}. \tag{15.47}$$

The stiffness matrix can be obtained in the same way as for the Timoshenko beam element, with the difference that in the Kirchoff formulation only the potential energy linked with bending deformation is taken into account.

The generalized strain vector $\boldsymbol{\epsilon}$ and the stress vector $\boldsymbol{\sigma}$ in this case contain, respectively, the curvatures of the mid-plane of the plate and the moments per unit length

$$\boldsymbol{\epsilon} = \left\{\begin{array}{c} -\dfrac{\partial^2 u_z}{\partial x^2} \\[2ex] -\dfrac{\partial^2 u_z}{\partial y^2} \\[2ex] 2\dfrac{\partial^2 u_z}{\partial x \partial y} \end{array}\right\}, \quad \boldsymbol{\sigma} = \left\{\begin{array}{c} \mathbf{M}_x \\ \mathbf{M}_y \\ \mathbf{M}_{xy} \end{array}\right\}. \tag{15.48}$$

Equation (15.3) still holds, provided that matrix $\mathbf{B}$ is

$$\mathbf{B} = \left[\begin{array}{ccc} -\dfrac{\partial^2 \mathbf{N}_1^T}{\partial x^2} & -\dfrac{\partial^2 \mathbf{N}_1^T}{\partial y^2} & -\dfrac{\partial^2 \mathbf{N}_1^T}{\partial x \partial y} \end{array}\right]^T. \tag{15.49}$$

The stress–strain relationship (15.4) holds, provided that matrix $\mathbf{E}$ is defined as

$$\mathbf{E} = \frac{Eh^3}{12(1-\nu^2)} \begin{bmatrix} 1 & \nu & 0 \\ \nu & 1 & 0 \\ 0 & 0 & \dfrac{1-\nu}{2} \end{bmatrix}^T , \tag{15.50}$$

as it can be immediately obtained from Eqs. (13.7) and (13.9).

Once that matrices $\mathbf{B}$ and $\mathbf{E}$ have been defined, the stiffness matrix is readily obtained from Eq. (15.5).

If the rotational inertia of the cross section is neglected, only the shape functions related to displacement u_z enter Eq. (15.7) for the computation of the mass matrix

$$\mathbf{M} = \rho h \int_{-a/2}^{a/2} \int_{-b/2}^{b/2} \mathbf{N}_1^T(x,\, y) \mathbf{N}_1(x,\, y) \, dx \, dy . \tag{15.51}$$

The same holds for Eq. (15.8) allowing in this case the computation of the nodal forces corresponding to a surface force distribution on the plate element.

15.5   Plate element: Mindlin formulation

As stated in Section 13.2, the Mindlin plate formulation includes shear deformation. The perpendiculars to the mid-plane in the undeformed configuration remain straight but not necessarily perpendicular to the mid surface after deformation, unlike to what happens in the Kirchoff plate.

Equation (15.42) does not hold any more and the expressions for rotations must be defined independently from that for displacement u_z. In the case of a rectangular element with four nodes, the following polynomial expressions can be stated instead of Eq. (15.43):

$$\begin{aligned} u_z &= a_1 + a_2\xi + a_3\eta + a_4\xi\eta, \\ \phi_y &= a_5 + a_6\xi + a_7\eta + a_8\xi\eta, \\ \phi_x &= a_9 + a_{10}\xi + a_{11}\eta + a_{12}\xi\eta . \end{aligned} \tag{15.52}$$

Also in this case the 12 coefficients a_i can be easily computed as functions of the nodal displacements, obtaining the whole matrix of the shape functions $\mathbf{N}$ that has again 3 lines and 12 columns.

The mass and stiffness matrices and the nodal force vector can be obtained in the same way as seen in Section 15.1. The present formulation has however the tendency to *lock* when the thickness of the plate is small, i.e., when shear deformation becomes negligible. This can be easily shown by writing the expressions of the curvatures of the plate, which are assumed to be coincident with the derivatives of the rotations ϕ_x and ϕ_y owing to the small displacement assumption

$$\frac{\partial \phi_x}{\partial x} = \frac{1}{a}\left(a_6 + a_8 \eta\right) \ , \quad \frac{\partial \phi_x}{\partial x} = \frac{1}{a}\left(a_{11} + a_{12}\xi\right) \tag{15.53}$$

and the shear deformations

$$\gamma_z = -\frac{\partial u_z}{\partial x} - \phi_y = -\frac{a_2}{a} - \frac{a_4}{a}\eta - a_9 - a_{10}\xi - a_{11}\eta - a_{12}\xi\eta,$$

$$\gamma_y = \frac{\partial u_z}{\partial x} - \phi_x = \frac{a_3}{b} + \frac{a_4}{b}\eta - a_5 - a_6\xi - a_7\eta - a_8\xi\eta \ . \tag{15.54}$$

If the plate is very thin and shear deformations vanish, all coefficients except a_1 are also vanishingly small. This causes also the curvatures of the plate to be negligible, and the plate element results heavily overconstrained. Practically, this results in an overestimating of the elements of the stiffness matrix.

The most common procedures used in order to avoid, or at least to reduce, the tendency of Mindlin plate elements to lock are the so-called *reduced integration* and the *selective integration* practices. They are usually employed when the integration required to compute the stiffness matrix is performed numerically, mainly using Gauss technique.

The first method is simply based on the computation of the integrals using an integration order lower than needed. The values of the integrals are then underestimated and this balances, at least in an approximate way, the overestimating of the stiffness matrix typical of 'locked' elements. Its empirical nature and the possibility of giving way to other drawbacks, makes the use of reduced integration not advisable. When the selective integration is used, the various parts of the stiffness matrix are integrated using different integration orders, so that the parts of the integrals giving way to locking are underestimated. The results are usually acceptable and this practice is common.

The use of elements with a larger number of nodes is effective, particularly when some nodes are not on the border of the elements: selective integration on elements with quadratic shape functions with nine nodes (one internal) leads to good results, while using cubic shape functions and 16 nodes (four internal) good results can be obtained even with complete integration. Many alternative formulations for Mindlin plate elements have been developed in order to prevent the element from locking: a deeper discussion can be found in the books by Hinton and Owen and Bathe.[2]

---

[2]E. Hinton, D.R.J. Owen, *Finite Element Software for Plates and Shells*, Pineridge Press, Swansea, 1984; K.J. Bathe, *Finite Element Procedures in Engineering Analysis*, Prentice Hall, Englewood Cliffs, 1982.

15.6   Brick elements

As an example of solid elements, the regular hexahedron 8-node element will be described. In a similar way, a tetrahedron 4-node element or a prismatic element with triangular bases with six nodes can be defined. If other nodes are inserted at the center of the sides, the number of nodes becomes, respectively, 20, 10, and 15. More complex elements, with nodes also on the faces or even inside the element, have been formulated.

Consider the element shown in Fig. 15.4. Each node has 3 degrees of freedom, leading to a total number of 24 degrees of freedom for the whole element. Each internal point of the element has 3 degrees of freedom, the displacements of the internal points in the direction of the axes

$$
\begin{aligned}
u_z &= a_1 + a_2\xi + a_3\eta + a_4\zeta + a_5\xi\eta + a_6\xi\zeta + a_7\eta\zeta + a_8\xi\eta\zeta, \\
u_y &= a_9 + a_{10}\xi + a_{11}\eta + a_{12}\zeta + a_{13}\xi\eta + a_{14}\xi\zeta + a_{15}\eta\zeta + a_{16}\xi\eta\zeta, \\
u_z &= a_{17} + a_{18}\xi + a_{19}\eta + a_{20}\zeta + a_{21}\xi\eta + a_{22}\xi\zeta + a_{23}\eta\zeta + a_{24}\xi\eta\zeta\ ,
\end{aligned}
$$
$$(15.55)$$

where the nondimensional coordinates ξ and η are the same already defined for the plate element, while coordinate ζ is referred to the thickness c of the element in the z-direction.

With simple computations it is possible to show that the three rows of the shape functions matrix $\mathbf{N}_1$, $\mathbf{N}_2$, and $\mathbf{N}_3$ are

$$
\begin{bmatrix}
\mathbf{N}_1 \\
\mathbf{N}_2 \\
\mathbf{N}_3
\end{bmatrix}
= \begin{bmatrix} N_{11}\mathbf{I} & N_{12}\mathbf{I} & N_{13}\mathbf{I} & N_{14}\mathbf{I} & N_{15}\mathbf{I} & N_{16}\mathbf{I} & N_{17}\mathbf{I} & N_{18}\mathbf{I} \end{bmatrix},
$$
$$(15.56)$$

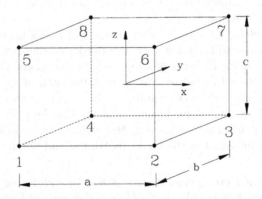

FIGURE 15.4. Regular hexahedron element with eight nodes.

where $\mathbf{I}$ is a 3×3 identity matrix and functions N_{1i} are

$$\begin{aligned}
N_{11} &= 1 - \xi - \eta - \zeta + \xi\eta + \xi\zeta + \eta\zeta - \xi\eta\zeta, \\
N_{12} &= \xi - \xi\eta - \xi\zeta + \xi\eta\zeta, \qquad N_{16} = \xi\zeta - \xi\eta\zeta, \\
N_{13} &= \eta - \xi\eta - \eta\zeta + \xi\eta\zeta, \qquad N_{17} = \eta\zeta - \xi\eta\zeta, \\
N_{14} &= \xi\eta - \xi\eta\zeta, \qquad\qquad\quad N_{18} = \xi\eta\zeta \\
N_{15} &= \zeta - \xi\zeta - \eta\zeta + \xi\eta\zeta
\end{aligned} \tag{15.57}$$

Since all generalized displacements are actual displacements, the general formulation described in Section 15.1 is directly applicable without any change. The strain vector is the usual column matrix representation of the three-dimensional strain tensor ϵ

$$\epsilon\left(x,y,z,t\right) = \begin{bmatrix} \epsilon_x & \epsilon_y & \epsilon_z & \gamma_{xy} & \gamma_{xz} & \gamma_{yz} \end{bmatrix}^T \tag{15.58}$$

and matrix $\mathbf{B}\left(x,y,z\right)$ in this case has 6 rows and 24 columns. Its various rows, indicated as $\mathbf{B}_i$, are

$$\mathbf{B}_1 = \frac{1}{a}\frac{\partial \mathbf{N}_1}{\partial \xi}, \qquad \mathbf{B}_2 = \frac{1}{b}\frac{\partial \mathbf{N}_2}{\partial \eta},$$

$$\mathbf{B}_3 = \frac{1}{a}\frac{\partial \mathbf{N}_3}{\partial \xi}, \qquad \mathbf{B}_4 = \left(\frac{1}{b}\frac{\partial \mathbf{N}_1}{\partial \eta} + \frac{1}{a}\frac{\partial \mathbf{N}_2}{\partial \xi}\right), \tag{15.59}$$

$$\mathbf{B}_5 = \left(\frac{1}{c}\frac{\partial \mathbf{N}_1}{\partial \zeta} + \frac{1}{a}\frac{\partial \mathbf{N}_3}{\partial \xi}\right), \qquad \mathbf{B}_6 = \left(\frac{1}{c}\frac{\partial \mathbf{N}_2}{\partial \zeta} + \frac{1}{b}\frac{\partial \mathbf{N}_3}{\partial \eta}\right).$$

Matrix $\mathbf{E}$ linking the stress vector $\boldsymbol{\sigma}$

$$\boldsymbol{\sigma}\left(x,y,z,t\right) = \begin{bmatrix} \sigma_x & \sigma_y & \sigma_z & \tau_{xy} & \tau_{xz} & \tau_{yz} \end{bmatrix}^T \tag{15.60}$$

with the already defined strain vector can be computed by inverting the compliance matrix of the material. If the material is isotropic, it follows that

$$\mathbf{E} = E \begin{bmatrix}
1 & -\nu & -\nu & 0 & 0 & 0 \\
-\nu & 1 & -\nu & 0 & 0 & 0 \\
-\nu & -\nu & 1 & 0 & 0 & 0 \\
0 & 0 & 0 & 2\left(1+\nu\right) & 0 & 0 \\
0 & 0 & 0 & 0 & 2\left(1+\nu\right) & 0 \\
0 & 0 & 0 & 0 & 0 & 2\left(1+\nu\right)
\end{bmatrix}^{-1}. \tag{15.61}$$

The computation of the stiffness and mass matrices and of the nodal force vector is straightforward and can be done directly using Eqs. (15.6)–(15.8).

Remark 15.8 *Solid elements allow a very detailed modeling of complex shapes but may lead a large number of degrees of freedom. R.H. Gallagher*[3]

---

[3]R.H. Gallagher, *Finite Element Analysis*, Prentice Hall, Englewood Cliffs, 1975.

speaks of a curse of dimensionality *to stress how quickly the number of degrees of freedom of a model grows from one-dimensional to two-dimensional and to three-dimensional representations.*

Consider as an example a one-dimensional structure, e.g., a bar, subdivided in 10 elements. It has 11 nodes, each having a single degree of freedom: the total number of degrees of freedom is 11.

Now consider a two-dimensional structure, e.g., a square planar element. If each side is subdivided into 10 parts, a total of 100 4-node rectangular elements with 121 nodes is required. Since each node has 2 degrees of freedom, the total number of degrees of freedom is 242.

If the corresponding three-dimensional structure is considered, for instance, a cube with each side divided into 10 parts, a total of 1,000 8-node brick elements with 1,331 nodes is required. Since each node has 3 degrees of freedom, the total number of degrees of freedom is 3,993.

The large number of degrees of freedom usually involved with models based on solid elements makes their use not very common for dynamic studies. Reduction techniques (see Chapter 4) are very effective and largely employed in connection with solid FEM modeling.

15.7   Isoparametric elements

The elements seen in the previous sections have all a regular shape and their study was performed using a rectangular coordinate frame directly applied to the element. This is however seldom applicable in actual cases, when the use of more or less distorted elements is mandatory in order to adequately model the geometry of the system. As an example, a 4-node rectangular element can be distorted into a quadrilateral element whose sides are not at right angle (Fig. 15.5a,b). The points of the latter, of coordinates (x, y), can be put in a one-to-one relationship with the points of the former, on which a set of non-dimensional coordinates (ξ, η) is stated. This mapping can be expressed by the functions

$$\left\{ \begin{array}{c} x \\ y \end{array} \right\} = \left\{ \begin{array}{c} f_1\,(\xi, \eta) \\ f_2\,(\xi, \eta) \end{array} \right\} . \qquad (15.62)$$

Functions f_1 and f_2 can be chosen in various ways, the most common approach being using the shape functions to express also the distortion of the element. In this case the element is said to be *isoparametric*.

In the general case of a brick element, Eq. (15.62) reduces to

$$\left\{ \begin{array}{c} x \\ y \\ z \end{array} \right\} = \left[\begin{array}{ccc} \mathbf{N}(\xi, \eta, \zeta) & \mathbf{0} & \mathbf{0} \\ \mathbf{0} & \mathbf{N}(\xi, \eta, \zeta) & \mathbf{0} \\ \mathbf{0} & \mathbf{0} & \mathbf{N}(\xi, \eta, \zeta) \end{array} \right] \left\{ \begin{array}{c} \mathbf{x}_i \\ \mathbf{y}_i \\ \mathbf{z}_i \end{array} \right\} . \qquad (15.63)$$

where $\mathbf{x}_i$, $\mathbf{y}_i$, and $\mathbf{z}_i$ are the coordinates of the nodes.

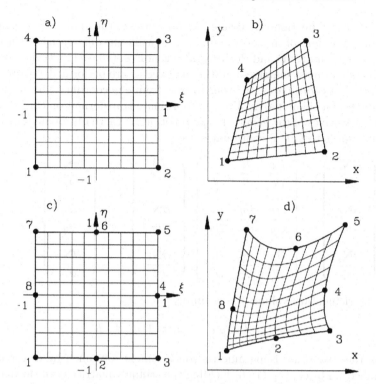

FIGURE 15.5. Isoparametric elements obtained by distorting a 4-node and an 8-node rectangular element.

If the functions used in order to define the distortion of the elements are polynomials whose degree is smaller than that of the shape functions, the element is said to be *subparametric*; in the opposite case it is said to be *superparametric*.

In the case of a general, 4-node quadrilateral isoparametric element whose nodal coordinates are (x_i, y_i) $(i = 1, ...4)$ mapped on a rectangular element extending from $\xi = -1$ to $\xi = 1$ and from $\eta = -1$ to $\eta = 1$, the coordinate transformation is

$$\left\{ \begin{array}{c} x \\ y \end{array} \right\} = \left[\begin{array}{c} N_1 x_1 + N_2 x_2 + N_3 x_3 + N_4 x_4 \\ N_1 y_1 + N_2 y_2 + N_3 y_3 + N_4 y_4 \end{array} \right], \qquad (15.64)$$

where

$$\begin{array}{ll} N_1 = (1 - \xi)(1 - \eta), & N_2 = (1 - \xi)(1 + \eta), \\ N_3 = (1 + \xi)(1 - \eta), & N_4 = (1 + \xi)(1 + \eta). \end{array} \qquad (15.65)$$

The mapping is represented in Fig. 15.5a and b. An 8-node rectangular element can be distorted with similar techniques into a curvilinear element, whose sides are arcs of parabola, as shown in Fig. 15.5c and d.

The shape functions are defined with respect to the undistorted element, i.e., are functions of the coordinates ξ, η, ζ, while the differentiations needed to compute matrix $\mathbf{B}$ and the integrations in Eqs. (15.6)–(15.8) yielding the stiffness and mass matrices and the load vector are performed with respect to x-, y-, z- coordinates. In order to pass from one coordinate frame to the other the Jacobian matrix $\mathbf{J}$ of the coordinate transformation can be used. The derivatives of the generic shape function N_i (actually, of any function) with respect to ξ, η, ζ coordinates are

$$
\left\{
\begin{array}{c}
\dfrac{\partial \mathbf{N}_i}{\partial \xi} \\[2ex]
\dfrac{\partial \mathbf{N}_i}{\partial \eta} \\[2ex]
\dfrac{\partial \mathbf{N}_i}{\partial \zeta}
\end{array}
\right\}
=
\left[
\begin{array}{ccc}
\dfrac{\partial x}{\partial \xi} & \dfrac{\partial y}{\partial \xi} & \dfrac{\partial z}{\partial \xi} \\[2ex]
\dfrac{\partial x}{\partial \eta} & \dfrac{\partial y}{\partial \eta} & \dfrac{\partial z}{\partial \eta} \\[2ex]
\dfrac{\partial x}{\partial \zeta} & \dfrac{\partial y}{\partial \zeta} & \dfrac{\partial z}{\partial \zeta}
\end{array}
\right]
\left\{
\begin{array}{c}
\dfrac{\partial \mathbf{N}_i}{\partial x} \\[2ex]
\dfrac{\partial \mathbf{N}_i}{\partial y} \\[2ex]
\dfrac{\partial \mathbf{N}_i}{\partial z}
\end{array}
\right\}
= \mathbf{J}
\left\{
\begin{array}{c}
\dfrac{\partial \mathbf{N}_i}{\partial x} \\[2ex]
\dfrac{\partial \mathbf{N}_i}{\partial y} \\[2ex]
\dfrac{\partial \mathbf{N}_i}{\partial z}
\end{array}
\right\} .
\tag{15.66}
$$

The elementary volume dV is simply

$$
dV = dx\, dy\, dz = \det(\mathbf{J})\, d\xi\, d\eta\, d\zeta .
\tag{15.67}
$$

In the case of an isoparametric solid element, obtained by distorting a hexahedron brick, Eq. (15.6) yielding the stiffness matrices transforms into

$$
\mathbf{K} = \int_{-1}^{1} \int_{-1}^{1} \int_{-1}^{1} \mathbf{B}^T \mathbf{E} \mathbf{B} \det(\mathbf{J})\; d\xi\, d\eta\, d\zeta .
\tag{15.68}
$$

Equations (15.7) and (15.8) yielding the mass matrix and the nodal force vector transform into

$$
\mathbf{M} = \int_{-1}^{1} \int_{-1}^{1} \int_{-1}^{1} \rho \mathbf{N}^T \mathbf{N} \det(\mathbf{J})\; d\xi\, d\eta\, d\zeta .
\tag{15.69}
$$

$$
\mathbf{f}(t) = \int_{-1}^{1} \int_{-1}^{1} \int_{-1}^{1} \mathbf{f}(\xi, \eta, \zeta, t) \mathbf{N} \det(\mathbf{J})\; d\xi\, d\eta\, d\zeta .
\tag{15.70}
$$

The integrations needed for the computation of the relevant matrices and vectors cannot usually be performed in closed form, since functions can be rather complex (the Jacobian matrix is included in matrix $\mathbf{B}$ and appears explicitly). This complicates only marginally the construction of the model, since the integration can be performed numerically element by element without any trouble. The integration method usually employed is Gauss method, which allows to obtain a good accuracy computing the function to be integrated only in a limited number of points.

Isoparametric elements are by far the most common in routine FEM computations.

15.8 Some considerations on the consistent mass matrix

As already stated, often the FEM is used only to compute the stiffness matrix to be used in the context of the lumped parameter approach. In this case the consistent mass matrix is not computed and a diagonal mass matrix obtained by lumping the mass at the nodes is used. When using beam elements, one half of the mass of each element can be lumped in each node. Transversal moments of inertia are completely neglected, while polar moments of inertia can be lumped in the same way as masses. If the degrees of freedom are ordered as in Eq. (15.10), the mass matrix of the element can thus be written in the form

$$\mathbf{M} = \frac{\rho l}{2}\mathrm{diag}\begin{bmatrix} A & A & A & 0 & 0 & I_p & A & A & A & 0 & 0 & I_p \end{bmatrix}. \quad (15.71)$$

It is also possible to insert transversal moments of inertia I_x and I_y into Eq. (15.71), but this does not usually increase the accuracy of the lumped parameter approach. The advantage is that of dealing with a diagonal mass matrix, whose inversion in order to compute the dynamic matrix is far simpler than that of the consistent mass matrix. As already stated, the accuracy is however reduced or, better, a greater number of elements is needed to reach the same accuracy and which formulation is more expedient must be assessed in each case.

It can be easily shown that the values of the total mass obtained through Eq. (4.55) using the two approaches are coincident.

If the beam undergoes a rigid body translational motion, vectors $\boldsymbol{\delta}_i$ are

$$\begin{aligned} \boldsymbol{\delta}_x &= \begin{bmatrix} 1 & 0 & 0 & 0 & 0 & 0 & 1 & 0 & 0 & 0 & 0 & 0 \end{bmatrix}^T \\ \boldsymbol{\delta}_y &= \begin{bmatrix} 0 & 1 & 0 & 0 & 0 & 0 & 0 & 1 & 0 & 0 & 0 & 0 \end{bmatrix}^T \\ \boldsymbol{\delta}_z &= \begin{bmatrix} 0 & 0 & 1 & 0 & 0 & 0 & 0 & 0 & 1 & 0 & 0 & 0 \end{bmatrix}^T \end{aligned} . \quad (15.72)$$

It is easy to verify that the value of the mass obtained by introducing Eq. (4.55) and both formulations of the mass matrix into Eq. (4.55) is the correct value of the mass of the beam

$$m_t = \rho A l .$$

Consider now a unit rigid body rotation of the beam element about an axis passing through the center of mass and parallel to y-axis. The displacement vector is

$$\mathbf{q} = \begin{bmatrix} -\frac{l}{2} & 0 & 0 & 0 & 1 & 0 & -\frac{l}{2} & 0 & 0 & 1 & 0 & 0 \end{bmatrix}^T . \quad (15.73)$$

By using a procedure similar to that used to obtain Eq. (4.55), it is possible to show that the expression

$$\mathbf{q}^T\mathbf{M}\mathbf{q} \qquad\qquad (15.74)$$

is nothing else than the moment of inertia J_y about the center of mass. By inserting the consistent mass matrix into expression (15.74), it follows that

$$J_y = \rho I_y l + \frac{\rho A l^3}{12} . \tag{15.75}$$

which coincides with the actual value.

By performing the same computation using the diagonal lumped mass matrix (15.71), in which also transversal moments of inertia of the cross sections are included, it follows that

$$J_y = \rho I_y l + \frac{\rho A l^3}{4} . \tag{15.76}$$

Clearly, the approximation which is obtainable using Eq. (15.76) instead of Eq. (15.75) is better if the length of the element is very small with respect to the length of the structure, i.e., the accuracy increases when a finer mesh is used.

Remark 15.9 *In the case of beam elements, the consistent approach allows to take into account exactly the inertia of the element in its rigid body motions. It constitutes however an approximation, since the shape functions used for the beam element yield the exact deflected shape (i.e., the deflected shape obtained through the continuum theory) only in the case of beams loaded at the ends. When inertia forces are applied along the element, the shape functions do not coincide with the results obtainable from the beam theory.*

15.9   Assembling the structure

The equations of motion of the element are written with reference to a local or element reference frame that has an orientation determined by the features of the element. In a beam element, for example, the z-axis can coincide with the axis of the beam, while the x- and y-axes are principal axes of inertia of the cross section. The various local reference frames of the elements have, in general, a different orientation in space.

To describe the behavior of the structure as a whole, another reference frame, namely the global or structure reference frame, is defined. The orientation in space of any local frame can be expressed, with reference to the orientation of the global frame, by a suitable rotation matrix

$$\mathbf{R} = \begin{bmatrix} l_x & m_x & n_x \\ l_y & m_y & n_y \\ l_z & m_z & n_z \end{bmatrix} , \tag{15.77}$$

where l_i, m_i, and n_i are the direction cosines of the axes of the former in the global frame. The expressions $\mathbf{q}_{il}$ and $\mathbf{q}_{ig}$ of the displacement vector

$\mathbf{q}_i$ of the ith node in the local and global reference frames are linked by the usual coordinate transformation

$$\mathbf{q}_{il} = \mathbf{R}\mathbf{q}_{ig} \, . \tag{15.78}$$

The generalized coordinates in the displacement vector of the element can be transformed from the local to the global reference frame using a similar relationship in which an expanded rotation matrix $\mathbf{R}'$ is used to deal with all the relevant generalized coordinates. It is essentially made by a number of matrices of the type of Eq. (15.77) suitably assembled together.

Remark 15.10 *The assumption of small displacements and rotations allows consideration of the rotations about the axes as the components of a vector, which can be rotated in the same way as displacements.*

The force vectors can also be rotated using the rotation matrix $\mathbf{R}'$, and the equation of motion of the element can be written with reference to the global frame and premultiplied by the inverse of matrix $\mathbf{R}'$, obtaining

$$\mathbf{R}'^{-1}\mathbf{M}\mathbf{R}'\ddot{\mathbf{q}}_g + \mathbf{R}'^{-1}\mathbf{K}\mathbf{R}'\mathbf{q}_g = \mathbf{f}_g \, . \tag{15.79}$$

Because the inverse of a rotation matrix is coincident with its transpose, the expressions of the mass and stiffness matrices of the element rotated from the local to the global frame are

$$\mathbf{M}_g = \mathbf{R}'^T \mathbf{M}_l \mathbf{R}' \tag{15.80}$$

and

$$\mathbf{K}_g = \mathbf{R}'^T \mathbf{K}_l \mathbf{R}' \, . \tag{15.81}$$

Similarly, the nodal load vector can be rotated using the obvious relationship

$$\mathbf{f}_g = \mathbf{R}'^T \mathbf{f}_l \, . \tag{15.82}$$

Once the mass and stiffness matrices of the various elements have been computed with reference to the global frame, it is possible to easily obtain the matrices of the whole structure. The n generalized coordinates of the structure can be ordered in a single vector $\mathbf{q}_g$. The matrices of the various elements can be rewritten in the form of matrices of order $n \times n$, containing all elements equal to zero except those that are in the rows and columns corresponding to the generalized coordinates of the relevant element. Because the kinetic and potential energies of the structure can be obtained simply by adding the energies of the various elements, it follows that

$$\begin{aligned}
\mathcal{T} &= \frac{1}{2}\sum_{\forall i} \dot{\mathbf{q}}_g^T \mathbf{M}_i \dot{\mathbf{q}}_g = \frac{1}{2}\dot{\mathbf{q}}_g^T \mathbf{M}\dot{\mathbf{q}}_g \, ; \\
\mathcal{U} &= \frac{1}{2}\sum_{\forall i} \mathbf{q}_g^T \mathbf{K}_i \mathbf{q}_g = \frac{1}{2}\mathbf{q}_g^T \mathbf{K}\mathbf{q}_g \, .
\end{aligned} \tag{15.83}$$

Matrices $\mathbf{M}$ and $\mathbf{K}$ are the mass and stiffness matrices of the whole structure and are obtained simply by adding all the mass and stiffness matrices of the elements. In practice, the various matrices of size $n \times n$ for the elements are never written. Each term of the matrices of all elements is just added into the global mass and stiffness matrices in the correct place.

The matrices of the structure are easily assembled, and this is one of the simplest steps. If the generalized coordinates are taken into a suitable order, the assembled matrices have a band structure; many general-purpose computer codes have a routine that reorders the coordinates in such a way that the bandwidth is the smallest possible.

In a similar way the nodal force vector can be easily assembled

$$\mathbf{f} = \sum_{\forall i} \mathbf{f}_i. \tag{15.84}$$

Remark 15.11 *The forces that are exchanged between the elements at the nodes cancel each other in this assembling procedure, and then the force vectors that must be inserted into the global equation of motion are only those related to the external forces applied to the structure.*

15.10   Constraining the structure

One of the advantages of the FEM is the ease with which the constraints can be defined. If the ith degree of freedom is rigidly constrained, the corresponding generalized displacement vanishes and, as a consequence, the ith column of the stiffness and mass matrices can be neglected, because they multiply a displacement and an acceleration, respectively, that are equal to zero.

Because one of the generalized displacements is known, one of the equations of motion can be neglected when solving for the deformed configuration of the system. The ith equation can thus be separated from the others, which amounts to canceling the ith row of all matrices and of the force vector. Note that the ith equation could, in this case, be used after all displacements have been computed to obtain the value of the ith generalized nodal force, which, in this case, is the unknown reaction of the constraint.

To rigidly constrain a degree of freedom, it is sufficient to cancel the corresponding row and column in all matrices and vectors. This approach allows simplification of the formulation of the problem, which is particularly useful in dynamic problems, but this simplification is often marginal, as the number of constrained degrees of freedom is very small, compared with the total number of degrees of freedom.

To avoid restructuring the whole model and rewriting all the matrices, rigid constraints can be transformed into very stiff elastic constraints.

If the ith degree of freedom is constrained through a linear spring with stiffness k_i, the potential energy of the structure is increased by the potential energy of the spring

$$\mathcal{U} = \frac{1}{2} k_i q_i^2 \,. \tag{15.85}$$

To take the presence of the constraint into account, it is sufficient to add the stiffness k_i to the element in the ith row and ith column of the global stiffness matrix. This procedure is very simple, which explains why a very stiff elastic constraint is often added instead of canceling a degree of freedom in the case of rigid constraints.

An additional advantage is that the reaction of the constraint can be obtained simply by multiplying the large generalized stiffness k_i by the correspondingly small generalized displacement q_i.

Example 15.1 *Write the mass and stiffness matrices for the study of the flexural behavior of the transmission shaft with a central joint sketched in Fig. 15.6. The joint, whose mass is m_G, is supported by an elastic constraint with stiffness k_G. Use the FEM, neglecting shear deformation and rotational inertia of the cross sections.*

A two-beam element model is sketched in Fig. 15.6b. There are 4 degrees of freedom of each element involved in flexural behavior in the xz-plane. Using the Euler–Bernoulli approach, the stiffness matrices of the elements are

$$\mathbf{M} = \frac{EI_y}{l^3} \begin{bmatrix} 12 & 6l & -12 & 6l \\ & 4l^2 & -6l & 2l^2 \\ & & 12 & -6l \\ symm. & & & 4l^2 \end{bmatrix} \,.$$

The total number of degrees of freedom of the unconstrained system is 7, because the connection between the elements is performed through a joint that allows different rotations of the two elements at node 2. The two constraints lock translational degrees of freedom at nodes 1 and 3, reducing the number of degrees of freedom to 5. Directly considering the constraints, the map of the degrees of freedom, i.e., the table stating the correspondence between the degrees of freedom of the elements and those of the structure is as follows:

Element	d.o.f.	1	2	3	4			
1	type	transl.	rot.	transl.	rot.			
Element	d.o.f.			1		2	3	4
2	type			transl.		rot.	transl.	rot.
Global	d.o.f.	constr.	1	2	3	4	constr.	5

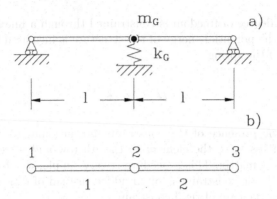

FIGURE 15.6. (a) Sketch of the system and (b) model based on two beam elements.

Noting that the stiffness k_G and the mass m_G must be added to element 22, the stiffness and mass matrices are

$$\mathbf{K} = \frac{EI_y}{l^3} \begin{bmatrix} 4l^2 & -6l & 2l^2 & 0 & 0 \\ & 24+k^* & -6l & 6l & 6l \\ & & 4l^2 & 0 & 0 \\ & & & 4l^2 & 2l^2 \\ & symm. & & & 4l^2 \end{bmatrix},$$

$$\mathbf{M} = \frac{\rho A}{420} \begin{bmatrix} 4l^2 & 13l & -3l^2 & 0 & 0 \\ & 312+m^* & -22l & 22l & -13l \\ & & 4l^2 & 0 & 0 \\ & & & 4l^2 & -3l^2 \\ & symm. & & & 4l^2 \end{bmatrix},$$

where

$$k^* = \frac{k_G l^3}{EI_y}, \qquad m^* = \frac{420 m_G}{\rho Al}.$$

Example 15.2 *Compute the natural frequencies of the beam of Example 14.4 using the FEM.*
Repeat the computations using both the lumped-parameters and the consistent approach and compare the results with those obtained using the Myklestadt method. Regard the supports as both rigid and very stiff elastic constraints.

The simplest approach is the lumped-parameters one. The model follows closely what was seen in Example 14.4: The stiffness matrices of the six Timoshenko beam elements are readily computed (they are all equal) and assembled. The assembly procedure is so simple that a map is not even needed. Because there are seven nodes and the element has 2 degrees of freedom per node (bending in one of the lateral planes), the order of the matrices is 14.

The stiffness of the constraints is assumed to be 1×10^{11} N/m. This number is then added to the elements 3,3 and 11,11 of the stiffness matrix. The mass matrix is diagonal, with the lumped masses in the positions with odd subscript and zeros in those with even subscripts, because the moments of inertia are neglected. Because the mass matrix is singular, the stiffness matrix is inverted to obtain the dynamic matrix D.

The natural frequencies, up to the fifth, are as follows:

$\omega_1 = 2,158$ *rad/s* $= 343.4$ *Hz,* $\omega_2 = 5,091$ *rad/s* $= 810.6$ *Hz,*
$\omega_3 = 6,780$ *rad/s* $= 1,079.1$ *Hz,* $\omega_4 = 10,469$ *rad/s* $= 1,666.2$ *Hz,*
$\omega_5 = 15,858$ *rad/s* $= 2,523.9$ *Hz.*

The values coincide up to the last digit reported with those computed using the transfer-matrix approach.

By canceling the third and eleventh row and column of the stiffness and mass matrices and again solving the eigenproblem, the following values of the first five natural frequencies can be found:

$\omega_1 = 2,158$ *rad/s* $= 343.4$ *Hz,* $\omega_2 = 5,093$ *rad/s* $= 343.4$ *Hz,*
$\omega_3 = 6,790$ *rad/s* $= 1,080.7$ *Hz,* $\omega_4 = 10,488$ *rad/s* $= 1,669.2$ *Hz,*
$\omega_5 = 15,869$ *rad/s* $= 2,525.6$ *Hz.*

The results are close to the ones obtained earlier, showing that a stiffness of 1×10^{11} N/m for the constraints is high enough to model rigid supports.

The consistent mass matrices of the elements were then built and the eigenanalysis was repeated. The natural frequencies, computed assuming elastic constraints with $k_c = 1 \times 10^{11}$ N/m, are as follows:

$\omega_1 = 2,239$ *rad/s* $= 356.3$ *Hz,* $\omega_2 = 6,031$ *rad/s* $= 959.8$ *Hz,*
$\omega_3 = 8,241$ *rad/s* $= 1,311.6$ *Hz,* $\omega_4 = 11,989$ *rad/s* $= 1,980.4$ *Hz,*
$\omega_5 = 20,552$ *rad/s* $= 3,271.0$ *Hz.*

The values of natural frequencies resulting from the consistent approach are higher than those obtained from the lumped model, as would be expected. The difference is not small, particularly for the higher-order modes, and this shows that the modelization with only six elements is too rough.

15.11 Dynamic stiffness matrix

As already stated, the finite element method yields approximate results even in the case of beam elements, because the shape functions cannot give way to the exact inflected shape. To overcome this limitation, which is more apparent than real, an exact method has been developed for beam elements, in which the expression of the deflected shape obtained through the continuous model is assumed.

In the case of a homogeneous, prismatic Euler–Bernoulli beam performing harmonic oscillations with frequency ω in the xz-plane, for example, the general solution of Eq. (12.60) yielding the deflected shape with any boundary condition is

$$q(z) = C_1 \sin(az) + C_2 \cos(az) + C_3 \sinh(az) + C_4 \cosh(az), \qquad (15.86)$$

where

$$a = \sqrt{\omega} \sqrt[4]{\frac{\rho A}{E I_y}}.$$

Constants C_i can be easily obtained from the boundary conditions, i.e., from the nodal generalized displacements. Note that in this case the shape functions that can be obtained from Eq. (15.86) depend on the frequency ω of the harmonic motion. They can be used to compute the mass and stiffness matrices of the element in the usual way. In this case, these matrices contain the frequency ω of the harmonic motion, and it is possible to directly write the dynamic stiffness matrix

$$\mathbf{K}_{dyn} = \mathbf{K} - \omega^2 \mathbf{M} \,,$$

which for the Euler–Bernoulli beam, is:[4]

$$\mathbf{K}_{dyn} = \frac{E I_y a}{\cos(al)\cosh(al) - 1} \begin{bmatrix} a^2 k_1 & ak_2 & a^2 k_3 & ak_4 \\ & -k_1 & ak_4 & k_5 \\ & & a^2 k_1 & ak_2 \\ \text{symm.} & & & -k_1 \end{bmatrix}, \qquad (15.87)$$

where

$$k_1 = \sin(al)\cosh(al) - \cos(al)\sinh(al),$$

$$k_2 = \sin(al)\sinh(al), \qquad k_3 = \sin(al) + \sinh(al),$$

$$k_4 = \cos(al) - \cosh(al), \qquad k_5 = \sinh(al) - \sin(al).$$

A similar approach can be used, at least in principle, for any type of element, provided that an expression of the type of Eq. (15.86) can be obtained from a continuous model. The dynamic matrices of the various elements can be assembled in the same way seen for mass and stiffness matrices. When the response to harmonic excitation has to be computed, the frequency of the motion is known and the computation is straightforward.

---

[4]See, for example, R.D. Henshell, G.B. Warburton, 'Transmission of vibration in beam systems', *Int. J. Numer. Meth. Eng.*, Vol. 1, 47–66, 1969. The dynamic matrices for both Euler–Bernoulli and Timoshenko beams are reported in detail.

On the contrary, when the natural frequencies of the system have to be computed, the solution of the eigenproblem is quite complicated because it is impossible to express it in standard form and then apply the usual numerical methods. The solution follows the lines seen for methods based on transfer matrices: A value of the frequency is assumed and the determinant of the dynamic stiffness matrix is computed. The determinant is not equal to zero, unless a natural frequency has been chosen, and new values of the frequency are assumed. A plot of the value of the determinant as a function of the frequency is then drawn. From the plot, the values of the frequency that cause the determinant to vanish can be obtained. The problem here is that the determinant of a fairly large matrix has to be computed many times and this is, from a computational point of view, far more involving than solving an eigenproblem in standard form using standard techniques.

The dynamic stiffness matrix, which is now a function of the frequency, can be expressed as a power series in the frequency ω:

$$\mathbf{K}_{dyn} = \mathbf{K}_{d_0} + \omega^2 \mathbf{K}_{d_2} + \omega^4 \mathbf{K}_{d_4} + \dots .$$

It is possible to show that only even powers of ω are present and that the first term $\mathbf{K}_{d_0}$ depends only on the elastic properties. In the case of beam elements it is coincident with the stiffness matrix. The second term $\mathbf{K}_{d_2}$ depends only on the inertial properties and is coincident, except for the sign, which is obviously changed, with the mass matrix. All other matrices depend on both inertial and stiffness properties and represent a correction of the potential and kinetic energies of the element due to a more precise formulation of the displacement field. The standard approach can thus be thought of as a truncation at the second term of the series for the exact dynamic stiffness matrix.

The drawbacks of the dynamic stiffness matrix approach, mainly those of leading to long and costly computations and of being applicable only to harmonic motion, are greater, in the opinion of the author, than its advantage of giving a better approximation, which can be obtained simply by using a finer mesh, with far less computational difficulty.

Remark 15.12 *The term exact used in this context means simply that it leads to the same results of the beam theory, i.e., of the continuous model. All the approximations linked to that model itself are, however, present.*

15.12 Damping matrices

It is possible to take into account the damping of the structure in a way that closely follows what has been said for the stiffness. If elements that can be modeled as viscous dampers are introduced into the structure between two nodes or between a node and the ground, a viscous damping matrix can be obtained using the same procedures used for the stiffness matrix of spring elements or elastic constraints. Actually, the relevant equations are equal, once the damping coefficient is substituted for the stiffness. If the damping of some of the elements can be modeled as hysteretic damping, within the limits of validity of the complex stiffness model, an imaginary part of the element stiffness matrix can be obtained by simply multiplying its real part by the loss factor.

Viscous or structural damping matrices are then assembled following the same rules seen for mass and stiffness matrices.

Remark 15.13 *If there is a geometric stiffness matrix, the imaginary part of the stiffness matrix must be computed before adding the geometric matrix to the stiffness matrix.*

Remark 15.14 *The real and imaginary parts of the stiffness matrices must be assembled separately, because, when the loss factor is not constant along the structure, they are not proportional to each other.*

15.13   Finite elements in time

An alternative to the conventional finite element discretization as described in Section 15.1 is the space–time approach in which time is considered as an added dimension and the four-dimensional space–time is meshed in a way similar to the tridimensional meshing of the standard FEM.

Equation (15.1) defining the shape functions now becomes

$$\mathbf{u}(x, y, z, t) = \mathbf{N}(x, y, z, t)\mathbf{q}, \tag{15.88}$$

where the elements of vector q are constants expressing the values of the generalized coordinates of the element at different times and the shape functions are functions of time and space coordinates.

The equation of motion of the element can be directly obtained from Hamilton's principle, stating that

$$\int_{t_1}^{t_2} [\delta(\mathcal{T} - \mathcal{U}) + \delta\mathcal{L}] \, dt = 0 \tag{15.89}$$

where $\delta(T - U)$ and δL are, respectively, the variation of the Lagrangian of the system and the virtual work of nonconservative forces corresponding to the virtual displacement δu. The generalized velocities, the strains, and the virtual displacements can be expressed as

$$\dot{\mathbf{u}}(x, y, z, t) = \left[\frac{\partial \mathbf{N}}{\partial t}\right]\mathbf{q}, \qquad \epsilon = \mathbf{B}\mathbf{q}, \qquad \delta\mathbf{u} = \mathbf{N}\delta\mathbf{q}, \tag{15.90}$$

where matrix $\mathbf{B}$ is of the same type as the matrix defined by Eq. (15.3), the only difference being that in this case it is also a function of time.

In the case where all generalized coordinates are related to displacements, the Lagrangian of the element can be expressed as

$$T - U = \frac{1}{2}q^T \left(\int_V \rho \left[\frac{\partial \mathbf{N}}{\partial t}\right]^T \left[\frac{\partial \mathbf{N}}{\partial t}\right] dV \right) q - \frac{1}{2}q^T \left(\int_V \mathbf{B}^T \mathbf{E} \mathbf{B} dV \right) q.$$

(15.91)

If a generalized force $f(x, y, z, t)$ is acting on the element, Hamilton's principle can be written in the form

$$\int_{t_1}^{t_2} \int_V \left\{ \delta q^T \left[\rho \left[\frac{\partial \mathbf{N}}{\partial t}\right]^T \left[\frac{\partial \mathbf{N}}{\partial t}\right] - \mathbf{B}^T \mathbf{E} \mathbf{B} \right] q + \delta q^T \mathbf{N}^T \mathbf{f} \right\} dV dt = 0,$$

(15.92)

i.e.,

$$\mathbf{K}^* q = \mathbf{f}^*,$$

(15.93)

where matrix K^* and vector f^* are

$$\mathbf{K}^* = \int_{V^*} \left\{ -\rho \left[\frac{\partial \mathbf{N}}{\partial t}\right]^T \left[\frac{\partial \mathbf{N}}{\partial t}\right] + \mathbf{B}^T \mathbf{E} \mathbf{B} \right\} dV^*,$$

(15.94)

$$\mathbf{f}^* = \int_{V^*} \mathbf{N}^T \mathbf{f} dV^*.$$

and volume V^* is the four-dimensional volume of the element.

The element matrices K^* and force vectors f^* are then assembled in the usual way, and an equation of the same type as Eq. (15.93) for the whole structure is written.

Remark 15.15 *The number of degrees of freedom needed when using the space–time approach is far larger than that resulting from conventional FEM modeling.*

An example of space–time modeling of a rectangular plate is shown in Fig. 15.7. Because the structure is two dimensional in space, the model has three dimensions. In the case shown in the figure, the elements are rectangular in time and are arranged in regular layers along the time coordinate. This is not necessarily the case, and the present method demonstrates its greatest advantages when the mesh is tailored in time to suit the local needs and triangular (hypertetrahedral) elements are used.

In the case of rectangular elements, the equations can be assembled layer by layer first. In this way, the generalized coordinates can be partitioned according to the time to which they refer, and the equation of the ith layer can be written as

$$\begin{bmatrix} \mathbf{A}_i & \mathbf{B}_i \\ \mathbf{C}_i & \mathbf{D}_i \end{bmatrix} \left\{ \begin{array}{c} q_{i-1} \\ q_i \end{array} \right\} = \left\{ \begin{array}{c} f_{1_i} \\ f_{2_i} \end{array} \right\},$$

(15.95)

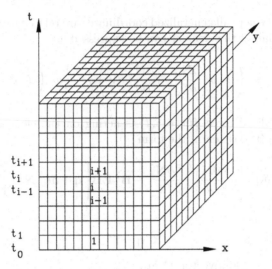

FIGURE 15.7. Space–time model of a rectangular plate using *rectangular* elements arranged in regular layers in time.

where matrices A_i, B_i, C_i, and $D_i{}^5$ are submatrices of K^* for the ith layer and the subscript in vector q_i refers to time t_i and not to the ith layer.

Once the matrices for the layers are assembled, the solution can be obtained starting from the first layer, in which the initial conditions are usually stated, and proceeding layer by layer in a way similar to conventional step-by-step time integration. The equation for the ith layer can be written in the form

$$\mathbf{B}_{i+1} q_{i+1} = -\left(\mathbf{D}_i + \mathbf{A}_{i+1}\right) q_i - \mathbf{C}_i q_{i-1} + \frac{1}{2}\left(\mathbf{f}_{2_{i-1}} + \mathbf{f}_{1_i}\right). \qquad (15.96)$$

Equation (15.96) directly yields the values of the generalized coordinates at time t_{i+1} from the values they take at time t_i and t_{i-1}.

The simplest type of shape functions for elements that are rectangular in time is

$$\mathbf{N}(x, y, z, t) = \left[\ \mathbf{N}_1(x, y, z)(1 - \tau) \quad \mathbf{N}_1(x, y, z)\tau\ \right], \qquad (15.97)$$

where $\tau = t/h$ is the nondimensional time obtained by dividing the time t by the time step h and N_1 is the matrix of the shape functions of the conventional FEM approach. With simple computations, it is possible to show that the integration in space can thus be performed independently

---

[5]Matrices $\mathbf{A}$, $\mathbf{B}_i$, $\mathbf{C}_i$, and $\mathbf{D}_i$ obviously have nothing to do with the matrices with the same name referred to as the quadruple of the system.

from the integration in time. By also taking into account the presence of a possible Rayleigh dissipation function, Eq. (15.96) reduces to

$$
\left(\mathbf{M} + \frac{h^2}{6}\mathbf{K} + \frac{h}{2}\mathbf{C}\right)\mathbf{q}_{i+1} = 2\left(\mathbf{M} - \frac{h^2}{3}\mathbf{K}\right)\mathbf{q}_i +
$$
$$
-\left(\mathbf{M} + \frac{h^2}{6}\mathbf{K} - \frac{h}{2}\mathbf{C}\right)\mathbf{q}_{i-1} + \frac{h^2}{2}\left(\mathbf{f}_{i-1} + \mathbf{f}_i\right), \tag{15.98}
$$

where M, C, and K are the matrices computed using the conventional FEM approach. Equation (15.98) is close to that obtained from the central difference algorithm for step-by-step integration in time (see Section A.5).

However, as already stated, the space–time finite element approach gives better results if the elements are not rectangular in time and the shape functions cannot be expressed as the product of a function of space by a function of time. There are some limitations on the ratio between the space and the time dimensions of the elements, which should approach the propagation velocity of the elastic waves in the material.[6] This consideration can cause a large increase is the number of degrees of freedom.

15.14 Exercises

Exercise 15.1 *Repeat the computations of Exercise 12.1 using the FEM. Use three, five, and seven Timoshenko beam elements.*

Exercise 15.2 *Compute the natural frequencies of the connecting rod of Exercise 12.7 using the FEM. Use the geometric matrix to take into account the compressive axial force. First use a single beam element and then repeat the computation using a finer subdivision.*

Exercise 15.3 *Consider the prismatic homogeneous beam sketched Fig. 15.8. It is clamped at one end and supported at the other end. In its midpoint it is connected to a frictionless prismatic slide constraining rotations. A mass m is located at the midpoint where also a force F(t) is acting. Write the equation of motion of the system using the FEM. Neglect the mass of the beam and its shear deformation and keep the model as simple as possible.*

Exercise 15.4 *Repeat the computation of the first natural frequency of the system studied in Example 9.1, assuming that during lift-off an acceleration of 6 g acts on the beam in axial direction. Use the FEM and model the beam with a single element.*

---

[6]Z. Kaczkowski, 'The method of finite space-time elements in dynamics of structures', *J. Tech. Phys.*, Vol. 16, No. 1, 69–84, 1975. See also C. Bajer, C. Bonthoux, "State-of-the-art in the space–time element method", *The Shock and Vibration Digest*, Vol. 23, 3–9, May 1991.

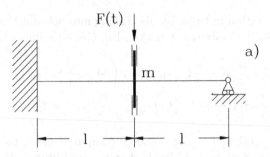

FIGURE 15.8. Exercise 5.4: sketch of the system.

Exercise 15.5 *Consider the transmission shaft of Example 15.1. Compute the numerical value of the natural frequency using the Raleigh method. Compute the first three natural frequencies and the mode shapes of the flexural vibrations using a four-element subdivision and compare the result obtained for the first mode with that previously obtained. Data: inner diameter = 70 mm, outer diameter = 90 mm, length of each part of the shaft = 500 mm, $E = 2.1 \times 10^{11}$ N/m^2, $\nu = 0.3$, $\rho = 7,810$ kg/m^3, $m_G = 2$ kg, $k_G = 10^5$ N/m.*

Exercise 15.6 *Repeat the dynamic analysis of the previous exercise by using the component-mode synthesis. Subdivide the system into two substructures coinciding with two spans of the shaft and take into account the first mode of each part.*

Exercise 15.7 *Consider a frame made by three beams arranged as an equilateral triangle. The frame is clamped at one of the vertices and carries two concentrated masses at the other vertices. Perform the dynamic analysis of the frame using one beam element for each side. Data: annular cross section with inner diameter = 50 mm, outer diameter = 70 mm, length of each side = 500 mm, $E = 2.1 \times 10^{11}$ N/m^2, $\nu = 0.3$, concentrated masses at the vertices $m_c = 10$ kg.*

Exercise 15.8 *Compute the field transfer matrix for a Timoshenko beam from the stiffness matrix of an element of the same type.*

Exercise 15.9 *Consider the beam supported by an actuator studied in Example 12.4. Plot the roots locus at varying proportional gain as in Fig. 12.10b, for the two cases of the sensor at the end of the beam and at the 90% of the length (the actuator is always at the end of the beam). Use the same relationship between the proportional and the derivative gain as in Example 12.4. Solve the problem using the FEM, with 10 Euler–Bernoulli elements. Data: length $l = 1$ m, rectangular cross section with thickness $t = 50$ mm and width $w = 100$ mm; density $\rho = 7,810$ kg/m^3, $E = 2.1 \times 10^{11}$ N/m^2.*

16

Dynamics of Multibody Systems

The previous chapters were mostly devoted to the study of elastic and visco-elastic systems, dealt with under the assumptions of small displacements and rotations. When some of the concentrated masses or deformable solids are replaced by rigid bodies, possibly connected to each other not only by springs and dampers but also by constraints of various nature, things become more complicated. In most cases, when dealing with mechanisms, displacements, and above all rotations, are not small and thus the equations of motion cannot be reduced to a set of linear equations. Today the most common approach is the use of multibody dynamics codes, operating in the time domain through numerical integration of the equations.

16.1 General considerations

Consider a system made by a number of rigid bodies, connected to each other by linkages that constrain a number of relative motions and by springs and dampers operating in the same way as seen for the discrete systems studied in the previous chapters.

The most general approach for dealing with systems of this kind is to build a multibody model, i.e., by writing the equations of motion of the various bodies and then adding the constraint equations that model the connections between them. Most commercial computer codes widely used in dynamic analysis operate along these lines.

The number of degrees of freedom resulting from this approach may be quite large: in tridimensional space each rigid body has 6 degrees of freedom (see below) and thus a number of $6n$ second-order differential equations must be written, if there are n rigid bodies.

Owing to the internal constraints, however, the number of degrees of freedom of the system, i.e., the minimum number of parameters needed to define the configuration of the system, may be much smaller. For instance, if two of the bodies are connected with each other by a cylindrical hinge a single parameter (the angle of rotation about the axis of the hinge) is needed to define completely their relative position: the cylindrical hinge thus constrains 5 degrees of freedom and can be modeled by a set of 5 algebraic equations.

The complexity of the study depends also on the configuration of the system. If the various bodies constitute an open chain, or a tree system, things are much simpler than in the case of a closed chain, or multiply connected system. In many cases the same system can be set in open and closed chain configurations, so it must be studied in different layouts.

An example of an open chain system is the robotic arm sketched in Fig. 16.1. It can be modeled as made of six rigid bodies: the support, the two beams, the wrist, and the two parts of the gripper. The number of degrees of freedom of these bodies is thus 36 and the dynamic model of the system consists of 36 second-order differential equations. The support is constrained by a cylindrical hinge, allowing just a rotation about the vertical axis. Also the two beams are constrained to the support and to each other by cylindrical hinges, as is the wrist to the second beam and the two parts of the gripper to the wrist: a total of six cylindrical hinges, constraining 30 degrees of freedom. A further constraint is supplied by the

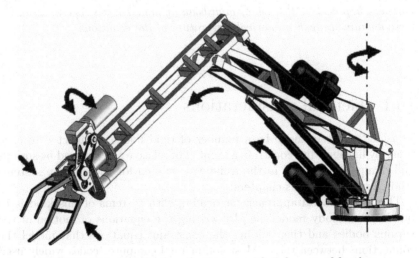

FIGURE 16.1. Sketch of a robotic arm with 5 degrees of freedom.

two gear wheels that force the two parts of the gripper to move together, by eliminating a further degree of freedom.

The constraint equations are thus 31. They can be used to eliminate 31 out of the 36 generalized coordinates, obtaining a set of five differential equations in the five independent generalized coordinates. It must be emphasized that in the 36 equations of motion originally written, the forces the various bodies exchange at the constraints are included; these forces are then eliminated when the constraint equations are introduced. The 31 equations so eliminated can be used to compute the constraint forces.

If the internal constraints are holonomic and the system is of the tree type, it is possible to resort to a simpler approach. One of the bodies may be chosen as *main body*, to which a number of secondary, *first-level* bodies are attached. Other secondary bodies, considered as *second-level* bodies, are then attached and so on. Secondary bodies have only the degrees of freedom allowed by the constraints. In this way, the minimum number of equations needed for the study is directly obtained. Such equations are all differential equations, usually of the second order. The forces exchanged at the constraints between the bodies do not appear explicitly in the equations and need not be computed in the dynamic study of the system as a whole.

The equations of motion of multibody systems may be obtained following two different approaches that are usually widely defined as Eulerian and Lagrangian. The first approach is based on writing the equations of motion of the various bodies by stating directly the dynamic equilibrium equations, usually in the form stating that the (generalized) external forces equal the time derivatives of the (generalized) momenta. This approach leads to a full set of differential equations to be combined with the constraint equations.

The Lagrangian approach is based on writing the kinetic and potential energies of the system and then applying the methods of analytical dynamics, Lagrange equations being one of the most common approach. As already stated, they can be written in two forms: a full set of equations, whose number is equal to all the generalized coordinates, to which the constraint equations must be added, possibly using the Lagrange multipliers approach, or a reduced set of equations, written in the independent generalized coordinates only.

Only this last approach will be followed in the present book.

It must be observed that multibody modeling is usually considered as a branch of mechanics having little to do with vibration mechanics and is not dealt with in books on vibration. This is done for a very good reason: Classical multibody models deal mostly with rigid bodies connected through linkages which behave following kinematic rules and thus are not subjected to vibration, not to mention that adding also this aspect of dynamics to a book on vibration would end in too wide a scope to be pursued in a reasonable way.

However, the trend in multibody dynamics is to include also deformable bodies, usually modeled through the finite element method, and springs and dampers, giving this kind of models the ability do deal with vibration problems.

Here a compromise is searched: In this short chapter the basics of the motion of rigid bodies is dealt with, without any attempt at delving in detail into multibody modeling, a field that has been the subject of many specialized books. To be consistent with the previous chapters, mainly the Lagrangian approach will be followed and no constraint equation is explicitly written.

Example 16.1 *Study the linearized dynamics of the planar system shown in Fig. 16.2. It is made of two rigid bodies, the first suspended on two elastic supports, having different horizontal and vertical stiffness and the second one connected to the first by two bars and forming a bi-filar pendulum. In the figure an inertial reference frame OXY centered in the static equilibrium position O of the center of mass G_1 and a body-fixed frame G_1xy are shown.*

Since the system is planar, each rigid body has a total or 3 degrees of freedom, so that, considering also the two bar connecting the bodies, the total number of degrees of freedom is $4 \times 3 = 12$. The four cylindrical hinges constrain 2 degrees of freedom each, for a total of 8 degrees of freedom. The number of independent degrees of freedom is thus 4.

Note that it is not a branched system, but includes a closed loop.

If the Lagrangian approach is followed, a minimum number of four generalized coordinates is needed: a simple choice is to chose the horizontal and vertical displacements X and Y of the first body and angles θ and ϕ that axis y and line BG_2 make with the vertical direction.

The position of the various points of interest in the inertial frame are

$$(\overline{G_1 - O}) = \left\{ \begin{array}{c} X \\ Y \end{array} \right\} \quad , \quad (\overline{A_i - O}) = \left\{ \begin{array}{c} X \pm b\cos(\theta) + a\sin(\theta) \\ Y \pm b\sin(\theta) - a\cos(\theta) \end{array} \right\} .$$

$$(\overline{B_i - O}) = \left\{ \begin{array}{c} X \pm c\cos(\theta) + d\sin(\theta) \\ Y \pm c\sin(\theta) - d\cos(\theta) \end{array} \right\} ,$$

$$(\overline{C_i - O}) = \left\{ \begin{array}{c} X \pm b\cos(\theta) + a\sin(\theta) + l\sin(\phi) \\ Y \pm b\sin(\theta) - a\cos(\theta) - l\cos(\phi) \end{array} \right\} .$$

$$(\overline{G_2 - O}) = \left\{ \begin{array}{c} X + (d+e)\sin(\theta) + l\sin(\phi) \\ Y - (d+e)\cos(\theta) - l\cos(\phi) \end{array} \right\} .$$

The velocities of the centers of mass and the angular velocities of the two bodies are

$$V_1 = \left\{ \begin{array}{c} \dot{X} \\ \dot{Y} \end{array} \right\} , \quad V_2 = \left\{ \begin{array}{c} \dot{X} + (d+e)\dot{\theta}\cos(\theta) + l\dot{\phi}\cos(\phi) \\ \dot{Y} + (d+e)\dot{\theta}\sin(\theta) + l\dot{\phi}\sin(\phi) \end{array} \right\} , \quad \Omega_1 = \Omega_2 = \dot{\theta}.$$

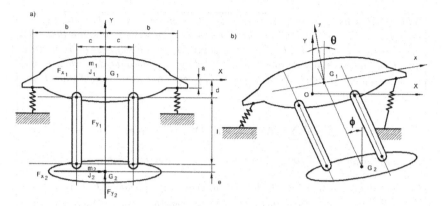

FIGURE 16.2. Two-body planar system. (a) Static equilibrium configuration; (b) deformed configuration.

Note that the two bodies have the same angular velocity, since the bi-filar pendulum keeps them parallel to each other.

Neglecting the mass of the two beams, the kinetic energy of the system is thus

$$\mathcal{T} = m_1 V_1^2 + m_2 V_2^2 + (J_1 + J_2)\dot{\theta}^2 \; ,$$

i.e.,

$$\mathcal{T} = \frac{1}{2}m\left(\dot{X}^2 + \dot{Y}^2\right) + \frac{1}{2}J\dot{\theta}^2 + \frac{1}{2}m_2 l^2 \dot{\phi}^2 + m_2 fl\dot{\theta}\dot{\phi}\cos(\theta - \phi) + \\ + m_2 f\dot{\theta}\left[\dot{X}\cos(\theta) + \dot{Y}\sin(\theta)\right] + m_2 l\dot{\phi}\left[\dot{X}\cos(\phi) + \dot{Y}\sin(\phi)\right] \; ,$$

where

$$m = m_1 + m_2 \; , \quad J = J_1 + J_2 + m_2 f^2 \; , \; f = d + e \; .$$

Its linearized expression is

$$\mathcal{T} = \frac{1}{2}m\left(\dot{X}^2 + \dot{Y}^2\right) + \frac{1}{2}J\dot{\theta}^2 + \frac{1}{2}m_2 l^2 \dot{\phi}^2 + m_2 fl\dot{\theta}\dot{\phi} + \\ + m_2 f\dot{\theta}\dot{X} + m_2 l\dot{\phi}\dot{X} \; .$$

The gravitational potential energy is simply

$$\mathcal{U}_g = mgY - m_2 gf\cos(\theta) - m_2 gl\cos(\phi) \; .$$

By linearizing and neglecting the constant terms, it follows that

$$\mathcal{U}_g = mgY + \frac{1}{2}m_2 gf\theta^2 + \frac{1}{2}m_2 gl\phi^2 \; .$$

Some further assumptions are needed for the computation of the potential energy due to the suspension springs. To simplify the matter, assume that they are linear, reacting to displacements of points A_i with a stiffness k_v in vertical direction and k_h in horizontal direction. Taking into account that when $Y = 0$ the springs are already deformed, since they carry the static weight, the elastic potential energy is

$$\mathcal{U}_e = \frac{1}{2} \sum_{i=1}^{2} \left[K_h X_{A_i}^2 + K_v \left(Y_{A_i} - Y_{0_i} + a \right)^2 \right] \ ,$$

i.e., neglecting the constant terms that do not enter the equations of motion,

$$\begin{aligned} \mathcal{U}_e \ = \ & K_h X^2 + K_v Y^2 + \left(K_h b^2 + K_v a^2 \right) \cos^2(\theta) + \left(K_h a^2 + K_v b^2 \right) \sin^2(\theta) + \\ & + 2 K_h a X \sin(\theta) - 2 a K_v \left(Y - Y_0 + a \right) \cos(\theta) - 2 K_v Y \left(Y_{0_i} - a \right) \ . \end{aligned}$$

Linearizing the expression of the elastic potential energy, it follows that

$$\mathcal{U}_e = K_h X^2 + K_v Y^2 + K_\theta \theta^2 + 2 K_h a X \theta - 2 K_v Y Y_{0_i},$$

where

$$K_\theta = \left(K_h - K_v \right) \left(a^2 - b^2 \right) - a K_v \left(Y_0 - a \right) \ .$$

The external forces are applied to the two centers of mass. Given a virtual displacement

$$\delta s = [\delta X, \ \delta Y, \ \delta \theta, \ \delta \phi]^T \ ,$$

the virtual displacements of G_1 and G_2 are

$$\delta s_{G_1} = \left\{ \begin{array}{c} \delta X \\ \delta Y \end{array} \right\} \ , \quad \delta s_{G_2} = \left\{ \begin{array}{c} \delta X + (d + e) \cos(\theta) \delta \theta + l \cos(\phi) \delta \phi \\ \delta Y + (d + e) \sin(\theta) \delta \theta + l \sin(\phi) \delta \phi \end{array} \right\} \ .$$

The virtual work of the external forces is

$$\begin{aligned} \delta \mathcal{L} \ = \ & \left(F_{x_1} + F_{x_2} \right) \delta X + \left(F_{y_1} + F_{y_2} \right) \delta Y + \left[F_{x_2} \cos(\theta) + F_{y_2} \sin(\theta) \right] f \delta \theta \\ & + \left[F_{x_2} \cos(\phi) + F_{y_2} \sin(\phi) \right] l \delta \phi \ . \end{aligned}$$

When the equations of motion are linearized, the external forces must not contain the generalized coordinates, so

$$\delta \mathcal{L} = \left(F_{x_1} + F_{x_2} \right) \delta X + \left(F_{y_1} + F_{y_2} \right) \delta Y + F_{x_2} f \delta \theta + F_{x_2} l \delta \phi \ .$$

The equations of motion can thus be obtained by introducing the relevant derivatives into the Lagrange equations. The second equation uncouples from the other ones, yielding

$$m \ddot{Y} + 2 K_v Y = -mg - 2 K_v Y_{0_i} + F_{y_1} + F_{y_2}.$$

Assuming that $Y = 0$ and $Q_y = 0$, the static equilibrium condition is readily obtained:

$$Y_{0_i} = \frac{mg}{2K_v} \quad for \ i = 1, \ 2,$$

and thus the equation for the motion about the equilibrium position reduces to that of a mass–spring system

$$m\ddot{Y} + K_v Y = F_{y_1} + F_{y_2} \ .$$

Substituting for Y_{0_i} its value, the other equations can be summarized as

$$\begin{bmatrix} m & m_2 f & m_2 l \\ m_2 f & J & m_2 fl \\ m_2 l & m_2 fl & m_2 l^2 \end{bmatrix} \begin{Bmatrix} \ddot{X} \\ \ddot{\theta} \\ \ddot{\phi} \end{Bmatrix} +$$

$$+ \begin{bmatrix} 2K_h & 2K_h a & 0 \\ 2K_h a & K'_\theta & 0 \\ 0 & 0 & m_2 gl \end{bmatrix} \begin{Bmatrix} X \\ \theta \\ \phi \end{Bmatrix} = \begin{Bmatrix} F_{x_1} + F_{x_2} \\ F_{x_2} f \\ F_{x_2} l \end{Bmatrix} ,$$

where

$$K'_\theta = 2 \left(K_h - K_v \right) \left(a^2 - b^2 \right) + 2a^2 K_v a + g \left(m_2 f - ma \right) \ .$$

The motion of the bi-filar pendulum is inertially coupled with that of the suspending mass. The linearized equations of motion allow to study completely the motion in the small about the static equilibrium position.

Remark 16.1 *The previous example is in a way misleading: It deals with a planar system and thus is too simple to represent the complexity of the multibody approach. In particular, owing to both the two-dimensional nature of the system and the small-displacement assumption, it was possible to linearize the equations of motion, obtaining a formulation suitable for a frequency-domain study. This is usually not the case when motion in tridimensional space is possible.*

16.2 Lagrange equations in terms of pseudo-coordinates

Often the equations of motion in the state space are written with reference to generalized velocities that are not simply the derivatives of the generalized coordinates. In particular, often it is expedient to use as generalized velocities w_i suitable combinations of the derivatives of the coordinates $v_i = \dot{x}_i$:

$$\{w_i\} = \mathbf{A}^T \{\dot{x}_i\} \ . \tag{16.1}$$

Remark 16.2 *Equation (16.1) that defines the generalized velocities as functions of the derivatives of the generalized coordinates is usually said to be the kinematic equation of the system.*

The coefficients of the linear combinations included in matrix A^T may be constant, but in general are themselves functions of the generalized coordinates.

Equation (16.1) may in general be inverted, obtaining[1]

$$\{\dot{x}_i\} = \mathbf{B}\{w_i\} \ , \tag{16.2}$$

where

$$\mathbf{B} = \mathbf{A}^{-T} \tag{16.3}$$

and symbol A^{-T} indicates the inverse of the transpose of matrix A.

In some cases matrix A^T is a rotation matrix and its inverse coincides with its transpose. In such cases

$$\mathbf{B} = \mathbf{A}^{-T} = \mathbf{A} \ .$$

However, in general this does not occur and

$$\mathbf{B} \neq \mathbf{A} \ .$$

While v_i are the derivatives of the coordinates x_i, in general it is not possible to express w_i as the derivatives of suitable coordinates. Equation (16.1) can be written in the infinitesimal displacements dx_i

$$\{d\theta_i\} = \mathbf{A}^T\{dx_i\} \ , \tag{16.4}$$

obtaining a set of infinitesimal displacements $d\theta_i$, corresponding to the velocities w_i. Equations (16.4) can be integrated, yielding displacements θ_i corresponding to the velocities w_i, only if

$$\frac{\partial a_{js}}{\partial x_k} = \frac{\partial a_{ks}}{\partial x_j} \ .$$

If this is not the case, Eq. (16.4) cannot be integrated and velocities w_i cannot be considered as the derivatives of true coordinates. In such a case they are said to be the derivatives of pseudo-coordinates.

As a first consequence of the nonexistence of coordinates corresponding to velocities w_i, Lagrange equation (3.37) cannot be written directly using velocities w_i, but must be modified to allow using velocities and coordinates that are not directly one the derivative of the other.

The use of pseudo-coordinates is fairly common, in particular when studying the dynamics of structures that perform also rigid-body motions (robots, spacecraft, etc.). If, for instance, in the dynamics of a rigid body the generalized velocities in a reference frame following the body in its motion are used, while the coordinates x_i are the displacements in an inertial

---

[1]Matrices $\mathbf{A}$ and $\mathbf{B}$ here defined must not be confused with the dynamic matrix $\mathcal{A}$ and with the input gain matrix $\mathcal{B}$ or the compliance matrix $\mathbf{B}$.

frame, matrix $\mathbf{A}^T$ is simply the rotation matrix allowing to pass from one reference frame to the other. In this case matrix $\mathbf{B}$ coincides with $\mathbf{A}$, but both are not symmetrical and then the velocities in the body-fixed frame cannot be considered as the derivatives of true coordinates.

Remark 16.3 *The body-fixed frame rotates and then it is not possible to integrate the velocities along the body-fixed axes to obtain the displacements along the same axes. That notwithstanding, it is possible to use the components of the velocity along the body-fixed axes to write the equations of motion.*

The kinetic energy can be written in general in the form

$$T = T(w_i, x_i, t) .$$

The derivatives $\partial T / \partial \dot{x}_i$ included in the equations of motion are

$$\frac{\partial T}{\partial \dot{x}_k} = \sum_{i=1}^{n} \frac{\partial T}{\partial w_i} \frac{\partial w_i}{\partial \dot{x}_k} , \tag{16.5}$$

i.e., in matrix form,

$$\left\{ \frac{\partial T}{\partial \dot{x}} \right\} = \mathbf{A} \left\{ \frac{\partial T}{\partial w} \right\} , \tag{16.6}$$

where

$$\left\{ \frac{\partial T}{\partial \dot{x}} \right\} = \left[\begin{array}{ccc} \dfrac{\partial T}{\partial \dot{x}_1} & \dfrac{\partial T}{\partial \dot{x}_2} & \cdots \end{array} \right]^T ,$$

$$\left\{ \frac{\partial T}{\partial w} \right\} = \left[\begin{array}{ccc} \dfrac{\partial T}{\partial w_1} & \dfrac{\partial T}{\partial w_2} & \cdots \end{array} \right]^T .$$

By differentiating with respect to time, it follows that

$$\frac{\partial}{\partial t} \left(\left\{ \frac{\partial T}{\partial \dot{x}} \right\} \right) = \mathbf{A} \frac{\partial}{\partial t} \left(\left\{ \frac{\partial T}{\partial w} \right\} \right) + \dot{\mathbf{A}} \left\{ \frac{\partial T}{\partial w} \right\} . \tag{16.7}$$

The generic element $\dot{a}_{jk}$ of matrix $\dot{A}$ is

$$\dot{a}_{jk} = \sum_{i=1}^{n} \frac{\partial a_{jk}}{\partial x_i} \dot{x}_i = \dot{\mathbf{x}}^T \left\{ \frac{\partial a_{jk}}{\partial x} \right\} \tag{16.8}$$

and then

$$\dot{a}_{jk} = \mathbf{w}^T \mathbf{B}^T \left\{ \frac{\partial a_{jk}}{\partial x} \right\} . \tag{16.9}$$

The various $\dot{a}_{jk}$ so computed can be written in matrix form

$$\dot{\mathbf{A}} = \left[\mathbf{w}^T \mathbf{B}^T \left\{ \frac{\partial a_{jk}}{\partial x} \right\} \right] . \tag{16.10}$$

The computation of the derivatives of the generalized coordinates $\{\partial T / \partial x\}$ is usually less straightforward. The generic derivative

$$\frac{\partial T}{\partial x_k}$$

is

$$\frac{\partial T^*}{\partial x_k} = \frac{\partial T}{\partial x_k} + \sum_{i=1}^{n} \frac{\partial T}{\partial w_i} \frac{\partial w_i}{\partial x_k} = \frac{\partial T}{\partial x_k} + \sum_{i=1}^{n} \frac{\partial T}{\partial w_i} \sum_{j=1}^{n} \frac{\partial a_{ij}}{\partial x_k} \dot{x}_j , \qquad (16.11)$$

where T^* is the kinetic energy expressed as a function of the generalized coordinates and their derivatives (the expression to be introduced into Lagrange equation in its usual form), while T is expressed as a function of the generalized coordinates and of the velocities w_i. Equation (16.11) can be written as

$$\frac{\partial T^*}{\partial x_k} = \frac{\partial T}{\partial x_k} + \mathbf{w}^T \mathbf{B}^T \frac{\partial \mathbf{A}}{\partial x_k} \left\{ \frac{\partial T}{\partial w} \right\} , \qquad (16.12)$$

where product

$$\mathbf{w}^T \mathbf{B}^T \frac{\partial \mathbf{A}}{\partial x_k}$$

yields a row matrix with n elements and that multiplied by the column matrix $\{\partial T / \partial w\}$ yields the required number.

By combining those row matrices, a square matrix is obtained:

$$\left[\mathbf{w}^T \mathbf{B}^T \frac{\partial \mathbf{A}}{\partial x_k} \right] , \qquad (16.13)$$

and thus the column containing the derivatives with respect to the generalized coordinates is

$$\left\{ \frac{\partial T^*}{\partial x} \right\} = \left\{ \frac{\partial T}{\partial x} \right\} + \left[\mathbf{w}^T \mathbf{B}^T \frac{\partial \mathbf{A}}{\partial x} \right] \left\{ \frac{\partial T}{\partial w} \right\} . \qquad (16.14)$$

By definition, the potential energy does not depend on the generalized velocities and then the term $\partial \mathcal{U} / \partial x_i$ is not influenced by the way the generalized velocities are written.

The Rayleigh dissipation function F can be written in terms of the velocities w. Its derivatives are

$$\left\{ \frac{\partial \mathcal{F}}{\partial \dot{x}} \right\} = \mathbf{A} \left\{ \frac{\partial \mathcal{F}}{\partial w} \right\} . \qquad (16.15)$$

The equation of motion (3.37) is thus

$$\mathbf{A} \frac{d}{dt} \left(\left\{ \frac{\partial T}{\partial w} \right\} \right) + \mathbf{\Gamma} \left\{ \frac{\partial T}{\partial w} \right\} - \left\{ \frac{\partial T}{\partial x} \right\} + \left\{ \frac{\partial \mathcal{U}}{\partial x} \right\} + \mathbf{A} \left\{ \frac{\partial \mathcal{F}}{\partial w} \right\} = \mathbf{Q} , \qquad (16.16)$$

where

$$\boldsymbol{\Gamma} = \left[\mathbf{w}^T \mathbf{B}^T \left\{ \frac{\partial a_{jk}}{\partial x} \right\} \right] - \left[\mathbf{w}^T \mathbf{B}^T \frac{\partial \mathbf{A}}{\partial x_k} \right] \qquad (16.17)$$

and Q is a vector containing the n generalized forces Q_i.

By premultiplying all terms by matrix $B^T = A^{-1}$ and attaching the kinematic equations to the dynamic equations, the final form of the state-space equations is obtained:

$$\begin{cases} \dfrac{d}{dt} \left(\left\{ \dfrac{\partial \mathcal{T}}{\partial w} \right\} \right) + \mathbf{B}^T \boldsymbol{\Gamma} \left\{ \dfrac{\partial \mathcal{T}}{\partial w} \right\} - \mathbf{B}^T \left\{ \dfrac{\partial \mathcal{T}}{\partial x} \right\} + \mathbf{B}^T \left\{ \dfrac{\partial \mathcal{U}}{\partial x} \right\} + \left\{ \dfrac{\partial \mathcal{F}}{\partial w} \right\} = \mathbf{B}^T \mathbf{Q} \\[2mm] \{\dot{q}_i\} = \mathbf{B} \{w_i\} \ . \end{cases}$$

$$(16.18)$$

16.3   Motion of a rigid body

16.3.1   Generalized coordinates

Consider a rigid body free in tridimensional space. Define an inertial reference frame $OXYZ$ and a frame $Gxyz$ fixed to the body and centered in its center of mass. The position of the rigid body is defined once the position of frame $Gxyz$ is defined with respect to $OXYZ$, that is, once the transformation leading $OXYZ$ to coincide with $Gxyz$ is defined. It is well known that the motion of the second frame can be considered as the sum of a displacement plus a rotation and then the parameters to be defined are six: three components of the displacement, two of the components of the unit vector defining the rotation axis (the third component need not be defined and may be computed from the condition that the unit vector has unit length) and the rotation angle. A rigid body has thus six degrees of freedom in the tridimensional space.

There is no problem in defining the generalized coordinates for the translational degrees of freedom, since the coordinates of any point fixed to the body expressed in any inertial reference frame (in particular, in frame $OXYZ$) can be used, with the simplest, and the most obvious, choice being usually the coordinates of the center of mass G. For the other generalized coordinates the choice is much more complicated. It is possible to resort, for instance, to two coordinates of a second point and to one of the coordinates of a third point, not on a straight line through the other two, but this choice is far from being the most expedient.

An obvious way to define the rotation of frame $Gxyz$ with respect to $OXYZ$ is to express directly the rotation matrix linking the two reference frames. It is a square matrix of size 3×3 (in tridimensional space) and then has nine elements. Three of them are independent, while the other six may be obtained from the first three using suitable equations.

Alternatively, the position of the body-fixed frame can be defined stating a sequence of three rotations about the axes. Since rotations are not vectors, the order in which they are performed must be specified.

Start rotating, for instance, the inertial frame about the X-axis. The second rotation may be performed about axes Y or Z (obviously in the position they take after the first rotation), but not about the X-axis, because in the latter case the two rotations would simply add to each other and would amount to a single rotation. Assume for instance to rotate the frame about the Y-axis.

The third rotation may occur about either X-axis or Z-axis (in the new position, taken after the second rotation), but not about the Y-axis.

The possible rotation sequences are 12, but may be subdivided into two types: those like $X \rightarrow Y \rightarrow X$ or $X \rightarrow Z \rightarrow X$, where the third rotation occurs about the same axis as the first one, and those like $X \rightarrow Y \rightarrow Z$ or $X \rightarrow Z \rightarrow Y$, where the third rotation is performed about a different axis.

In the first case the angles are said to be Euler angles, since they are of the same type of the angles Euler proposed to study the motion of gyroscopes (precession ϕ about Z-axis, nutation θ about X-axis, and rotation ψ, again about Z-axis). In the second case they are said to be Tait–Bryan angles.[2]

The possible rotation sequences are reported in the following table:

First	X				Y				Z			
Second	Y		Z		X		Z		X		Y	
Third	X	Z	X	Y	Y	Z	Y	X	Z	Y	Z	X
Type	E	TB	E	TB	E	TB	E	TB	E	TB	E	TB

Euler angles are indeterminate when the plane containing the axes of the first two rotations (xy in the first case) of the body-fixed frame is parallel to the plane with the same name of the inertial frame, i.e., when the second rotation angle vanishes.

In the dynamics of moving structures (road vehicles, aircraft, spacecraft, etc.) the most common approach is using Tait–Bryan angles of the type $Z \rightarrow Y \rightarrow X$ so defined (Fig. 16.3):

- Rotate frame XYZ (whose XY plane is usually horizontal or parallel to the ground) about Z-axis until axis X coincides with the projection of x-axis on plane XY (Fig. 16.3a). Such a position of X-axis can be indicated as x^*. The rotation angle between axes X and x^* is the *yaw* angle ψ. The rotation matrix allowing to pass from the x^*y^*Z frame that will be referred to as the intermediate frame to the inertial frame

---

[2]Sometimes all sets of three ordered angles are said to be Euler angles. With this wider definition also Tait–Bryan angles are considered as Euler angles.

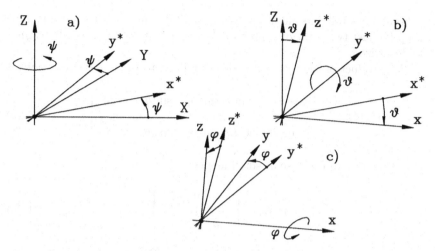

FIGURE 16.3. Definition of angles: yaw ψ (**a**), pitch θ (**b**), and roll ϕ (**c**).

XYZ is

$$\mathbf{R}_1 = \begin{bmatrix} \cos(\psi) & -\sin(\psi) & 0 \\ \sin(\psi) & \cos(\psi) & 0 \\ 0 & 0 & 1 \end{bmatrix} . \tag{16.19}$$

- The second rotation is the *pitch* rotation θ about y^*-axis, leading axis x^* in the position of x-axis (Fig. 16.3b). The rotation matrix is

$$\mathbf{R}_2 = \begin{bmatrix} \cos(\theta) & 0 & \sin(\theta) \\ 0 & 1 & 0 \\ -\sin(\theta) & 0 & \cos(\theta) \end{bmatrix} . \tag{16.20}$$

- The third rotation is the *roll* rotation ϕ about x-axis, leading axes y^* and z^* to coincide with axes y and z (Fig. 16.3c). The rotation matrix is

$$\mathbf{R}_3 = \begin{bmatrix} 1 & 0 & 0 \\ 0 & \cos(\phi) & -\sin(\phi) \\ 0 & \sin(\phi) & \cos(\phi) \end{bmatrix} . \tag{16.21}$$

The rotation matrix allowing to rotate any vector from the body-fixed frame xyz to the inertial frame XYZ is clearly the product of the three matrices

$$\mathbf{R} = \mathbf{R}_1 \mathbf{R}_2 \mathbf{R}_3 . \tag{16.22}$$

Performing the product of the rotation matrices, it follows

$$\mathbf{R} = \begin{bmatrix} c(\psi)c(\theta) & c(\psi)s(\theta)s(\phi) - s(\psi)c(\phi) & c(\psi)s(\theta)c(\phi) + s(\psi)s(\phi) \\ s(\psi)c(\theta) & s(\psi)s(\theta)s(\phi) + c(\psi)c(\phi) & s(\psi)s(\theta)c(\phi) - c(\psi)s(\phi) \\ -s(\theta) & c(\theta)s(\phi) & c(\theta)c(\phi) \end{bmatrix} , \tag{16.23}$$

where symbols cos and sin have been substituted by c and s.

Sometimes roll and pitch angles are small. In this case it is expedient to keep the last two rotations separate from the first one that usually cannot be linearized.

The product of the rotation matrices related to the last two rotations is

$$\mathbf{R}_2\mathbf{R}_3. = \begin{bmatrix} \cos(\theta) & \sin(\theta)\sin(\phi) & \sin(\theta)\cos(\phi) \\ 0 & \cos(\phi) & -\sin(\phi) \\ -\sin(\theta) & \cos(\theta)\sin(\phi) & \cos(\theta)\cos(\phi) \end{bmatrix} \tag{16.24}$$

that becomes, in the case of small angles,

$$R_2R_3 \approx \begin{bmatrix} 1 & 0 & \theta \\ 0 & 1 & -\phi \\ -\theta & \phi & 1 \end{bmatrix}. \tag{16.25}$$

The angular velocities $\dot{\psi}$, $\dot{\theta}$, and $\dot{\phi}$ are not applied along x-, y-, and z-axes, and then are not the components Ω_x, Ω_y, and Ω_z of the angular velocity in the body-fixed reference frame.[3] Their directions are those of axes Z, y^*, and x, and then the angular velocity in the body-fixed frame is

$$\left\{ \begin{array}{c} \Omega_x \\ \Omega_y \\ \Omega_z \end{array} \right\} = \dot{\phi}\mathbf{e}_x + \dot{\theta}\mathbf{R}_3^T\mathbf{e}_y + \dot{\psi}\left[\mathbf{R}_2\mathbf{R}_3\right]^T\mathbf{e}_z , \tag{16.26}$$

where the unit vectors are obviously

$$\mathbf{e}_x = \left\{ \begin{array}{c} 1 \\ 0 \\ 0 \end{array} \right\}, \ \mathbf{e}_y = \left\{ \begin{array}{c} 0 \\ 1 \\ 0 \end{array} \right\}, \ \mathbf{e}_z = \left\{ \begin{array}{c} 0 \\ 0 \\ 1 \end{array} \right\}. \tag{16.27}$$

By performing the products, it follows that

$$\left\{ \begin{array}{l} \Omega_x = \dot{\phi} - \dot{\psi}\sin(\theta) \\ \Omega_y = \dot{\theta}\cos(\phi) + \dot{\psi}\sin(\phi)\cos(\theta) \\ \Omega_z = \dot{\psi}\cos(\theta)\cos(\phi) - \dot{\theta}\sin(\phi) , \end{array} \right. \tag{16.28}$$

or, in matrix form

$$\left\{ \begin{array}{c} \Omega_x \\ \Omega_y \\ \Omega_z \end{array} \right\} = \begin{bmatrix} 1 & 0 & -\sin(\theta) \\ 0 & \cos(\phi) & \sin(\phi)\cos(\theta) \\ 0 & -\sin(\phi) & \cos(\phi)\cos(\theta) \end{bmatrix} \left\{ \begin{array}{c} \dot{\phi} \\ \dot{\theta} \\ \dot{\psi} \end{array} \right\}. \tag{16.29}$$

---

[3]Often symbols p, q, and r are used for the components of the angular velocity in the body-fixed frame.

If the pitch and roll angles are small enough to linearize the relevant trigonometric functions, the components of the angular velocity may be approximated as

$$\begin{cases} \Omega_x = \dot{\phi} - \theta\dot{\psi} \\ \Omega_y = \dot{\theta} + \phi\dot{\psi} \\ \Omega_z = \dot{\psi} - \phi\dot{\theta} \,. \end{cases} \tag{16.30}$$

16.3.2 Equations of motion – Lagrangian approach

Consider a rigid body in tridimensional space and chose as generalized coordinates the displacements X, Y, and Z of its center of mass and the set of three Tait–Bryan angles ψ, θ, and ϕ defined in the previous section. Assuming that the body axes xyz are principal axes of inertia, the kinetic energy of the rigid body is

$$\begin{aligned} T = {} & \tfrac{1}{2}m\left(\dot{X}^2 + \dot{Y}^2 + \dot{Z}^2\right) + \tfrac{1}{2}J_x\left[\dot{\phi} - \dot{\psi}\sin(\theta)\right]^2 + \\ & + \tfrac{1}{2}J_y\left[\dot{\theta}\cos(\phi) + \dot{\psi}\sin(\phi)\cos(\theta)\right]^2 + \\ & + \tfrac{1}{2}J_z\left[\dot{\psi}\cos(\theta)\cos(\phi) - \dot{\theta}\sin(\phi)\right]^2 . \end{aligned} \tag{16.31}$$

Introducing the kinetic energy into the Lagrange equations

$$\frac{d}{dt}\left(\frac{\partial T}{\partial \dot{q}_i}\right) - \frac{\partial T}{\partial q_i} = Q_i \,,$$

and performing the relevant derivatives, the six equations of motion are directly obtained. The three equations for translational motion are

$$\begin{cases} m\ddot{X} = Q_X \\ m\ddot{Y} = Q_Y \\ m\ddot{Z} = Q_Z \,. \end{cases} \tag{16.32}$$

The equations for rotational motion are much more complicated

$$\begin{aligned} & \ddot{\psi}\left[J_x\sin^2(\theta) + J_y\sin^2(\phi)\cos^2(\theta) + J_z\cos^2(\phi)\cos^2(\theta)\right] - \ddot{\phi}J_x\sin(\theta) + \\ & + \ddot{\theta}(J_y - J_z)\sin(\phi)\cos(\phi)\cos(\theta) + \dot{\phi}\dot{\theta}\cos(\theta)\left\{\left[1 - 2\sin^2(\phi)\right]\times \right. \\ & \left. \times(J_y - J_z) - J_x\right\} + 2\dot{\phi}\dot{\psi}(J_y - J_z)\cos(\phi)\cos^2(\theta)\sin(\phi) + \\ & + 2\dot{\theta}\dot{\psi}\sin(\theta)\cos(\theta)\left[J_x - \sin^2(\phi)J_y - \cos^2(\phi)J_z\right] + \\ & + \dot{\theta}^2(-J_y + J_z)\sin(\phi)\cos(\phi)\sin(\theta) = Q_\psi, \end{aligned}$$

$$\begin{aligned} & \ddot{\psi}(J_y - J_z)\sin(\phi)\cos(\theta)\cos(\phi) + \ddot{\theta}\left[J_y\cos^2(\phi) + J_z\sin^2(\phi)\right] + \\ & + 2\dot{\phi}\dot{\theta}(J_z - J_y)\sin(\phi)\cos(\phi) + \dot{\phi}\dot{\psi}(J_y - J_z)\cos(\theta)\left[1 - 2\sin^2(\phi)\right] + \\ & + \dot{\psi}\dot{\phi}J_x\cos(\theta) - \dot{\psi}^2\sin(\theta)\cos(\theta)\left[J_x - J_y\sin^2(\phi) - J_z\cos^2(\phi)\right] = Q_\theta, \end{aligned}$$

$$\tag{16.33}$$

$$J_x\ddot{\phi} - \sin(\theta)J_x\ddot{\psi} - \dot{\theta}\dot{\psi}J_z\sin^2(\phi)\cos(\theta)+$$
$$-\dot{\psi}\dot{\theta}\cos(\theta)\left\{J_x + J_y\left[1 - 2\sin^2(\phi)\right] - J_z\cos^2(\phi)\right\} +$$
$$+\dot{\theta}^2\left(J_y - J_z\right)\sin(\phi)\cos(\phi) - \dot{\psi}^2\left(J_y - J_z\right)\cos(\phi)\cos^2(\theta)\sin(\phi) = Q_\phi.$$

The generalized forces Q_X, Q_Y, and Q_Z are the actual forces applied in the direction of X-, Y-, and Z-axes, but Q_ψ, Q_θ, and Q_ϕ are not the moments applied about axes X, Y, and Z of the inertial frame or x, y, and z of the body-fixed frame.

Angle ψ does not appear explicitly in the equations of motion, and then if the roll and pitch angles are small all trigonometric functions can be linearized. If also the angular velocities are small, the equations of motion for rotations reduce to

$$\begin{cases} J_z\ddot{\psi} = Q_\psi \\ J_y\ddot{\theta} = Q_\theta \\ J_x\ddot{\phi} = Q_\phi. \end{cases} \tag{16.34}$$

In this case, the kinetic energy may be directly simplified, by developing the trigonometric functions in Taylor series and neglecting all terms containing products of three or more small quantities. For instance, the term

$$\left[\dot{\phi} - \dot{\psi}\sin(\theta)\right]^2$$

reduces to

$$\left[\dot{\phi} - \dot{\psi}\theta + \dot{\psi}\theta^3/6 + ...\right]^2 \approx \dot{\phi}^2,$$

since all other terms contain products of at least three small quantities. The kinetic energy thus reduces to

$$T \approx \frac{1}{2}m\left(\dot{X}^2 + \dot{Y}^2 + \dot{Z}^2\right) + \frac{1}{2}\left(J_x\dot{\phi}^2 + J_y\dot{\theta}^2 + J_z\dot{\psi}^2\right). \tag{16.35}$$

Remark 16.4 *This approach is simple only if the roll and pitch angles and their derivatives with respect to time are small. If not, the equations of motion obtained in terms of angular velocities $\dot{\phi}$, $\dot{\theta}$, and $\dot{\psi}$ are quite complicated and another approach is more expedient.*

16.3.3 *Equations of motion using pseudo-coordinates*

Since often the forces and moments applied to the rigid body are written with reference to the body-fixed frame, the equations of motion are best written with reference to the same frame. The kinetic energy can then be written in terms of the components v_x, v_y, and v_z (often referred to as u, v, and w) of the velocity and Ω_x, Ω_x e Ω_x (often referred to as p, q, and r) of the angular velocity.

If the body-fixed frame is a principal frame of inertia, the expression of the kinetic energy is

$$T = \frac{1}{2}m\left(v_x^2 + v_y^2 + v_z^2\right) + \frac{1}{2}\left(J_x\Omega_x^2 + J_y\Omega_y^2 + J_z\Omega_z^2\right). \tag{16.36}$$

The components of the velocity and of the angular velocity in the body-fixed frame are not the derivatives of coordinates, but are linked to the coordinates by the six kinematic equations

$$\mathbf{V} = \left\{ \begin{matrix} v_x \\ v_y \\ v_z \end{matrix} \right\} = \mathbf{R}^T \dot{\mathbf{X}} = \mathbf{R}^T \left\{ \begin{matrix} \dot{X} \\ \dot{Y} \\ \dot{Z} \end{matrix} \right\} , \tag{16.37}$$

$$\mathbf{\Omega} = \left\{ \begin{matrix} \Omega_x \\ \Omega_y \\ \Omega_z \end{matrix} \right\} = \mathbf{A}_2^T \dot{\boldsymbol{\theta}} = \begin{bmatrix} 1 & 0 & -\sin(\theta) \\ 0 & \cos(\phi) & \sin(\phi)\cos(\theta) \\ 0 & -\sin(\phi) & \cos(\theta)\cos(\phi) \end{bmatrix} \left\{ \begin{matrix} \dot{\phi} \\ \dot{\theta} \\ \dot{\psi} \end{matrix} \right\} , \tag{16.38}$$

that is, in more compact form,

$$\mathbf{w} = \mathbf{A}^T \dot{\mathbf{q}} , \tag{16.39}$$

where the vectors of the generalized velocities and of the derivatives of the generalized coordinates are

$$\mathbf{w} = \begin{bmatrix} v_x & v_y & v_z & \Omega_x & \Omega_y & \Omega_z \end{bmatrix}^T , \tag{16.40}$$

$$\dot{\mathbf{q}} = \begin{bmatrix} \dot{X} & \dot{Y} & \dot{Z} & \dot{\phi} & \dot{\theta} & \dot{\psi} \end{bmatrix}^T , \tag{16.41}$$

and matrix A is

$$\mathbf{A} = \begin{bmatrix} \mathbf{R} & \mathbf{0} \\ \mathbf{0} & \mathbf{A}_2 \end{bmatrix} . \tag{16.42}$$

The second submatrix is not a rotation matrix (the first one is such) and thus

$$\mathbf{A}^{-1} \neq \mathbf{A}^T ; \qquad \mathbf{B} \neq \mathbf{A} . \tag{16.43}$$

The inverse transformation is Eq. (16.2)

$$\dot{\mathbf{q}} = \mathbf{B}\mathbf{w} ,$$

where

$$\mathbf{B} = \mathbf{A}^{-T} = \begin{bmatrix} \mathbf{R} & \mathbf{0} \\ \mathbf{0} & \mathbf{A}_2^{-T} \end{bmatrix} . \tag{16.44}$$

None of the velocities included in vector w can be integrated to obtain a set of generalized coordinates, and thus all components of the velocities must be considered as derivatives of pseudo-coordinates.

The state-space equation, made of the six dynamic and the six kinematic equations, is then Eq. (16.18), simplified since in the present case no potential energy and Rayleigh dissipation function is present:

$$\begin{cases} \dfrac{d}{dt}\left(\left\{ \dfrac{\partial \mathcal{T}}{\partial w} \right\} \right) + \mathbf{B}^T \mathbf{\Gamma} \left\{ \dfrac{\partial \mathcal{T}}{\partial w} \right\} - \mathbf{B}^T \left\{ \dfrac{\partial \mathcal{T}}{\partial q} \right\} = \mathbf{B}^T \mathbf{Q} \\ \\ \{ \dot{q}_i \} = \mathbf{B}\{ w_i \} . \end{cases} \tag{16.45}$$

Here $\mathbf{B}^T\mathbf{Q}$ is just a column matrix containing the three components of the force and the three components of the moment applied to the body along the body-fixed axes x, y, and z.

The most difficult part of the computation is writing matrix $\mathbf{B}^T\mathbf{T}$. Performing somewhat difficult computations it follows[4] that

$$\mathbf{B}^T\mathbf{\Gamma} = \begin{bmatrix} \widetilde{\Omega} & \mathbf{0} \\ \widetilde{\mathbf{V}} & \widetilde{\Omega} \end{bmatrix}, \tag{16.46}$$

where $\widetilde{\Omega}$ and $\widetilde{\mathbf{V}}$ are skew-symmetric matrices containing the components of the angular and linear velocities

$$\widetilde{\Omega} = \begin{bmatrix} 0 & -\Omega_z & \Omega_y \\ \Omega_z & 0 & -\Omega_x \\ -\Omega_y & \Omega_x & 0 \end{bmatrix}, \quad \widetilde{\mathbf{V}} = \begin{bmatrix} 0 & -v_z & v_y \\ v_z & 0 & -v_x \\ -v_y & v_x & 0 \end{bmatrix}. \tag{16.47}$$

By separating the equations for translational and rotational degrees of freedom, the first equation (16.45) becomes

$$\begin{cases} \dfrac{\partial}{\partial t}\left(\left\{\dfrac{\partial\mathcal{T}}{\partial\mathbf{V}}\right\}\right) + \widetilde{\Omega}\left\{\dfrac{\partial\mathcal{T}}{\partial\mathbf{V}}\right\} - \mathbf{R}^T\left\{\dfrac{\partial\mathcal{T}}{\partial\mathbf{X}}\right\} = \mathbf{F} \\[3mm] \dfrac{\partial}{\partial t}\left(\left\{\dfrac{\partial\mathcal{T}}{\partial\mathbf{\Omega}}\right\}\right) + \widetilde{\mathbf{V}}\left\{\dfrac{\partial\mathcal{T}}{\partial\mathbf{V}}\right\} + \widetilde{\Omega}\left\{\dfrac{\partial\mathcal{T}}{\partial\mathbf{\Omega}}\right\} - \mathbf{A}_2^{-1}\left\{\dfrac{\partial\mathcal{T}}{\partial\theta}\right\} = \mathbf{M}, \end{cases}$$
$$\tag{16.48}$$

where $\boldsymbol{F}$ and $\boldsymbol{M}$ are the forces and moments in the body-fixed reference frame.

If the body-fixed axes are principal axes of inertia, the dynamic equations are simply

$$\begin{cases} m\dot{v}_x = m\Omega_z v_y - m\Omega_y v_z + F_x \\ m\dot{v}_y = m\Omega_x v_z - m\Omega_z v_x + F_y \\ m\dot{v}_z = m\Omega_y v_x - m\Omega_x v_y + F_z \\ J_x\dot{\Omega}_x = \Omega_y\Omega_z\left(J_y - J_z\right) + M_x \\ J_y\dot{\Omega}_y = \Omega_x\Omega_z\left(J_z - J_x\right) + M_y \\ J_z\dot{\Omega}_z = \Omega_x\Omega_y\left(J_x - J_y\right) + M_x. \end{cases} \tag{16.49}$$

Remark 16.5 *The equations so obtained are much simpler than Eq. (16.33). The last three equations are nothing else than Euler equations.*

---

[4]L. Meirovitch, *Methods of Analytical Dynamics*, McGraw-Hill, New York, 1970.

16.3.4   Eulerian approach

The linear and angular momentum of a rigid body can be written in the form

$$\mathbf{L} = m\mathbf{V}$$
$$\mathbf{H} = \mathbf{J}\boldsymbol{\Omega} .$$

(16.50)

The velocity $\mathbf{V}$ and the angular velocity $\boldsymbol{\Omega}$ may be written in any reference frame, but the latter must be consistent with the inertia tensor $\mathbf{J}$. Since $\mathbf{J}$ is constant only in a body-fixed frame, it is expedient to write the angular velocity in the same frame, while the velocity $\mathbf{V}$ may be written both in the inertial or in the body-fixed frame. The latter alternative will be followed here to be consistent with what was done in the previous section.

If the body-fixed frame is the principal inertia frame of the rigid body, the expression of the momenta is

$$\mathbf{L} = \left\{ \begin{array}{c} mv_x \\ mv_y \\ mv_z \end{array} \right\} , \quad \mathbf{H} = \left\{ \begin{array}{c} J_x\Omega_x \\ J_y\Omega_y \\ J_z\Omega_z \end{array} \right\} .$$

(16.51)

The equations of motion are thus

$$\frac{d\mathbf{L}}{dt} = \mathbf{F} , \quad \frac{d\mathbf{H}}{dt} = \mathbf{M} ,$$

(16.52)

where the forces and moments $\mathbf{F}$ and $\mathbf{M}$ must be expressed in the same frame as $\mathbf{L}$ and $\mathbf{H}$. The derivative must be computed with reference to an inertial frame: The general expression for the derivative of a general vector $\mathbf{A}$ in a non-inertial frame, having an absolute angular velocity $\boldsymbol{\Omega}$, is

$$\left.\frac{d\mathbf{A}}{dt}\right|_i = \left.\frac{d\mathbf{A}}{dt}\right|_{ni} + \boldsymbol{\Omega}\Lambda\mathbf{A} = \left.\frac{d\mathbf{A}}{dt}\right|_{ni} + \tilde{\boldsymbol{\Omega}}\mathbf{A} ,$$

(16.53)

where matrix $\tilde{\boldsymbol{\Omega}}$ is the skew-symmetric matrix defined by Eq. (16.47).

The equations of motion are thus

$$\left.m\frac{d\mathbf{V}}{dt}\right|_{ni} + m\tilde{\boldsymbol{\Omega}}\mathbf{V} = \mathbf{F},$$

(16.54)

$$\left.\mathbf{J}\frac{d\boldsymbol{\Omega}}{dt}\right|_{ni} + \tilde{\boldsymbol{\Omega}}\mathbf{J}\boldsymbol{\Omega} = \mathbf{M} ,$$

(16.55)

obtaining again Eq. (16.49):

$$\left\{ \begin{array}{l} m\dot{v}_x = m\Omega_z v_y - m\Omega_y v_z + F_x \\ m\dot{v}_y = m\Omega_x v_z - m\Omega_z v_x + F_y \\ m\dot{v}_z = m\Omega_y v_x - m\Omega_x v_y + F_z \\ J_x\dot{\Omega}_x = \Omega_y\Omega_z\left(J_y - J_z\right) + M_x \\ J_y\dot{\Omega}_y = \Omega_x\Omega_z\left(J_z - J_x\right) + M_y \\ J_z\dot{\Omega}_z = \Omega_x\Omega_y\left(J_x - J_y\right) + M_x. \end{array} \right.$$

16.4   Exercises

Exercise 16.1 *Consider the system of Example 16.1. Compute the natural frequencies and the corresponding eigenvectors. A hammer with a mass $m_h = 5$ kg hits the pendulum horizontally with a speed $V_h = 20$ m/s. Assuming that all the momentum of the hammer is transferred to the system with an inelastic shock, and that the excitation can be considered as impulsive, compute the time history of the system.*

Numerical data: $m_1 = 20$ kg, $m_2 = 15$ kg, $J_1 = 0.6$ kg m^2, $J_2 = 0.3$ kg m^2, $k_h = 1000$ kN/m, $k_v = 10,000$ kN/m, $a = 50$ mm, $b = 250$ mm, $c = 150$ mm, $d = 80$ mm, $e = 20$ mm, and $l = 500$ mm.

Exercise 16.2 *Repeat the computations of Exercise 16.1, with springs 1,000 times softer and a pendulum with length $l = 100$ mm. Add an hysteretic damping with a loss factor $\eta = 0.01$ applied to the elastic supports. Transform the hysteretic damping into an equivalent viscous damping with constant damping coefficient, computed for each mode.*

Exercise 16.3 *Write the nonlinear equations of motion for the system of Example 16.1 and repeat the computation of the response to the impulse excitation by numerical integration of the equations of motion. Compare the results with those obtained in the previous section. Divide by a factor of 10 the impulse and repeat the computation of the response.*

Exercise 16.4 *Repeat the computation of the response of the nonlinear system of the previous exercise, assuming that at the moment of the shock the constraints are severed and the system flies away falling down under the effect of gravity.*

17

Vibrating Systems in a Moving Reference Frame

In the preceding chapters vibrating systems were studied with reference to an inertial frame or to a frame that only has a translational motion with respect to an inertial frame. There are however cases in which a vibrating system is located on a rigid body (in the following defined 'carrier') that can undergo complex motion: In the case the carrier moves in a known way the problem is easily (at least in theory) solved by adding the necessary inertia forces. If however the motion of the carrier is influenced by the motion of the system on board, the coupled problem is much more complex.

17.1   General considerations

The vibrational dynamics of discrete systems was studied up to now under the assumption that the system is, on an average, stationary with reference to an inertial frame. When the motion was studied in a moving reference frame, its motion was assumed to be a translational one (since its velocity has not been assumed to be constant in time, the reference frame is non-inertial). The results so obtained may be used also to study cases where the vibrating system is on board of an object that moves in a general way, provided the dynamics of the *carrier* is uncoupled from the vibrational dynamics and the inertial effects are generally small.

There are however cases when this simplification cannot be used. A spacecraft with large flexible appendages, such as solar panels or antennas, whose natural frequencies are close to the frequencies involved in attitude control,

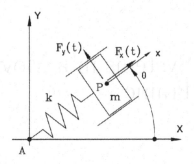

FIGURE 17.1. Spring–mass system on a carrier rotating about point A.

or the turbojet rotor installed on an aircraft performing quick manoeuvres, may be examples of systems in which uncoupling is impossible.

The coupled study of the motion of the carrier and the flexible system carried on board may be fairly complex and usually yields nonlinear equations. When they can be linearized in an approximate way it is possible to obtain closed-form solution and reach frequency-domain formulations yielding natural frequencies, mode shapes, etc. When, on the contrary, no linearization is possible the only approach is numerical integration in time.

A few very simple examples can show some peculiar characteristics of moving vibrating systems.

Example 17.1 *Write the equation of motion of a spring-mass system located on a massless carrier that can rotate about point A (Fig. 17.1) with a given law $\theta(t)$. The system has a single degree of freedom, and coordinate x, measured from the fixed Point A, may be used to define the position of P. The position of point P, referred to the inertial frame AXY, is*

$$\left\{ \begin{array}{c} X \\ Y \end{array} \right\} = \overline{(P-A)} = \mathbf{R} \left\{ \begin{array}{c} x \\ 0 \end{array} \right\},$$

where

$$\mathbf{R} = \left[\begin{array}{cc} \cos(\theta) & -\sin(\theta) \\ \sin(\theta) & \cos(\theta) \end{array} \right].$$

It follows that

$$\left\{ \begin{array}{c} X \\ Y \end{array} \right\} = \left\{ \begin{array}{c} x\cos(\theta) \\ x\sin(\theta) \end{array} \right\}.$$

By differentiating the coordinates of point P, its velocity is obtained:

$$\left\{ \begin{array}{c} \dot{X} \\ \dot{Y} \end{array} \right\} = \left\{ \begin{array}{c} \dot{x}\cos{(\theta)} - x\dot{\theta}\sin{(\theta)} \\ \dot{x}\sin{(\theta)} + x\dot{\theta}\cos{(\theta)} \end{array} \right\} \, .$$

The kinetic energy of the system is thus

$$T = \frac{1}{2}m\left(\dot{X}^2 + \dot{Y}^2 \right) = \frac{1}{2}m\left[\dot{x}^2 + \dot{\theta}^2 x^2 \right] \, .$$

The potential energy due to the spring is

$$\mathcal{U} = \frac{1}{2}k\,(x - x_0)^2 \, ,$$

where x_0 is the length at rest of the spring.
The external forces are assumed to act on point P in the direction of the rotating reference frame. The virtual displacement and the forces acting on point P are

$$\delta\mathbf{x} = \left[\begin{array}{cc} \delta x & 0 \end{array} \right]^T \, , \quad \mathbf{f}(t) = \left[\begin{array}{cc} f_x & f_y \end{array} \right]^T \, .$$

The virtual work due to the external forces is thus

$$\delta\mathcal{L} = f_x\delta x \, .$$

By performing the relevant derivatives

$$\frac{d}{dt}\left[\frac{\partial(T - \mathcal{U})}{\partial\dot{x}} \right] = m\ddot{x} \, , \quad \frac{\partial(T - \mathcal{U})}{\partial x} = m\dot{\theta}^2 x - k\,(x - x_0) \, , \quad \frac{\partial\delta\mathcal{L}}{\partial\delta x} = f_x \, ,$$

the equation of motion is found:

$$m\ddot{x} + \left(k - m\dot{\theta}^2 \right)x = k\,x_0 + f_x \, .$$

The equation of motion is thus the same as for the standard spring–mass system, with the centrifugal term $m\dot{\theta}^2 x$ added at the right side.
The system is stable only if the angular velocity $\dot{\theta}$ is small enough: If

$$\dot{\theta} > \sqrt{\frac{k}{m}}$$

the stiffness term becomes negative and no stable equilibrium position is found.

Remark 17.1 *The angular velocity $\dot{\theta} = \sqrt{k/m}$ is what will be defined as the* critical speed *in Chapter 23. The conclusion that above this speed the system becomes unstable is strictly linked with the presence of the guide constraining point P to move along x-axis.*

Example 17.2 *Write the equations of motion of the system of Fig. 17.1 in which the guide has been removed and point P is drawn toward both x- and y- axes (with an offset x_0) by a spring whose stiffness is k. This amounts to state that point P is drawn toward the point with coordinates $(x_0, 0)$ by a spring with the same stiffness in both x and y directions.*

The law $\theta(t)$ is stated and thus the system has two degrees of freedom. The displacements x and y may be used as generalized coordinates.

The position of point P, referred to the inertial frame OXY, is now

$$\left\{ \begin{array}{c} X \\ Y \end{array} \right\} = (\overline{P - O}) = \mathbf{R} \left\{ \begin{array}{c} x \\ y \end{array} \right\} ,$$

i.e.,

$$\left\{ \begin{array}{c} X \\ Y \end{array} \right\} = \left\{ \begin{array}{c} x \cos(\theta) - y \sin(\theta) \\ x \sin(\theta) + y \cos(\theta) \end{array} \right\} .$$

By differentiating the coordinates of point P, its velocity is obtained:

$$\left\{ \begin{array}{c} \dot{X} \\ \dot{Y} \end{array} \right\} = \left\{ \begin{array}{c} \dot{x} \cos(\theta) - \dot{y} \sin(\theta) - x\dot{\theta} \sin(\theta) - y\dot{\theta} \cos(\theta) \\ \dot{x} \sin(\theta) + \dot{y} \cos(\theta) + x\dot{\theta} \cos(\theta) - y\dot{\theta} \sin(\theta) \end{array} \right\} .$$

The kinetic energy of the system is thus

$$\mathcal{T} = \frac{1}{2}m\left(\dot{X}^2 + \dot{Y}^2\right) = \frac{1}{2}m\left\{\dot{x}^2 + \dot{y}^2 + \dot{\theta}^2\left[x^2 + y^2\right]\right\} + \\ + m\dot{\theta}\left(\dot{y}x - \dot{x}y\right) .$$

The potential energy due to the spring is

$$\mathcal{U} = \frac{1}{2}k\left[(x - x_0)^2 + y^2\right] .$$

The external forces are assumed to act in the direction of the rotating reference frame. The virtual displacement and the forces acting on point P are

$$\delta\mathbf{x} = \left[\begin{array}{cc} \delta x & \delta y \end{array} \right]^T , \quad \mathbf{f}(t) = \left[\begin{array}{cc} f_x & f_y \end{array} \right]^T .$$

The virtual work due to the external forces is

$$\delta\mathcal{L} = f_x\delta x + f_y\delta y .$$

By performing the relevant derivatives, the Lagrange equations yield the equations of motion

$$m\left[\begin{array}{cc} 1 & 0 \\ 0 & 1 \end{array} \right]\left\{ \begin{array}{c} \ddot{x} \\ \ddot{y} \end{array} \right\} + 2m\dot{\theta}\left[\begin{array}{cc} 0 & -1 \\ 1 & 0 \end{array} \right]\left\{ \begin{array}{c} \dot{x} \\ \dot{y} \end{array} \right\} + \\ \left[\begin{array}{cc} k - m\dot{\theta}^2 & -m\ddot{\theta} \\ m\ddot{\theta} & k - m\dot{\theta}^2 \end{array} \right]\left\{ \begin{array}{c} x \\ y \end{array} \right\} = \left\{ \begin{array}{c} kx_0 + f_x \\ f_y \end{array} \right\} .$$

Again the equations of motion are linear, but now contain a gyroscopic matrix,
i.e., a skew-symmetric matrix multiplying the velocities. Neglecting the angular
acceleration (i.e., if the speed $\dot\theta$ is kept constant), the stiffness matrix now is
always positive semidefined and positive defined at all speeds except the critical
speed defined in the previous example. As it will be seen in Chapter 23, a stable
equilibrium position exists at all speeds.

Example 17.3 *Repeat Example 17.1, but now the motion of the rotating
frame is not stated, i.e., law $\theta(t)$ is one of the unknowns of the problem. The
system has two degrees of freedom, and x and θ can be chosen as generalized
coordinates.*
*The position of point P is the same as before, and so is its velocity and its
kinetic energy*

$$T = \frac{1}{2}m\left(\dot X^2 + \dot Y^2\right) = \frac{1}{2}m\left[\dot x^2 + \dot\theta^2 x^2\right] .$$

Also the potential energy is the same:

$$\mathcal{U} = \frac{1}{2}k\left(x - x_0\right)^2 .$$

The virtual displacements now are

$$\delta\mathbf{x} = \left[\begin{array}{cc} \delta x & \delta\theta \end{array}\right]^T,$$

and the virtual displacement of point P in the inertial frame is

$$\left\{\begin{array}{c} \delta X \\ \delta Y \end{array}\right\} = \left\{\begin{array}{c} \delta x\cos(\theta) - \delta\theta x\sin(\theta) \\ \delta x\sin(\theta) + \delta\theta x\cos(\theta) \end{array}\right\} .$$

The forces acting in the inertial frame are

$$\left\{\begin{array}{c} f_X \\ f_Y \end{array}\right\} = \mathbf{R}\left\{\begin{array}{c} f_x \\ f_y \end{array}\right\} .$$

If also a torque M_z acts on the rotating frame, the virtual work is

$$\delta\mathcal{L} = \left\{\begin{array}{c} \delta X \\ \delta Y \end{array}\right\}^T \mathbf{R}\left\{\begin{array}{c} f_x \\ f_y \end{array}\right\} + \delta\theta M_z ,$$

i.e.,

$$\delta\mathcal{L} = \delta x f_x + \delta\theta\left(x f_y + M_z\right).$$

By performing the relevant derivatives the equations of motion are found:

$$\left\{\begin{array}{l} m\ddot x + \left(k - m\dot\theta^2\right)x = f_x + kx_0 \\ m\ddot\theta x^2 + 2m\dot\theta\dot x x = x f_y + M_z . \end{array}\right.$$

Remark 17.2 *These equations of motion are strongly nonlinear: As it will be seen later this is typical when the dynamics of the 'carrier' (in this case the rotating frame) interacts with the dynamics of the vibrating system.*

The motion of the carrier is best studied modeling the system as a discrete system, while the compliant part may be modeled as an elastic continuum. This approach is usually referred to as *hybrid formulation* of the equations of motion and implies the mixed use of ordinary and partial differential equations.

It became common in spacecraft dynamics in the 1980s: The spacecraft body may often be considered as a rigid body, particularly when dealing with attitude control, since its own natural frequencies are much higher than the frequencies involved in attitude control, but some parts, like large solar panel arrays or long antennas, may be so flexible that their vibration modes affect the overall spacecraft dynamics.

Remark 17.3 *The hybrid approach suffers from the same limitations seen for continuous models: As soon as the configuration gets complicated, the analytical difficulties suggest to resort to discretization and the model returns to be completely made of ODEs.*

In most cases, the continuous parts of the system are thus discretized, by either using the FEM or resorting to modal coordinates. Actually, these two approaches can be used together: A FEM modelization is used to compute the eigenvalues and the eigenvectors and then a global dynamic model is obtained in terms of modal coordinates.

Since the trend nowadays is toward discretization, particularly in connection with the finite elements method, the hybrid formulation will be used here only for stating the problem; further development will be based on discretization techniques.

17.2   Vibrating system on a rigid carrier

Consider a compliant system attached to a rigid body (the *carrier*) free to move in space. The carrier may be an actual material part of the overall system or may simply be reduced to a massless body whose only aim is defining a body-fixed reference frame. A reference configuration of the compliant system is also defined. It may be the configuration 'at rest' of the compliant system, i.e., the configuration it takes when no force acts on it.

When the compliant system is in its reference position it is possible to define a global center of mass G and an inertia tensor $\mathbf{J}$ of the whole system that can be considered as a rigid body.

Point P_c is the center of mass of the carrier.

While the definition of the inertial frame is immaterial, two possible choices can be made for the frame following the carrier in its motion. The most obvious alternative is to locate its origin in the center of mass of the carrier G_c and to state that its axes coincide with the principal axes of inertia of the latter.

This, however, does not yield the simplest formulation of the equations of motion. A much better choice is to center the frame fixed to the carrier in the center of mass G of the whole system, frozen in a reference configuration. The body-fixed frame may be a principal frame of inertia of the whole system, again in the reference configuration, but this is actually not needed.

The position of a generic point belonging to the compliant system will be defined as P_0 when the body is in its undeformed position and as P when the configuration is deformed. As shown in Fig. 17.2, vectors $\mathbf{r}$ and $\mathbf{u}$ define the position of P_0 and the displacement $P-P_0$. The coordinates of the center of mass of the carrier P_c and of a generic point P of the elastic system in the inertial frame are

$$\overline{(P_c - O)} = \mathbf{X} + \mathbf{R}\mathbf{r}_c ,$$

$$\overline{(P - O)} = \mathbf{X} + \mathbf{R}(\mathbf{r} + \mathbf{u}) ,$$

(17.1)

where

- $\mathbf{X} = \begin{bmatrix} X & Y & Z \end{bmatrix}_G^T$ is the vector defining the position of the center of mass G of the system in the inertial frame,

- $\mathbf{R}$ is the rotation matrix of the carrier expressed by Eq. (16.23),

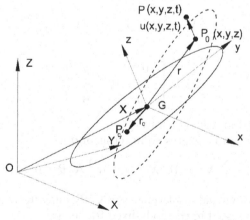

FIGURE 17.2. Sketch of a system made by a rigid carrier (*full line*) to which a continuous compliant system (*dashed line*) is attached.

- $\mathbf{r} = \begin{bmatrix} x & y & z \end{bmatrix}_P^T$ is the vector defining the position of point P in its reference position P_0 (usually corresponding to the undeformed configuration of the elastic system), expressed in the body-fixed frame, and

- $\mathbf{u} = \begin{bmatrix} u_x & u_y & u_z \end{bmatrix}_P^T$ is the displacement vector of the same point expressed in the same frame. In many cases, the displacement $\mathbf{u}$ may be considered as a small displacement.

The generalized coordinates of the system are thus

$$\mathbf{q} = \begin{bmatrix} \mathbf{X}^T & \boldsymbol{\theta}^T & \mathbf{u}^T \end{bmatrix}^T , \tag{17.2}$$

where $\boldsymbol{\theta}$ is the set of Tait–Bryan angles defined in Section 16.3.1.

The discrete part of the system has 6 degrees of freedom (the degrees of freedom of the carrier), while vector $\mathbf{u}\,(x, y, z, t)$ contains the generalized coordinates of the continuous part.

Using the same approach seen in the previous chapter (Eq. (16.37) and (16.38)), the generalized velocities are

$$\mathbf{w} = \begin{Bmatrix} \mathbf{V} \\ \boldsymbol{\Omega} \\ \dot{\mathbf{u}} \end{Bmatrix} = \mathbf{A}^T \begin{Bmatrix} \dot{\mathbf{X}} \\ \dot{\boldsymbol{\theta}} \\ \dot{\mathbf{u}} \end{Bmatrix} = \begin{bmatrix} \mathbf{R}^T & \mathbf{0} & \mathbf{0} \\ \mathbf{0} & \mathbf{A}_2^T & \mathbf{0} \\ \mathbf{0} & \mathbf{0} & \mathbf{I} \end{bmatrix} \dot{\mathbf{q}} , \tag{17.3}$$

where $\mathbf{I}$ is a suitable identity matrix.

The velocity of point P expressed in the body-fixed frame is

$$\mathbf{V}_P = \mathbf{V} + \boldsymbol{\Omega}\boldsymbol{\Lambda}(\mathbf{r} + \mathbf{u}) + \dot{\mathbf{u}} . \tag{17.4}$$

By expressing the vector product in matrix notation, it follows that

$$\mathbf{V}_P = \mathbf{V} + (\tilde{\mathbf{r}} + \tilde{\mathbf{u}})^T \boldsymbol{\Omega} + \dot{\mathbf{u}} , \tag{17.5}$$

where

$$\tilde{\mathbf{r}} = \begin{bmatrix} 0 & -z & y \\ z & 0 & -x \\ -y & x & 0 \end{bmatrix} , \tag{17.6}$$

and $\tilde{\mathbf{u}}$ is defined in a similar way.

The velocity can be expressed in terms of $\dot{\mathbf{q}}$ as

$$\mathbf{V}_P = \mathbf{R}^T \dot{\mathbf{X}} + (\tilde{\mathbf{r}} + \tilde{\mathbf{u}})^T \mathbf{A}_2^T \dot{\boldsymbol{\theta}} + \dot{\mathbf{u}} . \tag{17.7}$$

Given a set of virtual displacements $\delta\mathbf{X}$, $\delta\boldsymbol{\theta}$, and $\delta\mathbf{u}$, the virtual displacements δx_P, expressed in the body-fixed frame, are

$$\delta x_P = \mathbf{R}^T \delta\mathbf{X} + (\tilde{\mathbf{r}} + \tilde{\mathbf{u}})^T \mathbf{A}_2^T \delta\boldsymbol{\theta} + \delta\mathbf{u} . \tag{17.8}$$

The virtual work of a distributed force $\mathbf{f}(x, y, z, t)$ applied to the flexible body in the direction of the axes of the body-fixed frame is

$$\delta\mathcal{L} = \int_v \delta x_P^T \mathbf{f} \, dv = \int_v \left(\delta\mathbf{X}^T \mathbf{R}\mathbf{f} + \delta\boldsymbol{\theta}^T \mathbf{A}_2 (\tilde{\mathbf{r}} + \tilde{\mathbf{u}}) \, \mathbf{f} + \delta\mathbf{u}^T \mathbf{f} \right) \, dv \, , \quad (17.9)$$

where v is the volume occupied by the compliant part of the system.

If a force $\mathbf{F}_c$ and a moment $\mathbf{M}_c$ act on the carrier (the force is applied in its center of mass $\mathbf{P}_c$), the total virtual work acting on the system is thus

$$\delta\mathcal{L} = \delta\mathbf{X}^T \mathbf{R}\mathbf{F}_c + \delta\boldsymbol{\theta}^T \mathbf{A}_2 \left(\tilde{\mathbf{r}}_c \mathbf{F}_c + \mathbf{M}_c \right) + \int_v \left[\delta\mathbf{X}^T \mathbf{R}\mathbf{f} + \delta\boldsymbol{\theta}^T \mathbf{A}_2 (\tilde{\mathbf{r}} + \tilde{\mathbf{u}}) \, \mathbf{f} + \delta\mathbf{u}^T \mathbf{f} \right] dv \, .$$
$$(17.10)$$

The kinetic energy of the infinitesimal volume dv about point P is

$$d\mathcal{T} = \frac{1}{2} \rho \mathbf{v}_P^T \mathbf{v}_P dv = \frac{1}{2} \rho \left[\mathbf{V}^T \mathbf{V} + \boldsymbol{\Omega}^T \left(\tilde{\mathbf{r}}\tilde{\mathbf{r}}^T + 2\tilde{\mathbf{r}}\tilde{\mathbf{u}}^T + \tilde{\mathbf{u}}\tilde{\mathbf{u}}^T \right) \boldsymbol{\Omega} + \quad (17.11)\right.$$

$$\left. + \dot{\mathbf{u}}^T \dot{\mathbf{u}} + 2\mathbf{V}^T \left(\tilde{\mathbf{r}}^T + \tilde{\mathbf{u}}^T \right) \boldsymbol{\Omega} + 2\mathbf{V}^T \dot{\mathbf{u}} + 2\boldsymbol{\Omega}^T \left(\tilde{\mathbf{r}} + \tilde{\mathbf{u}} \right) \dot{\mathbf{u}} \right] dv \, .$$

The kinetic energy of the carrier is

$$\mathcal{T}_c = \frac{1}{2} m_c \left(\mathbf{V}^T \mathbf{V} + \boldsymbol{\Omega}^T \tilde{\mathbf{r}}_c \tilde{\mathbf{r}}_c^T \boldsymbol{\Omega} + 2\mathbf{V}^T \tilde{\mathbf{r}}_c^T \boldsymbol{\Omega} \right) +$$

$$+ \frac{1}{2} \boldsymbol{\Omega}^T \mathbf{J}_c \boldsymbol{\Omega} \quad (17.12)$$

The global inertia properties of the whole system in the reference configuration are

$$m = m_c + \int_v \rho \, dv \, , \quad \mathbf{J} = \mathbf{J}_c + \int_v \tilde{\mathbf{r}}_c \tilde{\mathbf{r}}_c^T \rho \, dv \, .$$

Moreover, since point G is the overall mass center of the system,

$$m_c \tilde{\mathbf{r}}_c^T + \int_v \rho \tilde{\mathbf{r}}^T \, dv = 0 \, .$$

The total kinetic energy of the system is thus

$$\mathcal{T} = \frac{1}{2} m \mathbf{V}^T \mathbf{V} + \frac{1}{2} \boldsymbol{\Omega}^T \mathbf{J} \boldsymbol{\Omega} + \frac{1}{2} \int_v \rho \dot{\mathbf{u}}^T \dot{\mathbf{u}} \, dv + \frac{1}{2} \boldsymbol{\Omega}^T \left(\int_v \rho \left(2\tilde{\mathbf{r}} + \tilde{\mathbf{u}} \right) \tilde{\mathbf{u}}^T \, dv \right) \boldsymbol{\Omega} +$$
$$(17.13)$$

$$+ \mathbf{V}^T \left(\int_v \rho \tilde{\mathbf{u}}^T \, dv \right) \boldsymbol{\Omega} + \mathbf{V}^T \int_v \rho \dot{\mathbf{u}} \, dv + \boldsymbol{\Omega}^T \int_v \rho \left(\tilde{\mathbf{r}} + \tilde{\mathbf{u}} \right) \dot{\mathbf{u}} \, dv \, .$$

Remark 17.4 *The kinetic energy contains terms that are products of more than two velocities or displacements, leading to nonlinear terms in the equations of motion.*

The elastic potential energy due to the deformation of the rigid body is not affected by generalized coordinates $\mathbf{X}$ and $\boldsymbol{\theta}$, but only by the coordinates $\mathbf{u}$ and their derivatives with respect to the spatial coordinates. Usually the first and the second derivatives $\mathbf{u}'$ and $\mathbf{u}''$ are included, but there may be cases in which also higher-order derivatives are present:

$$\mathcal{U} = \mathcal{U}\left(\mathbf{u},\, \mathbf{u}',\, \mathbf{u}''\right) . \qquad (17.14)$$

As it was seen in Section 12.8, it may be necessary to take into account also nonlinear terms in the strains to include effects like the influence of inertia forces on the elastic behavior of some parts of the system. A typical example is that of turbine blades: the rotation of the machine causes centrifugal accelerations that in turn cause axial loading of the blades and then an increase of their stiffness. In this case the angular velocity is usually assumed to be constant (or at least stated as a known function of time) and thus the problem may be linearized; in other cases true nonlinear effects need to be considered.

As seen in Section 15.2.6 this may be accounted for easily through a geometric matrix, but in general this matrix may be a function of the accelerations acting on the system.

Also a Rayleigh dissipation function, which is independent from generalized coordinates $\mathbf{X}$ and $\boldsymbol{\theta}$ and from velocities $\mathbf{V}$ and $\boldsymbol{\Omega}$, can be defined. It is a function only of $\dot{\mathbf{u}}$ (and possibly of $\mathbf{u}$) and of their derivatives with respect to space coordinates.

17.3 Lumped-parameters discretization

The compliant system is discretized as a number n of masses m_i, connected to each other and to the rigid body by springs and dampers. On each mass a force $\mathbf{F}_i$, expressed in the reference frame attached to the carrier, acts (Fig. 17.3)

Let P_i be the position of the point where mass m_i is located in the deformed position and P_{0i} the position of the same point in the reference (undeformed) position.

The discretized system has now $3\left(n+2\right)$ degrees of freedom.

The kinetic energy of the system is still expressed by Eq. (17.13), where the integrals are substituted by sums

$$\mathcal{T} = \frac{1}{2}m\mathbf{V}^T\mathbf{V} + \frac{1}{2}\boldsymbol{\Omega}^T\mathbf{J}\boldsymbol{\Omega} + \frac{1}{2}\sum_{i=1}^{n} m_i \left(\dot{\mathbf{u}}_i^T\dot{\mathbf{u}}_i + \boldsymbol{\Omega}^T\tilde{\mathbf{u}}_i\tilde{\mathbf{u}}_i^T\boldsymbol{\Omega} + 2\boldsymbol{\Omega}^T\tilde{\mathbf{r}}_i\tilde{\mathbf{u}}_i^T\boldsymbol{\Omega} + \right.$$

$$\left. +2\mathbf{V}^T\tilde{\mathbf{u}}_i^T\boldsymbol{\Omega} + 2\mathbf{V}^T\dot{\mathbf{u}}_i + 2\boldsymbol{\Omega}^T\tilde{\mathbf{r}}_i\dot{\mathbf{u}}_i + 2\boldsymbol{\Omega}^T\tilde{\mathbf{u}}_i\dot{\mathbf{u}}_i \right) . \qquad (17.15)$$

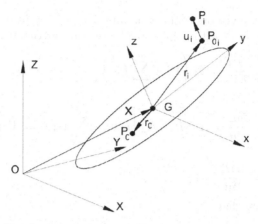

FIGURE 17.3. Sketch of a system made by a rigid carrier to which a discrete elastic system is attached.

Under the usual small-displacement assumption, the potential energy and the Rayleigh dissipation function may be written in the usual way

$$\mathcal{U} \approx \frac{1}{2}\mathbf{u}^T\mathbf{K}_e\mathbf{u} \ , \quad \mathcal{F} \approx \frac{1}{2}\dot{\mathbf{u}}^T\mathbf{C}_e\dot{\mathbf{u}} \ , \qquad (17.16)$$

where vector $\mathbf{u}$ is now a vector, containing $3n$ elements, in which the various vectors $\mathbf{u}_i$ are listed one after the other and $\mathbf{K}_e$ and $\mathbf{C}_e$ are the stiffness and damping matrices of the system when the carrier is fixed to the inertial reference frame.

As already stated, inertia forces can affect the stiffness of the system. To account for this it is possible to introduce a geometric matrix $\mathbf{K}_g$ that may depend on the acceleration $\dot{\mathbf{V}}$, the angular velocity $\boldsymbol{\Omega}$ or its derivative $\dot{\boldsymbol{\Omega}}$, but also on other effects like thermo-elastic stressing or any other load conditions. These effects must be introduced in each case.

By operating in the usual way, the equations of motion are

$$\begin{cases} \dfrac{\partial}{\partial t}\left(\left\{\dfrac{\partial \mathcal{T}}{\partial \mathbf{V}}\right\}\right) + \tilde{\boldsymbol{\Omega}}\left\{\dfrac{\partial \mathcal{T}}{\partial \mathbf{V}}\right\} - \mathbf{R}^T\left\{\dfrac{\partial \mathcal{T}}{\partial \mathbf{X}}\right\} = \mathbf{R}^T\mathbf{Q}_X \\[3mm] \dfrac{\partial}{\partial t}\left(\left\{\dfrac{\partial \mathcal{T}}{\partial \boldsymbol{\Omega}}\right\}\right) + \tilde{\mathbf{V}}\left\{\dfrac{\partial \mathcal{T}}{\partial \mathbf{V}}\right\} + \tilde{\boldsymbol{\Omega}}\left\{\dfrac{\partial \mathcal{T}}{\partial \boldsymbol{\Omega}}\right\} - \mathbf{A}_2^{-1}\left\{\dfrac{\partial \mathcal{T}}{\partial \theta}\right\} = \mathbf{A}_2^{-1}\mathbf{Q}_\theta \\[3mm] \dfrac{\partial}{\partial t}\left(\left\{\dfrac{\partial \mathcal{T}}{\partial \dot{\mathbf{u}}_i}\right\}\right) - \left\{\dfrac{\partial \mathcal{T}}{\partial \mathbf{u}_i}\right\} + \left\{\dfrac{\partial \mathcal{U}}{\partial \mathbf{u}_i}\right\} + \left\{\dfrac{\partial \mathcal{F}}{\partial \dot{\mathbf{u}}_i}\right\} = \mathbf{Q}_{u_i}, \end{cases}$$

$$(17.17)$$

where $\mathbf{Q}_i$ are the vectors of the generalized forces obtained by differentiating the virtual work with respect to the relevant virtual displacements:

$$
\begin{cases}
\mathbf{Q}_X = \dfrac{\partial \delta \mathcal{L}}{\partial \delta \mathbf{X}} = \mathbf{R}\left(\mathbf{F}_c + \displaystyle\sum_{i=1}^{n} \mathbf{F}_i\right) \\[3mm]
\mathbf{Q}_\theta = \dfrac{\partial \delta \mathcal{L}}{\partial \delta \theta} = \mathbf{A}_2 \mathbf{M}_c + \mathbf{A}_2 \left[\tilde{\mathbf{r}}_c \mathbf{F}_c + \displaystyle\sum_{i=1}^{n} \left(\tilde{\mathbf{r}}_i + \tilde{\mathbf{u}}_i\right)\mathbf{F}_i\right] \\[3mm]
\mathbf{Q}_{u_i} = \dfrac{\partial \delta \mathcal{L}}{\partial \delta \mathbf{u}_i} = \mathbf{F}_i \ .
\end{cases}
\tag{17.18}
$$

Remembering that

$$
\tilde{\mathbf{a}}^T \mathbf{b} = \tilde{\mathbf{b}} \mathbf{a}, \quad \mathbf{a}^T \tilde{\mathbf{b}} = \mathbf{b}^T \tilde{\mathbf{a}}^T, \quad \tilde{\mathbf{a}} \mathbf{a} = \mathbf{0},
$$

by introducing the expressions of the kinetic energy, the first three equations yield

$$
m\dot{\mathbf{V}} + m\tilde{\boldsymbol{\Omega}}\mathbf{V} + \sum_{\forall i} m_i \left(\ddot{\mathbf{u}}_i + 2\tilde{\boldsymbol{\Omega}}\dot{\mathbf{u}}_i + \dot{\tilde{\boldsymbol{\Omega}}}\mathbf{u}_i + \tilde{\boldsymbol{\Omega}}^2 \mathbf{u}_i\right) = \mathbf{F}_c + \sum_{i=1}^{n} \mathbf{F}_i \ . \tag{17.19}
$$

The three equations for the rotational degrees of freedom of the carrier are

$$
\mathbf{J}\dot{\boldsymbol{\Omega}} + \tilde{\boldsymbol{\Omega}}\mathbf{J}\boldsymbol{\Omega} + \sum_{\forall i} m_i \left(\tilde{\mathbf{r}}_i \ddot{\mathbf{u}}_i + \tilde{\mathbf{u}}_i \ddot{\mathbf{u}}_i + \tilde{\boldsymbol{\Omega}}\tilde{\mathbf{r}}_i \dot{\mathbf{u}} + 2\dot{\tilde{\mathbf{u}}}_i \tilde{\mathbf{u}}_i^T \boldsymbol{\Omega} + \right.
$$

$$
+ \tilde{\mathbf{u}}_i \tilde{\mathbf{u}}_i^T \dot{\boldsymbol{\Omega}} + 2\tilde{\mathbf{r}}_i \dot{\tilde{\mathbf{u}}}_i^T \boldsymbol{\Omega} + 2\tilde{\mathbf{r}}_i \tilde{\mathbf{u}}_i^T \dot{\boldsymbol{\Omega}} + \tilde{\mathbf{u}}_i \dot{\mathbf{V}} + \tilde{\mathbf{V}}\tilde{\mathbf{u}}_i^T \boldsymbol{\Omega} + \tilde{\boldsymbol{\Omega}}\tilde{\mathbf{u}}_i \mathbf{V} +
$$

$$
\left. + \tilde{\boldsymbol{\Omega}}\tilde{\mathbf{u}}_i \dot{\mathbf{u}}_i + \tilde{\boldsymbol{\Omega}}\tilde{\mathbf{u}}_i \tilde{\mathbf{u}}_i^T \boldsymbol{\Omega} + 2\tilde{\boldsymbol{\Omega}}\tilde{\mathbf{r}}_i \tilde{\mathbf{u}}_i^T \boldsymbol{\Omega}\right) = \mathbf{M}_0 + \tilde{\mathbf{r}}_c \mathbf{F}_c + \sum_{i=1}^{n} \left(\tilde{\mathbf{r}}_i + \tilde{\mathbf{u}}_i\right)\mathbf{F}_i \ .
$$
$$
\tag{17.20}
$$

The ith group of three equations yielding the behavior of the flexible part of the system is

$$
m_i \left(\ddot{\mathbf{u}}_i + \dot{\mathbf{V}} + \tilde{\mathbf{r}}_i^T \dot{\boldsymbol{\Omega}} + \tilde{\mathbf{u}}_i^T \dot{\boldsymbol{\Omega}} - \tilde{\boldsymbol{\Omega}}^T \tilde{\boldsymbol{\Omega}} \mathbf{u}_i - \tilde{\boldsymbol{\Omega}}^T \tilde{\mathbf{r}}_i^T \boldsymbol{\Omega} - \tilde{\boldsymbol{\Omega}}^T \mathbf{V} + 2\tilde{\boldsymbol{\Omega}}\mathbf{u}_i\right) + \dfrac{\partial \mathcal{U}}{\partial \mathbf{u}_i} = \mathbf{F}_i.
\tag{17.21}
$$

Centrifugal and Coriolis forces are immediately identified in the expressions of the equations of motion.

The equations can be written in matrix form as

$$
\begin{bmatrix}
m\mathbf{I} & \mathbf{0} & m_1\mathbf{I} & m_2\mathbf{I} & \cdots \\
 & \mathbf{J} & m_i\tilde{\mathbf{r}}_1 & m_i\tilde{\mathbf{r}}_2 & \cdots \\
 & & m_1\mathbf{I} & \mathbf{0} & \cdots \\
 & & & m_2\mathbf{I} & \cdots \\
\text{symm.} & \cdots & \cdots & \cdots & \cdots
\end{bmatrix}
\begin{Bmatrix}
\dot{\mathbf{V}} \\
\dot{\boldsymbol{\Omega}} \\
\ddot{\mathbf{u}}_1 \\
\ddot{\mathbf{u}}_2 \\
\cdots
\end{Bmatrix} +
\tag{17.22}
$$

$$
+ \left(\begin{bmatrix} m\tilde{\Omega} & 0 & 2m_1\tilde{\Omega} & 2m_2\tilde{\Omega} & \dots \\ 0 & \tilde{\Omega}J & C_{\theta 1} & C_{\theta 2} & \dots \\ -m_1\tilde{\Omega} & C_{1\theta} & 2\tilde{\Omega} & 0 & \dots \\ -m_2\tilde{\Omega} & C_{2\theta} & 0 & 2\tilde{\Omega} & \dots \\ \dots & \dots & \dots & \dots & \dots \end{bmatrix} + \begin{bmatrix} 0 & 0 \\ 0 & C_e \end{bmatrix} \right) \left\{ \begin{matrix} V \\ \Omega \\ \dot{u}_1 \\ \dot{u}_2 \\ \dots \end{matrix} \right\} +
$$

$$
\text{(17.23)}
$$

$$
+ \left(\begin{bmatrix} 0 & 0 & K_{X1} & K_{X2} & \dots \\ 0 & 0 & K_{\theta 1} & K_{\theta 2} & \dots \\ 0 & 0 & K_{11} & 0 & \dots \\ 0 & 0 & 0 & K_{22} & \dots \\ \dots & \dots & \dots & \dots & \dots \end{bmatrix} + \begin{bmatrix} 0 & 0 \\ 0 & (K_e + K_g) \end{bmatrix} \right) \left\{ \begin{matrix} X \\ \theta \\ u_1 \\ u_2 \\ \dots \end{matrix} \right\} =
$$

$$
= \left\{ \begin{matrix} F_c + \sum_{i=1}^{n} F_i \\ M_c + \tilde{r}_c F_c + \sum_{i=0}^{n} (\tilde{r}_i + \tilde{u}_i) F_i \cdot \\ F_1 \\ F_2 \\ \dots \end{matrix} \right\} ,
$$

where

$$
C_{\theta i} = m_i \left[\tilde{\Omega}\tilde{r}_i + 2\tilde{r}_i\tilde{\Omega} + \tilde{\Omega}\tilde{u}_i \right] ,
$$

$$
C_{i\theta} = m_i \tilde{\Omega}^T \tilde{r}_i , \quad K_{Xi} = K_{ii} = m_i \left(\tilde{\Omega}^2 + \dot{\tilde{\Omega}} \right) ,
$$

$$
K_{\theta i} = m_i \left(2\tilde{r}_i\dot{\tilde{\Omega}} + \dot{\tilde{V}}^T + \ddot{\tilde{u}}_i^T + \tilde{V}\tilde{\Omega} + \tilde{\Omega}\tilde{V}^T + 2\tilde{\Omega}\tilde{r}_i\tilde{\Omega} + \tilde{\Omega}\tilde{u}_i\tilde{\Omega} + \tilde{u}_i\dot{\tilde{\Omega}} + 2\dot{\tilde{u}}_i\tilde{\Omega} \right) .
$$

Example 17.4 *Consider a rigid body (mass m_{rb}, principal moments of inertia J_{rb_x}, J_{rb_y}, and J_{rb_z}). A mass m_1 is free to move along its x-axis, constrained by a spring with stiffness k. At rest, the distance between the mass center G of the whole system and points P and P_c (which lie on the x-axis) is r and r_c (Fig. 17.4).*

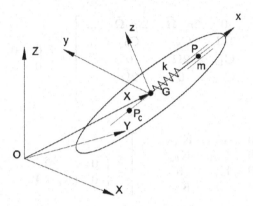

FIGURE 17.4. System made by a rigid body to which a mass, free to move along x-axis, is attached through a spring.

Vectors $\mathbf{u}_1$ and $\mathbf{r}_1$ are simply $u\mathbf{e}_x$ and $r\mathbf{e}_x$, where $\mathbf{e}_x$ is the unit vector of x-axis and the displacement u is the only deformation degree of freedom of the system. The force acting on point P is $\mathbf{F}_1$; the forces and moments acting on the carrier are reduced to a force $\mathbf{F}$ acting in point G and a moment $\mathbf{M}$. Matrix $\mathbf{J}$ reduces to

$$\mathbf{J} = \begin{bmatrix} J_{rb_x} & 0 & 0 \\ 0 & J_{rb_y} + m_1 r^2 + m_c r_c^2 & 0 \\ 0 & 0 & J_{rb_z} + m_1 r^2 + m_c r_c^2 \end{bmatrix}.$$

The first three equations reduce to

$$\begin{cases} m\dot{v}_x + m_1\ddot{u} - m\left(\Omega_z v_y - \Omega_y v_z\right) - m_1 u\left(\Omega_y^2 + \Omega_z^2\right) = F_x + F_{1_x} \\[2mm] m\dot{v}_y + m_1\dot{u}\Omega_z + m\left(\Omega_z v_x - \Omega_x v_z\right) + m_1 u\Omega_x\Omega_y + 2m_1\dot{u}\Omega_z = F_y + F_{1y} \\[2mm] m\dot{v}_z - m_1\dot{u}\Omega_y - m\left(\Omega_y v_x - \Omega_x v_y\right) + m_1 u\Omega_x\Omega_z - 2m_1\dot{u}\Omega_y = F_z + F_{1_z}. \end{cases}$$

The following three equations are

$$\begin{cases} J_x\dot{\Omega}_x + \Omega_z\Omega_y\left(J_z - J_y\right) = M_x \\[2mm] \dot{\Omega}_y\left[J_y + m_1 u\left(2r + u\right)\right] + \Omega_z\Omega_x\left[J_x - J_z' - m_1 u\left(2r + u\right)\right] - m_1 u\dot{v}_z + \\ \qquad + 2m_1\dot{u}\left(r + u\right)\Omega_y + m_1 u\left(v_x\Omega_y - v_y\Omega_x\right) = M_y - rF_z \\[2mm] \dot{\Omega}_z\left[J_z' + m_1 u\left(2r + u\right)\right] + \Omega_x\Omega_y\left[J_y' - J_x + m_1 u\left(2r + u\right)\right] + m_1 u\dot{v}_y + \\ \qquad + 2m_1\dot{u}\left(r + u\right)\Omega_z + m_1 u\left(v_x\Omega_z - v_z\Omega_x\right) = M_z + rF_y. \end{cases}$$

The last equation is

$$m_1\ddot{u} + m_1\dot{v}_x - m_1\left(\Omega_z v_y - \Omega_y v_z\right) - m_1\left(r + u\right)\left(\Omega_y^2 + \Omega_z^2\right) + ku = F_{1_x}.$$

17.3.1   Small elastic deformations

The displacements u_i of the compliant part of the system are usually quite small. In this condition some simplification can be introduced into the equations of motion.

The mass matrix is unchanged. In the damping matrix only term $\mathbf{C}_{\theta i}$ is changed:

$$\mathbf{C}_{\theta i} = m_i \left(\tilde{\mathbf{\Omega}} \tilde{\mathbf{r}}_i + 2 \tilde{\mathbf{r}}_i \tilde{\mathbf{\Omega}} \right).$$

Also in the stiffness matrix the only term to be modified is $\mathbf{K}_{\theta i}$:

$$\mathbf{K}_{\theta i} = m_i \left(2 \tilde{\mathbf{r}}_i \dot{\tilde{\mathbf{\Omega}}} + \dot{\tilde{\mathbf{V}}}^T + \tilde{\mathbf{V}} \tilde{\mathbf{\Omega}} + \tilde{\mathbf{\Omega}} \tilde{\mathbf{V}}^T + 2 \tilde{\mathbf{\Omega}} \tilde{\mathbf{r}}_i \tilde{\mathbf{\Omega}} \right).$$

Also the second term of the force vector should be changed

$$\mathbf{M}_0 + \sum_{i=0}^{n} \tilde{\mathbf{r}}_\mathbf{i} \mathbf{F}_i$$

so that the external forces act on the undeformed configuration, as usual when the small-displacement assumption is considered.

More extensive simplification can be performed in specific cases. In motor vehicle dynamics, for instance, if the x-axis lies in the symmetry plane of the vehicle and is more or less parallel to the road plane, only the component v_x of the velocity cannot be considered as small. The other components v_y and v_z, although not vanishing, in normal driving are small. Also the angular velocities, except that of the wheels about their axis, can be considered as small. The equations of motion can thus be reduced to a much simpler form.

Another well-known example is rotordynamics. If the axis of rotation of the rotor lies in z-direction, all velocities and angular velocities except Ω_z can be considered as small.

17.3.2   Fully linearized equations

A true simplification occurs when all rotations and generalized velocities are considered as small quantities too (no assumption needs to be done on rigid body displacements $\mathbf{X}$):

$$m_t \dot{\mathbf{V}} + \sum_{\forall i} m_\mathbf{i} \ddot{\mathbf{u}}_i = \mathbf{F} + \sum_{\forall i} \mathbf{F}_i \ , \tag{17.24}$$

$$\mathbf{J}_t \dot{\mathbf{\Omega}} + \sum_{\forall \mathbf{i}} m_\mathbf{i} \tilde{\mathbf{r}}_\mathbf{i} \ddot{\mathbf{u}}_\mathbf{i} = \mathbf{M} + \sum_{\forall i} \tilde{\mathbf{r}}_i \mathbf{F}_i, \tag{17.25}$$

$$m_1 \left(\ddot{\mathbf{u}}_i + \dot{\mathbf{V}} + \tilde{\mathbf{r}}_i^T \dot{\mathbf{\Omega}} \right) + \frac{\partial \mathcal{U}}{\partial \mathbf{u}_i} = \mathbf{F}_i, \tag{17.26}$$

i.e.,

$$
\begin{bmatrix}
m\mathbf{I} & \mathbf{0} & m_1\mathbf{I} & m_2\mathbf{I} & \dots \\
 & \mathbf{J} & m_i\tilde{\mathbf{r}}_1 & m_i\tilde{\mathbf{r}}_2 & \dots \\
 & & m_1\mathbf{I} & \mathbf{0} & \dots \\
 & & & m_2\mathbf{I} & \dots \\
\text{symm.} & \dots & \dots & \dots & \dots
\end{bmatrix}
\begin{Bmatrix}
\dot{\mathbf{V}} \\
\dot{\boldsymbol{\Omega}} \\
\ddot{\mathbf{u}}_1 \\
\ddot{\mathbf{u}}_2 \\
\dots
\end{Bmatrix}
+
\begin{bmatrix}
\mathbf{0} & \mathbf{0} \\
\mathbf{0} & \mathbf{C}_e
\end{bmatrix}
\begin{Bmatrix}
\mathbf{V} \\
\boldsymbol{\Omega} \\
\dot{\mathbf{u}}_1 \\
\dot{\mathbf{u}}_2 \\
\dots
\end{Bmatrix}
+
$$

$$
(17.27)
$$

$$
+
\begin{bmatrix}
\mathbf{0} & \mathbf{0} \\
\mathbf{0} & \mathbf{K}_e + \mathbf{K}_g
\end{bmatrix}
\begin{Bmatrix}
\mathbf{X} \\
\boldsymbol{\theta} \\
\mathbf{u}_1 \\
\mathbf{u}_2 \\
\dots
\end{Bmatrix}
=
\begin{Bmatrix}
\mathbf{F}_c + \sum_{i=1}^{n} \mathbf{F}_i \\
\mathbf{M}_c + \tilde{\mathbf{r}}_c \mathbf{F}_c + \sum_{i=0}^{n} \tilde{\mathbf{r}}_i \mathbf{F}_i \\
\mathbf{F}_1 \\
\mathbf{F}_2 \\
\dots
\end{Bmatrix}.
$$

Example 17.5 *Write the equations of motion of the previous example under the assumptions of partial and full linearization.*
If only displacement u is small, the equations of motion from 4 to 6 are

$$
\begin{cases}
J_x \dot{\Omega}_x + \Omega_z \Omega_y \left(J_z - J_y \right) = M_x \\[1mm]
\Omega_y \left[J_y' + m_1 ur \right] + \Omega_z \Omega_x \left[J_x - J_z' - 2m_1 ur \right] - m_1 \left(r + u \right) \dot{v}_z + \\
\quad + 2m_1 \dot{u} r \Omega_y + m_1 \left(r + u \right) \left(v_x \Omega_y - v_y \Omega_x \right) = M_y - r F_z \\[1mm]
\Omega_z \left[J_z' + m_1 ur \right] + \Omega_x \Omega_y \left[J_y' - J_x + 2m_1 ur \right] + m_1 \left(r + u \right) \dot{v}_y + \\
\quad + 2m_1 \dot{u} r \Omega_z + m_1 \left(r + u \right) \left(v_x \Omega_z - v_z \Omega_x \right) = M_z + r F_y.
\end{cases}
$$

If on the contrary it is possible to fully linearize the equations, it follows that

$$
\begin{cases}
m_t \dot{v}_x + m_1 \ddot{u} = F_x + F_{1x} \\[1mm]
m_t \dot{v}_y + m_1 r \dot{\Omega}_z = F_y + F_{1y} \\[1mm]
m_t \dot{v}_z - m_1 r \dot{\Omega}_y = F_z + F_{1z}
\end{cases}
\qquad
\begin{cases}
J_x \dot{\Omega}_x = M_x \\[1mm]
\dot{\Omega}_y J_y' - m_1 r \dot{v}_z = M_y - r F_z \\[1mm]
\dot{\Omega}_z J_z' + m_1 r \dot{v}_y = M_z + r F_y
\end{cases}
$$

$$
m_1 \ddot{u} + m_1 \dot{v}_x + k u = F_{1x}.
$$

17.4   Modal discretization

The displacements **u** of the continuous compliant parts of the system can be expressed in terms of modal coordinates (Eq. 12.26) once its eigenfunctions are known:

$$\mathbf{u}(x, y, z, t) = \sum_{i=0}^{\infty} \eta_i(t) \mathbf{q}_i(x, y, z),$$

or, by truncating the modal expansion after a number of terms,

$$\mathbf{u}(x, y, z, t) = \boldsymbol{\phi}(x, y, z) \boldsymbol{\eta}(t),$$

where $\boldsymbol{\eta}(t)$ is a column matrix containing the n modal coordinates retained and $\boldsymbol{\phi}(x, y, z)$ is a matrix containing n columns (in each column an eigenfunction) and three rows (the 3 components of the eigenfunction).

By introducing the modal coordinates into the expression for the kinetic energy, it follows that

$$\mathcal{T} = \frac{1}{2} m \mathbf{V}^T \mathbf{V} + \frac{1}{2} \boldsymbol{\Omega}^T \mathbf{J} \boldsymbol{\Omega} + \frac{1}{2} \dot{\boldsymbol{\eta}}^T \left(\int_v \rho \boldsymbol{\phi}^T \boldsymbol{\phi} dv \right) \dot{\boldsymbol{\eta}} + \left(\int_v \rho \mathbf{r}^T \tilde{\boldsymbol{\Omega}}^T \tilde{\boldsymbol{\Omega}} \boldsymbol{\phi} dv \right) \boldsymbol{\eta} +$$

$$+ \frac{1}{2} \boldsymbol{\eta}^T \left(\int_v \rho \boldsymbol{\phi}^T \tilde{\boldsymbol{\Omega}}^T \tilde{\boldsymbol{\Omega}} \boldsymbol{\phi} dv \right) \boldsymbol{\eta} + \boldsymbol{\Omega}^T \tilde{\mathbf{V}}^T \left(\int_v \rho \boldsymbol{\phi} dv \right) \boldsymbol{\eta} + \qquad (17.28)$$

$$+ \mathbf{V}^T \left(\int_v \rho \boldsymbol{\phi} dv \right) \dot{\boldsymbol{\eta}} + \boldsymbol{\Omega}^T \left(\int_v \rho \tilde{\mathbf{r}} \boldsymbol{\phi} dv \right) \dot{\boldsymbol{\eta}} + \boldsymbol{\eta}^T \left(\int_v \rho \boldsymbol{\phi}^T \tilde{\boldsymbol{\Omega}}^T \boldsymbol{\phi} dv \right) \dot{\boldsymbol{\eta}}.$$

One of the integrals is immediately computed:

$$\overline{\mathbf{M}} = \int_v \rho \boldsymbol{\phi}^T \boldsymbol{\phi} dv \qquad (17.29)$$

is the (diagonal) modal mass matrix of the compliant system.

Other integrals are straightforward:

$$\overline{\mathbf{M}}_1 = \int_v \rho \boldsymbol{\phi} dv, \qquad (17.30)$$

$$\overline{\mathbf{M}}_2 = \int_v \rho \tilde{\mathbf{r}} \boldsymbol{\phi} dv \qquad (17.31)$$

are constant matrices with three rows and n columns.

The four integrals containing $\tilde{\boldsymbol{\Omega}}$ must be further developed to perform the integration along the space coordinates.

Defining nine matrices $\mathbf{S}_{ij}$, whose size is $1 \times n$

$$\mathbf{S}_{ij} = \int_v \rho r_j \phi_i dv \quad \text{for } i, j = x, y, z,$$

it follows that

$$\int_v \rho \mathbf{r}^T \tilde{\boldsymbol{\Omega}}^T \tilde{\boldsymbol{\Omega}} \boldsymbol{\phi} dv = \Omega_x^2 \left(\mathbf{S}_{yy} + \mathbf{S}_{zz} \right) + \Omega_y^2 \left(\mathbf{S}_{xx} + \mathbf{S}_{zz} \right) + \Omega_z^2 \left(\mathbf{S}_{xx} + \mathbf{S}_{yy} \right) +$$

$$-\Omega_x\Omega_y\left(\mathbf{S}_{xy}+\mathbf{S}_{yx}\right)-\Omega_x\Omega_z\left(\mathbf{S}_{xz}+\mathbf{S}_{zx}\right)-\Omega_y\Omega_z\left(\mathbf{S}_{yz}+\mathbf{S}_{zy}\right).$$

A further set of nine matrices $\mathbf{T}_{ij}$, whose size is $n\times n$, can be defined as

$$\mathbf{T}_{ij}=\int_v\rho\phi_i^T\phi_j dv\quad\text{for }i,\,j=x,\,y,\,z\;.$$

Note that

$$\mathbf{T}_{ij}=\mathbf{T}_{ji}^T\quad\text{for }i,\,j=x,\,y,\,z\;.$$

Since in general the components along the axes of the eigenfunction do not have peculiar orthogonality properties, the nine matrices $\mathbf{T}_{ij}$ may be all present and different.

It thus follows that

$$\int_v\rho\phi^\mathbf{T}\tilde{\Omega}^\mathbf{T}\tilde{\Omega}\phi\mathbf{dv}=\Omega_x^2\left(\mathbf{T}_{yy}+\mathbf{T}_{zz}\right)+\Omega_y^2\left(\mathbf{T}_{xx}+\mathbf{T}_{zz}\right)+\Omega_z^2\left(\mathbf{T}_{xx}+\mathbf{T}_{yy}\right)+$$

$$-\Omega_x\Omega_y\left(\mathbf{T}_{xy}+\mathbf{T}_{yx}\right)-\Omega_x\Omega_z\left(\mathbf{T}_{xz}+\mathbf{T}_{zx}\right)-\Omega_y\Omega_z\left(\mathbf{T}_{yz}+\mathbf{T}_{zy}\right)$$

$$\int_v\rho\phi^T\tilde{\Omega}\phi dv=\Omega_x\left(\mathbf{T}_{zy}-\mathbf{T}_{yz}\right)+\Omega_y\left(\mathbf{T}_{xz}-\mathbf{T}_{zx}\right)+\Omega_z\left(\mathbf{T}_{yx}-\mathbf{T}_{xy}\right).$$

The expression of the kinetic energy can thus be expanded into the explicit form

$$\mathcal{T}=\frac{1}{2}m\mathbf{V}^T\mathbf{V}+\frac{1}{2}\Omega^T\mathbf{J}\Omega+\frac{1}{2}\dot{\eta}^T\overline{\mathbf{M}}\dot{\eta}+\Omega_x^2\left(\mathbf{S}_{yy}+\mathbf{S}_{zz}\right)\eta+\Omega_y^2\left(\mathbf{S}_{xx}+\mathbf{S}_{zz}\right)\eta+$$

$$+\Omega_z^2\left(\mathbf{S}_{xx}+\mathbf{S}_{yy}\right)\eta-\Omega_x\Omega_y\left(\mathbf{S}_{xy}+\mathbf{S}_{yx}\right)\eta-\Omega_x\Omega_z\left(\mathbf{S}_{xz}+\mathbf{S}_{zx}\right)\eta+$$

$$-\Omega_y\Omega_z\left(\mathbf{S}_{yz}+\mathbf{S}_{zy}\right)\eta+\frac{1}{2}\eta^\mathbf{T}\Omega_x^2\left(\mathbf{T}_{yy}+\mathbf{T}_{zz}\right)\eta+\frac{1}{2}\eta^\mathbf{T}\Omega_y^2\left(\mathbf{T}_{xx}+\mathbf{T}_{zz}\right)\eta+$$

$$(17.32)$$

$$+\frac{1}{2}\eta^\mathbf{T}\Omega_z^2\left(\mathbf{T}_{xx}+\mathbf{T}_{yy}\right)\eta-\frac{1}{2}\eta^\mathbf{T}\Omega_x\Omega_y\left(\mathbf{T}_{xy}+\mathbf{T}_{yx}\right)\eta+$$

$$-\frac{1}{2}\eta^\mathbf{T}\Omega_x\Omega_z\left(\mathbf{T}_{xz}+\mathbf{T}_{zx}\right)\eta-\frac{1}{2}\eta^\mathbf{T}\Omega_y\Omega_z\left(\mathbf{T}_{yz}+\mathbf{T}_{zy}\right)\eta+$$

$$+\Omega^\mathbf{T}\tilde{\mathbf{V}}^\mathbf{T}\overline{\mathbf{M}}_1\eta+\mathbf{V}^T\overline{\mathbf{M}}_1\dot{\eta}+\Omega^\mathbf{T}\overline{\mathbf{M}}_2\dot{\eta}+\eta^\mathbf{T}\Omega_x\left(\mathbf{T}_{zy}-\mathbf{T}_{yz}\right)\dot{\eta}+$$

$$+\eta^\mathbf{T}\Omega_y\left(\mathbf{T}_{xz}-\mathbf{T}_{zx}\right)\dot{\eta}+\eta^\mathbf{T}\Omega_z\left(\mathbf{T}_{yx}-\mathbf{T}_{xy}\right)\dot{\eta}\;.$$

The relevant derivatives can thus be performed:

$$\frac{\partial\mathcal{T}}{\partial\mathbf{V}}=m\mathbf{V}+\tilde{\Omega}\overline{\mathbf{M}}_1\eta+\overline{\mathbf{M}}_1\dot{\eta}\;,$$

$$\frac{d}{dt}\left(\frac{\partial\mathcal{T}}{\partial\mathbf{V}}\right)=m\dot{\mathbf{V}}+\overline{\mathbf{M}}_1\ddot{\eta}+\tilde{\Omega}\overline{\mathbf{M}}_1\dot{\eta}+\dot{\tilde{\Omega}}\overline{\mathbf{M}}_1\eta\;,$$

$$\frac{\partial \mathcal{T}}{\partial \mathbf{\Omega}} = \mathbf{J}\mathbf{\Omega} + \mathbf{U}_1\boldsymbol{\eta} + \frac{1}{2}\mathbf{U}_2\boldsymbol{\eta} + \tilde{\mathbf{V}}^{\mathbf{T}}\overline{\mathbf{M}}_1\boldsymbol{\eta} + \overline{\mathbf{M}}_2\dot{\boldsymbol{\eta}} + \mathbf{U}_3\dot{\boldsymbol{\eta}} \ ,$$

where

$$\mathbf{U}_1 = \left\{ \begin{array}{l} 2\Omega_x \left(\mathbf{S}_{yy} + \mathbf{S}_{zz}\right) - \Omega_y \left(\mathbf{S}_{xy} + \mathbf{S}_{yx}\right) - \Omega_z \left(\mathbf{S}_{xz} + \mathbf{S}_{zx}\right) \\ 2\Omega_y \left(\mathbf{S}_{xx} + \mathbf{S}_{zz}\right) - \Omega_x \left(\mathbf{S}_{xy} + \mathbf{S}_{yx}\right) - \Omega_z \left(\mathbf{S}_{yz} + \mathbf{S}_{zy}\right) \\ 2\Omega_z \left(\mathbf{S}_{xx} + \mathbf{S}_{yy}\right) - \Omega_x \left(\mathbf{S}_{xz} + \mathbf{S}_{zx}\right) - \Omega_y \left(\mathbf{S}_{yz} + \mathbf{S}_{zy}\right) \end{array} \right\} \ ,$$

$$\mathbf{U}_2 = \left\{ \begin{array}{l} \boldsymbol{\eta}^{\mathbf{T}} \left[2\Omega_x \left(\mathbf{T}_{yy} + \mathbf{T}_{zz}\right) - \Omega_y \left(\mathbf{T}_{xy} + \mathbf{T}_{yx}\right)\right] - \Omega_z \left(\mathbf{T}_{xz} + \mathbf{T}_{zx}\right) \\ \boldsymbol{\eta}^{\mathbf{T}} \left[2\Omega_y \left(\mathbf{T}_{xx} + \mathbf{T}_{zz}\right) - \Omega_x \left(\mathbf{T}_{xy} + \mathbf{T}_{yx}\right)\right] - \Omega_z \left(\mathbf{T}_{yz} + \mathbf{T}_{zy}\right) \\ \boldsymbol{\eta}^{\mathbf{T}} \left[2\Omega_z \left(\mathbf{T}_{xx} + \mathbf{T}_{yy}\right)\right] - \Omega_x \left(\mathbf{T}_{xz} + \mathbf{T}_{zx}\right) - \Omega_y \left(\mathbf{T}_{yz} + \mathbf{T}_{zy}\right) \end{array} \right\} \ ,$$

$$\mathbf{U}_3 = \left\{ \begin{array}{l} \boldsymbol{\eta}^{\mathbf{T}} \left(\mathbf{T}_{zy} - \mathbf{T}_{yz}\right) \\ \boldsymbol{\eta}^{\mathbf{T}} \left(\mathbf{T}_{xz} - \mathbf{T}_{zx}\right) \\ \boldsymbol{\eta}^{\mathbf{T}} \left(\mathbf{T}_{yx} - \mathbf{T}_{xy}\right) \end{array} \right\} \ .$$

$\mathbf{U}_1$ and $\mathbf{U}_2$ are functions of $\mathbf{\Omega}$ and $\mathbf{U}_2$ and $\mathbf{U}_3$ are functions of $\boldsymbol{\eta}$. The two terms $\mathbf{U}_2\boldsymbol{\eta}$ and $\mathbf{U}_3\dot{\boldsymbol{\eta}}$ can thus be regarded as small quantities when the modal displacements are small.

Proceeding with the derivatives, it follows that

$$\frac{d}{dt}\left(\frac{\partial \mathcal{T}}{\partial \mathbf{\Omega}}\right) = \mathbf{J}\dot{\mathbf{\Omega}} + \left(\overline{\mathbf{M}}_2 + \mathbf{U}_3\right)\ddot{\boldsymbol{\eta}} + \left(\mathbf{U}_1 + \frac{1}{2}\mathbf{U}_2 + \tilde{\mathbf{V}}^{\mathbf{T}}\overline{\mathbf{M}}_1\right)\dot{\boldsymbol{\eta}} +$$

$$+ \left(\dot{\mathbf{U}}_1 + \frac{1}{2}\dot{\mathbf{U}}_2 + \dot{\tilde{\mathbf{V}}}^{\mathbf{T}}\overline{\mathbf{M}}_1\right)\boldsymbol{\eta} + \dot{\mathbf{U}}_3\dot{\boldsymbol{\eta}} \ .$$

Neglecting small terms, it follows that

$$\frac{d}{dt}\left(\frac{\partial \mathcal{T}}{\partial \mathbf{\Omega}}\right) = \mathbf{J}\dot{\mathbf{\Omega}} + \overline{\mathbf{M}}_2\ddot{\boldsymbol{\eta}} + \left(\mathbf{U}_1 + \tilde{\mathbf{V}}^{\mathbf{T}}\overline{\mathbf{M}}_1\right)\dot{\boldsymbol{\eta}} + \left(\dot{\mathbf{U}}_1 + \dot{\tilde{\mathbf{V}}}^{\mathbf{T}}\overline{\mathbf{M}}_1\right)\boldsymbol{\eta} \ ,$$

$$\frac{\partial \mathcal{T}}{\partial \dot{\boldsymbol{\eta}}} = \overline{\mathbf{M}}\dot{\boldsymbol{\eta}} + \overline{\mathbf{M}}_1^T \mathbf{V} + \overline{\mathbf{M}}_2^T \mathbf{\Omega} + \mathbf{U}_4\boldsymbol{\eta} \ ,$$

where

$$\mathbf{U}_4 = \Omega_x \left(\mathbf{T}_{zy} - \mathbf{T}_{yz}\right)^T + \Omega_y \left(\mathbf{T}_{xz} - \mathbf{T}_{zx}\right)^T + \Omega_z \left(\mathbf{T}_{yx} - \mathbf{T}_{xy}\right)^T \ ,$$

$$\frac{d}{dt}\left(\frac{\partial \mathcal{T}}{\partial \dot{\boldsymbol{\eta}}}\right) = \overline{\mathbf{M}}\ddot{\boldsymbol{\eta}} + \overline{\mathbf{M}}_1^T \dot{\mathbf{V}} + \overline{\mathbf{M}}_2^T \dot{\mathbf{\Omega}} + \mathbf{U}_4\dot{\boldsymbol{\eta}} + \dot{\mathbf{U}}_4\boldsymbol{\eta} \ ,$$

$$\frac{\partial \mathcal{T}}{\partial \boldsymbol{\eta}} = \overline{\mathbf{M}}_1^{\mathbf{T}}\tilde{\mathbf{V}}\mathbf{\Omega} + \mathbf{U}_6\boldsymbol{\eta} + \mathbf{U}_5^T\mathbf{\Omega} + \mathbf{U}_4^T\dot{\boldsymbol{\eta}} \ ,$$

where

$$\mathbf{U}_5 = \left\{ \begin{array}{l} \Omega_x \left(\mathbf{S}_{yy} + \mathbf{S}_{zz}\right) - \Omega_y \mathbf{S}_{xy} - \Omega_z \mathbf{S}_{xz} \\ -\Omega_x \mathbf{S}_{yx} + \Omega_y \left(\mathbf{S}_{xx} + \mathbf{S}_{zz}\right) - \Omega_z \mathbf{S}_{yz} \\ -\Omega_x \mathbf{S}_{zx} - \Omega_y \mathbf{S}_{zy} + \Omega_z \left(\mathbf{S}_{xx} + \mathbf{S}_{yy}\right) \end{array} \right\} \ ,$$

$$\mathbf{U}_6 = \Omega_x^2 \left(\mathbf{T}_{yy} + \mathbf{T}_{zz}\right) + \Omega_y^2 \left(\mathbf{T}_{xx} + \mathbf{T}_{zz}\right) + \Omega_z^2 \left(\mathbf{T}_{xx} + \mathbf{T}_{yy}\right) +$$
$$-\Omega_x\Omega_y \left(\mathbf{T}_{xy} + \mathbf{T}_{yx}\right) - \Omega_x\Omega_z \left(\mathbf{T}_{xz} + \mathbf{T}_{zx}\right) - \Omega_y\Omega_z \left(\mathbf{T}_{yz} + \mathbf{T}_{zy}\right) \ .$$

The potential energy and the Rayleigh dissipation function may be written in the usual way

$$\mathcal{U} \approx \frac{1}{2}\boldsymbol{\eta}^T\overline{\mathbf{K}}\boldsymbol{\eta} \ , \quad \mathcal{F} \approx \frac{1}{2}\dot{\boldsymbol{\eta}}^T\overline{\mathbf{C}}\dot{\boldsymbol{\eta}} \ . \tag{17.33}$$

Remark 17.5 *While $\overline{\mathbf{K}}$ is diagonal, in general $\overline{\mathbf{C}}$ is not.*

Also the modal geometric matrix is generally not a diagonal matrix, and it couples the various modes.

The virtual work of the forces applied to the system is

$$\delta\mathcal{L} = \delta\mathbf{X}^T\mathbf{R}\mathbf{F} + \delta\boldsymbol{\theta}^T\mathbf{A}_2\mathbf{M} + \int_v \left[\delta\boldsymbol{\theta}^T\mathbf{A}_2\tilde{\mathbf{u}}\mathbf{f} + \delta\boldsymbol{\eta}^T\boldsymbol{\phi}^T\mathbf{f}\right] dv \ , \tag{17.34}$$

where

$$\mathbf{F} = \mathbf{F}_c + \int_v \mathbf{f}dv \ .$$

$$\mathbf{M} = \tilde{\mathbf{r}}_c\mathbf{F}_c + \mathbf{M}_c + \int_v \tilde{\mathbf{r}}\mathbf{f}dv \ .$$

Neglecting the term in $\tilde{\mathbf{u}}\mathbf{f}$, which is negligible if the modal coordinates are small, and introducing the modal forces

$$\overline{\mathbf{F}} = \int_v \boldsymbol{\phi}\mathbf{f}dv \ , \tag{17.35}$$

the forces applied to the system reduce to

$$\mathbf{Q}_X = \mathbf{F} \ , \quad \mathbf{Q}_\theta = \mathbf{M} \ , \quad \mathbf{Q}_\eta = \overline{\mathbf{F}} \ . \tag{17.36}$$

The equations of motion are still Eq. (17.17), where $\boldsymbol{\eta}$ has been substituted for $\mathbf{q}$.

The final form of the equations of motion is thus

$$\begin{bmatrix} m\mathbf{I} & 0 & \overline{\mathbf{M}}_1 \\ 0 & \mathbf{J} & \overline{\mathbf{M}}_2 \\ \overline{\mathbf{M}}_1^T & \overline{\mathbf{M}}_2^T & \overline{\mathbf{M}} \end{bmatrix} \begin{Bmatrix} \dot{\mathbf{V}} \\ \dot{\boldsymbol{\Omega}} \\ \ddot{\boldsymbol{\eta}} \end{Bmatrix} +$$

$$+ \left(\begin{bmatrix} m\,\tilde{\boldsymbol{\Omega}} & 0 & 2\tilde{\boldsymbol{\Omega}}\overline{\mathbf{M}}_1 \\ 0 & \tilde{\boldsymbol{\Omega}}\mathbf{J} & \mathbf{U}_1 + \tilde{\boldsymbol{\Omega}}\overline{\mathbf{M}}_2 \\ \overline{\mathbf{M}}_1^T\tilde{\boldsymbol{\Omega}} & \mathbf{U}_5^T & 2\mathbf{U}_4 \end{bmatrix} + \begin{bmatrix} 0 & 0 \\ 0 & \overline{\mathbf{C}} \end{bmatrix} \right) \begin{Bmatrix} \mathbf{V} \\ \boldsymbol{\Omega} \\ \dot{\boldsymbol{\eta}} \end{Bmatrix} + \tag{17.37}$$

$$+ \left(\begin{bmatrix} 0 & 0 & \left(\dot{\tilde{\boldsymbol{\Omega}}} + \tilde{\boldsymbol{\Omega}}^2\right)\overline{\mathbf{M}}_1 \\ 0 & 0 & \mathbf{K}_{\theta\eta} \\ 0 & 0 & \dot{\mathbf{U}}_4 - \mathbf{U}_6 \end{bmatrix} + \begin{bmatrix} 0 & 0 \\ 0 & \overline{\mathbf{K}} + \overline{\mathbf{K}}_g \end{bmatrix} \right) \begin{Bmatrix} \mathbf{X} \\ \boldsymbol{\theta} \\ \boldsymbol{\eta} \end{Bmatrix} = \begin{Bmatrix} \mathbf{F} \\ \mathbf{M} \\ \overline{\mathbf{F}} \end{Bmatrix} \ ,$$

where

$$\mathbf{K}_{\theta\eta} = \dot{\mathbf{U}}_1 + \dot{\mathbf{V}}^T\overline{\mathbf{M}}_1 + \left(\tilde{\mathbf{V}}\tilde{\boldsymbol{\Omega}} - \tilde{\boldsymbol{\Omega}}\tilde{\mathbf{V}}\right)\overline{\mathbf{M}}_1 + \tilde{\boldsymbol{\Omega}}\mathbf{U}_1 \ .$$

17.5   Planar systems

In case of a two-dimensional problem, the rigid body degrees of freedom are only 3. Using a pseudo-coordinates approach, it is possible to use the two components v_x and v_y of the velocity and the angular velocity about the z-axis, which reduces to $\dot{\psi}$. The displacement vector and the eigenfunctions have only two components and are functions of just two coordinates. The kinetic energy reduces to

$$\mathcal{T} = \frac{1}{2}m\mathbf{V}^T\mathbf{V} + \frac{1}{2}J\dot{\psi}^2 + \frac{1}{2}\dot{\boldsymbol{\eta}}^T\overline{\mathbf{M}}\dot{\boldsymbol{\eta}} + \dot{\psi}^2\left(\mathbf{S}_{xx} + \mathbf{S}_{yy}\right)\boldsymbol{\eta} +$$

$$+\frac{1}{2}\boldsymbol{\eta}^T\dot{\psi}^2\left(\mathbf{T}_{xx} + \mathbf{T}_{yy}\right)\boldsymbol{\eta} + \dot{\psi}\left[\begin{array}{cc} V_y & -V_x \end{array}\right]\overline{\mathbf{M}}_1\boldsymbol{\eta} + \mathbf{V}^T\overline{\mathbf{M}}_1\dot{\boldsymbol{\eta}} + \qquad (17.38)$$

$$+\dot{\psi}\overline{\mathbf{M}}_2\dot{\boldsymbol{\eta}} + \boldsymbol{\eta}^T\dot{\psi}\left(\mathbf{T}_{yx} - \mathbf{T}_{xy}\right)\dot{\boldsymbol{\eta}} .$$

Now

$$\overline{\mathbf{M}}_1 = \int_v \rho\phi dv \qquad (17.39)$$

has two rows and n columns and

$$\overline{\mathbf{M}}_2 = \int_v \rho\left(-y\phi_x + x\phi_y\right)dv \qquad (17.40)$$

is a matrix with one row and n columns.

The relevant derivatives can thus be performed:

$$\frac{\partial\mathcal{T}}{\partial\mathbf{V}} = m\mathbf{V} + \dot{\psi}\left\{\begin{array}{c} -\overline{\mathbf{M}}_{1y} \\ \overline{\mathbf{M}}_{1x} \end{array}\right\}\boldsymbol{\eta} + \overline{\mathbf{M}}_1\dot{\boldsymbol{\eta}},$$

$$\frac{d}{dt}\left(\frac{\partial\mathcal{T}}{\partial\mathbf{V}}\right) = m\dot{\mathbf{V}} + \overline{\mathbf{M}}_1\ddot{\boldsymbol{\eta}} + \dot{\psi}\left\{\begin{array}{c} -\overline{\mathbf{M}}_{1y} \\ \overline{\mathbf{M}}_{1x} \end{array}\right\}\dot{\boldsymbol{\eta}} + \ddot{\psi}\left\{\begin{array}{c} -\overline{\mathbf{M}}_{1y} \\ \overline{\mathbf{M}}_{1x} \end{array}\right\}\boldsymbol{\eta} .$$

Neglecting the small terms, the derivative with respect to $\dot{\psi}$ is

$$\frac{\partial\mathcal{T}}{\partial\dot{\psi}} = J\dot{\psi} + 2\dot{\psi}\left(\mathbf{S}_{xx} + \mathbf{S}_{yy}\right)\boldsymbol{\eta} + \left[\begin{array}{cc} V_y & -V_x \end{array}\right]\overline{\mathbf{M}}_1\boldsymbol{\eta} + \overline{\mathbf{M}}_2\dot{\boldsymbol{\eta}},$$

$$\frac{d}{dt}\left(\frac{\partial\mathcal{T}}{\partial\dot{\psi}}\right) = J\ddot{\psi} + \overline{\mathbf{M}}_2\ddot{\boldsymbol{\eta}} + \left[2\dot{\psi}\left(\mathbf{S}_{xx} + \mathbf{S}_{yy}\right) + \left[\begin{array}{cc} V_y & -V_x \end{array}\right]\overline{\mathbf{M}}_1\right]\dot{\boldsymbol{\eta}} +$$

$$+ \left(2\ddot{\psi}\left(\mathbf{S}_{xx} + \mathbf{S}_{yy}\right) + \left[\begin{array}{cc} \dot{V}_y & -\dot{V}_x \end{array}\right]\overline{\mathbf{M}}_1\right)\boldsymbol{\eta},$$

$$\frac{\partial\mathcal{T}}{\partial\dot{\boldsymbol{\eta}}} = \overline{\mathbf{M}}\dot{\boldsymbol{\eta}} + \overline{\mathbf{M}}_1^T\mathbf{V} + \overline{\mathbf{M}}_2^T\dot{\psi} + \dot{\psi}(\mathbf{T}_{yx} - \mathbf{T}_{xy})^T\boldsymbol{\eta},$$

$$\frac{d}{dt}\left(\frac{\partial\mathcal{T}}{\partial\dot{\boldsymbol{\eta}}}\right) = \overline{\mathbf{M}}\ddot{\boldsymbol{\eta}} + \overline{\mathbf{M}}_1^T\dot{\mathbf{V}} + \overline{\mathbf{M}}_2^T\ddot{\psi} + \dot{\psi}\left(\mathbf{T}_{yx} - \mathbf{T}_{xy}\right)^T\dot{\boldsymbol{\eta}} + \ddot{\psi}\left(\mathbf{T}_{yx} - \mathbf{T}_{xy}\right)^T\boldsymbol{\eta},$$

$$\frac{\partial T}{\partial \eta} = \dot{\psi} \overline{\mathbf{M}}_1^{\mathbf{T}} \left[\begin{array}{cc} V_y & -V_x \end{array} \right]^T + \dot{\psi}^2 \left(\mathbf{S}_{xx} + \mathbf{S}_{yy} \right)^T +$$

$$+ \dot{\psi}^2 \left(\mathbf{T}_{xx} + \mathbf{T}_{yy} \right) \boldsymbol{\eta} + \dot{\psi} \left(\mathbf{T}_{yx} - \mathbf{T}_{xy} \right) \dot{\boldsymbol{\eta}} .$$

The expressions for the potential energy and the Rayleigh dissipation function are the same as for the tridimensional case.

The final form of the equations of motion is thus

$$\begin{bmatrix} m & 0 & 0 & \overline{\mathbf{M}}_{1x} \\ 0 & m & 0 & \overline{\mathbf{M}}_{1y} \\ 0 & 0 & J_z & \overline{\mathbf{M}}_2 \\ \overline{\mathbf{M}}_{1x}^T & \overline{\mathbf{M}}_{1y}^T & \overline{\mathbf{M}}_2^T & \overline{\mathbf{M}} \end{bmatrix} \begin{Bmatrix} \dot{V}_x \\ \dot{V}_y \\ \ddot{\psi} \\ \ddot{\eta} \end{Bmatrix} +$$

$$+ \left(\begin{bmatrix} 0 & -m\dot{\psi} & 0 & -2\dot{\psi}\overline{\mathbf{M}}_{1y} \\ m\dot{\psi} & 0 & 0 & 2\dot{\psi}\overline{\mathbf{M}}_{1x} \\ 0 & 0 & 0 & 2\dot{\psi} \left(\mathbf{S}_{xx} + \mathbf{S}_{yy} \right) \\ \dot{\psi}\overline{\mathbf{M}}_{1y}^T & -\dot{\psi}\overline{\mathbf{M}}_{1x}^T & -\dot{\psi} \left(\mathbf{S}_{xx} + \mathbf{S}_{yy} \right)^T & -2\dot{\psi} \left(\mathbf{T}_{yx} - \mathbf{T}_{xy} \right) \end{bmatrix} + \right.$$

$$\text{(17.41)}$$

$$\left. + \begin{bmatrix} 0 & 0 \\ 0 & \mathbf{C} \end{bmatrix} \right) \begin{Bmatrix} V_x \\ V_y \\ \dot{\psi} \\ \dot{\eta} \end{Bmatrix} + \left(\begin{bmatrix} 0 & 0 & 0 & -\ddot{\psi}\overline{\mathbf{M}}_{1y} - \dot{\psi}^2\overline{\mathbf{M}}_{1x} \\ 0 & 0 & 0 & \ddot{\psi}\overline{\mathbf{M}}_{1x} - \dot{\psi}^2\overline{\mathbf{M}}_{1y} \\ 0 & 0 & 0 & K_{\theta\theta} \\ 0 & 0 & 0 & K_{\theta\eta} \end{bmatrix} + \right.$$

$$\left. + \begin{bmatrix} 0 & 0 \\ 0 & \overline{\mathbf{K}} + \overline{\mathbf{K}}_g \end{bmatrix} \right) \begin{Bmatrix} X \\ Y \\ \psi \\ \eta \end{Bmatrix} = \begin{Bmatrix} F_x \\ F_y \\ M \\ \overline{\mathbf{F}} \end{Bmatrix} ,$$

where

$$K_{\theta\theta} = 2\ddot{\psi}_z \left(\mathbf{S}_{xx} + \mathbf{S}_{yy} \right) + \dot{V}_y \overline{\mathbf{M}}_{1x} - \dot{V}_x \overline{\mathbf{M}}_{1y} + \dot{\psi} \left(V_y \overline{\mathbf{M}}_{1y} + V_x \overline{\mathbf{M}}_{1x} \right)$$

$$K_{\theta\eta} = \ddot{\psi} \left(\mathbf{T}_{yx} - \mathbf{T}_{xy} \right)^T - \dot{\psi}^2 \left(\mathbf{T}_{xx} + \mathbf{T}_{yy} \right) .$$

17.6 Beam attached to a rigid body: planar dynamics

Consider an Euler–Bernoulli beam attached to a rigid body (Fig. 17.5) and, to simplify the problem, assume also that

- only the bending behavior of the beam is important,

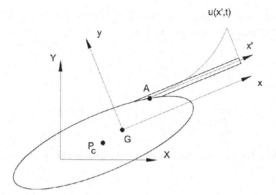

FIGURE 17.5. An Euler–Bernoulli beam attached to a rigid body performing a planar motion.

- the displacement u is small,

- the problem is planar in the xy plane.

The rigid body has then 3 degrees of freedom (X, Y, and ψ) and each point of the beam has an additional degree of freedom, displacement u. The only non-zero component of the angular velocity, Ω_z, reduces to $\dot\psi$.

Without any loss of generality, it is possible to chose the direction of the x-axis so that it is parallel to the axis of the beam. In this way vector u and the eigenfunctions ϕ have a single component, along the y-axis. Vector $\mathbf{r}$ reduces to

$$\mathbf{r} = \begin{bmatrix} x_A + x' & y_A \end{bmatrix}^T .$$

Assuming that the beam is clamped to the carrier and free at the other end and that the beam is prismatic, the eigenfunctions are (see Section 12.4)

$$q_i(\zeta) = \frac{1}{N_2} \left\{ \sin(\beta_i\zeta) - \sinh(\beta_i\zeta) - N_1 \left[\cos(\beta_i\zeta) - \cosh(\beta_i\zeta)\right] \right\} ,$$

where

$$\zeta = \frac{x'}{l} , \qquad N_1 = \frac{\sin(\beta_i) + \sinh(\beta_i)}{\cos(\beta_i) + \cosh(\beta_i)} ,$$

$$N_2 = \sin(\beta_i) - \sinh(\beta_i) - N_1 \left[\cos(\beta_i) - \cosh(\beta_i)\right]$$

and coefficients β_i are

$$\beta_1 = 1.875 , \ \beta_2 = 4.694 , \ \beta_3 = 7.855 , \ \beta_4 = 10.996 , \ \beta_i = \pi(i - 0.5) .$$

The modal masses and the modal stiffnesses are easily computed. Assuming that the beam is prismatic,

$$\overline{M}_i = \int_0^l \rho A \left[q_i(x')\right]^2 dx' = \frac{1}{4}\rho A l = \frac{1}{4}m, \qquad \forall i,$$

$$\overline{K}_i = \int_0^l EI \left[\frac{d^2 q_i(x')}{dx'^2}\right]^2 dx' = \frac{EI}{l^3}\frac{\beta_i^4}{4}, \qquad \forall i.$$

$\overline{M}_1$ and $\overline{M}_2$ have a single component, along the y- and z-axes, respectively:

$$\overline{M}_{1x} = m \int_0^1 q_i(\zeta)d\zeta , \qquad (17.42)$$

$$\overline{M}_2 = m \int_0^1 (l\zeta + x_a)\, q_i(\zeta)d\zeta . \qquad (17.43)$$

They must be computed numerically in each case.

Out of the nine matrices $\mathbf{S}_{ij}$ and $\mathbf{T}_{ij}$, only $\mathbf{S}_{yx}$, $\mathbf{S}_{yy}$, and $\mathbf{T}_{yy}$ are different from 0.

The modal geometric matrix is easily computed. The increase of the potential energy due to the axial stresses (see Section 12.8, Eq. (17.44))

$$\Delta\mathcal{U} = \frac{1}{2}\int_0^l A\sigma_x \left(\frac{\partial u}{\partial x'}\right)^2 dx' = \frac{1}{2}\boldsymbol{\eta}^T \left(\int_0^l A\sigma_x \frac{\partial q^T}{\partial x'}\frac{\partial q}{\partial x'}dx'\right)\boldsymbol{\eta} . \quad (17.44)$$

The axial stress due to rotation is

$$\sigma_{x\Omega} = \int_{x'}^l \rho\dot{\psi}^2 (u + x_a)\, du = \rho\dot{\psi}^2 l^2 \frac{1 - \zeta^2 + 2\zeta_A - 2\zeta_A\zeta}{2}. \qquad (17.45)$$

An axial stress may also be present due to the linear acceleration

$$\sigma_{x_V} = \int_{x'}^l \rho\dot{V} du = \rho\dot{V}l\,(1 - \zeta). \qquad (17.46)$$

The modal geometric matrix is thus

$$\overline{\mathbf{K}}_g = \rho A l \left(\dot{\psi}^2\overline{\mathbf{K}}_g^* + \frac{\dot{V}}{l}\overline{\mathbf{K}}_g^{**}\right) ,$$

where $\overline{K}_g^*$ and $\overline{K}_g^{**}$ must be computed numerically in each case from the equations

$$\overline{\mathbf{K}}_g^* = \frac{1}{2}\int_0^1 (1 - \zeta^2)\frac{\partial q^T}{\partial \zeta}\frac{\partial q}{\partial \zeta}d\zeta + \zeta_A \int_0^1 (1 - \zeta)\frac{\partial q^T}{\partial \zeta}\frac{\partial q}{\partial \zeta}d\zeta, \qquad (17.47)$$

$$\overline{\mathbf{K}}_g^{**} = \int_0^l (1 - \zeta) \frac{\partial q^T}{\partial \zeta} \frac{\partial q}{\partial \zeta} d\zeta \ . \tag{17.48}$$

The equations of motion are thus

$$\begin{Bmatrix} F_x \\ F_y \\ \mathbf{M} \\ \overline{\mathbf{F}} \end{Bmatrix} = \begin{bmatrix} m & 0 & 0 & \mathbf{0} \\ 0 & m & 0 & \overline{\mathbf{M}}_1 \\ 0 & 0 & J_z & \overline{\mathbf{M}}_2 \\ \mathbf{0} & \overline{\mathbf{M}}_1^T & \overline{\mathbf{M}}_2^T & \overline{\mathbf{M}} \end{bmatrix} \begin{Bmatrix} \dot{V}_x \\ \dot{V}_y \\ \ddot{\psi} \\ \ddot{\boldsymbol{\eta}} \end{Bmatrix} +$$

$$+ \left(\begin{bmatrix} 0 & -m\dot{\psi} & 0 & -2\dot{\psi}\overline{\mathbf{M}}_1 \\ m\dot{\psi} & 0 & 0 & \mathbf{0} \\ 0 & 0 & 0 & 2\dot{\psi}y_A\overline{\mathbf{M}}_1 \\ \dot{\psi}\overline{\mathbf{M}}_{1y}^T & 0 & -\dot{\psi}y_A\overline{\mathbf{M}}_1^T & 0 \end{bmatrix} + \begin{bmatrix} 0 & 0 \\ 0 & \overline{\mathbf{C}} \end{bmatrix} \right) \begin{Bmatrix} V_x \\ V_y \\ \dot{\psi} \\ \dot{\boldsymbol{\eta}} \end{Bmatrix} +$$

$$+ \left(\begin{bmatrix} 0 & 0 & 0 & -\ddot{\psi}\overline{\mathbf{M}}_1 \\ 0 & 0 & 0 & \dot{\psi}^2\overline{\mathbf{M}}_1 \\ 0 & 0 & 0 & \left(2\ddot{\psi}_z y_A - \dot{V}_x + \dot{\psi}V_y\right)\overline{\mathbf{M}}_1 \\ 0 & 0 & 0 & -\dot{\psi}^2\overline{\mathbf{M}} \end{bmatrix} + \begin{bmatrix} 0 & 0 \\ 0 & \mathbf{K} + \overline{\mathbf{K}}_g \end{bmatrix} \right) \begin{Bmatrix} X \\ Y \\ \psi \\ \boldsymbol{\eta} \end{Bmatrix}$$

(17.49)

17.7   The rotating beam

Consider the planar system of Fig. 17.5, but now assume that the angular velocity Ω of the carrier is constant and that point G is fixed. The motion of the carrier is fully determined, and the only degrees of freedom are those of the beam, i.e., the modal coordinates η_i.

The equations of motion are

$$\overline{\mathbf{M}}\ddot{\boldsymbol{\eta}} + \left(\overline{\mathbf{C}} - \Omega^2 y_A \overline{\mathbf{M}}_1^T\right)\dot{\boldsymbol{\eta}} + \left(-\Omega^2\overline{\mathbf{M}} + \overline{\mathbf{K}} + \overline{\mathbf{K}}_g\right)\boldsymbol{\eta} = \overline{\mathbf{F}} \ . \tag{17.50}$$

If just one mode is considered and the modal mass is normalized to one, the values of the parameters are

$$\overline{\mathbf{M}} = 1 \ , \quad \overline{\mathbf{M}}_1 = 1.5660 \ , \quad \overline{\mathbf{K}} = \frac{\beta^4 EI}{l^4 \rho A} = \frac{\beta^4 EI}{ml^3} \ , \tag{17.51}$$

$$\overline{\mathbf{K}}_g = 1.1933 + 1.5709\zeta_A \ .$$

If $\zeta_A = 0$ and $\zeta_A = 1$, the usual values for the upper bounds $\overline{\mathbf{K}}_g = 1.1933$ and $\overline{\mathbf{K}}_g = 2.7642$ found in the literature for the rotating beam[1] are obtained.

---

[1]See, for instance, G. Genta, *Dynamics of Rotating Systems*, Springer, New York, 2005, p. 483.

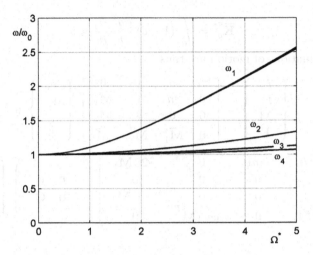

FIGURE 17.6. First four natural frequencies of a rotating beam as functions of the nondimensional rotational speed. The frequency has been computed setting the frequency of the nonrotating beam equal to one.

If the system is undamped and $y_A = 0$, and introducing the nondimensional frequency and speed

$$\Omega^* = \Omega \sqrt{\frac{ml^3}{EI}} \ , \qquad \omega^* = \omega \sqrt{\frac{ml^3}{EI}} \ , \tag{17.52}$$

the dependence of the natural frequencies on the speed is

$$\omega^* = \sqrt{\text{eig} \left[\Omega^{*2} \left(\overline{K}_g^* - I \right) + \beta_i^4 \right]} \ . \tag{17.53}$$

The first four natural frequencies in the rotation plane of a rotating beam with $\zeta_A = 1$ are plotted as functions of the nondimensional rotational speed in Fig. 17.6. In the plot the results obtained with the minimum number of modes (number of modes equal to the order of the natural frequency) and with 10 modes are superimposed. It is clear that, although the geometric matrix couples the various modes, the errors due to neglecting modal coupling are very small.

17.8   Exercises

Exercise 17.1 *Consider the system of Example 17.1 and add two equal viscous dampers with damping coefficient c in parallel to the springs. Write the equation of motion, using both the rotating coordinate frame Axy and the inertial frame OXY. Study the stability of the system.*

Exercise 17.2 *Consider the rigid body with a mass m_1 free to move along its x-axis, studied in Example 17.4 and shown in Fig. 17.4. At rest, the distance between the mass center G of the whole system and points P and P_c (which lie on the x-axis) is r and r_c.*

Compute the time history of the motion of the system with the following initial condition: $X_0 = Y_0 = Z_0 = u_0 = 0$, $\phi_0 = \theta_0 = \psi_0 = 0$, $\dot{X}_0 = 10$ m/s, $\dot{Y}_0 = 200$ m/s, $\dot{Z}_0 = 500$ m/s, $u_0 = \dot{X}_0$ (i.e., point P is initially at rest in the X-direction of the inertial frame), $\dot{\phi}_0 = \dot{\theta}_0 = \dot{\psi}_0 = 10$ rad/1. Use the fully nonlinear, the semilinearized, and the linearized approaches.

Repeat the computation with initial conditions on the angular velocities 100 times smaller.

Data $m_{rb} = 20$ kg, $J_{rb_x} = 4$ kg m^2, $J_{rb_y} = 4$ kg m^2, $J_{rb_z} = 6$ kg m^2, $m_1 = 5$ kg, $r_1 = 300$ mm, and $k = 20$ kN/m.

Exercise 17.3 *Consider an isotropic beam clamped radially to a rigid disc rotating at constant speed Ω_z on stiff bearings. Assume the rotation axis as Z-axis and the axis of the beam as x-axis. Compute the first natural frequency both in the rotation plane and in a plane containing the rotation axis using the modal approach (consider a single mode in each plane).*

Data radius of the disc $R = 300$ mm, length of the beam $l = 500$ mm, circular cross-section with radius $r = 20$ mm, $\rho = 7,810$ kg m^3, and $E = 2.1 \times 10^{11}$ N/m^2.

Exercise 17.4 *Consider a rigid body to which a beam is clamped (Fig. 17.7). The axis of the beam coincides with the x-axis of the body, which is a principal axis of inertia.*

Compute the time history of the motion of the system through modal approximation. Consider only the first mode of the beam in xy-plane and the first one in xz-plane.

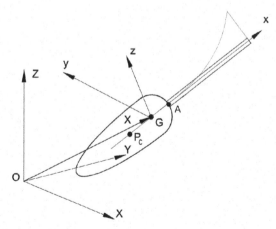

FIGURE 17.7. Beam clamped to a rigid body moving in the three-dimensional space.

Data $m_{rb} = 20$ *kg*, $J_{rb_x} = 4$ *kg* m^2, $J_{rb_y} = 4$ *kg* m^2, $J_{rb_z} = 6$ *kg* m^2, $x_A = 200$ *mm, length of the beam* $l = 500$ *mm, annular cross-section with inner radius* $r_i = 8$ *mm and outer radius* $r_0 = 9$ *mm*, $\rho = 7,810$ *kg* m^3, *and* $E = 2.1 \times 10^{11}$ N/m^2.

Initial condition: $X_0 = Y_0 = Z_0 = 0$, $\phi_0 = \theta_0 = \psi_0 = 0$, $\eta_{1_0} = \eta_{2_0} = 0$, $\dot{X}_0 = 10$ *m/s*, $\dot{Y}_0 = 15$ *m/s*, $\dot{Z}_0 = 20$ *m/s*, $\dot{\phi}_0 = \dot{\theta}_0 = \dot{\psi}_0 = 10$ *rad/s*, $\dot{\eta}_{1_0} = \dot{\eta}_{2_0} = 0$.

Part II

Dynamics of Nonlinear and time Variant Systems

18

Free Motion of Conservative Nonlinear Systems

The linearized approach seen in the previous chapters is just an approximation. If this simplifying assumption is removed and nonlinear modeling is pursued both the methods used in the study and the results obtained may change deeply, not only quantitatively but also qualitatively. While in some cases only small refinements are obtained, in others new phenomena enter the picture.

18.1   Linear versus nonlinear systems

Only linear systems were studied in the previous chapters. The real world is, however, generally nonlinear, and the use of linear models is always a source of approximation. In structural dynamics, the linearized approach is often justified by a number of considerations:

1. The study of linear systems can be performed using a set of second-order linear differential equations that yield a general solution by superimposing the solution of the homogeneous equations (defining the free behavior of the system) to a particular solution of the complete equations (defining the forced behavior). Above all, it is possible to resort to modal analysis and other techniques directly linked with the linearity of the model.

2. The behavior of structures is well approximated by linear models if

- Stresses and strains are low enough not to exceed the proportionality limit of the material, and displacements are small enough not to introduce geometrical nonlinearities.
- No inherently nonlinear element is present or, at least, influences their behavior in a significant way. Among the more common nonlinear devices, rolling element bearings (particularly ball bearings), elastomeric springs and dampers, and nonsymmetric shock absorbers can be mentioned.
- No inherently nonlinear mechanism considerably influences the behavior of the system. The main sources of nonlinearity are. play, Coulomb friction, and dependence of structural damping of most materials on the amplitude of the stress cycle.

3. The linearization of the equations of motion allows one to obtain interesting information on the behavior of nonlinear systems. The behavior of the system about an equilibrium position or, as is usually said, its behavior *in the small*, allows assessment of the stability of the equilibrium. Even if no linearization is, strictly speaking, possible, a linearized analysis may yield important results.

The behavior of a nonlinear system may be qualitatively different from that of the corresponding linearized system, and thus the behavior in the large can differ significantly from the behavior in the small.

At least a qualitative knowledge of the behavior of nonlinear systems is needed by the structural analyst, even if he plans to use, in most cases, linearized analysis: The differences between the actual behavior of a structure or machine and the predicted one are often explained by the departure from linearity of some of its components.

One of the phenomena that can occur in nonlinear systems is chaotic vibration: its occurrence is usually linked with situations in which small changes in the initial conditions lead to large variations of the behavior of the system. The study of chaotic vibration is usually based on the numerical integration in time of the equations of motion or on the use of experimental techniques. This field is still mainly studied by theoretical mechanicists, whose work have found, until now, limited practical application in structural analysis; only a short account on chaotic vibration will be given here.

18.2   Equation of motion

18.2.1   Configuration space

A simple model of a conservative nonlinear system with a single degree of freedom is shown in Fig. 18.1a. It is similar to the linear system shown

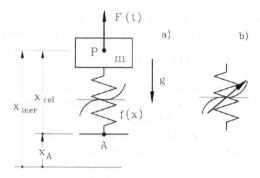

FIGURE 18.1. Conservative nonlinear system with a single degree of freedom.

in Fig. 1.1a, except that the force exerted by the nonlinear spring is not proportional to the displacement x but can be expressed by the general function $f(x)$, or, better, $f(x - x_A)$. The symbol used in Fig. 18.1a, although common, is not as widely used as the corresponding symbol for linear springs. The symbol shown in Fig. 18.1b is sometimes used when the force is also an explicit function of time.

The equation of motion of a nonlinear, conservative, time-invariant system is

$$m\ddot{x} + f(x - x_A) = F(t) - mg. \qquad (18.1)$$

Remark 18.1 *Since the system, although nonlinear, is time invariant, the equation describing its free behavior is autonomous.*

The absolute acceleration to which the inertia force is proportional is $\ddot{x}_{iner} = \ddot{x}_{rel} + \ddot{x}_A$, and Eq. (18.1), written in terms of relative displacements becomes

$$m\ddot{x}_{rel} + f(x_{rel}) = F(t) - mg - m\ddot{x}_A. \qquad (18.2)$$

In this case, the system is not insensitive to a time-independent translation of the reference frame and, consequently, the constant terms of Eq. (18.2) cannot be neglected or studied separately.

If function $f(x)$ can be subdivided into a linear part, kx, and a nonlinear part, $f'(x)$, and omitting the subscript *rel*, the equation of motion (18.2) can be written as

$$m\ddot{x} + kx + f'(x) = F(t) - mg - m\ddot{x}_A. \qquad (18.3)$$

The aforementioned equations can be extended to multi-degree-of-freedom systems. By separating the nonlinear part of the equations from the linear part, the equations of motion of a general conservative, time-invariant, nonlinear system can be written in the following matrix form:

$$\mathbf{M}\ddot{\mathbf{x}} + \mathbf{K}\mathbf{x} + \mu\mathbf{g} = \mathbf{f}(t) \,. \tag{18.4}$$

Matrices $\mathbf{M}$ and $\mathbf{K}$, all of order $n \times n$, define the behavior of the linear system that can be associated to the original system and are defined accordingly with what seen in Chapter 1.

Vector $\mathbf{g}$ contains the nonlinear part of the system. It consists of n functions g_i, which, in the most general case, are

$$g_i = g_i(\mathbf{x}, \mu, t) \,,$$

although it will be considered here as independent from time. Parameter μ is a number controlling the relative size of the nonlinear and linear parts of the system. With increasing μ, the behavior of the system gets less linear and, generally speaking, the solution of the equations of motion becomes increasingly difficult.

Another way of writing the equation of motion of a time-invariant conservative nonlinear system is

$$\mathbf{M}(\mathbf{x})\ddot{\mathbf{x}} + \mathbf{K}(\mathbf{x})\mathbf{x} = \mathbf{f}(t) \,, \tag{18.5}$$

where the matrices of the system are assumed to be functions of the generalized coordinates.

18.2.2 Modal coordinates

Also in the case of nonlinear systems it is possible to resort to the modal transformation using the modal coordinates of the linearized system. Any possible point in the configuration space can be thus defined and, hence, the description of any possible deformed shape of the nonlinear system can be described using modal coordinates. Equation (18.4) can thus be written in the form

$$\overline{\mathbf{M}}\ddot{\boldsymbol{\eta}} + \overline{\mathbf{K}}\boldsymbol{\eta} + \mu\boldsymbol{\Phi}^T\mathbf{g} = \overline{\mathbf{f}}(t) \,. \tag{18.6}$$

However the equations of motion (18.6) are not uncoupled owing to the nonlinearity of the system. All functions g_i, which are now functions of the modal coordinates η_i, can be present in each equation . If the system is only weakly nonlinear, coupling can be neglected and the behavior of the system can be approximated using a number of nonlinear equations describing systems with a single degree of freedom.

Remark 18.2 *Modal uncoupling is only an approximation in the case of a nonlinear system, even if no damping is present. Moreover, there is no way to estimate the approximation so introduced. In many cases modal coupling makes the modal formulation (18.6) of the equation of motion practically useless.*

18.2.3   State space

The motion can be studied with reference to the state space also in the case of nonlinear systems. Operating in the same way as seen in Section 1.6 and using the derivatives of the coordinates

$$\mathbf{v} = \dot{\mathbf{x}}$$

as auxiliary coordinates, the usual definition of the state vector

$$\mathbf{z} = \left\{ \begin{array}{c} \mathbf{v} \\ \mathbf{x} \end{array} \right\}$$

is obtained.

The equation of motion (18.2) of a system with a single degree of freedom can be written in the form

$$\dot{\mathbf{z}} = \left\{ \begin{array}{c} -\dfrac{1}{m} \left[f(x) - F(t) \right] \\ v \end{array} \right\}. \tag{18.7}$$

If Eq. (18.4) can be used and the linear part of the system can be separated from the nonlinear one, the state space equation of a nonlinear time-invariant system can be written in the form

$$\dot{\mathbf{z}} = \mathcal{A}\mathbf{z} + \left\{ \begin{array}{c} -\mu \mathbf{M}^{-1}\mathbf{g} \\ \mathbf{0} \end{array} \right\} + \mathcal{B}\mathbf{u}(t). \tag{18.8}$$

where the dynamic matrix, the input gain matrix, and the input vector are defined in Section 1.6.

Also in this case, nonlinearity can be introduced by writing matrices that depend on the state vector

$$\dot{\mathbf{z}} = \mathcal{A}(\mathbf{z})\,\mathbf{z} + \mathcal{B}(\mathbf{z})\,\mathbf{u}(t). \tag{18.9}$$

The two latter equations are not exactly equivalent, and in particular the last one can be used also for non-conservative systems.

18.2.4   Motion about an equilibrium position (in the small)

If the input acting on the system is constant in time, a static equilibrium position $\mathbf{x}_0$ can be easily obtained by assuming that both velocity and acceleration are equal to 0. In the case of a system with a single degree of freedom (Eq. 18.2), it follows that

$$f(x_0) = F_0 - mg, \tag{18.10}$$

where F_0 is a constant value of the force $F(t)$ and x_0 is the equilibrium position.

Remark 18.3 *The nonlinear equation (18.10) may have a number of different solutions, i.e., the system may have several equilibrium positions.*

If function $f(x)$ can be differentiated with respect to x in the point of the state plane with coordinates $x = x_0$ and $\dot{x} = 0$, the motion in the vicinity of an equilibrium position or motion in the small can be studied through the following linearized equation:

$$m\ddot{x} + \left(\frac{\partial f}{\partial x}\right)_{\substack{x = x_0 \\ \dot{x} = 0}} (x - x_0) = F(t) - mg - m\ddot{x}_A . \tag{18.11}$$

The linearized equation (18.11) allows the study of the stability of the static equilibrium position. If the derivative

$$(\partial f / \partial x)_{x=x_0, \ \dot{x}=0}$$

is positive, when point P is displaced from the equilibrium position, a force opposing this displacement is produced: The equilibrium position is statically stable. If it is negative, the equilibrium position is unstable; if it vanishes, the knowledge of the higher-order derivatives is needed to analyze the behavior of the system.

In the case of systems with many degrees of freedom, the state equation (18.8) can be linearized as

$$\dot{z} = \begin{bmatrix} 0 & -\mathbf{M}^{-1}\left(\mathbf{K} + \mu\left[\dfrac{\partial g_i}{\partial x_j}\right]\right) \\ \mathbf{I} & 0 \end{bmatrix} \mathbf{z} + \mathcal{B}\mathbf{u}(t) , \tag{18.12}$$

where $(\partial g_i / \partial x_j)$ is the Jacobian matrix containing the derivatives of functions $\mathbf{g}$ with respect to $\mathbf{x}$.

18.3   Free oscillations

Assume that x_0 is an equilibrium position and, without loss of generality, that $f(x_0) = 0$. The free oscillations about such a position can be studied using the equation

$$m\ddot{x} + f(x) = 0 . \tag{18.13}$$

Equation (18.13) is nonlinear but autonomous. If function $f(x)$ is substituted by its Taylor series about point x_0, since $f(x_0) = 0$, the equation of motion becomes

$$m\ddot{x} + (x - x_0)\left(\frac{\partial f}{\partial x}\right)_{x=x_0} + \frac{(x - x_0)^2}{2!}\left(\frac{\partial^2 f}{\partial x^2}\right)_{x=x_0} + \cdots = 0 . \tag{18.14}$$

There is no general closed-form solution to Eq. (18.13), so the study of the dynamic behavior of nonlinear systems is often performed by resorting to numerical experiments.

Remark 18.4 *Numerical simulation is a very powerful tool without which many phenomena could not even be discovered. However, it has the drawback of supplying general information on the behavior of the system only at the cost of performing a large quantity of experiments. In this, numerical simulation is very similar to physical experimentation.*

In many cases it is possible to resort to techniques yielding results that, although only being approximated, are valid in general, at least from a qualitative viewpoint.

To get a physical insight to the relevant phenomena, greater stress will be placed on analytical techniques, even when the equations they yield need, at any rate, to be solved numerically. The reader must, however, be aware that only the simultaneous use of analytical and numerical techniques yields useful information on the behavior of most nonlinear systems.

Consider a system whose behavior is modeled by Eq. (18.13). Assume that $f(x)$ is an odd function of $x - x_0$, and set the origin of the x-coordinate in the static equilibrium position ($x_0 = 0$). If the derivative $\partial f / \partial x$ in the origin is positive, force $-f(x)$ is a symmetrical restoring force keeping mass m in the static equilibrium position. The motion of the system consists of undamped oscillations centered about the position of static equilibrium. Since function $f(x)$ is odd, only the derivatives whose order is odd are present in Eq. (18.14).

Truncating the series in Eq. (18.14) after the second term, the characteristics of the nonlinear spring can be expressed by the law

$$f(x) = kx(1 + \mu x^2). \tag{18.15}$$

This type of restoring force is often referred to as a *Duffing-type* restoring force, because its general properties were first studied in detail by G. Duffing in 1918.[1]

Remark 18.5 *Equation (18.15) can be considered as the expression of Eq. (18.14) for a system with an odd restoring force (which makes even powers to disappear), truncated at the second term. It can thus be regarded as a second approximation, where the first approximation is linearization.*

Parameter μ has the dimensions of a length at the power -2. Constant k is usually positive, a condition necessary to lead to a stable static equilibrium position at $x = 0$. If μ is positive, the spring is said to be of the *hardening* type, because its stiffness increases with the displacement. On the contrary, if μ is negative, the spring is said to be of the *softening* type. Equation (18.15) is plotted in nondimensional form in Fig. 18.2.

---

[1]G. Duffing, *Erzwungene Schwingungen bei veränderlicher Eigenfrequenz*, F. Vieweg u. Sohn, Braunschweig, Germany, 1918.

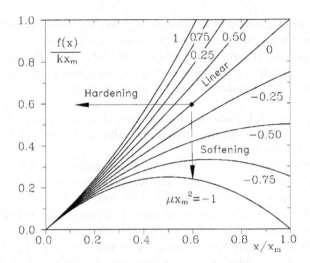

FIGURE 18.2. Law $f(x)$ expressed as a third-degree polynomial with two terms; nondimensional plot for both hardening and softening systems.

In the case of a 'Duffing-type' hardening system there is a single equilibrium position and it is stable. A softening system has three equilibrium positions, but two of them are unstable.

If both k and μ are negative, the equilibrium position in the origin is unstable, but two other stable equilibrium position exist. Systems of this type have been widely used in studies on the chaotic behavior of mechanical systems. If on the contrary k is negative and μ is positive no stable equilibrium positions exist.

The potential energy yielding a Duffing-type restoring force is

$$\mathcal{U} = \frac{1}{2}kx^2\left(1 + \mu\frac{x^2}{2}\right). \tag{18.16}$$

18.4   Direct integration of the equations of motion

Consider a system with $f(x)$ an odd function of x, positive when x is positive. The acceleration can be written as

$$\ddot{x} = \frac{1}{2}\frac{d(\dot{x}^2)}{dx}. \tag{18.17}$$

Equation (18.13) can be rearranged in the form

$$\frac{d(\dot{x}^2)}{dx} = -\frac{2}{m}f(x). \tag{18.18}$$

The physical meaning of Eq. (18.18) is clear: The variation of the kinetic energy of the system is, at any instant, equal to the work performed by the force $-f(x)$ in the displacement dx. By separating the variables and integrating, the following expression is obtained:

$$\dot{x}^2 = -\frac{2}{m} \int f(x)dx + C. \tag{18.19}$$

Since the system is symmetrical with respect to point $x = 0$, the undamped free oscillations are symmetrical about this point. During oscillation, the velocity vanishes at the extreme position x_m: $\dot{x} = 0$ for $x = x_m$. The constant of integration in Eq. (18.19) and the limits of integration can be computed. Equation (18.19) can then be integrated again, obtaining

$$t = \sqrt{\frac{m}{2}} \int_0^x \frac{du}{\sqrt{\int_u^{x_m} f(u')du'}}, \tag{18.20}$$

where the dummy variables u and u' have been introduced. Equation (18.20), obtained assuming that at time $t = 0$ mass m (i.e., point P) passes through the equilibrium position, is the law of motion of the system. Instead of being expressed as $x = x(t)$, it is in the form $t = t(x)$ and obviously applies only for one-quarter of the period of oscillation, with $0 \leq x \leq x_m$. Since all quarters of the period are identical, it describes completely the motion.

The period of oscillation is four times the time needed to reach the maximum amplitude x_m and can be easily computed

$$T = 4\sqrt{\frac{m}{2}} \int_0^{x_m} \frac{dx}{\sqrt{\int_x^{x_m} f(u)du}}. \tag{18.21}$$

Remark 18.6 *The period is a function of the amplitude of oscillation: This is a general feature of nonlinear systems.*

If the velocity at any position is known, Eq. (18.19) can be used to compute the value of the amplitude of the motion. As an example, if the initial velocity $(\dot{x})_0$ is known, the value of x_m, i.e., the amplitude of the oscillation, can be obtained from the equation

$$(\dot{x})_0^2 = \frac{2}{m} \int_0^{x_m} f(x)dx. \tag{18.22}$$

Remark 18.7 *Equations (18.20) to (18.22) allow the free motion of an undamped nonlinear autonomous system to be studied in detail. Unfortunately, the integral of Eq. (18.20) can be solved in closed form only in a few particular cases and, generally speaking, numerical integration is needed.*

This approach is, however, much simpler than numerical integration of the equation of motion, because only a simple integral has to be performed numerically.

Consider again a system with a Duffing-type restoring force. The relationship between the maximum elongation in free vibration and the speed at the equilibrium position is obtained by equating the potential energy (Eq. 18.16) to the kinetic energy at the equilibrium position

$$\frac{1}{2}m(\dot{x})_0^2 = \frac{1}{2}kx_m^2\left(1 + \mu\frac{x_m^2}{2}\right). \tag{18.23}$$

This equation can be solved in the velocity, yielding

$$\dot{x}_0 = x_m\sqrt{\frac{k}{m}\left(1 + \mu\frac{x_m^2}{2}\right)}, \tag{18.24}$$

or in the amplitude:

$$x_m = \sqrt{\frac{1 - \sqrt{1 + 2\frac{m}{k}\mu(\dot{x})_0^2}}{|\mu|}}. \tag{18.25}$$

From Eq. (18.25), it is clear that oscillatory motion of softening systems is possible only if the amplitude is not greater than $\sqrt{1/|\mu|}$. If this value is reached, the force is no more restoring and causes point P to move indefinitely away.

Remark 18.8 *All the equations reported so far must be used with care in the context of softening systems, always verifying that the displacements are small enough, or, which is the same, that the total energy of the system is not higher than the maximum potential energy it can store.*

The waveform of the free oscillations, obtained by numerically integrating Eq. (18.20), is reported in nondimensional form in Fig. 18.3 for different values of the nondimensional parameter μx_m^2. From Fig. 18.3, it is clear that all curves are quite close to the sine wave characterizing the behavior of the

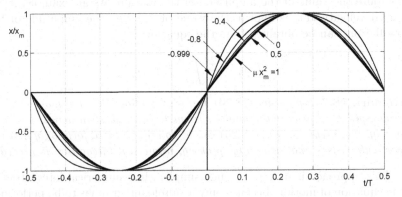

FIGURE 18.3. Time history of the free oscillations of a system with nonlinear restoring force of the type shown in Fig. 18.2.

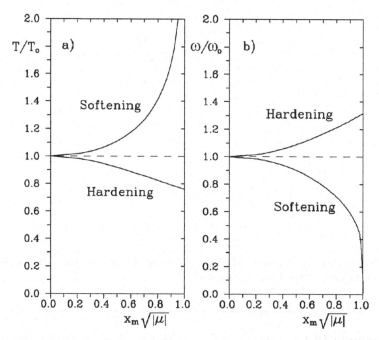

FIGURE 18.4. Frequency of the fundamental harmonic as a function of the amplitude of the free oscillations of a system with a law $f(x)$ of the type shown in Fig. 18.3. The period and frequency are made nondimensional by dividing by the values characterizing the linear system.

linear system ($\mu = 0$). Only in the case of softening systems oscillating with an amplitude close to the maximum possible amplitude is the waveform far from that of a harmonic oscillation. The period, obtained by numerical integration of Eq. (18.21), is plotted as a function of $x_m\sqrt{|\mu|}$ in Fig. 18.4a. The frequency of the fundamental harmonic is shown in Fig. 18.4b. The frequency increases with the amplitude (i.e., the period of oscillation gets shorter) in the case of hardening systems, while decreases in the case of softening ones. As already stated, the dependence of the period on the amplitude of oscillation is a general rule for nonlinear systems.

The free response $x(t)$ is not harmonic, because of the nonlinearity of function $f(x)$. It is, however, periodic with period T and can easily be transformed into a series of harmonic terms. Because function $f(x)$ was assumed to be odd, only terms whose circular frequency is an odd multiple of the fundamental frequency $2\pi/T$ are present. If at time $t = 0$ the system passes through the origin, the law $x(t)$ can be expressed as

$$x = x_m\left[a_1\sin\left(\frac{2\pi}{T}t\right) + a_3\sin\left(\frac{6\pi}{T}t\right) + \cdots + a_i\sin\left(\frac{2i\pi}{T}t\right) + \ldots\right].$$

$$(18.26)$$

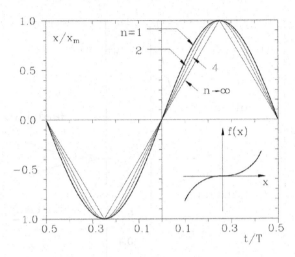

FIGURE 18.5. Time history of the free oscillations of a system with nonlinear restoring force of the type $f'(x) = kx^{2n-1}$.

Example 18.1 *Consider a hardening restoring force expressed by the odd function*

$$f(x) = kx^{(2n-1)}$$

with n an integer positive number. If $n > 1$ no linearized solution can be found, since $(\partial f/\partial x)_{x=0} = 0$.
The time history of the free oscillations and the maximum elongation are expressed by the equations

$$t = \sqrt{n}\sqrt{\frac{m}{2}}\int_0^x \frac{du}{\sqrt{x_m^{2n} - u^{2n}}}, \qquad x_m = \left[(\dot{x})_0\sqrt{n}\sqrt{\frac{m}{2}}\right]^{1/n}.$$

If $n = 1$ the system behaves linearly and the time history $x(t)$ is harmonic.
If $n \to \infty$ the system can be assimilated to two rigid walls in the points of coordinates $x = 1$ and $x = -1$, with point P bouncing between the walls, moving at a constant speed. If $n \neq 1$, no closed-form integration of the equation of motion is possible. When $n = 2$, i.e., the restoring force follows a cubic law in the displacement, the value of the period is

$$T = \frac{4}{x_m}\sqrt{\frac{2m}{k}}\int_0^1 \frac{d\chi}{\sqrt{1 - \chi^4}},$$

where $\chi = x/x_m$.
By using the approximate value of the elliptical integral, the period can be expressed as

$$T = \frac{7.4164}{x_m}\sqrt{\frac{m}{k}}.$$

As usual for nonlinear systems, the period is a function of the amplitude, in this case being proportional to the reciprocal of the latter. The law $x(t)$ for different values of n is plotted in nondimensional form for different values of n in Fig. 18.5.

The values of the first five coefficients of the series for the response were computed numerically and listed in the following table

	a_1	a_3	a_5	a_7	a_9
$n = 1$ (linear)	1	0	0	0	0
$n = 2$	0.9522	−0.0440	0.0025	−0.0004	0.0002
$n = 3$	0.9238	−0.0641	0.0094	−0.0015	0.0004
$n = 4$	0.9045	−0.0745	0.0152	−0.0036	0.0010
$n = 5$	0.8907	−0.0804	0.0195	−0.0058	0.0019
$n \to \infty$	0.8106	−0.0901	0.0324	−0.0165	0.0100

The coefficients were computed by numerically integrating equation (18.20) (Gauss method) in 400 steps in each quarter of period, and then using a discrete Fourier algorithm (1,600 points in the period).

The numerical errors can be significant in the higher-order harmonics, except for the case with $n \to \infty$, in which a closed-form solution exists: $a_i = 8/(i\pi)^2$

The table shows that the response of the system is not far from being harmonic, i.e., the contribution of all harmonics except the first is quite small, even if the nonlinearity is strong. This is due to the fact that function $f(x)$ is odd. If no symmetry was assumed, higher-order harmonics would have been more important, and all harmonics, including a constant term (zero-order harmonic), could have been present.

18.5   Harmonic balance

A different approach to the study of both free and forced oscillations of nonlinear systems is to approximately evaluate a few (or even just one) of the terms of the series (18.26). In the case of free oscillations of the undamped system, this approach yields an approximate relationship between the amplitude and the value of the fundamental frequency of the motion.

The simplest technique, based on the introduction of a series solution of the type of Eq. (18.26) into the equation of motion, is the so-called *harmonic balance*. Because all harmonics of the time history of the motion must be balanced, a set of nonlinear algebraic equations in the amplitudes of the various harmonics is obtained. The procedure is approximate for two reasons: The series (18.26) is truncated, usually after a few terms (or even after the first one), and some of the harmonic terms cannot be balanced, because the number of equations obtained is greater than the number of unknowns.

Remark 18.9 *Although it may seem too an empirical approach, if properly used with sound engineering common sense, this is a powerful tool for solving many nonlinear problems.*

Example 18.2 *Find an approximate solution for the fundamental harmonic of the free response of an undamped system whose nonlinear spring has a characteristic expressed by Eq. (18.15)*

$$f(x) = kx \left(1 + \mu x^2\right).$$

If only the fundamental harmonic is retained, the solution can be expressed in the form

$$x = x_m \sin(\omega t).$$

By introducing this time history into the equation of motion, it follows

$$(-m\omega^2 + k)x_m \sin(\omega t) + k\mu x_m^3 \sin^3(\omega t) = 0.$$

By remembering some trigonometric identities and introducing the natural frequency of the linearized system $\omega_0 = \sqrt{k/m}$, the latter equation yields

$$\left[1 - \left(\frac{\omega}{\omega_0}\right)^2\right] x_m \sin(\omega t) + \frac{3}{4}\mu x_m^3 \sin(\omega t) - \frac{1}{4}\mu x_m^3 \sin^3(3\omega t) = 0.$$

To satisfy this equation for all values of time, the coefficients of both $\sin(\omega t)$ and $\sin(3\omega t)$ must vanish. The first condition leads to the equation

$$\left[1 - \left(\frac{\omega}{\omega_0}\right)^2\right] + \frac{3}{4}\mu x_m^2 = 0,$$

which is the required relationship linking the frequency of the fundamental harmonic with the amplitude. Balancing the third harmonic would lead to another equation in the unknown x_m that cannot be satisfied, owing to the intrinsic approximations of the method.
The relationship between the amplitude and the frequency

$$\frac{\omega}{\omega_0} = \sqrt{1 + \frac{3}{4}\mu x_m^2}$$

is shown in nondimensional form in Fig. 18.6a, together with the solution already reported in Fig. 18.4b. The precision obtained through the harmonic balance technique is very good in the case of hardening systems and, for softening systems, when the amplitude of the oscillations is small or the system is mildly nonlinear, i.e., for small values of parameter $|\mu|x_m^2$. With increasing values of $|\mu|x_m^2$, the solution loses precision. In particular, the value of the amplitude at which the natural frequency vanishes, i.e., the period tends to infinity or the system no longer has an oscillatory behavior, is $\sqrt{4/3|\mu|}$, while the correct value is $\sqrt{1/|\mu|}$.

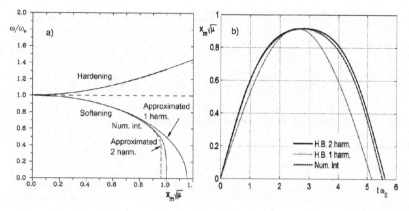

FIGURE 18.6. System with Duffing-type law $f(x)$ (Fig. 18.2). (a) Approximate values of the frequency of free oscillations as a function of the amplitude obtained using harmonic balance, compared with the results obtained through numerical integration. For hardening systems, the curves are completely superimposed. (b) Time history of the free oscillations computed using the harmonic balance technique (1 and 2 harmonics) and numerical integration $(x_m \sqrt{|\mu|} = 0.8788)$.

A better approximation can be obtained by assuming a solution in which two harmonics are included:

$$x = a_1 \sin(\omega t) + a_2 \sin(3\omega t) \,.$$

There are now two unknowns and two harmonics can be balanced.

By introducing the assumed time history of the solution into the equation of motion, it follows

$$(-m\omega^2 + k)a_1 \sin(\omega t) + (-9m\omega^2 + k)a_2 \sin(3\omega t)+$$

$$+k\mu \left[a_1 \sin(\omega t) + a_2 \sin(3\omega t)\right]^3 = 0 \,.$$

By remembering some trigonometric identities and introducing the natural frequency of the linearized system $\omega_0 = \sqrt{k/m}$, the latter equation yields

$$a_1 \left\{ \left[1 - \left(\frac{\omega}{\omega_0}\right)^2\right] + \frac{3}{4}\mu \left(a_1^2 + 2a_2^2 - a_1 a_2\right) \right\} \sin(\omega t)+$$

$$+ \left\{ a_2 \left[1 - 9\left(\frac{\omega}{\omega_0}\right)^2\right] + \frac{1}{4}\mu \left(-a_1^3 + 3a_2^3 + 6a_1^2 a_2\right) \right\} \sin(3\omega t)+$$

$$+ \frac{3}{4}\mu a_1 a_2 \left(-a_1 + a_2\right) \sin(5\omega t) - \frac{3}{4}\mu a_1 a_2^2 \sin(7\omega t) - \frac{1}{4}\mu a_2^3 \sin(9\omega t) = 0$$

Only the first two harmonics can be balanced, yielding the two equations

$$\begin{cases} 1 - \left(\dfrac{\omega}{\omega_0}\right)^2 + \dfrac{3}{4}\mu\left(a_1^2 + 2a_2^2 - a_1 a_2\right) = 0 \\ a_2\left[1 - 9\left(\dfrac{\omega}{\omega_0}\right)^2\right] + \dfrac{1}{4}\mu\left(-a_1^3 + 3a_2^3 + 6a_1^2 a_2\right) = 0 \end{cases}$$

The first equation can be solved in the frequency

$$\left(\dfrac{\omega}{\omega_0}\right)^2 = 1 + \dfrac{3}{4}\mu\left(a_1^2 + 2a_2^2 - a_1 a_2\right) \ .$$

Introducing it into the second equation and introducing the nondimensional amplitudes

$$a_i^* = a_i \sqrt{|\mu|} \ ,$$

it follows

$$51 a_2^{*3} - 27 a_1^* a_2^{*2} + \left(21 a_1^{*2} \pm 32\right) a_2^* + a_1^{*3} = 0 \ ,$$

where the upper sign holds for hardening and the lower one for softening systems.
It is then possible to state various values of a_1^* and to compute the corresponding values of a_2^* and of the frequency.
The amplitude of the poly-harmonic motion is

$$x_m = a_1 - a_2 \ .$$

The frequency computed in this way is plotted in Fig. 18.6a, dashed line. In the case of the hardening system there is little difference, while for the softening system the result is closer to that obtained by numerically integrating the equation of motion.
Consider the case of a softening system and assume that $a_1\sqrt{|\mu|} = 0.99$. The nondimensional frequency is $\omega/\omega_0 = 0.5583$, and the amplitude of the third harmonics is $a_2\sqrt{|\mu|} = 0.0740$. The amplitude of the oscillation is thus $x_m\sqrt{|\mu|} = 0.9160$, and the corresponding nondimensional frequency computed using a mono-harmonic solution is $\omega/\omega_0 = 0.6088$.
The time history during half a period is shown in nondimensional form in Fig. 18.6b, where the solutions obtained through harmonic balance are compared with that obtained by numerically integrating the equation of motion. From the plot it is clear that adding the third harmonics improves dramatically the results.

18.6   Ritz averaging technique

Other solution methods are the so-called averaging techniques. Into this category fall, among others, the method of Kirilov and Bogoljubov and that usually referred to as the Ritz averaging method, which will be described in detail. Practically, they are based on the substitution of some nonlinear functions with their average over one period, thus obtaining simplified expressions yielding approximate results. Usually, averaging is based on energy consideration.

The Ritz averaging technique is based on the observation that in free undamped vibrations the average value of the virtual work done in a complete cycle must vanish. The solution $x(t)$ is approximated with a linear combination of arbitrary time functions[2]

$$x(t) = \sum_{i=1}^{n} a_i \phi_i(t) , \qquad (18.27)$$

where a_i are constants and ϕ_i are arbitrary functions of time. If $f(x)$ is an odd function, functions ϕ_i can be the harmonic terms of the series (18.26).

When looking for a first approximation of the fundamental harmonic of the time history of the motion, only one function ϕ_i is used. As in the case of the harmonic balance technique, constants a_i can be computed by introducing the solution (18.27) into the equation of motion. The latter clearly cannot be satisfied exactly, because only an approximation of the true time history has been assumed. Instead of requiring that the equation of motion be satisfied at each instant, which is impossible, the equation of motion is satisfied as an average over one period of the motion.

The equation of motion (18.13) states that elastic and inertial forces must always be in equilibrium. If a virtual displacement δx is imposed on the system starting from an equilibrium condition characterized by given values of the position and velocity, the virtual work done by the applied forces must vanish, i.e.,

$$[m\ddot{x} + f(x)] \, \delta x = 0 . \qquad (18.28)$$

Using an averaged approach, this condition cannot be satisfied exactly. Although not vanishing in any instant, the virtual work must be equal to zero as an average over a period: The integral of Eq. (18.28) over one period is thus equal to zero. If a set of n virtual displacements are stated using the same functions ϕ_i assumed in the definition of the time history, it follows

---

[2]Equation (18.27) must not be confused with expression (14.1) used in connection with the assumed-modes method. In the latter, the unknowns $a_i(t)$ are functions of time, and the assumed functions are functions of the space coordinates. The problem is that of obtaining the deformed shape of the system, continuous or discrete, which is approximated by arbitrary functions, while the time history is not difficult to compute (in the case of free oscillations, it is harmonic). In this case, however, the system is nonlinear; then the difficulty lies in the determination of the time history, which is approximated by the arbitrary time functions. In the case of systems with a single degree of freedom, the unknowns are constants. They are functions of space coordinates if the method is applied to continuous systems or vectors for multi-degree-of-freedom discrete systems.

that

$$\delta x = \sum_{i=1}^{n} \delta a_i \phi_i(t). \tag{18.29}$$

The averaged expression of the virtual work is thus

$$\delta L = \sum_{i=1}^{n} \int_0^T \left[\ddot{x} + \frac{f(x)}{m} \right] \delta a_i \phi_i(t) dt = 0. \tag{18.30}$$

Because the virtual displacements δa_i are arbitrary, each term of the sum must vanish in order to satisfy Eq. (18.30)

$$\int_0^T \left[\ddot{x} + \frac{f(x)}{m} \right] \phi_i(t) dt = 0, \qquad \text{for } i = 1, 2, \ldots, n. \tag{18.31}$$

Remark 18.10 *The n equations (18.31) are a set of nonlinear equations in the n unknowns a_i. Their solution usually requires the use of numerical techniques and may constitute a difficult mathematical problem.*

Example 18.3 *Repeat the computation of Example 18.2 by resorting to the Ritz averaging technique. The same nonharmonic solution used in Example 18.2 is assumed. There is only one function of time*

$$\phi(t) = \sin(\omega t)$$

and, correspondingly, only one unknown coefficient, $a = x_m$. Equation (18.31), reduces to

$$\int_0^T \left[x_m \left(\frac{k}{m} - \omega^2 \right) \sin(\omega t) + \frac{k}{m} \mu x_m^3 \sin^3(\omega t) \right] \sin(\omega t) dt = 0.$$

Remembering that

$$\int_0^T \sin^2(\omega t) dt = \frac{\pi}{\omega}, \qquad \int_0^T \sin^4(\omega t) dt = \frac{3\pi}{4\omega},$$

and introducing the natural frequency of the linearized system $\omega_0 = \sqrt{k/m}$, the equation linking the fundamental frequency of the free vibration with the amplitude is

$$\left[1 - \left(\frac{\omega}{\omega_0} \right)^2 \right] + \frac{3}{4} \mu x_m^3 = 0,$$

This result is coincident with that obtained through the harmonic balance technique. Also in this case, a second-approximation solution could be obtained by using two arbitrary functions instead of one.

18.7   Iterative techniques

Iterative methods are restricted to weakly nonlinear and non-autonomous systems (or quasi-linear systems), i.e., systems whose equation of motion can be separated into a part containing only linear terms and a second one, relatively small with respect to the first, containing the nonlinear terms. What weakly actually means is difficult to assess, and the applicability of this approach depends on many factors, including the method used and the nature of the problem. If the equation of motion of a system with a single degree of freedom is written in the form of Eq. (18.4), the dynamic equilibrium between inertia and nonlinear restoring forces is

$$\ddot{x} = -\omega_0^2 x - \mu \frac{f'(x)}{m}\,, \tag{18.32}$$

where ω_0 is the natural frequency of the linearized system $\omega_0 = \sqrt{k/m}$. By adding a term $\omega^2 x$ at both sides, it follows

$$\ddot{x} + \omega^2 x = \left(\omega^2 - \omega_0^2\right) x - \mu \frac{f'(x)}{m}\,. \tag{18.33}$$

A zero-order approximation solution, with the same harmonic behavior in time than the solution of the linearized system, but with a different frequency, can be stated

$$x^{(0)} = x_m \sin(\omega t)\,. \tag{18.34}$$

The zero-order solution (Eq. (18.34)) is then introduced at right-hand side of the equation of motion (18.33) in order to obtain a new first order solution

$$\ddot{x}^{(1)} + \omega^2 x^{(1)} = \left(\omega^2 - \omega_0^2\right) x^{(0)} - \mu \frac{f'\left(x^{(0)}\right)}{m}\,. \tag{18.35}$$

If the system is mildly nonlinear (μ is small), ω cannot be much different from ω_0 and both the terms at right-hand side are small. The process is repeated until the required precision is obtained.

Example 18.4 *Compute the first-approximation solution for a system with a Duffing-type nonlinearity. With the present notation function $f'(x)$ is*

$$\frac{f'(x)}{m} = \omega_0^2 x^3\,,$$

Equation (18.35) then becomes

$$\ddot{x}^{(1)} + \omega^2 x^{(1)} = \left(\omega^2 - \omega_0^2\right) x_m \sin(\omega t) - \mu \omega_0^2\, x_m^3 \sin^3(\omega t)\,.$$

By remembering the identity

$$\sin^3 (\alpha) = \frac{3}{4} \sin (\alpha) - \frac{1}{4} \sin (3\alpha) \, ,$$

the equation

$$\ddot{x}^{(1)} + \omega^2 x^{(1)} = \left[\omega^2 - \omega_0^2 \left(1 + \frac{3}{4} \mu x_m^2 \right) \right] x_m \sin (\omega t) + \frac{1}{4} \mu \omega_0^2 \, x_m^3 \sin (3\omega t)$$

can be regarded as the equation of motion of a system whose natural frequency is ω, excited by two harmonic forces with frequencies ω and $3\,\omega$. The solution of the equation of motion of the system must have an oscillatory character, while, owing to the fact that the exciting frequency coincides with the natural frequency of the system, the amplitude builds up to infinity, growing linearly in time. The term in $\sin (\omega t)$ originates then a secular term, which must be eliminated:

$$\omega^2 - \omega_0^2 \left(1 + \frac{3}{4} \mu x_m^2 \right) = 0 \, .$$

This relationship between the frequency of the oscillation and the amplitude allows a first approximation of the frequency to be obtained:

$$\omega = \omega_0 \sqrt{1 + \frac{3}{4} \mu x_m^2} \, ,$$

which coincides with those obtained through harmonic balance and Ritz averaging technique.

The equation above can be considered the equation of motion of an undamped linear system with a harmonic excitation and its solution is the sum of the 'free oscillation' solution (harmonic with frequency ω) added to the solution regarding the forced oscillations (harmonic with frequency $3\,\omega$).

Using the equations developed in Chapter 6, the following solution is obtained

$$x^{(1)} = C_1 \sin (\omega t) + C_2 \cos (\omega t) - \frac{1}{32} \mu \, x_m^3 \sin (3\omega t) \, .$$

Constants C_1 and C_2 can be obtained from the initial conditions. If at time $t = 0$ the system passes through the origin, C_2 must vanish. As the maximum amplitude, reached at $\omega t = \pi/2$, is x_m, it follows

$$C_1 = x_m \left(1 - \frac{1}{32} \mu x_m^2 \right) \, .$$

The first-approximation solution is thus

$$x^{(1)} = x_m \left(1 - \frac{1}{32} \mu x_m^2 \right) \sin (\omega t) - \frac{1}{32} \mu \, x_m^3 \sin (3\omega t) \, .$$

The first-approximation solution obtained through the iterative technique shown is already better than that obtained from harmonic balance or Ritz averaging technique with a single harmonic, since it allows to obtain an estimate of the amplitude of the third harmonics.

18.8   Perturbation techniques

In perturbation techniques the solution is assumed to be a power series of a small parameter linked with the nonlinear part of the system, e.g., parameter μ of Eq. (18.4).

$$x = x^{(0)} + \mu x^{(1)} + \mu^2 x^{(2)} + \mu^3 x^{(3)} + \ldots \tag{18.36}$$

The equation of motion (18.32) can be written, for the sake of compactness, as

$$L(x) = N(x) \tag{18.37}$$

where

$$L(x) = \ddot{x} + \omega_0^2 x \quad, \quad N(x) = -\mu \frac{f'(x)}{m} \tag{18.38}$$

are, respectively, the linear and the nonlinear parts of the equation of motion. By introducing Eq. (18.36) into the equation of motion (18.37), it follows that

$$L\left(x^{(0)}\right) + \mu L\left(x^{(1)}\right) + \mu^2 L\left(x^{(2)}\right) + \ldots = \mu N\left(x^{(0)} + \mu x^{(1)} + \mu^2 x^{(2)} + \ldots\right) \tag{18.39}$$

The right-hand side can be expanded in a power series of μ, yielding

$$L\left(x^{(0)}\right) + \mu L\left(x^{(1)}\right) + \mu^2 L\left(x^{(2)}\right) + \ldots =$$
$$= \mu \left[N_0\left(x^{(0)}\right) + \mu N_1\left(x^{(0)}, x^{(1)}\right) + \mu^2 N_2\left(x^{(0)}, x^{(1)}, x^{(2)}\right) + \ldots \right] \tag{18.40}$$

In the case of an autonomous undamped system, the nonlinear operator $N(x)$ is proportional to $f'(x)$, whose power series expansion about point x_0 is simply

$$f'(x) = f'\left(x^{(0)}\right) + \mu x^{(1)} \frac{df'\left(x^{(0)}\right)}{dx} + \mu^2 \left(x^{(2)} \frac{df'\left(x^{(0)}\right)}{dx} + \frac{1}{2!} x^{(1)2} \frac{d^2 f'\left(x^{(0)}\right)}{dx^2} \right) + \ldots \tag{18.41}$$

By introducing Eq. (18.41) into Eq. (18.39) and equating the coefficients of the like terms in μ, the following set of equations can be obtained

$$\begin{cases} \ddot{x}^{(0)} + \omega_0^2 x^{(0)} = 0 \\[2mm] \ddot{x}^{(1)} + \omega_0^2 x^{(1)} = -\frac{1}{m} f'\left(x^{(0)}\right) \\[2mm] \ddot{x}^{(2)} + \omega_0^2 x^{(2)} = -\frac{1}{m} x^{(1)} \frac{df'\left(x^{(0)}\right)}{dx} \\[2mm] \ddot{x}^{(3)} + \omega_0^2 x^{(3)} = -\frac{1}{m} x^{(2)} \frac{df'\left(x^{(0)}\right)}{dx} - \frac{1}{2m} x^{(1)2} \frac{d^2 f'\left(x^{(0)}\right)}{dx^2} \\[2mm] \ldots\ldots\ldots\ldots\ldots \end{cases} \tag{18.42}$$

which can be solved recursively. The generating solution $x^{(0)}$ can be obtained from the first equation (18.42) and coincides with the solution of the linear equation. It is very easy to verify that if the solution

$$x^{(0)} = x_m \sin(\omega_0 t + \Phi) \qquad (18.43)$$

of the linear system is introduced as a generating solution, secular terms increasing to infinity are obtained, at least if only a limited number of terms in the series for $x(t)$ are retained. The term secular derives from the astronomical field, in which the perturbation technique was first applied. This difficulty is mainly due to the presence of the nonlinear term affecting the frequency of oscillation and not only their amplitude.

This could be taken into account by retaining enough terms of the series, but this leads to strong computational difficulties.[3] Here a solution based on Lindstedt's Method, which assumes that, if the solution of the equation of motion is oscillatory, the nonlinear term alters the period of an amount of the order of μ, will be shown. The frequency can thus be expressed by the power series in μ

$$\omega = \omega_0 + \mu\omega_1 + \mu^2\omega_2 + \mu^3\omega_3 + \dots \qquad (18.44)$$

where the frequency of the linear system is

$$\omega_0 = \sqrt{k/m} \ .$$

It is advisable to use a nondimensional time variable

$$\tau = \omega t \ , \qquad (18.45)$$

spanning from 0 to 2π in a period. All derivatives with respect to time can be expressed as functions of the derivatives with respect to τ

$$\dot{x} = \omega\frac{dx}{d\tau} = \omega x'., \quad \ddot{x} = \omega^2\frac{d^2x}{d\tau^2} = \omega^2 x''., \quad \dots, \qquad (18.46)$$

i.e.,

$$\dot{x} = \omega_0 x' + \mu\omega_1 x' + \mu^2\omega_2 x' + \dots$$

$$\ddot{x} = \omega_0^2 x'' + 2\mu\omega_0\omega_1 x'' + \mu^2\left(\omega_1^2 + 2\omega_0\omega_2\right) x'' + \dots \qquad (18.47)$$
....

---

[3]L. Meirovitch, *Methods of Analytical Dynamics*, McGraw-Hill, New York, 1970.

By introducing Eq. (18.47) into Eq. (18.40), the following set of equations can be obtained

$$
\begin{cases}
\omega_0^2 x''^{(0)} + \omega_0^2 x^{(0)} = 0 \\[2mm]
\omega_0^2 x''^{(1)} + \omega_0^2 x^{(1)} = -\frac{1}{m} f'\left(x^{(0)}\right) - 2\omega_0\omega_1 x''^{(0)} \\[2mm]
\omega_0^2 x''^{(2)} + \omega_0^2 x^{(2)} = -\frac{1}{m} x^{(1)} \frac{df'\left(x^{(0)}\right)}{dx} - \left(\omega_1^2 + 2\omega_0\omega_2\right) x''^{(0)} - 2\omega_0\omega_1 x''^{(1)} \\[2mm]
\quad \ldots\ldots\ldots\ldots
\end{cases}
$$

$$(18.48)$$

The components of the frequency ω_i can be computed by assuming that each function $x^{(i)}$ is periodical in τ with a period of 2π

$$x^{(i)}\left(\tau + 2\pi\right) = x^{(i)}\left(\tau\right) \tag{18.49}$$

Example 18.5 *Consider again a system with a Duffing-type nonlinearity and repeat the computation of the first-approximation solution.*
Introducing the relevant expression of function $f'(x)$, i.e., $f'(x) = m\omega_0^2 x^3$, Eq. (18.48) reduce to

$$
\begin{cases}
x''^{(0)} + x^{(0)} = 0 \\[2mm]
x''^{(1)} + x^{(1)} = -x^{(0)3} - 2\omega_1^* x''^{(0)} \\[2mm]
x''^{(2)} + x^{(2)} = -3x^{(1)} x^{(0)2} - \left(\omega_1^{*2} + 2\omega_2^*\right) x''^{(0)} - 2\omega_1^* x''^{(1)} \\[2mm]
\quad \ldots\ldots\ldots\ldots
\end{cases}
$$

where $\omega_i^* = \omega_i/\omega_0$.
As usual for autonomous systems, without any loss of generality, time $t = 0$ will be assumed in such a way that $x = 0$ for $t = 0$. The generating solution, with the mentioned initial condition, is simply

$$x^{(0)} = x_m \sin\left(\tau\right) \ .$$

Remembering the usual expression for $\sin^3\left(\tau\right)$, the second equation becomes

$$x''^{(1)} + x^{(1)} = -x_m \left(\frac{3}{4} x_m^2 - 2\omega_1^*\right) \sin\left(\tau\right) + \frac{1}{4} x_m^3 \sin\left(3\tau\right) \ .$$

Again the first term at right-hand side is a secular term, since the frequency of the excitation of the linear undamped system coincides with its natural frequency. In order to eliminate the secular term, the coefficient of $\sin\left(\tau\right)$ must vanish, as already seen in detail when dealing with the iterative method

$$\frac{3}{4}x_m^2 - 2\omega_1^* = 0 , \quad i.e \quad \omega_1^* = \frac{3}{8}x_m^2 .$$

Once the secular term has been eliminated, and proceeding as seen with iterative techniques, it follows that

$$x^{(1)} = C_1 \sin(\tau) + C_2 \cos(\tau) - \frac{1}{32}x_m^3 \sin(3\tau) .$$

Constants C_1 and C_2 can be obtained from the initial conditions. If $x = 0$ at time $t = 0$, C_2 vanishes. As the maximum amplitude, reached at $\tau = \pi/2$, is x_m, the value of constant C_1 is

$$C_1 = -\frac{x_m^3}{32} .$$

The final expression of solution $x^{(1)}$ is thus

$$x^{(1)} = -\frac{x_m^3}{32}\left[\sin(\tau) + \sin(3\tau)\right] .$$

This solution leads to a frequency and a time history of the free oscillations that coincide with those obtained by using the iterative technique if the series in μ is truncated after the first term.
The third equation (18.48) yields

$$x''^{(2)} + x^{(2)} = x_m\left(\frac{21}{128}x_m^4 + 2\omega_2^*\right)\sin(\tau) - \frac{3}{16}x_m^3\sin(3\tau) - \frac{3}{128}x_m^5\sin(5\tau) .$$

By equating to zero the secular term, the following value of ω_2^* can be obtained

$$\omega_2^* = -\frac{21}{256}x_m^4 .$$

Once that the secular term has been eliminated, the expression of $x^{(2)}$ becomes

$$x^{(2)} = C_1 \sin(\tau) + C_2 \cos(\tau) + \frac{3}{128}x_m^3\sin(3\tau) + \frac{1}{1024}x_m^5\sin(5\tau) .$$

The values of the constants of integration can be obtained by the usual procedure

$$C_1 = -\frac{23}{1024}x_m^5 , \qquad C_2 = 0$$

The complete solution, up to the third approximation, is thus

$$x = x_m\left(1 - \tfrac{1}{32}\mu x_m^2 + \tfrac{23}{1024}\mu^2 x_m^4\right)\sin(\omega t) +$$

$$-\tfrac{1}{32}\mu\, x_m^3\left(1 - \tfrac{24}{32}\mu x_m^2\right)\sin(3\omega t) + \tfrac{1}{1024}\mu^2 x_m^5\sin(5\omega t) ,$$

where the frequency is

$$\omega = \omega_0\left(1 + \frac{3}{8}\mu x_m^2 + \frac{21}{256}\mu^2 x_m^4\right) .$$

TABLE 18.1. Values of the nondimensional frequency and coefficients of the first three harmonics of the response of a system with a Duffing-type law $f(x)$ with different values of the nondimensional parameter μx_m^2.

μx_m^2		ω/ω_0	a_1/x_m	a_3/x_m	a_5/x_m
0	Num. Int.	1.0000	1.0000	0.00000	0.000000
	Exact	1	1	0	0
0.25	Num. Int.	1.0892	0.9934	−0.00658	0.000043
	Har. bal. (1h.)	1.0897	1		−
	Har. bal. (2h.)	1.0891	0.9934	−0.0066	−
	Iterative	1.0897	0.9922	−0.00781	−
	Perturbation	1.0886	0.9936	−0.00634	0.000061
1	Num. Int.	1.3178	0.9817	−0.01796	0.000321
	Har. bal. (1h.)	1.3229	1		−
	Har. bal. (2h.)	1.3197	0.9820	−0.0180	−
	Iterative	1.3229	0.9688	−0.03125	−
	Perturbation	1.2930	0.9912	−0.00781	000977
−0.25	Num. Int.	0.9004	1.0095	0.00963	0.000523
	Har. bal. (1h.)	0.9014	1		−
	Har. bal. (2h)	0.9004	1.0096	0.0096	−
	Iterative	0.9014	1.0078	0.00781	−
	Perturbation	0.9011	1.0092	0.00928	0.000610
−0.75	Num. Int.	0.6370	1.0536	0.05666	0.003230
	Har. bal. (1h)	0.6614	1		−
	Har. bal. (2h)	0.6331	1.0574	0.0574	−
	Iterative	0.6614	1.0234	0.02344	−
	Perturbation	0.6726	1.0361	0.03662	0.000549

By comparing the solutions obtained using iterative and perturbation techniques, it is clear that the two solutions (time histories and frequencies) are coincident up to the terms in μ.

The frequency and the coefficients of the series expressing the time history, computed using the perturbation and the iterative methods are compared in Table 18.1 with those obtained by integrating numerically the equation of motion and applying a Fourier transform to the solution. Note that systems with $\mu x_m^2 = 1$ or $\mu x_m^2 = -0.75$ cannot be defined as weakly nonlinear.

In this case, harmonic balance yields the best results (in the sense of closer to those obtained through numerical integration).

18.9   Solution in the state plane

By rewriting the state equation (18.7) for the free behavior of an undamped system and explicitly introducing the velocity $v = \dot{x}$, the following equation is obtained

$$\begin{cases} \dot{v} = -\dfrac{1}{m}f(x)\,, \\ \dot{x} = v\,, \end{cases} \qquad (18.50)$$

or, by eliminating time between the two equations,

$$\frac{dv}{dx} = -\frac{f(x)}{mv}\,. \qquad (18.51)$$

The last equation directly yields the slope of the trajectories of the system in the state plane and can be integrated without problems, at least numerically. When the system is linear, function $f(x)$ is simply kx, the integration yields the equation of elliptical trajectories

$$v^2 + \frac{k}{m}x^2 = C\,. \qquad (18.52)$$

Their size is determined by the value of the constant of integration C, which depends on the initial conditions.

Consider the system studied in Fig. 18.2, whose nonlinear part has a third power characteristic. Introducing Eq. (18.15) into Eq. (18.51), separating the variables, and integrating, it follows that

$$v^2 + \frac{k}{m}x^2\left(1 + \frac{\mu}{2}x^2\right) = C\,. \qquad (18.53)$$

Equation (18.53) is plotted in Fig. 18.7 in nondimensional form for various values of constant C. Defining the nondimensional states

$$v^* = v\sqrt{|\mu|}\frac{m}{k}\,, \qquad x^* = x\sqrt{|\mu|}\,,$$

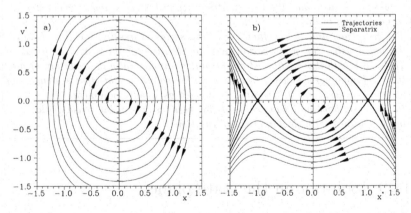

FIGURE 18.7. State portrait of a system whose restoring force is expressed by Eq. (18.15). Nondimensional velocity $v^* = v\sqrt{|\mu|m/k}$ as a function of the nondimensional displacement $x^* = x\sqrt{|\mu|}$ for (a) hardening and (b) softening systems.

the behavior of all possible systems whose function $f(x)$ is of the type of Eq. (18.15) can be summarized in just two plots: one for hardening (Fig. 18.7a) and one for softening systems (Fig. 18.7b).

Remark 18.11 *All trajectories run from left to right in the upper half-plane, where the velocity is positive, and from right to left in the lower half-plane.*

If the system is hardening there is a single equilibrium position and the trajectories are very similar to ellipses (circles, in the nondimensional plot) for small amplitudes. With growing amplitudes, the curves are elongated on the velocity axis and they lose the shape of ellipses. The equilibrium point is what is called a singular point of the center type. No trajectory can pass through that point, but all of them orbit around it.

In the case of softening systems there are three equilibrium points: one stable (the origin) and two unstable. The state portrait is divided into five domains or basins of attraction by a separatrix. The first domain of attraction is centered in the origin and is the basin of the stable solution. In it, the motion is periodic, with trajectories of elliptical (circular) form, if the amplitude is small, and an increasingly elongated form (along the displacement axis) with growing amplitude. In the two basins of attraction that lie above and below the first one, the motion is nonperiodic and characterized by high speed. Point P comes from the right (lower basin) or left (upper basin) and slows down, feeling a repulsive force unless the force changes sign in correspondence with the unstable equilibrium position. It is then caught by the central force field of the origin, accelerates again, and dashes past it. It then slows down again until the force changes sign at the other unstable equilibrium position and then accelerates away under the effect of the repulsive force field.

The other two domains of attraction, on the right and on the left, are characterized by large displacement and relatively small speed. The motion is again nonperiodic, and point P comes toward the origin, but the repulsive force prevents it from reaching the unstable equilibrium position and drives it back.

The separatrices pass through the unstable equilibrium points: They represent the trajectory of a point that has just enough energy to reach the unstable equilibrium position but gets there with vanishingly small speed. The two unstable equilibrium points are again singular, this time of the saddle type.

If the linear term is repulsive instead of attractive, i.e., k is negative, the state portraits are of the type shown in Fig. 18.8. When μ is negative, i.e., the coefficient of the term containing the third power of the displacement is positive, the behavior is of the hardening type (Fig. 18.8a). There are three static equilibrium positions, one unstable in the origin and two stable ones. The separatrix defines three domains of attraction. The motions taking

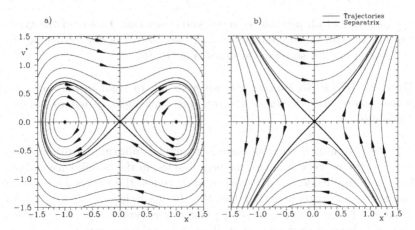

FIGURE 18.8. State portrait of a system whose restoring force is expressed by Eq. (18.15), as in Fig. 18.7, but with a negative value of the stiffness k. Nondimensional velocity $v^* = v\sqrt{|\mu|}m/k$ as a function of nondimensional displacement $x^* = x\sqrt{|\mu|}$ for (**a**) hardening (in this case with negative value of μ) and (**b**) softening systems.

place in the two domains centered in the two stable equilibrium positions, corresponding to two centers, are simple oscillatory motions about the equilibrium positions. If the amplitude is small, these motions are almost harmonic, while the trajectories depart from a circular shape when the amplitude grows. The domain of attraction lying outside the separatrix is that of oscillatory motions of large amplitude that cross all three equilibrium positions and are centered about the origin. The origin is a saddle point.

In the case of softening systems with repulsive linear stiffness, no oscillatory motion is possible and nonperiodic motions occur in the whole state space. The origin is again a saddle point.

Remark 18.12 *The approximate methods seen in the previous sections are applicable only within the domains of attraction of the stable equilibrium positions, while the state-space solution yields exact results in the whole state plane.*

A more detailed analysis of the state-plane approach and a review of the possible singular points will be seen in Chapter 20, when dealing with damped systems.

18.10   Exercises

Exercise 18.1 *Study the free oscillations of an undamped pendulum using the technique described in Section 18.4.*

Exercise 18.2 *Repeat the study of the undamped oscillations of a pendulum by approximating the restoring force as a Duffing-type force. Compare the re-*

sults with those obtained using the Ritz averaging technique and those obtained in Exercise 18.1.

Exercise 18.3 *Repeat the study performed in Exercise 18.1 by resorting to the state plane. Plot the state trajectories and compare the results obtained with those shown in Fig. 18.7b.*

Exercise 18.4 *Consider the double pendulum of Exercise 4.5. Write the nonlinear equations of motion and those obtained by retaining two terms of the series for the trigonometric functions of the generalized coordinates. Study by numerical integration of the equations of motion the free motion starting from a configuration with only angle θ_2 different from 0 and all velocities equal to 0. Perform the computation for two cases with $\theta_{2_0} = 2°$ and $60°$, respectively. Compare the results with those obtainable from the linearized system (analytical solution).*

19

Forced Response of Conservative Nonlinear Systems

When a linear system is excited by a harmonic forcing function, its response is harmonic. In the same conditions the response of a nonlinear system is more complex: not only the waveform is not harmonic, a thing that can be expressed by saying that it includes higher-order harmonics, but also other harmonic components, like subharmonics and combination tones, may appear. In case of conservative systems, a free response that does not decay quickly in time is also present: Owing to nonlinearity the free and forced responses interact in a complex way. Generally speaking, this complex motion can be studied only by numerical (or physical) experiments.

19.1 Approximate evaluation of the response to a harmonic forcing function

Forced oscillation of a conservative nonlinear system can be studied by using the non-autonomous equation obtained by adding a forcing function to Eq. (18.13). By separating the linear part of the restoring force $f(x)$ from the nonlinear part as in Eq. (18.3), the following nonlinear and non-autonomous equation is obtained:

$$m\ddot{x} + kx + f'(x) = F(t). \qquad (19.1)$$

The solution of Eq. (19.1) is, from the mathematical viewpoint, a formidable task, and no general solution can be obtained except by numerically integrating the equation of motion.

Among the many approximated approaches, the Ritz averaging technique can be used by introducing the virtual work due to the forcing function $F(t)$ into Eq. (18.31), which becomes

$$\int_0^T \left[\ddot{x} + \frac{k}{m}x + \frac{f'(x)}{m} + \frac{f(x)}{m} - \frac{F(t)}{m} \right] \phi_i(t)dt = 0, \qquad (19.2)$$

for $i = 1, 2, \ldots, n$.

If the number of functions $\phi_i(t)$ is large (usually when it is greater than 1) the computational difficulties can become overwhelming.

When the forcing function is harmonic, or at least periodic, the results obtained through harmonic balance are usually identical to those achievable through averaging techniques. The complexity of the analysis grows very rapidly with the number of harmonics considered.

If both the nonlinear part $f'(x)$ of $f(x)$ and the non-autonomous part of $F(t)$ are small, iterative or perturbation techniques can be used. In the latter case, Eq. (18.37) must be modified to include the forcing function:

$$L(x) = N(x) + F(t) . \qquad (19.3)$$

A set of equations similar to those seen for free vibration are obtained, with the difference that now the generating solution is the solution of the non-autonomous equation

$$\ddot{x}^{(0)} + \omega_0^2 x^{(0)} = \frac{F(t)}{m} . \qquad (19.4)$$

Remark 19.1 *Unfortunately, little general insight into the relevant phenomena can be gained through solutions of this type, due to the impossibility of resorting to the superimposition technique. The advantages of these complex analytical techniques over the numerical integration of the equation of motion are thus questionable.*

If the forcing function is harmonic in time with frequency ω, assuming that at time $t = 0$ it vanishes while increasing from negative values, it follows that

$$F(t) = f_0 \sin(\omega t) .$$

If the Ritz averaging technique is used, it is not necessary to assume that the nonlinear part of the restoring force $f'(x)$ is small. The accuracy of the results is, however, better in the case of mildly nonlinear systems.

To evaluate approximately the fundamental harmonic of the response, the steady-state response can be assumed to be periodic, with a period equal to the period of the forcing function, and in phase with the latter owing to the lack of damping. Reducing the response to its fundamental harmonic, a time history of the type

$$x = x_m \sin(\omega t) \qquad (19.5)$$

can be assumed. If function $f'(x)$ is not odd, even harmonics, including a constant term, are also present and Eq. (19.6) can lead to poor results, even when only an approximation of the fundamental harmonic is searched.

By applying the Ritz averaging technique in the form of Eq. (19.2), it follows that

$$\int_0^T [m\ddot{x} + kx + f'(x) - f_0 \sin(\omega t)] \sin(\omega t) dt = 0 . \qquad (19.6)$$

By defining a nondimensional time variable $\tau = \omega t$, whose values span from 0 to 2π in a period, introducing Eq. (19.5) into Eq. (19.6) and integrating, it follows that

$$x_m \left[1 - \left(\frac{\omega}{\omega_0}\right)\right]^2 - \frac{f_0}{k} + \frac{1}{k\pi} \int_0^{2\pi} f'(x) \sin(\tau) d\tau = 0 . \qquad (19.7)$$

Equation (19.7) allows the amplitude of the response x_m to be computed as a function of the amplitude and the frequency of the forcing function. Once function $f'(x)$ has been stated, it can easily be transformed through Eq. (19.5) into a function of time and then integrated, at least numerically. The frequency response, i.e., the amplitude of the response as a function of the forcing frequency, can thus be readily obtained.

Often, it is simpler to obtain the frequency response in the form of a function $\omega(x_m)$ instead of in the more common form $x_m(\omega)$ by solving Eq. (19.7) in the frequency

$$\frac{\omega}{\omega_0} = \sqrt{1 - \frac{f_0}{kx_m} + \frac{1}{k\pi x_m} \int_0^{2\pi} f'(x) \sin(\tau) d\tau} . \qquad (19.8)$$

19.2 Undamped Duffing's equation

19.2.1 First-approximation solution

Consider a system whose restoring force is that studied in Fig. 18.2, on which a harmonic forcing function is acting. The relevant equation of motion, often referred to as *undamped Duffing's equation*, is

$$m\ddot{x} + kx(1 + \mu x^2) = f_0 \sin(\omega t) . \qquad (19.9)$$

As usual with nonlinear systems, there is no general solution to Eq. (19.9), but different techniques can be used to obtain approximate solutions. As already stated, the solution of the equation of motion of a nonlinear system is, generally speaking, non-harmonic, even if the forcing function is harmonic. However, a good approximation, at least of the fundamental harmonic of the response, can be found using the Ritz averaging technique and assuming a mono-harmonic time history.

By introducing the nonlinear function $f'(x) = k\mu x^3$ into Eq. (19.8), it follows that

$$\frac{\omega}{\omega_0} = \sqrt{1 - \frac{f_0}{kx_m} + \frac{3}{4}\mu x_m^2} = \sqrt{1 - \left(\frac{x_m k}{f_0}\right)^{-1} + \frac{3}{4}\frac{\mu f_0^2}{k^2}\left(\frac{x_m k}{f_0}\right)^2}.$$

(19.10)

The second form of Eq. (19.10) evidences two nondimensional parameters, namely

$$x_m \frac{k}{f_0} \quad \text{and} \quad \sqrt{|\mu|}\frac{f_0}{k}.$$

The first is nothing other than the magnification factor defined in the way already seen for linear systems. It can be plotted as a function of the nondimensional frequency of the forcing function ω/ω_0. The second defines the nonlinearity of the system. If such a parameter vanishes, the response of the linear system is found. Otherwise, the dependence of the magnification factor on $\sqrt{|\mu|}f_0/k$ shows that the magnification factor is a function of the amplitude of the excitation as well as of the parameters of the system. This is typical of nonlinear systems.

The amplitude of the response x_m is plotted as a function of the forcing frequency, using the amplitude of the forcing function f_0/k as a parameter, in Fig. 19.1a and b, for the cases of hardening and softening systems, respectively. Note that the solution obtained here is only an approximation that gets worse when the value of the nondimensional frequency ω/ω_0 is far from unity. The approximation is particularly bad in static conditions, as can be immediately assessed by comparing the expression for the amplitude obtainable from Eq. (19.10) with ($\omega = 0$) and the expression obtainable

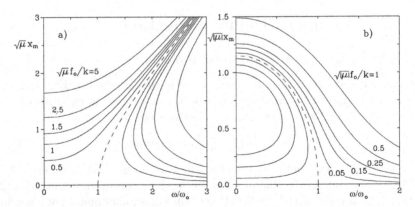

FIGURE 19.1. Response of an undamped system governed by Duffing's equation; amplitude of the response as a function of the frequency for various values of the amplitude of the harmonic forcing function; plot for (a) hardening and (b) softening systems in nondimensional form.

directly from the equation of motion (19.9) rewritten for static conditions (constant forcing function, $\ddot{x} = 0$).

The dashed curves have been obtained from Eq. (19.10) by setting to zero the amplitude of the forcing function. They yield the frequency of the free oscillations as a function of their amplitude and coincide with the curves of Fig. 18.6. Such curves are often referred to as the *backbone* or *skeleton* of the response and are very important in the evaluation of the response to harmonic excitation of nonlinear systems, both undamped and damped.

Remark 19.2 *The backbone curve expresses the resonance conditions for a nonlinear system, in the sense that defines the amplitude–frequency relationship at which the nonlinear restoring force balances the inertia force, at least as an average.*

Three zones can be identified in the frequency response, like in the case of linear systems:

1. At low frequency the behavior of the system is dominated by the stiffness, which, in this case, is nonlinear. The magnification factor depends on the force f_0; the phase is 0.

2. At high frequency the behavior is dominated by the inertia of mass m. Because inertia forces are linear, the response is very similar to that of the linear system; the phase is 180°.

3. When conditions approach those typical of free oscillations, i.e., when the curve $x_m(\omega/\omega_0)$ approaches the backbone, a sort of resonance occurs and the behavior of the system is dominated by damping. Little can be said by using an undamped model.

In the nonlinear case, however, the resonance peak is not straight, but rather it slopes to the right for hardening systems and to the left for softening ones. Consequently, there is no finite value of the frequency for which the amplitude of the response grows to infinitely large values when damping tends to zero.

Consider a hardening system excited at very low frequency. The response corresponds to point A in Fig. 19.2a. With increasing frequency, the amplitude of the response grows, following the line AFB, always remaining in phase with the excitation. After reaching point F, however, the solution is no longer the only possible one, and, at a certain frequency, mainly defined by the damping of the system (point B), the amplitude drops at the value corresponding to point C in the figure, and the phase lag shifts to 180°. At higher frequencies, line CD is followed.

If, on the contrary, the frequency is decreased from that of point D, line DCE is followed, with a phase of 180°. At frequencies lower than that of point E, no solution with a phase of 180° exists, and an increase in the amplitude of the response, accompanied by a change of phase to 0, occurs (line EF). The amplitude then decreases following line FA.

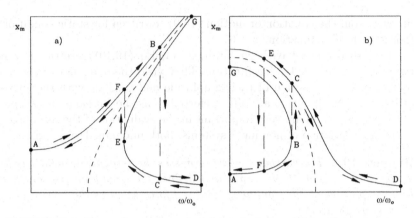

FIGURE 19.2. Jump phenomenon in (**a**) hardening and (**b**) softening systems.

Remark 19.3 *This behavior is often referred to as the* jump phenomenon *and is typical of nonlinear systems. It also holds, qualitatively, for systems governed by nonlinear equations other than Duffing's equation.*

A similar behavior is shown by softening systems, as illustrated in Fig. 19.2b. The path followed when the frequency of the excitation increases is AFBCD, with a jump at BC. When the frequency is reduced, the path is DCEFA, with a downward jump at EF. While the downward jumps (BC in hardening systems and EF in softening systems) occur at a frequency that is often not well determined, depending mainly on damping but also on external disturbances, the frequency at which the upward jumps occur (BC and EF, respectively) can be computed, at least within the first-approximation solution, using the undamped model.

Remark 19.4 *In case of softening systems the expression of the backbone gets less and less accurate with decreasing frequency and also the accuracy with which the higher branch of the response is obtained becomes worse. The whole part of the plot of Fig. 19.2b close to points G and E has little meaning.*

It is easy to verify that the frequency at which point E is located in Fig. 19.2a or point B is located in Fig. 19.2b is

$$\frac{\omega}{\omega_0} = \sqrt{1 \pm \sqrt[3]{\frac{81}{16}|\mu| \left(\frac{f_0}{k}\right)^2}}, \qquad (19.11)$$

where the upper sign holds for hardening systems and the lower one for softening systems.

It is possible to demonstrate that when three values of the amplitude are found, the greatest and the smallest correspond to stable conditions of the system, while the intermediate one is unstable and cannot actually be obtained. The unstable branch is the one indicated as EG in Fig. 19.2a (hardening system) and BG in Fig. 19.2b (softening system). The attraction domains of the solutions will be studied in detail when dealing with the damped Duffing's equation.

19.2.2 Higher harmonics

As already stated, the time history of the response is not harmonic; since the restoring force is an odd function of x, only odd harmonics, with frequencies 3ω, 5ω, ..., are present. These harmonics, whose frequency is higher than the forcing frequency, are usually referred to as *super-harmonics*. Although their amplitude is usually not great, their presence is a symptom of nonlinearity, like a sloping resonance peak.

A detailed study of the harmonic content of the solution of Duffing's equation, based on the perturbation technique, is reported in the mentioned book by L. Meirovitch, where the interested reader can find all the details.

Harmonic motion is used as a generating solution. The amplitude of the response is computed by stating that the secular terms, present in the equation yielding the second term of the power series $x^{(1)}$, vanish. The same value of the amplitude obtained using the Ritz averaging technique is so obtained. Solution $x^{(1)}$, which contains two terms with frequencies ω and 3ω, is thus obtained by stating that the secular terms that appear in the solution for $x^{(2)}$ vanish. Note that while obtaining the solution for the third harmonic, a new estimate for the amplitude of the fundamental harmonic is obtained, but it is usually close to that previously found. The analysis can continue obtaining new harmonics and corrections for the already computed results, but the relevant computations become very involving.

A different approach, based on harmonic balance, is shown here. As it was done for the free response, a second-approximation solution can be searched assuming that the response to a forcing function

$$f = f_0 \sin(\omega t) \tag{19.12}$$

is

$$x = a_1 \sin(\omega t) + a_2 \sin(3\omega t). \tag{19.13}$$

Operating as done in Example 18.2, by introducing the assumed time history of the solution into the equation of motion, and balancing the first two harmonics, it follows that

$$
\begin{cases}
a_1\left[1 - \left(\dfrac{\omega}{\omega_0}\right)^2 + \dfrac{3}{4}\mu\left(a_1^2 + 2a_2^2 - a_1 a_2\right)\right] = \dfrac{f_0}{k} \\[4mm]
a_2\left[1 - 9\left(\dfrac{\omega}{\omega_0}\right)^2\right] + \dfrac{1}{4}\mu\left(-a_1^3 + 3a_2^3 + 6a_1^2 a_2\right) = 0 \ .
\end{cases}
\tag{19.14}
$$

The fifth, seventh and ninth harmonics cannot be balanced.
By introducing the nondimensional parameters

$$\omega^* = \frac{\omega}{\omega_0}, \quad a_i^* = a_i \sqrt{|\mu|}, \quad f^* = \sqrt{|\mu|}\frac{f_0}{k} \qquad (19.15)$$

the equations reduce to

$$
\begin{cases}
3a_1^{*2} + 6a_2^{*2} - 3a_1^* a_2^* \pm 4\left(1 - \omega^{*2}\right) \mp 4\dfrac{f^*}{a_1} = 0 \\[2mm]
-a_1^{*3} + 3a_2^{*3} + 6a_1^2 a_2^* \pm 4a_2^* \left(1 - 9\omega^{*2}\right) = 0 \ ,
\end{cases}
\qquad (19.16)
$$

where the upper and lower signs refer to hardening and softening systems, respectively.

If instead of computing a_i^* as a function of ω^*, the frequency is obtained as a function of the amplitude, the first equation reduces to

$$\omega^* = \frac{1}{2}\sqrt{\pm 3a_1^{*2} \pm 6a_2^{*2} \mp 3a_1^* a_2^* + 4 - 4\frac{f^*}{a_1}} \ . \qquad (19.17)$$

Substituting the value of ω^* into the second equation, it follows that

$$51a_2^{*3} - 27a_1^* a_2^{*2} + \left(21a_1^{*2} \pm 32 \mp 36\frac{f^*}{a_1}\right)a_2^* + a_1^{*3} = 0 \ . \qquad (19.18)$$

Consider now an hardening system and drive it at a generic frequency ω with a nondimensional amplitude $f^* = 1$. The nondimensional frequency response is shown in Fig. 19.3, where the response obtained considering just the first harmonic is also reported.

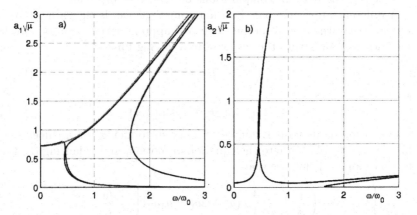

FIGURE 19.3. Nondimensional frequency response of a hardening Duffing-type system. Forcing amplitude $f^* = 1$. Amplitude of the first (**a**) and third (**b**) harmonics. The dashed line refers to the solution obtained by considering just the first harmonics.

Generally speaking, the difference between the solution with a single harmonic and that obtained by taking into account also the third harmonic is not large, but there is a small frequency range, not far from $\omega/\omega_0 = 1/3$, where a strong third harmonic is present. This is easily explained, since in that zone a sort of resonance between the third harmonics of the excitation and the free motion of the system occurs. Since the system is hardening, the resonance peak slopes to the right and this resonance occurs at a frequency in excess to $\omega_0/3$.

At frequencies higher than $\omega_0/3$ there is a possibility of a super-harmonic, with a large amplitude of the third harmonic and a correspondingly low value of the first harmonic. It is however questionable whether this type of motion can be sustained in an actual system, owing to damping (see next chapter).

Consider an excitation occurring at a frequency $\omega = \omega_0/2$. From Fig. 19.3, it is possible to obtain that the amplitude of the motion computed balancing just the first harmonics is

$$a_1^* = 0.8073.$$

Considering two harmonics, three solutions are obtained. The amplitudes are

$$a_1^* = 0.7307 , \qquad a_1^* = 0.3842 , \qquad a_1^* = 0.3023 ,$$
$$a_3^* = -0.2402 , \qquad a_3^* = 1.1779 , \qquad a_3^* = -1.2151 .$$

The relevant time histories are shown in Fig. 19.4, where they are compared with solutions obtained through numerical integration. To force the system to oscillate in the required way, the numerical integration was started with the initial conditions computed from the harmonic balance solution. The presence of a strong third harmonic is confirmed and the harmonic balance solution with two harmonic predicts quite accurately the motion of the system. The solution with a single harmonic on the contrary yields quite inaccurate results.

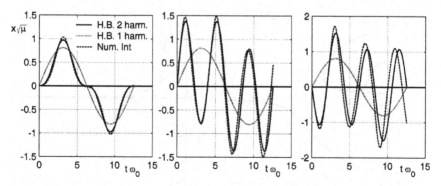

FIGURE 19.4. Time histories of the forced oscillations of the nonlinear system of Fig. 19.3 at a frequency $\omega = \omega_0/2$. The solutions obtained through harmonic balance are compared with those computed by numerically integrating the equation of motion.

Remark 19.5 *If the initial conditions corresponding to the solution with a single harmonics are used to start numerical integration, a solution that is anyway different from a mono-harmonic oscillation is obtained. This shows that no mono-harmonic solution is possible when the system is excited in a frequency range where a strong third harmonics is present.*

Some difference between the solution with two harmonics and that obtained by numerically integrating the equation of motion anyway exists: higher-order harmonics, although weak, are present. In a frequency range close to $\omega_0/5$ a strong fifth harmonic is expected; close to $\omega_0/7$ a seventh harmonics and so on.

The response of a softening system driven at a generic frequency ω with a nondimensional amplitude $f^* = 0.1$ is shown in Fig. 19.5. Owing to the greater complexity of the plot, the sign of the amplitude is also reported.

Here a strong third harmonics is found below the frequency $\omega/3$, since the resonance peaks is sloping to the left instead of the right. Consider an excitation occurring at a frequency $\omega = \omega_0/10$. From Fig. 19.5, it is possible to obtain that the amplitudes of the motion computed balancing just the first harmonics are

$$a_1^* = 1.0946 \,, \quad 0.1019 \,, \quad -1.1964 \,.$$

Considering two harmonics, nine solutions are obtained. The amplitudes are

$$
\begin{array}{llll}
a_1^* = 1.1731 \,, & a_3^* = 0.3271 \,; & a_1^* = 0.6386 \,, & a_3^* = -0.4545 \,; \\
a_1^* = 0.5919 \,, & a_3^* = 0.7757 \,; & a_1^* = 0.1019 \,, & a_3^* = -0.0003 \,; \\
a_1^* = -0.1120 \,, & a_3^* = 1.0899 \,; & a_1^* = -0.1525 \,, & a_3^* = -1.0807 \,; \\
a_1^* = -0.3749 \,, & a_3^* = -0.9748 \,; & a_1^* = -0.5901 \,, & a_3^* = 0.6403. \\
& a_1^* = -1.2759 \,, & a_3^* = -0.3325 \,.
\end{array}
$$

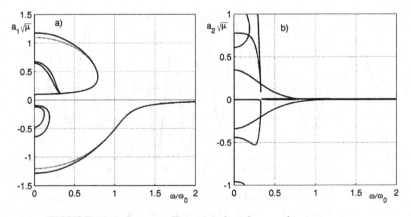

FIGURE 19.5. Same as Fig. 19.3, but for a softening system.

Six of the relevant time histories (in order, the first five and the last one in the list above) are shown in Fig. 19.6. The results contradict completely those of Fig. 19.4: now harmonic balance fails in obtaining acceptable results, except for the fourth case. The time histories obtained through numerical integration show a strong ninth harmonic that was clearly not included into the analysis. The fourth case is an exception: the amplitude is so low that the system behaves almost linearly, and the motion is practically harmonic: Just balancing the first harmonic is enough to get good results. The results obtained with a single harmonic are reported only for the branches of the response where they exist.

The poor behavior of the harmonic balance approach in this case is due to the low value of the forcing frequency at which the time history was computed ($\omega = \omega_0/10$). This leads to a strong response on high-order harmonics, with a strong fifth harmonic; close to $\omega_0/7$ a seventh harmonics is expected and so on. A completely different picture would have been obtained if a higher frequency were used. Such a low frequency was used to show that the upper branch of the response lying in the low-frequency region obtained using approximated approaches is much questionable in the case of softening systems.

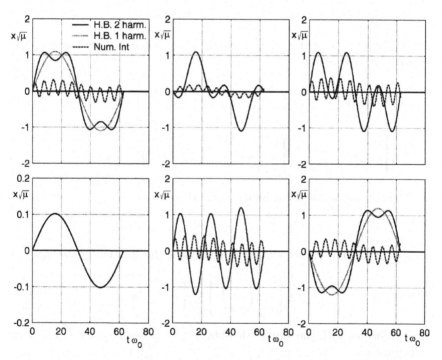

FIGURE 19.6. Time histories of the forced oscillations of the nonlinear system of Fig. 19.5 at a frequency $\omega = \omega_0/10$.

The results reported here were obtained using an undamped model. As it will be seen in the next chapter, damping can deeply modify this picture.

19.2.3 Subharmonics

Harmonics whose frequency is a submultiple of the forcing function (i.e., *subharmonics*) can also be excited. In linear systems, subharmonics can be present when the frequency of the forcing function is a multiple of the natural frequency; these subharmonic oscillations are, however, seldom observed because the damping of the system makes them disappear quite soon. The presence of sustained subharmonic oscillations is then another characteristic feature of nonlinear systems. In the case of a system governed by Duffing's equation, only odd subharmonics, i.e., subharmonics with frequencies $\omega/3$, $\omega/5$, $\omega/7$, ..., are found, because the restoring force is an odd function of the displacement.

A simple way of computing the subharmonic response to a forcing function

$$f = f_0 \sin(\omega t)$$

is assuming a response of the type

$$x = a_1 \sin\left(\frac{\omega}{3}t\right) + a_2 \sin(\omega t). \tag{19.19}$$

Operating as done for super-harmonics and balancing the first two harmonics, it follows that

$$\begin{cases} a_1\left[1 - \frac{1}{9}\left(\frac{\omega}{\omega_0}\right)^2 + \frac{3}{4}\mu\left(a_1^2 + 2a_2^2 - a_1 a_2\right)\right] = 0 \\[4mm] a_2\left[1 - \left(\frac{\omega}{\omega_0}\right)^2\right] + \frac{1}{4}\mu\left(-a_1^3 + 3a_2^3 + 6a_1^2 a_2\right) = f_0 \ . \end{cases} \tag{19.20}$$

The first equation yields immediately that

$$a_1 = 0$$

is a solution. The corresponding value of a_2 is the solution of the equation

$$a_2\left[1 - \left(\frac{\omega}{\omega_0}\right)^2\right] + \frac{3}{4}\mu a_2^3 = f_0 \tag{19.21}$$

and coincides with the amplitude obtained by assuming a mono-harmonic time history.

This fact is very significant: solutions in which no subharmonic is present are always possible.

By introducing the usual nondimensional parameters, the equations yielding the other solutions reduce to

$$
\begin{cases}
3a_1^{*2} + 6a_2^{*2} - 3a_1^*a_2^* \pm 4\left(1 - \dfrac{\omega^{*2}}{9}\right) = 0 \\[2mm]
-a_1^{*3} + 3a_2^{*3} + 6a_1^2 a_2^* \pm 4a_2^*\left(1 - \omega^{*2}\right) \mp 4f_0 = 0 \ ,
\end{cases}
\tag{19.22}
$$

where the upper signs refer to hardening systems and the lower ones to softening systems.

The nondimensional frequency is thus

$$
\omega^* = 3\sqrt{1 \pm \frac{3}{4}\left(a_1^{*2} + 2a_2^{*2} - a_1^*a_2^*\right)} \ .
\tag{19.23}
$$

Substituting the value of ω^* into the second equation, it follows that

$$
a_1^{*3} + 21a_2^*a_1^{*2} - 27a_2^{*2}a_1^* + 51a_2^{*3} \pm 32a_2^* \mp 4f^* = 0 \ .
\tag{19.24}
$$

Consider again a hardening system driven at a generic frequency ω with a nondimensional amplitude $f^* = 0.5$. The nondimensional frequency response computed using a time history containing the fundamental harmonic and the subharmonic of order $1/3$ is shown in Fig. 19.7.

The solution in which the subharmonic is not present is the same already studied. At high frequency, above $3\ \omega_0$, two responses containing a strong subharmonic are present. When this occurs, the amplitude of the fundamental harmonic is very small, and the motion is almost a pure subharmonic oscillation. To show the value of a_2^*, an enlargement of the zone of the plot for low amplitudes is reported in Fig. 19.7b.

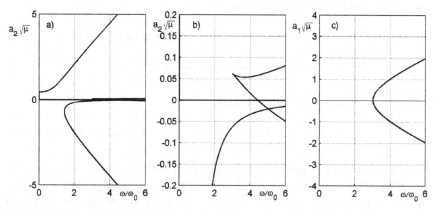

FIGURE 19.7. Subharmonic response of a hardening Duffing system. The zone of figure (a) close to the ω-axis is enlarged in (b) to show the very low amplitude of the synchronous component of the subharmonic response.

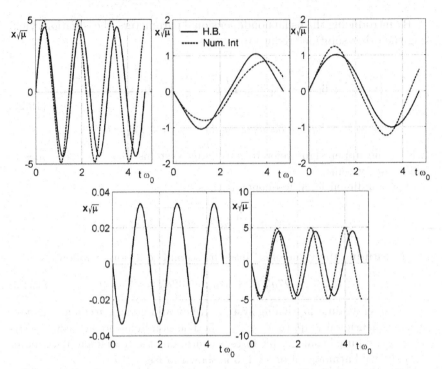

FIGURE 19.8. Time histories of the system studied in Fig. 19.7 at a nondimensional frequency $\omega^* = 4$. The responses related to five different solutions have been obtained using harmonic balance and numerical integration.

This is clearly seen in Fig. 19.8 where the time histories of the motion at a nondimensional frequency $\omega/\omega_0 = 4$ are shown. The amplitudes computed using the harmonic balance method are

- Responses with a single harmonic, i.e., with vanishing a_1^*:

$$a_2^* = 4.48871 , \quad a_2^* = -0.03334 , \quad a_2^* = -4.45538 ;$$

- Responses with a subharmonic:

$$a_1^* = -0.9898 , \quad a_2^* = -0.0548 ; \quad a_1^* = 1.0267, \quad a_2^* = 0.0171 .$$

The first and the last two plots deal with the three mono-harmonic responses. The one with the small amplitude is obtained so well with harmonic balance that it cannot be told from the solution obtained through numerical integration. The two with large amplitude show stronger nonlinear effects, as it should be expected, and the actual response is not so well computed assuming that the response is harmonic.

The motion was followed for 3 periods, i.e., for a period of the third subharmonic, to show that no subharmonic is present.

The two intermediate solutions are clearly dominated by the subharmonic. The solution obtained using harmonic balance gives fair results, even if the numerical integration shows some difference in both amplitude and frequency. At any rate it is clear that the period of the numerical solution is close to that of the third subharmonic.

A similar effect is usually reported for softening systems, with the difference that now the range at which the subharmonic resonance starts is below and not above the frequency $3\omega_0$. However, like in the case of the super-harmonics seen above, the motion of softening systems is much more complicated: By integrating numerically the equation of motion a mixture of subharmonics and super-harmonics is often found.

As already stated, while super-harmonics are nothing else than a distortion of the waveform and thus are inherent to any solution of a nonlinear system (they tend to disappear only at low amplitude, when the system tends to behave linearly), subharmonics are not necessarily present and can appear only if suitable conditions are given.

19.2.4 Combination tones

When a linear system is excited by a poly-harmonic forcing function (i.e., a forcing function that can be considered the sum of harmonic components), the response is also poly-harmonic and can be computed by adding the responses to each component of the excitation.

In the case of nonlinear systems, the response is much more complex, and if a Fourier analysis is performed, it contains harmonics that are not present in the excitation. This could be expected, because the effect of distorting the waveform, which is peculiar to nonlinear systems, is exactly that of introducing new harmonics.

Consider a system governed by Duffing's equation excited by a forcing function consisting of the sum of two harmonic components ω_1 and ω_2. By using perturbation techniques, a first-approximation solution containing harmonic components with frequencies ω_1, ω_2, $2\omega_1 + \omega_2$, $2\omega_1 - \omega_2$, $\omega_1 + 2\omega_2$, $\omega_1 - 2\omega_2$, $3\omega_1$, and $3\omega_2$ is obtained. The fundamental harmonics with frequencies ω_1 and ω_2, which would have been in the response of a linear system, are then present together with higher-order harmonics (third order, in this case), *combination tones*, and possibly subharmonics. If the solution is carried on to higher-order approximations, other harmonics are found.

If two forcing frequencies are close to each other, the response of a linear system gives way to a beat, as already seen. The period of the beat tends to infinity, when the two frequencies tend to coincide. This is not the case of nonlinear systems, where the phenomenon of *entrainment of frequency* takes place.

When the two forcing frequencies get close enough to each other, the beat disappears and the two excitations combine giving way to a single oscillation. The frequency range in which this occurs is referred to as the *interval of synchronization.*

19.3   Conclusions

No allowance for the presence of damping was taken in the present chapter devoted to the study of the forced response of a nonlinear system to a harmonic or periodic excitation. Energy dissipation can change deeply this picture, making it impossible for the system to oscillate at certain frequencies with certain amplitudes.

In the case of a hardening Duffing oscillator, for example, the oscillation with the largest amplitude is possible only up to a certain frequency due to damping, as it will be seen in the next chapter.

Subharmonic oscillation may also be prevented by damping.

Another point not considered is stability. The fact that a certain oscillation, with given values of amplitude and frequency, has been found using the methods seen above does not mean that the system necessarily can oscillate in a stable way in these conditions. As it will be seen further on, when different values of the amplitude of the oscillation are found at a certain frequency, not all of them correspond to stable motions. In general, listing the absolute values of the amplitudes in ascending order, they correspond alternatively to stable and unstable conditions.

The particular higher-order harmonics and combination tones found in connection with the Duffing equation are typical of systems governed by this equation, or more in general of systems with a symmetrical characteristic. Other frequencies would have been obtained if a different restoring force was assumed, and in particular even-order super- and subharmonics would have been found.

A non-symmetrical characteristic can be obtained by a simple translation of a symmetrical one. For instance, adding a static load to a system governed by the Duffing equation causes it to work about a different static equilibrium condition and thus to behave in a non-symmetrical way. Even-order harmonics may enter the picture.

Example 19.1 *Evaluate the amplitude of the fundamental harmonic of the response to a harmonic forcing function of a piecewise linear system, i.e., a system in which the characteristic of the restoring force can be represented by a number of straight lines. Assume that the restoring force is expressed by an odd function of the displacement and that there are only two values of the stiffness (Fig. 19.9a).*

Function $f'(x)$ expressing the nonlinear part of the restoring force is

$$
\begin{aligned}
f'(x) &= k(\alpha - 1)(x + x_1) & for \quad x < -x_1, \\
f'(x) &= 0 & for \quad -x_1 < x < x_1, \\
f'(x) &= k(\alpha - 1)(x - x_1) & for \quad x > x_1,
\end{aligned}
$$

where k is the linearized stiffness and coincides with k_1 and

$$
\alpha = \frac{k_2}{k_1} .
$$

Function $f'(x)$ must be integrated separately in each of the fields identified in Fig. 19.9b. The first expression for $f'(x)$ holds between τ_3 and τ_4, while the third equation holds between τ_1 and τ_2. Function $f'(x)$ vanishes between 0 and τ_1, τ_2 and τ_3, and τ_4 and 2π. The integral to be introduced into Eq. (19.8) is simply

$$
\int_0^{2\pi} f'(x)\sin(\tau)d\tau = k(\alpha - 1)\left\{ \int_{\tau_1}^{\pi - \tau_1} \left[x_m \sin^2(\tau) - x_1 \sin(\tau) \right] d\tau + \right.
$$

$$
\left. + \int_{\pi + \tau_1}^{2\pi - \tau_1} \left[x_m \sin^2(\tau) - x_1 \sin(\tau) \right] d\tau \right\} ,
$$

where the value of τ_1 is

$$
\tau_1 = \arcsin\left(\frac{x_1}{x_m} \right) .
$$

The response of the system can thus be easily computed through Eq. (19.8). By performing the integration and writing the response in nondimensional form, it follows that

$$
\frac{\omega}{\omega_0} = \sqrt{ 1 - \frac{f_0}{kx_1}\frac{x_1}{x_m} + (\alpha - 1)\left[1 + \frac{\sin(2\tau_1) - 2\tau_1}{\pi} - \frac{4x_1}{\pi x_m}\cos(\tau_1) \right] } .
$$

This expression holds only if $x_m > x_1$, otherwise the system behaves as if it were linear. The response is plotted in Fig. 19.9c in nondimensional form, for the case with $\alpha = 2$. If $\alpha > 1$, as in the case shown, the behavior is that of hardening systems, with a resonance curve leaning to the right.

It must be stressed that the solution so obtained is only an approximation, that is better if the response is close to a harmonic function. In the case studied, this happens if the amplitude is either only slightly higher than x_1 or very much higher than the same value. The limiting case with $x_m \gg x_1$ is that of a linear system with stiffness k_2, whose natural frequency is

$$
\omega = \omega_0\sqrt{\alpha} .
$$

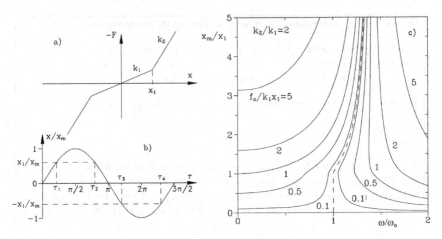

FIGURE 19.9. System with a piecewise linear characteristic: (a) force–displacement curve; (b) law $x(\tau)$ (only the fundamental harmonic is considered); (c) approximate response in nondimensional form.

19.4   Exercises

Exercise 19.1 *Consider the system of Example 15.1 and substitute for the linear support with stiffness k_G a nonlinear support with a hardening characteristic of the same type as seen for Duffing's equation. Compute the backbone of the response. Assume that a harmonic excitation with variable frequency and amplitude 20 N acts on the central joint. Compute the fundamental harmonic of the response in the range of frequencies between 1 and 100 Hz (only a first approximation is required).*

Data: inner diameter = 70 mm, outer diameter = 90 mm, length of each part of the shaft = 500 mm, $E = 2.1 \times 10^{11}$ N/m^2, $\nu = 0.3$, $\rho = 7{,}810$ kg/m^3, mass of the joint 2 kg, linear stiffness of the central support 105 N/m, $\mu = 5 \times 10^7$ $1/m^2$.

Exercise 19.2 *Consider again the transmission shaft studied in Example 15.1, excited by the same force as in Exercise 19.1. Instead of assuming that the support of the joint has a nonlinear elasticity, study the nonlinear effect due to the axial force when the constraints at the ends and all joints are infinitely stiff in axial direction. Assume that the bending compliance of the beams is negligible compared to that of the joint, and assume that the deformed shape is the same used for the solution with the Rayleigh method.*

Exercise 19.3 *Consider the system of Example 1.2. Replace the shaft connecting the first moment of inertia to point A with a nonlinear element with a cubic hardening characteristic. The linear part of the stiffness coincides with the stiffness in Example 1.2, while $\mu = 30$ $1/rad^2$. Compute a first approximation of the fundamental harmonic of the frequency response of the system when a moment with an amplitude of 0.5 N m and frequency varying between 0 and 6 rad/s acts on* the third inertia. Compare the results obtained with those shown in Fig. 7.11. Do not resort to modal coordinates, but use dynamic reduction to reduce the nonlinear problem to a single equation.

Exercise 19.4 *Repeat the computations of the preceding exercise by resorting to modal coordinates. State whether the coupling between the modes due to the nonlinear behavior of the system is large.*

Exercise 19.5 *A rotating machine is suspended through a system of elastomeric springs whose force–displacement characteristics can be approximated by the equation*

$$F = k_1 x - k_2 x^3 + k_3 x^5 .$$

where $k_1 = 10^5$ *N/m,* $k_2 = 4 \times 10^8$ *N/m^3, and* $k_3 = 10^{12}$ *N/m^5. The total mass of the system is* $m = 100$ *kg and the product of the mass of the rotor by its eccentricity is* $m_r \epsilon = 0.5$ *kg m. Neglecting the weight of the system and the damping of the suspension, compute a first approximation of the fundamental harmonic of the response of the system when the angular velocity of the machine spans between 100 and 1,000 rpm.*

Exercise 19.6 *Repeat the computations of the previous exercise without neglecting the weight of the machine and compare the results.*

20

Free Motion of Damped Nonlinear Systems

Damping causes the amplitude of free vibration to decay in time and limits the amplitude of forced vibration. In nonlinear systems however its effects are more complex and a closer mixing of restoring and damping forces may occur. While in linear systems stability is an intrinsic property of the system, when nonlinearities are present a system may be either stable or unstable depending on the working conditions. While the stability in the small is easily assessed, the general stability is much more difficult to study.

20.1 Nonlinear damping

If the force exerted by the nonlinear spring of Fig. 18.1a can be expressed by the general function $f(x, \dot{x})$, or, better, $f(x - x_A, \dot{x} - \dot{x}_A)$, instead of being function of the displacement only, the system becomes non-conservative, i.e., dissipates energy during the motion.

If the restoring force is also an explicit function of time, its equation of motion becomes

$$m\ddot{x} + f(x - x_A, \dot{x} - \dot{x}_A, t) = F(t) - mg, \qquad (20.1)$$

or in terms of relative coordinate (Eq. 18.2)

$$m\ddot{x}_{rel} + f(x_{rel}, \dot{x}_{rel}, t) = F(t) - mg - m\ddot{x}_A. \qquad (20.2)$$

In addition to being nonlinear, the system is non-autonomous, because time appears explicitly in the equation through the forcing function $F(t)$

and function $f(x, \dot{x}, t)$. Also the equation for the study of free vibrations is then non-autonomous.

If the restoring and damping forces are just $f(x, \dot{x})$ (i.e., is not an explicit function of time), a simple way to represent the characteristics of the system is to plot a tridimensional diagram of the type shown in Fig. 20.1, often referred to as *state force mapping*. The figure refers to the so-called Van der Pol oscillator, whose function $f(x, \dot{x})$ is

$$f(x, \dot{x}) = x - \mu \dot{x} \left(1 - x^2\right). \tag{20.3}$$

There are cases in which function $f(x, \dot{x})$ can be subdivided into a linear part, $c\dot{x} + kx$, and a nonlinear part, $f'(x, \dot{x})$, and the equation of motion, written in terms of relative coordinates (Fig. 18.1), becomes

$$m\ddot{x} + c\dot{x} + kx + f'(x, \dot{x}) = F(t) - mg - m\ddot{x}_A, \tag{20.4}$$

which corresponds to Eq. (18.3) with damping added.

As seen for the case of conservative systems, Eq. (20.4) can be extended to multi-degrees-of-freedom systems in the form

$$\mathbf{M}\ddot{\mathbf{x}} + \mathbf{C}\dot{\mathbf{x}} + \mathbf{K}\mathbf{x} + \mu\mathbf{g} = \mathbf{f}(t), \tag{20.5}$$

where matrix $\mathbf{C}$, of order $n \times n$, is the damping matrix of the linearized system, defined in Chapter 3. Vector $\mathbf{g}$ now consists of n functions g_i, which, in the most general case, are $g_i = g_i(\mathbf{x}, \dot{\mathbf{x}}, \mu, t)$.

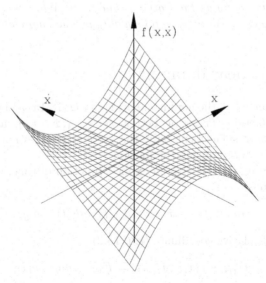

FIGURE 20.1. State force mapping for a nonlinear element characterized by the function $f(x, \dot{x}) = x - \mu \dot{x} \left(1 - x^2\right)$, with $\mu = 0.2$ (Van der Pol's equation).

The modal coordinates of the linearized undamped system allow to transform Eq. (20.5) in the modal form

$$\overline{\mathbf{M}}\ddot{\boldsymbol{\eta}} + \overline{\mathbf{C}}\dot{\boldsymbol{\eta}} + \overline{\mathbf{K}}\boldsymbol{\eta} + \mu \boldsymbol{\Phi}^T \mathbf{g} = \overline{\mathbf{f}}(t). \tag{20.6}$$

The modal equations of motion (20.6) are not uncoupled, not only owing to the nonlinear terms, but also because the modal-damping matrix is not diagonal, except in the case of proportional damping. As shown in Chapter 5, in many cases the modal coupling due to damping can be neglected and the situation is similar to that seen for undamped nonlinear systems.

Remark 20.1 *As seen for undamped nonlinear systems, a strong modal coupling can make the modal formulation (20.6) of the equation of motion practically useless.*

Also the state-space formulation of the equations of motion is similar to that seen in Chapter 18,

$$\dot{\mathbf{z}} = \left\{ \begin{array}{c} -\dfrac{1}{m}\left[f(x, \dot{x}, t) - F(t)\right] \\ \dot{x} \end{array} \right\}, \tag{20.7}$$

with the only difference that now function f depends also on the velocities $\dot{\mathbf{x}}$.

Again, if the linear part of the system can be separated from the nonlinear one, the state-space equation can be written in the form

$$\dot{\mathbf{z}} = \mathcal{A}\mathbf{z} + \left\{ \begin{array}{c} -\mu\mathbf{M}^{-1}\mathbf{g}(x, \dot{x}, t) \\ \mathbf{0} \end{array} \right\} + \mathcal{B}\mathbf{u}(t). \tag{20.8}$$

20.2 Motion about an equilibrium position (in the small)

The static equilibrium position is not affected by damping, and the motion in the small can be studied with reference to the same points in the state space as seen for the undamped system. If function $f(x, \dot{x})$ can be differentiated with respect to x and $\dot{x}$ in the point of the state plane with coordinates $x = x_0$, $\dot{x} = 0$, the motion in the vicinity of an equilibrium position or motion in the small can be studied through the linearized equation:

$$m\ddot{x} + \left(\frac{\partial f}{\partial \dot{x}}\right)_{\substack{x = x_0 \\ \dot{x} = 0}} \dot{x} + \left(\frac{\partial f}{\partial x}\right)_{\substack{x = x_0 \\ \dot{x} = 0}} (x - x_0) = F(t) - mg - m\ddot{x}_A, \tag{20.9}$$

which corresponds to Eq. (18.11), but has a damping term added.

The study of the static stability of the equilibrium position follows the same lines seen in Chapter 6, but the presence of damping allows the study of the dynamic equilibrium, i.e., to assess whether the system tends to return to the equilibrium position in time or tends to depart from it. If the derivative

$$\left(\frac{\partial f}{\partial \dot{x}}\right)_{x=x_0,\ \dot{x}=0}$$

is positive, the system shows a damped behavior, and if the position is statically stable, dynamic stability is also assured. If it is negative, self-excited divergent oscillations are to be expected. Finally, if it vanishes an undamped behavior can be predicted.

In the case of systems with many degrees of freedom, the state equation (20.8) can be linearized as

$$\dot{\mathbf{z}} = \left[\begin{array}{cc} -\mathbf{M}^{-1}\left(\mathbf{C} + \mu\left[\frac{\partial g_i}{\partial \dot{x}_j}\right]\right) & -\mathbf{M}^{-1}\left(\mathbf{K} + \mu\left[\frac{\partial g_i}{\partial x_j}\right]\right) \\ \mathbf{I} & \mathbf{0} \end{array} \right] \mathbf{z} + \mathcal{B}\mathbf{u}(t),$$

$$(20.10)$$

where $[\partial g_i/\partial \dot{x}_j]$ and $[\partial g_i/\partial x_j]$ are Jacobian matrices containing the derivatives of functions $\mathbf{g}$ with respect to $\dot{\mathbf{x}}$ and $\mathbf{x}$. The stability in the small can be studied as usual by computing the eigenvalues of the dynamic matrix of the linearized system.

Apart from studying the stability *in the small*, i.e., in the vicinity of an equilibrium point, in the case of nonlinear systems, it is possible to define a stability in the large. The stability of nonlinear systems will be dealt with in Section 20.6.

20.3   Direct integration of the equation of motion

Consider a system of the type shown in Fig. 18.1, in which the restoring force is a function not only of the position of point P but also of its velocity. If $f(x_0, 0)$ vanishes, the free oscillations about the equilibrium point $(x_0, 0)$ can be studied using a nonlinear autonomous equation that is similar to Eq. (18.13):

$$m\ddot{x} + f(x, \dot{x}) = 0. \qquad (20.11)$$

As already stated, if the derivative $(\partial f/\partial \dot{x})_{x=x_0, \dot{x}=0}$ is positive, the motion in the small is damped. In the opposite case, function $f(x, \dot{x})$ has an exciting effect on the motion.

Remark 20.2 *Generally speaking, it is not said that function $f(x, \dot{x})$ maintains its damping or exciting properties in the whole field in which it is defined, and the motion in the large may be complex.*

Function $f(x, \dot{x})$ may have, for example, a damping effect for large values of x and an exciting nature for values of x smaller than a given quantity, like in the case of the Van der Pol oscillator of Eq. (20.3). In this case, the motion in the large is self-excited at small amplitudes and damped in the case of large motions, and the existence of a limit cycle in the phase plane $x, \dot{x}$, to which all oscillations tend with time, can usually be demonstrated.

There are cases in which function $f(x, \dot{x})$ can be considered as the sum of a function of the displacement x and a function of the velocity $\dot{x}$, i.e., in which the restoring and the damping forces act in a separate way. Using the symbol $f(x)$ for the former and $\beta(\dot{x})$ for the latter, Eq. (20.11) reduces to

$$m\ddot{x} + \beta(\dot{x}) + f(x) = 0 . \tag{20.12}$$

As in the case of undamped systems, Eq. (20.11) can be rewritten as a first-order differential equation:

$$\frac{1}{2} m \frac{d\left(\dot{x}^2\right)}{dx} = -f(x, \dot{x}) . \tag{20.13}$$

In general, however, no separation of the variables is possible, and Eq. (20.13) cannot be integrated in the way seen for the undamped systems in Section 18.4. Each case has to be studied in a particular way, often yielding only qualitative results. To obtain quantitative results, it is usually necessary to resort to approximate methods, like iterative or perturbation procedures, or to the numerical integration of the equations of motion.

Remark 20.3 *Note that the motion is, strictly speaking, not periodic since the amplitude decreases (or increases) in time. The Ritz averaging technique cannot be used unless the system is lightly damped, and thus the reduction of amplitude in each period is small enough to be neglected.*

Example 20.1 *Consider a nonlinear system in which a damper following the Coulomb model is added to a device providing a restoring force having no damping component. The term Coulomb damping is commonly used for a drag force that is independent of the speed. It has important practical applications since it can be used as a first-approximation model for dry friction.*
Equation (20.12) can be used, and function $\beta(\dot{x})$ can be expressed in the form

$$\beta(\dot{x}) = F \frac{\dot{x}}{|\dot{x}|} ,$$

where F is the absolute value of the drag force. By introducing the afore-mentioned expression for $\beta(\dot{x})$ into Eq. (20.13), the following equation can be obtained:

$$\frac{1}{2}\frac{d(\dot{x})^2}{dx} = \frac{-f(x) \mp F}{m},$$

where the upper sign holds when the velocity is positive. It can be integrated as

$$(\dot{x})^2 = \frac{2}{m} \int \left[-f(x) \mp F\right] + C,$$

with the obvious limitation that when the sign of the velocity changes, the integration must be interrupted and resumed with a different sign and a different value of the constant C. By separating the variables and integrating, the following relationship between t and x can be obtained:

$$t = \pm\sqrt{\frac{m}{2}} \int \frac{du}{\sqrt{\int \left[-f(x) \mp F\right] dx + C}} + C_1.$$

This equation can be studied in a way not much different from that seen for undamped systems in Section 18.4.

Assume that force $f(x)$ is linear with displacement x ($f(x) = kx$). If at time $t = 0$ the mass is displaced in the position $x = x_0$ (with positive value of x_0) and then released with vanishingly small velocity, the lower sign in the equation must be chosen at the beginning, because the velocity is negative, at least for a while.

This holds only if the restoring force is strong enough to overcome dry friction, i.e., $x_0 > F/k$; otherwise no motion results. Actually a greater value of the friction is to be expected when the mass m is at standstill, but this effect will be neglected here. It is easy to perform the integration and to compute the value of constant C, obtaining

$$\begin{cases} (\dot{x})^2 = \dfrac{2}{m}\left(-\dfrac{kx^2}{2} + Fx + \dfrac{kx_0^2}{2} - Fx_0\right), \\[2mm] t = \pm\sqrt{\dfrac{m}{2}} \displaystyle\int \dfrac{du}{\sqrt{x_0^2 - 2\frac{F}{k}x_0 + 2\frac{F}{k}u - u^2}} + C_1, \end{cases}$$

where constant C_1 and the sign must be chosen so that $x = x_0$ for $t = 0$ and that x decreases with increasing time. By performing the integration it follows that

$$t = \pm\sqrt{\frac{m}{2}} \arcsin\left(\frac{kx - F}{kx_0 - F}\right) + C_1,$$

or, expressing x as a function of time and introducing a phase angle Φ,

$$x = \left(x_0 - \frac{F}{k}\right)\sin\left(\sqrt{\frac{k}{m}}t + \Phi\right) + \frac{F}{k} = \left(x_0 - \frac{F}{k}\right)\cos\left(\sqrt{\frac{k}{m}}t\right) + \frac{F}{k}.$$

The last expression is obtained considering that at time $t = 0$ the mass m is in x_0 and the phase angle Φ is equal to $\pi/2$. It holds for half a period ($0 < t < \pi\sqrt{m/k}$).

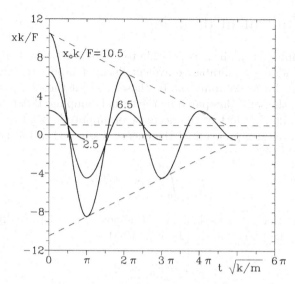

FIGURE 20.2. Free oscillations of a linear system with dry friction (Coulomb damping).

The motion in the first half-period follows a harmonic law, with the only difference with respect to the undamped case of being centered about the value F/k instead of being centered on the time axis. The study can then proceed in the same way for the second half-period, remembering that the sign of the velocity is changed and that the starting value of x is that obtained at the end of the first half-period.

At the end of a full period the position of mass m is $x_2 = x_0 - 4F/k$.

The amplitude decreases by the quantity $4F/k$ in a period. The motion then proceeds, as shown in Fig. 20.2, with half-periods of a cosine law, all with the same frequency (that of the undamped system), each one with an amplitude reduced by the amount $2F/k$. Eventually the mass stops within the band with half-width F/k and all motion is extinguished.

The motion so obtained has some similarities with that obtained for viscous damping, in particular where the frequency is concerned, but with the important difference that the law of motion is included, at least within the usual approximations, between two straight lines of equation

$$x = \pm \left(x_0 - \frac{2F}{\pi\sqrt{mk}} t \right) ,$$

and that the motion extinguishes in a finite time in a position that is different from the equilibrium position of the spring. The number n of half-periods needed to stop the system is

$$n = int\left(\frac{x_0 k}{2F} + \frac{1}{2} \right) .$$

20.4   Equivalent damping

When damping is small, it is possible to approximate the behavior of the system by adding a suitable equivalent linear damping to the undamped system. This linear damping can be chosen so that the energy it dissipates in a cycle is the same dissipated by the actual damping. If damping is small enough to allow neglect of the decrease of amplitude occurring in a cycle and neglecting also preloading, the energy dissipated in the period T can be computed as

$$\mathcal{L} = \int_0^T -f(x, \dot{x})\dot{x}\,dt \,. \tag{20.14}$$

Assuming that the time history of the motion is substantially harmonic, the law $x(t)$ can be expressed by Eq. (19.5) as

$$x = x_m \sin(\omega t) \,,$$

obtained by neglecting all harmonics except the fundamental. The expression of the energy dissipated is thus

$$\mathcal{L} = -x_m \int_0^{2\pi} -f(x, \dot{x})\cos(\omega t)d(\omega t) \,. \tag{20.15}$$

The energy dissipated in a cycle by the equivalent linear damping is

$$\mathcal{L}_{eq} = -\int_0^T c_{eq}\dot{x}^2 dt = -x_m^2 \omega c_{eq} \int_0^{2\pi} \cos^2(\omega t)d(\omega t) = x_m^2 c_{eq}\pi\omega \,. \tag{20.16}$$

By equating the two expressions for the energy dissipated in a cycle, the following value of the equivalent damping is obtained

$$c_{eq} = \frac{1}{x_m \pi \omega} \int_0^{2\pi} f(x, \dot{x})\cos(\omega t)d(\omega t) = \frac{1}{x_m \pi \omega} \int_0^{2\pi} \beta(\dot{x})\cos(\omega t)d(\omega t) \,; \tag{20.17}$$

the second expression holds if Eq. (20.12) can be used, i.e., if the restoring force and the damping force are independent of each other. The integral in Eq. (20.17) can be easily computed, at least numerically, once that law $f(x, \dot{x})$ (or $\beta(\dot{x})$) has been defined. The equivalent damping depends thus on both the amplitude and the frequency of the motion.

Remark 20.4 *This definition of the equivalent damping is consistent with that given in Section 3.4.3 for structural damping. In that case, however, the equivalent damping was independent from the amplitude of the motion, because structural damping is a form of linear damping.*

Example 20.2 *Compute the equivalent damping for aerodynamic damping. If the Reynolds number is high enough, aerodynamic forces are proportional to the square of the velocity; law $\beta(\dot{x})$ can thus be expressed as*

$$\beta(\dot{x}) = c\dot{x}|\dot{x}| \ .$$

By introducing this law into Eq. (20.17), it follows that

$$c_{eq} = \frac{cx_m\omega}{\pi}\left[\int_{-\pi/2}^{\pi/2}\cos^3(\omega t)d(\omega t) + \int_{\pi/2}^{3\pi/2}-\cos^3(\omega t)d(\omega t)\right] = \frac{8cx_m\omega}{3\pi} \ .$$

20.5   Solution in the state plane

The equation yielding the trajectories in the state plane is still Eq. (20.18), where function f now depends on both position and velocity:

$$\frac{dv}{dx} = -\frac{f(x,\dot{x})}{mv} \ . \tag{20.18}$$

In general, it is impossible to integrate it in closed form, but there is no difficulty in obtaining the state portrait by numerical integration.

For a system with a Duffing-type restoring force expressed by Eq. (18.15), linear damping, and no excitation, the equation of the trajectories in the state plane is

$$\frac{dv}{dx} = -\frac{kx(1 + \mu x^2) + cv}{mv} \ . \tag{20.19}$$

Stating a nondimensional velocity

$$v^* = v\sqrt{|\mu|}m/k$$

and a nondimensional coordinate

$$x^* = x\sqrt{|\mu|} \ ,$$

the behavior of the system depends on a single nondimensional parameter, the damping ratio

$$\zeta = \frac{c}{2\sqrt{km}} \ .$$

The state portrait is plotted in Fig. 20.3 in nondimensional form for a value of $\zeta = 0.1$. The two plots are related to hardening (Fig. 20.3a) and softening systems (Fig. 20.3b).

Remark 20.5 *The state portraits are qualitatively different from those related to the undamped system reported in Fig. 18.7.*

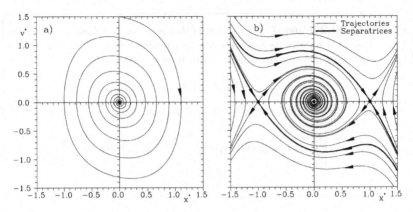

FIGURE 20.3. State portrait of a system whose restoring force is expressed by
Eq. (18.15). Nondimensional velocity $v^* = v\sqrt{|\mu|}\,m/k$ as a function of nondimen-
sional displacement $x^* = x\sqrt{|\mu|}$ for (a) hardening and (b) softening systems.
Nondimensional damping $\zeta = c/2\sqrt{km} = 0.1$.

If the system is hardening there is a single equilibrium position and the
trajectories are spirals that wind inward, toward the equilibrium point in
the origin. Because all trajectories tend to the singular point, this is an
attractor and its basin of attraction extends to the whole state plane. A
singular point of this type is said to be a *stable focus*.

The state portrait for softening systems (Fig. 20.3b) is more complex. In
this case there are three singular points, corresponding to the equilibrium
positions. The one in the center is again a stable focus, but now its basin
of attraction does not encompass the whole state plane, but only the zone
that lies within the two separatrices.

The other two singular points are saddle points and are repellors because
the state trajectories tend to depart from them, even if they can be initially attracted. The parts of the plane outside the basin of attraction of
the stable equilibrium positions are characterized by nonperiodic motions,
which are essentially not different from the ones seen for the undamped
system.

If the linear term is repulsive instead of attractive, i.e., the stiffness k
is negative, the state portraits are of the type shown in Fig. 20.4. When
μ is also negative, i.e., the coefficient of the term containing the third
power of the displacement is positive, the behavior is of the hardening type
(Fig. 20.4a). There are three static equilibrium positions: one unstable in
the origin and two stable ones. The corresponding singular points are a
saddle point and two stable foci.

The separatrix defines two domains of attractions, whose shape in this
case is quite intricate. The domain of the focus on the right has been

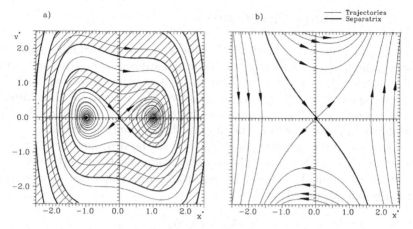

FIGURE 20.4. State portrait of a system whose restoring force is expressed by Eq. (18.15), as in Fig. 20.3, but with a negative value of the stiffness k. Nondimensional velocity $v^* = v\sqrt{|\mu|}m/k$ as a function of nondimensional displacement $x^* = x\sqrt{|\mu|}$ for (a) hardening (in this case with negative value of μ) and (b) softening systems. Nondimensional damping $\zeta = c/2\sqrt{|k|m} = 0.1$.

hatched in the figure to make the state portrait easier to understand. All motions end up in either of the foci, but the interwoven nature of the domains of attraction makes the final result quite sensitive to the initial condition. Consider, for example, a motion starting from rest with point P on the negative side of the x-coordinate. If the initial value of x is small, the motion is attracted toward the equilibrium position at the left. With increasing absolute value of x, however, motions that are attracted alternatively by the two different equilibrium positions are found.

In the case of softening systems with repulsive linear stiffness, no oscillatory motion is possible and nonperiodic motions occur in all domains of attraction. The origin is again a saddle point.

The types of singular points seen up to this point (centers, foci, and saddle points) are not the only possible ones. A *singular point* is an equilibrium position of the system; it can easily be found from Eq. (20.2), which, without loss of generality, can be written as

$$f(x_0, 0) = 0 ,$$

because static forces acting on the system can be included in function $f(x, \dot{x})$.

The free motion in the small in the vicinity of the singular point can easily be studied using the homogeneous linear equation associated with Eq. (20.9).

20.6   Stability in the small

The usual practical definitions of static and dynamic stabilities can be used. The first one defined a case when the stiffness is positive and point P is attracted toward the equilibrium position when displaced, while in the second case damping is zero or positive. If it is zero, the time history does not diverge, while if it is positive the time history actually reaches the equilibrium position, at least asymptotically (asymptotic stability).

More elaborate definitions for stability are needed, however. Consider a vector $\vec{X}(t)$ in the state plane, whose components are $x(t)$ and $v(t)$, and use the symbol $|\vec{X}(t)|$ for its euclidean norm. Following Liapunov, an equilibrium position defined by

$$\vec{X}_0 = \{x_0, 0\}$$

is stable if, for any arbitrarily small positive quantity ϵ, there exists a positive quantity δ such that the inequality

$$|\vec{X}(t) - \vec{X}_0| < \epsilon \qquad \text{for } 0 \leq t < \infty \tag{20.20}$$

holds if

$$|\vec{X}(0) - \vec{X}_0| < \delta \ .$$

This means that any trajectory starting within a circle of radius δ centered in the equilibrium point in the state plane remains within a circle of radius ϵ for all values of time. If to this condition it is added that

$$\lim_{t \to \infty} |\vec{X}(t) - \vec{X}_0| = 0 , \tag{20.21}$$

then the equilibrium position is asymptotically stable. This definition also holds for systems with many degrees of freedom, provided vector $\vec{X}(t)$ has as many components as the number of dimensions of the state space. The linearized homogeneous equation of motion associated with Eq. (20.9) has a solution of the type

$$x = x_0 + ae^{st} . \tag{20.22}$$

By introducing solution (20.22) into the equation of motion (20.9), the following characteristic equation yielding the value of s is readily obtained:

$$s^2 + \left(\frac{\partial f}{\partial \dot{x}}\right)_{\substack{x = x_0 \\ \dot{x} = 0}} s + \left(\frac{\partial f}{\partial x}\right)_{\substack{x = x_0 \\ \dot{x} = 0}} = 0 . \tag{20.23}$$

If the roots of the equation are real and have the same sign, the singular point is a node. If they are positive the node is unstable; if they are negative the node is stable. An example of a stable node is the equilibrium position of an overdamped linear system.

If the roots of the equation are real but their signs are different, the singular point is a saddle point. An example was shown in Fig. 18.7b. The motion in the vicinity of a node or saddle point is nonperiodic.

If the roots of the equation are imaginary and obviously have opposite signs, since they must be conjugate, the singular point is a center and the motion is an undamped oscillation. An example of a center is the equilibrium position of Fig. 18.6a.

If the roots of the equation are complex conjugate, the singular point is a focus. If their real part is positive, the focus is unstable; if it is negative, the focus is stable. An example of a stable focus is the equilibrium position of Fig. 20.3a. The motion is a damped or self-excited oscillation.

A stable node or stable focus is strictly attractors; centers can also be called attractors because all state trajectories lying within a certain basin orbit about them.

The roots of Eq. (20.23) can be reported on the complex plane, obtaining a plot of the same type as those shown in Table 20.1. As already stated

TABLE 20.1. Various types of singular points for a system with a single degree of freedom together with the corresponding position of the roots of Eq. (20.23) on the Argand plane. Note that the center has been considered as an intermediate situation between stable and unstable singular points.

for linear systems, this type of representation is very useful, particularly when the stability of a system with many degrees of freedom has to be studied: If any root lies at the right of the imaginary axis, i.e., it has a positive real part, the system is unstable. If the influence of the variation of a parameter on the stability of the system has to be studied, the plot can be drawn for different values of the relevant parameter. Each root then describes a curve on the complex plane (the root locus), and the onset of instability corresponds to the value of the parameter that causes one of the curves to cross the imaginary axis. Also, if stability in the large is considered, apart from singular points, there are attractors and repellors of another type: the limit cycles.

Remark 20.6 *If the number of dimensions of the state space is greater than two, other types of singular points exist, and other types of attractors, as toroidal attractors and strange attractors, can be present.*

20.7 The Van der Pol oscillator

The Van der Pol oscillator is described by the equation

$$m\ddot{x} + c(x^2 - 1)\dot{x} + kx = 0, \qquad \text{with } c > 0. \tag{20.24}$$

By introducing the nondimensional time, the damping ratio, and the natural frequency of the linearized system

$$\tau = \omega_0 t \, , \quad \zeta = \frac{c}{2\sqrt{mk}} \, , \quad \omega_0 = \sqrt{\frac{k}{m}} \, ,$$

the equation of motion can be written in the nondimensional form

$$\frac{d^2x}{d\tau^2} + 2\zeta(x^2 - 1)\frac{dx}{d\tau} + x = 0. \tag{20.25}$$

The state portrait, obtained by numerical integration with $\zeta = 0.1$, is shown in Fig. 20.5. The central singular point is an unstable focus. All trajectories tend to a stable limit cycle, which has a shape tending to a circle when $\zeta \rightarrow 0$. Within the limit cycle the behavior of the system is self-excited; outside, it is damped.

The time histories obtained for three different values of ζ ($\zeta = 0.05$, $\zeta = 0.5$, $\zeta = 50$) are shown in Fig. 20.6. In all three cases the starting point in the state plane is the origin, or, better, a point close to the origin since in the numerical simulation starting from an unstable point is not sufficient to allow self-excitation of the oscillation. The numerical integration was performed for 100 units of the nondimensional time to allow the amplitude to grow, and then other 100 units were recorded.

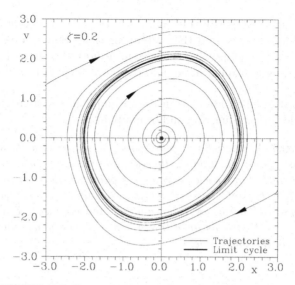

FIGURE 20.5. State portrait of the Van der Pol oscillator.

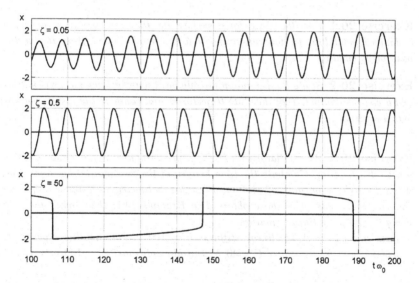

FIGURE 20.6. Time histories of the free oscillations of a Van der Pol oscillator with three different values of ζ ($\zeta = 0.05$, $\zeta = 0.5$, $\zeta = 50$).

For small values of ζ the waveform is almost harmonic and its frequency almost coincides with that of the linearized system (ω_0). If the value of ζ is large, the waveform is almost a square wave, i.e., the oscillator snaps between two positions. This behavior is often called *relaxation oscillations*.

Notice that a damper of this kind dissipates energy at large amplitudes and introduces energy into the system for small amplitudes. It must be able to exchange energy with the system in both directions (as is usually said, to work in four quadrants), and then must be considered as an active system.

The Van der Pol oscillator is just an example of a system giving way to a limit cycle surrounding an equilibrium point.

Remark 20.7 *If a single singular point is surrounded by several limit cycles nested within each other, they are alternatively stable and unstable, with the unstable limit cycles acting as separatrices between the attractors represented by the stable ones. Sometimes two limit cycles, a stable and an unstable one, can merge into one another and give way to a semi-stable limit cycle, which acts as an attractor on one side and as a repellor, or separatrix, on the other side.*

20.8 Exercises

Exercise 20.1 *Plot the state force surface for the restoring force acting on a damped pendulum without assuming that the displacement from the vertical position is small.*

Exercise 20.2 *Study the pendulum already dealt with in Exercise 18.1, but now add a viscous damper. Plot the state trajectories and compare the results obtained with Fig. 20.3b.*

Exercise 20.3 *An oscillator with Duffing-type hardening restoring force is damped by dry friction. Compute the time history of the free motion by numerically integrating the equations of motion. First use a linearized restoring force and compare the results with those obtained in Example 20.1; then repeat the study using the fully nonlinear equation.*

Data mass $m = 1$ kg, linear stiffness $k = 100$ kN/m; nonlinear parameter $\mu = 500$ $1/m^2$, friction force $F = 40$ N. Initial conditions: $x_0 = 100$ mm, $v_0 = 0$.

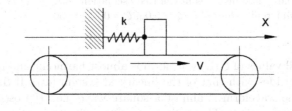

FIGURE 20.7. Spring–mass system sliding on a moving belt.

Exercise 20.4 *A mass m is linked by a spring with stiffness k to a fixed point and slides on a belt moving at a speed V (Fig. 20.7). Assuming that the friction coefficient between the belt and the mass is f when there is relative motion and f_s ($f_s > f$) when the mass is not moving with respect to the belt, compute the time history of the mass using the state-plane approach. Compute the time history also for the case with $f_s = f$.*

Data: $m = 10$ kg, $k = 500$ N/m, $f = 0.6$, $f_s = 1.2$, $V = 1$ m/s, $g = 9.81$ m/s^2. Initial condition: at time $t = 0$ the mass is in the equilibrium point ($x_0 = 0$) and does not slip on the belt ($v_0 = V$).

21

Forced Response of Damped Nonlinear Systems

Damped nonlinear systems may exhibit a wide variety of forced responses, even when they are excited by the simplest harmonic functions. These responses may span from quasi-harmonic or poly-harmonic to very complex nonperiodic. While approximated analytical techniques can yield approximations of the former, only experimental studies, either based on physical experiments or on numerical integration in time, can yield a complete picture. In particular, even very simple systems may exhibit complex chaotic behavior: this issue has not yet been fully understood.

21.1   Reduction of the size of the problem

There are cases where not all the equations of motion of the system are nonlinear, or at least the mathematical model can be reduced to a formulation of this kind. A case of practical interest is, for instance, that of a linear system constrained by a number of supports having nonlinear behavior.

Since the solution of a nonlinear set of equations becomes rapidly more difficult with increasing number of equations, it would be very expedient to reduce the size of the nonlinear set of equation to a minimum.

Consider a system with many degrees of freedom modeled by an equation of the type of Eq. (18.4), with a damping matrix added and functions g_i not depending explicitly on time. Assume also that some of the functions g_i vanish.

The degrees of freedom can be reordered, listing first those corresponding to equations containing a nonlinear part and then all the others. The equations of motion can thus be written in the form

$$
\begin{bmatrix} \mathbf{M}_{11} & \mathbf{M}_{12} \\ \mathbf{M}_{21} & \mathbf{M}_{22} \end{bmatrix} \left\{ \begin{array}{c} \ddot{\mathbf{x}}_1 \\ \ddot{\mathbf{x}}_2 \end{array} \right\} + \begin{bmatrix} \mathbf{C}_{11} & \mathbf{C}_{12} \\ \mathbf{C}_{21} & \mathbf{C}_{22} \end{bmatrix} \left\{ \begin{array}{c} \dot{\mathbf{x}}_1 \\ \dot{\mathbf{x}}_2 \end{array} \right\} +
$$
$$
+ \begin{bmatrix} \mathbf{K}_{11} & \mathbf{K}_{12} \\ \mathbf{K}_{21} & \mathbf{K}_{22} \end{bmatrix} \left\{ \begin{array}{c} \mathbf{x}_1 \\ \mathbf{x}_2 \end{array} \right\} + \mu \left\{ \begin{array}{c} \mathbf{g} \\ \mathbf{0} \end{array} \right\} = \left\{ \begin{array}{c} \mathbf{f}_1(t) \\ \mathbf{f}_2(t) \end{array} \right\} . \tag{21.1}
$$

A possible way to get rid of a number of linear equations, thus reducing the size of the problem, is by resorting to Guyan reduction. Some of the nonlinear degrees of freedom are regarded as master degrees of freedom while the others are considered as slave: The usual Guyan reduction technique allows thus to eliminate some of the equations, yielding a set of equations that are fewer in number than the original ones. Functions g_i may even contain slave coordinates, but this does not introduce difficulties because they are easily computed from the master coordinates through Eq. (10.2).

Remark 21.1 *As usual with Guyan reduction, the procedure yields only approximate results; if a large-scale condensation leading to the suppression of all linear degrees of freedom is performed, these approximations may well lead to unacceptable results.*

The extent of the reduction of the size depends on the nature of the system and a suitable strategy needs to be decided in each case. There are cases in which it is possible to eliminate all linear degrees of freedom, reaching a nonlinear system whose size is as small as possible, but usually a trade-off between complexity and precision must be attempted.

A completely different approach is resorting to modal reduction. Following that seen in Section 10.7, a coordinate transformation matrix

$$
\boldsymbol{\Psi} = \begin{bmatrix} \mathbf{I} & \mathbf{0} \\ -\mathbf{K}_{22}^{-1}\mathbf{K}_{21} & \boldsymbol{\Phi} \end{bmatrix} \tag{21.2}
$$

can be defined, where $\boldsymbol{\Phi}$ is the eigenvector matrix of the MK system (i.e., in this case, the undamped linearized system) whose matrices are $\mathbf{M}_{22}$ and $\mathbf{K}_{22}$.

As usual with coordinate transformations based on the eigenvector matrix, it is possible to use a reduced number of modes by resorting to the reduced eigenvector matrix $\boldsymbol{\Phi}^*$. Operating in this way matrix $\boldsymbol{\Psi}$ is no more square and the final number of degrees of freedom is reduced, but the nonlinear part of the equation is not affected.

The reduction is only approximated, not only owing to the coupling due to damping but also because the nonlinearities affect the behavior of the whole system, introducing a further coupling.

21.2   First approximation of the response to a harmonic forcing function

21.2.1   Systems with a single degree of freedom

Consider a nonlinear system with a single degree of freedom excited by a harmonic forcing function. If the equation of motion is written separating the linear part of the restoring and damping forces from the nonlinear part expressed by function $f'(x, \dot{x})$ and introducing the natural frequency of the linearized system ω_0 and its damping ratio ζ, the equation of motion is

$$\ddot{x} + 2\zeta\omega_0\dot{x} + \omega_0^2 x + \frac{\omega_0^2}{k} f(x, \dot{x}) = \frac{\omega_0^2}{k} f_0 \sin(\omega t) \,. \qquad (21.3)$$

As it was done when searching for a first approximation of the undamped system, assume that the response is harmonic in time. Due to damping, the response will not be in phase with the excitation, and when searching a first approximation of the fundamental harmonic, two unknowns must be stated: the amplitude and the phase of the response, or the amplitude in phase and the amplitude in quadrature. Instead of working in phase with the excitation, it is simpler to assume that time $t = 0$ is the instant when the response vanishes, i.e., to work in phase with the response. The response can thus be assumed to be of the type of Eq. (19.5)

$$x = x_m \sin(\omega t) \,,$$

and the unknown phase can be included in the expression of the forcing function, which becomes

$$f_0 \sin(\omega t + \Phi) \,.$$

Using the Ritz averaging technique, it follows that

$$\int_0^T \left[\ddot{x} + 2\zeta\omega_0\dot{x} + \omega_0^2 x + \frac{\omega_0^2}{k} f(x, \dot{x}) - \frac{\omega_0^2}{k} f_0 \sin(\omega t + \Phi) \right] \phi(t) dt = 0 \,,$$

$$\qquad (21.4)$$

where the arbitrary function $\phi(t)$ has been assumed as

$$\phi(t) = \sin(\omega t) \,,$$

and the unknown a is the amplitude x_m. By remembering that

$$\int_0^{2\pi} \sin(\alpha) \cos(\alpha) d\alpha = 0 \,,$$

and introducing the nondimensional time $\tau = \omega t$, it follows that

$$\left[1 - \left(\frac{\omega}{\omega_0} \right)^2 \right] x_m - \frac{f_0}{k} \cos(\Phi) + \frac{1}{k\pi} \int_0^{2\pi} f'(x, \dot{x}) \sin(\tau) d\tau = 0 \,. \quad (21.5)$$

Equation (21.5) contains the two unknowns, x_m and Φ, and another equation, which can be readily obtained by observing that the energy dissipated by the system must be equal to the work performed by the excitation, is needed. Again, instead of equating the relevant quantities in each instant, the equation is written as an average in a cycle. The energy dissipated in a cycle is expressed by the integral

$$\mathcal{L}_d = \int_0^T \left[\ddot{x} + 2\zeta\omega_0\dot{x} + \omega_0^2 x + \frac{\omega_0^2}{k} f'(x, \dot{x}) \right] \dot{x}\,dt . \qquad (21.6)$$

By introducing the expression of the velocity $\dot{x}$ into Eq. (21.6) and integrating, it follows that

$$\mathcal{L}_d = 2\pi\zeta\omega_0\omega^2 x_m + \frac{\omega\omega_0^2 x_m}{k} \left[\int_0^{2\pi} f'(x, \dot{x}) \cos(\tau)d\tau \right] . \qquad (21.7)$$

The work performed by the forcing function in a cycle is

$$\mathcal{L}_f = \frac{\omega_0^2}{k} \int_0^T f_0 \sin(\omega t - \Phi)\dot{x}\,dt = \frac{\pi x_m \omega \omega_0^2}{k} f_0 \sin(\Phi) . \qquad (21.8)$$

Equating expressions (21.7) and (21.8), the required equation is obtained:

$$2\zeta\frac{\omega}{\omega_0} x_m + \frac{1}{k\pi} \int_0^{2\pi} f'(x, \dot{x}) \cos(\tau)d\tau - \frac{f_0}{k} \sin(\Phi) = 0 . \qquad (21.9)$$

Equations (21.5) and (21.9) allow the phase and the amplitude of the response to be computed, once function $f'(x, \dot{x})$ has been stated.

Often, however, the problem can be solved in a simpler way. Assume that the nonlinear restoring force can be expressed by function $f'(x)$ while the nonlinear damping force can be expressed by function $\beta'(\dot{x})$ and that $f'(x)$ and $\beta'(\dot{x})$ are odd functions of x and $\dot{x}$, respectively. This means that $\beta'(-\dot{x}) = -\beta'(\dot{x})$ and $f'(-x) = -f'(x)$. Displacement $x(t)$ expressed by Eq. (19.5) is itself an odd function of time, while its derivative, the velocity, is an even function of time. From this, it follows that functions $\beta'(\tau)$ and $f(\tau)$ are even and odd, respectively. A consequence of this consideration is that

$$\int_0^{2\pi} \beta'(\dot{x}) \sin(\tau)d\tau = 0 ; \qquad \int_0^{2\pi} f'(x) \cos(\tau)d\tau = 0 . \qquad (21.10)$$

Note that if functions $\beta'(\dot{x})$ and $f'(x)$ are not odd, the center of the forced oscillations does not coincide with the static equilibrium position, and the assumption that the response can be approximated with a simple harmonic motion of the type of Eq. (19.5) becomes unacceptable. At least a constant term of the displacement must be included, and the whole procedure shown here must be modified.

Actually, if Eq. (21.10) is not satisfied, the restoring force $f'(x)$ would enter Eq. (21.9), i.e., would participate in the energy dissipation, which is clearly not the case.

By replacing function $f'(x, \dot{x})$ with functions $\beta'(\dot{x})$ and $f'(x)$ and remembering Eq. (21.10), Eqs. (21.5) and (21.9) reduce to

$$
\begin{cases}
\left[1 - \left(\dfrac{\omega}{\omega_0}\right)^2\right] x_m + \dfrac{1}{k\pi} \displaystyle\int_0^{2\pi} f'(x) \sin(\tau)d\tau = \dfrac{f_0}{k} \cos(\Phi), \\[4mm]
2x_m\zeta \dfrac{\omega}{\omega_0} + \dfrac{1}{k\pi} \displaystyle\int_0^{2\pi} \beta'(\dot{x}) \cos(\tau)d\tau = \dfrac{f_0}{k} \sin(\Phi).
\end{cases}
\tag{21.11}
$$

By adding the squares of the two equations (21.11), a relationship allowing the amplitude of the response to be computed as a function of the driving frequency is readily obtained:

$$
\left\{\left[1 - \left(\frac{\omega}{\omega_0}\right)^2\right] x_m + \frac{1}{k\pi} \int_0^{2\pi} f'(x) \sin(\tau)d\tau\right\}^2 +
$$
$$
+ \left[2x_m\zeta \frac{\omega}{\omega_0} + \frac{1}{k\pi} \int_0^{2\pi} \beta'(\dot{x}) \cos(\tau)d\tau\right]^2 = \left(\frac{f_0}{k}\right)^2.
\tag{21.12}
$$

The phase can be easily computed by dividing the second equation (21.11) by the first:

$$
\Phi = \arctan \left\{ \frac{2x_m\zeta \dfrac{\omega}{\omega_0} + \dfrac{1}{k\pi} \displaystyle\int_0^{2\pi} \beta'(\dot{x}) \cos(\tau)d\tau}{\left[1 - \left(\dfrac{\omega}{\omega_0}\right)^2\right] x_m + \dfrac{1}{k\pi} \displaystyle\int_0^{2\pi} f'(x) \sin(\tau)d\tau} \right\}.
\tag{21.13}
$$

21.2.2 Systems with many degrees of freedom

If the system has many degrees of freedom, but is excited by a harmonic forcing function, it is possible to proceed in the same way. If only a first approximation of the fundamental harmonic of the response to a harmonic forcing function

$$
\mathbf{f} = \mathbf{f}_{0s} \sin(\omega t) + \mathbf{f}_{0c} \cos(\omega t)
\tag{21.14}
$$

is to be obtained, a solution of the type

$$
\mathbf{x} = \mathbf{x}_{0s} \sin(\omega t) + \mathbf{x}_{0c} \cos(\omega t)
\tag{21.15}
$$

can be assumed.

If the excitation is not synchronized, it is impossible to state a single phase reference and thus it is advisable to use as unknowns directly the sine and cosine components of the response.

Operating in the same way seen for systems with a single degree of freedom, and applying the Ritz averaging technique, it follows that

$$\int_0^T [\mathbf{M}\ddot{x} + \mathbf{C}\dot{x} + \mathbf{K}x + \mu\mathbf{g}(\mathbf{x}, \dot{\mathbf{x}}) - \mathbf{f}_{0s}\sin(\omega t) - \mathbf{f}_{0c}\cos(\omega t)]\sin(\omega t)dt = 0,$$
(21.16)

$$\int_0^T [\mathbf{M}\ddot{x} + \mathbf{C}\dot{x} + \mathbf{K}x + \mu\mathbf{g}(\mathbf{x}, \dot{\mathbf{x}}) - \mathbf{f}_{0s}\sin(\omega t) - \mathbf{f}_{0c}\cos(\omega t)]\cos(\omega t)dt = 0.$$
(21.17)

Introducing the solution (21.15) and performing the integrations, the following set of equations is obtained

$$\begin{cases} \left(\mathbf{K} - \omega^2\mathbf{M}\right)\mathbf{x}_{0s} - \omega\mathbf{C}\mathbf{x}_{0c} + \mathbf{f}_{ns} = \mathbf{f}_{0s}, \\ \omega\mathbf{C}\mathbf{x}_{0s} + \left(\mathbf{K} - \omega^2\mathbf{M}\right)\mathbf{x}_{0c} + \mathbf{f}_{nc} = \mathbf{f}_{0c}, \end{cases}$$
(21.18)

where

$$\mathbf{f}_{ns}(\mathbf{x}_{0s}, \mathbf{x}_{0c}) = \frac{\omega}{\pi}\int_0^T \mu\mathbf{g}(\mathbf{x}, \dot{\mathbf{x}})\sin(\omega t)dt,$$
(21.19)

$$\mathbf{f}_{nc}(\mathbf{x}_{0s}, \mathbf{x}_{0c}) = \frac{\omega}{\pi}\int_0^T \mu\mathbf{g}(\mathbf{x}, \dot{\mathbf{x}})\cos(\omega t)dt.$$
(21.20)

By introducing a complex amplitude

$$\mathbf{x}_0 = \mathbf{x}_{0s} + i\mathbf{x}_{0c},$$
(21.21)

the equation reduces to

$$\mathbf{K}_{dyn}\mathbf{x}_0 + \mathbf{f}_{ns}(\mathbf{x}_{0s}, \mathbf{x}_{0c}) + i\mathbf{f}_{nc}(\mathbf{x}_{0s}, \mathbf{x}_{0c}) = \mathbf{f}_{0s} + i\mathbf{f}_{nc},$$
(21.22)

where

$$\mathbf{K}_{dyn} = \mathbf{K} - \omega^2\mathbf{M} + i\omega\mathbf{C}$$

is the usual dynamic stiffness matrix of the linearized system, which is generally complex.

If not all degrees of freedom are related to the nonlinear behavior of the system, the dynamic stiffness matrix can be partitioned in the same way as the matrices in Eq. (21.1) and subjected to *dynamic reduction* (i.e., to the usual procedure of static reduction but performed on the dynamic stiffness matrix). The fact that when the system is damped the dynamic stiffness matrix is complex does not complicate the relevant computations.

Remark 21.2 *The dynamic stiffness matrix is a function of the frequency of the forcing function ω and then when computing the frequency response of the system, the reduction procedure has to be repeated for each value of the frequency. Moreover, no result can be obtained at those frequencies for which matrix $\mathbf{K}_{dyn_{22}}(\omega)$ is singular.*

This technique allows the reduction of the number of nonlinear equations to be solved, but the computation of the response, although easier than that of the original system, is still a formidable problem. Since dynamic reduction does not introduce any error, the only source of approximation is having assumed that the response is harmonic, a thing that never holds exactly in the case of nonlinear vibrating systems.

If the system is undamped (i.e., with $\mathbf{C} = 0$ and functions g_i depending only on the coordinates) and all forcing functions are in phase, also all the responses, reduced to their fundamental harmonic, are in phase with each other and with the excitation. Their time histories can be written in the form

$$\mathbf{f}(t) = \mathbf{f}_0 \sin(\omega t) , \quad \mathbf{x}(t) = \mathbf{x}_0 \sin(\omega t) .$$

The equations of motion of the reduced system are then

$$\mathbf{K}_{dyn_{cond}} \mathbf{x}_0 + \frac{1}{\pi} \int_0^{2\pi} \mathbf{g} \sin(\omega t) d(\omega t) = \mathbf{f}_{0_{cond}} , \qquad (21.23)$$

which is the required set of algebraic nonlinear equations.

Generally speaking, the integration can be performed without much difficulty. The problem of solving the set of nonlinear algebraic equations can, however, be difficult, particularly in the fields of frequency where more than one solution exists. There is a number of computation procedures that can be used, but the Newton–Raphson method is usually the best choice, although it cannot guarantee convergence in all cases.

Also in the case of systems with many degrees of freedom, it is possible to plot the backbone of the response, which is usually made of a number of separate branches equal to the number of natural frequencies of the linearized system. The backbone can be obtained simply by computing the response to a vanishingly small excitation.

21.3   Duffing's equation with viscous damping

Consider a nonlinear system governed by Duffing's equation, to which a linear viscous damper has been added. The equation of motion is

$$m\ddot{x} + c\dot{x} + kx(1 + \mu x^2) = f_0 \sin(\omega t) . \qquad (21.24)$$

The procedure seen in the preceding section can be used provided that

$$f'(x) = k\mu x^3 ; \qquad \beta'(\dot{x}) = 0 . \qquad (21.25)$$

Remembering that

$$\int_0^{2\pi} f'(x) \sin(\tau) d\tau = k\mu \int_0^{2\pi} [x_m \sin(\tau)]^3 \sin(\tau) d\tau = \frac{3\pi}{4} k\mu x_m^3 , \quad (21.26)$$

Equations (21.12) and (21.13) allow the amplitude and the phase of the response to be computed

$$\begin{cases} \left[(k - m\omega^2)x_m + \dfrac{3}{4}k\mu x_m^3 \right]^2 + x_m^2 c^2 \omega^2 = f_0^2, \\[4mm] \Phi = \arctan\left(\dfrac{c\omega}{k - m\omega^2 + \frac{3}{4}k\mu x_m^2} \right). \end{cases} \qquad (21.27)$$

The phase so obtained is that of the exciting force with respect to the displacement and is positive, i.e., the force leads the displacement.

By introducing the nondimensional frequency and amplitude of the displacement and the forcing function

$$\omega^* = \frac{\omega}{\omega_0}, \quad x_0^* = x_0\sqrt{|\mu|}, \quad f^* = \sqrt{|\mu|}\frac{f_0}{k}, \qquad (21.28)$$

and the damping ratio, the first equation (21.27), can be expressed in the following nondimensional form:

$$\omega^{*4} - 2\omega^{*2}\left[1 + \frac{3}{4}x_0^{*2} - 2\zeta^2 \right] + \left[1 + \frac{3}{4}x_0^{*2} \right]^2 - \left(\frac{f^*}{x_0^*} \right)^2 = 0. \qquad (21.29)$$

Equation (21.29) can be solved in the frequency and function $x_0^*(\omega^*)$ can then be obtained from function $\omega^*(x_0^*)$. Unfortunately, because such a plot is a function of two parameters, namely, excitation and damping, it is not possible to summarize the general behavior of the system in a single chart.

The amplitude and phase, computed with a single value of the excitation $\sqrt{\mu}f_0/k = 1$ and some different values of ζ, are plotted in Fig. 21.1. The

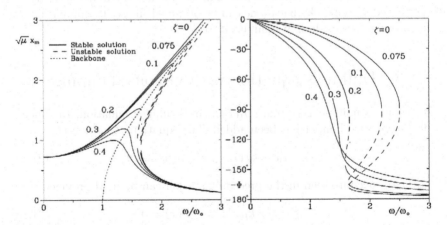

FIGURE 21.1. Amplitude and phase of the response of a system governed by Duffing's equation (*hardening spring*) with viscous damping under the effect of a harmonic excitation; $\sqrt{\mu}f_0/k = 1$. Phase $-\Phi$ of the displacement with respect to the force, which is negative, as in the case of linear systems, has been plotted.

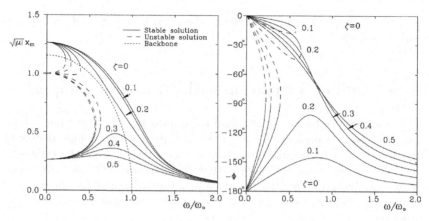

FIGURE 21.2. Same as Fig. 21.1, but for softening systems; $\sqrt{\mu}f_0/k = 0.25$.

plot refers to hardening systems. For $\zeta = 0$ the solution of the undamped system is obtained. With increasing values of the damping, the resonance peak gets smaller. The jump phenomenon is present only for small values of the damping, as can be easily inferred from the consideration that at high values of ζ only one possible value of the amplitude can be found at any value of the frequency. The same consideration can be drawn from the plot of the phase against the frequency. At high damping, the phase changes continuously from $0°$ to $180°$ with increasing frequency. If the damping is small enough, an abrupt variation of phase takes place when the system jumps from one configuration to the other. It is interesting to note that the frequency at which the phase takes the value of $90°$ depends on the damping. Such a value of the phase occurs when the amplitude curve crosses the resonance line, i.e., the backbone of the response.

The case of a softening system is shown in Fig. 21.2. At high values of damping the curves split into two parts; the lower one is that actually followed by the system and the upper one is merely a theoretical result, affected by strong approximations. A similar pattern is shown by the plot of the phase, in which the curves are constituted by two branches at low values of the damping, when the jump phenomenon is present. The phase takes a value of 0 at low frequency, to decrease slowly following the upper branch until the jump occurs. The phase then shifts beyond $-90°$, to decrease in absolute value slightly and then increases up to $-180°$ at very high frequency. At high values of damping, when no jump occurs, the phase follows the line on the right.

Remark 21.3 *The results here reported are approximate and refer only to the fundamental harmonic of the response. For a more complete evaluation of the response it is possible to resort to the harmonic balance approach or to perturbation techniques, which lead to long computations. The* *numerical integration of the equations of motion is nowadays the most common approach.*

21.4   Duffing's equation with structural damping

It is also possible to use the structural damping model in the case of non-linear systems. Consider, for example, a system with a single degree of freedom with a hardening spring whose characteristic is of the type seen in Duffing's equation, and assume that the response $x(t)$ can be approximated as a harmonic law. To introduce structural damping, use the complex stiffness model with regard to the linear part of the restoring force of the system

$$f(x) = kx_m \sin(\omega t + \eta) + \mu k x_m^3 \sin^3(\omega t). \qquad (21.30)$$

The linear part of the restoring force is out of phase with respect to the displacement, and a hysteresis cycle of the type shown in Fig. 21.3, curve a, is obtained (the phase is assumed to be equal to the loss factor η).

If the whole reaction $f(x)$ is assumed to lag the displacement (i.e., $\omega t + \eta$ is introduced also in the term in $\sin^3$), a hysteresis cycle of the type shown in Fig. 21.3, curve b, is obtained. The hysteresis cycles shown in Fig. 21.3 allow modeling the actual behavior of many machine elements as elastomeric springs and vibration insulators.

The response of a system governed by Duffing's equation with structural damping of the type of Eq. (21.30) can be easily computed:

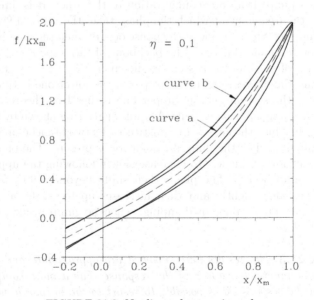

FIGURE 21.3. Nonlinear hysteresis cycles.

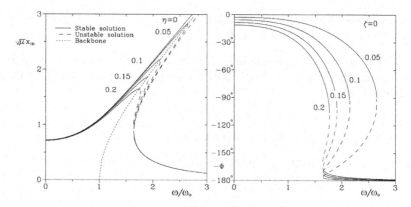

FIGURE 21.4. Amplitude and phase of the response of a system governed by Duffing's equation (*hardening spring*) with structural damping as defined by Eq. (21.30) under the effect of a harmonic excitation; $\sqrt{\mu}f_0/k = 1$.

$$
\begin{cases}
\omega^{*4} - 2\omega^{*2}\left[1 + \dfrac{3}{4}x_0^{*2}\right] + (1 + \eta^2)\left[1 + \dfrac{3}{4}x_0^{*2}\right]^2 - \left(\dfrac{f^*}{x_0^*}\right)^2 \,, \\[4mm]
\Phi = \arctan\left(\eta\dfrac{1 + \frac{3}{4}x_0^{*2}}{1 - \omega^{*2} + \frac{3}{4}x_0^{*2}}\right) .
\end{cases}
\tag{21.31}
$$

The response is plotted in Fig. 21.4 for $f^* = \sqrt{\mu}f_0/k = 1$ and various values of the loss factor η.

21.5   Backbone and limit envelope

Also in the case of damped systems it is possible to define the backbone of the response. If the amplitude of the forcing function vanishes, the first equation (21.11) yields the backbone of the response

$$
\frac{\omega}{\omega_0} = \sqrt{1 + \frac{1}{k\pi x_m}\int_0^{2\pi} f'(x)\sin(\tau)d\tau} \,.
\tag{21.32}
$$

The backbone depends neither on damping nor on the forcing function. Actually it is a feature of the conservative part of the system and is more correlated to its free behavior than to the response to a given excitation.

Remark 21.4 *The role of the backbone of a nonlinear system is the same that the natural frequency has for linear systems; on the backbone, inertia* *forces balance nonlinear restoring forces (as an average on a cycle) and the response lags the excitation by 90°.*

The value of the amplitude in such resonant conditions can be computed by stating that the phase lag is 90°. From the second equation (21.11), it follows that

$$\frac{\omega}{\omega_0} = \frac{f_0}{2\zeta k x_m} - \frac{1}{2\zeta k \pi x_m} \int_0^{2\pi} \beta'(\dot{x}) \cos(\tau) d\tau \,, \tag{21.33}$$

or in case of the Duffing's equation with linear damping,

$$\frac{\omega}{\omega_0} = \frac{f_0}{2\zeta k x_m} \,. \tag{21.34}$$

Equations (21.33) and (21.34) define a curve in the $x_m(\omega)$ plane that is usually referred to as a *limit envelope*. It divides the plane into two regions: The one spanning under the curve includes all possible working conditions, since the forcing function can supply the energy needed to sustain the oscillations. In the zone above the limit envelope, the energy dissipated by damping is greater than that supplied by the excitation and thus no steady oscillation is possible.

The presence of a limit envelope explains the downward jump phenomenon that could not be explained by the undamped model. What causes the amplitude to suddenly decrease is the impossibility of the forcing function to sustain the larger amplitude against energy dissipations.

Remark 21.5 *While the backbone defines the conditions in which there is equilibrium between inertia forces and restoring forces (resonance conditions), the limit envelope states the conditions for equilibrium between damping and external forces.*

Remark 21.6 *The limit envelope does not depend on the mass of the system or on the type of restoring force. In particular, it does not change whether the restoring force is linear, hardening, or softening, provided that damping and the forcing functions are not changed.*

The response of a damped nonlinear system with a single degree of freedom can easily be approximated by cutting the undamped response using the limit envelope (Fig. 21.5). This procedure is the generalization of that shown in Fig. 7.2d for linear systems. The properties of the backbone and the limit envelope are valid for any type of functions $f'(x)$ and $\beta'(\dot{x})$ (provided they are odd) and any law linking the amplitude of the forcing function f_0 to the frequency ω.

Remark 21.7 *In some cases the limit envelope can be completely above the backbone, with no intersection between them. The amplitude in this*

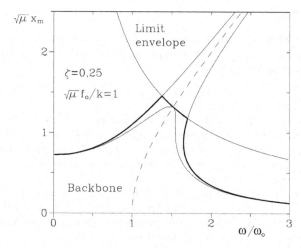

FIGURE 21.5. Response of a nonlinear damped system with a single degree of freedom approximated by cutting the undamped response with the limit envelope.

case grows indefinitely, with no downward jump with increasing frequency, since the forcing function is powerful enough to sustain the oscillation with the larger amplitude at any value of the frequency.

The limit envelope, in general, does not exist for multi-degrees-of-freedom systems. However, when it exists, it is a very powerful tool for understanding the system's behavior.

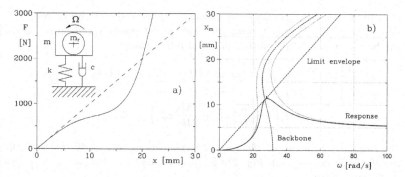

FIGURE 21.6. **(a)** Scheme of the system and force–displacement characteristics of the elastomeric springs; **(b)** first approximation of the fundamental harmonic of the response, backbone, and limit envelope. The response of the undamped system is also reported.

Example 21.1 *A rotating machine is suspended through a system of elastomeric springs whose force–displacement characteristics (Fig. 21.6a) can be approximated by the equation*

$$F = k_1 x - k_2 x^3 + k_3 x^5 \, ,$$

where $k_1 = 10^5$ N/m, $k_2 = 4 \times 10^8$ N/m^3, and $k_3 = 10^{12}$ N/m^5 (see Exercise 19.5). The damping of the suspension system will be assumed to be of the viscous type, with damping coefficient $c = 1{,}200$ Ns/m. The total mass of the system is $m = 100$ kg, and the product of the mass of the rotor by its eccentricity is $m_r \epsilon = 0.5$ kg m.

Compute a first approximation of the fundamental harmonic of the response when the angular velocity of the machine spans between 20 and 300 rpm.

The unbalance causes a force whose vertical component varies harmonically with frequency ω equal to the rotational speed Ω of the machine

$$F = m_r \epsilon \Omega^2 \sin(\Omega t).$$

The equation of motion of the system is

$$m\ddot{x} + c\dot{x} + k_1 x - k_2 x^3 + k_3 x^5 = m_r \epsilon \Omega^2 \sin(\Omega t) \,.$$

The natural frequency and the damping ratio of the linearized system are

$$\omega_0 = \sqrt{\frac{k_1}{m}} = 31.62 \ rad/s; \qquad \zeta = \frac{c}{2\sqrt{k_1 m}} = 0.19 \,.$$

The backbone and the limit envelope of the response can be obtained directly from Eqs. (21.32) and (21.33)

$$\frac{\omega}{\omega_0} = \sqrt{1 - \frac{3k_2}{4k_1}x_m^2 + \frac{15k_3}{24k_1}x_m^4} \,; \qquad \frac{\omega}{\omega_0} = \frac{m_r \epsilon \omega^2}{2\zeta k_1 x_m} \,.$$

The amplitude of the fundamental harmonic of the response can be computed using Eq. (21.12). By performing the relevant integrations and rewriting the equation as a quadratic equation in ω^2, it follows that

$$\left(\frac{\omega}{\omega_0}\right)^4 \left[1 - \left(\frac{m_r \epsilon}{m x_m}\right)^2\right] - 2\left(\frac{\omega}{\omega_0}\right)^2 \left(1 - \frac{3k_2}{4k_1}x_m^2 + \frac{15k_3}{24k_1}x_m^4 - 2\zeta^2\right) +$$

$$+ \left(1 - \frac{3k_2}{4k_1}x_m^2 + \frac{15k_3}{24k_1}x_m^4\right)^2 = 0 \,.$$

The frequency response of the system is plotted, together with the backbone and the limit envelope, in Fig. 21.6b. Also the undamped response is plotted in the same figure.

Example 21.2 *Consider a linear system with dry friction, modeled as shown in Example 20.1 using the Coulomb damping model, excited by a harmonic forcing function. The equation of motion is*

$$m\ddot{x} + F\frac{\dot{x}}{|\dot{x}|} + kx = f_0 \sin(\omega t).$$

The integral defining the equivalent damping must be computed separately for the parts of the cycle in which the velocity is positive $(-\pi/2 < \omega t < \pi/2)$ or negative $(\pi/2 < \omega t < 3\pi/2)$

$$c_{eq} = \frac{4F}{x_m \pi \omega}; \qquad \zeta_{eq} = \frac{2F\omega_0}{x_m \pi \omega k}.$$

The amplitude of the response of the linear system with equivalent damping can be computed using Eq. (7.6). Note that in this case damping depends on amplitude and a further rearranging of the equations is required to obtain the following frequency response:

$$x_m = \frac{f_0}{k}\sqrt{\frac{1 - \left(\dfrac{4F}{\pi f_0}\right)^2}{1 - \left(\dfrac{\omega}{\omega_0}\right)^2}}; \qquad \Phi = \arctan\left[\frac{-1}{\sqrt{\left(\dfrac{\pi f_0}{4F}\right)^2 - 1}}\right].$$

An apparently inconsistent result is thus obtained, because at resonance the amplitude tends to infinity, in spite of damping, at least if

$$\frac{F}{f_0} < \frac{\pi}{4}.$$

This can, however, be easily explained by the consideration that both the energy dissipated by damping and the energy input due to the exciting force are proportional to the amplitude. Thus, if dry friction is small enough to allow the system to oscillate, it cannot avoid the building up of an infinite amplitude at resonance. The case with viscous damping was different, as the drag force was proportional to the amplitude, and consequently, the energy dissipated was proportional to its square. No matter how low the damping of the system was, there was a finite value of the amplitude for which the energy dissipated by damping was equal to the work performed by the excitation.

The same conclusion can be immediately drawn from the study of the limit envelope. By introducing the equivalent damping into the equation of the limit envelope, it follows that

$$\frac{F}{f_0} = \frac{\pi}{4}.$$

This expression does not represent a line on the ω, x_m plane, i.e., no limit envelope actually exists, but it is nevertheless the condition for which motion is possible, as stated earlier.

The results here obtained are only approximated and are close to the correct results only if the motion of the system is not far from being a harmonic motion.

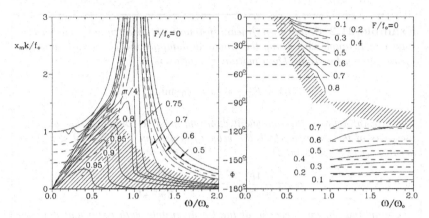

FIGURE 21.7. Response of a system with linear restoring force and Coulomb damping to a harmonic excitation: exact solution (*full lines*) against the current approximate solution (*dashed lines*). In the hatched region the motion is not continuous; in the zone in the lower left corner more than one stop occurs in each cycle.

An exact solution to the problem was obtained by Den Hartog (J.P. Den Hartog, "Forced vibrations with combined coulomb and viscous damping", Trans. ASME, Vol. 53, 1931.) and is reported in Fig. 21.7. In the part of Fig. 21.7 lying within the hatched region, the motion is not continuous but contains at least one stop at each cycle. In the part in the lower left corner, the motion stops at least twice each cycle, and no solution was found. The dashed lines are the solutions obtained from the current model using the equivalent damping concept. In the vicinity of the resonance peak, the approximate curves are very close to those obtained in the aforementioned paper, as could be expected. On the contrary, when the motion is intermittent, or, generally speaking, when damping is not small, the approximate solution is quite far from the correct one. Where motion is intermittent no phase has been computed, because the very concept of phase loses its meaning. Also, in cases where the approximate solution leads to the conclusion that no motion is possible, an exact solution exists.

21.6   Multiple Duffing equations

Consider an undamped structure whose behavior is linear, supported on a number m of elastic constraints with a force–elongation characteristic of the type used in Duffing's equation:

$$F_i = -k_i x_i (1 + \mu_i x_i^2).$$

Assume that the system is discretized in such a way that the generalized displacements of the supporting points are chosen as the first m generalized coordinates and the linear part of the characteristic of the supports is accounted for in the stiffness matrix of the system. Note that each function g_i contains only the generalized coordinate x_i. It is then possible to separate the nonlinear part of the problem from the linear part, eliminating the last $n - m$ equations by dynamic reduction. By using the Ritz averaging technique, Eq. (21.23) becomes

$$\mathbf{K}_{dyn_{cond}}\mathbf{x}_0 + \frac{1}{\pi}\left\{\int_0^{2\pi} k_i\mu_i x_i^3 \sin(\omega t)d(\omega t)\right\} = \mathbf{f}_{0_{cond}}, \qquad (21.35)$$

where

$$x_i = x_{0_i}\,\sin(\omega t)\,.$$

By performing the integrations, the following set of m algebraic equations is obtained:

$$\mathbf{K}_{dyn_{cond}}x_0 + \frac{3}{4}\{k_i\mu_i x_{0_i}^3\} = f_{0_{cond}}, \qquad (21.36)$$

where the condensed matrix and vector must be computed for each value of the frequency ω. A set of m cubic equations is so obtained.

In the particular case of a system with 2 degrees of freedom, a simple solution can be found by solving one of the unknowns from the equation in which it is at first power and then substituting it into the other equation. From the first equation the following value for x_{02} can be obtained

$$x_{02} = \frac{f_{01} - K_{11}x_{01} - \frac{3}{4}k_1\mu_1 x_{01}^3}{K_{12}}\,.$$

By introducing this value of x_{02} into the second equation of motion, it yields an equation in x_{01} that can be solved numerically.

If the system is damped things are much more complex. Assuming that no nonlinear damping is present, the same system seen above leads to the following equation:

$$\mathbf{K}_{dyn}\left(\mathbf{x}_{0s} + i\mathbf{x}_{0c}\right) + \mathbf{f}_{ns}\left(\mathbf{x}_{0s}, \mathbf{x}_{0c}\right) + i\mathbf{f}_{nc}\left(\mathbf{x}_{0s}, \mathbf{x}_{0c}\right) = \mathbf{f}_{0s} + i\mathbf{f}_{nc}, \qquad (21.37)$$

where

$$\mathbf{f}_{ns} = \frac{\omega}{\pi}\left(\int_0^T k_i\mu_i x_{0_i}^3\,\sin(t)dt\right), \qquad (21.38)$$

$$\mathbf{f}_{nc} = \frac{\omega}{\pi}\left\{\int_0^T k_i\mu_i x_{0_i}^3\,\cos(t)dt\right\}, \qquad (21.39)$$

and

$$x_0 = x_{0s}\sin(\omega t) + x_{0c}\cos(\omega t)\,. \qquad (21.40)$$

Since

$$[a\sin\left(\omega t\right) + b\cos\left(\omega t\right)]^3 = \frac{3}{4}a\left(b^2 + a^2\right)\sin\left(\omega t\right) + \qquad (21.41)$$

$$+\frac{3}{4}b\left(b^2+a^2\right)\cos\left(\omega t\right)+\frac{1}{4}a\left(3b^2-a^2\right)\sin\left(\omega t\right)+\frac{1}{4}b\left(b^2-3a^2\right)\cos\left(\omega t\right)\ ,$$

the integrals of Eqs. (21.38) and (21.39) are

$$\int_0^{2\pi}\left[x_{0s}\sin\left(\omega t\right)+x_{0c}\cos\left(\omega t\right)\right]^3\sin\left(\omega t\right)d\left(\omega t\right)=\frac{3\pi}{4}x_{0s}\left(x_{0s}^2+x_{0c}^2\right),$$

$$\tag{21.42}$$

$$\int_0^{2\pi}\left[x_{0s}\sin\left(\omega t\right)+x_{0c}\cos\left(\omega t\right)\right]^3\cos\left(\omega t\right)d\left(\omega t\right)=\frac{3\pi}{4}x_{0c}\left(x_{0s}^2+x_{0c}^2\right)\ .$$

The ith element of vector $\mathbf{f}_{ns}+i\mathbf{f}_{nc}$ is thus

$$f_{ns_i}+if_{nc_i}=\frac{3}{4}k_i\mu_i\left(x_{0s_i}^2+x_{0c_i}^2\right)\left(x_{0s_i}+ix_{0c_i}\right)\ ,\tag{21.43}$$

i.e.,

$$\mathbf{f}_{ns}+i\mathbf{f}_{nc}=\frac{3}{4}\mathrm{diag}\left\{k_i\mu_i\left(x_{0s_i}^2+x_{0c_i}^2\right)\right\}\left(\mathbf{x}_{0s}+i\mathbf{x}_{0c}\right)\ .\tag{21.44}$$

The equation allowing to compute the amplitude of the response from the amplitude of the excitation is

$$\left[\mathbf{K}_{dyn}+\frac{3}{4}\mathrm{diag}\left\{k_i\mu_i\left(x_{0s_i}^2+x_{0c_i}^2\right)\right\}\right]\left(\mathbf{x}_{0s}+i\mathbf{x}_{0c}\right)=\mathbf{f}_{0s}+i\mathbf{f}_{nc}\ ,\tag{21.45}$$

The reduced dynamic stiffness matrix is a function of the frequency that must be computed numerically for each value of ω. Only in the case where no dynamic reduction has been performed, the mass, stiffness and damping matrices of the linear part of the model can be introduced separately into the equation.

Example 21.3 *Consider the rigid beam on nonlinear springs with linear dampers in parallel sketched in Fig. 21.8a, excited by a harmonic forcing function applied to the center of mass. Assume the following data: $m = 10$ kg, $a = 50$ mm, $b = 120$ mm, radius of inertia $r = \sqrt{J/m} = 80$ mm, $\mu_1 = \mu_2 = \mu = 80,000\ 1/m^2$, $k_1 = k_2 = k = 10\ kN/m$, $c_1 = c_2 = c = 30$ Ns/m, $f_0 = 10\ N$. With simple computations the mass, damping and stiffness matrices, and the force vector can be shown to be*

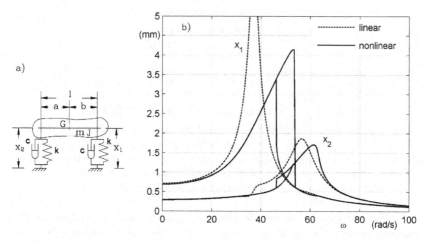

FIGURE 21.8. Sketch (**a**) and response (**b**) of a nonlinear system with 2 degrees of freedom, excited by a harmonic forcing function applied in the center of mass. Amplitude of the displacements of the supporting points.

$$\mathbf{M} = \frac{m}{l^2} \begin{bmatrix} a^2 + r^2 & ab - r^2 \\ ab - r^2 & b^2 + r^2 \end{bmatrix}, \qquad \mathbf{C} = c \begin{bmatrix} 1 & 0 \\ 0 & 1 \end{bmatrix},$$

$$\mathbf{K} = k \begin{bmatrix} 1 & 0 \\ 0 & 1 \end{bmatrix}, \qquad \mathbf{f}_0 = \frac{f_0}{l} \begin{Bmatrix} a \\ b \end{Bmatrix}.$$

Equation (21.37) is a nonlinear set of two complex equations and no closed form solution is possible. An efficient approach is to resort to the Newton–Raphson method and, in this context, it is expedient to separate the real part of the equations from their imaginary part. The four resulting equations of motion can be written in the form

$$\mathbf{p}(x_i) = 0,$$

where function $\mathbf{p}(x_i)$ *is*

$$\mathbf{p}(x_i) = \left\{ \begin{bmatrix} -\omega^2 \mathbf{M} + \mathbf{K} & -\mathbf{C} \\ \mathbf{C} & -\omega^2 \mathbf{M} + \mathbf{K} \end{bmatrix} + \mathbf{g}' \right\} \begin{Bmatrix} x_{1s} \\ x_{2s} \\ x_{1c} \\ x_{2c} \end{Bmatrix} - \begin{Bmatrix} \mathbf{f}_0 \\ 0 \end{Bmatrix},$$

and the nonlinear function $\mathbf{g}'$ *is*

$$\mathbf{g}' = \frac{3k\mu}{4} \, diag \begin{bmatrix} x_{1s}^2 + x_{1c}^2 & x_{2s}^2 + x_{2c}^2 & x_{1s}^2 + x_{1c}^2 & x_{2s}^2 + x_{2c}^2 \end{bmatrix}.$$

The vector of the unknowns at the $(i+1)$*th iteration is obtained from the vector at the* i*th iteration from the usual relationship*

$$\mathbf{x}^{(i+1)} = \mathbf{x}^{(i)} - \mathbf{S}^{-1}\left(\mathbf{x}^{(i)}\right) \mathbf{p}\left(\mathbf{x}^{(i)}\right),$$

where the Jacobian matrix $\mathbf{S}$ *is*

$$\mathbf{S} = \begin{bmatrix} -\omega^2\mathbf{M}+\mathbf{K} & -\mathbf{C} \\ \mathbf{C} & -\omega^2\mathbf{M}+\mathbf{K} \end{bmatrix} +$$

$$+\frac{3k\mu}{4}\begin{bmatrix} 3x_{1s}^2 + x_{1c}^2 & 0 & 2x_{1s}x_{1c} & 0 \\ 0 & 3x_{2s}^2 + x_{2c}^2 & 0 & 2x_{2s}x_{2c} \\ 2x_{1s}x_{1c} & 0 & x_{1s}^2 + 3x_{1c}^2 & 0 \\ 0 & 2x_{2s}x_{2c} & 0 & x_{2s}^2 + 3x_{2c}^2 \end{bmatrix}.$$

The response in terms of the modulus of the complex amplitude at the supports was computed twice, once starting from $\omega = 0$ and proceeding with small increases of frequency and each time using as starting solution that obtained in the previous computation, and once doing the same, but starting from a high value of frequency ($\omega = 100$ rad/s) and decreasing it to 0. The results are shown in Fig. 21.8b.

The first computation yields the downward jumps, while the second one the upward jumps. The results are compared with the results obtained from the linearized system.

This procedure yields only an approximation of the fundamental harmonic of the result, and even for that only some of the possible solutions are found. Other possible solutions that might exist are not found in this way.

The same results are converted to displacement at the center of gravity x_G and rotation θ in Fig. 21.9b using the relationship

$$\begin{Bmatrix} x_G \\ \theta \end{Bmatrix} = \frac{1}{a+b}\begin{bmatrix} a & b \\ 1 & -1 \end{bmatrix}\begin{Bmatrix} x_1 \\ x_2 \end{Bmatrix}.$$

It is clear that the second mode is much less excited than the first one and thus it is much less affected by nonlinearities.

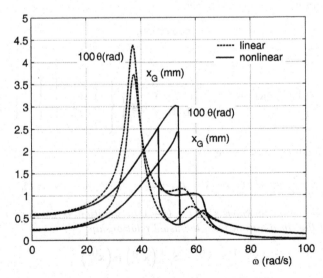

FIGURE 21.9. Same as Fig. 21.8, but in terms of amplitude of the displacement of the center of mass x_G and of rotation θ.

21.7   Approximated sub- and super-harmonic response

To compute the response including other harmonics it is possible to resort to the Ritz averaging technique, the harmonic balance, or other approaches. In general, if the forcing function is harmonic, it is possible to write

$$\mathbf{f}\,(t) = \mathbf{f}_{0s}\sin(\omega t) + \mathbf{f}_{0c}\cos(\omega t)\,. \tag{21.46}$$

The response can be written as

$$\mathbf{x}\,(t) = \sum_{i=1}^{m}\mathbf{x}_{0s}\sin(k_i\omega t) + \sum_{i=1}^{m}\mathbf{x}_{0c}\cos(k_i\omega t)\,, \tag{21.47}$$

where m is the number of harmonics of the response that are considered and k_i is the order or the various harmonics. For instance, $k_i = 1/3$ for the third subharmonics and 3 for the third super-harmonics.

If the system has n degrees of freedom and m harmonics are taken into account, the unknowns sine and cosine components of the amplitude are $2\,mn$.

An exception is when one of the harmonic is a zero-order harmonic, i.e., a term constant in time. When $k_i = 0$, the component in sine does not exist and the unknowns are $2\,mn - 1$.

By introducing the response and the excitation into the equation of motion (20.5)

$$\mathbf{M}\ddot{\mathbf{x}} + \mathbf{C}\dot{\mathbf{x}} + \mathbf{K}\mathbf{x} + \mu\mathbf{g} = \mathbf{f}(t)\,,$$

it can be transformed into a set of algebraic nonlinear equations.

It can be shown that the nonlinear function $\mathbf{g}$ can be transformed into

$$\mathbf{g} = \sum_{i=1}^{m}\mathbf{g}_{is}\sin(k_i\omega t) + \sum_{i=1}^{m}\mathbf{g}_{ic}\cos(k_i\omega t) + \dots\,.$$

Expressions $\mathbf{g}_{is}$ and $\mathbf{g}_{ic}$ are now nonlinear functions of the sine and cosine amplitudes of the various harmonics, and the (...) show that there are additional terms that cannot be balanced and thus must be dropped. Harmonic balance is anyway an approximated procedure.

By balancing the harmonics, the following set of nonlinear equations is obtained

$$\mathbf{p}(\mathbf{x}_0) = \mathbf{K}_d\mathbf{x}_0 + \mu\mathbf{g}(\mathbf{x}_0) - \mathbf{f}_0 = \mathbf{0}\,,$$

where

$$\mathbf{K}_d = \begin{bmatrix} \mathbf{K} - k_1^2\omega^2\mathbf{M} & -k_1\omega\mathbf{C} & \mathbf{0} & \mathbf{0} & \dots \\ k_1\omega\mathbf{C} & \mathbf{K} - k_1^2\omega^2\mathbf{M} & \mathbf{0} & \mathbf{0} & \dots \\ \mathbf{0} & \mathbf{0} & \mathbf{K} - k_2^2\omega^2\mathbf{M} & -k_2\omega\mathbf{C} & \dots \\ \mathbf{0} & \mathbf{0} & k_2\omega\mathbf{C} & \mathbf{K} - k_2^2\omega^2\mathbf{M} & \dots \\ \dots & \dots & \dots & \dots & \dots \end{bmatrix}\,,$$

$$\mathbf{x}_0 = \begin{Bmatrix} x_{1s} \\ x_{1c} \\ x_{2s} \\ x_{2c} \\ \cdots \end{Bmatrix}, \quad \mathbf{g}_0 = \begin{Bmatrix} g_{1s} \\ g_{1c} \\ g_{2s} \\ g_{2c} \\ \cdots \end{Bmatrix}, \quad \mathbf{f}_0 = \begin{Bmatrix} f_{1s} \\ f_{1c} \\ f_{2s} \\ f_{2c} \\ \cdots \end{Bmatrix}.$$

If the forcing function is mono-harmonic, as in Eq. (21.46), only the first two components of vector $\mathbf{f}_0$ are present. If it is poly-harmonic, also combination tones should be introduced. It is possible, but the number of harmonics so introduced becomes large.

The difficulty lies usually in defining functions g_1. Even using symbolic computer codes, it is quite a long and difficult process.

If the equation is solved using the Newton–Raphson procedure, the solution at the $(i+1)$th iteration is obtained from that at the ith iteration using the equation

$$\mathbf{x}_0^{(i+1)} = \mathbf{x}_0^{(i)} - \mathbf{S}^{-1}(\mathbf{x}_0^{(i)})\mathbf{p}(\mathbf{x}_0^{(i)}),$$

where the Jacobian matrix is

$$\mathbf{S} = \mathbf{K}_d + \mu \left[\frac{\partial \mathbf{g}_{0i}}{\partial \mathbf{x}_{0j}} \right].$$

21.7.1 Damped Duffing equation

The first super-harmonic of the Duffing equation is the third harmonics. The solution is thus

$$x = x_{1s} \sin(\omega t) + x_{1c} \cos(\omega t) + x_{2s} \sin(3\omega t) + x_{2c} \cos(3\omega t).$$

Since it is possible to balance only the first two harmonics

$$g_{1s} = \frac{3}{4}k \left[x_{1c}^2 x_{2s} + x_{1c}^2 x_{1s} - x_{1s}^2 x_{2s} + 2x_{1s} x_{2s}^2 + 2x_{1s} x_{2c}^2 - 2x_{1c} x_{1s} x_{2c} + x_{1s}^3 \right],$$

$$g_{1c} = \frac{3}{4}k \left[x_{1c} x_{1s}^2 + 2x_{1c} x_{2c}^2 + 2x_{1c} x_{2s}^2 + x_{1c}^2 x_{2c} - x_{1s}^2 x_{2c} + 2x_{1c} x_{1s} x_{2s} + x_{1c}^3 \right],$$

$$g_{2s} = \frac{1}{4}k \left[6x_{1s}^2 x_{2s} + 6x_{1c}^2 x_{2s} + 3x_{1c}^2 x_{1s} + 3x_{2c}^2 x_{2s} - x_{1s}^3 + 3x_{2s}^3 \right],$$

$$g_{2c} = \frac{1}{4}k \left[3x_{2c} x_{2s}^2 + 6x_{1c}^2 x_{2c} - 3x_{1c} x_{1s}^2 + 6x_{1s}^2 x_{2c} + x_{1c}^3 + 3x_{2c}^3 \right].$$

The system has a single degree of freedom and thus a set of four algebraic equations is needed to obtain the amplitudes of the sine and cosine components of the two harmonics. The nondimensional results in terms of the moduli a_1 and a_2 of the amplitudes of the fundamental and third harmonics for a case with $f_0\sqrt{\mu}/k = 1$ (the same as Fig. 19.3) and $\zeta = 0.1$ are reported in Fig. 21.10. The solution obtained taking into account the

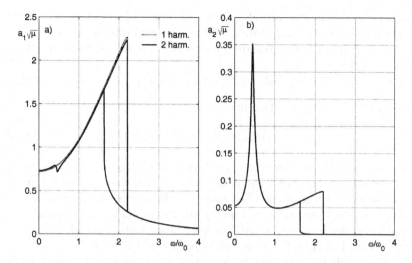

FIGURE 21.10. Response of a hardening system with $f_0\sqrt{\mu}/k = 1$ and $\zeta = 0.1$.
(a) Amplitude of the fundamental harmonic; (b) amplitude of the third harmonic.

super-harmonic is compared with that obtained by approximating the time
history with a harmonic law. As was seen in Fig. 19.3 for the undamped
system, the range where a strong super-harmonic response is found is quite
limited and damping prevents from reaching very large amplitudes. The
jump phenomenon is not affected by the presence of the third harmonic.

To obtain a subharmonic response is much more difficult. If a solution
of the type

$$x = x_{1s} \sin\left(\frac{\omega}{3}t\right) + x_{1c} \cos\left(\frac{\omega}{3}t\right) + x_{2s} \sin\left(\omega t\right) + x_{2c} \cos\left(\omega t\right)$$

is assumed, it is easy to check that also in the damped case a solution in
which the amplitude subharmonic vanishes exists. While it is impossible to
rule out that other solutions exist, it is difficult to obtain that the solution
algorithm converges on it. Moreover, it is impossible to state, using this
simplified approach, whether and with which initial conditions the subharmonic oscillation is actually started. Only numerical or physical experiment
can shed light on the matter, and only for the specific condition in which
the experiment is performed.

21.8   Van der Pol method: stability of the steady-state solution

The approach seen in Section 21.2 allows a first approximation of the
fundamental harmonic of the steady-state response to a harmonic forcing function to be obtained. The so-called Van der Pol method relies on

similar approximations but also allows the study of solutions that are not restricted to steady-state motions; it is a powerful tool in the study of the stability of the motion. Consider a nonlinear non-autonomous equation of motion of the type

$$m\ddot{x} + f(x, \dot{x}) = f_0 \cos(\omega t), \qquad (21.48)$$

and assume that the solution has the form

$$x(t) = x_1(t) \cos(\omega t) + x_2(t) \sin(\omega t) + \bar{x}, \qquad (21.49)$$

where x_1 and x_2 are slowly varying functions of time and a constant displacement $\bar{x}$ is taken into account for considering the case of non-symmetrical damping or restoring force. The motion is then a harmonic motion with slowly varying amplitude. x_1 and x_2 are the in-phase and in-quadrature amplitudes, respectively, while $\bar{x}$ is usually vanishingly small because, in many cases, function $f(x, \dot{x})$ is odd in both x and $\dot{x}$.

It is possible to show that by introducing the solution (21.49) into Eq. (21.48) and neglecting the second derivatives of functions x_1 and x_2 with respect to time, which are negligible due to the assumption that the amplitude varies slowly in time, the problem reduces to

$$\begin{cases} \dfrac{dx_1}{dt} = F_1(x_1, x_2, \bar{x}), \\[2mm] \dfrac{dx_2}{dt} = F_2(x_1, x_2, \bar{x}), \\[2mm] 0 = F_3(x_1, x_2, \bar{x}). \end{cases} \qquad (21.50)$$

In the case where $\bar{x}$ is vanishingly small, the last equation is not present. Only this case will be studied in detail here. Equation (21.50) can be dealt with using the same methods as for the study of Eq. (18.50), and the motion can be studied in the (x_1, x_2)-plane. This plane is sometimes referred to as the Van der Pol plane, but the terms phase and state plane are also common. The author thinks that the latter terms are better reserved to the $(x, \dot{x})$-plane, to avoid confusion. Neglecting $\bar{x}$, Eq. (21.50) can be reduced to

$$\frac{dx_2}{dx_1} = \frac{F_2(x_1, x_2)}{F_1(x_1, x_2)}, \qquad (21.51)$$

which allows plotting the trajectories of the system in the Van der Pol plane and obtaining the steady-state solution as fixed points in the same plane. The stability of the steady-state solutions can be studied in the same way seen for the stability of equilibrium positions in the phase plane.

Consider, for example, Duffing's equation. By introducing the solution (21.49) into the equation of motion and neglecting the terms containing the second derivatives of x_1 and x_2 with respect to time, those containing the

product of their derivatives for the damping coefficient, and those containing the trigonometric functions of 3 ωt (as seen for the harmonic balance
method) and then equating the coefficients of the sine and cosine of ωt, it
follows that

$$
\begin{cases}
\dfrac{dx_1}{dt} = \dfrac{1}{2\omega}\left[(\omega_0^2 - \omega^2)x_2 - 2\zeta\omega\omega_0 x_1 + \dfrac{3}{4}\mu\omega_0^2 x_2(x_1^2 + x_2^2)\right], \\[4mm]
\dfrac{dx_2}{dt} = \dfrac{1}{2\omega}\left[-(\omega_0^2 - \omega^2)x_1 - 2\zeta\omega\omega_0 x_2 + \dfrac{f_0}{k}\omega_0^2 - \dfrac{3}{4}\mu\omega_0^2 x_1(x_1^2 + x_2^2)\right].
\end{cases}
$$
$$(21.52)$$

Equation (21.52) is just Eq. (21.50) with the right-hand sides defining
functions F_1 and F_2. The trajectories obtained by numerically integrating
Eq. (21.51) for $\zeta = 0.1$, $f_0\sqrt{\mu}/k = 1$, and two values of ω/ω_0 are plotted
in nondimensional form in Fig. 21.11.

The three singular points are the steady-state solutions that coincide
with those obtained using the Ritz averaging or the harmonic balance
techniques. The solutions with the maximum and minimum amplitudes
correspond to stable foci, while the intermediate one is a saddle point. The
stability of the former two solutions and the instability of the last one are
so demonstrated.

The basins of attraction of the two stable solutions have a complex shape
and are quite interwoven. This means that starting from a certain position
in the Van der Pol plane (i.e., starting from an oscillation with a certain amplitude and phase), the system settles to the oscillation with larger
amplitude; starting from an oscillation with larger amplitude, the steady-
state condition with smaller amplitude is obtained. Also the phase can be

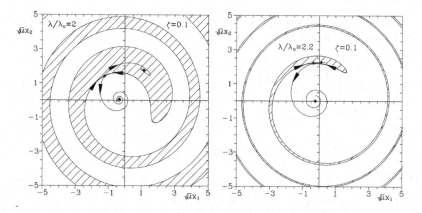

FIGURE 21.11. Trajectories in the Van der Pol plane for a system governed by
Duffing's equation, with $\zeta = 0.1$ and $f_0\sqrt{\mu}/k = 1$; nondimensional plots drawn for
$\omega/\omega_0 = 2.0$ and 2.2, respectively. The hatched zone is the basin of attraction of
the stable solution with larger amplitude.

important in deciding which solution the system will tend to, although the importance of phase decreases with growing amplitude, as the separatrices tend to assume a more circular form.

Consider a system governed by Duffing's equation, excited by a forcing function with a frequency lower than that at which three solutions can exist. There is only one solution, a stable focus in the Van der Pol plane. If the forcing frequency is increased, at a certain moment two more solutions appear and the plane splits into two domains of attraction. The domain of the larger solution is very large and that of the smaller one is very narrow.

With increasing frequency the smaller domain of attraction grows while the other one reduces. In the center of the frequency range in which three solutions exist, the two domains have comparable extensions (left-hand side of Fig. 21.11). The domain of the larger solution reduces its size, to disappear at the frequency at which the jump takes place and three solutions no longer exist. This explains why it can be difficult to sustain an oscillation with the larger amplitude near the jump frequency: When the domain of attraction of this solution is very small, as in the right-hand side of Fig. 21.11, and the solution is very near the separatrix, any small change of amplitude makes the system shift toward the other solution.

To study the stability of a singular point x_{1_0}, x_{2_0}, Eq. (21.50) can be easily linearized about it. Because it is a singular point, functions F_1 and F_2 vanish and the linearized equation reduces to

$$
\begin{cases}
\dfrac{dx_1}{dt} = \left(\dfrac{\partial F_1}{\partial x_1}\right)_{x_{1_0},x_{2_0}} x_1 + \left(\dfrac{\partial F_1}{\partial x_2}\right)_{x_{1_0},x_{2_0}} x_2 \, ; \\[4mm]
\dfrac{dx_2}{dt} = \left(\dfrac{\partial F_2}{\partial x_1}\right)_{x_{1_0},x_{2_0}} x_1 + \left(\dfrac{\partial F_2}{\partial x_2}\right)_{x_{1_0},x_{2_0}} x_2 \, .
\end{cases}
\tag{21.53}
$$

By assuming a solution of the type $\mathbf{x} = \mathbf{a}e^{st}$, the following characteristic equation is obtained

$$
\det \begin{bmatrix}
\left(\dfrac{\partial F_1}{\partial x_1}\right)_{x_{1_0},x_{2_0}} - s & \left(\dfrac{\partial F_1}{\partial x_2}\right)_{x_{1_0},x_{2_0}} \\[4mm]
\left(\dfrac{\partial F_2}{\partial x_1}\right)_{x_{1_0},x_{2_0}} & \left(\dfrac{\partial F_2}{\partial x_2}\right)_{x_{1_0},x_{2_0}} - s
\end{bmatrix} = 0 \, ,
\tag{21.54}
$$

i.e.,

$$
s^2 + \beta s + \gamma = 0 \, ,
$$

where

$$
\begin{cases}
\beta = -\left(\dfrac{\partial F_1}{\partial x_1}\right)_{x_{1_0},x_{2_0}} - \left(\dfrac{\partial F_2}{\partial x_2}\right)_{x_{1_0},x_{2_0}} \, , \\[4mm]
\gamma = \left(\dfrac{\partial F_1}{\partial x_1}\right)_{x_{1_0},x_{2_0}} \left(\dfrac{\partial F_2}{\partial x_2}\right)_{x_{1_0},x_{2_0}} - \left(\dfrac{\partial F_1}{\partial x_2}\right)_{x_{1_0},x_{2_0}} \left(\dfrac{\partial F_2}{\partial x_1}\right)_{x_{1_0},x_{2_0}} \, .
\end{cases}
$$

From the signs of the real and imaginary parts of the solutions s it is possible to state whether the point is a saddle, a node, a focus, or a center and to study its stability, following the same rules seen in Section 3.5.3.

Remark 21.8 *The Van der Pol method is as approximate as are the harmonic balance method and Ritz averaging technique, both used with only the fundamental harmonic: It is practically an application of harmonic balance to slowly varying nonstationary motions.*

Its main advantage is that of eliminating time by working in a plane characterized by a reference frame rotating with angular velocity equal to the forcing frequency ω in the complex plane of Fig. 7.1, the only difference being that the direction of the ordinate axis is reversed. In that plane, the equation of motion is autonomous, and a bi-dimensional state plane can be used. In this sense, the Van der Pol plane is a state plane.

21.9   Strongly nonlinear systems

All the methods seen in the previous sections lead to approximated solutions of the equation of motion. In most of them the time history of the response is assumed to be harmonic or, at best, poly-harmonic. In others the solution of the linearized system is assumed as a first approximation and then is corrected in steps leading to better and better approximations.

These approaches however work only if the system behaves as a weakly nonlinear system, meaning either that the system is weakly nonlinear in itself or that it operates in conditions (i.e., small amplitude oscillations) preventing its nonlinear nature from strongly affecting its behavior.

Moreover, with increasing importance of nonlinearities, these methods become computationally heavier and heavier. If the response of the system is almost harmonic, the results obtained by using the harmonic balance method with just the fundamental harmonic are good and easily obtained. If two harmonics are used the computations start becoming time consuming. If many harmonics are required for obtaining a good approximation, the mentioned methods become soon overwhelmingly complex, even if symbolic computer codes are used to obtain the nonlinear set of algebraic equations. Moreover, once the nonlinear algebraic equations are obtained, their solution may be a formidable problem, and above all it may be very difficult to obtain all significant solutions.

If the system is strongly nonlinear, its behavior may be qualitatively different from what can be found using these approximated methods and can be studied only through numerical or physical experiments. Clearly also numerical integration of the equations of motion yield approximated results, but this term has a different meaning. As a first point, all solutions obtained from mathematical models are only approximations of the behavior of actual systems, but this is obvious. Numerical integration introduces approximations owing to the discretization in time on which it is based, but usually the related errors can be reduced to a minimum, even if at the cost of increasing computer time.

The problem linked with this numerical approach to nonlinear dynamics is that the resulting time history depends not only quantitatively but also qualitatively on the forcing function and the initial conditions. Even in case of a model as simple as the Duffing equation the number of simulations required to be reasonably confident that all possible solutions have been explored may be large. And at the end one can never be sure that all possible outcomes have been investigated.

A problem is also to resort to representations that allow to put in a single chart (or a limited number of charts) all significant results. One of the most common of such representations is the *Poincaré map*. Since it is based on the state space, this map can be actually plotted in the case of systems with a single degree of freedom. When the number of degrees of freedom is equal to 2 or larger, only bi- or tridimensional sections of the state space can be plotted.

21.10   Poincaré mapping

As already stated in Section 1.6, in the case of non-autonomous systems the state space must have one added dimension: time. The trajectories related to a system with a single degree of freedom are thus tridimensional curves. The equation of motion of the system can be easily written in the form of a set of three autonomous equations in the variables x, v, and t

$$\begin{cases} \dot{v} = F(x, v, t), \\ \dot{x} = v, \\ \dot{t} = 1, \end{cases} \tag{21.55}$$

where function $F(x, v, t)$ also includes the forcing function. In the case of periodic motion, the trajectories wind around the time axis, assuming the shape of a helix whose pitch is equal to the period of the response.

Consider, for example, a linear damped system with $\zeta = 0.2$, excited at a frequency slightly smaller than the natural frequency ($\omega_n/\omega = 1.5$). Assume that at time $t = 0$ the system is at a standstill and the forcing function is at its maximum positive value. The tridimensional phase trajectory, computed using Eq. (7.11), is plotted in Fig. 21.12a. The choice of a linear system for this example is due to the fact that an exact solution for the time history exists. The time history is plotted in Fig. 21.12b. Clearly, it is the projection of the phase trajectory on the (x, t)-plane. The state projection, i.e., the projection of the tridimensional trajectory on the (x, v)-plane, is plotted in Fig. 21.12c; the projected trajectory intersects at many points.

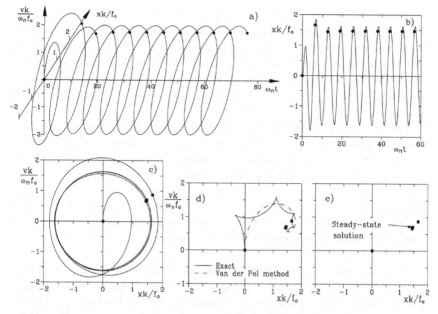

FIGURE 21.12. Behavior of a linear damped system excited by a harmonic forcing function in the state space. The *dots* are the positions at time $t = 0$ and after a number of periods of the forcing function: (**a**) tridimensional state trajectories; (**b**) time history as the projection of the state trajectory on (x, t)-plane; (**c**) state projection on (x, v)-plane; (**d**) trajectories in the Van der Pol plane; (**e**) Poincaré map. Nondimensional plot with $\zeta = 0.2$ and $\omega_n/\omega = 1.5$.

The dots in the figure are the positions at time $t = 0$ and after 1, 2, ... periods of the forcing function.

The Van der Pol plane was seen in Section 21.8 with the Van der Pol approximate method. Actually, it can be defined by the following coordinate transformation:

$$\begin{cases} x(t) = x_1(t)\cos(\omega t) + x_2(t)\sin(\omega t)\,, \\[2mm] \dfrac{1}{\omega}v(t) = -x_1(t)\sin(\omega t) + x_2(t)\cos(\omega t)\,. \end{cases} \qquad (21.56)$$

The first equation (21.56) coincides with Eq. (21.49) when the constant term in the latter has been neglected: Coordinate transformation (21.56) defines the Van der Pol plane as well as Eq. (21.49). The inverse transformation is obviously

$$\begin{cases} x_1(t) = x(t)\cos(\omega t) - \dfrac{1}{\omega}v(t)\sin(\omega t)\,, \\[2mm] x_2(t) = x(t)\sin(\omega t) + \dfrac{1}{\omega}v(t)\cos(\omega t)\,. \end{cases} \qquad (21.57)$$

No assumption on functions $x_1(t)$ and $x_2(t)$, such as they are slowly varying in time, has been made and, consequently, the exact trajectory can be represented. If ω is the fundamental frequency of the response, the Van der Pol plane can be considered a plane that travels along the time axis while rotating about the origin with angular velocity ω. The trajectories then unwind in the Van der Pol plane, yielding a sort of state portrait with non-intersecting lines.

The trajectory in the Van der Pol plane for the system of Fig. 21.12 has been plotted in Fig. 21.12d. In the same figure the trajectory obtained using the Van der Pol method, i.e., assuming that functions $x_1(t)$ and $x_2(t)$ are slowly varying in time, has been plotted with a dashed line.

Remark 21.9 *The approximations linked with the Van der Pol solution in this case do not affect the steady-state solution, because it is exactly harmonic, but only the way in which it is reached.*

A more common way of representing the tridimensional trajectory on a plane is the so-called *Poincaré mapping* or *Poincaré section*. A set of points on the tridimensional trajectory is sampled with an interval equal to the period of the forcing function and then projected on the (x, v)-plane. When studying the steady-state motion with a period equal to the period of the forcing function, all points on the Poincaré map are superimposed on each other. If the system starts from a nonstationary motion, a set of points that gets closer and closer to the one related to the stationary condition is obtained. The attractor on the Poincaré map for a system that undergoes a steady-state oscillation with a period equal to the period of the forcing function is thus a single point. The Poincaré mapping of the tridimensional trajectory shown in Fig. 21.12a is plotted in Fig. 21.12e.

As already stated, a nonlinear system excited by a harmonic forcing function with frequency ω can, in certain cases, exhibit what is usually called a *subharmonic oscillation*, i.e., an oscillatory motion whose fundamental frequency is smaller than that of the forcing function, usually a submultiple of it. A subharmonic of order n is a solution with fundamental frequency equal to ω/n, i.e., with a period equal to n times the period of the forcing function. Because the points used for the construction of the Poincaré mapping are sampled with a period equal to that of the excitation, the map for steady-state oscillations is made of n distinct points. The attractor is then a set of n points, usually referred to as *periodic points*, which are cyclically touched at regular intervals by the tridimensional state trajectory.

In the Poincaré map different attractors may exist, separated by lines that define their basins of attraction. Also, simple point attractors can coexist with n-point attractors: This simply means that depending on the initial conditions, i.e., depending on the basin of attraction in which the first point is, the motion can evolve into a periodic motion with fundamental frequency ω or into a subharmonic motion with frequency ω/n. To

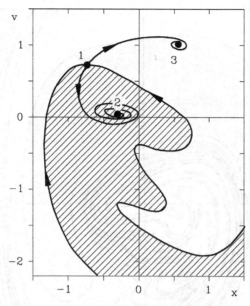

FIGURE 21.13. Domains of attraction in the Poincaré map for Duffing's equation. From C. Hayashi, *Nonlinear Oscillations in Physical Systems*, Princeton University Press, Princeton, N.J., 1985.

show the basins of attraction in the Poincaré map, some plots obtained by Hayashi from Duffing's equation with vanishing linear stiffness, using a specially designed analog computer, are reported in Figs. 21.13 and 21.14. The domains of attraction reported in Fig. 21.13 refer to a particular set of parameters for which two stable steady-state solutions exist (points 2 and 3). Point 1 is a saddle point corresponding to an unstable solution and lies on the separatrix of the domains of attraction.

The whole picture is not very different from those reported in Fig. 21.11, although the latter was only approximate, having been obtained with the Van der Pol method.

A far more complicated situation is reported in Fig. 21.14. Working with the same equation, but with different values of the parameters, subharmonic solutions of the third order have also been obtained, and the domains of attraction become very intricate. Again solutions 2 and 3 correspond to periodic attractors with fundamental frequency equal to the frequency of the forcing function, and solution 1 is a saddle point, as in Fig. 21.13. In this case, however, two subharmonic solutions of the third order also exist. The one represented by periodic points 4, 5, and 6 is unstable and the points lay on the separatrix, while the solution represented by points 7, 8, and 9 is stable. Note that all domains of attraction of the subharmonic solution lie within the domain of attraction of solution 2.

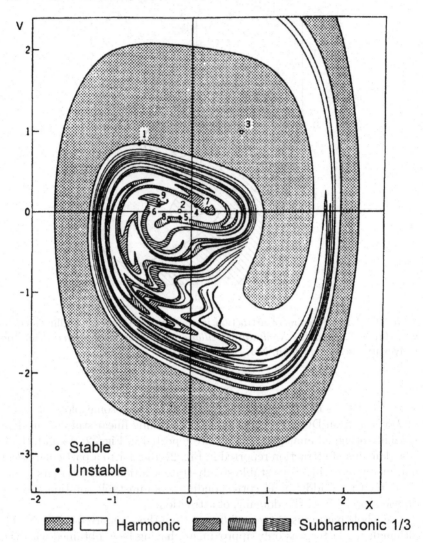

FIGURE 21.14. Domains of attraction in the Poincaré map for Duffing's equation, as in Fig. 21.13, but with different values of the parameters. From C. Hayashi, *Nonlinear Oscillations in Physical Systems*, Princeton University Press, Princeton, N.J., 1985.

21.11   Chaotic vibrations

The possibility of performing numerical experiments on nonlinear differential equations that cannot be integrated analytically allowed the discovery that a system modeled using a standard differential equation, and hence a fully deterministic model, can give way to a seemingly random motion.

Actually, the possibility of obtaining irregular time histories from a deterministic system excited by a regular forcing function had been observed for a long time, but was usually ascribed to noise or the influence of some external disturbances rather than the intrinsic behavior of the system. However, it is clear that such irregular behavior may take place in many systems due to their own characteristics and that very simple models can also simulate this behavior.

The study of the forced response of the Van der Pol equation did show that in the cases in which two stable attractors exist, when starting the motion from certain positions of the state space the initial transient can hesitate to choose the attractor to converge to. An arbitrarily long transient shifting from the region around one attractor to the region around the other can be present before the motion eventually settles out. This behavior is now known as *transient chaos*.

Another form of chaotic behavior, perhaps the true one, is *persistent chaos*. When a system undergoes such a process, the seemingly random motion will not settle out in time, and no attractor, in the conventional sense, is present. Because chaotic motion does not die out, it is possible to identify a different sort of attractor, typical of chaotic motion, which is called a *strange* or *chaotic attractor*.

To give way to chaotic motion, the state space must have at least three dimensions, and the system must be quite sensitive to the initial conditions. The latter condition seems to be the key factor for chaotic behavior. The first deep study on systems that are very sensitive to initial conditions was performed by Lorentz in 1963 on a three-equation model of atmospheric dynamics for weather predictions. He found that very slight changes in some parameters lead, after time, to large variations in the results.

Consider, for example, the following damped Duffing equation with vanishing linear stiffness:

$$\ddot{x} + 0.05\dot{x} + x^3 = 7.5\cos(t).$$

This very equation has been studied by Ueda,[1] and the results have also been reported by many other authors. To show how much the time history of the system can depend on the initial conditions, two time histories starting from the conditions $x = 3$, $v = 4$ and $x = 3.1$, $v = 4.1$ for $t = 0$ are plotted in Fig. 21.15a. The time histories are initially very similar but, as times goes on, the differences quickly build up, and after time they do not look related anymore. Of course, the system is fully deterministic, and every time a computation is started with exactly the same initial conditions, identical results are found. However, if the initial conditions are known in

---

[1]Y. Ueda, 'Steady motion exhibited by Duffing's equation: A picture book of regular and chaotic motions', *New Approaches in Nonlinear Problems in Dynamics*, P.J. Holmes (Editor), SIAM, Philadelphia, 1980, 311–322.

an approximate way, completely different results are obtained depending on the approximation chosen.

A system that shows this behavior is said to be structurally unstable. To define structural stability, it is possible to verify whether two trajectories starting in the state space within a sphere of small radius δ tend to remain close to each other within a sphere with a limited radius or separate. To show the same behavior in the state projection, the projection of the state trajectories on the (x,v)-plane of the system studied in Fig. 21.15a is reported in Fig. 21.15b. The heavy dots denote the position at the ends of each cycle, i.e., the points that enter the Poincaré map.

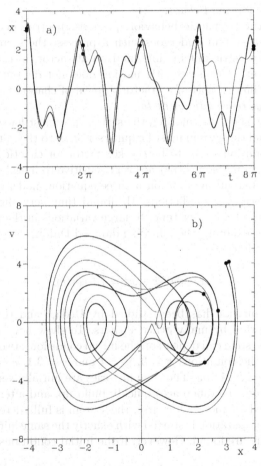

FIGURE 21.15. (a) Time history and (b) state projection for a system modeled by the equation $\ddot{x} + 0.05\dot{x} + x^3 = 7.5\cos(t)$ with two slightly different initial conditions.

The time history and state projection seem quite irregular, but if the system is followed for a longer period of time, a particular form of regularity emerges. After the transient dies out, although the time history does not show any regularity or periodicity, all points of the Poincaré map settle on a very complicated geometrical figure that is the attractor of the system. The attractor resulting from the aforementioned equation, usually referred to as the *Ueda attractor*, is shown in Fig. 21.16.

The Ueda attractor is one of the better-studied chaotic attractors. Its shape is quite complex, and its structure follows a fractal geometry. This is a general feature of strange attractors. To show the fractal structure, a very detailed representation must be obtained, which means that the numerical integration must be carried on for many cycles with sufficient precision. A great deal of computer time is needed to perform detailed studies on chaotic vibrations.

In addition to a strange attractor, another characteristic feature of chaotic motion is the spectrum of the time history. Although the response of a linear system to a harmonic excitation is mono-harmonic and that of a nonlinear non-chaotic system contains a number of harmonics (usually very few), when chaotic motion is present a continuous spectrum, like those encountered in random vibrations, is found.

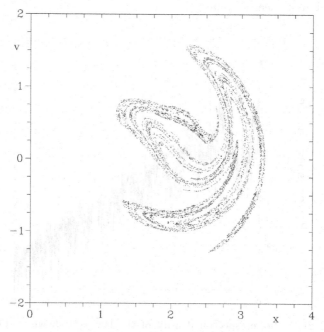

FIGURE 21.16. Ueda attractor represented as a Poincaré map; picture obtained following the system through 8,000 cycles.

Obviously the phenomena are different, because there is nothing random in chaotic motion. If the same numerical experiment is repeated several times with exactly the same initial conditions, the same outputs are obtained. And even if the initial conditions are different, once steady-state chaotic motion is achieved, the same attractor is found. The presence of a well-defined attractor makes it clear that under the random-like appearance, there is an underlying order.

The power spectral density of the law $x(t)$ shown in Fig. 21.15a is shown in Fig. 21.17. The motion was followed for many periods, and then a fast Fourier algorithm was applied on a set of 8,192 points. The fundamental frequency and some higher harmonics are clearly visible, but the spectrum is continuous as if broadband noise were present, as is typical for chaotic vibrations.

The Duffing equation can give way to chaotic response to harmonic excitation not only with the values of the various parameters considered here. Actually, chaotic motion has been observed with both hardening and softening systems, with positive, negative, or vanishing linear parts. The presence of multiple equilibrium positions is then not needed to enable the occurrence of chaotic vibrations.

What is actually needed for chaotic motion is strong nonlinearity. The more experimental work, physical and numerical, is performed on chaotic vibration, the more this type of behavior seems to be a common possibility for heavily nonlinear systems.

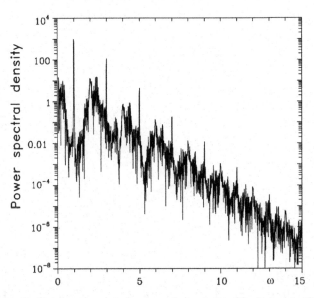

FIGURE 21.17. Power spectral density of the law $x(t)$ shown in Fig. 21.15a obtained using a fast Fourier algorithm on a set of 8,192 points and Hanning windowing.

Much theoretical and experimental work is still needed before chaotic behavior of mechanical systems can be fully understood and design methods based on it developed. For now, the study of chaotic systems is mostly a theoretical issue, pursued by mathematicians and mechanicists with strong theoretical interests. The author, however, thinks that these studies may, in the future, supply the basis for applicative studies that will influence design and structural analysis methods.

The interested reader can find more details on these topics in the many specialized texts that have been published in the last 20 years. Some of them are listed in the bibliography.

21.12 Exercises

Exercise 21.1 *Consider the quarter-car model studied in Example 9.2. Substitute the linear spring with a spring with cubic characteristics with the same stiffness and a nonlinear parameter* $\mu = 400 \ m^{-2}$. *Plot the frequency response of the system for amplitudes of the excitation equal to 10 and 25 mm.*

Exercise 21.2 *Add a linear viscous damper with damping coefficient* $c = 200$ *Ns/m located on the central support to the system studied in Exercise 19.5. Plot the limit envelope and compute the forced response of the system, using the same excitation of Exercise 19.5. State whether the system shows the jump phenomenon.*

Exercise 21.3 *Repeat the computations of Exercise 19.3 without neglecting damping.*

Exercise 21.4 *Consider an oscillator governed by the equation*

$$\ddot{x} + 0.2\dot{x} - x\left(1 - x^2\right) = 0.3\cos\left(1.29\ t\right) \ .$$

Compute the forced response starting from the initial condition $x = 0$, $\dot{x} = 0$ *for* $t = 0$ *using numerical integration. Plot the time history and state-space trajectory for 20 cycles and plot the Poincarè map for a larger number of cycles.*

22

Time Variant and Autoparametric Systems

A particular class of time-variant systems that has many interesting applications is that in which the parameters of the system are periodic in time. In this case the periodic variation of the characteristics of the system can act as an excitation, the so-called parametric excitation. Time-variant systems may be linear, and then a solution can be demonstrated to exist and be unique, or nonlinear. In the latter case their solution is even more complex than that of the nonlinear systems studied in the previous chapters. Another class of systems is that of autoparametric systems; they have some characteristics in common with parametrically excited systems, but their behavior is usually strictly linked with the nonlinear nature of their mathematical models.

22.1   Linear time-variant systems

All systems studied in the previous sections were modeled using equations whose homogeneous part did not contain functions of time. There are, however, many systems of practical importance for which models of this type do not hold. The simplest time-variant model is a linear system in which the mass, damping, or stiffness are functions of time. By dividing the equation by the mass, it is possible to eliminate the dependence on time of the first term, obtaining an equation of motion of the type

$$\ddot{x} + p_1(t)\dot{x} + p_2(t)x = f(t). \tag{22.1}$$

Up to this point, only two types of excitation were considered, namely *external excitation* and *self-excitation*. The first, studied in several sections of this book, is due to a forcing function $f(t)$ applied to the system. At least in linear systems, it does not depend on the generalized coordinates. An external exciting force supplies energy to the system, from some external energy source.

Self-excitation is linked with the homogeneous equation of motion and thus with the very nature of the system. It involves some sort of instability and requires that either the system is provided with energy that can be used to excite vibration or it is able to extract energy from the environment.

The simplest case occurs when the stiffness is negative or when the stiffness matrix is negative defined, and an example is the inverted pendulum. Here the energy is supplied by the gravitational field. Other possibilities are the presence of a circulatory matrix or of negative damping.

Equation (22.1) may be excited in another way: Even if the external excitation is set to zero, i.e., if $f(t) = 0$, the presence of the functions of time in the homogeneous equation can act as an excitation. Because this type of excitation acts from within the parameters of the system, it is usually referred to as *parametric excitation*.

It can coexist with external excitation and self-excitation.

Equation (22.1) is linear, although its coefficients are not constant, and its general solution can be obtained by adding a particular solution of the complete equation to the general solution of the homogeneous equation. Moreover, any linear combination of the solutions of the homogeneous equation is itself a solution.

A particular case of time-dependent systems that has a great practical importance is that of models expressed by linear second-order differential equation of the type of Eq. (22.1) (or a set of such equations) in which functions $p_1(t)$ and $p_2(t)$ are periodic functions of time with period T.

The study of homogeneous equations of this type was published by Floquet in 1883 and hence is usually referred to as *Floquet theory*.

Under the only added condition that $p_1(t)$ is differentiable with respect to time, it is possible to demonstrate that the homogeneous equation associated to the homogeneous of motion (22.1) can be reduced to the form

$$\ddot{x}^* + p(t)x^* = 0\,, \qquad \text{with } p(t) = p(t+T)\,, \tag{22.2}$$

where

$$x = x^* e^{-\frac{1}{2}\int p_1(t)dt}\,, \qquad p(t) = p_2 - \frac{1}{4}p_1^2 - \frac{1}{2}\dot{p}_1\,.$$

This means that the free behavior of the damped system can be obtained from that of an undamped system by multiplying the time history of the latter by an appropriate decaying factor and slightly modifying the frequency by a change of the stiffness. This clearly holds also for linear systems with constant parameters.

In the latter case,

$$p_1(t) = \frac{c}{m}\,, \quad p_2(t) = \frac{k}{m}\,, \quad p(t) = \frac{k}{m}\left(1 - \zeta^2\right) \tag{22.3}$$

and

$$x = x^* e^{-\frac{1}{2}\frac{c}{m}t} = x^* e^{-\zeta\sqrt{\frac{k}{m}}t}\,.$$

It is easy to verify that Eq. (22.2)

$$\ddot{x}^* + p(t)x^* = \left(\ddot{x} + 2\zeta\sqrt{\frac{k}{m}}\dot{x} + \frac{k}{m}x\right)e^{\zeta\sqrt{\frac{k}{m}}t} = 0 \tag{22.4}$$

coincides with the usual equation of motion of a linear system multiplied by $e^{\zeta\sqrt{\frac{k}{m}}t}$. Since the latter factor is always positive, the two equations coincide.

22.2   Hill's equation

Equation (22.2)

$$\ddot{x} + p(t)x = 0\,, \qquad \text{with } p(t) = p(t + T)\,, \tag{22.5}$$

is usually referred to as *Hill's equation*, because it was first studied by Hill is 1886 in the determination of the perigee of lunar orbit. From Floquet theory, it follows that the two independent solutions $x_i(t)$, whose linear combinations yield all possible solutions of the homogeneous equation, can be written as

$$x_i(t) = e^{\alpha_i t}\phi_i(t) \qquad i = 1, 2\,, \tag{22.6}$$

where functions $\phi_i(t)$ are periodic with period T and constants α_i can be complex. From Eq. (22.6) it immediately follows that if the real parts of constants α_i are positive, the solution is unbounded and the behavior of the system is unstable. If the imaginary part of the same constants is not vanishingly small, the result is the product of a function with period T by another function whose period can be different, and then the solution can be an oscillation, damped or self-excited, whose period can be different from that of function $p(t)$.

Since function $p(t)$ is periodic with period T, it can be written as a Fourier series:

$$p(t) = p_0 + \sum_{i=1}^{\infty}\left[p_{1i}\cos\left(i\omega t\right) + p_{2i}\sin\left(i\omega t\right)\right]\,, \tag{22.7}$$

where

$$\omega = \frac{2\pi}{T}\,. \tag{22.8}$$

If only the constant term is considered, the Hill's equation reduces to the usual equation of an undamped system with constant parameters.

If the first terms only are considered, stating $t = 0$ at the instant when p reaches its maximum, the equation reduces to the *Mathieu equation*. Using symbols δ and 2ϵ for the coefficients of the series for $p(t)$, it can be written as

$$\ddot{x} + \left[\delta + 2\epsilon \cos(\omega t)\right] x = 0 . \tag{22.9}$$

Considering the first two terms in cosine (plus the constant term), the Whittaker-Hill equation is obtained:

$$\ddot{x} + \left[A + B \cos(\omega t) + C \cos(2\omega t)\right] x = 0 . \tag{22.10}$$

Approximated solutions can be obtained by writing the solution (22.6) as a Fourier series truncated at the first n terms:

$$x(t) = e^{\alpha t} \left\{ a_0 + \sum_{k=1}^{n} \left[a_k \cos\left(k\omega t\right) + b_k \sin\left(k\omega t\right)\right] \right\} \tag{22.11}$$

or

$$x(\tau) = e^{\alpha t} \sum_{k=-n}^{n} \phi_k e^{ikt} \tag{22.12}$$

and then obtaining the $2n + 1$ unknowns a_k and b_k (or ϕ_k) from a set of $2n + 1$ linear algebraic equations.

22.3   Pendulum on a moving support: Mathieu equation

Consider a pendulum suspended to a point A that can move with a prescribed law (Fig. 22.1). To obtain the equation of motion through the Lagrange equation, the position and the velocity of point P must be computed:

$$\left\{ \begin{matrix} x \\ y \end{matrix} \right\}_P = \left\{ \begin{matrix} x_A + l\sin(\theta) \\ y_A - l\cos(\theta) \end{matrix} \right\} \qquad \left\{ \begin{matrix} \dot{x} \\ \dot{y} \end{matrix} \right\}_P = \left\{ \begin{matrix} \dot{x}_A + l\dot{\theta}\cos(\theta) \\ \dot{y}_A + l\dot{\theta}\sin(\theta) \end{matrix} \right\} . \tag{22.13}$$

The kinetic and potential energies are immediately obtained:

$$\begin{aligned} \mathcal{T} &= \tfrac{1}{2}m \left\{ \dot{x}_A^2 + \dot{y}_A^2 + l^2\dot{\theta}^2 + 2\dot{\theta}l\left[\dot{x}_A \cos(\theta) + \dot{y}_A \sin(\theta)\right] \right\} , \\ \mathcal{U} &= mg\left[y_A - l\cos(\theta)\right] . \end{aligned} \tag{22.14}$$

Lagrange equation thus yields the equation of motion

$$\ddot{\theta} + \frac{\ddot{x}_A}{l}\cos(\theta) + \frac{g + \ddot{y}_A}{l}\sin(\theta) = 0 . \tag{22.15}$$

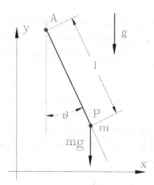

FIGURE 22.1. Pendulum on a moving support: sketch of the system.

Equation (22.15) is a nonlinear equation with parametric excitation. If the amplitude of the motion is small, it can be linearized, obtaining

$$\ddot{\theta} + \frac{g + \ddot{y}_A}{l}\theta = -\frac{\ddot{x}_A}{l}. \tag{22.16}$$

If the supporting point moves only in the x-direction, i.e., in the same direction in which the pendulum moves for small movements about the equilibrium (vertical) position, the equation of motion becomes

$$\ddot{\theta} + \omega_p^2\theta = -\frac{\ddot{x}_A}{l}, \tag{22.17}$$

where

$$\omega_p = \sqrt{\frac{g}{l}} \tag{22.18}$$

is the natural frequency of the small oscillations of the pendulum. The excitation due to motion of the supporting point has the usual effects that can be easily computed using the techniques seen in Chapter 6.

If the supporting point moves only in the y-direction the equation of motion is

$$\ddot{\theta} + \frac{g + \ddot{y}_A}{l}\theta = 0. \tag{22.19}$$

Remark 22.1 *The direction in which the supporting point moves is at right angle to the motion of the pendulum in the equilibrium position. If the pendulum could move along a straight line, the two motions are uncoupled and no excitation would occur. The motion of the pendulum occurs however along an arc of a circle and thus it moves slightly also in y-direction, so that there is some excitation. From the equation it is clear that this excitation grows with increasing θ, or, better, sin(θ) if no linearization is performed.*

The motion of the supporting point in y-direction thus provides a parametric excitation. The homogeneous equation (22.19) is a Hill's equation.

If the motion in the y-direction of the supporting point is harmonic with amplitude A and frequency ω

$$y_A = A\cos(\omega t) \,,$$

remembering that the amplitude of the acceleration is equal to that of the displacement multiplied by $-\omega^2$, without loss of generality the equation of motion can be written as

$$\ddot{\theta} + \left[\omega_p^2 - \frac{A}{l}\omega^2\cos(\omega t)\right]\theta = 0 \,. \tag{22.20}$$

Introducing the nondimensional parameters

$$\tau = \frac{\omega t}{2} \,, \qquad \delta = \left(\frac{2\omega_p}{\omega}\right)^2 \,, \qquad \epsilon = -\frac{2A}{l} \,,$$

it becomes a Mathieu equation in its standard form (22.9)

$$\frac{d^2\theta}{d\tau^2} + \left[\delta + 2\epsilon\cos(2\tau)\right]\theta = 0. \tag{22.21}$$

Its solution is of the type of equation (22.6) where function $\phi(\tau)$ is periodic with period equal to π and, consequently, can be expressed by the following Fourier series

$$\phi(\tau) = \sum_{k=-\infty}^{\infty} \phi_k e^{2ik\tau} \tag{22.22}$$

and then

$$\theta(\tau) = e^{\alpha\tau}\sum_{k=-\infty}^{\infty} \phi_k e^{2ik\tau} = \sum_{k=-\infty}^{\infty} \phi_k e^{(\alpha+2ik)\tau}. \tag{22.23}$$

Remembering that

$$2\cos(2\tau) = e^{2i\tau} + e^{-2i\tau}$$

and introducing the series expressing the solution into the equation of motion (22.21), the latter yields

$$\sum_{k=-\infty}^{\infty} \phi_k\left[(\alpha+2ik)^2 + \delta\right]e^{(\alpha+2ik)\tau} +$$

$$+\epsilon\sum_{k=-\infty}^{\infty} \phi_k\left\{e^{[\alpha+2i(k+1)]\tau} + e^{[\alpha+2i(k-1)]\tau}\right\} = 0 \,. \tag{22.24}$$

By separately equating the various terms of Eq. (22.24), a set made of an infinity of equations is obtained:

$$
\begin{bmatrix}
\cdots & \cdots & \cdots & \cdots & \cdots & \cdots & \cdots \\
\cdots & (\alpha-4i)^2+\delta & \epsilon & 0 & 0 & 0 & \cdots \\
\cdots & \epsilon & (\alpha-2i)^2+\delta & \epsilon & 0 & 0 & \cdots \\
\cdots & 0 & \epsilon & \alpha^2+\delta & \epsilon & 0 & \cdots \\
\cdots & 0 & 0 & \epsilon & (\alpha+2i)^2+\delta & \epsilon & \cdots \\
\cdots & 0 & 0 & 0 & \epsilon & (\alpha+4i)^2+\delta & \cdots \\
\cdots & \cdots & \cdots & \cdots & \cdots & \cdots & \cdots
\end{bmatrix} \times
$$

$$
\times \left\{ \begin{matrix} \cdots \\ \phi_{-2} \\ \phi_{-1} \\ \phi_0 \\ \phi_1 \\ \phi_2 \\ \cdots \end{matrix} \right\} = \mathbf{0}. \tag{22.25}
$$

Equation (22.25) can be used to compute the exponent α of Eq. (22.6) by equating to zero the determinant of the matrix of the coefficients. Such a determinant, which has an infinity of rows and columns, is usually referred to as Hill's infinite determinant. Although it is impossible to obtain an exact solution of this eigenproblem, approximate solutions are readily obtainable by only considering the central part of the determinant. For example, if only the central part with dimension 3×3 is retained, the constant term and the fundamental harmonic of the series in Eq. (22.22) are computed. This will be referred to as *first approximation*.

The second approximation, allowing the computation of the second harmonic of the series, comes from the 5×5 determinant obtained considering all terms written explicitly in Eq. (22.25).

Hill's determinant can be also used to assess the stability of the motion. The motion is unstable, i.e., the amplitude grows indefinitely in time, if the real part of exponent α is positive. The motion can be stable or unstable depending on the values of parameters δ and ϵ: It is possible to plot the boundary of the regions in which an unstable behavior is present on the (δ,ϵ)-plane simply by setting to zero the real part of α. When $\alpha = 0$ an undamped oscillation with nondimensional period equal to π is obtained, while when $\alpha = \pm i$ the undamped motion has a period 2π.

By introducing these values of α, the equation stating that Hill's determinant is equal to zero becomes an eigenproblem in δ, if ϵ is stated, or in ϵ if δ is stated. The hatched regions in Fig. 22.2a are the regions of instability in the (δ,ϵ)-plane obtained from the fourth-order approximation (determinant whose size is 9×9). These results are obviously only approximate, because a reduced form of Hill's determinant has been used.

When δ is negative the motion is almost always unstable, except for a few combinations of δ and ϵ. A negative δ physically means that the pendulum is inverted and the values of ϵ leading to stability correspond to values of

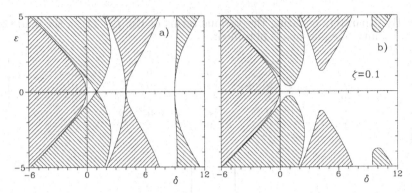

FIGURE 22.2. Regions of instability in the (δ,ϵ)-plane for Mathieu equation: **(a)** without damping; **(b)** with damping.

the parametric excitation causing the pendulum to be stable in its inverted position.

Remark 22.2 *High values of δ correspond to low forcing frequencies: to stabilize the inverted pendulum δ must be quite low and hence high forcing frequencies are required. Also the values of ϵ required for stabilization are high, i.e., the supporting point must move at fairly high frequency with large amplitudes. The lower the frequency, the larger is the required amplitude.*

When δ is positive, i.e., the pendulum is in its normal position, the fields of instability are very narrow unless the parametric excitation ϵ is quite large. There are however values of δ for which the instability conditions reach the ϵ axis: Instability occurs for very small values of the excitation, i.e., the system becomes self-excited (*autoparametric excitation*).

The first value of δ for which this occurs is $\delta = 1$, i.e.,

$$\omega = 2\omega_p .$$

The frequency of the excitation is twice the natural frequency of the pendulum, i.e., the pendulum performs two oscillations in a cycle of the forcing frequency.

Remark 22.3 *This kind of resonance could be expected, since the pendulum moves up and down twice for each oscillation cycle; thus at $\omega = 2\omega_p$ there is a sort of resonance between the vertical motion of the supporting point and the oscillation of the pendulum.*

Other resonant conditions occur when

$$\delta = 4, 9, \ldots = i^2 \text{ for } i = 2, 3, \ldots .$$

Including also the previously seen value, the values of ω for autoparametric instability are

$$\omega = \frac{2\omega_p}{i} \text{ for } i = 1, 2, 3, \ldots . \tag{22.26}$$

If damping is not neglected, the linearized homogeneous equation of motion of the pendulum in nondimensional form is

$$\frac{d^2\theta}{d\tau^2} + 2\zeta\frac{d\theta}{d\tau} + [\delta + 2\epsilon\cos(2\tau)]\theta = 0, \tag{22.27}$$

where

$$\zeta = \frac{\Gamma}{ml^2\omega} = \zeta^*\sqrt{\delta} \text{ and } \zeta^* = \frac{c}{2ml^2\omega_p}.$$

By using the transformation seen for Eq. (22.2), Eq. (22.27) can be reduced to a Mathieu equation in standard form. The presence of damping reduces the growth rate of the oscillations from α to $\alpha - \zeta$ and modifies the natural frequency from $\sqrt{\delta}$ to $\sqrt{\delta - \zeta^2}$. The regions of instability for $\zeta = 0.1$, computed using the fourth approximation of Hill's determinant, are plotted in Fig. 22.2b. The effect of damping is that of reducing the instability regions, particularly for low forcing frequencies, i.e., high values of δ.

There are many different approaches, all approximate, to the study of the Mathieu and Hill equations. Only that based on Hill's determinant is shown here, because it is a very suitable tool for the study of systems with many degrees of freedom and is very easily implemented on digital computers.

Example 22.1 *Consider a pendulum whose length is 0.5 m, connected to a support that can move in a vertical direction (along the y-axis in Fig. 22.1). Using the results obtained from the Mathieu equation, choose the values of the frequency and the amplitude of the harmonic time history of the displacement of the supporting point so that the inverted vertical position of the pendulum is stable. Verify this stability by numerically integrating the equation of motion of the pendulum with initial conditions corresponding to a displacement from the vertical of 0.05 rad (about 2.86°).*

At first the relevant portion of the stability map of Fig. 22.2a is plotted in Fig. 22.3a. A point lying within the stability zone is chosen (point A, with $\delta = -0.1$, $\epsilon = 0.75$). The frequency and the amplitude of the motion of the supporting point are immediately computed (note that $g = -9.81$ m/s², as the pendulum is inverted):

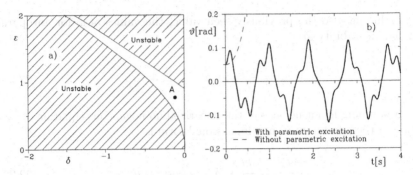

FIGURE 22.3. (a) Boundary of the stability zone for the inverted pendulum in the (δ, ϵ)-plane (Fig. 22.2a); (b) time histories of the response of the inverted pendulum with and without parametric excitation supplied by the vertical motion of the supporting point.

$$\omega = \sqrt{\frac{4g}{l\delta}} = 28 \ rad/s = 4.45 \ Hz; \qquad |A| = \frac{l\epsilon}{2} = 0.188 \ m.$$

The equation of motion of the pendulum, without making any small oscillations assumption, is then

$$\ddot{\theta} - \left[19.62 + 294.78\cos(28t)\right]\sin(\theta) = 0.$$

The result of the numerical integration of the equation of motion of the system, with initial conditions $\theta = 0.05$, $\dot{\theta} = 0$, is reported in Fig. 22.3b, with the time history computed without taking into account parametric excitation. From the figure, it is clear that the prescribed motion of the supporting point allows the inverted pendulum to oscillate about the vertical position while it falls off if the supporting point is fixed.

22.4   The elastic pendulum

The pendulum on a moving support is an interesting example of a system in which the excitation, which is essentially an external excitation since it is due to a motion of the supporting point imposed from outside, acts through the variability of the coefficients of the equation of motion. If the amplitude of the oscillations is small, the equation of motion is linear and can be written in the form of a Mathieu equation in its standard form.

In many cases a system can undergo parametric excitation without the need of external excitation. The simplest of these *autoparametric systems*[1] is the elastic pendulum, i.e., a pendulum in which the suspension wire has been replaced by a spring.

---

[1]See A. Tondl, T. Ruijgrok, F. Verhulst, R. Nabergoj, *Autoparametric Resonance in Mechanical Systems*, Cambridge University Press, 2000.

Assume that the length at rest of the pendulum under the effect of the weight of mass m is l and that the stiffness of the spring is k.

The position of point P (Fig. 22.4a) is

$$(\overline{P - O}) = \left\{ \begin{array}{c} (l+x)\sin(\theta) \\ -(l+x)\cos(\theta) \end{array} \right\} , \qquad (22.28)$$

where x is the stretching of the spring and θ is the rotation of the pendulum.

The velocity of point P is

$$\left\{ \begin{array}{c} \dot{x} \\ \dot{y} \end{array} \right\}_P = \left\{ \begin{array}{c} \dot{x}\sin(\theta) + \dot{\theta}(l+x)\cos(\theta) \\ -\dot{x}\cos(\theta) + \dot{\theta}(l+x)\sin(\theta) \end{array} \right\} . \qquad (22.29)$$

The kinetic energy of the system is thus

$$T = \frac{1}{2}m\left[\dot{x}^2 + \dot{\theta}^2(l+x)^2\right] . \qquad (22.30)$$

Since in the static equilibrium conditions used to define the stretching of the spring x, the spring has already a deformation equal to mg/k, the sum of the gravitational and elastic potential energies is

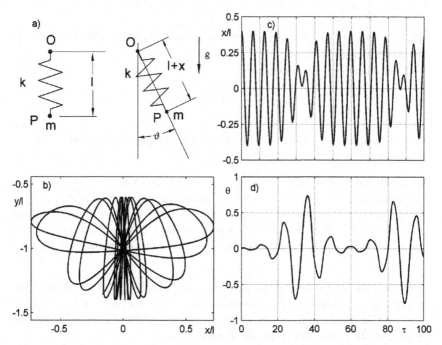

FIGURE 22.4. Elastic pendulum: sketch of the system: (a) time histories (c) and (d) trajectory (b) of point P for a system with $\alpha = 0.5$, starting from a position with $x/l = 0.4$ and $\theta = 0.01$ with zero speed.

$$\mathcal{U} = -mg\left(l + x\right)\cos\left(\theta\right) + \frac{1}{2}k\left(x + \frac{mg}{k}\right)^2 . \qquad (22.31)$$

By introducing the kinetic and potential energies into the Lagrange equations and performing the relevant derivatives, the equations of motion are found:

$$\begin{cases} \ddot{x} - \dot{\theta}^2\left(l + x\right) + g\left[1 - \cos\left(\theta\right)\right] + \dfrac{k}{m}x = 0 \\[2mm] \ddot{\theta}\left(l + x\right)^2 + 2\dot{\theta}\dot{x}\left(l + x\right) + g\left(l + x\right)\sin\left(\theta\right) = 0 . \end{cases} \qquad (22.32)$$

By introducing the natural frequencies of the spring and the pendulum, the nondimensional displacement and time and the ratio between the natural frequencies

$$\omega_s = \sqrt{\frac{k}{m}}, \quad \omega_p = \sqrt{\frac{g}{l}}, \quad x^* = \frac{x}{l}, \quad \tau = \omega_s t, \quad \alpha = \frac{\omega_p}{\omega_s}, \qquad (22.33)$$

the equations of motion become

$$\begin{cases} x^{*\prime\prime} - \theta^{\prime 2}\left(1 + x^*\right) + \alpha^2\left[1 - \cos\left(\theta\right)\right] + x^* = 0 \\[2mm] \theta^{\prime\prime} + 2\theta^{\prime}x^{*\prime}\dfrac{1}{1 + x^{*\prime}} + \dfrac{\alpha^2}{1 + x^{*\prime}}\sin\left(\theta\right) = 0 , \end{cases} \qquad (22.34)$$

where $(\prime)$ and $(\prime\prime)$ indicate the first and second derivatives with respect to the nondimensional time τ.

If the equations are linearized

$$\begin{cases} x^{*\prime\prime} + x^* = 0 \\ \theta^{\prime\prime} + \alpha^2\theta = 0 \end{cases} \qquad (22.35)$$

they uncouple and reduce to the those of a spring–mass system and of a pendulum.

Remark 22.4 *The reason of this uncoupling is clear: in the linear case the spring is involved in the motion of point P which is always in a perpendicular direction to that of the pendular motion: the two motions cannot influence each other. This is different from what is seen for the pendulum on a moving support, where the two directions are perpendicular only for* $\theta = 0$.

Even if the amplitude of the oscillations is small, the interaction is thus essentially a nonlinear phenomenon, due to the terms in $\theta^{\prime 2}$ and $\theta^{\prime}x^{*\prime}$. Stating that θ is small, it is possible to obtain a semi-linearized equation of motion: It is however a nonlinear equation.

Like all homogeneous equations, Eq. (22.34) has a trivial solution, with $x = 0$ and $\theta = 0$.

There is also a semitrivial solution, with $\theta = 0$ (for all values of time, and then also $\theta' = 0$ and $\theta'' = 0$) and $x \neq 0$. Stating $\theta = 0$ in the equations of motion, the first reduces to

$$x^{*\prime\prime} + x^* = 0 \qquad (22.36)$$

while the second one is always satisfied. The motion is that of a spring–mass system that moves vertically, without oscillating.

On the contrary a semitrivial solution with $x = 0$ and $\theta \neq 0$, i.e., a pendular oscillation of the system without deformations of the spring, does not exist. Introducing $x = 0$ into the equation of motion it follows that

$$\begin{cases} -\theta'^2 + \alpha^2 \left[1 - \cos(\theta) \right] = 0 \\ \\ \theta'' + \alpha^2 \sin(\theta) = 0 \ . \end{cases} \qquad (22.37)$$

This equation has no solution, except for the trivial one $\theta = 0$.

Apart from the trivial and the semitrivial solutions, there are other solutions that cannot be obtained in closed form since the equations are essentially nonlinear.

Of particular interest is the case with $\alpha = 1/2$, i.e, with a natural frequency of the spring system equal to twice that of the pendulum. A sort of nonlinear beat takes place: The energy swaps from the pendulum to the spring, as shown in Fig. 22.4b–d. The waveforms show clearly that the system behaves in a nonlinear way.

22.5   Autoparametric systems

Usually vibrating systems constituted by two or more subsystems which can influence each other to the point that one may destabilize the other are called *autoparametric systems*. The conditions for this destabilization to occur are referred to as *autoparametric resonance*. In most cases they are nonlinear, and hence the term 'destabilize' must be interpreted with caution: The system may be unstable in the small, but with growing amplitude a stable limit cycle or even a subsequent decrease of the amplitude may occur.

This is the case of the elastic pendulum studied in the previous section: When the autoparametric resonance occurs, the motion of the spring (involving mostly coordinate x) destabilizes the pendular motion (involving coordinate θ), causing its amplitude to grow at the expense of the spring motion, as seen in Fig. 22.4b and c. With growing amplitude this instability is reduced and then the pendular motion starts reducing again, giving way to what has been called a nonlinear beat.

Since it must be made of at least two subsystem, an autoparametric system must have at least two degrees of freedom. Like in the case of the elastic pendulum, the coupling between the two subsystems is strictly linked with nonlinearities.

22.5.1 Pendulum on elastic support: free response

A typical autoparametric system is the pendulum on a mass–spring–damper system shown in Fig. 22.5. Here the damped mechanical oscillator is the principal system, while the pendulum is the secondary system. The first moves in y-direction, while for small amplitudes the pendulum moves in x-direction. As typical for autoparametric systems, the coupling is caused by the nonlinearities and may have a large effect on its behavior.

The position and velocity of point P are

$$(\overline{P - O}) = \left\{ \begin{array}{c} l \sin(\theta) \\ y - l \cos(\theta) \end{array} \right\}, \tag{22.38}$$

$$\left\{ \begin{array}{c} \dot{x} \\ \dot{y} \end{array} \right\}_P = \left\{ \begin{array}{c} \dot{\theta} l \cos(\theta) \\ \dot{y} + \dot{\theta} l \sin(\theta) \end{array} \right\}, \tag{22.39}$$

where y is the y-coordinate of point C.

Operating as usual, the Lagrangian of the system is

$$\mathcal{T} - \mathcal{U} = \frac{1}{2} m_t \dot{y}^2 + \frac{1}{2} m \left[l^2 \dot{\theta}^2 + 2l \dot{y} \dot{\theta} \sin(\theta) \right] - m_t g y +$$

$$+ mgl \cos(\theta) - \frac{1}{2} k \left(y - \frac{m_t g}{k} \right)^2, \tag{22.40}$$

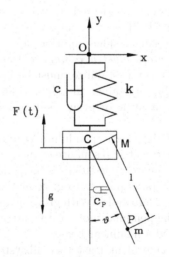

FIGURE 22.5. Pendulum on an elastic support. Sketch of the system.

where
$$m_t = m + M \ .$$

Owing to the presence of the two dampers, a Rayleigh dissipation function can be defined
$$\mathcal{F} = \frac{1}{2}c\dot{y}^2 + \frac{1}{2}c_p\dot{\theta}^2 \ . \tag{22.41}$$

The equation for free motion is
$$\begin{cases} m_t\ddot{y} + ml\ddot{\theta}\sin(\theta) + ml^2\dot{\theta}^2\cos(\theta) + c\dot{y} + ky = 0 \\ ml^2\ddot{\theta} + ml\ddot{y}\sin(\theta) + mgl\sin(\theta) + c_p\dot{\theta} = 0 \ . \end{cases} \tag{22.42}$$

Again, all coupling disappears if the system is linearized. The nonlinear system has a trivial solution, with $y = 0$ and $\theta = 0$, and a semitrivial solution, with $\theta = 0$ and y the solution of the equation
$$m_t\ddot{y} + c\dot{y} + ky = 0 \ , \tag{22.43}$$

i.e., the usual equation of motion of a spring–mass–damper system.

Nontrivial solutions can be studied by numerical integration, as seen for the elastic pendulum.

Remark 22.5 *No free unstable motion is possible, since no energy source able to compensate for energy dissipation is present.*

22.5.2 Pendulum on elastic support: forced response

Assume that a force
$$F = F_0 \cos(\omega t)$$

acts on mass M in y-direction. By introducing the following nondimensional parameters
$$\omega_1 = \sqrt{\frac{k}{m_t}} \ , \quad \omega_p = \sqrt{\frac{g}{l}} \ , \quad \alpha = \frac{\omega_p}{\omega_1} \ ,$$
$$\mu = \frac{m}{m_t} \ , \quad \zeta_1 = \frac{c}{2\sqrt{km_t}} \ , \quad \zeta_p = \frac{c_p}{2\mu l^2\sqrt{km_t}}$$
$$\tau = \omega t \ , \quad f_0 = \frac{F_0}{kl} \ , \quad \beta = \frac{\omega_1}{\omega} \ , \quad y^* = \frac{y}{l}$$

(now the nondimensional time is related to the frequency of the excitation and not to the natural frequency of the system) the equations reduce to the nondimensional form:
$$\begin{cases} y^{*\prime\prime} + \mu\theta^{\prime\prime}\sin(\theta) + \mu\theta^{\prime 2}\cos(\theta) + 2\zeta\beta y^{*\prime} + \beta^2 y^* = \beta^2 f_0 \cos(\tau) \\ \theta^{\prime\prime} + y^{*\prime\prime}\sin(\theta) + 2\zeta_p\beta\theta^{\prime} + \alpha^2\beta^2\sin(\theta) = 0 \ . \end{cases} \tag{22.44}$$

The pendulum acts as a sort of dynamic vibration absorber and, when tuned, can effectively reduce the amplitude of vibration of the main system. Tuning here occurs when $\alpha = 1/2$, i.e., when the pendulum performs one oscillation in the time the main systems perform two oscillations. This is fairly obvious, since the pendulum performs two up–down cycles in each oscillation, while the main system performs only one.

However, here the coupling is nonlinear and thus the effectiveness of the pendulum depends on the amplitude. When the motion of the mass is small, the pendulum does not move, and the solution is again a semitrivial solution with $\theta = 0$ (although $y^* \neq 0$). At a certain amplitude the pendulum starts being effective and absorbs a large part of the vibration energy, subtracting it from the main system. This behavior is often referred to as *vibration quenching*.

The stability of the trivial, semitrivial, and nontrivial solutions is studied in the book by A. Tondl et al.[2] mainly on the base of small displacements assumptions.

Here only an example is reported to show the phenomenon; owing to the strong nonlinearity of the system, the study was performed using numerical integration in time.

Assume the following values of the relevant parameters:

$$\alpha = \frac{1}{2} , \quad \mu = 0.1 , \quad \zeta = 0.02 , \quad \zeta_p = 0.04 , \quad f = 0.02 .$$

The forcing function is applied for 500 units of the nondimensional time, i.e., for about 80 cycles of the forcing function, starting from

$$y^* = 0.01 , \quad \theta = 0.02 , \quad y^{*\prime} = 0 , \quad \theta' = 0 .$$

The position is slightly deviated from the position at rest, since starting from exactly $\theta = 0$ would cause a motion following the semitrivial solution (in the numerical simulation no perturbation is present if not explicitly introduced).

Various values of β spanning from 10 to 0.5 (i.e., of the forcing frequency, from $\omega/\omega_1 = 0.1$ to $\omega/\omega_1 = 2$) were introduced. The results for $\beta = 1$, i.e., for a resonant excitation on the main system, are reported in Fig. 22.6a, b, and d. At resonance, the amplitude starts growing toward the amplitude of the system without pendulum ($y^* = 0.5$), but after a number of periods (about 8) the pendulums kicks in and starts absorbing (and dissipating) energy reducing the amplitude of the main system. A steady-state solution with an amplitude about 1/5 of that of the system without pendulum is eventually reached.

---

[2]A. Tondl, T. Ruijgrok, F. Verhulst, R. Nabergoj, *Autoparametric Resonance in Mechanical Systems*, Cambridge University Press, Cambridge, 2000.

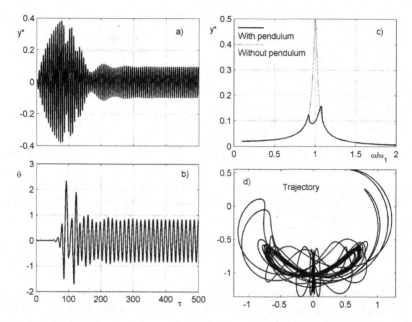

FIGURE 22.6. Pendulum on a spring–mass–damper system. Time history of the oscillations of the main system (a) and of the pendulum (b), trajectory of point P (d) and dependence of the amplitude of the motion of the main system on the forcing frequency (c). Data: $\alpha = 0.5$, $\mu = 0.1$, $\zeta = 0.02$, $\zeta_p = 0.04$, $f = 0.02$. Resonant conditions, except for (c).

Remark 22.6 *Very large oscillations of the pendulum are obtained; in some cases the pendulum performs even full revolutions about its suspension point. Clearly in this case no semi-linearized solutions could be used.*

The steady-state amplitude for different values of the forcing frequency (or better, of ω/ω_1) is reported in Fig. 22.6c. Outside resonance the amplitude of the response is that of the system without pendulum: The semitrivial solution is stable and the presence of the pendulum has no effect. In quasi-resonant or resonant conditions the quenching phenomenon occurs: the pendulum absorbs and then dissipates much of the energy (thanks to its damper) reducing the amplitude of oscillation of the main system.

This effect is quite sensitive to the parameters of the system. The results obtained for $\beta = 1$ with the same values of the parameters except for ζ_p are shown in Fig. 22.7a.

Quenching occurs with moderate damping of the pendulum. If it is increased the motion of the pendulum is reduced and less energy is transferred from the main system with a resulting larger amplitude of the motion of the latter. With very little pendulum damping the amplitude of its motion

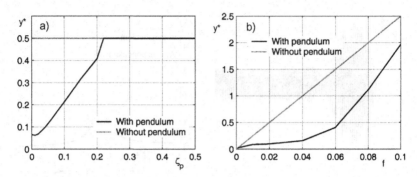

FIGURE 22.7. Dependence of the amplitude of the system of Fig. 22.7 on the damping of the pendulum (**a**) and on the amplitude of the forcing function (**b**). Nondimensional plot.

is increased, but the amplitude of the vibration of the main system remains low.

To show the strong nonlinearity of the phenomenon, the dependence of the amplitude of the response on the amplitude of the force is shown in Fig. 22.7b. In the case of the linear system the plot is a straight line. In this case, the quality factor is 25 and hence the nondimensional displacement is 25 times the nondimensional force. With the pendulum, at low amplitudes quenching does not occur, since the pendulum does not move (semitrivial solution). At a certain amplitude quenching starts, and the amplitude of the response is almost independent from the amplitude of the excitation: by increasing the energy introduced into the system, the energy supplied to the pendulum increases, but the energy ending in the main system is almost constant. At a certain point the pendulum is unable to absorb more energy, and the amplitude of the response of the main system starts growing again. At very large amplitudes, quenching becomes again marginal.

Remark 22.7 *What has been shown for a pendulum on a single-degree-of-freedom system can be extended to multi-degrees-of-freedom systems, provided the oscillation of the pendulum occurs in a direction perpendicular to that of the vibration of the supporting point.*

It is possible to conclude that a pendulum oscillating in the same direction of the vibration of the supporting point is a standard dynamic vibration absorber, and if the amplitude of the oscillations is small, a linear phenomenon occurs. Tuning occurs when the two frequencies are equal.

On the contrary, if the oscillation is at right angles with the vibration of the supporting point, an autoparametric resonance occurs and the phenomenon is strongly nonlinear even in the case of small oscillations. Tuning occurs when the frequency of the pendulum is half the frequency of the main system.

22.6   Exercises

Exercise 22.1 *A dynamometric test rig for rockets has a stiffness $k = 800$ kN/m and a damping $c = 100$ Ns/m. A rocket with a mass of $m_0 = 2,000$ kg is fired on it. The rate at which the fuel is burnt is assumed to be constant $\dot{m} = 10$ kg/s, the total amount of fuel is $m_f = 1,600$ and the exhaust velocity is $v = 2,000$ m/s. Compute the time history of the system for a time equal to 120% of the burn time of the rocket.*

Exercise 22.2 *Plot the map of Fig. 22.2b, for $\zeta = 0.05$.*

Exercise 22.3 *Repeat the computation of the time history of the system studied in Example 22.1 (Fig. 22.3b), comparing the results obtained using the linearized and nonlinear equation of motion. Repeat the comparison assuming an initial displacement 10 times larger.*

Exercise 22.4 *Consider the system studied in Example 22.1 working at point A in Fig. 22.3a. Compute the eigenvalues α of the system taking into account a different number of harmonics, from 1 to 4.*

Part III

Dynamics of Rotating and Reciprocating Machinery

23

Elementary Rotordynamics: The Jeffcott Rotor

The simplest model for understanding the vibrational behavior of rotating machinery is the so-called Jeffcott rotor, consisting of a point mass located on a massless elastic, and possibly damped, shaft. Several peculiar features of the vibrational behavior of rotors, like the unbalance response, self-centering, and the roles of rotating and nonrotating damping, can be studied using this much simplified model.

23.1 Elementary rotordynamics

While usually structures are stationary with respect to an inertial frame of reference, apart from the vibratory motion studied by structural dynamics, many machines contain rotating elements that may vibrate owing to their elasticity and inertia. Rotating bodies and structures are usually defined rotors.

Following the ISO definition, a *rotor* is a body suspended through a set of cylindrical hinges or bearings that allow it to rotate freely about an axis fixed in space. Transmission shafts, parts of reciprocating machines that have only rotational motion, and many other rotating machine elements can thus be considered rotors. If no reference is made to the type of supports or their existence, space vehicles or celestial bodies rotating about an axis whose direction is constant can also be regarded as rotors, at least for some features of their behavior.

Remark 23.1 *While rotors on fixed bearings, or* fixed *rotors, are studied under the assumption that the spin speed is imposed (usually is constant), isolated, or* free, *rotors are governed by the conservation of linear and angular momenta. It is possible to show that under the assumption of small displacements and rotations the two assumptions coincide.*

The parts of the machine that do not rotate will be referred to with the general definition of *stator*.

Some simplifications that allow the linearization of the mathematical model of rotors can be made. The rotor has, in the undeformed configuration, a well-defined and fixed rotation axis, which coincides with one of the baricentrical principal axes of inertia, if the rotor is perfectly balanced. Actually, this is only approximately true, but the *unbalance*, i.e., the deviation from this ideal condition, is usually small. The displacement of the rotational axis from its nominal position due to the deformations of the system is also assumed to be small. The two assumptions of small unbalance and small displacement allow the linearization of the equations of motion in a way consistent with what was seen for the case of the dynamics of structures where similar small-displacement assumptions were required to obtain linear equations of motion.

Some cases that, strictly speaking, could not be studied using the aforementioned assumptions can still be dealt with in the same way. Consider, for example, the rotor of an aircraft turbojet during maneuvered flight. The direction of the axis of the rotor changes continuously in time, and no small-angle assumption can be considered for this motion. However, the motion of the rotor can be studied in a reference frame that is fixed to the aircraft, provided that the motion of the latter can be considered independent of the dynamic behavior of the first and the related inertia forces are added. This way of separating the problem into its dynamic and quasi-static parts is possible if the characteristic times of the different phenomena under study are widely different. In the example given earlier, this is clearly the case if the frequencies that characterize the motion of the rotor of the turbine with respect to the aircraft are of several Hertz (periods of fractions of seconds), while the rotations of the airframe have characteristic times of the order of several seconds. On the contrary, the seismic actions on the rotor of a machine in a building can have frequencies of the same order of magnitude as those that characterize the rotor itself, and the problem may have to be studied without any uncoupling being possible.

Another common assumption is that of axial symmetry. The dynamic study of an axi-symmetrical rotor is greatly simplified and is usually performed using a nonrotating reference frame. If, on the contrary, the rotor cannot be considered axially symmetrical, the study becomes very complicated, unless an axial symmetry assumption can be made on the nonrotating parts of the system. In the latter case, a reference frame that rotates at the angular velocity of the rotor can be used and simplified equations can be obtained. If both stator and rotor are isotropic with respect to the rotation axis, very simple models can be devised. In these simple models the rotor is usually modeled as a beam-like structure, on which concentrated mass elements, sometimes with their moments of inertia, are located.

23.2   Vibrations of rotors: the Campbell diagram

As already stated, a rotor is a body that is free to rotate about a well-defined axis. Usually the dynamic study of rotors is performed under the assumption that the angular velocity about the axis of rotation, usually referred to as *spin speed*, is constant, at least in its average value. Because the natural frequencies of a deformable rotor, or more generally, of a machine that contains a rotor, can depend on the spin speed, the dynamic behavior of such systems is usually summarized by a plot of the natural frequencies as functions of the rotational speed.

Because in many cases the frequencies of the exciting forces also depend on the speed, they can be reported on the same plot, obtaining what is generally known as a *Campbell diagram*. If the dynamic behavior of the system can be described in terms of complex frequencies, the *decay rate plot*, in which the decay rates are reported as functions of the spin speed, can be obtained together with the Campbell diagram. Alternatively, the frequency can be plotted against the decay rate, obtaining what is generally referred to as a *roots locus*.

There are cases in which, in spite of what was said earlier, the natural frequencies are constant with respect to the rotational speed: The Campbell diagram in this case is made of horizontal straight lines.

Remark 23.2 *The Campbell diagram can be plotted only in the case of linear systems, because only in this case does the very concept of natural frequencies apply. However, in the case of nonlinear systems, the Campbell diagram of the linearized system may yield important information on the behavior of the system.*

Example 23.1 *A case in which the natural frequencies of the system are strongly influenced by the spin speed is that of the rotating pendulum, a pendulum attached to the outer radius of a disc rotating at a constant angular velocity (Fig. 23.1a).*
As the angular velocity Ω of the disc is imposed, the system only has 2 degrees of freedom, and angles θ (between the projection of line PC on the plane of the disc and radius OC) and ϕ (between line PC and the mentioned plane) can be assumed to be generalized coordinates.

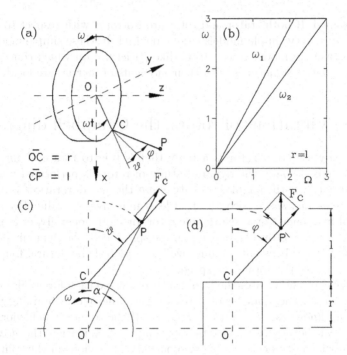

FIGURE 23.1. Rotating pendulum. (a) Sketch of the system and generalized coordinates, (b) Campbell diagram, (c) situation in the xy-plane, (d) situation in a plane containing the z-axis.

All other force fields except the centrifugal field are neglected.
The position of point P is

$$(\overline{P\text{-}O}) = \left\{ \begin{array}{c} r\cos(\Omega t) + l\cos(\phi)\cos(\Omega t + \theta) \\ r\sin(\Omega t) + l\cos(\phi)\sin(\Omega t + \theta) \\ l\sin(\phi) \end{array} \right\}.$$

By differentiating the expressions of the coordinates with respect to time, the velocity of point P is readily obtained

$$\vec{V}_P = \left\{ \begin{array}{c} -\Omega r\sin(\Omega t) - \dot{\phi}l\sin(\phi)\cos(\Omega t + \theta) - l(\Omega + \dot{\theta})\cos(\phi)\sin(\Omega t + \theta) \\ \Omega r\cos(\Omega t) - \dot{\phi}l\sin(\phi)\sin(\Omega t + \theta) + l(\Omega + \dot{\theta})\cos(\phi)\cos(\Omega t + \theta) \\ \dot{\phi}l\cos(\phi) \end{array} \right\}.$$

The kinetic energy of the mass located in point P is simply

$$\mathcal{T} = \frac{1}{2}m|\vec{V}_P|^2 = \frac{1}{2}m\left[\Omega^2 r^2 + \dot{\phi}^2 l^2 + l^2(\Omega + \dot{\theta})^2\cos^2(\phi) + \right.$$

$$\left. -2\Omega r l\dot{\phi}\sin(\phi)\sin(\theta) + 2\Omega r l(\Omega + \dot{\theta})\cos(\phi)\cos(\theta)\right].$$

The equations of motion can be easily obtained by resorting to Lagrange equations:

$$\begin{cases} l\ddot{\theta}\cos^2(\phi) - 2l(\Omega + \dot{\theta})\dot{\phi}\cos(\phi)\sin(\phi) + \Omega^2 r\cos(\phi)\sin(\theta) = 0, \\ l\ddot{\phi} + l(\Omega + \dot{\theta})^2\cos(\phi)\sin(\phi) + \Omega^2 r\sin(\phi)\cos(\theta) = 0. \end{cases}$$

The equations of motion are nonlinear, but can be linearized in the study of the small oscillations of the pendulum about the static equilibrium position

$$\begin{cases} l\ddot{\theta} + \Omega^2 r\theta = 0, \\ l\ddot{\phi} + \Omega^2(r + l)\phi = 0. \end{cases}$$

The linearized equations can also be obtained directly from an expression of the kinetic energy truncated after quadratic terms. By introducing the series for the sine and cosine and neglecting products of the generalized coordinates in which terms of order greater than two are contained, the kinetic energy can be written as

$$T = \frac{1}{2}m\left[\Omega^2(r + l)^2 + \dot{\phi}^2 l^2 + \dot{\theta}^2 l^2 - \Omega^2 l(r + l)\phi^2 - \Omega^2 r l\theta^2 + 2\Omega l(r + l)\dot{\theta}\right].$$

The expression of the kinetic energy can be subdivided into three terms:

$$\mathcal{T}_0 = \frac{1}{2}m\left[\Omega^2(r + l)^2 - \Omega^2 l(r + l)\phi^2 - \Omega^2 r l\theta^2\right]$$

is independent of the generalized velocities.

Apart from a constant term, whose derivatives are nil, it yields the so-called geometric stiffness terms in the equation of motion. As usual in rotating systems, they constitute a centrifugal stiffening and are proportional to the square of the spin speed:

$$\mathcal{T}_1 = m\Omega l(r + l)\dot{\theta}$$

is linear in the generalized velocities. However, this term is independent of the displacements, and its derivatives in the equations of motion are nil; there is no gyroscopic term in the equations of the rotating pendulum:

$$\mathcal{T}_2 = \frac{1}{2}m\left[\dot{\phi}^2 l^2 + \dot{\theta}^2 l^2\right]$$

is quadratic in the generalized velocities and yields the inertia terms of the equations of motion, which are also present in natural systems.

The motion in the rotation plane xy is uncoupled, within the validity of the linearization of the equations of motion, from the motion in axial direction z. The former is the equation of motion of a pendulum whose length is l within a constant force field whose acceleration is $r\Omega^2$, and the latter is the equation motion of the same pendulum within a constant force field whose acceleration is $(r + l)\Omega^2$. The natural frequencies of the motions outside and within the rotation planes are, respectively,

$$\omega_1 = \Omega\sqrt{1 + \frac{r}{l}}, \qquad \omega_2 = \Omega\sqrt{\frac{r}{l}}.$$

The Campbell diagram, shown in Fig. 23.1b for the case in which $r = l$, is then made of two straight lines. A simple explanation of the different behaviors of the system in a plane containing the axis of rotation and a plane perpendicular to it (xy-plane) is shown in Fig. 23.1c and d. In the former, the restoring force acting on the pendulum in a direction perpendicular to line PC is

$$F_c \sin(\phi) \approx m\Omega^2(r + l)\phi.$$

In the xy-plane the restoring force is

$$F_c \sin(\theta - \alpha) \approx m\Omega^2(r + l)(\theta - \alpha).$$

As $\alpha \approx \theta l/(r + l)$, the restoring force is $\approx m\Omega^2 r\theta$.
The analysis reported here shows a simple case in which the natural frequency of the system depends on the angular velocity. It also shows how the cylindrical symmetry of the centrifugal field induces a sort of anisotropy into the system: The behavior in the xy-plane is different from that in any plane containing the axis of rotation. The rotating pendulum also has some important technological applications and will be studied again when dealing with damping of torsional vibration.

23.3   Forced vibrations of rotors: critical speeds

Often rotors are subjected to forces that vary in time, and sometimes their time history is harmonic. This is the case, for example, of forces due to the unbalance of the rotor itself, which can be described as a vector rotating with the same angular speed as the rotor and whose components in the fixed reference frame vary harmonically in time with circular frequency equal to the rotational speed Ω. In other cases the time history is less regular but, if it is periodical, can always be represented as the sum of harmonic components. In the case of rotors there are also, however, instances in which the forcing functions can be described only in statistical terms.

   In the first two cases the frequency of the forcing function or of its harmonic components is often linked with the spin speed of the rotor and can be plotted on the Campbell diagram. In the case of the excitation due to unbalance, for example, the forcing frequency can be represented on the $(\omega\Omega)$-plane by the straight line $\omega = \Omega$, i.e., by the bisector of the first quadrant. In this case, the excitation is said to be *synchronous*. The relationship linking the frequency of the forcing function to the spin speed is often of simple proportionality and can be represented on the Campbell diagram by a straight line through the origin.

   The spin speeds at which one of the forcing functions has a frequency coinciding with one of the natural frequencies of the system are usually

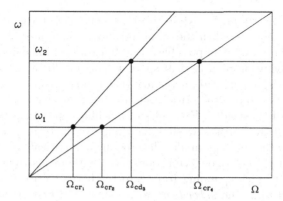

FIGURE 23.2. Graphical representation of the critical speeds in the case in which the forcing frequencies are proportional to Ω while the natural frequencies are constant.

referred to as *critical speeds* and can be identified on the Campbell diagram by intersecting the curves related to the natural frequencies with those related to the forcing frequencies. A case in which the natural frequencies are independent of the speed and the forcing frequencies are proportional to Ω is reported in Fig. 23.2.

Not all the intersections on the Campbell diagram are equally dangerous. If the frequency of a forcing function coincides with the natural frequency of a mode that is completely uncoupled with it (or, better, if the modal force corresponding to the forcing function and the resonant mode is vanishingly small), no resonance actually occurs. For example, if the frequency of the driving torque (i.e., of the torsional moment on the rotor) is coincident with a flexural natural frequency of the rotor and torsional and flexural behaviors are completely uncoupled, no resonance takes place. In other cases the resonance can be very weak and the damping of the system can be sufficient to avoid any measurable effect.

There are, however, cases in which a very strong resonance takes place and the rotor cannot operate at or near a critical speed without strong vibrations and even a catastrophic failure. In particular, the resonances due to the coincidence of one of the flexural natural frequencies with the spin speed are particularly dangerous. They can be detected on the Campbell diagram by the intersection of the curves related to the natural frequencies with the straight line $\omega = \Omega$. They are usually referred to as *flexural critical speeds*, without further indications, while other critical speeds related to bending behavior, which are usually less dangerous, are often said to be *secondary critical speeds*.

When a rotor operates at a critical speed, with the meaning just given, the amplitude of the vibration grows linearly in time and only the damping of the stator and the supports (as will be shown later, the damping of the rotor is completely ineffective in this case) and the unavoidable nonlinearities, which show up when the amplitude grows, can prevent the failure of the rotor. Actually, it is possible to design rotating machinery in such a way that operation at a critical speed is possible for a limited period of time, but it is, at any rate, necessary that the normal operating range be either below or above the critical speeds and that sustained operation at critical speed be avoided. Examples of machines operating above the first critical speed are domestic washing machines: In the transition between the washing and the spinning modes the crossing of a critical speed is generally easily detected from strong vibrations, even without instruments.

The speed range spanning from zero to the first critical speed is usually referred to as the *subcritical range*; above the first critical speed the *supercritical range* starts. A growing number of machines work in the supercritical range, and then at least one of the critical speeds must be crossed during start-up and shut-down procedures.

If the Campbell diagram related to flexural vibrations is made by straight lines parallel to the Ω-axis, i.e., if the natural frequencies are independent of the speed, the numerical values of the critical speeds coincide with those of the natural frequencies at standstill, as can be seen from Fig. 23.2.

Remark 23.3 *Some confusion between the concepts of critical speed and natural frequency that can still be found can probably be ascribed to this. Even if the numerical values are coincident, the two physical phenomena are different, particularly where the stressing of the rotor is concerned.*

The force due to unbalance, which can be expressed in a stationary reference frame as a vector rotating with angular velocity Ω, has a constant direction if seen in a reference frame fixed to the rotor. If the machine is axially symmetrical, as will be seen in the following sections, at the critical speed the rotor rotates in a deflected configuration but does not vibrate, in the sense that there is no cyclic stressing.

Remark 23.4 *The flexural critical speed can be defined as the speed at which the centrifugal forces due to the bending of the rotor are in indifferent equilibrium with the elastic restoring forces, and from this point of view the situation is more similar to that characterizing elastic instability than that typical of vibratory phenomena. A rotor operating at a critical speed is not subject to vibration but is a source of periodic excitation that can cause vibration, often very strong, in the nonrotating parts of the machine.*

The very concept of critical speed has been defined with reference to a linear system, and it is impossible to define critical speeds in this sense in the case of nonlinear rotors. However, a more general definition of critical speed, as a speed in which strong vibrations are encountered, is often used. This definition, which also holds in the case of nonlinear rotors, has a certain degree of arbitrarity, because the amplitude of the vibration depends on the cause that produces it. In the case of nonlinear rotors the speed at which the maximum amplitude is reached, i.e., the critical speed following the last definition, also depends on the entity of the exciting causes (for example, the unbalance in the case of flexural critical speeds). The critical speeds of linear systems are, on the contrary, characteristic of the system and are independent of the excitations.

23.4   Fields of instability

Rotors can develop an unstable behavior in well-defined velocity ranges. The term *unstable* can have several meanings, and different definitions of stability exist, one of the most common being that introduced by Liapunov and reported in Section 20.6.

However, this theoretical definition of stability may be difficult to apply in many engineering situations, and a technical definition of stability can be used: The behavior of a machine is considered *stable* when the amplitude of vibration in normal operation does not exceed a value considered acceptable. For rotating machinery, it was stated by A. Muszynska that[1]

a rotating machine is stable if its rotor performs a pure rotational motion around an appropriate axis at a required rotational speed and this motion is not accompanied by other modes of vibrations of the rotor, its elements or other stationary parts of the machine, or, if such vibrations take place, their amplitudes do not exceed admitted, acceptable values. The stable rotating machine is immune to external perturbing forces, i.e., any random perturbation cannot drastically change its behavior. Such a perturbation causes only a transient decaying process leading to a previous regime of performance, or to a new one, which is included in the acceptable limits.

When studying the behavior of damped linear systems, the amplitude of free vibration was seen to decay exponentially in time, owing to the energy dissipation due to damping. In the case of rotors, however, there is a source of energy, the centrifugal field, that may in some cases cause an unbounded growth in time of the amplitude of free vibrations. The frequency ranges in which this growth occurs, i.e., in which self-excited vibrations can develop, are usually called *instability fields* or *instability ranges*, and the speed at which the first such field starts is the *threshold of instability*.

Remark 23.5 *Instability ranges must not be confused with critical speeds: Critical speeds are a sort of resonance between a natural frequency and a forcing function acting on the rotor, while in instability ranges true self-excited vibrations occur.*

---

[1]A. Muszynska, *Rotor Instability*, Senior Mechanical Engineering Seminar, Carson City, June 1984.

For an instability range to occur, some source of energy to sustain the vibration with increasing amplitude is needed; in the present case the energy can be supplied by the kinetic energy linked with rotation at the spin speed Ω. It is easy to verify that the kinetic energy stored in the rotor is greater by some orders of magnitude than the elastic potential energy the rotor can store without failure.

Consider, for example, a thin ring with radius r and material density ρ rotating at speed Ω. As the hoop stress in the ring is simply

$$\sigma_h = \rho\Omega^2 r^2 \, ,$$

the relationship linking the kinetic energy with the stress is

$$\mathcal{T} = \frac{1}{2}mr^2\Omega^2 = \frac{1}{2}\frac{m\sigma_h}{\rho} \, .$$

The maximum potential energy the ring can store in an axi-symmetrical tensile deformation is

$$\mathcal{U} = \frac{1}{2}\frac{V\sigma_U^2}{E} = \frac{1}{2}\frac{m\sigma_U^2}{E\rho} \, ,$$

where V and σ_U are the material volume and the ultimate strength, respectively. The ratio between the kinetic energy and the potential energy corresponding to a deformation causing the failure of the ring (at failure $\sigma_h = \sigma_U$) is

$$\frac{\mathcal{T}}{\mathcal{U}} = \frac{E}{\sigma_U} \, .$$

By introducing the expression of the hoop stress into the last equation, it is easy to see that ratio E/σ_U is nothing other than the ratio v_s/V_U between the speed of sound in the material and the peripheral velocity at failure of the ring, whose value is usually far greater than 10. The kinetic energy stored in the rotor is then at least one or two orders of magnitude greater than the energy needed to deform the rotor until failure occurs. Similar considerations would hold for other geometrical configurations or deformation patterns.

Remark 23.6 *It is then sufficient that a small portion of the kinetic energy of the rotor is transformed into deformation potential energy to cause failure. This situation is typical of structural elements that are in close contact with an energy source that can excite and sustain vibration, another case being that of aeroelastic vibrations.*

It is easy, at least from a theoretical point of view, to predict the onset of unstable working conditions in a linear system. If the time history of the system is expressed in the form

$$\mathbf{x} = \mathbf{x}_0 e^{i\omega t} \, , \tag{23.1}$$

the sign of the decay rate, i.e., of the imaginary part of the complex frequency ω, gives directly the stability condition: If it is positive the amplitude decays exponentially in time (stable condition), while unstable operation is characterized by the negative imaginary part of the complex frequency. The plot of the decay rate as a function of the spin speed must then be drawn with the Campbell diagram. At vanishingly small speeds all decay rates are obviously positive, because there is no external source of energy that can excite vibration. With increasing speed the decay rate of some modes can decrease, showing a reduction of stability. If at a certain value of the spin speed Ω one of them vanishes and then becomes negative, that speed is the threshold of instability of the system.

It is clear that when the time history is written in the form

$$\mathbf{x} = \mathbf{x}_0 e^{st} , \qquad (23.2)$$

the system is stable if the real parts of all complex eigenvalues are negative.

Actually, it can be difficult to evaluate the decay rate with enough precision, because it is influenced by many factors that are difficult to evaluate, one of them being damping. Often, it is only possible to perform a first-approximation theoretical or numerical study, whose results must be verified experimentally.

As a general rule, if the conditions for uncoupling of flexural, axial, and torsional behaviors are met, only the first can give way to self-excited vibration. There are many mechanisms that can cause unstable conditions, including internal rotor damping due to material damping and friction between the various components assembled by bolting, riveting, shrink fitting, and so on; rubbing between stator and rotor; and fluid viscosity in journal bearings and seals. All the mentioned mechanisms are potentially dangerous, but they do not necessarily always cause instability. The better known destabilizing effects are those due to the material damping in the rotor and viscosity in lubricated journal bearings. The latter can cause the well-known oil whip phenomenon, which consists in very strong vibrations starting from a speed that is, in many cases, close to twice the first flexural critical speed.

To make it easier to distinguish between critical speeds and fields of instability, the following features can be listed:

Critical speeds

- They occur at well-defined values of the spin speed.
- The amplitude grows linearly in time if no damping is present. It can be maintained within reasonable limits and, as a consequence, a critical speed can be passed.
- The value of the speed is fixed, but that of the maximum amplitude depends on the amplitude of the perturbation. In particular, the main flexural critical speeds do not depend on the amount of unbalance, but the amplitude increases with increasing unbalance.

Fields of instability

- Their span is usually quite large. Often all speeds in excess of the threshold of instability give way to unstable behavior.
- If a threshold of instability exists, it is usually located in the super-critical range.
- The amplitude grows exponentially in time. It grows in an uncontrollable way, and then working above the threshold of instability is impossible. When it falls within the working range, the system must be modified to raise it well above the maximum operating speed. Only possible nonlinearities of the system can maintain the amplitude within a limit, giving way to a limit cycle.

23.5   The undamped linear Jeffcott rotor

23.5.1   Free whirling

The simplest model that can be used to study the flexural behavior of rotors is the so-called *Jeffcott rotor*, consisting of a point mass attached to a massless shaft. The only force acting on the mass m is that due to the elasticity of the shaft. The weight of the rotor is thus neglected, but this does not detract much from the validity of the model. The stiffness k providing the restoring force can be considered the stiffness of the shaft, the supporting structure, or a combination of the two. The two schemes sketched in Fig. 23.3 yield the same results, as long as the system is undamped and axially symmetrical.

Point P, in which mass m is fixed, is always contained in the xy-plane. This statement is justified by the uncoupling between axial and radial motions and relies on the small-displacement assumptions that are at the base of linear structural analysis.

A model with 2 degrees of freedom can thus be used for the study of the flexural behavior. The equations of motion of mass m are simply

$$\begin{cases} m\ddot{x} + kx = 0, \\ m\ddot{y} + ky = 0, \end{cases} \tag{23.3}$$

where x and y are the coordinates of point P at the generic time t.

The equations of motion along each axis are coincident with the equation of the free motion of a system with a single degree of freedom and their solution is a harmonic motion with frequency

$$\omega_n = \sqrt{\frac{k}{m}}.$$

The motion of point P can thus be thought of as the combination of two harmonic motions taking place along axes x and y with the same frequency

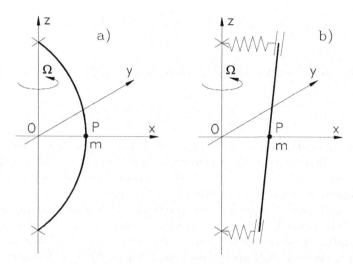

FIGURE 23.3. Sketch of a Jeffcott rotor. The model is sketched in its deformed configuration at the time in which point P crosses the xz-plane; (a) flexible shaft on stiff supports, (b) stiff shaft on compliant supports.

ω, coinciding with the natural frequency of the nonrotating shaft. They can add to each other giving way to a trajectory of point P that can be circular, elliptical, or straight in any direction in the xy-plane, depending on the initial conditions.

The same result could be obtained using the complex coordinate

$$z = x + iy \tag{23.4}$$

in the xy-plane. By multiplying the second part of Eq. (23.3) by the imaginary unit i and adding the two equations, it follows that

$$m\ddot{z} + kz = 0. \tag{23.5}$$

The solution of this homogeneous differential equation is

$$z = z_0 e^{i\omega t},$$

where z_0 is, generally speaking, a complex number

$$z_0 = x_0 + iy_0.$$

By introducing this solution into Eq. (23.5), the latter yields a homogeneous algebraic equation that has solutions other than the trivial solution $z_0 = 0$ only if

$$\omega = \pm\sqrt{\frac{k}{m}}\,.$$

The general solution of Eq. (23.5) is thus

$$z = Z_1 e^{i\sqrt{\frac{k}{m}}t} + Z_2 e^{-i\sqrt{\frac{k}{m}}t}\,. \qquad (23.6)$$

The physical meaning of Eq. (23.6) is obvious: z is a vector that rotates in the xy-plane with angular velocity ω. If the amplitude z_0 is real, point P crosses the x-axis at time $t = 0$. The motion expressed by Eq. (23.6) is the superimposition of a circular forward or direct motion (i.e., occurring in the same direction as the spin speed) and a circular backward motion. They both occur at an angular velocity, often called *whirl speed*, equal to the natural frequency of the nonrotating system. In the following study, the spin speed Ω will always be considered positive: A forward motion will then be characterized by positive whirl speed ω (first quadrant of the Campbell diagram), while a backward motion is characterized by a negative value of ω (fourth quadrant of the Campbell diagram). The result of the superimposition of the two motions depends on the initial conditions, i.e., on the values of complex constants Z_1 and Z_2. If, for example, Z_2 is equal to 0 a circular forward whirling occurs, while if the two constants are equal and real a harmonic vibration along the x-axis takes place.

The result obviously cannot depend on which solution is used, either that based on the study of the motion in the xz- and yz-planes (Eq. 23.4) or that based on the use of complex coordinates (Eq. 23.5). However, they enlighten different aspects of the phenomenon, and their results are not exactly equivalent. In the first, ω is the frequency of two harmonic motions in two planes and its sign has no physical meaning. Vector $e^{i\omega t}$, which can be used to express the motion in the form

$$x = x_0 e^{i\omega t}\,, \quad y = y_0 e^{i\omega t}\,,$$

is a rotating vector in the complex plane and only the real components of $x_0 e^{i\omega t}$ and $y_0 e^{i\omega t}$ have a physical meaning.

In the solution of Eq. (23.5) ω is, however, a true angular velocity, and the relevant vector rotates in the physical space and not in the complex plane. The deflected shape rotates about the undeformed configuration with angular velocity ω, although no rotation of a material object with that speed takes place. As a consequence, its sign states the direction of rotation of the deflected configuration.

At any rate, the natural frequency of the rotor or the whirl speed does not depend on the spin speed Ω: The Campbell diagram of a Jeffcott rotor is then made of straight lines, as shown in Fig. 23.2. The flexural critical speed, defined as the speed at which the natural frequency of the system is coincident with the frequency of rotation, coincides with the natural frequency of the nonrotating system:

$$\Omega_{cr} = \sqrt{\frac{k}{m}} \,. \tag{23.7}$$

23.5.2   Unbalance response

In real life, the center P of mass m cannot be exactly coincident with the center C of the cross-section of the shaft. The distance between the two points may be small, but the presence of the eccentricity ϵ (Fig. 23.3) causes a static unbalance $m\epsilon$ that can strongly affect the behavior of the system.

If the angle between line CP and the x-axis is imposed by the driving system, i.e., function $\theta(t)$ can be regarded as a known function of time, the number of degrees of freedom of the system is again equal to 2. Assuming the displacements x and y of point C as generalized coordinates, the position and velocity of point P and the Lagrangian function are, respectively,

$$(\overline{\text{P-O}}) = \left\{ \begin{matrix} x_P \\ y_P \end{matrix} \right\} = \left\{ \begin{matrix} x + \epsilon \cos(\theta) \\ y + \epsilon \sin(\theta) \end{matrix} \right\},$$

$$V_P = \left\{ \begin{matrix} \dot{x}_P \\ \dot{y}_P \end{matrix} \right\} = \left\{ \begin{matrix} \dot{x} - \epsilon\dot{\theta}\sin(\theta) \\ \dot{y} + \epsilon\dot{\theta}\cos(\theta) \end{matrix} \right\},$$

$$\mathcal{T} - \mathcal{U} = \frac{1}{2}m\left\{ \dot{x}^2 + \dot{y}^2 + \epsilon^2\dot{\theta}^2 + 2\epsilon\dot{\theta}\left[-\dot{x}\sin(\theta) + \dot{y}\cos(\theta) \right] \right\} - \frac{1}{2}k\left(x^2 + y^2 \right). \tag{23.8}$$

The equation of motion of the shaft can easily be obtained through the Lagrange equation. By performing the relevant derivatives and remembering that no assumption of constant spin speed $\dot{\theta}$ has been made, the following equations of motion are obtained:

$$\begin{cases} m\left[\ddot{x} - \epsilon\dot{\theta}^2\cos(\theta) - \epsilon\ddot{\theta}\sin(\theta) \right] + kx = 0, \\ m\left[\ddot{y} - \epsilon\dot{\theta}^2\sin(\theta) + \epsilon\ddot{\theta}\cos(\theta) \right] + ky = 0\,. \end{cases} \tag{23.9}$$

By multiplying the second equation of motion (23.9) by the imaginary unit i and adding the two equations, the following equation of motion in terms of the complex coordinate z is readily obtained:

$$m\ddot{z} + kz = m\epsilon\left(\dot{\theta}^2 - i\ddot{\theta} \right)e^{i\theta}\,. \tag{23.10}$$

The homogeneous equation is the same Eq. (23.5) already studied and does not contain the angular velocity of the rotor.

If the angular velocity of the rotor $\Omega = \dot{\theta}$ is assumed to be constant, angle θ can be expressed as Ωt, and Eq. (23.10) reduces to

$$m\ddot{z} + kz = m\epsilon\Omega^2 e^{i\Omega t}, \qquad (23.11)$$

whose particular integral is simply

$$z = z_0 e^{i\Omega t}.$$

By introducing it into Eq. (23.11) the latter transforms into the algebraic equation

$$(-m\Omega^2 + k)z_0 = m\epsilon\Omega^2, \qquad (23.12)$$

which yields

$$z_0 = \epsilon \frac{m\Omega^2}{k - m\Omega^2} = \epsilon \frac{\Omega^2}{\Omega_{cr}^2 - \Omega^2}, \qquad (23.13)$$

where Ω_{cr} is the critical speed defined by Eq. (23.7). The value of the amplitude z_0 so obtained is real and, as a consequence, vector $(\overline{\text{C-O}})$, i.e., z, rotates with velocity Ω in the xy-plane remaining in line with vector $(\overline{\text{P-C}})$. The value of Ω that causes the denominator of the expression for z_0 in Eq. (23.13) to vanish, i.e., that causes the amplitude to reach an infinite value, is coincident with the flexural critical speed of the rotor.

The amplitude of motion of point C due to the presence of the unbalance $m\epsilon$, i.e., the unbalance response, is reported as a function of the speed in the nondimensional plot of Fig. 23.4a. In the subcritical range, the amplitude grows from zero to a value tending to infinity at the critical speed, always remaining positive. In the supercritical range, however, the value of the amplitude z_0 is negative, and its absolute value decreases monotonically with the speed. When the speed tends to infinity the amplitude tends to $-\epsilon$.

The sign of the solution determines the equilibrium configurations as shown in Fig. 23.4b and c. When the solution is positive, in the subcritical field, points O, C, and P are aligned in the mentioned order and the center of mass of the rotor lies outside the deformed configuration of the shaft. In the supercritical field, however, point P lies between points C and O, and when the speed tends to infinity the amplitude z_0 tends to $-\epsilon$, or point P tends to point O. This phenomenon is usually referred to as *self-centering* because the rotor tends to rotate about its center of mass instead of its geometrical center. When the behavior is controlled by the stiffness, rotation takes place about a point close to the geometrical center, while when it is dominated by the inertia it occurs about a point close to the mass center.

The motion of point C can thus be expressed as the superimposition of a free motion that can be circular, elliptical, or even rectilinear occurring with frequency

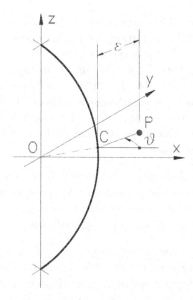

FIGURE 23.4. Unbalanced Jeffcott rotor sketched at time t.

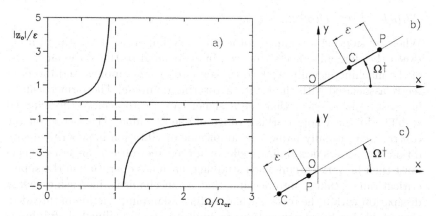

FIGURE 23.5. (a) Unbalance response of an undamped Jeffcott rotor. Nondimensional amplitude as a function of the nondimensional spin speed. Configuration of the system in (b) subcritical conditions and (c) supercritical conditions.

$$\omega = \sqrt{\frac{k}{m}}$$

and a circular motion with angular speed Ω. The amplitude of the first is controlled by the initial conditions and is independent of the speed. The presence of damping will cause the free motion to decay in time or, in some cases, to increase in time as will be shown in the following section. The amplitude of the second is strongly influenced by the speed and is constant in time. The unbalance response has been computed assuming constant spin speed: Fig. 23.5 gives the amplitude of the motion at the various speeds and cannot be used to describe the time history of the amplitude during the acceleration of the rotor.

Remark 23.7 *The Jeffcott rotor is obviously an oversimplification of real-world rotors but, nevertheless, allows understanding and modeling, at least qualitatively, some of the most important phenomena typical of rotor dynamics, namely critical speeds and self-centering.*

Remark 23.8 *The coincidence of the critical speed, computed from the unbalance response or from the free behavior of the system, with the natural frequency of the undamped system is a peculiar characteristic of the Jeffcott model or, more generally, of all those rotors in which the natural frequency does not depend on the speed, and must not be considered a general feature.*

23.6   Jeffcott rotor with viscous damping

23.6.1   Free whirling

When considering a damped rotor it is very important to distinguish between the damping effects that can be associated to the stationary parts of the machine, usually referred to as nonrotating damping, and those directly associated with the rotor, or rotating damping. The former usually has a stabilizing effect that the designer can use to achieve the required stability in the whole working range of the machine. The latter, on the contrary, can usually reduce the amplitude of vibration in subcritical conditions but shows destabilizing effects in the supercritical range. Designers must then be very careful when studying machines operating in the supercritical range, taking into account that all mechanisms increasing energy dissipation within the rotor, such as material damping, friction in threaded, riveted, shrink-fitted connections in built-up rotors, splined shafts, intershaft dampers in multi-shaft machines, etc., may cause severe instability problems.

The model of Fig. 23.3 can be extended to the damped system simply by adding the generalized forces due to damping to the right-hand side of Eq. (23.10). Because the generalized coordinates are the displacements in the x- and y-directions of point C, such forces are the components in the directions of the x- and y-axes of damping forces applied to the point mass in point C.

Assuming a viscous damping model and using the complex notation, the force due to nonrotating damping is

$$\vec{F}_n = F_{n_x} + i F_{n_y} = -c_n \dot{x} - i c_n \dot{y} = -c_n \dot{z}, \qquad (23.14)$$

where c_n is the nonrotating damping coefficient.

Remark 23.9 *In the model of Fig. 23.3b the nonrotating damping force may be due to the supports. Otherwise it may be due to any other damping device, provided that energy is dissipated in an element that does not rotate.*

A rotating reference frame $O\xi\eta z$ (Fig. 23.6) must be introduced for the study of rotating damping. The origin and z-axis of the rotating frame are the same as those of the fixed reference frame $Oxyz$ of Fig. 23.3, but axes ξ and η rotate in the xy-plane at the same speed of the rotor. When the rotational speed is constant, the angle between the two reference frames is simply $\theta = \Omega t$. Let a complex coordinate ζ be defined in the $\xi\eta$-plane

$$\zeta = \xi + i\eta = z e^{-i\theta}. \qquad (23.15)$$

From Eq. (23.15), the derivative of the complex coordinate ζ is readily obtained:

$$\dot{\zeta} = (\dot{z} - i\dot{\theta}z)e^{-i\theta}.$$

The force due to rotating viscous damping, with damping coefficient c_r, can be expressed in the $O\xi\eta z$ frame by the simple equation

$$\vec{F}_{r_{\xi\eta}} = -c_r \dot{\zeta} = -c_r(\dot{z} - i\dot{\theta}z)e^{-i\theta}. \qquad (23.16)$$

The expression of the force $F_{r_{xy}}$ in the $Oxyz$ frame is readily obtained from that of vector $F_{r_{\xi\eta}}$, expressed in the $O\xi\eta z$ frame

$$\vec{F}_{r_{xy}} = \vec{F}_{r_{\xi\eta}} e^{i\theta} = -c_r(\dot{z} - i\dot{\theta}z). \qquad (23.17)$$

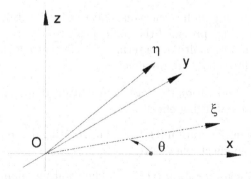

FIGURE 23.6. Reference frames $Oxyz$ and $O\xi\eta z$ at time t.

Introducing the expressions (23.14) and (23.17) for the forces due to nonrotating and rotating damping on the right-hand side of the equation of motion (23.10), it follows that

$$m\ddot{z} + (c_r + c_n)\dot{z} + (k - ic_r\dot{\theta})z = m\epsilon(\dot{\theta}^2 - i\ddot{\theta})e^{i\theta} \,. \tag{23.18}$$

Remark 23.10 *Rotating damping has been assumed here to be synchronous, i.e., to rotate at the same speed as the rotor. It is provided by any damping device that dissipates energy in an element rotating at the angular velocity* Ω.

If the spin speed is constant, Eq. (23.18) is easily modified by neglecting the term in $\ddot{\theta}$. In the latter case, the equation of motion is formally identical to the equation of motion of a mass m suspended on a spring with complex stiffness

$$k - i\Omega c_r$$

and a viscous damper with damping coefficient $c_n + c_r$ on which a force with harmonic time history with frequency $\Omega = \dot{\theta}$ and amplitude $m\epsilon\Omega^2$ is acting.

Equation (23.18) can be solved, as usual, by adding a particular integral to the complementary function. The solution of the homogeneous equation yielding the behavior of a perfectly balanced rotor is the usual one,

$$z = z_0 e^{i\omega t} \,,$$

where both the amplitude z_0 and the frequency ω are expressed by complex numbers. By introducing this solution into Eq. (23.18), the following characteristic equation is obtained:

$$m\omega^2 + i(c_r + c_n)\omega + k - i\Omega c_r = 0 \,. \tag{23.19}$$

The imaginary part is

$$c_n\omega + c_r(\omega - \Omega) \,. \tag{23.20}$$

Nonrotating damping has the same effect as in vibrating structures, and is always stabilizing (provided that c_n is positive).

Rotating damping multiplies a term, $(\omega - \Omega)$, that may be either positive or negative. Three cases are possible:

- Subcritical operation ($\Omega < \omega$). The rotating damping term is positive and has a stabilizing effect.

- Synchronous operation ($\Omega = \omega$). The rotating damping term vanishes and has no effect whatsoever.

- Supercritical operation ($\Omega > \omega$). The rotating damping term is negative and has a destabilizing effect.

This is also true for other forms of damping, like hysteretic damping: all forms of rotating damping are stabilizing in subcritical operation and destabilizing above the critical speed.[2]

The roots of this quadratic equation with complex coefficients are

$$\omega = i\frac{c_r + c_n}{2m} \pm \sqrt{\frac{-(c_r + c_n)^2 + 4m(k - i\Omega c_r)}{4m^2}}. \tag{23.21}$$

The real and imaginary parts of the complex frequency can be easily separated by resorting to the well-known formula

$$\sqrt{a \pm ib} = \sqrt{\frac{\sqrt{a^2 + b^2} + a}{2}} \pm i\sqrt{\frac{\sqrt{a^2 + b^2} - a}{2}}, \tag{23.22}$$

which yields

$$\omega = \pm\frac{1}{\sqrt{2}}\sqrt{\Gamma + \sqrt{\Gamma^2 + \left(\frac{\Omega c_r}{m}\right)^2}} + i\left[\frac{c_r + c_n}{2m} \mp \frac{1}{\sqrt{2}}\sqrt{-\Gamma + \sqrt{\Gamma^2 + \left(\frac{\Omega c_r}{m}\right)^2}}\right], \tag{23.23}$$

where

$$\Gamma = \frac{k}{m} - \frac{(c_r + c_n)^2}{4m^2}. \tag{23.24}$$

Two values of the complex whirl frequency or whirl speed ω can be found for each value of the spin speed Ω. They are not conjugate, because Eq. (23.19) has complex coefficients. Without loss of generality, because it is sufficient to assume that at time $t = 0$ point C crosses the x-axis, the amplitude z_0 can be assumed to be real. Separating the real part ω_R of the complex frequency ω from the imaginary part ω_I, the time history can then be written as

$$\begin{cases} x = z_0 e^{-\omega_I t} \cos(\omega_R t), \\ \\ y = z_0 e^{-\omega_I t} \sin(\omega_R t). \end{cases} \tag{23.25}$$

A logarithmic spiral is thus obtained as the result of two damped or amplified harmonic motions. The real part of ω has the meaning of a true angular velocity: It is the angular velocity at which the deflected shape rotates about the undeformed configuration. The imaginary part is a decay rate: If it is positive the amplitude decays in time and point C tends to point O. The rotor has a stable behavior as whirl motion tends to reduce its amplitude.

If ω_I is negative the amplitude grows exponentially in time. The motion is unstable, as any small perturbation can trigger this self-excited whirling.

---

[2]G. Genta , *On a Persisting Misunderstanding on the Role of Hysteretic Damping in Rotordynamics*, Journal of Vibration and Acoustics, Vol. 126, July 2004, 469–471, G. Genta , *On the Stability of Rotating Blade Arrays*, Journal of Sound and Vibration, 273, 2004, pp. 805–836.

The first of the two values of ω obtained in Eq. (23.23) (the one with the upper signs) has a positive real part and an imaginary part that may be either positive or negative. It corresponds to a forward whirl mode, which may be either damped or self-excited, depending on the sign of ω_I. With simple computations, the condition for stability can be shown to be

$$\Omega < \sqrt{\frac{k}{m}}\left(1 + \frac{c_n}{c_r}\right). \tag{23.26}$$

The rotor is then stable in the subcritical range, i.e., when the speed is lower than the critical speed $\sqrt{k/m}$ of the undamped system. In the supercritical field, stability depends on the value of ratio c_n/c_r between the nonrotating and the rotating damping. If there is no nonrotating damping, the motion is unstable in the whole supercritical range. Increasing nonrotating damping, the threshold of instability becomes higher.

The second value of ω (lower signs in Eq. (23.23)) has a negative real part and a positive imaginary part. It corresponds to a damped backward whirl mode that usually damps out quite fast and has little practical interest.

Equation (23.23) can be rewritten in the following nondimensional form:

$$\begin{cases} \omega_R^* = \pm\sqrt{\Gamma^* + \sqrt{\Gamma^{*2} + \Omega^{*2}\zeta_r^2}}, \\ \omega_I^* = \zeta_r + \zeta_n \mp \sqrt{-\Gamma^* + \sqrt{\Gamma^{*2} + \Omega^{*2}\zeta_r^2}}, \end{cases} \tag{23.27}$$

where

$$\omega^* = \omega\sqrt{\frac{m}{k}}, \quad \Omega^* = \Omega\sqrt{\frac{m}{k}}, \quad \zeta_r = \frac{c_r}{2\sqrt{km}},$$

$$\zeta_n = \frac{c_n}{2\sqrt{km}}, \quad \Gamma^* = \frac{1 - (\zeta_n + \zeta_r)^2}{2}.$$

The nondimensional whirl frequency ω^* is then a function of the spin speed Ω^* and of only two parameters ζ_n and ζ_r. The real and imaginary parts of ω^* are plotted as functions of the spin speed in Fig. 23.7. Four values of ζ_r have been considered, and rotating and nonrotating damping have been assumed to be equal ($\zeta_n = \zeta_r$).

It is clear that the real part of ω is little affected by the presence of damping: A value of ζ_n equal to 0.1 is already quite high and the curve $\omega_R(\Omega)$ is very close to that of the undamped case. The first and fourth quadrants of the (ω_R, Ω)-plane have been shown in the Campbell diagram. When using the complex notation, the sign of ω_R has an important physical meaning and it is advisable to plot at least two quadrants of the ω_R, Ω-plane: the first and second, or, as in the figure, the first and fourth.

Another way of representing the real and imaginary parts of the complex whirl frequency is the roots locus (Fig. 23.8), in which the imaginary parts of eigenvalues

$$s = i\omega$$

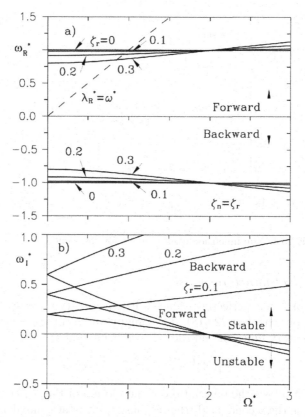

FIGURE 23.7. (a) Real and (b) imaginary parts of the complex whirl frequency as functions of the spin speed in nondimensional form. The curves for four different values of damping have been plotted. Rotating and nonrotating damping have been assumed to be equal: The threshold of instability is then twice the critical speed.

are plotted as functions of the real parts. As usual for root loci, a solution of the type of Eq. (23.2) is assumed: The imaginary part of s has the meaning of whirl speed, while the real part is the decay rate (negative for stability):

$$\Re(s) = -\Im(\omega) , \quad \Im(s) = \Re(\omega) .$$

Remark 23.11 *A difference between roots loci in general dynamics (e.g., that of Fig. 23.8) and those encountered in rotordynamics is that the latter are not symmetrical with respect to the real axis. This is because the eigenvalues are not conjugate or, in physical terms, the forward and backward complex whirl speeds are not equal (in the present case, the absolute values of ω_R are equal, while those of ω_I are not).*

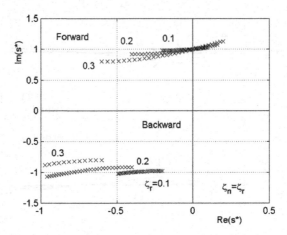

FIGURE 23.8. Nondimensional roots locus for the same system as Fig. 23.7

23.6.2   Unbalance response

If the rotor is not perfectly balanced, it is necessary to resort to the non-homogeneous equation (23.18). If the angular acceleration is neglected, the complementary function and the particular integral are, respectively,

$$z = Z_1 e^{i\omega_1 t} + Z_2 e^{i\omega_2 t} ; \qquad z = z_0 e^{i\Omega t} . \qquad (23.28)$$

The first allows the description of the motion of the perfectly balanced rotor, while the second yields the response to the static unbalance $m\epsilon$. It is a synchronous whirling, i.e., a whirling with $\omega = \Omega$. The amplitude of the unbalance response is then obtained by introducing the particular integral into the equation of motion, obtaining

$$z_0(-m\Omega^2 + i\Omega c_n + k) = m\epsilon\Omega^2 . \qquad (23.29)$$

As already seen, rotating damping does not enter Eq. (23.29): Unbalance produces a synchronous excitation, i.e., an excitation that rotates in the xy-plane at the same speed Ω as the rotor, and the latter rotates in the deflected configuration but is not subject to deformations that change in time.

The situation occurring in this condition is sketched in Fig. 23.9: In (a), a whirling shaft is shown in its deflected configuration, and in (b), the situation occurring in the xy-plane is sketched. As the spin speed is equal to the whirl speed, the zone of the cross-section of the shaft subjected to tensile stresses (shaded part close to point B) remains always under tensile loading, whereas that subjected to compression is always compressed. The situation is similar to that of the Moon, which always shows the same side to the Earth. In this condition, the material constituting the rotor is subjected to a constant stress state, which is good on one side as the rotor

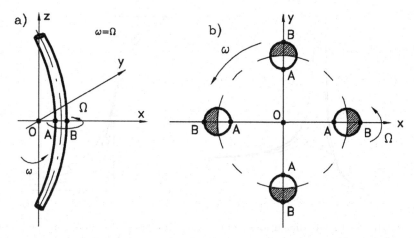

FIGURE 23.9. (a) Sketch of a shaft undergoing circular synchronous whirling, i.e., with whirl speed ω equal to the spin speed Ω; (b) situation on the xy-plane: The shaded area of the cross-section is under tensile stress.

is not subjected to fatigue, but on the other side, the internal damping of the material cannot dissipate energy and hence limit the amplitude, particularly when the system is working at a speed close to a critical speed.

The flexural critical speeds can, in fact, be defined as the speeds at which the centrifugal forces due to the bending of the rotor are in indifferent equilibrium with the elastic restoring forces, and from this point of view, the situation is more similar to that characterizing elastic instability than that typical of vibratory phenomena. A rotor operating at a critical speed is then not subject to vibration but is a source of periodic excitation that can cause vibrations, often very strong, in the nonrotating parts of the machine.

The amplitude z_0 is obviously expressed by a complex number. By separating the real from the imaginary part, it follows that

$$\Re(z_0) = \frac{m\epsilon\Omega^2(k - m\Omega^2)}{(k - m\Omega^2)^2 + \Omega^2 c_n^2} = \epsilon\frac{\Omega^{*2}(1 - \Omega^{*2})}{(1 - \Omega^{*2})^2 + 4\zeta_n^2\Omega^{*2}},$$

$$\Im(z_0) = -\frac{m\epsilon\Omega^3 c_n}{(k - m\Omega^2)^2 + \Omega^2 c_n^2} = -\epsilon\frac{2\Omega^{*3}\zeta_n}{(1 - \Omega^{*2})^2 + 4\zeta_n^2\Omega^{*2}},$$

$$|z_0| = \epsilon\frac{\Omega^{*2}}{\sqrt{(1 - \Omega^{*2})^2 + 4\zeta_n^2\Omega^{*2}}},$$

$$\Phi = \arctan\left(\frac{-2\Omega^*\zeta_n}{1 - \Omega^{*2}}\right). \tag{23.30}$$

The amplitude and phase of z_0 are plotted in nondimensional form as functions of the speed in Fig. 23.10. The different curves have been obtained with different values of the nonrotating damping ζ_n. The equation yielding the unbalance response is identical to that yielding the amplitude of the

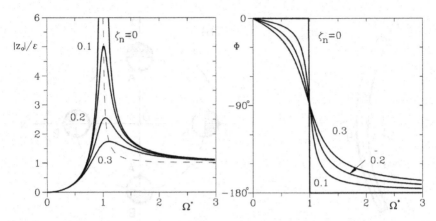

FIGURE 23.10. Nondimensional amplitude and phase of the unbalance response for three different values of nonrotating damping. Full line indicates curves for different values of damping; dashed line indicates line connecting the peaks.

response of a vibrating system with a single degree of freedom to a harmonic excitation whose amplitude is proportional to the square of the frequency. Fig. 23.10 is then identical to Fig. 7.6b. The damped resonance peak lies at the right of the undamped resonance, as was the case for Fig. 7.6b, and not on the left, as in Fig. 7.2.

The situation in the xy-plane is shown in Fig. 23.11b. Since angle Φ is always negative, point C always lags the line forming an angle of Ωt with the x-axis, i.e., ξ-axis. The delay is exactly 90° at the critical speed. The

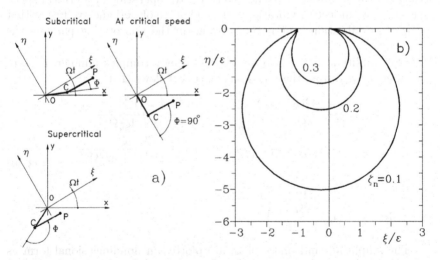

FIGURE 23.11. (a) Situation in the xy-plane in subcritical conditions, at the critical speed, and in supercritical conditions; (b) trajectories of point C in the $\xi\eta$-plane expressed in nondimensional form.

plot of the real and imaginary parts of the complex amplitude z_0 yields directly the trajectory of point C in the $\xi\eta$-plane (Fig. 23.11b). By adding vector ϵ to vector z_0, the trajectory of point P is obtained.

The plot of the trajectory of C is very similar to a Nyquist diagram, with the important difference that the latter is plotted in the complex plane, while the former gives the actual position of point C in the $\xi\eta$-plane in the physical space.

Remark 23.12 *The term* trajectory *has been used with an extended meaning: All these figures were obtained neglecting the angular acceleration, and the points of the trajectories refer to steady-state conditions at different speeds and are not successive positions during an acceleration of the rotor. However, if the acceleration is very slow and dynamic effects linked to it can be neglected, the curves of Fig. 23.11b can be assumed to be at least a good approximation for the actual trajectories in the $\xi\eta$-plane.*

Remark 23.13 *Self-centering is strictly linked with an increase of phase Φ from $0°$ to $-180°$, i.e., to a rotation of point C in the $\xi\eta$-plane.*

If point C were constrained to remain on the ξ-axis, like in Example 17.1, no self-centering would occur, as in the case of the system sketched in Fig. 23.12, where mass m is allowed to move along a rotating guide and is constrained to the center of rotation O by a spring of stiffness k. The system is only apparently similar to the Jeffcott rotor of Fig. 23.3: If the length of the guide were infinite, it would behave similar to a Jeffcott rotor in the subcritical field, but in supercritical conditions point P would not come back to the self-centered position and would remain at infinity (or, in real life, at the end of the guide).

In the damped Jeffcott rotor of Fig. 23.3, the motion of point C can be considered as the superimposition of a backward inward-spiral motion, which usually decays very quickly in time, a forward spiral motion, whose

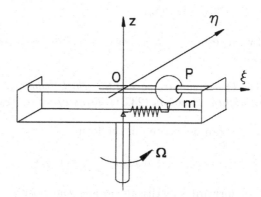

FIGURE 23.12. A system in which mass m is constrained to move along the ξ-axis of the rotating reference frame. No self-centering is possible.

amplitude can be decreasing, constant, or increasing in time depending on the stability of the system, and a circular synchronous motion, with constant amplitude. The amplitude of the latter depends on the spin speed and the eccentricity ϵ.

23.7   Jeffcott rotor with structural damping

Rotating and nonrotating damping have been supposed to be of the viscous type. This can be a realistic model for nonrotating damping, particularly when dampers of the squeeze-film type are used, but a structural damping model is usually better suited for rotating damping. There is no difficulty in introducing the complex stiffness model described in Section 3.4.1. Equation (3.62) yielding the equivalent damping must, however, be modified into

$$c_{eq} = \frac{\eta k}{|\omega_m|} \,,$$

where ω_m is the frequency at which the material goes through the hysteresis cycle. It coincides with $|\omega_R|$, modulus of the real part of the whirl frequency ω, in the case of nonrotating damping, while it takes the value $|\omega_R - \Omega|$ for rotating damping. If they are both of the structural type, by neglecting the imaginary part of frequency ω in the computation of ω_m, Eq. (23.19) for the study of free whirling becomes

$$-m\omega^2 + k + i\left(\frac{\omega_R}{|\omega_R|}\eta_n k_n + \frac{\omega_R - \Omega}{|\omega_R - \Omega|}\eta_r k_r\right) = 0\,. \qquad (23.31)$$

The terms neglected operating in this way

$$-\eta_n k_n \frac{\omega_I}{|\omega_R|} - \eta_r k_r \frac{\omega_I}{|\omega_R - \Omega|}$$

are very small with respect to k, at least if the loss factor is small and ω_R is different enough from Ω.

Remark 23.14 *The imaginary part of the whirl frequency vanishes at the threshold of instability. Equation (23.31) yields exact results in this case.*

Equation (23.31) can be written in the form

$$-m\omega^2 + k + i(\pm\eta_n k_n \pm \eta_r k_r) = 0\,. \qquad (23.32)$$

In the case of forward subcritical motion, the positive signs hold; in supercritical conditions the sign of η_n is positive and that of η_r is negative. They are both negative in the case of backward motion (Fig. 23.13).

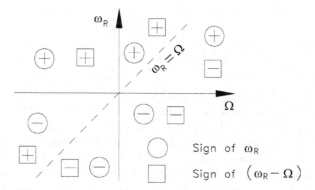

FIGURE 23.13. Signs of ω_R and $\omega_R - \Omega$ in the various zones of the Campbell diagram.

The characteristic equation can be solved in the same way as in the case of viscous damping. The real and imaginary parts of the whirl frequency are

$$
\begin{cases}
\omega_R = \pm \dfrac{1}{\sqrt{2m}} \sqrt{k + \sqrt{k^2 + (k_n \eta_n \pm k_r \eta_r)^2}}\,, \\[2mm]
\omega_I = \dfrac{1}{\sqrt{2m}} \sqrt{-k + \sqrt{k^2 + (k_n \eta_n + k_r \eta_r)^2}}\,,
\end{cases}
\tag{23.33}
$$

in the case of backward or forward subcritical whirl; in the supercritical field, the imaginary part is

$$
\begin{aligned}
\omega_I &= \tfrac{1}{\sqrt{2m}} \sqrt{-k + \sqrt{k^2 + (k_n \eta_n - k_r \eta_r)^2}} && \text{if } \eta_n k_n > \eta_r k_r, \\[2mm]
\omega_I &= -\tfrac{1}{\sqrt{2m}} \sqrt{-k + \sqrt{k^2 + (k_n \eta_n - k_r \eta_r)^2}} && \text{if } \eta_n k_n < \eta_r k_r\,.
\end{aligned}
\tag{23.34}
$$

The condition for stability in the supercritical field is then $\eta_n k_n > \eta_r k_r$. If this condition is satisfied, no threshold of instability exists and the system is stable at any speed. If, on the contrary, it is not satisfied, the threshold of instability coincides with the critical speed and no supercritical running is possible.

The case in which there are different types of damping can be studied in the same way. If rotating damping is of the structural type and nonrotating damping is viscous, the condition for stability is

$$
c_n > \eta_r k_r \sqrt{\frac{m}{k}}\,.
$$

In the case of the unbalance response, rotating damping has no influence on the behavior of the system. If nonrotating damping is of the structural type, the amplitude and phase of the unbalance response are

$$|z_0| = \epsilon \frac{\Omega^{*2}}{\sqrt{(1 - \Omega^{*2})^2 + \eta_n^2}} \, ,$$

$$\Phi = \arctan\left(\frac{-\eta_n}{1 - \Omega^{*2}}\right) .$$

(23.35)

The assumption of constant loss factor is only an approximation, particularly for conditions close to synchronous whirling. If $\omega = \Omega$, ω_m vanishes, and the very concept of hysteretic damping loses meaning.

If the dependence of the loss factor from the frequency is known, the free whirling frequencies can be computed iteratively. Because the whirl speed is little influenced by the value of the damping, the procedure can converge quickly. For the computation of the unbalance response, however, the dependence of damping from the frequency introduces no computational difficulty. Other effects, such as the dependence of damping from the value of the maximum stress, introduce nonlinearities into the model and make the solution very difficult.

23.8   Equations of motion in real coordinates

Equation (23.18) can be written using the real coordinates x and y

$$\begin{bmatrix} m & 0 \\ 0 & m \end{bmatrix} \begin{Bmatrix} \ddot{x} \\ \ddot{y} \end{Bmatrix} + \begin{bmatrix} c_n + c_r & 0 \\ 0 & c_n + c_r \end{bmatrix} \begin{Bmatrix} \dot{x} \\ \dot{y} \end{Bmatrix} +$$

(23.36)

$$+ \begin{bmatrix} k & \Omega c_r \\ -\Omega c_r & k \end{bmatrix} \begin{Bmatrix} x \\ y \end{Bmatrix} = m\epsilon \begin{Bmatrix} \dot{\theta}^2 \cos(\theta) + \ddot{\theta} \sin(\theta) \\ \dot{\theta}^2 \sin(\theta) - \ddot{\theta} \cos(\theta) \end{Bmatrix} .$$

Equation (23.36) has the form of Eq. (3.41), where the skew-symmetric gyroscopic matrix $\mathbf{G}$ vanishes while the skew-symmetric circulatory matrix $\mathbf{H}$ is present and contains rotating damping. As it will be shown better for gyroscopic systems, imaginary terms in the equation based on complex coordinates correspond to skew-symmetric terms when real coordinates are used. The presence of a circulatory matrix is linked with destabilizing effects, like in the current case rotating damping.

23.9   Stability in the supercritical field

The steady-state unbalance response of an undamped or damped Jeffcott rotor was computed in the preceding sections. However, no conclusion about the stability of the equilibrium position of the system was reached. A simple way to state the stability of the equilibrium position is by observing that the complete solution of the equation of motion can be obtained by adding the solution for free whirling to the unbalance response. When the first one leads to a stable behavior, the overall behavior of the system is stable. The equilibrium position is then stable in the whole supercritical range up to the previously computed threshold of instability.

However, better evidence can be obtained by writing the equation of motion with reference to the rotating reference frame $O\xi\eta$ shown in Fig. 23.6. If the angular velocity Ω is constant, the position, velocity, and acceleration of point C can be expressed as functions of complex coordinate ζ

$$\begin{cases} z = \zeta e^{i\Omega t}, \\ \dot{z} = \left(\dot{\zeta} + i\Omega\zeta\right) e^{i\Omega t}, \\ \ddot{z} = \left(\ddot{\zeta} + 2i\Omega\dot{\zeta} - \Omega^2\zeta\right) e^{i\Omega t}. \end{cases} \tag{23.37}$$

By introducing Eq. (23.37) into the equation of motion of the damped system (23.18) written for constant speed and rearranging the various terms, it follows that

$$m\ddot{\zeta} + (c_n + c_r + 2im\Omega)\dot{\zeta} + (k - m\Omega^2 + i\Omega c_n)\zeta = m\epsilon\Omega^2. \tag{23.38}$$

Note that the main differences between Eq. (23.18) written in the inertial frame and Eq. (23.38) are the presence of the terms $2im\Omega\dot{\zeta}$ linked with Coriolis acceleration and $m\Omega^2\zeta$ due to centrifugal forces and the fact that in the rotating frame unbalance forces are constant, in both direction and modulus. From Eq. (23.38), the static equilibrium equation is readily obtained. By equating to zero all derivatives of the generalized coordinates with respect to time, the equilibrium position ζ_0 is obtained:

$$\zeta_0 = \frac{m\epsilon\Omega^2}{k - m\Omega^2 + i\Omega c_n}. \tag{23.39}$$

Equation (23.39) coincides with Eq. (23.30). The stability of the equilibrium position is easily studied by assuming that the motion of point C occurs in the vicinity of the position expressed by Eq. (23.39)

$$\zeta(t) = \zeta_1(t) + \zeta_0. \tag{23.40}$$

By introducing Eq. (23.40) into the equation of motion (23.38) and remembering the expression for the equilibrium position (23.39), the following equation for the motion about the equilibrium position is obtained:

$$m\ddot{\zeta}_1 + (c_n + c_r + 2im\Omega)\dot{\zeta}_1 + (k - m\Omega^2 + i\Omega c_n)\zeta_1 = 0. \tag{23.41}$$

The solution of Eq. (23.41) is of the usual type

$$\zeta_1 = \zeta_{1_0} e^{i\omega' t},$$

yielding a spiral motion about the equilibrium position whose amplitude can be either decreasing or increasing in time depending on the sign of the imaginary part of the complex frequency ω'. By introducing this solution into Eq. (23.41), the following characteristic equation allowing the computation of the complex frequency is readily obtained:

$$-m\omega'^2 + \left[i(c_n + c_r) - 2m\Omega\right]\omega' + k - m\Omega^2 + i\Omega c_n = 0. \qquad (23.42)$$

The solution of the characteristic equation is

$$\omega' = -\Omega + i\frac{c_r + c_n}{2m} \pm \sqrt{\frac{-(c_r + c_n)^2 + 4m(k - i\Omega c_r)}{4m^2}}. \qquad (23.43)$$

By comparing Eq. (23.43) with Eq. (23.21), it is immediately clear that the imaginary parts of the two expressions are equal and the real parts differ by a term equal to $-\Omega$. This result is expected because the former is expressed in a frame of reference that rotates at an angular velocity equal to Ω with respect to the one in which Eq. (23.21) is expressed. Because the imaginary part of the complex frequency is the same as that given by Eq. (23.21), when condition (23.26) for stability is satisfied, the motion about the equilibrium position is, in the rotating reference frame, a decaying spiral and the equilibrium position is stable.

The equation of motion in the rotating frame (23.38) can be written using the real coordinates ξ and η instead of the complex coordinate ζ, obtaining

$$\begin{bmatrix} m & 0 \\ 0 & m \end{bmatrix}\begin{Bmatrix} \ddot{\xi} \\ \ddot{\eta} \end{Bmatrix} + \begin{bmatrix} c_n + c_r & -2m\Omega \\ 2m\Omega & c_n + c_r \end{bmatrix}\begin{Bmatrix} \dot{\xi} \\ \dot{\eta} \end{Bmatrix} + \qquad (23.44)$$

$$+ \begin{bmatrix} k - m\Omega^2 & -\Omega c_n \\ \Omega c_n & k - m\Omega^2 \end{bmatrix}\begin{Bmatrix} \xi \\ \eta \end{Bmatrix} = \begin{Bmatrix} m\epsilon\Omega^2 \\ 0 \end{Bmatrix}.$$

Remark 23.15 *In this case both the gyroscopic* **G** *and the circulatory matrix* **H** *are present, the first being due to Coriolis forces. A sort of centrifugal stiffening (the terms* $-m\Omega^2$ *in* **K***) is present but is again due to the non-inertial reference frame rather than a true centrifugal stiffening.*

23.10   Acceleration through the critical speed

If the angular velocity of the rotor is not constant, the equation of motion for the lateral behavior of the damped system is Eq. (23.18). A third equation must be added to express the dependence of the angular displacement θ on the driving torque M_z. Because the third degree of freedom of the system is linked with rotation about the z-axis, the polar moment of inertia J_z cannot be neglected. No torsional elasticity of the shaft is taken into account, because the driving torque is assumed to be directly applied to a torsionally stiff rotor and the Lagrangian of the system is expressed by Eq. (23.8), to which a term

$$\frac{1}{2}J_z\dot{\theta}^2$$

must be added.

By performing the relevant derivatives of the Lagrangian with respect to the angular coordinate θ and its derivative $\dot{\theta}$, the equation of motion for rotations about the z-axis is

$$(J_z + me^2)\ddot{\theta} + me\left[-\ddot{x}\sin(\theta) + \ddot{y}\cos(\theta)\right] = M_z. \tag{23.45}$$

Equation (23.45), which can be expressed in terms of the complex coordinate z, and Eq. (23.18) are the equations of motion of the system

$$\begin{cases} m\ddot{z} + (c_r + c_n)\dot{z} + (k - i\dot{\theta}c_r)z = me(\dot{\theta}^2 - i\ddot{\theta})e^{i\theta}, \\ (J_p + me^2)\ddot{\theta} + me\Im\left(\ddot{z}e^{-i\theta}\right) = M_z. \end{cases} \tag{23.46}$$

The same equations can be written in a rotor-fixed reference frame, which in this case does not rotate at constant speed. Equations (23.37) expressing the absolute position, velocity, and acceleration as functions of the same characteristics expressed in the rotating frame must be modified to take into account the fact that the rotational speed is not constant

$$\begin{cases} z = \zeta e^{i\theta}, \\ \dot{z} = \left(\dot{\zeta} + i\dot{\theta}\zeta\right)e^{i\theta}, \\ \ddot{z} = \left(\ddot{\zeta} + 2i\dot{\theta}\dot{\zeta} + i\ddot{\theta}\zeta - \dot{\theta}^2\zeta\right)e^{i\theta}. \end{cases} \tag{23.47}$$

The equations of motion written with reference to the rotating frame are thus

$$\begin{cases} m\ddot{\zeta} + (c_n + c_r + 2im\Omega)\dot{\zeta} + (k - m\Omega^2 + i\Omega c_n + im\ddot{\theta})\zeta = me(\dot{\theta}^2 - i\ddot{\theta}), \\ (J_p + me^2)\ddot{\theta} + me\Im\left(\ddot{\zeta} + 2i\dot{\theta}\dot{\zeta} + i\ddot{\theta}\zeta - \dot{\theta}^2\zeta\right) = M_z, \end{cases} \tag{23.48}$$

or, using real coordinates,

$$\begin{bmatrix} m & 0 \\ 0 & m \end{bmatrix}\left\{\begin{array}{c} \ddot{\xi} \\ \ddot{\eta} \end{array}\right\} + \begin{bmatrix} c_n + c_r & -2m\dot{\theta} \\ 2m\dot{\theta} & c_n + c_r \end{bmatrix}\left\{\begin{array}{c} \dot{\xi} \\ \dot{\eta} \end{array}\right\} +$$

$$+ \begin{bmatrix} k - m\dot{\theta}^2 & -\dot{\theta}c_n - m\ddot{\theta} \\ \dot{\theta}c_n + m\ddot{\theta} & k - m\dot{\theta}^2 \end{bmatrix}\left\{\begin{array}{c} \xi \\ \eta \end{array}\right\} = me\left\{\begin{array}{c} \dot{\theta}^2 \\ -\ddot{\theta} \end{array}\right\}, \tag{23.49}$$

$$(J_p + me^2)\ddot{\theta} + me\left(\ddot{\eta} + 2\dot{\theta}\dot{\xi} + \ddot{\theta}\xi - \dot{\theta}^2\eta\right) = M_z. \tag{23.50}$$

The study of an accelerating Jeffcott rotor can be performed following two different schemes: Either the time history of the driving torque or that of the angular displacement can be stated. In the first case, the two equations are coupled and the solution can be very complicated. In the second case, however, they uncouple and the first equation directly yields the flexural behavior of the system, while the second allows the computation of the time history of the driving torque needed to follow the assumed acceleration law. The laws $\theta(t)$, $\Omega(t) = \dot{\theta}(t)$, and $a(t) = \ddot{\theta}(t)$ are known and there is no difficulty in numerically integrating the equations of motion.

Even if in the literature some solutions for the constant acceleration case can be found,[3] they are so complicated that today it is easier to perform the numerical integration of the equations of motion. The amplitude of the motion of a Jeffcott rotor with $\zeta_n = 0.1$ and $\zeta_r = 0.01$ accelerating from standstill to a speed equal to three times the critical speed with constant acceleration is plotted in Fig. 23.14a. The plot has been obtained in nondimensional form, using the following nondimensional parameters:

$$\xi^* = \frac{\xi}{\epsilon} , \quad \eta^* = \frac{\eta}{\epsilon} , \quad \tau = \frac{t}{t_1} = \frac{ta}{\Omega_{cr}} , \quad a^* = \frac{a}{\Omega_{cr}^2} .$$

Remark 23.16 *The nondimensional time τ and the nondimensional velocity Ω^* coincide, because time has been made nondimensional with reference to time t_1 needed to reach the critical speed.*

With low values of the acceleration, the motion follows a pattern similar to that seen for the steady-state case, as was easily predicted. With increasing values of the acceleration, the peak amplitude is reduced and the self-centered conditions are reached after some oscillations are damped out. The oscillations become stronger with increasing acceleration.

The trajectory of point C in the rotating plane is shown in Fig. 23.14b. From this plot, it is clear that the oscillations are actually the result of a spiral motion of the system taking place about the self-centered position.

Remark 23.17 *In the case here studied the threshold of instability is never exceeded and the motion is stable. When the speed becomes higher than the threshold of instability, the spiral stops decaying and starts growing as unstable behavior develops.*

The driving torque needed to perform the acceleration is shown in Fig. 23.15. Actually, the nondimensional torque plotted is only the amount of torque that exceeds the value aJ_z needed to accelerate a perfectly balanced rotor. Also, strong oscillations are visible here in the supercritical field. In Fig. 23.15 the curve related to a vanishingly small acceleration has also been plotted. The equation of this curve can be obtained in the closed

---

[3]See, for example, F.M. Dimentberg, *Flexural vibrations of rotating shafts*, Butterworths, London, 1961.

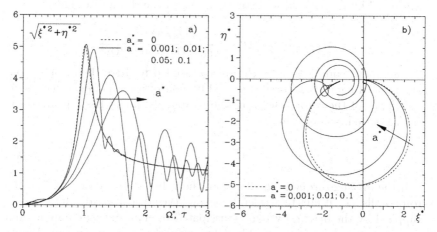

FIGURE 23.14. Motion of a Jeffcott rotor with $\zeta_n = 0.1$ and $\zeta_r = 0.01$ crossing a critical speed with constant acceleration. Nondimensional plot for some values of the angular acceleration; (a) time history of the amplitude; (b) trajectory in the rotating plane $\xi^* \eta^*$.

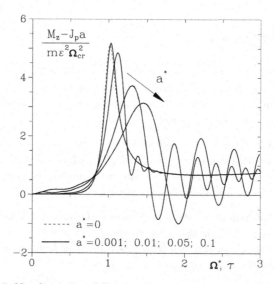

FIGURE 23.15. Nondimensional driving torque needed to perform the acceleration through the critical speed. Same system as in Fig. 23.14.

form by imposing that the displacement ζ is constant in time at the value expressed by Eq. (23.39) and stating $a = 0$ in the second Eq. (23.48)

$$M_z = m\epsilon\Im(-\zeta\Omega^2) = \frac{2k\epsilon^2\Omega^{*^5}\zeta_n}{(1-\Omega^{*2})^2 + 4\zeta_n^2\Omega^{*2}} .$$

The above equation yields the torque needed to drive the rotor at constant speed against the drag provided by nonrotating damping due to the unbalance of the rotor. The drag due to rotordynamics has a peak at the critical speed, whose value is

$$M_{z_{max}} = \frac{k\epsilon^2}{2\zeta_n} . \tag{23.51}$$

If the driving torque is smaller than the value given by Eq. (23.51), the rotor stalls, i.e., it fails to accelerate beyond the critical speed. All the power supplied by the driving system is dissipated by nonrotating damping, and acceleration is no longer possible. Obviously, the value so computed must be added to that required to overcome all other forms of drag, such as aerodynamic or bearing drag. In the high supercritical range, the torque needed to operate at constant speed grows linearly with speed. From the above equation, it follows that

$$\lim_{\Omega\to\infty} M_z = 2k\epsilon^2\Omega^*\zeta_n . \tag{23.52}$$

23.11   Exercises

Exercise 23.1 *Consider a rotor with a mass of $m = 20$ kg, running on two bearings with a stiffness $k_b = 5$ MN/m. Assuming that the center of mass is at midspan, compute the critical speed and the response to an unbalance $m\epsilon = 400$ gmm using the Jeffcott rotor model.*

Exercise 23.2 *Consider the Jeffcott rotor of the previous exercise, but add a rotating damper with damping $c_r = 3.5$ kNs/m and assume that the damping of the bearings is $c_{nb} = 1.2$ kNs/m. Compute the Campbell diagram, the decay rate plot, the roots locus, and the unbalance response.*

Exercise 23.3 *Repeat the study performed in Exercise 23.2, but assuming that damping can be modeled as hysteretic damping. The loss factor of the supports is $\eta_n = 0.15$. Rotating damping can be modeled as a loss factor $\eta_r = 0.2$, applied to a spring having the same stiffness k of the system.*

Exercise 23.4 *Consider the damped Jeffcott rotor of Exercise 23.2. Compute the response to an acceleration from standstill to 12000 rpm in 0.4 s at constant acceleration and then follow the motion for a further 0.4 s. Plot the amplitude of the response, the trajectory in the rotating plane, and the driving torque.*

24

Dynamics of Multi-Degrees-of-Freedom Rotors

Real-life rotors are much more complex than Jeffcott rotors and more realistic models are needed. The rotor must be modeled at least as a rigid body (versus the point mass of the Jeffcott rotor) and its moments of inertia, which can produce gyroscopic effects, must be accounted for. If the rotor is flexible, multi-degrees-of-freedom models are needed.

24.1 Model with 4 degrees of freedom: gyroscopic effect

24.1.1 Generalized coordinates and equations of motion

In the Jeffcott model the rotor was assumed to be a point mass. When studying the acceleration through the critical speed, the moment of inertia about the axis of rotation was introduced, but this had no effect on the flexural behavior. Actually, the rotor's moments of inertia can considerably influence its dynamic behavior since they are responsible for the gyroscopic moments that cause the natural frequencies of bending modes to depend on the spin speed and the Campbell diagram to be different from a number of straight lines running in a horizontal direction.

The simplest model to evaluate this effect is still that in Fig. 23.3, where a rigid body is located in P instead of a point mass. It has nonvanishing moments of inertia and one of its principal axes of inertia coincides, in the undeformed position, with the z-axis.

Its ellipsoid of inertia has axial symmetry with respect to the same axis. The principal moments of inertia of the rigid body will be referred to as the *polar* moment of inertia J_p about the rotation axis and the *transversal* moment of inertia J_t about any axis in the rotation plane.

If $J_p > J_t$, the body is usually referred to as a *disc*; the limiting case is that of an infinitely thin disc in which $J_p = 2J_t$. If $J_p = J_t$, the inertia ellipsoid degenerates into a sphere and the situation is, in many aspects, similar to that of the Jeffcott rotor. If $J_p < J_t$, the rotor is usually referred to as a *long rotor*.

Assume also that, owing to small errors, the position of point P, in which the center of mass of the rigid body is located, does not coincide with that of point C, the center of the shaft. The distance between the two points is the eccentricity ϵ. Moreover, the axis of symmetry of the rigid body does not coincide exactly with the rotation axis. The angle between them is the angular error χ. These two errors, which are assumed to be small, cause a static unbalance and a couple unbalance, respectively.

Strictly speaking, the system has 6 degrees of freedom, and six generalized coordinates must be defined for the study of its dynamic behavior. The uncoupling between axial, flexural, and torsional behaviors seen for beams will be assumed to hold for the elastic part of the system. The same uncoupling also occurs for the inertial part of the model and a model with 4 degrees of freedom is sufficient for the study of the flexural behavior at constant speed, at least under wide simplifying assumptions.

The generalized coordinates will be defined with reference to the frames shown in Fig. 24.1a.

- Frame $OXYZ$: Inertial frame, with its origin in O and its Z-axis coinciding with the rotation axis of the rotor.

- Frame $O\Xi HZ$ with its origin in O and its Z-axis coinciding with that of the preceding frame: Axes Ξ, H rotate in the XY-plane with angular velocity Ω, in the case of constant speed operation. In the case of variable speed, angle θ between the two systems coincides

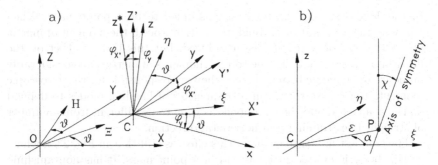

FIGURE 24.1. (a) Reference frames and (b): unbalance conditions in the $C\xi\eta\zeta$ reference frame.

with the same angle defined in Fig. 23.4. It will be referred to as a *rotating frame*.

- Frame $CX'Y'Z'$ with origin in C: Its axes remain parallel with those of frame $OXYZ$. The $X'Y'$-plane remains parallel to the XY-plane.

- Frame $Cxyz$ with origin in C: Its z-axis coincides with the rotation axis of the rigid body in the deformed position and the x- and y-axes are defined by the following rotations:

  - Rotate the axes of $CX'Y'Z'$ frame about the X'-axis by an angle $\phi_{X'}$ until the Y'-axis enters the rotation plane of the rigid body in its deformed configuration. Let the axes so obtained be the y- and z^*-axes. The rotation matrix allowing one to express the components of a vector in $CX'yz^*$ frame from those referred to $CX'Y'Z'$ frame (or to the inertial frame, as the directions of their axes coincide) is

$$\mathbf{R}_1 = \begin{bmatrix} 1 & 0 & 0 \\ 0 & \cos(\phi_{X'}) & \sin(\phi_{X'}) \\ 0 & -\sin(\phi_{X'}) & \cos(\phi_{X'}) \end{bmatrix} .$$

  - Rotate the frame so obtained about the y-axis until the X'-axis enters the rotation plane of the rigid body in its deformed configuration. Let the axis so obtained be the x-axis and the rotation angle be ϕ_y. Let the matrix expressing this second rotation be

$$\mathbf{R}_2 = \begin{bmatrix} \cos(\phi_y) & 0 & -\sin(\phi_y) \\ 0 & 1 & 0 \\ \sin(\phi_y) & 0 & \cos(\phi_y) \end{bmatrix} .$$

After the two mentioned rotations, the z-axis coincides, apart from the angular error χ, with the symmetry axis of the rigid body in its deformed configuration. Frame $Cxyz$ is centered in the center of the shaft of the rigid body and follows it in its whirling motion. However, it does not rotate with the spin speed Ω. It will be referred to as *whirling frame*.

- Frame $C\xi\eta z$ with origin in C: It is obtained from frame $Cxyz$ by rotating the x- and y-axes in the xy-plane by an angle equal to rotation angle θ of the rotor corresponding to the spin speed. If rotation occurs with constant spin speed Ω, angle θ is equal to Ωt. Frame $C\xi\eta z$ is actually fixed to the rigid body, although not being centered in its center of mass owing to the eccentricity ϵ, and not being principal of inertia due to the angular error χ. It will be referred to as *rotating*

618   24. Dynamics of Multi-Degrees-of-Freedom Rotors

and whirling frame. The matrix allowing one to express a vector in the $C\xi\eta z$-frame from the components in the $Cxyz$-frame is

$$\mathbf{R}_3 = \begin{bmatrix} \cos(\theta) & \sin(\theta) & 0 \\ -\sin(\theta) & \cos(\theta) & 0 \\ 0 & 0 & 1 \end{bmatrix}.$$

As already stated, the rotor is assumed to be slightly unbalanced. As the position of the rotor in the $C\xi\eta z$-frame is immaterial, the principal axis of inertia corresponding to the moment of inertia J_p will be assumed to be parallel to the ξz-plane. Because the static unbalance cannot be assumed to lie in the same plane as the couple unbalance, the eccentricity cannot be assumed to lie along the ξ-axis, as was the case of the Jeffcott rotor. The conditions of unbalance are summarized in Fig. 24.1b, where the static unbalance is shown to lead the couple unbalance of a phase angle α. A fourth rotation matrix $\mathbf{R}_4$ allowing passage from the rotor system of reference $C\xi\eta z$ to the principal axes of the rigid body is thus defined as

$$\mathbf{R}_4 = \begin{bmatrix} \cos(\chi) & 0 & -\sin(\chi) \\ 0 & 1 & 0 \\ \sin(\chi) & 0 & \cos(\chi) \end{bmatrix}.$$

Take the X-, Y-, and Z-coordinates of point C and angles $\phi_{X'}$, ϕ_y, and θ as generalized coordinates of the rigid body. A small displacement assumption on coordinates X, Y, Z, $\phi_{X'}$, and ϕ_y will allow great simplification of the problem. Coordinate θ, on the contrary, cannot be considered small.

To compute the kinetic energy of the rigid body, the velocity of its center of mass (point P) and its angular velocity expressed in the principal system of inertia must be computed. The position of point P is easily obtained

$$(\overline{\text{P}-\text{O}}) = \begin{Bmatrix} X \\ Y \\ Z \end{Bmatrix} + \mathbf{R}_1^T \mathbf{R}_2^T \mathbf{R}_3'^T \begin{Bmatrix} \epsilon \\ 0 \\ 0 \end{Bmatrix}, \tag{24.1}$$

where matrix $\mathbf{R}_3'$ is a matrix identical to $\mathbf{R}_3$, but obtained substituting angle $\theta + \alpha$ to angle θ. The small-displacement assumption allows the linearization of the trigonometric functions of angles $\phi_{X'}$ and ϕ_y and the neglect of some terms, which are of the same order of magnitude as those that are neglected when truncating the series for the trigonometric functions after the first term. Equation (24.1) then reduces to

$$(\overline{\text{P}-\text{O}}) = \begin{Bmatrix} X + \epsilon\cos(\theta + \alpha) \\ Y + \epsilon\sin(\theta + \alpha) \\ Z + \epsilon\left[\phi_{X'}\sin(\theta + \alpha) - \phi_y\cos(\theta + \alpha)\right] \end{Bmatrix}. \tag{24.2}$$

The velocity of the center of mass or the rotor is easily computed by performing the derivatives of vector $(\overline{P - O})$ with respect to time. As angle α is constant, it follows that

$$V_P = \left\{ \begin{array}{c} \dot{X} - \epsilon\dot{\theta}\sin(\theta + \alpha) \\ \dot{Y} + \epsilon\dot{\theta}\cos(\theta + \alpha) \\ \dot{Z} + \epsilon\left[(\dot{\theta}\phi_{X'} - \dot{\phi}_y)\cos(\theta + \alpha) + (\dot{\theta}\phi_y + \dot{\phi}_{X'})\sin(\theta + \alpha)\right] \end{array} \right\} . \tag{24.3}$$

In the third line of Eq. (24.3) there are two terms: The first is the velocity in the axial direction due to the displacement of point C in the same direction, and the second is the velocity in the axial direction due to the eccentricity and rotations of the cross section of the shaft. It is easy to verify that the last term causes a coupling between bending and axial behaviors of the rotor; however, if the eccentricity is small, it is negligible compared to the first one. In the following developments, all terms containing the product of the eccentricity or the angular error by a small quantity will be neglected and no axial–flexural coupling will be obtained. The translational kinetic energy is thus

$$\mathcal{T}_t = \frac{1}{2}mV_P^2 = \frac{1}{2}m\left\{\dot{X}^2 + \dot{Y}^2 + \dot{Z}^2 + \right. \tag{24.4}$$

$$\left. +\epsilon^2\dot{\theta}^2 + \epsilon\dot{\theta}\left[-\dot{X}\sin(\theta + \alpha) + \dot{Y}\cos(\theta + \alpha)\right]\right\} .$$

For the computation of the rotational kinetic energy, the angular velocity must be expressed with reference to a frame coinciding with the principal axes of the rotor. The three components of the angular velocity can be considered vectors acting in different directions: $\dot{\phi}_{X'}$ along the X'-axis, $\dot{\phi}_y$ along the y-axis, and $\dot{\theta}$ along the z-axis. Using the relevant rotation matrices, the components of the angular velocity along the principal axes of inertia of the rotor $\boldsymbol{\Omega}$ are

$$\boldsymbol{\Omega} = \mathbf{R}_4\mathbf{R}_3\mathbf{R}_2\left\{\begin{array}{c}\dot{\phi}_{X'}\\0\\0\end{array}\right\} + \mathbf{R}_4\mathbf{R}_3\left\{\begin{array}{c}0\\\dot{\phi}_y\\0\end{array}\right\} + \mathbf{R}_4\left\{\begin{array}{c}0\\0\\\dot{\theta}\end{array}\right\} . \tag{24.5}$$

By remembering the small-displacement assumptions, Eq. (24.5) reduces to

$$\boldsymbol{\Omega} = \left\{\begin{array}{c}\dot{\phi}_{X'}\cos(\theta) + \dot{\phi}_y\sin(\theta) - \chi\dot{\theta}\\-\dot{\phi}_{X'}\sin(\theta) + \dot{\phi}_y\cos(\theta)\\\dot{\phi}_{X'}[\chi\cos(\theta) + \phi_y] + \dot{\phi}_y\chi\sin(\theta) + \dot{\theta}\end{array}\right\} . \tag{24.6}$$

Since the components of Ω are referred to the principal axes of inertia, neglecting the small terms, the rotational kinetic energy can be easily computed as

$$\mathcal{T}_r = \frac{1}{2} \left\{ J_t(\dot{\phi}_{X'}^2 + \dot{\phi}_y^2 + \chi^2\dot{\theta}^2) + J_t(\dot{\theta}^2 + 2\dot{\theta}\dot{\phi}_{X'}\phi_y) + \right. \tag{24.7}$$

$$\left. +2\dot{\theta}\chi(J_p - J_t)\left[\dot{\phi}_{X'}\cos(\theta) + \dot{\phi}_y\sin(\theta)\right]\right\}.$$

The equations of motion can be obtained as usual through the Lagrange equation. In this case it is expedient to introduce into the Lagrange equations only the expression of the kinetic energy and then to add the elastic reaction of the shaft directly, as external generalized forces. The generalized forces Q_i that must be included in the equations of motion related to the translational degrees of freedom are simply the forces F_X, F_Y, and F_Z applied on the rigid body in the direction of the axes of the inertial frame. By performing the relevant derivatives, the first three equations of motion are obtained with simple computations

$$\begin{cases} m\ddot{X} = m\epsilon\left[\ddot{\theta}\sin(\theta + \alpha) + \dot{\theta}^2\cos(\theta + \alpha)\right] + F_X, \\ m\ddot{Y} = m\epsilon\left[-\ddot{\theta}\cos(\theta + \alpha) + \dot{\theta}^2\sin(\theta + \alpha)\right] + F_Y, \\ m\ddot{Z} = F_Z. \end{cases} \tag{24.8}$$

The generalized forces to be introduced into the three equations of motion related to the rotational degrees of freedom are not exactly the moments about the axes of the $Cxyz$ frame because the first generalized coordinate for rotation is referred to the X'-axis. It is easy to show that the generalized forces are related to the components of the moment acting on the rigid body in the directions of the axes of $Cxyz$ frame by the equations

$$Q_{\phi_{X'}} = M_x\cos(\phi_y) + M_z\sin(\phi_y), \qquad Q_{\phi_y} = M_y, \qquad Q_\theta = M_z. \tag{24.9}$$

By using expressions (24.9) for the generalized forces, performing all the relevant derivatives and linearizing again, the following equations of motion can be obtained:

$$\begin{cases} J_t\ddot{\phi}_{X'} + J_p\ddot{\theta}\phi_y + J_p\dot{\theta}\dot{\phi}_y = \chi(J_t - J_p)[\ddot{\theta}\cos(\theta) - \dot{\theta}^2\sin(\theta)] + \\ \qquad + M_x + \phi_y M_z, \\ J_t\ddot{\phi}_y - J_p\dot{\theta}\dot{\phi}_{X'} = \chi(J_t - J_p)[\ddot{\theta}\sin(\theta) + \dot{\theta}^2\cos(\theta)] + M_y, \\ (J_p + J_t\chi^2 + m\epsilon^2)\ddot{\theta} + m\epsilon\left[\ddot{Y}\cos(\theta) - \ddot{X}\sin(\theta)\right] + \\ \qquad + \chi(J_p - J_t)[\ddot{\phi}_{X'}\cos(\theta) + \ddot{\phi}_y\sin(\theta)] = M_z. \end{cases} \tag{24.10}$$

Moment M_z can be obtained from the last equation and substituted into the first. Since M_z is multiplied by the small quantity ϕ_y, in the first equation only the term $J_p\ddot{\theta}$ needs to be considered. By introducing the value of the moment about the z-axis into the first equation, it reduces to the form

$$J_t \ddot{\phi}_{X'} + J_p \dot{\theta} \dot{\phi}_y = \chi(J_t - J_p)[\ddot{\theta} \cos(\theta) - \dot{\theta}^2 \sin(\theta)] + M_x. \qquad (24.11)$$

Equation (24.11) and the last two Eq. (24.10) are the three equations of motion for the rotational degrees of freedom.

24.1.2 Behavior of the system at constant speed

The equation related to the translational degree of freedom along the z-axis uncouples from the others. If the angular velocity of the rotor is constant, the last equation also uncouples and the flexural behavior can be studied separately from axial and torsional behaviors. By stating that $\theta = \Omega t$, the four relevant equations of motion reduce to

$$
\begin{cases}
m\ddot{X} = m\epsilon\Omega^2 \cos(\Omega t + \alpha) + F_X, \\
m\ddot{Y} = m\epsilon\Omega^2 \sin(\Omega t + \alpha) + F_Y, \\
J_t \ddot{\phi}_{X'} + J_p \Omega \dot{\phi}_y = -\chi\Omega^2(J_t - J_p) \sin(\Omega t) + M_x, \\
J_t \ddot{\phi}_y - J_p \Omega \dot{\phi}_{X'} = \chi\Omega^2(J_t - J_p) \cos(\Omega t) + M_y.
\end{cases}
\qquad (24.12)
$$

The only forces and moments acting on the rigid body in P that will be considered are those due to the elastic reaction of the shaft. Because the behavior of the shaft is assumed to be linear, they are linked to the generalized coordinates by the stiffness matrix of the shaft. The situation in the xz-plane is similar to that in the yz-plane, but if the same elements of the stiffness matrix are used, owing to the axial symmetry of the shaft, the different sign conventions in the two coordinate planes compel to use opposite signs for the elements with subscripts 12 and 21

$$
\begin{Bmatrix} F_X \\ M_y \end{Bmatrix} = - \begin{bmatrix} K_{11} & K_{12} \\ K_{12} & K_{22} \end{bmatrix} \begin{Bmatrix} X \\ \phi_y \end{Bmatrix},
$$

$$
\begin{Bmatrix} F_Y \\ M_x \end{Bmatrix} = - \begin{bmatrix} K_{11} & -K_{12} \\ -K_{12} & K_{22} \end{bmatrix} \begin{Bmatrix} Y \\ \phi_{X'} \end{Bmatrix}.
\qquad (24.13)
$$

Note that while forces are defined with reference to the undeflected configuration, as is typical of the linear theory of elasticity, moments are, on the contrary, referred to the xyz-axes, which follow the deflected configuration. This is essentially equivalent, because rotations are assumed to be small, and amounts to release one of the usual simplifications of the theory of elasticity. By introducing the expressions (24.13) of the forces and moments into the equations of motion (24.12), it follows that

$$\begin{cases} m\ddot{X} + K_{11}X + K_{12}\phi_y = m\epsilon\Omega^2\cos(\Omega t + \alpha)\,, \\ m\ddot{Y} + K_{11}Y - K_{12}\phi_{X'} = m\epsilon\Omega^2\sin(\Omega t + \alpha)\,, \\ J_t\ddot{\phi}_{X'} + J_p\Omega\dot{\phi}_y - K_{12}Y + K_{22}\phi_{X'} = -\chi\Omega^2(J_t - J_p)\sin(\Omega t)\,, \\ J_t\ddot{\phi}_y - J_p\Omega\dot{\phi}_{X'} + K_{12}X + K_{22}\phi_y = \chi\Omega^2(J_t - J_p)\cos(\Omega t)\,. \end{cases}$$

(24.14)

Equations (24.14) can be written in the following matrix form:

$$\begin{bmatrix} m & 0 & 0 & 0 \\ 0 & J_t & 0 & 0 \\ 0 & 0 & m & 0 \\ 0 & 0 & 0 & J_t \end{bmatrix} \begin{Bmatrix} \ddot{X} \\ \ddot{\phi}_y \\ \ddot{Y} \\ \ddot{\phi}_{X'} \end{Bmatrix} + \Omega \begin{bmatrix} 0 & 0 & 0 & 0 \\ 0 & 0 & 0 & -J_p \\ 0 & 0 & 0 & 0 \\ 0 & J_p & 0 & 0 \end{bmatrix} \begin{Bmatrix} \dot{X} \\ \dot{\phi}_y \\ \dot{Y} \\ \dot{\phi}_{X'} \end{Bmatrix} +$$

$$+ \begin{bmatrix} K_{11} & K_{12} & 0 & 0 \\ K_{12} & K_{22} & 0 & 0 \\ 0 & 0 & K_{11} & -K_{12} \\ 0 & 0 & -K_{12} & K_{22} \end{bmatrix} \begin{Bmatrix} X \\ \phi_y \\ Y \\ \phi_{X'} \end{Bmatrix} = \Omega^2 \begin{Bmatrix} m\epsilon\cos(\Omega t + \alpha) \\ \chi(J_t - J_p)\cos(\Omega t) \\ m\epsilon\sin(\Omega t + \alpha) \\ -\chi(J_t - J_p)\sin(\Omega t) \end{Bmatrix}.$$

(24.15)

If $-\phi_{X'}$ is used instead of $\phi_{X'}$ as the generalized coordinate for rotations about the X-axis, the stiffness matrix assumes a more regular pattern with all terms positive and the skew-symmetric gyroscopic matrix, containing the polar moments of inertia, is replaced by its transpose.

It is possible to define a set of complex coordinates that allow the equations of motion to be written in a more compact form

$$\begin{cases} z = X + iY\,, \\ \phi = \phi_y - i\phi_{X'}\,. \end{cases}$$

(24.16)

Multiplying the second Eq. (24.15) by the imaginary unit i and adding it to the first one, and multiplying the third equation by $-i$ and adding it to the fourth one, reduces the equations of motion to

$$\begin{cases} m\ddot{z} + K_{11}z + K_{12}\phi = m\epsilon\Omega^2 e^{i(\Omega t + \alpha)}\,, \\ J_t\ddot{\phi} - i\Omega J_p\dot{\phi} + K_{12}z + K_{22}\phi = \chi\Omega^2(J_t - J_p)e^{i\Omega t}\,, \end{cases}$$

(24.17)

or, in a more compact form,

$$\mathbf{M}\ddot{\mathbf{q}} - i\Omega\mathbf{G}\dot{\mathbf{q}} + \mathbf{K}\mathbf{q} = \Omega^2\mathbf{f}e^{i\Omega t}\,.$$

(24.18)

The vectors of the complex coordinates and of the unbalances and the mass, gyroscopic, and stiffness matrices in Eq. (24.18) are

$$\mathbf{q} = \begin{Bmatrix} z \\ \phi \end{Bmatrix}, \quad \mathbf{M} = \begin{bmatrix} m & 0 \\ 0 & J_t \end{bmatrix}, \quad \mathbf{G} = \begin{bmatrix} 0 & 0 \\ 0 & J_p \end{bmatrix},$$

$$\mathbf{K} = \begin{bmatrix} K_{11} & K_{12} \\ K_{12} & K_{22} \end{bmatrix}, \quad \mathbf{f} = \begin{Bmatrix} m\epsilon e^{i\alpha} \\ \chi(J_t - J_p) \end{Bmatrix}.$$

Remark 24.1 *All matrices are symmetric when using the complex coordinate notation, while when using real coordinates the gyroscopic matrix is skew symmetric.*

The presence of damping can be accounted for in a very easy way if the viscous or hysteretic damping models can be accepted. Like in the case of the Jeffcott model, it is important to distinguish between nonrotating and rotating damping and to introduce separately the two damping matrices. In the case of viscous damping, the equation of motion of the damped system is

$$\mathbf{M\ddot{q}} + (\mathbf{C}_n + \mathbf{C}_r - i\Omega\mathbf{G})\mathbf{\dot{q}} + (\mathbf{K} - i\Omega\mathbf{C}_r)\mathbf{q} = \Omega^2\mathbf{f}e^{i\Omega t}, \qquad (24.19)$$

where the structure of matrices $\mathbf{C}_n$ and $\mathbf{C}_r$ is similar to that of the stiffness matrix.

Remark 24.2 *By linearizing the equation of motion of a rotating gyroscopic system a linear equation very similar to that of a discrete vibrating system has been obtained. The main differences are the presence of the gyroscopic matrix and, in the case of damped systems, of a term $i\Omega\mathbf{C}_r$ linked to rotating damping. Obviously, when the spin speed Ω tends to zero, the equation of motion of a nonrotating system is obtained.*

24.1.3 Free whirling of the undamped system

The free whirling of the undamped system can be studied using the homogeneous equation associated with Eq. (24.17). Introducing a solution of the type

$$z = z_0 e^{i\omega t}, \qquad \phi = \phi_0 e^{i\omega t}$$

into the equation of motion, the following algebraic linear equations are readily obtained:

$$\begin{cases} z_0 \left(-m\omega^2 + K_{11}\right) + \phi_0 K_{12} = 0, \\ z_0 K_{12} + \phi_0 \left(-J_t\omega^2 + J_p\omega\Omega + K_{22}\right) = 0. \end{cases} \qquad (24.20)$$

The characteristic equation allowing computation of the whirl frequency is

$$\det \begin{bmatrix} -m\omega^2 + K_{11} & K_{12} \\ K_{12} & -J_t\omega^2 + J_p\omega\Omega + K_{22} \end{bmatrix} = 0, \qquad (24.21)$$

which yields

$$\omega^4 - \Omega\omega^3 \frac{J_p}{J_t} - \omega^2 \left(\frac{K_{11}}{m} + \frac{K_{22}}{J_t}\right) + \Omega\omega \frac{K_{11}J_p}{mJ_t} + \frac{K_{11}K_{22} - K_{12}^2}{mJ_t} = 0. \quad (24.22)$$

Equation (24.22) has four real roots, two of which are positive. The Campbell diagram of the system is of the type shown in Fig. 24.2: At each speed Ω, four whirl modes, occurring at different frequencies, are possible. Two of them occur in the forward direction and two in the backward direction.

Because all solutions of Eq. (24.22) are real, the corresponding eigenvectors $\mathbf{q}_i$ are also real. At time $t = 0$, both Y and $\phi_{X'}$ vanish: the z-axis is contained in a plane also containing the Z-axis and rotating about the latter with a constant angular velocity equal to the whirl speed ω. The axis of the rotor describes a cone whose axis is the rotor axis in its undeformed position. Because the deflected shape is contained in a plane, the motion could have been studied in such a plane from the beginning, using a model with only 2 degrees of freedom. This model, however, cannot be used to study the damped system, and a more general model was preferred from the beginning. Actually, the free whirling of the system can be more complex because the motion is the combination of all four circular whirling motions that occur at different frequencies

$$\mathbf{q} = \sum_{k=1}^{4} \mathbf{q}_k e^{i\omega_k t}. \tag{24.23}$$

The gyroscopic terms couple the behavior in the planes passing through the rotation axis and make it impossible for the system to perform elliptical or rectilinear motions. An equation of the type of Eq. (24.23) can yield elliptical motions only in the case in which two of the eigenfrequencies ω_k have the same modulus and opposite sign, as in the case of the Jeffcott rotor. The introduction of gyroscopic moments causes forward whirl frequencies to be different from backward whirl frequencies and all whirl motions to be circular. Obviously, the four circular modes can add to each other, yielding Lissajous curves. Another effect of gyroscopic moments is causing the whirl frequencies to be functions of the spin speed. Generally speaking, the frequency of forward modes increases with speed, while the frequency of backward modes decreases.

To avoid solving Eq. (24.22) in ω, the Campbell diagram can be obtained solving the same equation in Ω. The equation is linear in this unknown and a closed-form solution can be easily found:

$$\Omega = \frac{mJ_t\omega^4 - (J_tK_{11} + mK_{22})\omega^2 + K_{11}K_{22} - K_{12}^2}{\omega J_p(m\omega^2 - K_{11})}. \tag{24.24}$$

From Fig. 24.2 it is clear that the Campbell diagram has three horizontal asymptotes; two for backward motions and one for forward whirling. Another asymptote has the equation $\omega = \Omega J_p/J_t$. The intersections between the curve $\omega(\Omega)$ and the bisector of the first quadrant $\omega = \Omega$ yield the conditions for forward synchronous whirling, i.e., the critical speeds. By introducing condition $\Omega = \omega$ into Eq. (24.22) the following quadratic equation in Ω^2 is obtained:

$$\Omega^4 m(J_p - J_t) - \Omega^2[(J_p - J_t)K_{11} - mK_{22}] - K_{11}K_{22} + K_{12}^2 = 0. \tag{24.25}$$

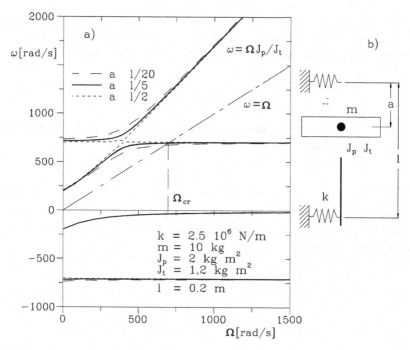

FIGURE 24.2. Campbell diagram of a system made by a rigid rotor on elastic supports.

By neglecting the negative solutions corresponding to the intersections of curve $\omega(\Omega)$ with the straight line $\omega = \Omega$ lying in the third quadrant of the $\omega\Omega$-plane, the following values of the critical speed are obtained:

$$\Omega_{cr} = \sqrt{\dfrac{K_{11}(J_p - J_t) - mK_{22} \pm \sqrt{[K_{11}(J_p - J_t) + mK_{22}]^2 - 4m(J_p - J_t)K_{12}^2}}{2m(J_p - J_t)}}.$$

$$(24.26)$$

Introducing the ratio

$$\delta = \dfrac{J_p - J_t}{m},$$

Eq. (24.26) can be written in the more compact form

$$\Omega_{cr} = \dfrac{1}{\sqrt{2m}} \sqrt{K_{11} - \dfrac{K_{22}}{\delta} \pm \sqrt{\left(K_{11} + \dfrac{K_{22}}{\delta}\right)^2 - 4\dfrac{K_{12}^2}{\delta}}}. \qquad (24.27)$$

If $J_p < J_t$ or $\delta < 0$, as happens in the case of long rotors, there are two real solutions and, as a consequence, two values of the critical speed.

If, on the contrary, $J_p > J_t$ or $\delta > 0$, as happens in the case of discs, one of the solutions is imaginary and only one critical speed exists.

If $K_{12} = 0$, i.e., if translational and rotational degrees of freedom are uncoupled, the two equations can be studied separately and the Campbell diagram degenerates into two straight lines with equations

$$\omega = \pm \sqrt{\frac{K_{11}}{m}},$$

as in the case of Jeffcott rotors, a curve that has for asymptotes the Ω-axis in backward whirling, and a curve with the asymptote

$$\omega = \Omega \frac{J_p}{J_t}$$

in forward whirling. The straight lines and the curves can cross either in the first quadrant, as in Fig. 24.2, or in the fourth quadrant. The values of the critical speed are

$$
\begin{cases}
\Omega_{cr_I} = \sqrt{\dfrac{K_{11}}{m}}, \\[3mm]
\Omega_{cr_{II}} = \sqrt{-\dfrac{K_{22}}{m\delta}} = \sqrt{-\dfrac{K_{22}}{J_p - J_t}}.
\end{cases}
\tag{24.28}
$$

The second value is real only if $\delta < 0$, i.e., $J_t > J_p$.

Consider a rotor made by a rigid gyroscopic body attached to a flexible uniform shaft running on rigid bearings (Fig. 24.3). The stiffness matrix of the system is

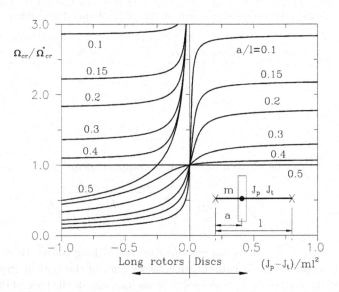

FIGURE 24.3. Influence of the gyroscopic moment on critical speeds.

$$\mathbf{K} = \frac{3EIl}{a(l-a)} \begin{bmatrix} \dfrac{l^2 - 3al + 3a^2}{a^2(l-a)^2} & \dfrac{2a-l}{a(l-a)} \\ \dfrac{2a-l}{a(l-a)} & 1 \end{bmatrix} . \qquad (24.29)$$

The influence of the gyroscopic moment on the critical speeds can be shown by plotting the values of the critical speed or, better, of ratio $\Omega_{cr}/\Omega_{cr}^*$, where Ω_{cr}^* is the critical speed computed neglecting gyroscopic effects, as functions of δ or, better, of the nondimensional parameter

$$\frac{\delta}{l^2} = \frac{J_p - J_t}{ml^2} .$$

The graph, plotted for various values of a/l, is shown in Fig. 24.3. If the rotor is at midspan, i.e., if $a/l = 0.5$, the gyroscopic moment has no effect on critical speed in the case of a disc rotor while causing a second critical speed in the case of long rotors.

In all other cases, a disc-type rotor causes an increase in the critical speed, which is sometimes explained by saying that the gyroscopic moment causes a stiffening of the system. It is only a phenomenological explanation because no actual stiffening takes place.

In the case of a long rotor, the critical speed decreases and a second critical speed, usually quite higher than the first, is present.

Remark 24.3 *The values of*

$$\frac{|J_p - J_t|}{ml^2}$$

are in practice quite small, particularly in the case of long rotors, and Fig. 24.3 holds only for a narrow zone about the ordinate axis. On the right side of this narrow zone the discs are too thin and they can no longer be considered as rigid bodies, and on the left the rotor becomes too slender to be considered as rigid again.

The considerations drawn from Fig. 24.3, although obtained for a particular case, are, however, qualitatively applicable in general.

24.1.4 Unbalance response of the undamped system

The unbalance response can be easily studied starting from Eq. (24.18), whose particular integral can be expressed in the form

$$z = z_0 e^{i\Omega t} , \qquad \phi = \phi_0 e^{i\Omega t} ,$$

where the amplitudes z_o and ϕ_o are, in general, complex numbers because static and couple unbalances do not lie in the same plane, i.e., the phase angle α is, in general, not equal to zero. The effects of the two types of unbalance can be studied separately, and then the relevant solutions can be added to each other.

In the case of a static unbalance with vanishing phase α, the amplitude of the response is

$$\begin{cases} z_0 = m\epsilon\Omega^2 \dfrac{(J_p - J_t)\Omega^2 + K_{22}}{\Delta}, \\ \phi_0 = -m\epsilon\Omega^2 \dfrac{K_{12}}{\Delta}, \end{cases} \tag{24.30}$$

where

$$\Delta = -m(J_p - J_t)\Omega^4 + [K_{11}(J_p - J_t) + mK_{22}]\,\Omega^2 + K_{11}K_{22} - K_{12}^2\,.$$

In the case of couple unbalance, the response is

$$\begin{cases} z_0 = \chi\Omega^2(J_p - J_t)\dfrac{K_{12}}{\Delta}, \\ \phi_0 = \chi\Omega^2(J_p - J_t)\dfrac{m\Omega^2 - K_{11}}{\Delta}. \end{cases} \tag{24.31}$$

Equations (24.30) and (24.31) are plotted in Fig. 24.4 for the same rotor as in Fig. 24.2 with $l/a = 4$. As is clear from the figure, in the current case, self-centering occurs at speeds in excess of the critical speed in the case of static unbalance, while a certain self-centering also occurs in the subcritical field in the case of couple unbalance.

This happens because the system has only one critical speed ($J_p > J_t$), and the natural frequency at standstill related to the rotational mode is lower than that linked with the translational mode.

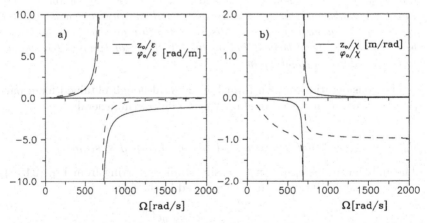

FIGURE 24.4. Amplitude of the unbalance response for the same rotor of Fig. 24.2 with $l/a = 4$: (a) response to static unbalance and (b) response to couple unbalance.

Example 24.1 *The schematic cross section of a turbomolecular pump is reported in Fig. 24.5, together with a few basic dimensions. Knowing that the operating speed is 30,000 rpm, the first critical speed must be lower than 15,000 rpm and the second must be above the working range, state the diameter of the shaft.*

The rotor of the pump can be modeled as a rigid body attached to a massless shaft, with its center of mass at the connection point. The inertia characteristics of the rigid body are $m = 6$ kg, $J_p = 0.035$ kg m^2, and $J_t = 0.055$ kg m^2. The material of the shaft has a Young's modulus $E = 2.1 \times 10^{11}$ N/m^2. The mass of the shaft has been neglected, and the model with 2 complex degrees of freedom can be used. The model is sketched in Fig. 24.5. Choose as generalized coordinates the displacement and rotation at the center of mass of the rigid body, neglect the compliance of the bearings and of the stator of the machine, and assume that the shaft has constant diameter.

The stiffness matrix can be computed using the FEM. However, due to the simple geometry, it is possible to compute the compliance matrix by applying a unit force and a unit moment in point P and computing the displacement and rotation in the same point (for simple geometries like the current one, the relevant formulas are reported in many handbooks), the compliance matrix $\boldsymbol{B}$ is obtained:

$$\boldsymbol{B} = \frac{1}{6EI} \begin{bmatrix} 2l_2^2(l_1+l_2) & l_2(2l_1+3l_2) \\ l_2(2l_1+3l_2) & 2(l_1+3l_2) \end{bmatrix}.$$

The stiffness matrix $\boldsymbol{K}$ is obtained by inverting the compliance matrix β

$$\boldsymbol{K} = \frac{6EI}{l_2^3(4l_1+3l_2)} \begin{bmatrix} 2(l_1+3l_2) & -l_2(2l_1+3l_2) \\ -l_2(2l_1+3l_2) & 2l_2^2(l_1+l_2) \end{bmatrix} =$$

$$= I \begin{bmatrix} 1.3696 & -0.0445 \\ -0.0445 & 0.001644 \end{bmatrix} \times 10^{16}.$$

The critical speeds can be computed using Eq. (24.26):

$$\Omega_{cr_I} = 0.8611 \times 10^7 \sqrt{I} \ rad/s = 0.88223 \times 10^8 \sqrt{I} \ rpm,$$

$$\Omega_{cr_{II}} = 5.504 \times 10^7 \sqrt{I} \ rad/s = 5.2566 \times 10^8 \sqrt{I} \ rpm.$$

The large difference between the first and the second critical speeds allows the choice of a low value of the first critical speed without any danger that the second is located within the working range. By assuming that $\Omega_{cr_I} = 10,000$ rpm, a value of the moment of inertia of the shaft $I = 1.479 \times 10^{-8}$ m^4 and then a shaft diameter of 23.4 mm is obtained. Choosing a value of 24 mm, the moment of inertia is $I = 1.628 \times 10^{-8}$ m^4 and the critical speeds are $\Omega_{cr_I} = 10,490$ rpm, $\Omega_{cr_{II}} = 67,080$ rpm, respectively. The second value is quite high, and it is likely that other critical speeds are located between the computed values. A more realistic model, which also takes into account the mass of the shaft and the compliance of the bearings, must, however, be used to obtain them.

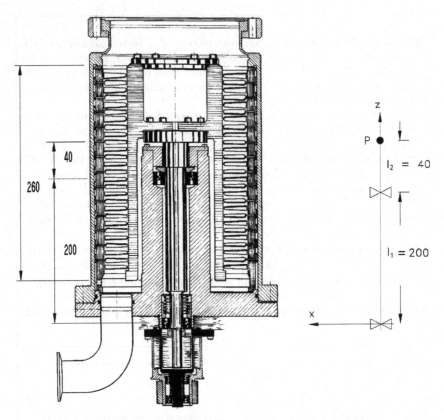

FIGURE 24.5. Schematic drawing of a turbomolecular pump (*Vacuum Technology; Its Foundation, Formulae and Tables*, Leybold-Heraeus, Köln, Germany) and model of the rotor. The drawing refers to a machine no longer in production.

24.2   Rotors with many degrees of freedom

The Jeffcott rotor and the model with 4 degrees of freedom allow at least a qualitative understanding of the dynamic behavior of linear axi-symmetrical rotors, modeled as a point mass or a rigid body connected to a massless elastic shaft running in massless elastic bearings. These models are, however, inadequate to supply quantitative information on the behavior of complex rotating systems, and more accurate models are needed in the design stage.

The next step is modeling rotors as beam-like objects: The degrees of freedom are the lateral displacements and the rotations of the cross sections. This approach is usually referred to as one-dimensional modeling, since the properties of the rotor depend on one coordinate only, the axial coordinate z.

In the present text the study is limited to this simple approach; for more detailed modeling, like the one-and-a-half-dimensional approach, in which the rotor is still basically a beam-like object, but the flexibility of the disks is accounted for in a simplified way, or the full tridimensional approach the reader can refer to specialized texts.[1]

Also, in the case of rotors, both the continuum and the discretized approaches are possible, at least in theory.

24.2.1   Continuum models

Consider a rotor that can be modeled as a beam slender enough to use the standard beam theory. If the supports can be assumed to be rigid, there is no difficulty computing the natural frequencies at standstill. If the polar moment of inertia of the cross-sections can be neglected, gyroscopic moments do not significantly affect the dynamic behavior of the system when rotating and the natural frequencies are independent of the spin speed. The values computed for the nonrotating beam hold for any value of the speed Ω, the Campbell diagram is made by straight lines parallel to the Ω-axis, and the values of the critical speeds coincide with those of the natural frequencies.

Transmission shafts with a simple geometrical shape are usually modeled as cylindrical Euler–Bernoulli beams. There is, however, no difficulty studying the dynamic behavior of a cylindrical shaft, taking into account the rotational inertia of the cross-section and shear deformation. Also, the gyroscopic effect can be accounted for, and it is possible to show that the three effects are of the same order of magnitude: If one of them is considered, it is advisable to include them all. Gyroscopic effects cause a stiffening of the system, which increases with increasing speed. In critical speed conditions, usually the combined effect of the three phenomena is that of increasing the critical speed, while at standstill they cause a decrease of the natural frequency.[2]

The continuum approach is, however, feasible only in the case of very simple geometrical shapes, such as shafts with constant cross-section. Dealing with more complicated shapes using the continuum approach leads to analytical difficulties, and today the most popular methods are based on some discretization technique. In particular, methods based on transfer matrices were, and still are, widely used, even if they are now yielding to the FEM.

Remark 24.4 *Very often, rotors are modeled using beam elements, with inertial properties modeled using the lumped or the consistent approach. Although there is no difficulty using other types of elements, the relevant*

---

[1]See, for instance, G. Genta, *Dynamics of Rotating Systems*, Springer, New York, 2005.

[2]See, for example, F.M. Dimentberg, *Flexural vibrations of rotating shafts*, Butterworths, London, 1961; G. Genta, "Consistent matrices in rotor dynamics", *Meccanica*, Vol. 20, (1985), 235–248.

gyroscopic matrices are usually not available to users of standard finite element codes.

24.2.2 Lumped-parameters models: Myklestadt–Prohl method

Among lumped-parameters methods, the transfer matrix approach, namely, the Myklestadt method was, and still is, widely used in rotor dynamics, mainly for critical speed prediction. The formulation seen in Section 14.6 holds with the only difference of adding the gyroscopic moment in the node transfer matrices expressed by Eq. (14.23). As it can be easily obtained from the second Eq. (24.20), the gyroscopic moment due to the ith rigid body located in the ith node can be accounted for by replacing

$$\omega \Omega J_p - \omega^2 J_t$$

for

$$-\omega^2 J_t$$

in the expression of the generalized inertia force related to the rotational degree of freedom. By performing the mentioned substitution in Eq. (14.23), the Myklestadt method is easily converted to the study of the dynamic behavior of rotors. If the aim is only to evaluate the critical speeds, it is possible to introduce $\omega = \Omega = \Omega_{cr}$ into the node transfer matrix, which becomes

$$\mathbf{T}_{i_n} = \begin{bmatrix} 1 & 0 & 0 & 0 \\ 0 & 1 & 0 & 0 \\ -m_i \Omega_{cr}^2 + k_c & 0 & 1 & 0 \\ 0 & -(J_{t_i} - J_{p_i})\Omega_{cr}^2 + \chi_c & 0 & 1 \end{bmatrix}. \qquad (24.32)$$

The computation follows the guidelines already seen in Section 14.6 without modifications.

When using the stiffness or compliance approach, the whole structure is modeled as for nonrotating systems in Section 14.6. If complex coordinates are used, the relevant equations are either Eq. (24.18) or (24.19), depending on whether or not damping is neglected. The size of all matrices and vectors is equal to the number of complex degrees of freedom, which coincides with the number of degrees of freedom related to bending behavior in the xz- and yz-planes. Mass, stiffness, and damping matrices $\mathbf{M}$, $\mathbf{K}$, and $\mathbf{C}$ are those related to the flexural behavior of the nonrotating system in the xz-plane, although the choice of the relevant plane depends on the way in which the complex coordinates were defined. This is obvious because the equation of motion must become the same that describes the free behavior of the nonrotating system when the spin speed Ω tends to zero. Also, in this case the stiffness matrix can be directly obtained using the FEM or from the matrix of the coefficients of influence.

If the complex coordinates are ordered as follows:

$$\mathbf{q} = \begin{bmatrix} z_1 & \phi_1 & z_2 & \phi_2 & \cdots & z_n & \phi_n \end{bmatrix}^T, \qquad (24.33)$$

matrices $\mathbf{M}$ and $\mathbf{G}$ and vector $\mathbf{f}$ of the unbalances are, respectively,

$$\mathbf{M} = \mathrm{diag} \begin{bmatrix} m_1 & J_{t_1} & m_2 & J_{t_2} & \cdots & m_n & J_{t_n} \end{bmatrix},$$

$$\mathbf{G} = \mathrm{diag} \begin{bmatrix} 0 & J_{p_1} & 0 & J_{p_2} & \cdots & 0 & J_{p_n} \end{bmatrix}, \qquad (24.34)$$

$$\mathbf{f} = \begin{bmatrix} m_1 \epsilon_1 e^{i\alpha_1} & (J_{t_1} - J_{p_1})\chi_1 e^{i\beta_1} & \cdots & m_n \epsilon_n e^{i\alpha_n} & (J_{t_n} - J_{p_n})\chi_n e^{i\beta_n} \end{bmatrix}^T.$$

The phases α_i and β_i are needed to take into account the possible different orientations in space of the vectors expressing static and couple unbalances. Rotating damping matrix $\mathbf{C}_r$ can be built in the same way as for general damping matrices, taking into account only the contribution to the overall energy dissipation due to rotating parts of the machine.

24.2.3   Models with consistent inertial properties

There is no difficulty in adding the contribution of rotation to the kinetic energy of an element of any type into the FEM. The mass matrix that results from this approach is coincident with that related to the nonrotating model. Also, the stiffness and damping matrices are not affected by the fact that the system rotates, apart from a possible geometrical effect due to the stiffening of the element that can be ascribed to centrifugal stressing. This effect, which is usually neglected in the formulation of the beam elements used to model rotating shafts, must be considered when dealing with beam elements in a direction perpendicular to the rotation axis, such as those used to model turbine blades, propellers, and similar structural members. If centrifugal stiffening must be considered, it can be generally taken into account by adding a term of the type $\Omega^2 \mathbf{K}_\Omega$ to the stiffness matrix, where $\mathbf{K}_\Omega$ is a matrix of constants that can be computed at the element level and then assembled in the usual way.

The effect of axial forces acting on the rotor can be accounted for by using the standard formulation of the geometric matrix.

The greatest difference between a finite element model for a rotor and that for the corresponding nonrotating structure is the presence in the first of a gyroscopic term. It is possible to show that the consistent gyroscopic matrix of a beam element is identical to the part of the mass matrix related to the rotational inertia of the cross-sections in which the polar moment of the cross-section I_z is substituted for the transversal moment of inertia I_y. As in the case of an axi-symmetrical beam $I_z = 2I_y$, the gyroscopic matrix is equal to twice the rotational contribution to the mass matrix. If there is damping in the system, the damping matrices of the nonrotating and rotating elements can be assembled separately, directly yielding the two matrices $\mathbf{C}_n$ and $\mathbf{C}_r$ to be introduced into the equations of motion. The consistent unbalance vectors for a linear distribution of static and couple unbalance are

$$\mathbf{f} = \mathbf{a}_1 \left\{ \begin{array}{c} \epsilon_{\xi_1} + i\epsilon_{\eta_1} \\ \epsilon_{\xi_2} + i\epsilon_{\eta_2} \end{array} \right\} + \mathbf{a}_2 \left\{ \begin{array}{c} \chi_{\eta_1} - i\chi_{\xi_1} \\ \chi_{\eta_2} + i\chi_{\xi_2} \end{array} \right\}, \qquad (24.35)$$

where

$$\mathbf{a}_1 = \frac{\rho Al}{120(1 - \Phi)} \begin{bmatrix} 42 + 40\Phi & 18 + 20\Phi \\ l(6 + 5\Phi) & l(4 + 5\Phi) \\ 18 + 20\Phi & 42 + 40\Phi \\ -l(4 + 5\Phi) & -l(6 + 5\Phi) \end{bmatrix},$$

$$\mathbf{a}_2 = \frac{\rho I_y}{12(1 - \Phi)} \begin{bmatrix} -6 & -6 \\ l(1 + 4\Phi) & l(-1 + 2\Phi) \\ 6 & 6 \\ l(-1 + 2\Phi) & l(1 + 4\Phi) \end{bmatrix}$$

and the other symbols have the same meaning seen in Section 15.2. The eccentricity of the elements has been defined by the components along the ξ- and the η-axes in correspondence to the two end nodes of the element. The couple unbalance has been defined in the same way, even if it is difficult to correctly define a distributed couple unbalance.

The matrices of the elements can be assembled and condensed in the same way as for nonrotating structures. Gyroscopic matrices are condensed using the same algorithm as for mass matrices.

The model can also include the stator of the machine as well as various rotors with different rotating speeds, provided that the whole system is axially symmetrical, as in the case of multishaft turbines.

24.3   Real versus complex coordinates

The equations of motion of axi-symmetrical rotors were obtained in the preceding sections with reference to a set of complex coordinates of the type defined by Eq. (24.16). The use of complex coordinates is very convenient in the case of axi-symmetrical systems, particularly if damping is neglected, because it allows to study the system using a model whose size is half the size of the same problem expressed in real coordinates.

Remark 24.5 *Even if the coordinates are complex, all relevant matrices are real and the computation of the critical speeds, the undamped Campbell diagram, and the undamped unbalance response does not involve actual working with complex numbers, as will be shown in detail in the following sections.*

Another advantage of the use of complex coordinates is that of obtaining the response in terms of rotating vectors that rotate in the physical space, directly giving complete information about the orbits of the various nodes and the direction of the whirling motions. Using real coordinates the rotating vectors are defined with reference to the complex plane and the whirling motion is obtained as the composition of harmonic vibrations in the directions of the x- and y-axes.

The main limitation of the complex coordinate approach is that it deals with difficulty with elliptical orbits, as those occurring in the case of non-axisymmetrical machines, as will be seen in Chapter 25.

Equation (24.19) can be written by separating the real part of each equation from the imaginary part, obtaining

$$\begin{bmatrix} \mathbf{M} & \mathbf{0} \\ \mathbf{0} & \mathbf{M} \end{bmatrix} \ddot{x} + \left(\Omega \begin{bmatrix} \mathbf{0} & \mathbf{G} \\ -\mathbf{G} & \mathbf{0} \end{bmatrix} + \begin{bmatrix} \mathbf{C} & \mathbf{0} \\ \mathbf{0} & \mathbf{C} \end{bmatrix} \right) \dot{x} + \qquad (24.36)$$

$$+ \left(\begin{bmatrix} \mathbf{K} & \mathbf{0} \\ \mathbf{0} & \mathbf{K} \end{bmatrix} + \Omega \begin{bmatrix} \mathbf{0} & \mathbf{C}_r \\ -\mathbf{C}_r & \mathbf{0} \end{bmatrix} \right) x = \Omega^2 \left\{ \begin{array}{c} \Re(\mathbf{f} e^{i\Omega t}) \\ \Im(\mathbf{f} e^{i\Omega t}) \end{array} \right\},$$

where

$$\mathbf{x} = [\Re(\mathbf{q})^T, \Im(\mathbf{q})^T]^T$$

and $\mathbf{C}$ is the total damping matrix,

$$\mathbf{C} = \mathbf{C}_n + \mathbf{C}_r .$$

Remark 24.6 *In the case of complex coordinates all matrices were symmetrical, but the use of real coordinates results in skew-symmetrical gyroscopic and rotating damping (circulatory) matrices.*

The real coordinates $\mathbf{x}$ defined earlier differ from the standard real coordinates used for the study of nonrotating systems for what the sign of the rotational degrees of freedom related to rotation about the x-axis is concerned. In many cases, a generalized form of Eq. (24.15) is used, in which the sign conventions for rotational coordinates are those standard in the FEM. The equation of motion of the damped system is thus

$$\begin{bmatrix} \mathbf{M}_x & \mathbf{0} \\ \mathbf{0} & \mathbf{M}_y \end{bmatrix} \ddot{x}^* + \left(\Omega \begin{bmatrix} \mathbf{0} & \mathbf{G}_{xy} \\ -\mathbf{G}_{yx} & \mathbf{0} \end{bmatrix} + \begin{bmatrix} \mathbf{C}_x & \mathbf{0} \\ \mathbf{0} & \mathbf{C}_y \end{bmatrix} \right) \dot{x}^* +$$

$$+ \left(\begin{bmatrix} \mathbf{K}_x & \mathbf{0} \\ \mathbf{0} & \mathbf{K}_y \end{bmatrix} + \Omega \begin{bmatrix} \mathbf{0} & \mathbf{C}_{r_{xy}} \\ -\mathbf{C}_{r_{yx}} & \mathbf{0} \end{bmatrix} \right) x^* \qquad (24.37)$$

$$= \Omega^2 \left\{ \begin{array}{c} \mathbf{f}_x \cos(\Omega t) - \mathbf{f}_y \sin(\Omega t) \\ \mathbf{f}_x \sin(\Omega t) + \mathbf{f}_y \cos(\Omega t) \end{array} \right\},$$

where matrices with subscript x are the same as for complex coordinates, matrices with subscript y are similar, except for the sign of elements with subscripts made by two numbers whose sum is odd. Matrices with subscripts xy and yx have signs that differ from those of the corresponding matrices in Eq. (24.36) and are such that the global gyroscopic and rotating damping matrices are skew-symmetrical. Vector $\mathbf{x}^*$ is a vector containing the generalized coordinates and is of the type of the vector of the generalized coordinates of Eq. (24.15).

The order of the degrees of freedom shown in Eqs. (24.36) and (24.37), with all coordinates related to the xz-plane listed one after the other followed by those related to the other plane, is just an indication because it would lead to matrices with a very large band. Actually, the degrees of freedom are mixed at the element level and the structure of the matrices shown here holds for the matrices of the elements and not for those related to the whole structure.

24.4 Fixed versus rotating coordinates

Although the flexural behavior of rotating axi-symmetrical systems can be studied using an inertial coordinate frame as seen in the preceding sections, it is possible to use a set of coordinates based on the rotating frame $O\,\Xi HZ$.

A set of complex coordinates $\mathbf{r}$, related to the coordinates $\mathbf{q}$, can thus be obtained using the relationship

$$\mathbf{r} = \mathbf{q}e^{-i\Omega t} \tag{24.38}$$

and the equation of motion (24.19) transforms into the following equation:

$$\mathbf{M}\ddot{\mathbf{r}} + \left[\mathbf{C}_n + \mathbf{C}_r + i\Omega(2\mathbf{M} - \mathbf{G})\right]\dot{\mathbf{r}} +$$
$$+ \left[\mathbf{K} - \Omega^2(\mathbf{M} - \mathbf{G}) + i\Omega\mathbf{C}_n\right]\mathbf{r} = \Omega^2\mathbf{f}. \tag{24.39}$$

The same considerations seen for Eq. (23.38) obtained for the Jeffcott model also hold for Eq. (24.39). Also in the case of rotating coordinates it is possible to chose between the complex coordinates approach seen earlier and an alternative formulation of the problem based on real coordinates. The relevant equation can be easily obtained by separating the real and imaginary parts of Eq. (24.39). The gyroscopic, Coriolis, and nonrotating damping terms give way to skew-symmetrical matrices. The use of rotating coordinates, which is not very convenient in the case of axi-symmetrical systems, becomes advisable when dealing with machines including a nonisotropic rotor (see Chapter 25).

24.5   State-space equations

All the preceding equations can be written with reference to the state space. If complex coordinates are used, the state vector is

$$\mathbf{z} = \left\{ \begin{array}{c} \dot{\mathbf{q}} \\ \mathbf{q} \end{array} \right\}. \tag{24.40}$$

The corresponding state matrix and product $\mathcal{B}\mathbf{u}$ due to unbalance are, respectively,

$$\mathcal{A} = \left[\begin{array}{cc} -\mathbf{M}^{-1}\left(\mathbf{C}_n + \mathbf{C}_r - i\Omega\mathbf{G}\right) & -\mathbf{M}^{-1}\left(\mathbf{K} + \Omega^2\mathbf{K}_\Omega - i\Omega\mathbf{C}_r\right) \\ \mathbf{I} & \mathbf{0} \end{array} \right], \tag{24.41}$$

$$\mathcal{B}\mathbf{u} = \Omega^2 e^{i\Omega t} \left\{ \begin{array}{c} \mathbf{f} \\ \mathbf{0} \end{array} \right\}. \tag{24.42}$$

Note that the dynamic matrix is complex, due to the presence of the gyroscopic and rotating damping (circulatory) terms and, as a consequence, its eigenvalues are non-conjugated. When the roots locus, with the spin speed used as a parameter, is plotted, it is not symmetrical with respect to the real axis. This was also the case for the Jeffcott rotor, as shown in Fig. 23.8, but in the current case it also occurs for the undamped system owing to the gyroscopic term.

State-space equations can also be written with reference to real coordinates, in which case a real dynamic matrix is always obtained. Rotor-fixed coordinates can also be used.

24.6   Static solution

In this section, the term *static solution* will be referred to as the computation of the deformed shape and of the stress and strain fields of a rotor under the effects of constant forces. The relevant equation yielding the static displacement is easily obtained by adding a vector of static nonrotating forces $\mathbf{f}_n$ on the right-hand side of Eq. (24.19) and neglecting all terms containing the time derivatives of the generalized coordinates. There is no difficulty in introducing the hysteretic damping matrix $\mathbf{K}_r''$, i.e., the imaginary part of the complex stiffness matrix related to the rotating part of the system into the equation of motion, remembering that in the case of a static solution the whirl speed Ω vanishes. Taking into account centrifugal stiffening, it follows that

$$\left(\mathbf{K} + \Omega^2\mathbf{K}_\Omega - i\Omega\mathbf{C}_r - i\mathbf{K}_r''\right)\mathbf{q} = \mathbf{f}_n. \tag{24.43}$$

Neglecting centrifugal stiffening, from Eq. (24.43) it is clear that the static solution for an undamped rotating system coincides with that of the corresponding nonrotating structure but, in case of a damped rotor, the deflected shape is influenced by the spin speed. This effect is easily understood if the force vector is assumed to be real, i.e., if all forces and moments act in the same plane, e.g., in the xz-plane, as in the case of a rotor loaded by self-weight. The presence of imaginary terms in the matrix of the coefficients of Eq. (24.43) results in a complex displacement vector $\mathbf{q}$. The presence of damping causes a lateral deviation of the deformed shape of the rotor from the plane in which all loads are acting. In the case of viscous damping, this deviation depends on the spin speed. Its measurement in a rotating–bending fatigue test has been suggested several times and was actually used to measure the internal damping of materials.

24.7   Critical-speed computation

Critical speeds can easily be computed from the homogeneous equation of motion (24.18), yielding free whirling, by adding the condition $\omega = \Omega$. The following eigenproblem is obtained:

$$\left(-\Omega^2(\mathbf{M} - \mathbf{G} - \mathbf{K}_\Omega) + \mathbf{K} \right)\mathbf{q} = \mathbf{0}. \tag{24.44}$$

Note that the size of the eigenproblem, in which only real quantities are involved, is equal to the number of complex degrees of freedom. Mathematically, the problem is that already seen for the computations of the natural frequencies of an undamped vibrating system whose mass matrix is $\mathbf{M} - \mathbf{G} - \mathbf{K}_\Omega$. The only difference is that in the current case the mass matrix can be nonpositive definite. All properties of eigenvectors seen for vibrating systems still hold because they were related only to the symmetry of the relevant matrices: If matrix $\mathbf{M} - \mathbf{G} - \mathbf{K}_\Omega$ is nonpositive definite, the only consequence is that some of the modal masses are negative and the corresponding eigenvalues are imaginary.[3] The eigenvectors are m-orthogonal only if complex coordinates are used: The real coordinates approach yields a skew-symmetrical gyroscopic matrix and makes it necessary to resort to a set of left and right eigenvectors.

In many cases excitations whose frequency is a multiple of the spin speed can be present. As will be seen in Chapter 25, often the excitation provided by constant forces on slightly nonsymmetrical rotors is accounted for by intersecting the Campbell diagram with the straight line $\omega = 2\Omega$, while the excitation due to lubricated journal bearings is studied using the straight

---

[3]For a detailed discussion of the meaning of the negative modal masses, see, for example, G. Genta and F. De Bona, "Unbalance response of rotors: A modal approach with some extensions to damped natural systems", *Journal of Sound and Vibrations*, 140(1), (1990), 129–153.

line $\omega = \Omega/2$. The secondary critical speeds occurring at the intersection of the generic straight line $\omega = h\Omega$ can be computed from the eigenproblem

$$\left(-\Omega^2(h^2\mathbf{M} - h\mathbf{G} - \mathbf{K}_\Omega) + \mathbf{K} \right)\mathbf{q}_0 = \mathbf{0}. \tag{24.45}$$

Also, in this case the use of complex coordinates allows the performance of the modal analysis of the system using mode shapes that are k- and m-orthogonal.

24.8 Unbalance response

The response to a generic unbalance distribution can be easily obtained by computing a particular integral of Eq. (24.18) or (24.19), which has the form

$$\mathbf{q} = \mathbf{q}_0 e^{i\Omega t}.$$

The algebraic equation yielding the response of the damped system is readily obtained

$$\left(-\Omega^2(\mathbf{M} - \mathbf{G} - \mathbf{K}_\Omega) + i\Omega\mathbf{C}_n + \mathbf{K} \right)\mathbf{q}_0 = \Omega^2\mathbf{f}. \tag{24.46}$$

In the case of undamped systems, Eq. (24.46) has no solution when the spin speed coincides with a critical speed and the matrix of the coefficients is singular. If vectors ϵ_i and angles χ_i are not all contained in the same plane, i.e., all phases α_i and β_i are not equal, vector $\mathbf{f}$ is complex and the solution $\mathbf{q}$ is also complex. Physically, this means that the deflected shape is a skew line. However, the matrix of the coefficients of Eq. (24.46) remains real and no actual working with complex numbers is needed. The real and imaginary parts of the solution can be obtained separately, solving two sets of real linear equations with the same matrix of the coefficients.

The computation of the response to an arbitrary unbalance distribution is formally similar to the computation of the response of a vibrating system excited by a harmonic forcing function, the differences being that the rotational speed Ω is used in the equation instead of the frequency ω, the excitation vector $\Omega^2\mathbf{f}$ is proportional to Ω^2, and the mass matrix of the system

$$\mathbf{M} - \mathbf{G} - \mathbf{K}_\Omega$$

can be nonpositive definite. Some of its eigenvalues may be negative, in terms of Ω^2, and then some of the critical speed can be imaginary, as was seen in detail in Section 24.1.3. The response in terms of displacement is more similar to an inertance than to a dynamic compliance.

When damping is not neglected, the response to unbalance is very similar to the response of a damped vibrating system to harmonic excitation, but only nonrotating damping enters the equation of motion. The matrix of the coefficients never becomes singular, because it has no real eigenvalue. It is, however, complex, and, if software for use with complex numbers is not available, the solution of Eq. (24.46) involves the solution of a set of equations whose size must be doubled in order to work with real numbers: The number of equations is then equal to the number of real coordinates, and using complex coordinates leads to little computational advantage.

The response to unbalance of an axi-symmetrical system could also be easily studied using rotating coordinates. In this case, the response is nothing other than the static response to forces that are constant in both modulus and direction, and the critical speeds are a sort of elastic instability condition, in which the stiffness matrix of the system becomes singular.

24.9   Campbell diagram and roots locus

To plot the Campbell diagram the whirl frequencies of the system must be computed as functions of the spin speed. If the stability of the system has to be studied, damping must also be taken into account. This does not make the problem conceptually more complicated but computations become much longer. In the case of viscous damping, the basic equation is the homogeneous equation associated with Eq. (24.19)

$$\mathbf{M}\ddot{\mathbf{q}} + (\mathbf{C}_n + \mathbf{C}_r - i\Omega\mathbf{G})\dot{\mathbf{q}} + (\mathbf{K} + \Omega^2\mathbf{K}_\Omega - i\Omega\mathbf{C}_r)\mathbf{q} = \mathbf{0}. \qquad (24.47)$$

The roots locus is directly obtained by computing the eigenvalues of the dynamic matrix (24.41); to take into account structural damping, it is sufficient to add to the stiffness matrix $\mathbf{K}$ the expression

$$\pm i\mathbf{K}_n'' \pm i\mathbf{K}_r'' \,,$$

where the double signs have the same meaning seen in Fig. 23.13.

The Campbell diagram is often referred to a solution of the type

$$\mathbf{q} = \mathbf{q}_0 e^{i\omega t}$$

instead of

$$\mathbf{q} = \mathbf{q}_0 e^{st} \,.$$

In this case, it is possible to find the complex whirl frequencies ω as eigenvalues of the matrix

$$\begin{bmatrix} \mathbf{M}^{-1}(\Omega\mathbf{G} + i\mathbf{C}_n + i\mathbf{C}_r) & \mathbf{M}^{-1}(\mathbf{K} + \Omega^2\mathbf{K}_\Omega \pm i\mathbf{K}_n'' \pm i\mathbf{K}_r'' - i\Omega\mathbf{C}_r) \\ \mathbf{I} & \mathbf{0} \end{bmatrix} . \qquad (24.48)$$

Operating in this way, a real eigenproblem must be solved in the case of undamped systems.

If the Campbell diagram has to be plotted by scanning the $\omega\Omega$-plane using m values of the spin speed Ω, an eigenproblem of order $2n$, where n

is the number of complex degrees of freedom, must be solved m times. The computation is thus time consuming, and large-scale condensation may be necessary to keep computer time within reasonable limits.

If the real-coordinate approach is used, the size of the problem doubles and the information on the direction of the whirl motion included in the eigenvalues is lost: To distinguish between forward and backward modes the eigenvectors must be studied.

In the case of the damped system, the solution of a complex eigenproblem requires a further doubling of the size of the relevant matrices, when no software for complex algebra is available. Moreover, if there is structural damping, several eigenproblems with the different signs present in Eq. (24.48) must be solved for each value of the spin speed. A maximum of four eigenproblems need to be solved, in the case where all forms of damping are present. After obtaining the solutions, there is no difficulty checking where the real part of the various solutions is located in the Campbell diagram and then in choosing which eigenvalues are to be discarded, following the scheme of Fig. 23.13. The presence of damping makes the computation of the Campbell diagram much heavier. In most cases, however, the presence of damping has little effect on the values of the whirl frequencies, while deeply affecting the decay rate. A good strategy is first studying the undamped system to locate the critical speeds and to define the general pattern of the whirl frequencies and then computing the complex whirl frequencies at some selected values of the speed, mainly in the supercritical range, where the occurrence of instability can be suspected.

In the case of undamped systems, a solution of the type

$$\mathbf{q} = \mathbf{q}_0 e^{i\omega t}$$

allows the obtaining of an eigenproblem with real matrices while assuming that

$$\mathbf{q} = \mathbf{q}_0 e^{st}$$

the dynamic matrix is complex. On the contrary, if damping is not neglected, the two notations lead to similar complexities and it may be expedient to use the second solution.

24.10   Acceleration of a torsionally stiff rotor

If the elastic behavior of the system is such that axial, torsional, and flexural behaviors are uncoupled, the same considerations seen in Section 23.10 for the simple model with 6 degrees of freedom also hold for more complex models. In particular, within the frame of the linearized theory, the axial degrees of freedom are uncoupled from flexural behavior. If the rotor is torsionally stiff, i.e., the torsional rotations of all cross sections are equal, and the acceleration is performed with an imposed law $\theta(t)$, the flexural behavior can also be studied independently, using equations of the type of the first two Eq. (24.8), the second Eq. (24.9), and Eq. (24.10). The only equation for the rotational degree of freedom is an equation of the type of the third Eq. (24.9). By introducing the usual complex coordinates, it follows that

$$
\begin{cases}
\mathbf{M\ddot{q}} + (\mathbf{C}_n + \mathbf{C}_r - i\Omega\mathbf{G})\mathbf{\dot{q}} + (\mathbf{K} - i\Omega\mathbf{C}_r)\mathbf{q} = (\Omega^2 - i\dot{\Omega})\mathbf{f}e^{i\theta}, \\[2mm]
M_z = J_{p_{tot}}\dot{\Omega} + \Im\left(\mathbf{\overline{f}}^T \mathbf{\ddot{q}}e^{-i\theta}\right),
\end{cases}
\tag{24.49}
$$

where centrifugal stiffening has been neglected and $J_{p_{tot}}$ is the total moment of inertia of the rotor about the z-axis, also taking into account static and couple unbalances. By introducing $\ddot{\theta} = 0$ and the solution for steady-state whirling into the last Eq. (24.49), the driving torque needed to maintain a constant angular velocity is readily obtained.

To perform the numerical integration in time of Eq. (24.49) it is easier to resort to rotating coordinates, as seen for the case of the Jeffcott rotor. The relevant equations of motion are thus

$$
\mathbf{M\ddot{r}} + \left[\mathbf{C}_n + \mathbf{C}_r - i\Omega(2\mathbf{M} - \mathbf{G})\right]\mathbf{\dot{r}} + \left[\mathbf{K} - \Omega^2(\mathbf{M} - \mathbf{G}) + \right.
$$
$$
\left. + i(\dot{\Omega}\mathbf{M} + \Omega\mathbf{C}_n)\right]\mathbf{r} = (\Omega^2 - i\dot{\Omega})\mathbf{f},
\tag{24.50}
$$
$$
M_z = J_{p_{tot}}\dot{\Omega} + \Im\left[\mathbf{\overline{f}}^T\left(\mathbf{\ddot{r}} + i\dot{\Omega}\mathbf{r} + 2i\Omega\mathbf{\dot{r}} - \Omega^2\mathbf{r}\right)\right].
$$

The rotating reference frame used for Eq. (24.50) rotates at a variable spin speed $\Omega(t)$, as defined by Eq. (23.47).

Example 24.2 *Study the dynamic behavior of the rotor of a small gas turbine sketched in Fig. 24.6a using both the FEM and the Myklestadt–Prohl method. The finite element model, which can be used for the lumped parameters approach, is sketched in Fig. 24.6b. It consists of 10 beam and 2 mass elements. The compressor and turbine rotors can be considered rigid bodies; the stiffness of beam elements 2,3,7, and 8 is thus large (large diameter) and their mass is zero (vanishing material density) because the relevant mass is introduced as concentrated mass elements at nodes 3 and 8. The main characteristics of the elements are as follows:*

El. number	1	2	3	4	5	6	7	8	9	10
ϕ_i [mm]	15	0	0	55	60	60	0	0	30	20
ϕ_o [mm]	30	120	120	63	68	68	150	150	40	30
l [mm]	30	68.3	34.2	30	88	88	25	30	40	20
ρ [kg/m²]	7810	0	0	7810	7810	7810	0	0	7810	7810
E [MN/m²]					210 000					

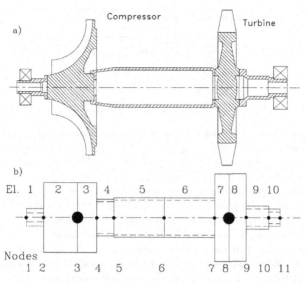

FIGURE 24.6. Sketch of the rotor of a small gas turbine rotor (**a**) and model for dynamic analysis (**b**).

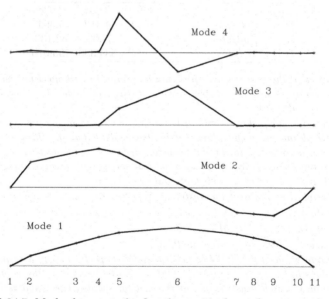

FIGURE 24.7. Mode shapes at the first four critical speeds computed using the model with 20 degrees of freedom.

The characteristics of the mass elements are as follows:

Element number	1	2
Node number	3	8
m [kg]	20.81	18.2
J_p [kgm^2]	0.285	0.269
J_t [kgm^2]	0.174	0.142

The bearings, located at nodes 1 and 10, are modeled as rigid supports. If, in the construction of the global stiffness matrix, they must be considered elastic bearings, a very high value of the stiffness, $k = 10^{15}$ N/m, can be used.

Solution using the FEM – consistent mass matrices.

If the bearings are modeled as very stiff elastic constraints, the total number of complex degrees of freedom of the model is 22; there are only 20 degrees of freedom if the constraints are considered rigid. The critical speeds are computed using all degrees of freedom and through Guyan reduction.

In the latter case two condensation schemes are considered; a very strong reduction in which only 2 degrees of freedom are retained (translations at nodes 3 and 8) and a more detailed scheme, in which 6 master degrees of freedom are considered (translations at nodes 3, 4, 6, 7, 8, and 9). The results are reported in the following table:

Master degrees of freedom	2	6	20
Ω_{cr_I} [rad/s]	2,080	2,032	2,028
$\Omega_{cr_{II}}$ [rad/s]	4,445	4,312	4,304
$\Omega_{cr_{III}}$ [rad/s]	–	34,062	32,130
$\Omega_{cr_{IV}}$ [rad/s]	–	534,380	795,010

It is clear that the first critical speed can be computed with good precision even with the simplest model, and its precision is still acceptable when searching the second critical speed.

The first four mode shapes obtained from the complete model are reported in Fig. 24.7. From the mode shapes, it is also evident that the third and fourth modes are mainly due to the deformation of the shaft connecting the turbine to the compressor: To compute such critical speeds with greater accuracy, the use of a finer model in the relevant zones is advisable.

The Campbell diagram of the system is reported in Fig. 24.8a. The intermediate reduction scheme has been used, since the diagram extends to speeds in excess of the second critical speed.

Solution with Myklestadt–Prohl method.

The computation of the first two critical speeds is performed by assuming various values of the speed from 0 to 5,000 rad/s and computing the value of the determinant, which must vanish at the critical speeds. The result is reported in Fig. 24.8b. The values of the first two critical speeds are

$$\Omega_{cr_I} = 1,888 \ rad/s \ and \ \Omega_{cr_{II}} = 3,833 \ rad/s \ .$$

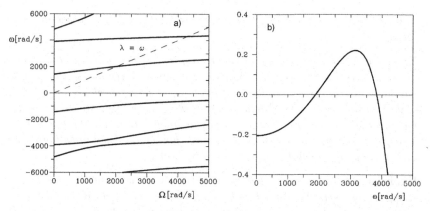

FIGURE 24.8. **(a)**: Campbell diagram of the rotor in Figure 24.6. **(b)**: Graph for the computation of the critical speeds of the same system, using Myklestadt-Prohl method.

The results so obtained are lower than those previously computed. This is consistent with the consideration that the computation has been performed by lumping the mass of the shaft at the nodes and neglecting the gyroscopic effects of the shafts. Note that higher values would have been obtained if the shear deformation of the shaft were neglected:

$$\Omega_{cr_I} = 2,115 \ rad/s \ and \ \Omega_{cr_{II}} = 4,919 \ rad/s \ .$$

Shear deformation is important in this case because the shaft is not slender and the results obtainable from a Euler–Bernoulli approach can be affected by large errors.

24.11 Exercises

Exercise 24.1 *Consider the transmission shaft of Exercise 15.5. Plot the Campbell diagram and compute the critical speeds.*

Exercise 24.2 *A flywheel must store 1 kW h at 17,000 rpm, with a working range between 8,500 and 17,000 rpm. Knowing that the energy density at the maximum speed of the rotor is 8 W h/kg and that the transversal moment of inertia is equal to 65% of the polar moment of inertia, compute its mass and moments of inertia.*

The flywheel runs on a pair of ball bearings. Knowing that the span between the bearings is 400 mm, the center of mass is at 30% of the span, and the stiffness of each bearing is 10^8 N/m, compute the Campbell diagram and the critical speeds. If a critical speed falls within a range spanning between 60% of the minimum operating speed and 130% of the maximum operating speed, change the stiffness *of the bearings in such a way that it is moved either below or above the stated range.*

Exercise 24.3 *Compute the unbalance response of the system of the previous exercise. If two solutions were obtained, compute both responses. Compute the unbalance mε needed to keep the total force due to static unbalance on the stator of the machine within 200 N in the whole working range (compute the forces from the amplitude of the orbits at the bearing locations). Compute the forces due to a couple unbalance due to two opposite static unbalances equal to half of the one earlier computed, applied in the planes of the bearings.*

Exercise 24.4 *Assume that the bearings of the flywheel of the preceding exercise are characterized by a loss factor $\eta = 0.08$. Compute the unbalance response from standstill to the maximum operating speed with the static unbalance computed and, if a critical speed must be passed, the maximum forces exerted on the stator. Compute the driving torque and power needed to overcome the drag due to nonrotating damping and the minimum torque and power needed to avoid stalling when crossing the critical speed.*

Exercise 24.5 *The flywheel of the preceding exercise is operated by an electric motor whose maximum power is 20 kW. Assuming that the controller maintains constant the driving torque during acceleration from standstill, compute the time needed to reach the maximum operating speed, neglecting the flexural behavior of the rotor. By integrating numerically in time the equations of motion, compute the unbalance response during an acceleration with the constant rate computed earlier. Compute the driving torque actually needed to follow the stated velocity time history and compare it with the torque supplied by the motor.*

Exercise 24.6 *Compute the critical speeds and plot the Campbell diagram of a Stodola–Green rotor (a rigid body attached at the end of a prismatic beam which is clamped at the other end) using both the Jeffcott rotor model and the model with 2 complex degrees of freedom.*
    Data. Beam: outer diameter $d_o = 40$ mm, inner diameter $d_i = 30$ mm, length $l = 500$ mm, and Young's modulus $E = 2.1 \times 10^{11}$ N/m^2. Inertial properties: $m = 10$ kg, $J_p = 1$ kgm^2, and $J_t = 0.6$ kgm^2.

Exercise 24.7 *Consider the small gas turbine of Example 24.2. Study the dependence of the first two critical speeds with the bearing stiffness, in the field $10^5 < k < 10^{10}$ N/m.*

Exercise 24.8 *Consider again the same small turbine of the previous exercise, assuming that the stiffness of the bearings is $k = 10^8$ N/m. Assume that the material constituting the shaft has a loss factor $\eta_r = 0.01$, while the loss factor of the bearings is $\eta_n = 0.02$. Compute the Campbell diagram, the decay rate plot and the roots locus. If an instability range is found, compute the threshold of instability.*
    Repeat the analysis assuming that the loss factor of the bearings is $\eta_n = 0.005$.

25

Nonisotropic Rotating Machines

The rotors studied in the previous chapters were assumed to be axially symmetrical. In the present chapter this assumptions will be dropped, and more general models will be introduced to understand the effects of the lack of axial symmetry of the stator, the rotor, or both.

The assumption of axial symmetry of the whole machine is usually not verified exactly in real life, and sometimes an axi-symmetrical model can be only a very rough approximation of the actual system.

If a rotating machine is not isotropic about the rotation axis, the deviation from symmetry can concern only the rotor, only the stator, or both. In the first two cases, the model is not exceedingly complicated, but in the third case no exact solution of the relevant equations of motion can be found.

In some cases the system displays a nonisotropic behavior, even if all the parts of the machine are geometrically axi-symmetrical. This occurs particularly when the rotor runs on lubricated journal bearings: Under the effect of external forces the journal takes an off-center position within the bearing and reacts in different ways to the forces acting in the various planes including the rotation axis.

In the following sections the study of unsymmetrical rotors will be dealt with in subsequent steps. At first a very simple configuration, based on the Jeffcott rotor, will be studied. After the relevant phenomena have been qualitatively understood using this simplified model, a more complete study, allowing quantitative results to be obtained even for complex systems, will be expounded.

25.1   Jeffcott rotor on nonisotropic supports

Consider the case of the rotor in Fig. 23.3b but assume that the stiffness of the supports is not isotropic in the xy-plane. All other assumptions made in Section 23.5.1, in particular the linearity of the system and the assimilation of the rotor to a point mass, will be retained. The motion will be studied in the xy-plane. In this plane the polar diagram of the stiffness of the supports is an ellipse, the so-called *ellipse of elasticity*.

Without any loss of generality, axes x and y will be assumed to coincide with the axes of the ellipse of elasticity, i.e., to be the principal axes of elasticity of the supporting structure and the stiffness along the x-direction to be lower than that along the y-direction. The elastic reaction of the shaft in this case is

$$F_x = -k_x x, \qquad\qquad F_y = -k_y y. \tag{25.1}$$

By introducing the two different values of the stiffness into the equation of motion (23.9), the latter transforms into

$$\begin{cases} m\ddot{x} + k_x x = m\epsilon[\dot{\theta}^2 \cos(\theta) + \ddot{\theta}\sin(\theta)], \\ m\ddot{y} + k_y y = m\epsilon[\dot{\theta}^2 \sin(\theta) - \ddot{\theta}\cos(\theta)]. \end{cases} \tag{25.2}$$

In the case of constant spin speed $\dot{\theta} = \Omega$, Eq. (25.2) with $\ddot{\theta} = 0$ still holds. From the homogeneous equation, it is clear that there are two natural frequencies, one (the lower) related to the motion in the xz-plane and the other related to the motion in the yz-plane:

$$\omega_{n_1} = \sqrt{\frac{k_x}{m}}, \qquad \omega_{n_2} = \sqrt{\frac{k_y}{m}}. \tag{25.3}$$

They are not influenced by the spin speed, and then the Campbell diagram is made by two straight lines.

Remark 25.1 *The motions in the two planes occur at different frequencies, so the two harmonic motions cannot combine to make circles or ellipses.*

Remark 25.2 *The fact that the two natural frequencies are independent of the spin speed causes the two critical speeds to coincide with the natural frequencies:* $\Omega_{cr_1} = \omega_{n_1}$ *and* $\Omega_{cr_2} = \omega_{n_2}$.

At the first critical speed the motion reduces to a straight vibration along the x-axis, and at the other critical speed it reduces to a straight motion along the y-axis.

The unbalance response can be obtained directly from Eq. (25.2) with $\ddot{\theta} = 0$. The response is in each plane equal to the response of the Jeffcott rotor, computed using the stiffness related to that plane

$$x_0 = \frac{m\epsilon\Omega^2}{k_x - m\Omega^2} \quad , \quad y_0 = \frac{m\epsilon\Omega^2}{k_y - m\Omega^2} \,. \qquad (25.4)$$

The unbalance response can be subdivided into three speed ranges (Fig. 25.1):

- From standstill to the first critical speed: The responses in the two planes have the same sign and are out of phase from each other by 90°, as clearly seen from Eq. (25.2), where the excitation is expressed by a sine and a cosine function. The orbit grows mainly along the x-axis and has the shape of an elongated ellipse. Approaching the first critical speed, the axis of the orbit along the x-axis tends to infinity.
- From the first to the second critical speed: The response along the x-axis is negative, having already crossed the critical speed; that along

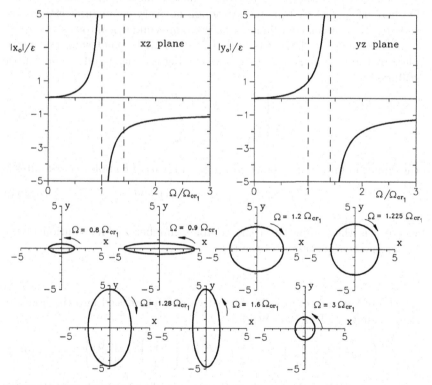

FIGURE 25.1. Unbalance response of a Jeffcott rotor on anisotropic supports. Amplitude of the motion along the x - and y-axes as a function of the speed and orbits at some selected values of Ω; ratio between the stiffnesses in the two planes $\alpha = k_y/k_x = 2$.

the y-axis is positive. When they combine, they give way to an elliptical motion in the backward direction. Close to the first critical speed the ellipse is elongated along the x-axis, and close to the second it is elongated in the other direction. There is an intermediate speed at which the amplitudes in the two planes are equal and the orbit is circular. Unbalance, an excitation that by definition is applied in the forward direction, can thus excite a backward synchronous whirling of the rotor.

- Above the second critical speed: The amplitudes in both planes are of the same sign, and they both tend to the same value, namely, to $-\epsilon$. An elliptic forward whirling that tends to become circular with increasing speed is so obtained; self-centering takes place normally. It could easily be expected that in the high supercritical field the elastic anisotropy has little influence on the dynamic behavior of the rotor: The behavior of the system is dominated by inertia forces that are clearly isotropic.

The behavior of the Jeffcott rotor on anisotropic supports has been studied here using real coordinates. Although not very common, it is also possible to use complex coordinates to study nonsymmetrical rotors. As usual, it is possible to add the first Eq. (25.2) to the second equation multiplied by the imaginary unit i. By introducing the mean and the deviatoric stiffness

$$\begin{cases} k_m = \dfrac{1}{2}(k_x + k_y) \\[2mm] k_d = \dfrac{1}{2}(k_x - k_y) \,, \end{cases} \tag{25.5}$$

the equation for the unbalance response in terms of complex coordinates is

$$m\ddot{z} + k_m z + k_d \bar{z} = m\epsilon\Omega^2 e^{i\Omega t} \,, \tag{25.6}$$

where $\bar{z}$ is the conjugate of the complex number z. The solution of the homogeneous equation is of the type

$$z = z_1 e^{i\omega t} + z_2 e^{-i\bar{\omega} t} \,,$$

which gives way to elliptical orbits. It can be introduced into the homogeneous equation of motion, yielding

$$\left(-\omega^2 \begin{bmatrix} m & 0 \\ 0 & m \end{bmatrix} + \begin{bmatrix} k_m & k_d \\ k_d & k_m \end{bmatrix} \right) \begin{Bmatrix} z_1 \\ \bar{z}_2 \end{Bmatrix} = \begin{Bmatrix} 0 \\ 0 \end{Bmatrix} \,. \tag{25.7}$$

Equation (25.7) can easily be solved in ω, yielding the values of the whirl frequencies coinciding with those expressed by Eq. (25.3).

The particular integral that allows the unbalance response to be computed is

$$z = z_1 e^{i\Omega t} + z_2 e^{-i\Omega t} .$$

By introducing it into the equation of motion and solving for the amplitudes of the forward and backward components z_1 and z_2, the following unbalance response is obtained:

$$z = \frac{m\epsilon\Omega^2}{(k_x - m\Omega^2)(k_y - m\Omega^2)} \left[(k_m - m\Omega^2)e^{i\Omega t} - k_d e^{-i\Omega t} \right] . \qquad (25.8)$$

It is easy to demonstrate that the orbits expressed by Eq. (25.8) coincide with those expressed by Eq. (25.4). The possibility of backward whirling to be excited by unbalance is, however, more obvious when complex coordinates are used: At the speed

$$\Omega_b = \sqrt{\frac{k_m}{m}}$$

the amplitude of the forward component vanishes, and the orbit is a circular backward whirl with amplitude

$$z_{0_b} = \epsilon \frac{k_d}{k_m} .$$

If the system is damped, the field in which backward whirling occurs is reduced and can, if damping is large enough, disappear altogether. When reaching the first critical speed, the amplitude of the elliptical orbits remains limited while the axis of the ellipse is not exactly aligned with one of the axes of elasticity of the supports. The orbit starts then to rotate in the xy-plane and to become thinner: if at a certain speed it reduces to a line, a reversal of the direction of whirling occurs.

The elliptical backward orbit then continues its rotation, with increasing and then decreasing width, until it again becomes a straight line and a new reversal of the whirling direction takes place. When forward whirling occurs, the direction of the larger axis of the ellipse has a direction close to the other axis of elasticity of the supports and the speed is close to the second critical speed.

If damping is large enough, the orbit never reduces to a line and the reversal of its direction does not occur.

The unbalance response of a Jeffcott rotor with rotating and nonrotating damping ratio $\zeta_r = 0.01$ and $\zeta_n = 0.1$ is shown in Fig. 25.2. With these values of the damping ratios, backward whirling actually occurs. Due to the complexity of the behavior, a new representation is used: The orbits are plotted in a tridimensional graph, stacked along the spin-speed axis. It was proposed to designate this representation as an orbital tube.[1] The tube starts as a point at zero speed and then enlarges, taking an elliptical

---

[1]C. Delprete, G. Genta, S. Carabelli, *Orbital tubes*, Meccanica, 32, 1997, pp. 485–492.

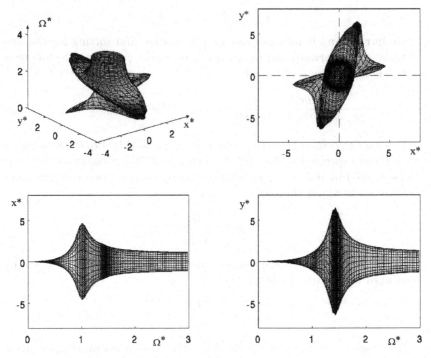

FIGURE 25.2. Nondimensional unbalance response of a damped Jeffcott rotor on anisotropic supports; (**a**) orbital tube, i.e., representation of the orbits at different speeds as a tri-dimensional plot; (**b**) orbital view; (**c**) and (**d**) projections on the Ωx- and Ωy-planes, respectively.

cross-section. At very high speed, due to self-centering, it tends to a circular cylinder with radius equal to the eccentricity.

The projection on the xy-plane (orbital view) directly gives the orbits at various speeds, superimposed on each other. The projections on the Ωx- and Ωy-planes give the peak-to-peak amplitude as a function of the spin speed.

25.2   Nonisotropic Jeffcott rotor

Now consider a Jeffcott rotor of the type shown in Fig. 23.3a, in which the stiffness of the shaft is not isotropic. The polar diagram of the stiffness is now an ellipse whose axes can be assumed, without loss of generality, to lie along the ξ- and η-axes. It is then possible to write an equation similar to Eq. (25.1) but referred to the $O\xi\eta z$-frame:

$$F_\xi = -k_\xi \xi, \qquad F_\eta = -k_\eta \eta.$$

When dealing with rotating asymmetry, it is advisable to write the equation of motion with reference to the rotating frame. By introducing the mean and deviatoric stiffness, defined as in the preceding section but with reference to the $O\xi\eta z$-frame and neglecting damping, the equation of motion (23.38) becomes

$$m\ddot{\zeta} + 2i\Omega m\dot{\zeta} - m\Omega^2\zeta + k_m\zeta + k_d\bar{\zeta} = m\epsilon\Omega^2 e^{i\alpha}, \qquad (25.9)$$

where α is the angle between the ξ-axis and the direction of the unbalance vector in the $\xi\eta$-plane (Fig. 24.1b). Using real coordinates, the same equation transforms into

$$\begin{bmatrix} m & 0 \\ 0 & m \end{bmatrix} \left\{ \begin{matrix} \ddot{\xi} \\ \ddot{\eta} \end{matrix} \right\} + \begin{bmatrix} 0 & -2m\Omega \\ 2m\Omega & 0 \end{bmatrix} \left\{ \begin{matrix} \dot{\xi} \\ \dot{\eta} \end{matrix} \right\} + \qquad (25.10)$$

$$+ \begin{bmatrix} k_\xi - m\Omega^2 & 0 \\ 0 & k_\eta - m\Omega^2 \end{bmatrix} \left\{ \begin{matrix} \xi \\ \eta \end{matrix} \right\} = m\epsilon\Omega^2 \left\{ \begin{matrix} \cos(\alpha) \\ \sin(\alpha) \end{matrix} \right\}.$$

The unbalance response is easily obtained as a steady-state solution of Eq. (25.10)

$$\xi = \frac{m\epsilon\Omega^2 \cos(\alpha)}{k_\xi - m\Omega^2}, \qquad \eta = \frac{m\epsilon\Omega^2 \sin(\alpha)}{k_\eta - m\Omega^2}, \qquad (25.11)$$

which represents, in a fixed reference frame, a circular whirling. The denominators of Eq. (25.11) vanish for two values of the spin speed, the critical speeds of the system

$$\Omega_{cr_I} = \sqrt{\frac{k_\xi}{m}}, \qquad \Omega_{cr_{II}} = \sqrt{\frac{k_\eta}{m}}. \qquad (25.12)$$

The free whirling of the system can easily be obtained from the homogeneous equation (25.9) or (25.10), being immaterial whether real or complex coordinates are used. In the first case the solution of the homogeneous equation (25.10) is

$$\xi = \xi_0 e^{i\omega' t}, \qquad \eta = \eta_0 e^{i\omega' t},$$

and ω' is a complex whirl speed in the $\xi\eta$-plane. It does not coincide with the whirl speed ω in the xy-plane but is linked with the latter by the obvious relationships

$$\Re(\omega) = \Re(\omega') + \Omega, \qquad \Im(\omega) = \Im(\omega').$$

The homogeneous equation (25.10) then yields an eigenproblem in ω'

$$\begin{bmatrix} \Omega_{cr_I}^2 - \omega'^2 - \Omega^2 & -2i\Omega\omega' \\ 2i\Omega\omega' & \Omega_{cr_{II}}^2 - \omega'^2 - \Omega^2 \end{bmatrix} \left\{ \begin{matrix} \xi_0 \\ \eta_0 \end{matrix} \right\} = \mathbf{0}. \qquad (25.13)$$

By introducing the nondimensional speeds and the stiffness ratio

$$\omega'^* = \frac{\omega'}{\Omega_{cr_I}} \ , \quad \omega^* = \frac{\omega}{\Omega_{cr_I}} \ , \quad \Omega^* = \frac{\Omega}{\Omega_{cr_I}} \ , \quad \alpha^* = \frac{k_\eta}{k_\xi} \ ,$$

the characteristic equation can be written in nondimensional form

$$\omega'^{*^4} - \omega'^{*^2}(1 + \alpha^* + 2\Omega^{*^2}) + (1 - \Omega^{*^2})(\alpha^* - \Omega^{*^2}) = 0 \ . \qquad (25.14)$$

The stiffness ratio α^* is greater than unity if $k_\xi < k_\eta$. By first solving Eq. (25.14) in ω'^{*^2} it follows that

$$\omega'^{*^2} = \Omega^{*^2} + \frac{1 + \alpha^*}{2} \pm \sqrt{2\Omega^{*^2}(1 + \alpha^*) + \frac{(1 - \alpha^*)^2}{4}} \ . \qquad (25.15)$$

The expression under the square root in Eq. (25.15) is always positive: The two solutions for ω'^{*^2} are then always real. The one with the upper sign $(+)$ is always positive and yields two real solutions in ω'^*, one positive and one negative. The solution with the lower sign $(-)$ is positive only if

$$\Omega^{*^4} - \Omega^{*^2}(1 + \alpha^*) + \alpha^* > 0 \ .$$

As α^* was assumed to be greater than unity, the last condition can be written in the form

$$\Omega^* < 1 \ ; \quad \alpha^* < \Omega^* \ , \text{ i.e.,} \quad \Omega < \sqrt{\frac{k_\xi}{m}} \ ; \quad \sqrt{\frac{k_\eta}{m}} < \Omega \ . \qquad (25.16)$$

If condition (25.16) is satisfied, the characteristic equation (25.14) has four real roots – two positive and two negative. The behavior of the system is stable. If, however, the value of the spin speed Ω lies between the two critical speeds of the system, as shown by condition (25.16), two imaginary roots are found. Because the equation of motion has real coefficients, the solutions are conjugate numbers, or, if they have vanishing real parts, they have opposite signs. A negative imaginary solution then exists, which corresponds to an unstable behavior of the system, with amplitude growing indefinitely with exponential law. The presence of an elastic anisotropy of the rotating parts of the system causes the occurrence of an instability range that spans from the lowest to the highest critical speed.

The Campbell diagram $\omega(\Omega)$ for a system with

$$\alpha^* = \frac{k_\eta}{k_\xi} = 2$$

obtained from Eq. (25.15) is plotted in Fig. 25.3. All four quadrants of the ω, Ω-plane have been reported, even if just two of them, the first and fourth, for example, give a complete picture of the situation.

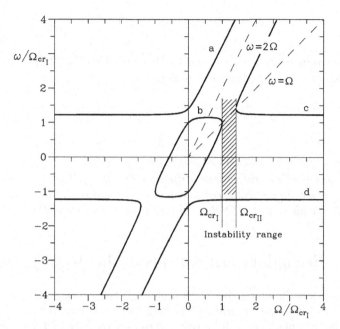

FIGURE 25.3. Campbell diagram for an anisotropic Jeffcott rotor; ratio between the stiffnesses in the two planes $\alpha^* = k_\eta/k_\xi = 2$.

The same conclusions already seen can be drawn from the figure. At low speed, up to the first critical speed, there are four solutions. Those on branches a and d of the curve come from the solution (25.15) with the upper sign $(+)$ and are a forward and a backward whirl. Those on branch b come from the solution with the lower sign $(-)$ and are one backward and one forward or both forward, depending on the value of Ω. At high speed, above the second critical speed, the situation is similar, the difference being that the solutions coming from the expression with the lower sign $(-)$ are both forward motions and lie on branch c of the curve. If the value of the speed lies in the instability range spanning between the critical speeds, there are two real solutions, on branches a and d of the curve, and two imaginary ones, not reported in the figure. The negative imaginary solution causes the behavior of the system to become unstable.

Rotating damping, which can cause the system to become unstable at high speed, reduces the instability range between the critical speeds. A similar effect is due to nonrotating damping, which, however, is stabilizing at any speed. If damping is high enough, the instability range gets smaller and can disappear altogether. The homogeneous equation of motion of the damped system in the rotating frame is

$$\begin{bmatrix} m & 0 \\ 0 & m \end{bmatrix} \left\{ \begin{array}{c} \ddot{\xi} \\ \ddot{\eta} \end{array} \right\} + \begin{bmatrix} c_n + c_r & -2m\Omega \\ 2m\Omega & c_n + c_r \end{bmatrix} \left\{ \begin{array}{c} \dot{\xi} \\ \dot{\eta} \end{array} \right\} + \qquad (25.17)$$

$$+ \begin{bmatrix} k_\xi - m\Omega^2 & -\Omega c_n \\ \Omega c_n & k_\eta - m\Omega^2 \end{bmatrix} \begin{Bmatrix} \xi \\ \eta \end{Bmatrix} = \{0\} \, .$$

By introducing the solution for free whirling into Eq. (25.17), the following characteristic equation is obtained

$$\det \begin{bmatrix} -m\omega'^2 + i\omega'(c_n + c_r) - m\Omega^2 + k_\xi & -m\Omega(2i\omega' + c_n) \\ m\Omega(2i\omega' + c_n) & -m\omega'^2 + i\omega'(c_n + c_r) - m\Omega^2 + k_\eta \end{bmatrix} = 0 \, .$$
$$(25.18)$$

Remark 25.3 *The system of Fig. 23.12 can be considered a limiting case of an asymmetrical rotor: The stiffness along the η-axis is infinitely high, causing a second critical speed that tends to infinity. The field of instability extends for all values of Ω that are above the critical speed.*

25.3   Secondary critical speeds due to rotor weight

All conditions in which there is resonance between one of the natural frequencies of the system and an exciting force different from the rotating force due to unbalance will be referred to as *secondary critical speeds*. It is well known that when constant bending forces, such as the self-weight of a rotor whose axis is horizontal, act on the rotor, the critical speeds and the Campbell diagram are not influenced by the presence of such forces. Whirling takes place about the deflected configuration of the rotor, but, due to linearity, the two effects, namely, static bending and whirling, do not interact. It is, however, well known that the weight of a rotor with a horizontal axis can cause the occurrence of secondary critical speeds, whose values are about half of those of primary critical speeds, or, more exactly, are located at the intersections on the Campbell diagram of the curves for free whirling with the straight line $\omega = 2\Omega$. The presence of these secondary critical speeds is linked to the deviations from a perfect axial symmetry of the rotor.

In the case of the Jeffcott rotor studied in the preceding section, these secondary critical speeds are easily deduced from Fig. 25.3. In the figure an intersection of branch b of the curve with the Ω-axis is clearly visible. The system has, at a well-determined speed, a natural frequency that is vanishingly small and then, at that speed, a sort of resonance with a static force, i.e., with a force constant in modulus and direction, is possible.

The phenomenon may be easier to understand with reference to the rotating frame $O\xi\eta z$, in which condition $\omega = 0$ is seen as $\omega' = -\Omega$. A constant force in the xy-plane, as self-weight of a horizontal rotor, is seen in the $\xi\eta$-plane as a force rotating with speed $-\Omega$, which can cause resonance when the natural frequency of the system has the same frequency.

The same phenomenon can also be seen in a different way. When the stiffness of the shaft is not isotropic in the $\xi\eta$-plane, its polar diagram is an ellipse, the ellipse of elasticity. The functions of time expressing the stiffnesses $k_x(t)$ and $k_y(t)$ are periodic in time, if the angular speed Ω is constant, and their period is equal to half a revolution (frequency equal to 2Ω). The conditions for resonance occur when the curves on the Campbell diagram intersect the straight line $\omega = 2\Omega$.

The two ways of seeing the same phenomenon are equivalent: From Fig. 25.3 it is clear that at the same value of the speed at which a branch of the curve intersects the Ω-axis, another intersects the line $\omega = 2\Omega$. From the characteristic equation, if

$$\Re(\omega) = \Re(\omega') + \Omega$$

is the real part of a solution for free whirling,

$$\Re(\omega) = -\Re(\omega') + \Omega$$

is also the real part of another solution. The value of the secondary critical speed for the Jeffcott rotor of Fig. 25.3 can be easily computed by introducing the condition $\omega' = -\Omega$ into Eq. (25.14) and solving it in Ω. It follows that

$$\Omega^* = \sqrt{\frac{\alpha^*}{2(1+\alpha^*)}} \text{ , i.e., } \qquad \Omega^* = \sqrt{\frac{k_\xi k_\eta}{2(k_\xi + k_\eta)}} \,. \qquad (25.19)$$

If k_ξ and k_η tend to a single value k, $\alpha^* \to 1$, the value of the secondary critical speed tends to

$$\Omega_{cr_s} = \frac{1}{2}\sqrt{\frac{k}{m}} = \frac{1}{2}\Omega_{cr} \,. \qquad (25.20)$$

Remark 25.4 *Equation (25.20) holds only in the case of a Jeffcott rotor; in all other cases the secondary critical speeds can be found at the intersections of the curve $\omega(\Omega)$ with the line $\omega = 2\Omega$.*

Remark 25.5 *All secondary critical speeds characterized by the condition $\omega > \Omega$, as is the case for those excited by self-weight, occur in the subcritical region and usually cannot trigger unstable behavior. The internal damping of the rotor is in these conditions stabilizing and, generally speaking, no unstable behavior is expected in the subcritical region.*

Also inertial anisotropy, i.e., difference in the moments of inertia about transversal baricentrical axes, has effects similar to those seen for elastic anisotropy.

25.4 Equation of motion for an anisotropic machine with many degrees of freedom

Consider a beam element of the type studied in Section 15.2, for which all the assumptions for uncoupling among flexural, torsional, and axial behavior hold. If the principal axes of inertia and elasticity lie in the xz-and yz-planes (hereafter designated by subscripts x and y), the mass and stiffness matrices related to flexural behavior can be written in the form

$$\mathbf{M} = \begin{bmatrix} \mathbf{M}_x & \mathbf{0} \\ \mathbf{0} & \mathbf{M}_y \end{bmatrix}, \qquad \mathbf{K} = \begin{bmatrix} \mathbf{K}_x & \mathbf{0} \\ \mathbf{0} & \mathbf{K}_y \end{bmatrix}, \qquad (25.21)$$

where matrices related to the two bending planes are shown in Section 15.2. In the following analytical development, the generalized coordinate for rotation in the yz-plane will be $-\phi_x$ instead of ϕ_x, in such a way that matrices related to the xz- and yz-planes are equal if an element is axially symmetrical. In this way, the introduction of complex coordinates will be straightforward.

When assembling the structure, assume that the global reference frame has the same z-axis as those of each element, but that the x-axes of the elements are rotated by an angle α with respect to the global reference frame. The rotation matrix is

$$\mathbf{R'} = \begin{bmatrix} \cos(\alpha)\,\mathbf{I} & \sin(\alpha)\,\mathbf{I} \\ -\sin(\alpha)\,\mathbf{I} & \cos(\alpha)\,\mathbf{I} \end{bmatrix}, \qquad (25.22)$$

and the stiffness matrix in the global reference frame is

$$\mathbf{K}_g = \begin{bmatrix} \cos^2(\alpha)\mathbf{K}_x + \sin^2(\alpha)\mathbf{K}_y & \sin(\alpha)\cos(\alpha)(\mathbf{K}_x - \mathbf{K}_y) \\ \sin(\alpha)\cos(\alpha)(\mathbf{K}_x - \mathbf{K}_y) & \sin^2(\alpha)\mathbf{K}_x + \cos^2(\alpha)\mathbf{K}_y \end{bmatrix}. \qquad (25.23)$$

By introducing the mean and deviatoric stiffness matrices of the elements

$$\mathbf{K}_m = \frac{1}{2}(\mathbf{K}_x + \mathbf{K}_y), \qquad \mathbf{K}_d = \frac{1}{2}(\mathbf{K}_x - \mathbf{K}_y),$$

Eq. (25.23) can be written as

$$\mathbf{K}_g = \begin{bmatrix} \mathbf{K}_m + \mathbf{K}_d \cos(2\alpha) & \mathbf{K}_d \sin(2\alpha) \\ \mathbf{K}_d \sin(2\alpha) & \mathbf{K}_m - \mathbf{K}_d \cos(2\alpha) \end{bmatrix}. \qquad (25.24)$$

If the element belongs to the rotor, angle α must be substituted by $\alpha + \theta$, or, in the case of constant spin speed Ω, by $\alpha + \Omega t$.

All the aforementioned considerations hold for mass and damping matrices and for elements other than beam elements. Note that, in general, all mean and deviatoric matrices of structural elements are symmetrical, with the exception of the stiffness and damping matrices of the elements used to model lubricated journal bearings in linearized theories (see Section 4.10). Once all matrices of the elements have been obtained and expressed in the global reference frame, it is possible to assemble the various elements to obtain the matrices related to the whole structure. Obviously, the rotating elements must be assembled separately from the nonrotating elements.

Due to the presence of the deviatoric matrices (it is sufficient that a single element has a nonvanishing deviatoric matrix), the structure of the assembled matrices is more complex than that of Eq. (25.21). For the stiffness matrix it follows that

$$\mathbf{K} = \begin{bmatrix} \mathbf{K}_x & \mathbf{K}_{xy} \\ \mathbf{K}_{yx} & \mathbf{K}_y \end{bmatrix} . \tag{25.25}$$

Due to the presence of the coupling terms with xy- and yx-subscripts, a new definition of the mean and deviatoric matrices for the whole structure is needed

$$\begin{cases} \mathbf{K}_m = \dfrac{1}{2}(\mathbf{K}_x + \mathbf{K}_y) + i\dfrac{1}{2}(\mathbf{K}_{yx} - \mathbf{K}_{xy}) \\[3mm] \mathbf{K}_d = \dfrac{1}{2}(\mathbf{K}_x - \mathbf{K}_y) + i\dfrac{1}{2}(\mathbf{K}_{yx} + \mathbf{K}_{xy}) . \end{cases} \tag{25.26}$$

Note that, except in the mentioned case of elements used for the linearized modeling of hydrodynamic bearings, the matrices with subscripts xy and yx are equal and the mean matrices are real. On the contrary, deviatoric matrices are, in general, complex. Using the complex-coordinate approach and the definitions of mean and deviatoric matrices given by Eq. (25.26), the equation of motion describing the flexural behavior of a general system containing stationary elements and elements rotating at constant spin speed Ω can be shown to be[2]

$$\mathbf{M}_m\ddot{\mathbf{q}} + (\mathbf{C}_m - i\Omega\mathbf{G})\dot{\mathbf{q}} + (\mathbf{K}_m - i\Omega\mathbf{C}_{r_m})\mathbf{q} + \mathbf{M}_{n_d}\ddot{\overline{\mathbf{q}}} +$$

$$+\mathbf{M}_{r_d}e^{2i\Omega t}(\ddot{\overline{\mathbf{q}}} + 2i\Omega\dot{\overline{\mathbf{q}}}) + \mathbf{C}_{n_d}\dot{\overline{\mathbf{q}}} + \mathbf{C}_{r_d}e^{2i\Omega t}\dot{\overline{\mathbf{q}}} + \tag{25.27}$$

$$+\mathbf{K}_{n_d}\overline{\mathbf{q}} + (\mathbf{K}_{r_d} - i\Omega\mathbf{C}_{r_d})e^{2i\Omega t}\overline{\mathbf{q}} = \mathbf{F}_n + \Omega^2\mathbf{F}_r e^{i\Omega t} .$$

Matrices and vectors with subscript r are related to the rotating elements, and those with subscript n are related to the stator of the machine. The mean matrices without r or n subscripts are related to the whole system and are obviously the sum of the corresponding nonrotating and rotating mean matrices. The nonrotating force vector is related to static forces, while the rotating vector is related to forces that are stationary in a reference frame rotating at the spin speed Ω. Because the latter are usually unbalance forces, their magnitude is proportional to the square of the spin speed.

Remark 25.6 *All terms containing the deviatoric matrices of rotating elements have coefficients that are periodic functions of time, with periods equal to half of the period of rotation. Moreover, the complex conjugate of the vector of the generalized coordinates is present in the terms related to deviatoric matrices.*

---

[2]G. Genta, "Whirling of unsymmetrical rotors: A finite element approach based on complex coordinates", *Journal of Sound and Vibration*, 124(1), (1988), 24–53.

Equation (25.27) is a linear differential equation with periodic coefficients of the type studied by the Floquet theory. If all deviatoric matrices vanish, as with axi-symmetrical systems, Eq. (25.27) reduces to the constant coefficient equation (24.19) already studied for isotropic rotors.

25.4.1 General system with anisotropic stator

Consider a system whose behavior is modeled by Eq. (25.27), but with an isotropic rotor. The simplest of the systems of this type is the Jeffcott rotor on nonisotropic supports studied in Section 4.8.2. Because all deviatoric matrices related to the rotor vanish, Eq. (25.27) reduces to the following differential equation with constant coefficients:

$$\mathbf{M}_m\ddot{\mathbf{q}} + (\mathbf{C}_m - i\Omega\mathbf{G})\dot{\mathbf{q}} + (\mathbf{K}_m - i\Omega\mathbf{C}_{r_m})\mathbf{q} +$$
$$+\mathbf{M}_{n_d}\ddot{\overline{\mathbf{q}}} + \mathbf{C}_{n_d}\dot{\overline{\mathbf{q}}} + \mathbf{K}_{n_d}\overline{\mathbf{q}} = \mathbf{F}_n + \Omega^2\mathbf{F}_r e^{i\Omega t}. \tag{25.28}$$

Equation (25.28) involves actually working with complex coordinates, because deviatoric matrices are generally complex, and, consequently, the advantage of resorting to complex coordinates depends on the time and cost-effectiveness of the available subroutines for computations involving complex numbers. Alternatively, it is possible to use the same equation of motion written in terms of real coordinates

$$\begin{bmatrix} \mathbf{M}_x & \mathbf{M}_{xy} \\ \mathbf{M}_{yx} & \mathbf{M}_y \end{bmatrix}\ddot{\mathbf{x}} + \left(\Omega\begin{bmatrix} 0 & \mathbf{G} \\ -\mathbf{G} & 0 \end{bmatrix} + \begin{bmatrix} \mathbf{C}_x & \mathbf{C}_{xy} \\ \mathbf{C}_{yx} & \mathbf{C}_y \end{bmatrix}\right)\dot{\mathbf{x}} +$$

$$+ \left(\begin{bmatrix} \mathbf{K}_x & \mathbf{K}_{xy} \\ \mathbf{K}_{yx} & \mathbf{K}_y \end{bmatrix} + \Omega\begin{bmatrix} 0 & \mathbf{C}_r \\ -\mathbf{C}_r & 0 \end{bmatrix}\right)\mathbf{x} = \tag{25.29}$$

$$= \Omega^2\left\{ \begin{array}{c} \Re(\mathbf{f}_r e^{i\Omega t}) \\ \Im(\mathbf{f}_r e^{i\Omega t}) \end{array} \right\} + \left\{ \begin{array}{c} \Re(\mathbf{f}_n) \\ \Im(\mathbf{f}_n) \end{array} \right\},$$

where the real-coordinates vector is defined as in Eq. (24.36).

The solution for static loading is similar to the corresponding solution for axi-symmetrical systems, i.e., a constant vector $\mathbf{q} = \mathbf{q}_0$, leading to the equation

$$(\mathbf{K}_m - i\Omega\mathbf{C}_{r_m})\mathbf{q}_0 + \mathbf{K}_{n_d}\overline{\mathbf{q}}_0 = \mathbf{f}_n, \tag{25.30}$$

or, if real coordinates are used,

$$\left(\begin{bmatrix} \mathbf{K}_x & \mathbf{K}_{xy} \\ \mathbf{K}_{yx} & \mathbf{K}_y \end{bmatrix} + \begin{bmatrix} 0 & \mathbf{C}_r \\ -\mathbf{C}_r & 0 \end{bmatrix}\right)\{x_0\} = \left\{ \begin{array}{c} \Re(\mathbf{f}_n) \\ \Im(\mathbf{f}_n) \end{array} \right\}. \tag{25.31}$$

The inflected shape, then, is a line (generally a skew line) that remains fixed in space. Rotating damping couples the behavior in the xz- and yz-planes, even if the coordinate planes are planes of symmetry for the stator.

The unbalance response is a synchronous elliptical whirling. The solution of the equation of motion can be expressed in the form

$$\mathbf{q} = \mathbf{q}_1 e^{i\Omega t} + \mathbf{q}_2 e^{-i\Omega t},$$

i.e., as the sum of two circular whirling motions taking place at speed Ω in opposite directions. Both $\mathbf{q}_1$ and $\mathbf{q}_2$ are, generally speaking, complex vectors, that physically correspond to elliptical orbits not having axes x and y as axes of symmetry. The unknowns of the problem are then $4n$ in number, i.e., the imaginary and real parts of two vectors of size n. By introducing this solution into the equation of motion (25.31), the latter yields

$$\begin{bmatrix} \mathbf{A}_{11} & \mathbf{A}_{12} \\ \mathbf{A}_{21} & \mathbf{A}_{22} \end{bmatrix} \begin{Bmatrix} \mathbf{q}_1 \\ \mathbf{q}_2 \end{Bmatrix} = \Omega^2 \begin{Bmatrix} \mathbf{f}_r \\ \mathbf{0} \end{Bmatrix}, \tag{25.32}$$

where

$$\begin{aligned} \mathbf{A}_{11} &= -\Omega^2 (\mathbf{M}_m - \mathbf{G}) + i\Omega \mathbf{C}_{n_m} + \mathbf{K}_m, \\ \mathbf{A}_{12} &= -\Omega^2 \mathbf{M}_{n_d} + i\Omega \mathbf{C}_{n_d} + \mathbf{K}_{n_d}, \\ \mathbf{A}_{21} &= -\Omega^2 \mathbf{M}_{n_d} - i\Omega \mathbf{C}_{n_d} + \mathbf{K}_{n_d}, \\ \mathbf{A}_{22} &= -\Omega^2 (\mathbf{M}_m + \mathbf{G}) - i\Omega (\mathbf{C}_{n_m} + 2\mathbf{C}_{r_m}) + \mathbf{K}_m. \end{aligned} \tag{25.33}$$

Rotating damping now enters the equation yielding the unbalance response: The shaft no longer rotates in the deformed configuration but actually vibrates, in the sense that each part of it experiences stresses that vary with time.

In the case of undamped systems, Eq. (25.32) reduces to

$$\left(\begin{bmatrix} \mathbf{K}_m & \mathbf{K}_{n_d} \\ \mathbf{K}_{n_d} & \mathbf{K}_m \end{bmatrix} - \Omega^2 \begin{bmatrix} \overline{\mathbf{M}}_m - \mathbf{G} & \mathbf{M}_{n_d} \\ \overline{\mathbf{M}}_{n_d} & \mathbf{M}_m + \mathbf{G} \end{bmatrix} \right) \begin{Bmatrix} \mathbf{q}_1 \\ \mathbf{q}_2 \end{Bmatrix} = \Omega^2 \begin{Bmatrix} \mathbf{f}_r \\ \mathbf{0} \end{Bmatrix}, \tag{25.34}$$

which is real if the stator is symmetrical with respect to the coordinate planes. Note that the mean mass matrix is always real and coincides with its conjugate.

By equating the determinant of the matrix of the coefficients of Eq. (25.34) to zero, an eigenproblem in Ω^2 allowing the critical speeds to be computed, is obtained. At certain speeds, vector $\mathbf{q}_1$ vanishes; this physically corresponds to a circular backward whirling motion due to unbalance, as was shown in Section 25.1 for the Jeffcott rotor.

The equation corresponding to Eq. (25.34) but obtained using real coordinates, and the complexity of the relevant computations, is very similar.

In the case of free whirling, the orbits of the system are elliptical. The relevant solution of the homogeneous equation of motion is of the type

$$\mathbf{q} = \mathbf{q}_1 e^{i\omega t} + \mathbf{q}_2 e^{-i\omega t},$$

which leads to the following algebraic equation

$$\left(-\omega^2 \begin{bmatrix} \mathbf{M}_m & \mathbf{M}_{n_d} \\ \overline{\mathbf{M}}_{n_d} & \mathbf{M}_m \end{bmatrix} + \omega\Omega \begin{bmatrix} \mathbf{G} & \mathbf{0} \\ \mathbf{0} & -\mathbf{G} \end{bmatrix} + i\omega \begin{bmatrix} \mathbf{C}_m & \mathbf{C}_{n_d} \\ \overline{\mathbf{C}}_{n_d} & \overline{\mathbf{C}}_m \end{bmatrix} + \right.$$

$$\left. + \begin{bmatrix} \mathbf{K}_m & \mathbf{K}_{n_d} \\ \mathbf{K}_{n_d} & \mathbf{K}_m \end{bmatrix} - i\Omega \begin{bmatrix} \mathbf{C}_r & \mathbf{0} \\ \mathbf{0} & -\mathbf{C}_r \end{bmatrix} \right) \begin{Bmatrix} \mathbf{q}_1 \\ \mathbf{q}_2 \end{Bmatrix} = \mathbf{0}. \tag{25.35}$$

The corresponding solution in terms of real coordinates is

$$\mathbf{x} = \Re(\mathbf{x}_0 e^{st}) \, ,$$

which yields the following algebraic equation

$$\left(s^2 \begin{bmatrix} \mathbf{M}_x & \mathbf{M}_{xy} \\ \mathbf{M}_{yx} & \mathbf{M}_y \end{bmatrix} + s\Omega \begin{bmatrix} \mathbf{0} & \mathbf{G} \\ -\mathbf{G} & \mathbf{0} \end{bmatrix} + s \begin{bmatrix} \mathbf{C}_{n_x} + \mathbf{C}_r & \mathbf{C}_{n_{xy}} \\ \overline{\mathbf{C}}_{n_{yx}} & \mathbf{C}_{n_y} + \mathbf{C}_r \end{bmatrix} + \right.$$

$$\left. + \begin{bmatrix} \mathbf{K}_x & \mathbf{K}_{xy} \\ \mathbf{K}_{yx} & \mathbf{K}_y \end{bmatrix} + \Omega \begin{bmatrix} \mathbf{0} & \mathbf{C}_r \\ -\mathbf{C}_r & \mathbf{0} \end{bmatrix} \right) \mathbf{x}_0 = \mathbf{0}. \tag{25.36}$$

In the case of an undamped system whose stator is symmetrical with respect to the coordinate planes, the two approaches are exactly equivalent, as both lead to a set of $2n$ real algebraic equations. The eigenvalues from Eq. (25.35) are real and those from Eq. (25.36) are imaginary. The eigenvectors of the former are real, and those of the latter are made up of real and imaginary terms, depending on the phasing of the various motions added to give the various orbits.

In the most general case, Eq. (25.35) yields a set of $2n$ complex equations, and Eq. (25.36) are always real. In the author's opinion, however, the physical interpretation is more straightforward in the case of the former equation, even if it is not sufficient to find out the sign of the eigenvalue ω to assess whether the whirl motion occurs in the forward or backward direction. In fact, if ω is a solution of the eigenproblem, $-\overline{\omega}$ is also a solution and, consequently, each mode is found twice, with opposite signs of $\Re(\omega)$.

Remark 25.7 *Whirling may occur, in some modes, in the forward direction at certain points of the rotor and in the backward direction at other points. They are sometimes referred to as* mixed modes.

Remark 25.8 *Although only the study of the eigenvectors can make clear which mode occurs in the forward or backward direction, the author feels that the physical interpretation of the solution is somehow more clear when using complex coordinates.*

Example 25.1 *Consider the rotor of the small gas turbine studied in Example 24.2. Compute the critical speeds and plot the Campbell diagram assuming that the rigid bearings are substituted by nonisotropic supports, whose stiffness is 1.2×10^7 N/m in the vertical plane and 8×10^6 N/m in the horizontal plane. The dynamic study will be performed using the same FEM model seen in Example 24.2, with the only difference that when performing matrix condensation, the translational coordinates of nodes 1 and 11 (i.e., the displacements*

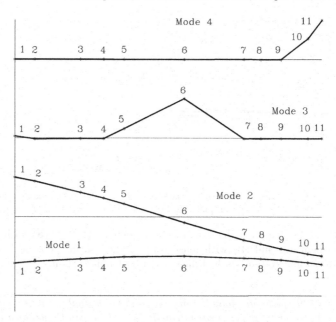

FIGURE 25.4. Mode shapes at the first four critical speeds computed using the model with eight degrees of freedom.

at the supports) are retained. The complex-coordinates approach will be followed, and the mean and deviatoric stiffness of the supports is first computed:
$k_m = 1 \times 10^7$ N/m, $k_d = 2 \times 10^6$ N/m.
A first computation is then run, neglecting the deviatoric stiffness, by simply
assembling the mean stiffness of the bearings in the stiffness matrix of the
rotor, in position 1, 1 and 11, 11.
The values of the critical speeds obtained from a condensation scheme with
eight master degrees of freedom (corresponding to the scheme with six master
degrees of freedom of Example 24.2) are 652.9, 1,352, 34,029, and 57,920 rad/s.
Note that, compared with the case with stiff bearings, the first two critical
speeds are reduced, and the third is almost unchanged. The fourth is completely
different, as can be explained by plotting the mode shapes (Fig. 25.4): In the
fourth mode the deformation is mainly localized at the bearing on the turbine
side, which in the previous model was assumed to be stiff.
The Campbell diagram is plotted in Fig. 25.5a.
If the anisotropy of the supports is not neglected, the deviatoric stiffness matrix
must be built: It is a matrix with almost all elements equal to zero, except for
two on the main diagonal that are equal to k_d. The other deviatoric matrices
are equal to zero. Eq. (25.34) can be used to compute the following values of
the first 10 critical speeds (in rad/s):

$$588.2, \quad 698.0, \quad 768.3, \quad 1,362, \quad 1,463,$$
$$3,530, \quad 33,617, \quad 34,029, \quad 51,632, \quad 57,938.$$

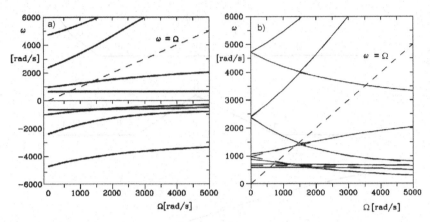

FIGURE 25.5. Campbell diagram of the rotor of Fig. 24.6 (a) with nonrigid isotropic bearings and (b) with anisotropy of the bearings not neglected. Full lines, anisotropic system; dashed lines, mean isotropic system.

> *Many critical speeds are found; from the undamped analysis it is impossible to state the severity of such critical conditions, which can be obtained only by plotting an unbalance response of the damped system.*
>
> *The Campbell diagram has been reported in Fig. 25.5b. Only one quadrant of the diagram has been plotted, because it contains all the information. From the diagram, it is impossible to state which modes are related to forward or backward whirling; a clue, however, is that the branches sloping upward are related to forward whirling. The curves computed for the isotropic system are also reported on the same plot (dotted lines).*
>
> *Note that the effects of the anisotropic bearings are quite limited, particularly where the higher-order modes are concerned.*

25.4.2   General system with anisotropic rotor

Consider a system modeled by Eq. (25.27), but whose stator is isotropic. The simplest system of this type is the anisotropic Jeffcott rotor studied in Section 4.8.3. All deviatoric matrices related to the stator vanish, and Eq. (25.27) reduces to

$$
\mathbf{M}_m\ddot{\mathbf{q}} + (\mathbf{C}_m - i\Omega\mathbf{G})\dot{\mathbf{q}} + (\mathbf{K}_m - i\Omega\mathbf{C}_{r_m})\mathbf{q} +
$$
$$
+\mathbf{M}_{r_d}e^{2i\Omega t}(\ddot{\bar{\mathbf{q}}} + 2i\Omega\dot{\bar{\mathbf{q}}}) + \mathbf{C}_{r_d}e^{2i\Omega t}\dot{\bar{\mathbf{q}}} + \qquad (25.37)
$$
$$
+(\mathbf{K}_{r_d} - i\Omega\mathbf{C}_{r_d})e^{2i\Omega t}\bar{\mathbf{q}} = \mathbf{F}_n + \Omega^2\mathbf{F}_r e^{i\Omega t}.
$$

Equation (25.37) can be transformed into an equation with constant coefficients by resorting to the rotating coordinates defined by Eq. (24.38)

$$
\mathbf{M}_m\ddot{\mathbf{r}} + [\mathbf{C}_m + i\Omega(2\mathbf{M}_m - \mathbf{G})]\dot{\mathbf{r}} +
$$
$$
+ \left[-\Omega^2(\mathbf{M}_m - \mathbf{G}) + \mathbf{K}_m + i\Omega\mathbf{C}_n\right]\mathbf{r} + \qquad (25.38)
$$

$$+\mathbf{M}_{r_d}\ddot{\bar{\mathbf{r}}} + \mathbf{C}_{r_d}\dot{\bar{\mathbf{r}}} + \left(\Omega^2\mathbf{M}_{r_d} + \mathbf{K}_{r_d}\right)\bar{\mathbf{r}} = \mathbf{F}_n e^{-i\Omega t} + \Omega^2\mathbf{F}_r\ .$$

Alternatively, it is possible to use the same equation of motion written in terms of real coordinates

$$
\begin{bmatrix} \mathbf{M}_x & \mathbf{M}_{xy} \\ \mathbf{M}_{yx} & \mathbf{M}_y \end{bmatrix} \ddot{\mathbf{x}} + \left(-\Omega \begin{bmatrix} \mathbf{M}_{yx} - \mathbf{M}_{xy} & \mathbf{M}_x + \mathbf{M}_y - \mathbf{G} \\ -\mathbf{M}_x - \mathbf{M}_y + \mathbf{G} & \mathbf{M}_{yx} - \mathbf{M}_{xy} \end{bmatrix} + \right.
$$

$$
\left. + \begin{bmatrix} \mathbf{C}_{r_x} + \mathbf{C}_n & \mathbf{C}_{r_{xy}} \\ \mathbf{C}_{r_{yx}} & \mathbf{C}_{r_y} + \mathbf{C}_n \end{bmatrix} \right) \dot{\mathbf{x}} + \qquad (25.39)
$$

$$
+ \left(\begin{bmatrix} \mathbf{K}_x & \mathbf{K}_{xy} \\ \mathbf{K}_{yx} & \mathbf{K}_y \end{bmatrix} - \Omega^2 \begin{bmatrix} \mathbf{M}_y - \mathbf{G} & -\mathbf{M}_{yx} \\ -\mathbf{M}_{xy} & \mathbf{M}_x - \mathbf{G} \end{bmatrix} + \right.
$$

$$
\left. +\Omega \begin{bmatrix} 0 & -\mathbf{C}_n \\ \mathbf{C}_n & 0 \end{bmatrix} \right) \mathbf{x} = \left\{ \begin{array}{c} \Re(\mathbf{f}_n e^{-i\Omega t}) \\ \Im(\mathbf{f}_n e^{-i\Omega t}) \end{array} \right\} + \Omega^2 \left\{ \begin{array}{c} \Re(\mathbf{f}_r) \\ \Im(\mathbf{f}_r) \end{array} \right\},
$$

where the real-coordinates vector is

$$\mathbf{x} = [\Re(\mathbf{r})^T, \Im(\mathbf{r})^T]^T\ .$$

The study of the unbalance response is easier than that of a static loading because the deformed configuration is stationary with respect to the reference frame. The solution of Eq. (25.39) for static loading is of the type

$$\mathbf{r} = \mathbf{r}_1 e^{-i\Omega t} + \mathbf{r}_2 e^{i\Omega t}\ ,$$

leading to the equation

$$
\begin{bmatrix} \mathbf{A}_{11} & \mathbf{A}_{12} \\ \mathbf{A}_{21} & \mathbf{A}_{22} \end{bmatrix} \left\{ \begin{array}{c} \mathbf{q}_1 \\ \overline{\mathbf{q}}_2 \end{array} \right\} = \left\{ \begin{array}{c} \mathbf{f}_n \\ 0 \end{array} \right\}, \qquad (25.40)
$$

where

$$\mathbf{A}_{11} = -i\Omega\mathbf{C}_{r_m} + \mathbf{K}_m\ ,$$
$$\mathbf{A}_{12} = -i\Omega\mathbf{C}_{r_d} + \mathbf{K}_d\ ,$$
$$\mathbf{A}_{21} = -i\Omega\overline{\mathbf{C}}_{r_d} + \overline{\mathbf{K}}_d\ ,$$
$$\mathbf{A}_{22} = -2\Omega^2(2\mathbf{M}_m - \mathbf{G}) - i\Omega(\overline{\mathbf{C}}_{r_m} + 2\overline{\mathbf{C}}_n) + \overline{\mathbf{K}}_m\ .$$

The same solution can be written with reference to the fixed frame as

$$\mathbf{q} = \mathbf{r}e^{i\Omega t} = \mathbf{r}_1 + \mathbf{r}_2 e^{2i\Omega t}.$$

The obvious meaning of $\mathbf{r}_1$ is the mean inflected shape, which is fixed in space, while $\mathbf{r}_2$ is a component of the deflected shape, which rotates at a speed equal to 2Ω. Static loading then causes the onset of vibrations, seen by the stator as occurring with a frequency 2Ω and by the rotor with a frequency Ω.

By equating to zero the matrix of the coefficients of Eq. (25.40), an eigenproblem in Ω is obtained. It yields the values of the secondary critical speeds due to a constant load distribution, as seen in Section 25.3 for the Jeffcott rotor.

The solution of the problem related to a given unbalance distribution is straightforward, leading to a synchronous circular whirling. The solution of the equation of motion is constant, $\mathbf{r} = \mathbf{r}_0$, leading to the equation

$$\left[-\Omega^2(\mathbf{M}_m - \mathbf{G}) + \mathbf{K}_m + i\Omega\mathbf{C}_n\right]\mathbf{r}_0 + \left(\Omega^2\mathbf{M}_{r_d} + \mathbf{K}_{r_d}\right)\bar{\mathbf{r}}_0 = \Omega^2\mathbf{f}_r . \quad (25.41)$$

Once written in the fixed reference frame, the solution is

$$\mathbf{q} = \mathbf{r}_0 e^{i\Omega t} .$$

The response to unbalance is, consequently, a pure circular synchronous whirling, and rotating damping has no effect on the behavior of the system because the rotor does not vibrate but merely rotates in the deflected configuration.

The solution for the free whirling of the system is of the type

$$\mathbf{r} = \mathbf{r}_1 e^{i\omega' t} + \mathbf{r}_2 e^{-i\bar{\omega}' t} ,$$

i.e., an elliptical whirling with reference to the rotating frame $\xi\eta z$. From the equation of motion (25.38), an algebraic equation is obtained:

$$\left(-\omega'^2 \begin{bmatrix} \mathbf{M}_m & \mathbf{M}_d \\ \mathbf{M}_d & \mathbf{M}_m \end{bmatrix} + \omega'\Omega \begin{bmatrix} 2\mathbf{M}_m + \mathbf{G} & 0 \\ 0 & 2\mathbf{M}_m - \mathbf{G} \end{bmatrix} + \right.$$

$$+ i\omega' \begin{bmatrix} \mathbf{C}_m + \mathbf{C}_{r_m} & \mathbf{C}_{r_d} \\ \overline{\mathbf{C}}_{r_d} & \overline{\mathbf{C}}_n + \overline{\mathbf{C}}_{r_m} \end{bmatrix} + i\Omega \begin{bmatrix} \mathbf{C}_n & 0 \\ 0 & -\overline{\mathbf{C}}_n \end{bmatrix} +$$

$$\left. + \begin{bmatrix} -\mathbf{M}_m + \mathbf{G} & \mathbf{M}_d \\ \overline{\mathbf{M}}_d & -\mathbf{M}_m + \mathbf{G} \end{bmatrix} + \begin{bmatrix} \mathbf{K}_m & \mathbf{K}_{n_d} \\ \overline{\mathbf{K}}_{n_d} & \overline{\mathbf{K}}_m \end{bmatrix} \right) \begin{Bmatrix} \mathbf{r}_1 \\ \bar{\mathbf{r}}_2 \end{Bmatrix} = 0. \quad (25.42)$$

By expressing the same solution in the fixed frame, it yields

$$\mathbf{q} = \mathbf{r}_1 e^{i\omega t} + \mathbf{r}_2 e^{i(2\Omega - \bar{\omega})t}.$$

The orbits are then elliptical when seen in the rotating frame, but become Lissajous curves when seen in the fixed frame. The equation of motion could be written directly in terms of fixed coordinates, obtaining an equation in which the whirl speed ω is present instead of ω'

$$\left(-\omega^2 \begin{bmatrix} \mathbf{M}_m & \mathbf{M}_d \\ \mathbf{M}_d & \mathbf{M}_m \end{bmatrix} + \omega\Omega \begin{bmatrix} \mathbf{G} & 2\mathbf{M}_d \\ 2\overline{\mathbf{M}}_d & 4\mathbf{M}_m - \mathbf{G} \end{bmatrix} + \right.$$

$$+ \Omega^2 \begin{bmatrix} 0 & 0 \\ 0 & -4\mathbf{M}_m + 2\mathbf{G} \end{bmatrix} + i\omega \begin{bmatrix} \mathbf{C}_m + \mathbf{C}_{r_m} & \mathbf{C}_{r_d} \\ \overline{\mathbf{C}}_{r_d} & \overline{\mathbf{C}}_n + \overline{\mathbf{C}}_{r_m} \end{bmatrix} +$$

$$\left. - i\Omega \begin{bmatrix} \mathbf{C}_{r_m} & \mathbf{C}_{r_d} \\ \overline{\mathbf{C}}_{r_d} & 2\overline{\mathbf{C}}_n + \overline{\mathbf{C}}_{r_m} \end{bmatrix} + \begin{bmatrix} \mathbf{K}_m & \mathbf{K}_{n_d} \\ \overline{\mathbf{K}}_{n_d} & \overline{\mathbf{K}}_m \end{bmatrix} \right) \begin{Bmatrix} \mathbf{r}_1 \\ \bar{\mathbf{r}}_2 \end{Bmatrix} = 0. \quad (25.43)$$

It is then not necessary, even in this case, to resort to the rotating frame. The problem of finding the fields of instability of systems with many degrees of freedom can easily be solved by obtaining the eigenvalues of Eq. (25.43) and then studying the sign of the imaginary part of the complex frequency.

25.4.3   General system with anisotropic stator and rotor

If neither the stator nor the rotor have axial symmetry, the linear equation of motion has periodic coefficients in both the rotating and nonrotating reference frames. As seen in Section 22.1, no general method exists to solve equations of this kind. Because the coefficients are periodic functions of time with period π/Ω, the solution of the homogeneous equation associated with Eq. (25.27) is of the type

$$\mathbf{q} = \mathbf{q}_1(t)e^{i\omega t} + \mathbf{q}_2(t)e^{-i\overline{\omega}t},$$

where both vectors $\mathbf{q}_1$ and $\mathbf{q}_2$ are periodic functions of time with the same period, i.e., with fundamental frequency equal to 2Ω. The general solution is then the sum of a number of terms of the type mentioned earlier, each with its value of ω, plus a solution of the complete equation. The study of the stability of the system can thus be performed in the same way as in the case with constant $\mathbf{q}_1$ and $\mathbf{q}_2$.

As seen in Section 22.3 for Hill's equation, unknown functions $\mathbf{q}_1$ and $\mathbf{q}_2$ can be expressed by trigonometric series

$$\mathbf{q} = \sum_{j=-\infty}^{\infty} \left(\mathbf{q}_{1_j} e^{i(\omega+2j\Omega)t} + \mathbf{q}_{2_j} e^{-i(\overline{\omega}+2j\Omega)t} \right). \qquad (25.44)$$

By introducing Eq. (25.44) into the homogeneous equation associated to Eq. (25.27), the following algebraic equation is readily obtained

$$\sum_{j=-\infty}^{\infty} \left(\mathbf{A}_j\mathbf{q}_{1_j} e^{i(\omega+2j\Omega)t} + \mathbf{B}_j\mathbf{q}_{2_j} e^{-i(\overline{\omega}+2j\Omega)t} + \overline{\mathbf{C}}_j\overline{\mathbf{q}}_{1_j} e^{-i(\overline{\omega}+2j\Omega)t} + \right.$$

$$\left. + \mathbf{D}_j\overline{\mathbf{q}}_{2_j} e^{i(\omega+2j\Omega)t} + \mathbf{E}_j\overline{\mathbf{q}}_{1_j} e^{-i[\overline{\omega}+2(j-1)\Omega]t} + \mathbf{E}_j\overline{\mathbf{q}}_{2_j} e^{i[\omega+2(j+1)\Omega]t} \right) = 0.$$

$$(25.45)$$

By separately equating to zero the various terms of Eq. (25.45) and stating

$$\mathbf{A}_j = -(\omega + 2j\Omega)^2\mathbf{M}_m + \Omega(\omega + 2j\Omega)\mathbf{G} + i(\omega + 2j\Omega)\mathbf{C}_m + \mathbf{K}_m - i\Omega\mathbf{C}_{r_m},$$

$$\mathbf{B}_j = -(\omega + 2j\Omega)^2\mathbf{M}_m - \Omega(\omega + 2j\Omega)\mathbf{G} + i(\omega + 2j\Omega)\overline{\mathbf{C}}_m + \overline{\mathbf{K}}_m + i\Omega\overline{\mathbf{C}}_{r_m},$$

$$\mathbf{C}_j = -(\omega + 2j\Omega)^2\overline{\mathbf{M}}_{n_d} + i(\omega + 2j\Omega)\overline{\mathbf{C}}_{n_d} + \overline{\mathbf{K}}_{n_d}, \qquad (25.46)$$

$$\mathbf{D}_j = -(\omega + 2j\Omega)^2\mathbf{M}_{n_d} + i(\omega + 2j\Omega)\mathbf{C}_{n_d} + \mathbf{K}_{n_d},$$

$$\mathbf{E}_j = [-\omega^2 - 2\omega\Omega(2j-1) - 4j\Omega^2(j-1)]\overline{\mathbf{M}}_{r_d} + i(\omega + 2j\Omega)\overline{\mathbf{C}}_{r_d} + \overline{\mathbf{K}}_{r_d} - i\Omega\overline{\mathbf{C}}_{r_d},$$

$$\mathbf{F}_j = [-\omega^2 - 2\omega\Omega(2j+1) - 4j\Omega^2(j+1)]\mathbf{M}_{r_d} + i(\omega + 2j\Omega)\mathbf{C}_{r_d} + \mathbf{K}_{r_d} + i\Omega\mathbf{C}_{r_d},$$

the following infinite set of algebraic equations is readily obtained

$$
\begin{bmatrix}
\ddots & \ddots & \ddots & & & & & \\
\mathbf{F}_{-2} & \mathbf{A}_{-1} & \mathbf{D}_{-1} & & & & & \\
 & \mathbf{C}_{-1} & \mathbf{B}_{-1} & \mathbf{E}_0 & & & & \\
 & & \mathbf{F}_{-1} & \mathbf{A}_0 & \mathbf{D}_0 & & & \\
 & & & \mathbf{C}_0 & \mathbf{B}_0 & \mathbf{E}_1 & & \\
 & & & & \mathbf{F}_0 & \mathbf{A}_1 & \mathbf{D}_1 & \\
 & & & & & \mathbf{C}_1 & \mathbf{B}_1 & \mathbf{E}_2 \\
 & & & & & & \mathbf{F}_1 & \mathbf{A}_2 & \mathbf{D}_2 \\
 & & & & & & \ddots & \ddots & \ddots
\end{bmatrix}
\times
\left\{
\begin{array}{c}
\vdots \\
\mathbf{q}_{1_{-1}} \\
\overline{\mathbf{q}}_{2_{-1}} \\
\mathbf{q}_{1_0} \\
\overline{\mathbf{q}}_{2_0} \\
\mathbf{q}_{1_1} \\
\overline{\mathbf{q}}_{2_1} \\
\mathbf{q}_{1_2} \\
\vdots
\end{array}
\right\}
= \{0\}.
$$

$$(25.47)$$

The homogeneous equation (25.47) has solutions different from the trivial one only if the matrix of the coefficients is singular. An eigenproblem, similar to the one linked with Hill's infinite determinant (Eq. (22.25)), but with a bandwidth $3n$, is obtained. Approximate solutions can be obtained by considering a limited array of $2m$ rows by $2m$ columns, in terms of matrices (the actual number of rows and columns is $2\ m \times n$), centered on the matrices with subscript 0. Let these limited matrices be denoted by $\mathbf{Z}_m$ and let matrix $\mathbf{A}_0$ be $\mathbf{Z}_0$.

The zero-order approximation involving matrix $\mathbf{Z}_0$ deals with a symmetrical system having the mean properties of the actual system.

Matrices $\mathbf{A}_j$ and $\mathbf{B}_j$ contain the mean properties of the system, matrices $\mathbf{C}_j$ and $\mathbf{D}_j$ are related to the unsymmetrical characteristics of the stator, and matrices $\mathbf{E}_j$ and $\mathbf{F}_j$ are related to the unsymmetrical properties of the rotor.

The first-order approximation obtained by considering matrix

$$\mathbf{Z}_1 = \begin{bmatrix} \mathbf{A}_0 & \mathbf{B}_0 \\ \mathbf{C}_0 & \mathbf{D}_0 \end{bmatrix}$$

coincides with the solution of a system with isotropic rotor (with mean properties) running on an asymmetrical stator (with the actual properties).

The equation allowing the study of a system with a symmetrical stator and an asymmetrical rotor is

$$\begin{bmatrix} \mathbf{B}_0 & \mathbf{E}_1 \\ \mathbf{F}_0 & \mathbf{A}_1 \end{bmatrix} \left\{ \begin{array}{c} \overline{\mathbf{q}}_{2_0} \\ \mathbf{q}_{1_1} \end{array} \right\} = \mathbf{0}. \qquad (25.48)$$

The infinite set of equations (25.47) and the corresponding nonhomogeneous set that includes the forcing functions, represent a general model for rotating machinery. The solution of the eigenproblem, however, is not easy. The equations must be rearranged to obtain an eigenproblem in standard form, which results in a further doubling of the size of the problem, becoming $4mn$ and then a problem of a large order must be faced.

If $\Re(\omega)$ is the real part of the solution of the eigenproblem, also

$$\Re(\omega) + 2j\Omega , \quad \Re(\omega) - 2j\Omega \;\; \forall j$$

are real parts of other solutions. This explains the apparent inconsistency of the number of the eigenvalues, which tend to infinity when increasing the size of the matrix $\mathbf{Z}_m$: While obtaining a better precision, solutions whose real parts are equal to those already obtained plus a multiple of 2Ω are obtained. Vectors $\mathbf{q}_j$ with j greater than 1 usually add only a small ripple on the basic solution, which is given by the vectors with $j \leq 1$. Third-order approximations should, consequently, give results that accurately simulate the behavior of the actual system.

Solutions for nonhomogeneous problems, like those related to the response to static loading and unbalance, can be obtained in a similar way. In the former case the solution can be expressed by the series

$$\mathbf{q} = \mathbf{q}_0 + \sum_{j=1}^{\infty} \mathbf{q}_{1_j} e^{2ij\Omega t} + \sum_{j=1}^{\infty} \mathbf{q}_{2_j} e^{-2ij\Omega t}. \tag{25.49}$$

The mean response, the forward components, and the backward components are written separately in Eq. (25.49). By using this expression of the deflected shape, the following infinite set of equations can be obtained from the equation of motion (25.27)

$$\begin{cases} \mathbf{A}_0\mathbf{q}_0 + \mathbf{C}_0\bar{\mathbf{q}}_0 + \mathbf{E}_1\bar{\mathbf{q}}_{1_1} = \mathbf{f}_n \\ \mathbf{F}_0\bar{\mathbf{q}}_0 + \mathbf{A}_1\mathbf{q}_{1_1} + \mathbf{D}_1\bar{\mathbf{q}}_{2_1} = \{0\} \\ \mathbf{C}_1\mathbf{q}_{1_1} + \mathbf{B}_1\bar{\mathbf{q}}_{2_1} + \mathbf{E}_2\mathbf{q}_{1_2} = \{0\} \\ \mathbf{F}_1\bar{\mathbf{q}}_{2_1} + \mathbf{A}_2\mathbf{q}_{1_2} + \mathbf{D}_1\bar{\mathbf{q}}_{2_2} = \{0\} \\ \cdots\cdots\cdots\cdots\cdots\cdots\cdots\cdots\cdots \end{cases} \tag{25.50}$$

where the relevant matrices can be obtained from Eq. (25.47) by setting $\omega = 0$. Here, again, a good approximation can be obtained by considering only the components $\mathbf{q}_0$, $\mathbf{q}_{11}$, and $\mathbf{q}_{21}$, i.e., by resorting to a reduced set formed by the first three equations (25.50) in which the term in $\mathbf{E}_2$ has been neglected. The size of the problem is thus reduced to $3n$ equations.

The response to an arbitrary unbalance distribution is similarly obtained by assuming a solution of the type

$$\mathbf{q} = \sum_{j=1}^{\infty} \mathbf{q}_{1_j} e^{i(2j+1)\Omega t} + \sum_{j=1}^{\infty} \mathbf{q}_{2_j} e^{-i(2j+1)\Omega t}. \tag{25.51}$$

By using the expressions (25.51) for the deflected shape, the following infinite set of equations can be obtained from the equation of motion (25.27)

$$\begin{bmatrix} \mathbf{A}_0 & \mathbf{D}_0 & & & \\ \mathbf{C}_0 & \mathbf{B}_0 & \mathbf{E}_1 & & \\ & \mathbf{F}_0 & \mathbf{A}_1 & \mathbf{D}_1 & \\ & & \mathbf{C}_1 & \mathbf{B}_1 & \mathbf{E}_2 \\ & \cdots & \cdots & \cdots & \end{bmatrix} \begin{Bmatrix} \mathbf{q}_{1_0} \\ \bar{\mathbf{q}}_{2_0} \\ \mathbf{q}_{1_1} \\ \bar{\mathbf{q}}_{2_1} \\ \cdots \end{Bmatrix} = \Omega^2 \begin{Bmatrix} \mathbf{f}_r \\ \{0\} \\ \{0\} \\ \{0\} \\ \cdots \end{Bmatrix}. \tag{25.52}$$

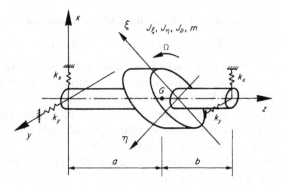

FIGURE 25.6. Sketch of an asymmetric rigid rotor on asymmetric elastic supports.

Here the relevant equations can be obtained from Eq. (25.47) by setting $\omega = \Omega$. In this case a good approximation can be obtained by considering only the first four equations in which the term in $\mathbf{E}_2$ has been neglected. The size of the problem thus reduces to $4n$ equations.

25.5   Exercises

Exercise 25.1 *Plot the Campbell diagram and the decay rate plot of a nonisotropic Jeffcott rotor running in isotropic supports. Study first the undamped system and then consider also damping. State whether an instability range exists.*
Data: $m = 10$ kg, $k_\xi = 10^5$ N/m, $k_\eta = 2 \times 10^5$ N/m, $c_n = 400$ Ns/m, $c_r = 40$ Ns/m.

Exercise 25.2 *Consider the gas turbine of Example 24.2. Repeat the dynamic analysis substituting the rigid bearings with isotropic supports with a radial stiffness of 10^7 N/m, with unsymmetrical supports, with a stiffness of 1.2×10^7 N/m in the vertical plane and 8.9×10^6 N/m in the horizontal plane.*

Exercise 25.3 *Consider a rigid unsymmetrical rotor running on two identical anisotropic bearings (Fig. 25.6) and model it as a 4 degrees of freedom (two complex) system. Assuming that the principal directions of both supports are the same, so that matrices K_{xy} and K_{yx} vanish, compute the Campbell diagram and the decay rate plot. Data: $m = 10$ kg, $J_\xi = 0.312$ kg m^2, $J_\eta = 0.648$ kg m^2, $J_p = 0.56$ kg m^2, $k_x = 1.2 \times 10^5$ N/m, $k_y = 0.8 \times 10^5$ N/m, $a = 100$ mm, $b = 300$ mm.*

Exercise 25.4 *Consider the same system of the previous exercise. Compute the unbalance response due to an eccentricity of 10 μm in the direction of the ξ-axis. Plot the orbit of the center of mass at a speed of 76 rad/s.*

26
Nonlinear Rotors

When rotors contain nonlinear element or work outside their linearity range, phenomena typical of nonlinear systems, like the dependence of the frequency of free vibration from the amplitude, the jump, or even deterministic chaos, may be present. Unlike nonlinear vibrating systems, a harmonic solution (circular whirling) is still possible, even if it is no more a unique solution. Circular whirling can be studied through closed-form solutions of the equation of motion, but for studying the other solution numerical integration in time must be used.

26.1 General considerations

It is well known that the behavior of both lubricated journal bearings and rolling-element bearings is strongly nonlinear and can cause rotors to behave in a nonlinear way. However, many other mechanisms may have similar effects, including nonlinear elasticity or dry friction. Often they can be neglected and nonlinear parts can be modeled as linear, or even rigid, elements. This is usually the case, for example, of bearings when they are much stiffer than other parts of the machine. This simple approach, however, may not be suitable in some cases and, consequently, the nonlinear problem must be faced. The determination of the critical speed, at least in the usual terms, and the plotting of the Campbell diagram become thus impossible, and the study is generally limited to the computation of the response of the system to given forcing causes, usually unbalance. A definition of critical speeds as those speeds at which the response of the system becomes very large, is possible, but the fact that they depend on the particular unbalance distribution makes their computation of limited use.

Remark 26.1 *Nonlinearity makes it impossible to study separately free whirling and the effects of static and dynamic loads because the dynamic behavior can be strongly influenced by the presence of static loads.*

In the case of nonlinear rotors there is a deep difference between the behavior of axi-symmetrical and nonisotropic systems. If the system is axially symmetrical, circular whirling is an exact solution of the problem, although nonlinearity makes other solutions possible.

Remark 26.2 *Contrary to what is seen for general nonlinear vibrating systems, in this case it is possible to obtain an exact solution of the equation of motion in closed form. This is generally impossible in the case of anisotropic systems, for which the usual approximation techniques seen in Part II must be used.*

26.2   Nonlinear Jeffcott rotor: equation of motion

Consider a Jeffcott rotor of the type seen in Section 23.5.1, but with restoring and damping forces that are nonlinear functions of the displacement and velocity, respectively. The axial symmetry of the system, however, causes the restoring force to have the same direction as the displacement and the damping force to have the same direction as the velocity. By resorting to the complex coordinate z, defined by Eq. (23.4), the restoring force and the forces due to nonrotating and rotating damping can be expressed in the form

$$\begin{cases} \vec{F}_e = -kz\,[1 + f(|z|)] \;, \\ \vec{F}_{dn} = -c_n\dot{z}\,[1 + \beta_n(|\dot{z}|)] \;, \\ \vec{F}_{dr} = -c_r(\dot{z} - i\Omega z)\,[1 + \beta_r(|\dot{z} - i\Omega z|)] \;. \end{cases} \tag{26.1}$$

Note that no interaction between restoring and damping force has been assumed: The current model is not the most general nonlinear Jeffcott rotor.

By introducing the expressions of the nonlinear forces into the equation of motion of the Jeffcott rotor (Eq. (23.18)) written for constant spin speed, it follows

$$m\ddot{z} + (c_n + c_r)\dot{z} + (k - i\Omega c_r)z + [c_n\beta_n(|\dot{z}|) + c_r\beta_r(|\dot{z} - i\Omega z|)]\,\dot{z} +$$

$$+kf(|z|)z - i\Omega c_r\beta_r(|\dot{z} - i\Omega z|)z = m\epsilon\Omega^2 e^{i\Omega t} + F_n \;.$$

$$\tag{26.2}$$

Equation (26.2) can be used to study the free behavior of the system (free circular whirling), the effect of a static load F_n (of weight, for example), or that of an eccentricity of mass m. The presence of nonlinear terms, however, makes it impossible to perform a general study by superimposing the various solutions and to obtain non-circular whirling by adding forward and backward whirling of different amplitudes.

26.3 Unbalance response

Consider the undamped system whose model can be obtained from Eq. (26.2) by neglecting all terms related to damping, linear, and nonlinear, and the nonrotating force. A possible solution allowing the response to the static unbalance $m\epsilon$ to be computed is

$$z = z_0 e^{i\Omega t} .$$

Remark 26.3 *Circular whirling is an exact solution of the nonlinear equation of motion and not just an approximation of the fundamental harmonic of the response. As a consequence, there are no higher-order harmonics in the response. This does not mean that there cannot be higher-order harmonics, because circular whirling is only one of the possible solutions; however, it can be demonstrated to be stable, at least in some conditions, and has been found both experimentally and by numerical experimentation.*

Introducing this solution into the equation of motion, the following algebraic equation is readily obtained:

$$\left[k - m\Omega^2 + kf(|z_0|)\right] z_0 = m\epsilon\Omega^2 . \tag{26.3}$$

The backbone of the response can be obtained by assuming that the unbalance is vanishingly small, i.e., by solving the homogeneous equation associated with Eq. (26.3).

Because the backbone defines, in the case of vibrating systems, the conditions for a sort of nonlinear resonance, here the situation on the backbone is similar to that occurring at a critical speed.

Example 26.1 *Compute the unbalance response of an undamped rotor characterized by a function $f(|z|)$ of the same type as that seen for Duffing's equation: $f(|z_0|) = \mu|z_0|^2$.*
The algebraic equations allowing computation of the amplitude of the circular whirling and the backbone are

$$\left(1 - \frac{m}{k}\Omega^2 + \mu|z_0|^2\right) z_0 = \frac{m}{k}\epsilon\Omega^2 ,$$

$$1 - \frac{m}{k}\Omega^2 + \mu|z_0|^2 = 0 .$$

The equation yielding the backbone is

$$z_0 = \sqrt{\frac{m\Omega^2 - k}{k\mu}} .$$

A result that is quite similar to that obtained for Duffing's equation is obtained. Some important differences, however, must be mentioned: Ω here is the spin speed and not a circular frequency, and the excitation is proportional to Ω^2. Moreover, since whirling is circular, the nonlinear element does not oscillate along the force–displacement characteristic but actually rotates, maintaining a given deformation. Most considerations regarding multiple solutions and the jump phenomenon, however, hold in general, regardless of the particular law $f(|z|)$.

The same circular whirling solution also holds in the case of damped systems. The algebraic equation that can be obtained in the usual way is

$$\left[1 - \frac{m}{k}\Omega^2 + f(|z_0|) + i\Omega\frac{c_n}{k} + i\Omega\frac{c_n}{k}\beta_n(|\Omega z_0|) \right] z_0 = \frac{m}{k}\epsilon\Omega^2 . \qquad (26.4)$$

As easily predictable, rotating damping plays no role in determining the conditions for synchronous whirling. The amplitude z_0 is, in this case, expressed by a complex number, as already seen in the linear case. By separating the real and imaginary parts of Eq. (26.4) and introducing the phase Φ of the response with respect to the unbalance vector, which is, as usual, assumed to be along the ξ-axis, it follows that

$$\begin{cases} z_0 \left[1 - \frac{m}{k}\Omega^2 + f(|z_0|) \right] = \frac{m}{k}\epsilon\Omega^2 \cos(\Phi) , \\[2mm] z_0 c_n \left[1 + \beta_n(|\Omega z_0|) \right] = -m\epsilon\Omega\sin(\Phi) . \end{cases} \qquad (26.5)$$

The amplitude and phase of the response are thus

$$\begin{cases} |z_0|^2 \left\{ \left[1 - \frac{m}{k}\Omega^2 + f(|z_0|) \right]^2 + \left(\frac{\Omega c_n}{k} \right)^2 [1 + \beta_n(|\Omega z_0|)]^2 \right\} = \left(\frac{m}{k}\epsilon \right)^2 \Omega^4 , \\[3mm] \Phi = -\arctan\left[\Omega c_n \dfrac{1 + \beta_n(|\Omega z_0|)}{k - m\Omega^2 + kf(|z_0|)} \right] . \end{cases}$$
$$(26.6)$$

Also, in the case of the nonlinear Jeffcott rotor, it is possible to obtain a limit envelope of the response. By stating that the phase Φ is equal to $-90°$, the second Eq. (26.5) directly yields the equation of the limit envelope

$$z_0 c_n \left[1 + \beta_n(|\Omega z_0|) \right] = m\epsilon\Omega . \qquad (26.7)$$

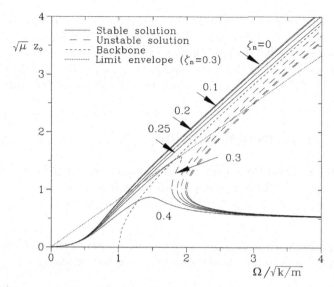

FIGURE 26.1. Unbalance response for a nonlinear Jeffcott rotor with linear damping. The characteristic of the nonlinear spring is assumed to be of the hardening Duffing type.

If damping is linear, the limit envelope is just a straight line whose equation is

$$z_0 = \frac{\Omega m \epsilon}{c_n} \ .$$

Some results obtained for a restoring force of the Duffing type with hardening characteristic and linear damping are shown in nondimensional form in Fig. 26.1. The plot has been obtained for a nondimensional nonlinear parameter

$$\mu^* = \mu \epsilon^2 = 0.25$$

and various values of the nondimensional damping

$$\zeta_n = \frac{c_n}{2\sqrt{km}} \ .$$

The differences with the usual results obtained for damped Duffing's equation are that the peak amplitude occurs now at the right of the backbone instead of on the left (the same thing happens in linear systems) and the curves do not close for low values of the damping. This suggests a greater difficulty in experiencing the jump phenomenon and, consequently, self-centering.

Remark 26.4 *Self-centering is much more difficult in the case of non-linear rotors than in the linear case, and higher nonrotating damping is needed to work in the supercritical range. The jump, needed to obtain the* *self-centered configuration, takes place when the energy supplied by the forcing function is not sufficient to sustain the motion with higher amplitude. In rotating systems the forcing function is due to unbalance, and the amount of energy supplied by the centrifugal field is large. This can be seen from the shape of the limit envelope, which may not cross the backbone.*

26.4 Free circular whirling

A possible solution of the homogeneous equation associated with Eq. (26.2), i.e., of the model for a perfectly balanced Jeffcott rotor, is

$$z = z_0 e^{i\omega t}.$$

Remark 26.5 *This solution, representing a circular free whirling of the system, is again an exact solution of the equation of motion.*

To avoid greater analytical complexities without losing generality of the results, damping will be assumed to be linear. By introducing the solution for circular free whirling into the homogeneous equation of motion, it follows that

$$\left\{ 1 - \frac{m}{k}\omega^2 + f(|z_0|) + i\left[\omega\Omega\frac{c_n}{k} + (\omega - \Omega)\frac{c_r}{k}\right] \right\} z_0 = 0. \qquad (26.8)$$

Equation (26.8) has a solution different from the trivial solution $z_0 = 0$ only if the expression in braces is equal to zero. An equation in ω is so obtained, which yields

$$\begin{cases} \omega_R^* = \pm\sqrt{\Gamma^* + \sqrt{\Gamma^{*2} + \Omega^{*2}\zeta_r^2}}\,, \\ \omega_I^* = \zeta_n + \zeta_r \mp \sqrt{-\Gamma^* + \sqrt{\Gamma^{*2} + \Omega^{*2}\zeta_r^2}}\,, \end{cases} \qquad (26.9)$$

where the nondimensional speeds Ω^* and ω^* and the damping ratios ζ_n and ζ_r are defined in the usual way, with reference to the linearized system, and

$$\Gamma^* = \frac{1 + f(|z_0|) - (\zeta_n + \zeta_r)^2}{2}.$$

Equation (26.9) is very similar to the corresponding expression for the linear case, with the difference that in the current case the whirl speed is a function of the amplitude of the motion. As is obvious for a nonlinear system, the Campbell diagram loses any meaning. A three-dimensional plot in which the amplitude $|z_0|$ is reported as a function of Ω and ω_R can, however, be introduced. The Ω,ω_R-plane of the tridimensional plot coincides with the Campbell diagram of the linearized system. The first Eq. (26.9) defines a surface expressing all the possible conditions of free whirling; it

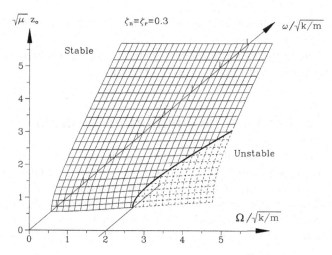

FIGURE 26.2. Tridimensional plot for the study of the free whirling of the non-linear Jeffcott rotor studied in Fig. 26.1.

can be considered the backbone of the system at varying spin speed Ω. The intersection of the surface with the $|z_0|, \omega_R$-plane expresses the relationship linking the natural frequency with the amplitude at standstill: It is then the backbone of the nonrotating system for a whirling mode in the xy-plane.

One of the mentioned plots has been reported in nondimensional form in Fig. 26.2. The figure has been plotted for a system with $\zeta_n = \zeta_r = 0.3$. The Campbell diagram of the linearized system coincides with that plotted in Fig. 23.7 (curve for the corresponding value of the damping ratio). The stability threshold of the linearized system is at $\Omega^* = 2$.

The intersection of the surface with the plane of equation $\omega = \Omega$ gives the conditions for free synchronous whirling: It then coincides with the backbone of the unbalance response shown in Fig. 26.1.

From the second Eq. (26.9) it is possible to obtain the condition for stability

$$\Omega < \sqrt{1 + f(|z_0|)}\sqrt{\frac{k}{m}}\left(1 + \frac{c_n}{c_r}\right). \tag{26.10}$$

The threshold of stability depends on the amplitude, as shown in Fig. 26.2, where the unstable part of the surface is dashed. If the spin speed Ω is lower than the threshold of instability of the linearized system, the motion is always stable and the amplitude of free whirling decays to zero. On the contrary, when the linearized analysis shows an unstable behavior in the small, the nonlinear effects reduce the instability, and the result is a motion with growing amplitude until the border separating the full lines from the dotted lines is reached. The amplitude corresponding to these conditions is that of a sort of limit cycle that constitutes an attractor for all free whirling motions. The amplitude of the limit cycle is a function of the speed and grows with increasing Ω.

This situation holds for hardening systems and qualitatively can be extended to systems with many degrees of freedom.

26.5   Stability of the equilibrium position

The stability of the equilibrium position for the unbalance response of the system can be obtained in the same way as for linear systems. Consider a damped nonlinear Jeffcott rotor whose restoring and damping forces have a cubic (Duffing type) and linear characteristic, respectively, and neglect nonrotating forces. The equation of motion in the rotating $\xi\eta$ reference frame can be easily obtained:

$$\begin{cases} m\ddot{\xi} + (c_n + c_r)\dot{\xi} - 2m\Omega\dot{\eta} + \left[k - m\Omega^2 + k\mu(\xi^2 + \eta^2)\right]\xi - c_n\Omega\eta = m\epsilon\Omega^2\,, \\[2mm] m\ddot{\eta} + (c_n + c_r)\dot{\eta} + 2m\Omega\dot{\xi} + \left[k - m\Omega^2 + k\mu(\xi^2 + \eta^2)\right]\eta + c_n\Omega\xi = 0\,. \end{cases}$$
$$(26.11)$$

The behavior of the system depends on three nondimensional parameters: the damping ratios ζ_n and ζ_r and the nonlinearity parameter $\epsilon^2\mu$. Equation (26.11) can also be used to study the motion of the system in nonstationary conditions with reference to the rotating frame. The unbalance response can be immediately computed by assuming a stationary solution in the $\xi\eta$-plane. This result coincides with the solution already obtained in Section 23.9. The study of the stability in the small about the equilibrium position can be performed in the usual way. If $\xi_1(t)$ and $\eta_1(t)$ are small displacements from the equilibrium position ξ_0, η_0, the equation of motion can be linearized, obtaining, in nondimensional terms,

$$\begin{bmatrix} 1 & 0 \\ 0 & 1 \end{bmatrix} \begin{Bmatrix} \xi_1^{*''} \\ \eta_1^{*''} \end{Bmatrix} + \begin{bmatrix} 2(\zeta_n + \zeta_r) & -2\Omega^* \\ 2\Omega^* & 2(\zeta_n + \zeta_r) \end{bmatrix} \begin{Bmatrix} \xi_1^{*'} \\ \eta_1^{*'} \end{Bmatrix} +$$

$$+ \begin{bmatrix} 1 - \Omega^{*2} + \mu^*\left(3\xi_0^{*2} + \eta_0^{*2}\right) & 2\left(-\Omega^*\zeta_n + \mu^*\xi_0^*\eta_0^*\right) \\ 2\left(\Omega^*\zeta_n + \mu^*\xi_0^*\eta_0^*\right) & 1 - \Omega^{*2} + \mu^*\left(\xi_0^{*2} + 3\eta_0^{*2}\right) \end{bmatrix} \begin{Bmatrix} \xi_1^* \\ \eta_1^* \end{Bmatrix} = \mathbf{0},$$
$$(26.12)$$

where, apart from the damping ratios and the usual nondimensional speed, the other nondimensional quantities are defined as

$$\xi^* = \frac{\xi}{\epsilon}\,, \quad \eta^* = \frac{\eta}{\epsilon}\,, \quad \tau = t\sqrt{\frac{k}{m}}\,, \quad \mu^* = \mu\epsilon^2\,,$$

and prime denotes differentiation with respect to τ.

The stability of the motion in the vicinity of an equilibrium position can be assessed by studying the sign of the decay rate of the small oscillations. Because the solution of a simple fourth-degree algebraic equation is required, the study can be carried on in closed form. Proceeding in this way it follows that, in the case of hardening systems, when three equilibrium positions are found, the solution with the largest amplitude is always stable. The intermediate solution is found to be unstable (saddle point), while the smallest solution (the self-centered one) is stable only up to a certain speed, which depends essentially on the ratio between the nonrotating and the rotating damping. Note that the nonlinearity of the system ensures that the stability condition for the unbalance response will not coincide with the stability condition of the free whirling, i.e., of the perfectly balanced system.

Equation (26.11) can be used to study numerically the behavior of the system in nonstationary conditions with the aim of assessing the stability in the large and the possibility of the existence of steady-state solutions different from those already obtained. An example of trajectories is shown in Fig. 26.3. The figure has been obtained in conditions yielding three stationary solutions, two of which are stable.

The trajectories tend to the stable solutions, even if a certain attraction is also felt toward the saddle point, which is unstable. By integrating the

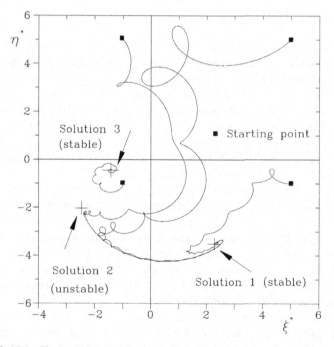

FIGURE 26.3. Trajectories in the rotating plane starting from four different points, one belonging to the domain of attraction of the lowest solution and three to the other domain. Values of parameters: $\Omega^* = 2$, $\mu^* = 0.2$, $\zeta_n = 0.2$, $\zeta_r = 0.03$.

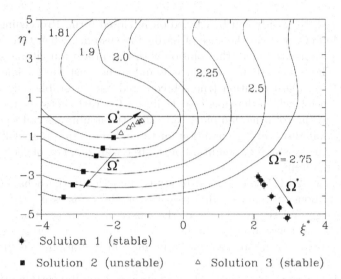

FIGURE 26.4. Basins of attraction of the equilibrium positions at different values of the speed computed by numerically integrating the equations of motion; same values of the parameters as in Fig. 26.3.

equations of motion of the system for different values of the speed, using as starting positions different pairs of values of coordinates ξ, η, the domains of attraction of the various solutions at different speeds can be obtained. The results obtained for some values of the speed are reported in Fig. 26.4. The figure was obtained using the same values of the parameters as for Fig. 26.3, numerically integrating the equations of motion 81,000 times for each value of the speed until a stable equilibrium position was found. In some cases, a few hundred steps were sufficient; in others, the motion had to be followed for thousands of steps.

The two domains of attraction are clearly well defined, and the unstable solution lies on the separatrix of the two attraction domains. At low speed, when the self-centered solution just starts existing, its domain of attraction is very small. By increasing the speed, the domain of attraction of the self-centered solution grows, and at the speed at which the jump occurs it extends for the whole plane. The physics of the phenomenon consequently does not show any critical dependence on the initial conditions, and chaotic behavior was never encountered.

Some solutions found in the literature in which chaotic behavior of an axi-symmetrical rotor has been found are available. However, a key factor that can trigger chaotic behavior of rotating systems seems to be the lack of axial symmetry, either due to geometric or material anisotropy or due to the presence of mechanisms such as bearing clearance or static loading, which make an isotropic system to operate in an offset position. At any rate, it must be stressed that the domains of attraction should be plotted in the state space, which in the current case has four dimensions (two positions and two velocities) and not in the space of the configurations as in Fig. 26.4.

26.5.1 Systems with many degrees of freedom

A very general mathematical model of a nonlinear rotor can be obtained from Eq. (26.11) by adding a generic vector function $\mathbf{f}(q_i, \dot{q}_i, \theta, t)$ to take into account the behavior of the nonlinear part of the system and by introducing the term $(\Omega^2 - ia)$ instead of Ω^2 to allow us to take into account angular accelerations

$$
\begin{aligned}
&+\mathbf{M}_m\ddot{\mathbf{q}} + \left(\mathbf{C}_m - i\Omega\mathbf{G}\right)\dot{\mathbf{q}} + \left(\mathbf{K}_m - i\Omega\mathbf{C}_{r_m}\right)\mathbf{q} + \mathbf{M}_{n_d}\ddot{\overline{\mathbf{q}}} + \\
&+\mathbf{M}_{r_d}e^{2i\theta}\ddot{\overline{\mathbf{q}}} + \mathbf{C}_{n_d}\dot{\overline{\mathbf{q}}} + \left(\mathbf{C}_{r_d} + 2i\Omega\mathbf{M}_{r_d}\right)e^{2i\theta}\dot{\overline{\mathbf{q}}} + \mathbf{K}_{n_d}\overline{\mathbf{q}} + \qquad (26.13)\\
&+\left(\mathbf{K}_{r_d} + i\Omega\mathbf{C}_{r_d}\right)e^{2i\theta}\overline{\mathbf{q}} + \mathbf{f}(q_i, \dot{q}_i, \theta, t) = \left(\Omega^2 - ia\right)\mathbf{f}_r e^{i\theta} + \mathbf{f}_n .
\end{aligned}
$$

Equation (26.13) has been obtained with only the assumption of uncoupling among flexural, axial, and torsional behavior and neglecting centrifugal stiffening. The latter can be accounted for simply by introducing appropriate stiffness matrices proportional to Ω^2.

The equation that allows us to describe the rotational degree of freedom of the system, which is just one due to the assumption of a torsionally rigid rotor, is again Eq. (24.49), which is not affected by either nonlinearities or deviations from axial symmetry.

In general, it is reasonable to expect that the time history of the accelerating system has a simpler expression in the rotating frame, where a slow variation of generalized coordinates in time should occur, than in the fixed frame, where the relevant quantities vary with a frequency equal to the rotational speed. However, if the system is nonisotropic, the unbalance response is at least poly-harmonic and then fast variation can occur in both the rotating and nonrotating frames. By introducing the rotating coordinates expressed by Eq. (24.38) into Eq. (26.13), it yields

$$
\begin{aligned}
&\mathbf{M}_m\ddot{\mathbf{r}} + \left[\mathbf{C}_m + i\Omega\left(2\mathbf{M}_m - \mathbf{G}\right)\right]\dot{\mathbf{r}} + \left[\mathbf{K}_m - \Omega^2\left(\mathbf{M}_m - \mathbf{G}\right) + \right.\\
&+i\left(a\mathbf{M}_m + \Omega\mathbf{C}_{r_m}\right)\bigg]\mathbf{r} + \mathbf{M}_{n_d}e^{-2i\theta}\ddot{\overline{\mathbf{r}}} + \left(\mathbf{C}_{n_d} + 2i\Omega\mathbf{M}_{n_d}\right)e^{-2i\theta}\dot{\overline{\mathbf{r}}} + \\
&+\left[\mathbf{K}_{n_d} - \Omega^2\mathbf{M}_{n_d} - i\left(a\mathbf{M}_{n_d} + \Omega\mathbf{C}_{n_d}\right)\right]e^{-2i\theta}\overline{\mathbf{r}} + \mathbf{M}_{r_d}\ddot{\overline{\mathbf{r}}} + \mathbf{C}_{r_d}\dot{\overline{\mathbf{r}}} +
\end{aligned}
$$

$$+\left(\mathbf{K}_{r_d} + \Omega^2 \mathbf{M}_{r_d} - ia\mathbf{M}_{r_d}\right)\bar{\mathbf{r}} + \mathbf{f}'(r_i, \dot{r}_i, \theta, t)\}e^{-i\theta} =$$

$$= \left(\Omega^2 - ia\right)\mathbf{f}_r + \mathbf{f}_n e^{-i\theta} . \tag{26.14}$$

The equation yielding the driving torque is still the second Eq. (24.50).

If the angular acceleration is not vanishingly small, Eq. (26.14) must be integrated numerically in time, and as a consequence there is no conceptual difficulty taking into account both nonlinearities and asymmetry. If software allowing us to deal with complex quantities is available, there is no need to split the equation into its real and imaginary parts before integration: Once laws $a(t)$, $\Omega(t)$, and $\theta(t)$ are stated, the numerical integration is straightforward. In general, Eq. (26.14) is a powerful tool to study by numerical integration in time many difficult rotordynamics problems, like constant speed whirling of unsymmetrical rotors or chaotic behavior due to the simultaneous presence of nonlinearities and asymmetry.

A simple example of the behavior of a nonlinear rotor performing a constant rate acceleration–deceleration cycle is shown in Fig. 26.5. The system is a nonlinear Jeffcott rotor of the same type already studied in Fig. 26.1, and the results are shown in nondimensional form, as in Fig. 23.14. The values of the nondimensional parameters used for the simulation are

$$a^* = 0.01 , \quad \zeta_r = 0.01 , \quad \zeta_n = 0.10 , \quad \mu\epsilon^2 = 0.03 .$$

In the same figure, the solutions obtained from the usual steady-state approach are reported (dashed lines). Note that the values of the parameters are such that the system can experience a downward jump and achieve

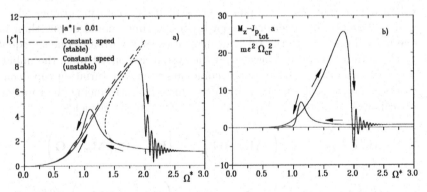

FIGURE 26.5. Acceleration and deceleration of a nonlinear Jeffcott rotor through the critical speed; (a) amplitude of the motion, compared with the amplitude in steady-state operation; (b) driving torque. ($a^* = 0.01$, $\zeta_r = 0.01$, $\zeta_n = 0.10$ and $\mu\epsilon^2 = 0.03$; Ω_{cr} is the critical speed of the linearized system.)

a self-centered condition. As in the linear case, the presence of the angular acceleration reduces the peak amplitude; however, in the linear case the displacement peak occurs at a higher speed than that characterizing the maximum steady-state amplitude, but the jump here occurs at a lower speed. The curve related to spin down is different from that related to the acceleration phase. This also occurs for linear rotors, when angular acceleration is accounted for, but here this effect is far greater, because both acceleration and nonlinearity contribute to it.

If the system is axially symmetrical and the spin speed is constant, Eq. (26.13) can be simplified as

$$\mathbf{M\ddot{q}} + \left(\mathbf{C}_n + \mathbf{C}_r - i\Omega\mathbf{G}\right)\mathbf{\dot{q}} + \left(\mathbf{K} - i\Omega\mathbf{C}_r\right)\mathbf{q} +$$

$$+\mathbf{f}(q_{0_i}, \dot{q}_{0_i}) = \Omega^2\mathbf{f}_r e^{i\Omega t} + \mathbf{f}_n .$$

(26.15)

If no nonrotating force acts on the structure, the unbalance response is a circular whirling that can be expressed by the solution

$$\mathbf{z} = \mathbf{z}_0 e^{i\Omega t} .$$

It must be stressed again that circular whirling is an exact solution of the equation of motion, not just an approximation of the fundamental harmonic of the response, as is customary in nonlinear vibrating systems. Introducing this solution into Eq. (26.15) and considering the presence of a nonrotating structural damping matrix $\mathbf{K}_n''$, the unbalance response of a nonlinear rotor with many degrees of freedom is readily obtained

$$\left(\mathbf{K} - \Omega^2(\mathbf{M} - \mathbf{G}) + i(\Omega\mathbf{C}_n + \mathbf{K}_n'')\right)\mathbf{z}_0 + \mathbf{f}_i(z_{0_i}) = \Omega^2\mathbf{f}_r. \quad (26.16)$$

To reduce the size of the problem, the condensation techniques described in Chapter 10 can be used, particularly when fewer generalized coordinates are directly linked with nonlinearities, as when dealing with a rotor that is linear in itself but runs on nonlinear bearings. A first reduction of the number of degrees of freedom is performed by eliminating those degrees of freedom not directly involved in nonlinearities with which a small fraction of the total mass is associated. This procedure, basically a Guyan reduction, introduces approximations, which are usually very small if the degrees of freedom to be dropped are chosen with care. The equations of motion are then divided into two groups: a linear and a nonlinear set of equations. A second dynamic condensation procedure, that must be applied at each value of the spin speed at which the unbalance response is to be obtained, can then follow to eliminate all the linear degrees of freedom.

In some particular cases the solution of the nonlinear problem can easily be performed, because it reduces to a single nonlinear equation. This includes the case of an undamped linear rotor supported by two nonlinear bearings, one of which has a behavior that can be approximated by a cubic characteristic. For the solution of the general problem, in which the nonlinear functions can have complicated expressions and there are many degrees of freedom, the Newton–Raphson method seems to be the most appropriate choice. In this case it is advisable to write the equation of motion using real coordinates instead of complex ones.

The iterative algorithm allowing the computation of the solution of Eq. (26.16) at the $(i+1)$th iteration from that at the ith iteration is

$$\mathbf{x}_{i+1} = \mathbf{x}_i - h\mathbf{S}(x_i)^{-1}\mathbf{p}(x_i), \qquad (26.17)$$

where h is a relaxation constant, and the vector of the unknown $\mathbf{x}$, the elements of Jacobian matrix $\mathbf{S}(x)$, and the functions $\mathbf{p}(x)$ are defined as

$$\mathbf{x} = \left\{ \begin{array}{c} \Re(\mathbf{z}_0) \\ \Im(\mathbf{z}_0) \end{array} \right\} = \left\{ \begin{array}{c} \mathbf{x}_\xi \\ \mathbf{x}_\eta \end{array} \right\}, \qquad S_{ij} = \frac{\partial p_i(x)}{\partial x_j},$$

$$\mathbf{p}(\mathbf{x}) = \left[\begin{array}{cc} \mathbf{K} - \Omega^2(\mathbf{M} - \mathbf{G}) & -\Omega(\mathbf{C}_n + \mathbf{K}_n'') \\ \Omega(\mathbf{C}_n + \mathbf{K}_n'') & \mathbf{K} - \Omega^2(\mathbf{M} - \mathbf{G}) \end{array} \right] \mathbf{x} +$$

$$+ \{f(\mathbf{x})\} - \Omega^2 \left\{ \begin{array}{c} \Re(\mathbf{f})_r \\ \Im(\mathbf{f})_r \end{array} \right\}. \qquad (26.18)$$

If there is no damping in the system and the unbalance distribution is contained in a plane (i.e., vector $\mathbf{f}_r$ is real), the computation of the response is much simpler because vector $\mathbf{z}_0$ is also real. All the aforementioned equations still hold, but they can be written directly using the unknowns $\mathbf{z}_0$. The number of equations is, consequently, halved, and, in some particular cases, the solution of the nonlinear part can be reduced to the solution of a single nonlinear equation.

Because the backbone curves are related to the undamped system, they can be computed using the same approach as for the latter, simply by neglecting the forcing vector $\mathbf{f}_r$ in the definition of functions $\mathbf{p}(\mathbf{z}_0)$.

The convergence of the generalized Newton–Raphson method can be, in some cases, a source of potential problems. In many cases, the basins of attraction of the solutions take very complicated shapes, or the computation can lock itself into a cycle without reaching any actual solution of the basic equation. For example, the domains of attraction of the various solutions plotted for the Jeffcott rotor already studied in Fig. 26.4 are reported in Fig. 26.6. The map was plotted for a value of the nondimensional speed $\Omega^* = 2$, at which three solutions (two stable and an unstable one) exist.

The structure of the map is fractal, as can be seen by enlarging selected zones, and the unstable solution also has a nonvanishing attraction domain. A consequence of the fractal structure of the map is the fact that there are zones in which very small changes of the initial values assumed for

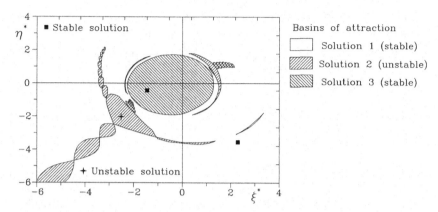

FIGURE 26.6. Basins of attraction of the solutions for the equilibrium positions of the system of Fig. 26.1 at a speed $\Omega^* = 2$, computed through the Newton–Raphson iterative technique.

the computation cause large differences on the solution obtained. When comparing the maps of Figs. 26.3 and 26.5 it is clear that the fractal nature of the second is strictly linked with the mathematical procedure used for the solution of the equation (i.e., the Newton–Raphson technique) and has nothing to do with the actual physical behavior of the system.

No improvement of the mathematical behavior of the equations has been obtained by introducing a relaxation factor. The map of the domains of attraction changes but retains its fractal structure and the characteristics of showing a domain of attraction of the unstable solution; however, the number of iterations required often increases.

The results obtained for the Jeffcott rotor show the fairly good convergence characteristics of the Newton–Raphson technique. Obviously this does not guarantee an equally well-behaved nature of the general mathematical model for systems with many degrees of freedom. The following considerations, however, can be extrapolated and used as guidelines in the solution of more complex problems:

- It seems that numerical damping is not of great use in avoiding non-convergence.
- Attractive stable cycles can be found, particularly in the fields just above a critical speed of the linearized system. After each iteration, it is necessary to check not only whether convergence has been obtained, but also whether the computation has been locked in a stable cycle.
- If this circumstance occurs, the computation can be started again using a different set of initial values. When operating in a self-centered branch, a good guess could be trying to start from a vector **x** obtained by multiplying the one of the preceding attempt by a constant smaller than one; in a high branch of the response a constant greater than 1 can be used.

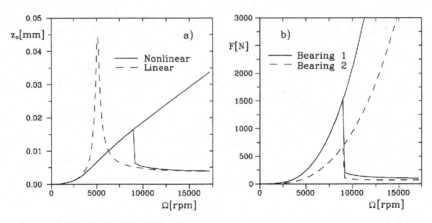

FIGURE 26.7. Amplitude of the orbit of the center of gravity of the (a) nonlinear and linearized rotors and (b) bearing forces as functions of the speed.

- The fractal nature of the domains of attraction can cause large differences in the results to be produced by small changes in the trial vector $\mathbf{x}$. The solution can also converge on an unstable branch of the response.

- The stable branches should be obtained using as a trial vector the result obtained in the previous computation.

- The aforementioned considerations also hold for the computation of the various branches of the backbone. Here, however, a trivial solution with all the elements of $\mathbf{x}$ vanishingly small exists and, therefore, the procedure must be started outside the basin of attraction of the trivial solution. A suggestion is to start at a speed just above a critical speed of the linearized system (for hardening systems) with a trial vector proportional to the eigenvector of the linearized system corresponding to the mentioned critical speed.

Example 26.2 *A flywheel, with a working range between 8,500 and 17,000 rpm, has a mass $m = 125$ kg and moments of inertia $J_p = 2.272$ kg m^2 and $J_t = 1.477$ kg m^2.*
It runs on a pair of ball bearings whose nonlinear characteristic can be approximated by the expression

$$F = -2 \times 10^7 (1 + 10^{10}(|z|)^2)z$$

and whose damping properties can be modeled as hysteretic damping applied only to the linear part of the stiffness with loss factor $\eta = 0.08$.

Knowing that the center of mass is at 30% of the span between the bearings (400 mm), study the response to a static unbalance $m\epsilon = 438 \times 10^{-6}$ kg m and state whether self-centering is possible. Compute the forces exerted on the stator.

Because the distances of the center of mass of the rotor from the bearings are $a = 120$ mm and $b = 280$ mm, respectively, the matrices entering into the mathematical model of the linearized system are

$$\mathbf{M} = \begin{bmatrix} 125 & 0 \\ 0 & 1.477 \end{bmatrix}, \quad \mathbf{G} = \begin{bmatrix} 0 & 0 \\ 0 & 2.272 \end{bmatrix},$$

$$\mathbf{K} = k \begin{bmatrix} 2 & a-b \\ a-b & a^2+b^2 \end{bmatrix} = \begin{bmatrix} 40 & -3.2 \\ -3.2 & 1.856 \end{bmatrix} \times 10^6.$$

In this case it is expedient to use as generalized coordinates the displacements at the bearing locations. The complex coordinates z and ϕ can be obtained from the complex coordinates z_1 and z_2 using a transformation matrix $\mathbf{T}$

$$\left\{ \begin{array}{c} z \\ \phi \end{array} \right\} = \mathbf{T} \left\{ \begin{array}{c} z_1 \\ z_2 \end{array} \right\} = \frac{1}{l} \begin{bmatrix} b & a \\ 1 & -1 \end{bmatrix} \left\{ \begin{array}{c} z_1 \\ z_2 \end{array} \right\}.$$

The equation of motion of the undamped system is thus

$$\mathbf{M}^* \left\{ \begin{array}{c} \ddot{z}_1 \\ \ddot{z}_2 \end{array} \right\} - i\Omega \mathbf{G}^* \left\{ \begin{array}{c} \dot{z}_1 \\ \dot{z}_2 \end{array} \right\} + \mathbf{K}^* \left\{ \begin{array}{c} z_1 \\ z_2 \end{array} \right\} + k\mu \left\{ \begin{array}{c} |z_1|^2 z_1 \\ |z_2|^2 z_2 \end{array} \right\} = \Omega^2 \mathbf{f}^* e^{i\Omega t},$$

where

$$\mathbf{M}^* = \mathbf{T}^T \mathbf{M} \mathbf{T}, \quad \mathbf{G}^* = \mathbf{T}^T \mathbf{G} \mathbf{T}$$

$$\mathbf{K}^* = \mathbf{T}^T \mathbf{K} \mathbf{T} = \begin{bmatrix} k & 0 \\ 0 & k \end{bmatrix}, \quad \mathbf{f}^* = \mathbf{T}^T \mathbf{f} = \frac{m\epsilon}{l} \left\{ \begin{array}{c} b \\ a \end{array} \right\}.$$

The solution for circular whirling can be obtained from Eq. (26.16):

$$\left(-\Omega^2 (\mathbf{M}^* - \mathbf{G}^*) + \mathbf{K}^*(1 + i\eta) \right) \left\{ \begin{array}{c} z_{1_0} \\ z_{2_0} \end{array} \right\} + k\mu \left\{ \begin{array}{c} |z_{1_0}|^2 z_{1_0} \\ |z_{2_0}|^2 z_{2_0} \end{array} \right\} = \Omega^2 \mathbf{f}^*.$$

The solution can be performed using the Newton–Raphson algorithm. The expression of the Jacobian matrix is simply

$$\mathbf{S} = \left(-\Omega^2 (\mathbf{M}^* - \mathbf{G}^*) + \mathbf{K}^*(1 + i\eta) \right) + 3k\mu \begin{bmatrix} |z_{1_0}|^2 & 0 \\ 0 & |z_{2_0}|^2 \end{bmatrix}.$$

Note that all matrices and vectors are complex and, if software allowing the use of complex numbers is available, there is no need to double the number of equations to work with real numbers.

The solution is performed twice. By increasing the speed from zero, the higher branch is obtained, and by reducing the speed from the maximum value the lower, self-centered branch is computed. At each speed, the solution obtained in the previous computation is chosen as the starting solution. The results are plotted in Fig. 26.7. From the figure it is clear that self-centering is not possible, due to the low value of nonrotating damping. Correspondingly, the forces on the bearings are very high.

26.6   Exercises

Exercise 26.1 *Consider the Jeffcott rotor with nonlinear Duffing-type restoring force and linear damping studied in Fig. 26.1 and following. Compute the unbalance response, the backbone, and the limit envelope for the case with $c_n = 400$ Ns/m. Repeat the computation for $c_n = 600$ Ns/m. Data: $m = 10$ kg, $k = 10^5$ N/m, $\mu = 2 \times 10^9$ $1/m^2$, $\epsilon = 10^{-5}$ m, $\Omega = 200$ rad/s.*

Exercise 26.2 *Consider the same Jeffcott rotor of Exercise 26.1. Choose four values of the initial conditions belonging to different domains of attraction and integrate numerically the equation of motion until the system settles in a circular whirling. Verify whether the steady-state solution is one of the exact solutions of the equation of motion and whether the solution corresponds to that obtained through Newton–Raphson technique. Data: $m = 10$ kg, $k = 10^5$ N/m, $\mu = 2 \times 10^9$ $1/m^2$, $c_n = 400$ Ns/m, $c_r = 60$ Ns/m, $\epsilon = 10^{-5}$ m, $\Omega = 200$ rad/s. Suggestion: perform the computations in the rotating frame with a nondimensional formulation of the equation of motion and use the starting conditions shown in Fig. 26.3.*

Exercise 26.3 *Compute the unbalance response of an undamped Jeffcott rotor with Duffing-type nonlinearity by using Newton–Raphson technique and plot a map showing, for each value of the speed, toward which solution the computation converges with the various value of the starting amplitudes (consider values of z_0^* within the range $[-5, 5]$). Does a set of points exist from which no convergence is obtained?*

Exercise 26.4 *Compute the response of the system of the previous exercise without neglecting damping and using Newton–Raphson algorithm. Plot a map of the basins of attractions of the solution in the $\xi^* \eta^*$ plane (within the range $[-6, 4]$) at constant speed. Choose a value $\Omega^* = 2$ for the nondimensional speed and $\zeta_n = 0.2$ for the nonrotating damping.*

Exercise 26.5 *Compute the unbalance response of the flywheel of Exercise 24.5 assuming that the bearings have the hardening characteristic $F = 2 \times 10^7(1 + 10^{10}z^2)z$ and that the hysteretic damping, with $\eta = 0.2$, applies only to the linear part of the stiffness.*

Study the response obtained and state whether self-centering is possible. Compute the forces exerted on the stator.

27

Dynamic Problems of Rotating Machines

The behavior of a rotor may be strongly influenced by that of its bearings and dampers, particularly when they are intrinsically nonlinear. The main characteristics of lubricated and magnetic bearings and whirl dampers are here summarized. The important topic of rotating machinery diagnostic is also touched upon.

27.1   Rotors on hydrodynamic bearings

27.1.1   Oil whirl and oil whip

It is very well known that rotors running on journal bearings show particular dynamic problems due to the fluid film, usually referred to as *oil whirl* and *oil whip*.

The first is a whirling of the rotor, taking place at a frequency of about half the speed of rotation (sometimes referred to as *half-frequency whirl*) and is superimposed to the other whirling motions, particularly the synchronous whirling due to unbalance. Its amplitude is usually not large and does not constitute an actual problem. At a speed that is usually not far from twice the first critical speed, the motion becomes more severe and can rapidly degenerate in a very violent, sometimes destructive, whirling that takes place at a frequency almost independent of the speed and coincident with the first natural whirl frequency of the rotor. This motion is usually referred to as *oil whip*.

A simple heuristic explanation of the phenomenon states that the whirling takes place at about half the rotational frequency by noting that the oil in the bearing moves around at a speed that is about half the peripheral velocity of the journal, providing a sort of rotating damping whose speed of rotation is close to half the spin speed. The behavior of the system is unstable in the region of the Campbell diagram lying below the straight line $\omega = \Omega/2$ and the threshold of instability can be found easily by intersecting it with the lowest branch of the Campbell diagram (Fig. 27.1). If the lowest branch of the Campbell diagram is a horizontal straight line, as is the case of the Jeffcott rotor, the threshold of instability occurs at twice the critical speed of the rotor.

The mentioned heuristic explanation also accounts for the fact that the frequency of the oil whirl is slightly lower than half the rotational speed, usually in the range of 0.45–0.48 Ω: The average velocity of the oil film is slightly lower than half the peripheral speed of the journal, depending on the clearance and the exact velocity profile.

However, the phenomenon is more complex, and the actual behavior of the lubricant film must be modeled in some detail. In particular, the intrinsic nonlinear nature of the bearing cannot be neglected, which makes the study of the dynamic behavior of the system more complex than the simple linear rotordynamic study that is often sufficient when no allowance is taken for the compliance of journal bearings. In particular, the whirl–whip transition is not as abrupt as shown in Fig. 27.1.

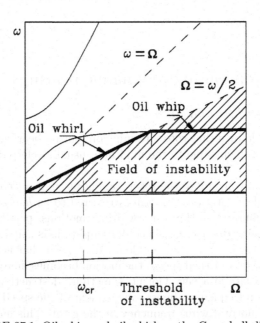

FIGURE 27.1. Oil whip and oil whirl on the Campbell diagram.

27.1.2   Forces exerted by the oil film on the journal

Consider the journal bearing sketched in Fig. 27.2a. Assume that the bearing is perfectly aligned, i.e., the axes of the bearing and of the journal are parallel. The nonrotating reference frame $Oxyz$ is centered in the center of the bearing and has the directions of its axes fixed in space, while the directions of the axes of reference frame $Ox'y'z$, whose axis x' contains the center of the journal C, are not fixed if point C moves about point O. Initially, only *stationary* motion will be considered: the coordinates x and y of the center of the journal are independent from time t and, consequently, reference frame $Ox'y'z$ is fixed in space and angle β is constant.

Assume that the pressure is linked to the thickness of the fluid film by the well-known Reynolds equation[1]

$$\frac{1}{6}\left[\frac{1}{R_j^2}\frac{\partial}{\partial\theta}\left(\frac{h^3}{\mu}\frac{\partial p}{\partial\theta}\right)+\frac{\partial}{\partial z}\left(\frac{h^3}{\mu}\frac{\partial p}{\partial z}\right)\right]=\Omega\frac{\partial h}{\partial\theta}+2\frac{\partial h}{\partial t}, \qquad (27.1)$$

where μ is the viscosity of the lubricant and

$$\Omega=\Omega_j-\Omega_b,$$

where subscripts b and j refer to the bearing and the journal, respectively. Usually, the bearing does not rotate; hereafter, it will be assumed that

$$\Omega_b=0, \quad \Omega_j=\Omega.$$

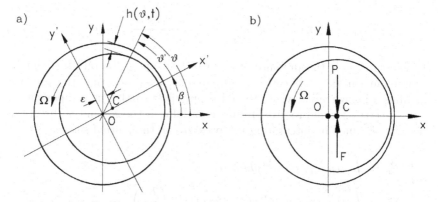

FIGURE 27.2. Lubricated journal bearing: **(a)** geometrical definitions, **(b)** position of the journal when a load P acts along the y-axis; solution obtained for stationary conditions from Eq. (27.5).

---

[1] O. Reynolds, 'On the theory of lubrication and its applications to Mr. Towers' experiments, *Philosophical Transaction of the Royal Society of London*, Vol. 177, (1886), 154–234.

The film thickness h is easily expressed as a function of the coordinates of the center of the journal with respect to the center of the bearing

$$h = c[1 - x^* \cos(\theta) - y^* \sin(\theta)] = c[1 - \epsilon^* \cos(\theta')], \qquad (27.2)$$

where the clearance c is simply given by the difference of the radii

$$c = R_b - R_j,$$

and the nondimensional eccentricity and coordinates are

$$\epsilon^* = \frac{\epsilon}{c}, \quad x^* = \frac{x}{c}, \quad y^* = \frac{y}{c}.$$

Equation (27.1) can be used to obtain the pressure distribution in the fluid film. However, if a numerical solution is considered as not general enough, some simplifications must be introduced to allow the pressure to be computed in closed form. If the bearing is assumed to be very long, it is possible to neglect the fluid flow and pressure gradient in axial direction, obtaining the so-called long-bearing approximation, often associated with the name of Sommerfeld. Equation (27.1) reduces to

$$\frac{1}{6R_j^2} \frac{\partial}{\partial \theta} \left(\frac{h^3}{\mu} \frac{\partial p}{\partial \theta} \right) = (\Omega x - 2\dot{y}) \sin(\theta) - (\Omega y + 2\dot{x}) \cos(\theta). \qquad (27.3)$$

In stationary conditions, the pressure distribution around the journal can be expressed by the relationship

$$p - p_0 = -6\mu\Omega \left(\frac{R_j}{c} \right)^2 \frac{\epsilon^*}{2 + \epsilon^{*2}} \frac{2 - \epsilon^* \cos(\theta')}{[1 - \epsilon^* \cos(\theta')]^3} \sin(\theta'). \qquad (27.4)$$

Pressure p_0 is the pressure at $\theta' = 0$, i.e., where the oil film is at its minimum thickness. It can be computed by assuming that the pressure attains a known value where the oil supply is located. The components of the force the journal receives from the oil film in the x'- and y'-directions can be obtained by integrating the pressure on the journal surface

$$\begin{cases} F_{x'} = \displaystyle\int_0^{2\pi} (p - p_0) \cos(\theta') d\theta = 0, \\[2mm] F_{y'} = \displaystyle\int_0^{2\pi} (p - p_0) \sin(\theta') d\theta = 12\pi\mu\Omega R_j l \left(\frac{R_j}{c} \right)^2 \frac{\epsilon^*}{(2 + \epsilon^{*2})\sqrt{1 - \epsilon^{*2}}}, \end{cases}$$
$$(27.5)$$

where l is the length of the bearing in the axial direction.

The force exerted by the oil film on the journal is thus directed along the y'-axis and the displacement occurs in the x'-direction. The journal is displaced in a direction perpendicular to the direction of the load, as shown in Fig. 27.2b, in which the load P acts in the vertical direction.

By linearizing the expression of the forces about the central position, the following expression is obtained:

$$\left\{ \begin{array}{c} F_x \\ F_y \end{array} \right\} = -\mathbf{K} \left\{ \begin{array}{c} x \\ y \end{array} \right\} = -6\pi\mu\Omega l \left(\frac{R_j}{c} \right)^3 \left[\begin{array}{cc} 0 & 1 \\ -1 & 0 \end{array} \right] \left\{ \begin{array}{c} x \\ y \end{array} \right\}. \qquad (27.6)$$

Remark 27.1 *The stiffness matrix obtained from the linearization of the bearing has vanishing elements on the main diagonal and is skew-symmetrical. Although the first feature is linked to the particular oversimplified formulation used, the non-symmetrical structure of the stiffness matrix is general.*

In a similar way, if the center of the journal C moves, i.e., if its x- and y-coordinates are not constant in time, it is possible to compute the forces due to this motion and then, by linearizing about the central position of the journal, to obtain a damping matrix

$$\left\{ \begin{array}{c} F_x \\ F_y \end{array} \right\} = -\mathbf{C} \left\{ \begin{array}{c} \dot{x} \\ \dot{y} \end{array} \right\} = -12\pi\mu l \left(\frac{R_j}{c} \right)^3 \left[\begin{array}{cc} 1 & 0 \\ 0 & 1 \end{array} \right] \left\{ \begin{array}{c} \dot{x} \\ \dot{y} \end{array} \right\}. \qquad (27.7)$$

Consider a circular whirling with very small amplitude ϵ and whirl speed ω. The position and velocity of the center of the journal are

$$\left\{ \begin{array}{c} x \\ y \end{array} \right\} = \epsilon \left\{ \begin{array}{c} \cos(\omega t) \\ \sin(\omega t) \end{array} \right\}, \qquad \left\{ \begin{array}{c} \dot{x} \\ \dot{y} \end{array} \right\} = \epsilon \left\{ \begin{array}{c} -\sin(\omega t) \\ \cos(\omega t) \end{array} \right\}, \qquad (27.8)$$

and the force that the journal receives from the oil film is

$$\left\{ \begin{array}{c} F_x \\ F_y \end{array} \right\} = -12\pi\mu l\epsilon \left(\frac{R_j}{c} \right)^3 \left(\frac{\Omega}{2} - \omega \right) \left\{ \begin{array}{c} \sin(\omega t) \\ -\cos(\omega t) \end{array} \right\}. \qquad (27.9)$$

This force is directed tangentially to the orbit of point C and, if $\omega < \Omega/2$, its direction is the same as the velocity of point C. The force drives the shaft along its motion with destabilizing effects.

If, however, $\omega > \Omega/2$, the direction of the force is opposite that of the velocity and it opposes the whirling motion with stabilizing effects. The same results already seen and sketched in Fig. 27.1 are then obtained. They are, however, affected by the assumptions used, mainly by the linearization about the central position, which is acceptable only in the case of very lightly loaded bearings, and by the pressure distribution expressed by Eq. (27.4) and plotted as a function of angle θ' in Fig. 27.3 for some selected values of the eccentricity ϵ^*.

From the figure, it is clear that when the eccentricity is high and the inlet pressure low, very low values of the absolute pressure can be reached in some parts of the oil film, or even negative absolute pressures, which is physically without meaning. When the pressure becomes lower than the vapor pressure of the lubricant at the relevant temperature, cavitation occurs and the oil film ruptures.

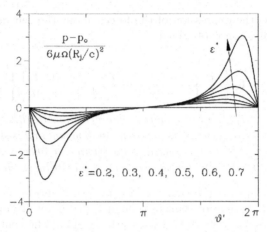

FIGURE 27.3. Pressure distribution on the journal along angle θ' for various values of the nondimensional eccentricity ϵ^*. Long-bearing assumption, Eq. (27.4).

A usual approach for the study of bearings with partially cavitated oil film is assuming that the pressure in the non-cavitated part of the bearing is equal to that computed by assuming a complete oil film. Because the pressure in the cavitated part of the bearing can be neglected, the same formulas already seen for the computation of the forces can be used, provided that the integration is performed between angles θ_1 and θ_2, defining the region on which the oil film extends.

A simple approach, usually referred to as *fully cavitated bearing*, is assuming that the oil film extends between $\theta' = \pi$ and $\theta' = 2\pi$, i.e., in the region in which the value of the pressure is higher than p_0. Once angles θ_1 and θ_2 have been defined, the static forces corresponding to a given displacement of the journal, i.e., to a pair of values x_{st} and y_{st} of the coordinates of point C, can be computed. For small motions about the static equilibrium position it is possible to linearize the expressions of the forces and to compute the stiffness and damping matrices of the bearing. In the case of the long-bearing assumption, the relevant equations are

$$\mathbf{f} = \frac{6\mu R_j^3 l\Omega}{c^2} \left\{ \begin{array}{c} I_7 \\ I_8 \end{array} \right\},$$

$$\mathbf{C} = -\left[\left(\frac{\partial F_i}{\partial \dot{x}_j} \right)_{\substack{x_1 = x_{st} \\ x_2 = y_{st}}} \right] = \frac{6\mu R_j^3 l}{n^2 c^3} \left[\begin{array}{cc} I_9 & I_{10} \\ I_{11} & I_{12} \end{array} \right], \qquad (27.10)$$

$$\mathbf{K} = -\left[\left(\frac{\partial F_i}{\partial x_j} \right)_{\substack{x_1 = x_{st} \\ x_2 = y_{st}}} \right] = -\frac{6\mu R_j^3 l\Omega}{(2+n^2)c^3} \left[\begin{array}{cc} I_1 + I_2 & -I_4 + I_5 \\ I_6 + I_5 & -I_1 + I_3 \end{array} \right],$$

where

$$I_1 = \int_{\theta_1}^{\theta_2} \frac{1+h_{st}^*}{h_{st}^{*3}} \sin(\theta)\cos(\theta)d\theta\,, \qquad I_2 = \int_{\theta_1}^{\theta_2} (2+h_{st}^*)\frac{\omega}{h_{st}^*}\cos^2(\theta)d\theta\,,$$

$$I_3 = \int_{\theta_1}^{\theta_2} (2+h_{st}^*)\frac{\omega}{h_{st}^*}\sin^2(\theta)d\theta\,, \qquad I_4 = \int_{\theta_1}^{\theta_2} \frac{1+h_{st}^*}{h_{st}^{*3}}\cos^2(\theta)d\theta\,,$$

$$I_5 = \int_{\theta_1}^{\theta_2} (2+h_{st}^*)\frac{\omega}{h_{st}^*}\sin(\theta)\cos(\theta)d\theta\,, \qquad I_6 = \int_{\theta_1}^{\theta_2} \frac{1+h_{st}^*}{h_{st}^{*3}}\sin^2(\theta)d\theta\,,$$

$$I_7 = \int_{\theta_1}^{\theta_2} \omega\cos(\theta)d\theta\,, \qquad I_8 = \int_{\theta_1}^{\theta_2} \omega\sin(\theta)d\theta\,,$$

$$I_9 = \int_{\theta_1}^{\theta_2} (Ry_{st}^* + Sx_{st}^*)\cos(\theta)d\theta\,, \qquad I_{10} = \int_{\theta_1}^{\theta_2} (-Rx_{st}^* + Sy_{st}^*)\cos(\theta)d\theta\,,$$

$$I_{11} = \int_{\theta_1}^{\theta_2} (Ry_{st}^* + Sx_{st}^*)\sin(\theta)d\theta\,, \qquad I_{12} = \int_{\theta_1}^{\theta_2} (-Rx_{st}^* + Sy_{st}^*)\sin(\theta)d\theta\,,$$

$$h_{st}^* = 1 - x_{st}^*\cos(\theta) + y_{st}^*\sin(\theta)\,, \qquad \omega = \frac{x_{st}^*\sin(\theta) - y_{st}^*\cos(\theta)}{h_{st}^{*3}}\,,$$

$$R = 2\omega h_{st}^* \frac{1+h_{st}^*}{2+\epsilon^{*2}}\,, \qquad S = \frac{1}{(1+\epsilon^*)^2} - \frac{1}{h_{st}^{*2}}\,.$$

The integrals in Eq. (27.10) must be solved numerically, but this does not imply long and costly computations.

If the bearing is relatively short and the long-bearing approach does not seem applicable, the flow in the circumferential direction can be neglected obtaining the so-called short-bearing approximation. By neglecting the term linked with the circumferential pressure gradients in Reynolds equation (27.1) and introducing into the latter the expression of the film thickness, it follows that

$$\frac{1}{6}\frac{\partial}{\partial z}\left(\frac{h^3}{\mu}\frac{\partial p}{\partial z}\right) = (\Omega x - 2\dot{y})\sin(\theta) - (\Omega y + 2\dot{x})\cos(\theta)\,. \qquad (27.11)$$

By operating in the same way as for the long-bearing Sommerfeld approximation, the following expressions for the forces in static conditions and the stiffness and damping matrices are obtained:

$$\mathbf{f} = \frac{\mu R_j l^3 \Omega}{2c^2}\left\{\begin{array}{c} I_7 \\ I_8 \end{array}\right\}\,,$$

$$\mathbf{C} = -\left[\left(\frac{\partial F_i}{\partial \dot{x}_j}\right)_{\substack{x_1 = x_{st} \\ x_2 = y_{st}}}\right] = \frac{\mu R_j l^3}{c^3}\left[\begin{array}{cc} I_4 & I_1 \\ I_1 & I_6 \end{array}\right]\,, \qquad (27.12)$$

$$\mathbf{K} = -\left[\left(\frac{\partial F_i}{\partial x_j}\right)_{\substack{x_1 = x_{st} \\ x_2 = y_{st}}}\right] = -\frac{\mu R_j l^3 \Omega}{2c^3}\left[\begin{array}{cc} I_1 + I_2 & -I_4 + I_5 \\ I_6 + I_5 & -I_1 + I_3 \end{array}\right]\,,$$

where

$$I_1 = \int_{\theta_1}^{\theta_2} \frac{\sin(\theta)\cos(\theta)}{h_{st}^{*3}}d\theta\,, \qquad I_2 = 3\int_{\theta_1}^{\theta_2} \frac{\omega}{h_{st}^*}\cos^2(\theta)d\theta\,,$$

$$I_3 = 3\int_{\theta_1}^{\theta_2} \frac{\omega}{h_{st}^*}\sin^2(\theta)d\theta\,, \qquad I_4 = \int_{\theta_1}^{\theta_2} \frac{\cos^2(\theta)}{h_{st}^{*3}}d\theta\,,$$

$$I_5 = 3\int_{\theta_1}^{\theta_2} \frac{\omega}{h_{st}^*}\sin(\theta)\cos(\theta)d\theta\,, \qquad I_6 = \int_{\theta_1}^{\theta_2} \frac{\sin^2(\theta)}{h_{st}^{*3}}d\theta\,,$$

$$I_7 = \int_{\theta_1}^{\theta_2} \omega\cos(\theta)d\theta\,, \qquad I_8 = \int_{\theta_1}^{\theta_2} \omega\sin(\theta)d\theta\,,$$

$$h_{st}^* = 1 - x_{st}^*\cos(\theta) - y_{st}^*\sin(\theta)\,, \qquad \omega = \frac{x_{st}^*\sin(\theta) - y_{st}^*\cos(\theta)}{h_{st}^{*3}}\,.$$

For the intermediate case of a bearing with non-negligible length, several authors have suggested approximate solutions, usually obtained from the Sommerfeld solution, multiplied by a function of the axial coordinate. Such approximate pressure distributions are usually more satisfactory than either of the two solutions mentioned earlier, because they take into account both axial and circumferential flows.

A solution of this type was proposed by Warner.[2] The pressure distribution is obtained by multiplying the pressure computed for an infinitely long-bearing p_∞ (Sommerfeld solution) by a coefficient

$$C_W = 1 - \frac{\cosh\left[\left(\frac{2z}{l}\right)\left(\frac{\gamma l}{2R_j}\right)\right]}{\cosh\left(\frac{\gamma l}{2R_j}\right)}\,, \tag{27.13}$$

where

$$\gamma^2 = \frac{\int_{\theta_1}^{\theta_2} [1 + \epsilon^*\cos(\theta)]^3 \left(\frac{dp_\infty}{d\theta}\right)^2 d\theta}{\int_{\theta_1}^{\theta_2} [1 + \epsilon^*\cos(\theta)]^3 p_\infty^2 d\theta}\,.$$

In the case of the Warner solution, if the derivatives of coefficient C_W with respect to the position and velocity are neglected, the expressions of the forces and of the stiffness and damping matrices can be directly obtained by multiplying those obtained for the long-bearing case by the factor

$$C_W' = 1 - \frac{2R_j}{\gamma l}\tanh\left(\frac{\gamma l}{2R_j}\right)\,, \tag{27.14}$$

obtained by integrating coefficient C_W in the axial direction.

---

[2]P.C. Warner, 'Static and dynamic properties of partial journal bearings', *Journal of Basic Engineering*, Trans. ASME, Series D, 85, (1963), 244.

Many other bearing types were developed with the aim of reducing the inherent instability of plain bearings. In the case of complex geometries there is no chance of obtaining a closed-form solution, or at least a solution computed with simple numerical integrations. Either experimental results or a more detailed numerical modeling of the fluid film are thus needed. They are presented in the form of general charts applicable to a given family of bearings. Nondimensional parameters, related to the average pressure

$$p_m = \frac{F}{2R_j l},$$

are commonly used. One of them is the Sommerfeld number, defined as

$$S = \frac{\mu \Omega}{2\pi p_m} \left(\frac{R_j}{c} \right)^2 = \frac{\mu \Omega R_j l}{\pi F} \left(\frac{R_j}{c} \right)^2. \tag{27.15}$$

The Sommerfeld number is well suited to the study of long bearings. For the short-bearing model, the load factor

$$O = \frac{2p_m}{\mu \Omega} \left(\frac{c}{R_j} \right)^2 \left(\frac{2R_j}{l} \right)^2 = \frac{F}{\mu \Omega l R_j} \left(\frac{c}{R_j} \right)^2 \left(\frac{2R_j}{l} \right)^2 \tag{27.16}$$

is commonly used. Barwell in 1956 proposed calling the load factor the Ocvirk number, and here it will be referred to with the symbol O. The two nondimensional parameters are linked by the relationship

$$O = \frac{1}{\pi S} \left(\frac{2R_j}{l} \right)^2. \tag{27.17}$$

The position of the center of the journal in stationary conditions is univocally determined, given a certain type of bearing, once the Sommerfeld number (or any other relevant nondimensional parameter) is stated. Charts giving the nondimensional coordinates of the center of the bearing x^* and y^*, or better, the eccentricity ϵ^* and the attitude angle β, as functions of the Sommerfeld number, summarize the static behavior of the bearing. The attitude angle defines the direction of the displacement of the center of the journal with respect to the direction of the force F, as shown in Fig. 27.4a. In the case of a non-cavitated long bearing, the attitude angle is equal to 90° for any value of the Sommerfeld number.

In many cases, the minimum thickness of the oil film

$$h^*_{min} = 1 - \epsilon^*$$

is given instead of the eccentricity. The plot of the eccentricity and the attitude angle as functions of the load factor for a fully cavitated short bearing is reported in Fig. 27.4b.

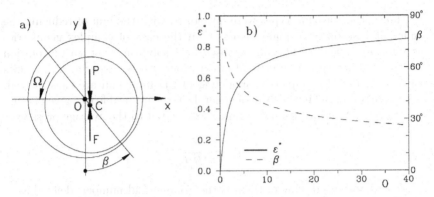

FIGURE 27.4. Lubricated journal bearing: (**a**) definition of the attitude angle β (load P acting along y-axis); (**b**) eccentricity and attitude angle as functions of the load factor for a fully cavitated short bearing.

In Fig. 27.5 the elements of the stiffness and damping matrices for the same bearing in Fig. 27.4b are reported in nondimensional form as functions of the load factor. Similar graphs for a grooved bearing with ratio

$$\frac{l}{2R_j} = 0.5$$

are reported in Fig. 27.6. Graphs of the type shown in Fig. 27.4, 27.5, and 27.6 are reported in the literature for different types of lubricated bearings.[3]

27.1.3   Interaction between the behavior of the oil film and that of the structure

The behavior of lubricated journal bearings was studied first by Robertson in 1933,[4] who investigated the stability of the ideal 360° infinitely long journal bearing. Using the expressions for the film forces obtained by Harrison in 1913,[5] he concluded that the rotor will be unstable at all speeds rather than at speeds above twice the first critical speed. This is easily ascribed to the fact that with a non-cavitated oil film the attitude angle, i.e., the angle between the direction of the load and that of the displacement of the journal, is 90° and the radial stiffness of the bearing vanishes.

Once a model of the bearing is introduced into the model of the rotor, a nonlinear problem results. The static equilibrium position, i.e., the

---

[3]Many charts and tables for bearings of different types are reported in T. Someya, *Journal-Bearing Databook*, Springer, Berlin, 1989.

[4]D. Robertson, 'Whirling of a journal in a sleeve bearing', *Philosophical Magazime.*, Series 7, Vol. 15, (1933), 113–130.

[5]W.J. Harrison, 'The hydrodynamical theory of lubrication with special reference to air as lubricant', *Transaction of Cambridge Philosophical Society*, Vol. 22, (1913), 39.

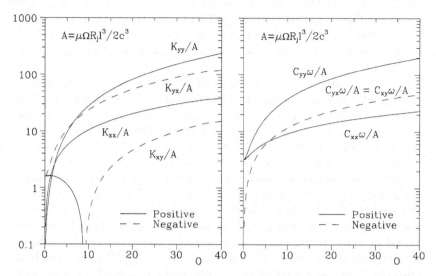

FIGURE 27.5. Elements of the stiffness and damping matrices for the same bearing as Fig. 27.4b as functions of the load factor. The static force is assumed to act in the y-direction.

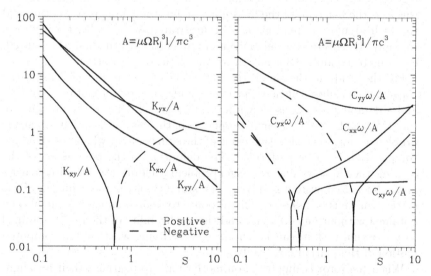

FIGURE 27.6. Elements of the stiffness and damping matrices for a grooved bearing as functions of the Sommerfeld number: ratio $l/2R_j = 0.5$. (J.S. Rao, *Rotor dynamics*, Wiley Eastern, New Delhi, 1983, 110–111.) The plots have been transformed to report the curves in a way that is more consistent with Fig. 27.5.

position the rotating shaft takes under the effects of static loads, is first obtained. The motion about this equilibrium position is then studied using a linearized model. This approach gives a general picture of the behavior of the rotor where the small oscillations are concerned and allows to study its stability in the small.

If a small unbalance is present, provided it gives way to displacements small enough not to go beyond the field in which the linearization holds, the unbalance response can be studied and the relevant results can be superimposed on those related to the free whirling. The well-known statement that the whirling (and eventually whipping) due to the lubricant does not interact with unbalance response holds exactly in this condition.

In this study the rotor is assumed to be axially symmetrical, while no similar assumption is made with regard to the stator. The intrinsic asymmetry of a loaded journal bearing makes it impossible to exploit the symmetrical characteristics of the stator, and on the contrary, an asymmetrical rotor loaded by static forces will produce an elliptical whirling, which is incompatible with the assumption of the existence of a stationary equilibrium position. If the rotor is not isotropic, it must be replaced by a symmetrical rotor with average properties in the computation of the static equilibrium position.

When the amplitude grows larger the nonlinear behavior cannot be neglected also in the dynamic study. The usual approach is numerically integrating the equations of motion to study the stability in the large. In this case the behavior of the fluid film can be studied in greater detail, and the hydrodynamic equations can be integrated numerically, for instance, to study the effects of misalignments between the bearing and the journal, e.g., due to bending deformations of the shaft, on cavitation.

For the study in the small, the journal bearings are modeled using the conventional eight-coefficient linearized model, i.e., by computing a stiffness and a damping matrix as shown in Section 27.1.2. Due to the influence of the static loads on the dynamic properties of the bearings, a nonlinear static problem is first solved for each value of the spin speed at which the dynamic computation is to be performed. The definition of the parameters of the bearing then follows and the eigenproblem yielding the eigenfrequencies of the linearized system is solved. The stability of the free whirling of the system can thus be assessed. The response to a small unbalance can also be obtained without any difficulty and the applicability of the result obtained to the actual system can be checked by verifying whether the computed motion of the journal is actually small.

When the rotor is supported on more than two bearings, their misalignment can have a large effect on the dynamic behavior of the machine by inducing preloads that affect the stiffness of the bearings. This can also be used by inducing by purpose such preloads through a control system to tailor the dynamic characteristics of the machine. The displacements of the bearing center with respect to the nominal position can be introduced into the model to also allow the study of this effect.

If the system is statically determined, the static loads on the bearings can be computed. Once the forces are known, it is possible to compute the two components of the relative displacement of the journals (with respect to the bearings) and then the eight coefficients characterizing their dynamic behavior. If graphs of the type of Fig. 27.5 are not available, it is possible, at least when the long- or the short-bearing assumptions can be made, to resort to an iterative technique such as the Newton–Raphson method. Note that the solution of the static problem is usually unique, and severe convergence problems are not expected.

If the system is not statically determinate, however, the loads on the bearings depend on the deformation of both the stator and the rotor and on possible misalignments. A coupled problem that is far more complex must be solved. The equation allowing the static deflected configuration of the rotor to be studied is Eq. (24.43). By also introducing structural damping and writing explicitly the forces due to weight, it yields

$$\left[\mathbf{K} - i(\Omega\mathbf{C} + \mathbf{K}'')\right]\mathbf{q} = \mathbf{f}_n + g\mathbf{M}(\boldsymbol{\delta}_x + i\boldsymbol{\delta}_y), \qquad (27.18)$$

where vector $\mathbf{f}_n$ contains static forces, not including weight, while the elements of vector $\boldsymbol{\delta}_x + i\boldsymbol{\delta}_y$ vanish for the rotational degrees of freedom and are equal to the cosines of the angle between the vertical direction and the x- and y-axes for translational coordinates.

Remark 27.2 *The deformed equilibrium position depends on the angular velocity only if there is viscous damping.*

The internal damping of the rotor will be assumed to be of the structural (hysteretic) type, because it allows the computations referred to the rotor to be performed only once. However, there is no difficulty modifying the equations to also take into account viscous damping.

Since when the Newton–Raphson technique is used for the solution of the nonlinear set of equations there is some advantage to resorting to real coordinates, Eq. (27.18) can be rewritten in the form

$$\mathbf{K}^*\left\{\begin{array}{c}\mathbf{x}_r \\ \mathbf{y}_r\end{array}\right\} = \mathbf{f}_r^*, \qquad (27.19)$$

where

$$\mathbf{K}^* = \left[\begin{array}{cc}\mathbf{K} & \mathbf{K}'' \\ -\mathbf{K}'' & \mathbf{K}\end{array}\right], \quad \mathbf{f}_r^* = \left\{\begin{array}{c}\mathbf{f}_{x_n} \\ \mathbf{f}_{y_n}\end{array}\right\} + g\left\{\begin{array}{c}\mathbf{M}\boldsymbol{\delta}_x \\ \mathbf{M}\boldsymbol{\delta}_y\end{array}\right\},$$

and vectors $\mathbf{x}_r$ and $\mathbf{y}_r$ contain the real and imaginary parts, respectively, of the complex coordinates of the rotor. The stiffness matrix is singular because the rotor has been considered unsupported and this part of the system is underconstrained.

It is possible to separate the vectors of the generalized coordinates into two subsets: the first, labeled with subscript 1, containing the displacements at the supporting points (i.e., the centers of the journals), and the second, with subscript 2, containing all other generalized displacements. The hydrodynamic bearings will be assumed to react only to translations of the journal, so the first set contains a number of elements equal to twice the number of the bearings. If the model of the bearing is modified to also include the moment due to angular displacements, the inclusion of the rotational degrees of freedom in this set of displacements is straightforward. By partitioning accordingly the relevant matrices and vectors, it follows that

$$
\begin{bmatrix} \mathbf{K}_{11}^* & \mathbf{K}_{12}^* \\ \mathbf{K}_{21}^* & \mathbf{K}_{22}^* \end{bmatrix} \left\{ \begin{array}{c} \left\{ \begin{array}{c} \mathbf{x} \\ \mathbf{y} \end{array} \right\}_1 \\ \left\{ \begin{array}{c} \mathbf{x} \\ \mathbf{y} \end{array} \right\}_2 \end{array} \right\} = \left\{ \begin{array}{c} \mathbf{f}_1^* \\ \mathbf{f}_2^* \end{array} \right\} . \tag{27.20}
$$

By applying the usual techniques of static reduction and assuming the generalized coordinates of the first group as master degrees of freedom, Eq. (27.18) reduces to

$$
\mathbf{K}_{cond_r} \left\{ \begin{array}{c} \mathbf{x} \\ \mathbf{y} \end{array} \right\}_1 = \mathbf{f}_{cond_r} , \tag{27.21}
$$

where the expressions of the condensed matrices are the usual ones and the internal generalized coordinates of the rotor are expressed by Eq. (10.2).

The model of the stator can be built in a way similar to that seen for the rotor, with two important differences. The stator does not need to be axially symmetrical and, because it is stationary, its static deformation is not affected by its damping. The equilibrium equation for the stator is Eq. (27.19) where subscript r has been replaced by s. Matrices $\mathbf{M}^*$ and $\mathbf{K}^*$ are, in general, symmetrical and contain the coupling terms between the behavior in the xz- and yz-planes. Also, in this case it is possible to separate the vectors of the generalized coordinates into two subsets: the first, labeled with subscript 3, containing the displacements at the supporting points (i.e., the center of the bearings) and the second, with subscript 4, containing all other generalized displacements, and then to resort to static reduction techniques.

The equilibrium equation referred to the displacements of the bearings is then Eq. (27.21) where subscripts r and 1 have been substituted with s and 3, respectively. It can be reduced through the usual algorithm.

Note that because the interface between stator and rotor is represented by the bearings, which react only to translations, the sets of generalized coordinates with subscripts 1 and 3 contain only displacements and no rotations. This allows the use of conventions for the rotations of the stator and the rotor that are not consistent and the use of any standard FEM code to build the model of the stator even if the conventions for rotations about the x-axis are different from those used in the model of the rotor. Any type of element can be used in both models, provided that the displacements at the interface are measured with reference to the same axes.

The relative displacements x_i and y_i of the center of the ith journal with respect to the center of the ith bearing can be easily obtained from the displacements x_{i_1} and y_{i_1} of the former and the displacements x_{i_3} and y_{i_3} of the latter, where subscripts 1 and 3 refer to the partitioning of the generalized coordinates. Taking into account that the center of the ith bearing may be displaced by the quantities Δ_{i_x} and Δ_{i_y}, with respect to the nominal position when no force acts on the bearing, the relative displacement of the journal with respect to the bearing can be expressed as

$$\left\{ \begin{array}{c} x_i \\ y_i \end{array} \right\} = \left\{ \begin{array}{c} x_{i_1} - x_{i_3} - \Delta_{i_x} \\ y_{i_1} - y_{i_3} - \Delta_{i_y} \end{array} \right\}. \tag{27.22}$$

Equation (27.22) allows the computation of the relative displacement of the bearing and then of the forces F_x and F_y that the journal receives from the oil film through numerical integration of the bearing model or the use of experimental graphs. The interaction between stator and rotor can thus be expressed by adding the forces due to the oil film to Eq. (27.21) for the rotor and the corresponding one for the stator

$$\left\{ \begin{array}{c} \mathbf{K}_{cond_r} \left\{ \begin{array}{c} x \\ y \end{array} \right\}_1 + \left\{ \begin{array}{c} \mathbf{f}_x \\ \mathbf{f}_y \end{array} \right\} = \mathbf{f}_{cond_r}, \\ \mathbf{K}_{cond_s} \left\{ \begin{array}{c} x \\ y \end{array} \right\}_3 - \left\{ \begin{array}{c} \mathbf{f}_x \\ \mathbf{f}_y \end{array} \right\} = \mathbf{f}_{cond_s}. \end{array} \right. \tag{27.23}$$

If the number of bearings is m, Eq. (27.23) is a set of $4m$ nonlinear equations with the $4m$ unknowns representing the displacements of the journals and the bearings in the x- and y-directions. The number of nonlinear equations can, however, be reduced because the actual unknowns of the nonlinear part of the equation are the differences between the displacements of the rotor and the stator, which are only $2m$ in number. Because the reduced stiffness matrix of the rotor is generally singular and cannot be inverted, the second set of Eq. (27.23) can be multiplied by $\mathbf{K}_{cond_r}\mathbf{K}_{cond_s}^{-1}$ and added to the first, obtaining the following nonlinear equation, which can be solved by resorting to the Newton–Raphson iterative technique

$$\mathbf{K}_{cond_r}\mathbf{x} + \mathbf{A} \left\{ \begin{array}{c} \mathbf{f}_x(\{\mathbf{x} - \Delta\}) \\ \mathbf{f}_y(\{\mathbf{x} - \Delta\}) \end{array} \right\} = \mathbf{f}^{**}, \tag{27.24}$$

where

$$\mathbf{x} = \left\{ \begin{array}{c} \mathbf{x}_1 - \mathbf{x}_3 \\ \mathbf{y}_1 - \mathbf{y}_3 \end{array} \right\}, \qquad \mathbf{A} = \mathbf{I} + \mathbf{K}_{cond_r}\mathbf{K}_{cond_s}^{-1},$$

$$\mathbf{f}^{**} = \mathbf{f}_{cond_r} - \mathbf{K}_{cond_r} \mathbf{K}_{cond_s}^{-1} \mathbf{f}_{cond_s} \, .$$

Once the values of the relative displacements $\mathbf{x}$ of the bearings in the static equilibrium position have been computed, the eight coefficients that characterize the dynamic behavior of the bearing can be easily obtained. Note that the solution of the static problem must be repeated for each value of the speed at which the natural frequencies are to be computed. Even if no viscous damping is considered and the relevant matrices can be computed only once, the forces in the bearing are strongly influenced by the speed.

The dynamic study in the small can be performed using the model described in Section 25.4.1, because the lubricated journal bearings cause the nonrotating parts of the machine to behave in an anisotropic way. Both real and complex coordinates can be used. Free whirling and unbalance response can be studied, the latter only if the amplitude of the response is small enough not to exceed the linearity limits.

Remark 27.3 *If the static forces acting on the bearings are vanishingly small, as in the case of a vertical bearing without static loading, the linearized model can supply only a very rough approximation of the actual behavior of the system. The system is generally unstable in the small and only a true nonlinear analysis allows the limit cycle to be found.*

To study the motion in the large and to find the limit cycles that can occur when the motion in the small is unstable, it is necessary to resort to numerical integration of the equations of motion. This can be performed using the same model described earlier, even if more detailed models of the oil film, also including angular misalignments of the bearings, are described in the literature. The inherent instability of plain journal bearings makes them unsuitable for high-speed supercritical machinery and many other bearing configurations have been developed with the aim of overcoming this difficulty. In particular, tilting pad bearings allow the instability problem to be solved completely at the cost of a reduction of damping at low speed and of added overall complexity. Among the many papers and books existing on this subject, those by Tondl and Muszynska[6] are worth mentioning.

Problems similar to those linked with lubricated journal bearings are also encountered in other cases in which a fluid is interposed between the stator and the rotor, as, for example, in labyrinth and liquid seals. In all cases, one of the most effective measures aimed at reducing the instability problems is the decrease of the peripheral velocity of the fluid around the

---

[6]See, for example, A. Tondl, *Some problems in rotor dynamics*, Czechoslovak Academy of Sciences, Prague, Czechoslovakia, 1965, and A. Muszynska, *Rotor instability*, Senior Mechanical Engineering Seminar, Carson City, June 1984.

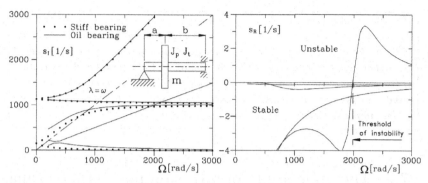

FIGURE 27.7. Sketch of the system used as example, together with the plot of the real and imaginary parts of the eigenvalues as functions of the spin speed.

shaft. This can be done using anti-swirl vanes, by roughening the stator walls, or by injecting the fluid in the tangential, backward direction.

Example 27.1 *Consider the rotor sketched in Fig. 27.7. It is made of a massless flexible shaft on which a disk with mass m and moments of inertia J_p and J_t is fitted.*

The shaft is supported on one side by a rigid bearing and on the other side by a hydrodynamic journal bearing. Due to the nonlinearity of the system, it is impossible to perform a general study and actual data must be stated: rotor: $a = 85$ mm, $b = 255$ mm, diameter of shaft = 25.4 mm, $m = 20$ kg, $J_p = 1$ kgm², and $J_t = 0.7$ kgm²; bearing: $l = 16$ mm, $c = 35.2$ μm, and $\mu = 0.02$ Ns/m².

Let the bearing be modeled using the short-bearing assumption, with a fully cavitated film, i.e., with the oil film extending only for 180°.

The imaginary and real parts of the eigenvalues, i.e., the actual whirl frequency and the decay rate, are plotted in Fig. 27.7. The whirl frequency is compared to the values obtained assuming that both bearings are rigid.

Due to the use of real coordinates (the size of the relevant matrices is 6, after reduction), it is not possible to distinguish immediately between forward and backward whirl motions without analyzing the eigenvectors. However, a clue is that the modes related to branches that slope downward on the Campbell diagram are backward modes.

From the plot it is clear that the presence of the hydrodynamic bearing has little influence on the whirl frequencies of the system but causes an added frequency to be present. This new mode, clearly an oil whirl, follows almost exactly the line $\omega = \Omega/2$. The decay rate is always negative, i.e., the motion is stable, but the oil whirl mode becomes unstable at about 2,000 rad/s.

> *The transition between oil whirl and oil whip occurs exactly when the line $\omega = \Omega/2$ crosses the first forward whirl mode, as predicted by the usual approximate criteria. As a conclusion, the behavior of the system is practically coincident with what could be expected using the approximate criteria, and the picture of the phenomenon shown in Fig. 27.1 holds. This is, however, not a general result, because in more complicated cases the presence of the bearings can have a more important effect on the dynamic behavior.*

27.2   Dynamic study of rotors on magnetic bearings

Magnetic suspension has been proposed for many applications, from vehicles to models in wind tunnels. In the field of rotating machinery, magnetic bearings can drastically reduce bearing drag, while completely avoiding the presence of lubricant and wear. The greatest advantages of magnetic bearings are, however, related to the dynamic behavior of the rotor: The stiffness and damping of the bearing system can be tailored for the application and can be adjusted following the operating conditions. In the case of active bearings, the control system can be used not only to maintain the rotor in the required position as a rigid body, minimizing the effects of unbalance and reducing the needs of strict balancing tolerances, but also to control the deformation modes.

The main goal of magnetic bearings is to keep the suspended body in the proper position and avoid rigid-body modes: A complete suspension must then constrain the six modes of a rigid body in space. In the case of a vehicle or rotor, one of the rigid-body modes must be left unconstrained (rotation about the axis, for a rotor) and a five-axis suspension is needed. Electromagnetic forces can be exerted by a passive device, based on permanent magnets or by uncontrolled electromagnets,[7] or by an active device suitably controlled.

A consequence of Earnshaw theorem is that a magnetic suspension based on a five-passive-axes layout is unstable except if diamagnetic materials or an electrodynamic layout are used. The choice is then between the use of a hybrid suspension system, in which at least one of the degrees of freedom of the rotor is constrained by a mechanical bearing, and that of a device that is at least partially active (or better controlled). The solutions span from that of a one-active-axis suspension, in which two passive radial bearings control the four degrees of freedom linked with the lateral behavior of the rotor and one axial active bearing restrains axial motion, to a fully active five-axis suspension, with active radial and axial bearings.

---

[7]It would be better to use the terms *uncontrolled* and *controlled* instead of *passive* and *active*: A bearing using uncontrolled electromagnets is actually active since it uses external power.

The acronym AMB is commonly used for active magnetic bearing.

In this section the equations of motion for the lateral behavior of a linear axi-symmetrical rotor running on a set of active radial bearings will be obtained.

By resorting to the complex coordinates defined in Section 24.1.2 (Eq. (24.16)), the equation of motion of the rotor–stator system is

$$\mathbf{M\ddot{q}} + (\mathbf{C}_n + \mathbf{C}_r - i\Omega\mathbf{G})\dot{\mathbf{q}} + (\mathbf{K} - i\Omega\mathbf{C}_r)\mathbf{q} = \mathbf{f}_c + \mathbf{f}_n + \Omega^2\mathbf{f}_r e^{i\Omega t}, \quad (27.25)$$

where vector $\mathbf{f}_c$ includes the forces exerted by the magnetic bearings on the rotor and the stator, and the stiffness matrix $\mathbf{K}$ is singular with four vanishing eigenvalues because the four rigid-body motions of the rotor are unconstrained. The real part of vector $\mathbf{f}_c$ relates to the forces due to the magnetic bearings in the xz-plane, while its imaginary part is linked with the forces exerted in the yz-plane.

Assume that each actuator is made by four electromagnets as shown in Fig. 27.8: Coils 1 and 3 are responsible for the force in the x-direction and coils 2 and 4 for the force in the y-direction. As a first approximation, the force exerted by a single coil depends on the current i and the air gap t between stator and rotor following the law

$$F = K\left(\frac{i}{t}\right)^2, \quad (27.26)$$

where K is a constant that includes all design parameters of the actuator.

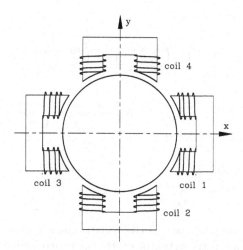

FIGURE 27.8. Sketch of an electromagnetic actuator of a magnetic bearing.

Usually, to linearize the overall behavior of the bearing, the coils are fed with a constant current, referred to as *bias current*, on which the control current is superimposed (see Example 27.3). A so-called static compensation current i_0' is superimposed to the bias current i_0 of the two coils acting in the same direction to withstand any static load. The bias and compensation currents in coils 1 and 3 acting in the x-direction can be written as $i_0 + i_0'$ and $i_0 - i_0'$. The total x-component of the force F_c exerted by the two identical coils 1 and 3 is

$$F_x = K \left[\left(\frac{i_0 + i_0' + i_x}{c - d - u_x} \right)^2 - \left(\frac{i_0 - i_0' - i_x}{c + d + u_x} \right)^2 \right], \qquad (27.27)$$

where i_x, c, d, and u_x are, respectively, the control current, the radial clearance, the static offset, and the radial displacement.

Remark 27.4 *Equation (27.27) holds only if $i_0 - i_0' - i_x > 0$: In the opposite case, the controller should switch off one of the coils, and the system works in nonlinear conditions.*

An actuator that works with a bias current high enough to allow all coils to be always energized is said to work as a 'class A' actuator while a 'class B' actuator works with only one of the two coils on, the one that carries the load. The intermediate condition, in which at least one coil is switched on and off, can be referred to as 'mixed-mode' working.

The force expressed by Eq. (27.27) can be linearized about the condition with $i_x = 0$, $u_x = 0$ as

$$F_x = F_0 + K_i i_x + K_u u_x, \qquad (27.28)$$

where

$$F_0 = K \left[\left(\frac{i_0 + i_0'}{c - d} \right)^2 - \left(\frac{i_0 - i_0'}{c + d} \right)^2 \right],$$

$$K_i = \left(\frac{\partial F_x}{\partial i_x} \right)_{\substack{i_x = 0 \\ u_x = 0}} = 2K \left[\frac{i_0 + i_0'}{(c - d)^2} + \frac{i_0 - i_0'}{(c + d)^2} \right], \qquad (27.29)$$

$$K_u = \left(\frac{\partial F_x}{\partial u_x} \right)_{\substack{i_x = 0 \\ u_x = 0}} = 2K \left[\frac{(i_0 + i_0')^2}{(c - d)^3} + \frac{(i_0 - i_0')^2}{(c + d)^3} \right].$$

If the sensor–actuator pair is co-located (the sensor reads the displacement at the location the actuator exerts the control force) and the control system is an ideal decentralized, proportional derivative (PD) controller, the magnetic bearing can be modeled as a spring–damper system. The control current the amplifiers supply to the actuators is in this case

$$i_x = -K_a \left(K_{cp} o_s + K_{cd} \dot{o}_s \right) , \tag{27.30}$$

where K_a, K_{cp}, and K_{cd} are the gain of the power amplifier and the proportional and derivative gains of the controller, respectively, and o_s is the output of the sensor. Since

$$o_s = K_s u_x , \tag{27.31}$$

where K_s is the gain of the sensor, the force exerted by the actuator is

$$F_x = F_0 + K_i K_a K_s \left(K_{cp} u_x + K_{cd} \dot{u}_x \right) + K_u u_x , \tag{27.32}$$

i.e.,

$$F_x = F_0 + \left(-K_i K_a K_s K_{cp} + K_u \right) u_x - K_i K_a K_s K_{cd} \dot{u}_x . \tag{27.33}$$

The equivalent stiffness k and damping c of the controlled magnetic bearing are

$$k = -K_u + K_i K_a K_s K_{cp} , \quad c = K_i K_a K_s K_{cd}. \tag{27.34}$$

The stiffness depends on the gains of the control loop but also on K_u, which is the absolute value of the negative electromechanical stiffness, expressing the instability of the bearing in open-loop conditions predicted by the Earnshaw theorem.

Remark 27.5 *Co-location of sensors and actuators simplifies the analysis and greatly reduces the possibility of unstable behavior but is seldom a realistic assumption. More realistic models in which the sensors and the actuators are located at different positions are needed.*

If the actuator operates in a central position, i.e., if $d = 0$, Eq. (27.34) reduces to

$$F_0 = 4K \frac{i_0 i_0'}{c^2} , \quad K_i = 4K \frac{i_0}{c^2} , \quad K_u = 4K \frac{i_0^2 + i_0'^2}{c^3}. \tag{27.35}$$

The force F_0 acting on the bearing determines the value of the current i_0' and hence affects the stiffness of the bearing. As a consequence, if the static forces acting in the x- and y-directions are not equal, as in the case of horizontal rotors, the behavior is anisotropic even if the geometrical and electrical characteristics are equal in the two directions. This effect is strong if the bearing operates with low bias currents and can be accounted for by using different gains of the controller for the two directions but, generally speaking, complete compensation is not possible.

When this occurs, the rotor exhibits the usual behavior of a rotor running on anisotropic bearings and, if damping is low enough, backward whirling occurs in the speed range spanning between the two rigid-body critical speeds in the two reference planes. The behavior of a spindle running on a five-active-axes magnetic suspension with decentralized PID controller is summarized in Fig. 27.9.[8] The figure has been obtained by recording the amplitude and phase of the synchronous components of the displacements in the x- (horizontal) and y- (vertical) directions during a spin-down test.

The greater stiffness in vertical direction is clear from the figure and occurs albeit the controller has a higher gain in horizontal direction to partially compensate for the anisotropy. A very simple PID decentralized controller has been used; if a more sophisticated control technique were

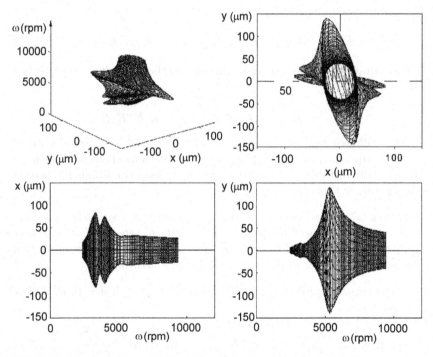

FIGURE 27.9. Dynamic behavior of a spindle running on a five-active-axes magnetic suspension with decentralized PID controller: same type of representation of the orbits at different speeds as in Fig. 25.2, except that in this case the experimental values of the synchronous components of the displacements measured during a spin-down test have been used.

[8]S. Carabelli, C. Delprete, G. Genta, 'Control strategies for decentralized control of active magnetic bearings', 4th International Symposium on Magnetic Bearings, Zurich, August 1994.

implemented, both anisotropy and amplitude at the critical speeds could have been controlled better.

Alternatively, a bias current larger than that required to withstand the static forces acting on the bearing can be used to obtain an almost isotropic behavior.

Equation (27.25) can be written with reference to the state space as

$$\dot{\mathbf{z}} = \mathcal{A}\mathbf{z} + \mathcal{B}_c\mathbf{u}_c(t) + \mathcal{B}_n\mathbf{u}_n(t) + \mathcal{B}_r\mathbf{u}_r(t)e^{i\Omega t}, \qquad (27.36)$$

where

- vector $\mathbf{z}$ contains the n complex state variables $\dot{\mathbf{q}}$ and $\mathbf{q}$;

- vector $\mathbf{u}_c(t)$ contains the r control input functions. It can be written in complex form, and the real and imaginary parts of its components are related to the forces in the two coordinate planes xz and yz. Vectors $\mathbf{u}_n(t)$ and $\mathbf{u}_r(t)$ are input vectors related to nonrotating and rotating forces, which, in the most general case, can be functions of time. They can be real vectors, in some cases simply scalar quantities, but it is possible to formulate the equations in such a way that they are complex;

- $\mathcal{A}$ is the complex dynamic matrix of the system,

$$\mathcal{A} = \begin{bmatrix} -\mathbf{M}^{-1}(\mathbf{C}_n + \mathbf{C}_r - i\Omega\mathbf{G}) & -\mathbf{M}^{-1}(\mathbf{K} - i\Omega\mathbf{C}_r) \\ \mathbf{I} & \mathbf{0} \end{bmatrix}; \qquad (27.37)$$

- matrices $\mathcal{B}_i$ are real or complex matrices that link the forces due to the control devices, nonrotating and rotating forces with the various inputs.

The equations linking the outputs of the system with the state vectors and control equations are the same as those already seen in Section 11.3 and need not be repeated here. The only difference between a controlled structure and a controlled rotor is the presence of the imaginary terms due to the gyroscopic matrix added to the damping matrix and those due to rotating damping added to the stiffness matrix. The gain matrices and vector of the state variables are complex.

Consider the rigid rotor sketched in Fig. 27.10a suspended on a four-active-axes magnetic suspension. Assume that the study of the lateral behavior of the system can be performed separately from the axial behavior and that the sensor–actuator pairs are co-located.

Remark 27.6 *The assumption of co-located sensors and actuators can be dropped when a model with many degrees of freedom is used and the displacements at the location of both sensors and actuators enter the equation of motion.*

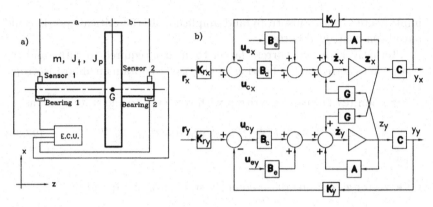

FIGURE 27.10. Rigid rotor on a four-axis active magnetic suspension: (a) sketch of the system; (b) block diagram, evidencing gyroscopic coupling between the xz- and yz-planes.

The rotor is assumed to be affected by static and couple unbalance and a constant lateral force due to weight acts in the xz-plane. There are two complex degrees of freedom, and, using the complex-coordinate notation seen in Section 8.3.2, the dynamic matrix and the state vector are

$$
\mathcal{A} = \begin{bmatrix} 0 & 0 & 0 & 0 \\ 0 & i\Omega\delta & 0 & 0 \\ 1 & 0 & 0 & 0 \\ 0 & 1 & 0 & 0 \end{bmatrix} , \qquad \mathbf{z} = \begin{Bmatrix} \dot{x} + i\dot{y} \\ \dot{\phi}_y - i\dot{\phi}_x \\ x + iy \\ \phi_y - i\phi_x \end{Bmatrix} , \qquad (27.38)
$$

where

$$
\delta = \frac{J_p}{J_t} .
$$

Both the stiffness and damping matrices vanish due to the fact that no mechanical support of the rotor exists.

The nonrotating forces due to weight, the rotating forces due to unbalance, and the control forces can be summarized in the following input vectors

$$
\mathcal{B}_n \mathbf{u}_n = \begin{Bmatrix} -g \\ 0 \\ 0 \\ 0 \end{Bmatrix} , \qquad \mathcal{B}_r \mathbf{u}_r e^{i\Omega t} = \Omega^2 \begin{Bmatrix} \epsilon e^{i\alpha} \\ \chi(1-\delta) \\ 0 \\ 0 \end{Bmatrix} e^{i\Omega t} , \qquad (27.39)
$$

$$
\mathcal{B}_c \mathbf{u}_c = \begin{bmatrix} 1/m & 1/m \\ -a/J_t & b/J_t \\ 0 & 0 \\ 0 & 0 \end{bmatrix} \begin{Bmatrix} F_{c_{x1}} + iF_{c_{y1}} \\ F_{c_{x2}} + iF_{c_{y2}} \end{Bmatrix} .
$$

In a similar way, the output of the system, i.e., the complex displacements and velocities at the transducer location, is

$$
\mathbf{y} = \left\{ \begin{array}{c} \dot{x}_1 + i\dot{y}_1 \\ \dot{x}_2 + i\dot{y}_2 \\ x_1 + iy_1 \\ x_2 + iy_2 \end{array} \right\} = \left[\begin{array}{cccc} 1 & -a & 0 & 0 \\ 1 & b & 0 & 0 \\ 0 & 0 & 1 & -a \\ 0 & 0 & 1 & b \end{array} \right] \mathbf{z} .
\tag{27.40}
$$

Assume that a decentralized control strategy, in which each axis is controlled independently, is implemented. This simplifies the control unit design, which reduces to a set of simpler units, each acquiring the information regarding the displacements and velocities in one radial direction at the bearing location and supplying the control input to the actuators that provide the control force in the corresponding direction. Each controller is then a SISO control system. To provide the required damping, the information regarding velocities is also required, possibly using velocity sensors, but usually by differentiating in time the displacements. In the case of deformable rotors, however, a state feedback performed through a suitable observer seems to be more appropriate, particularly where the response to external disturbances is concerned.[9]

Assuming a complete axial symmetry, which is consistent only with the case of a bearing completely unloaded from static forces, using a simple proportional–derivative control such as the one summarized by Eq. (11.27), and assuming that the reference position is characterized by a zero value of the displacements, the control forces for the rotor of Fig. 27.10a can be expressed as

$$
\left\{ \begin{array}{c} F_{c_{x_1}} + iF_{c_{y_1}} \\ F_{c_{x_2}} + iF_{c_{y_2}} \end{array} \right\} = \left\{ \begin{array}{c} F_{c_{x_{1_0}}} + iF_{c_{y_{1_0}}} \\ F_{c_{x_{2_0}}} + iF_{c_{y_{2_0}}} \end{array} \right\} +
\tag{27.41}
$$

$$
+ \left[\begin{array}{cccc} K_{d_1} & 0 & K_{p_1} - K_{u_1} 0 & 0 \\ 0 & K_{d_2} & 0 & K_{p_2} - K_{u_2} \end{array} \right] \mathbf{z} .
$$

where

$$
K_p = K_i K_a K_s K_{cp} , \quad K_d = K_i K_a K_s K_{cd}
$$

are the overall closed-loop proportional and derivative gain (Eq. (27.34)).

If the system is not axi-symmetrical, the use of complex coordinates implies a form of the relevant equations that include the conjugate of the state vector (see Section 9.1). The complex-coordinate approach is, however, not the only possible one and is best suited to cases in which the characteristics of the magnetic bearings are isotropic in the xy-plane. If this is not the

---

[9]P. Larocca, D. Fermental, E. Cusson, 'Performance comparison between centralized and decentralized control of the Jeffcott rotor', 2nd International Symposium on Magnetic Bearings, Tokyo, July 1990.

case, owing to bearing or control system anisotropy, the real and imaginary parts of equations from (27.36) to (27.41) can be separated in the same way seen in Chapter 24 for general rotordynamics. Real equations with double size are then obtained. The dynamic matrix $\mathcal{A}$ is in this case

$$\mathcal{A} = \begin{bmatrix} -\mathbf{M}^{-1}(\mathbf{C}_n + \mathbf{C}_r) & -\Omega\mathbf{M}^{-1}\mathbf{G} & -\mathbf{M}^{-1}\mathbf{K} & -\Omega\mathbf{M}^{-1}\mathbf{C}_r \\ \Omega\mathbf{M}^{-1}\mathbf{G} & -\mathbf{M}^{-1}(\mathbf{C}_n + \mathbf{C}_r) & \Omega\mathbf{M}^{-1}\mathbf{C}_r & -\mathbf{M}^{-1}\mathbf{K} \\ \mathbf{I} & 0 & 0 & 0 \\ 0 & \mathbf{I} & 0 & 0 \end{bmatrix},$$

(27.42)

where the various matrices are referred to flexural behavior in the xz-plane. The relevant matrices for the rigid rotor of Fig. 27.9, studied using real coordinates, are the following:

Open-loop dynamic matrix

$$\mathcal{A} = \begin{bmatrix} \begin{bmatrix} 0 & 0 & 0 & 0 \\ 0 & 0 & 0 & \Omega\delta \\ 0 & 0 & 0 & 0 \\ 0 & -\Omega\delta & 0 & 0 \end{bmatrix} & 0 \\ \mathbf{I} & 0 \end{bmatrix}.$$

(27.43)

State vector

$$\mathbf{z} = \begin{bmatrix} \dot{x} & \dot{\phi}_y & \dot{y} & -\dot{\phi}_x & x & \phi_y & y & \phi_x \end{bmatrix}^T.$$

Nonrotating input gain matrix and input vector

$$\mathcal{B}_n \mathbf{u}_n = \left\{ \begin{Bmatrix} -g \\ 0 \\ 0 \\ 0 \\ 0 \end{Bmatrix} \right\}.$$

Unbalance input gain matrix and input vector

$$\mathcal{B}_r \mathbf{u}_r = \left\{ \begin{Bmatrix} \epsilon \cos(\alpha + \Omega t) \\ \chi(1-\delta)\cos(\Omega t) \\ \epsilon \sin(\alpha + \Omega t) \\ \chi(1-\delta)\sin(\Omega t) \\ 0 \end{Bmatrix} \right\}.$$

(27.44)

Control input gain matrix and input vector

$$\mathcal{B}_c \mathbf{u}_c = \begin{bmatrix} \begin{bmatrix} 1/m & 0 & 1/m & 0 \\ -a/J_t & 0 & b/J_t & 0 \\ 0 & 1/m & 0 & 1/m \\ 0 & -a/J_t & 0 & b/J_t \end{bmatrix} \\ 0 \end{bmatrix} \left\{ \begin{matrix} F_{c_{x1}} \\ F_{c_{y1}} \\ F_{c_{x2}} \\ F_{c_{y2}} \end{matrix} \right\}.$$

Output vector

$$
\mathbf{y} = \left\{ \begin{array}{c} \dot{x}_1 \\ \dot{y}_1 \\ \dot{x}_2 \\ \dot{y}_2 \\ x_1 \\ y_1 \\ x_2 \\ y_2 \end{array} \right\} = \left[\begin{array}{cccccccc} 1 & 0 & -a & 0 & 0 & 0 & 0 & 0 \\ 0 & 1 & 0 & -a & 0 & 0 & 0 & 0 \\ 1 & 0 & b & 0 & 0 & 0 & 0 & 0 \\ 0 & 1 & 0 & b & 0 & 0 & 0 & 0 \\ 0 & 0 & 0 & 0 & 1 & 0 & -a & 0 \\ 0 & 0 & 0 & 0 & 0 & 1 & 0 & -a \\ 0 & 0 & 0 & 0 & 1 & 0 & b & 0 \\ 0 & 0 & 0 & 0 & 0 & 1 & 0 & b \end{array} \right] \mathbf{z}. \qquad (27.45)
$$

Controller gain matrix

$$
\mathbf{K}_y = \left[\begin{array}{cccccccc} k_{d_{x_1}} & 0 & 0 & 0 & k_{p_{x_1}} & 0 & 0 & 0 \\ 0 & k_{d_{y_1}} & 0 & 0 & 0 & k_{p_{y_1}} & 0 & 0 \\ 0 & 0 & k_{d_{x_2}} & 0 & 0 & 0 & k_{p_{x_2}} & 0 \\ 0 & 0 & 0 & k_{d_{y_2}} & 0 & 0 & 0 & k_{p_{y_2}} \end{array} \right]. \qquad (27.46)
$$

Note that the coupling between the xz- and yz-planes is due only to the gyroscopic terms. This is clearly shown in the block diagram of Fig. 27.10b, where matrices $\mathcal{A}$ are those of the corresponding non-gyroscopic system in each of the two planes xz and yz. The block $\mathbf{G}$ supplies the gyroscopic coupling between the two planes, and its gain is proportional to the spin speed Ω. Note that in Fig. 27.10 an output control acting separately in the xz- and yz-planes is used. On the contrary, no assumption on a control system acting separately on the two bearings has been made.

A simple PD or, better, a PID control can be used but more complex control laws are common. As a first thing, actual controller–actuator systems have a cut-off frequency that is influenced more by the characteristics of the actuator than by the controller.

A more complex dependence of the gain from the frequency can be used to obtain better performance: A high static stiffness could be useful in many applications, as machine-tool spindles, while low stiffness at high frequency can allow the system to work in supercritical conditions, using self-centering to reduce balancing requirements. Compensators are usually introduced into the control system to perform these tasks.

Remark 27.7 *Active magnetic bearings are usually unsuitable for applications where high static stiffness is required, while are at their best for soft-mounted rotors.*

Apart from the dependence of the gain on the frequency, it is also possible to introduce a dependence of the control law on the spin speed, producing a Campbell diagram in which the dependence of the various natural frequencies from the spin speed is different from that due to the gyroscopic effect alone. Also, the unbalance response can be strongly affected by the dependence of the gains of the control system on the speed, and smooth running can be achieved without the need of strict balancing tolerances. While the dependence of the stiffness on the frequency can shift the critical speeds, its dependence on the speed can make some critical speeds disappear altogether.

The separate control of each bearing can be suitable if only rigid-body behavior is to be controlled. When the control system is aimed at controlling the deformation modes, a true multivariable control system, based on an observer to perform state feedback, can be used, even if at the cost of added complexity.

Remark 27.8 *The modal approach is still possible, but the modal gyroscopic matrix is, generally speaking, non-diagonal, as is the modal damping matrix. However, while the latter is usually small and the errors due to neglecting the out-of-diagonal terms are negligible, the same does not hold for the first.*

The modal gyroscopic matrix can be reduced to a diagonal matrix to uncouple the modes even in this case, but errors due to spillover can be large. Their amount depends on the actual rotor configuration and must be checked in each case.

Using either the complex- or real-coordinates approach, it is possible to study the closed-loop dynamics of the system. Although the results obtainable are not in principle dissimilar from those typical of rotors running on bearings of other type, the particular features of magnetic bearings can produce a different behavior. The control system can introduce a much higher damping than that achievable with mechanical means. This is favorable in both fighting high-speed instability and reducing unbalance response when crossing critical speeds.

The whole pattern of the Campbell diagram can be strongly affected: The usual consideration that damping affects the decay rate but has little influence on the natural frequencies is no longer valid at the high damping levels obtainable in controlled systems. Some modes can become, in the whole speed range or only a part of it, overdamped, and the relevant branch of the Campbell diagram can go to zero and then disappear. The cut-off frequency of the control system and actuators may be lower than the maximum rotational frequency of the rotor, reducing both the stiffness and the damping of the supports at the frequencies of synchronous whirling. This effect can affect favorably the unbalance response, particularly in the case of low stiffness.

Many strategies aimed at obtaining a sort of self-balancing have been attempted and used. The simplest is that of using a notch filter centered on a frequency that follows the spin speed and remains synchronized. Although this is effective in achieving a self-centered configuration, it has poor rejection characteristics for disturbances at the synchronous frequency and is not advisable. A feedforward strategy in which the synchronous disturbance

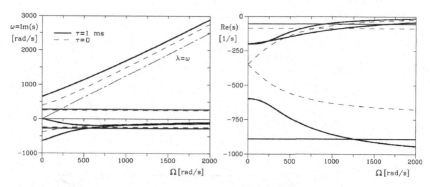

FIGURE 27.11. Campbell diagram of a rigid rotor on magnetic bearings: (a) imaginary part and (b) real part of the Laplace variable s.

due to unbalance is measured by averaging in time the outputs of the sensors at that frequency and then is fed to the controller, forcing the rotor to spin about its principal axis of inertia, yields very interesting results.

A last dynamic problem to be studied in the case of a rotor on magnetic bearings is the dynamic behavior in the case of failure of the control system. Usually rotors running on magnetic bearings are provided with a set of emergency touch-down bearings, usually of the rolling-element type, that can perform the spin-down task in the case of high-speed failure of the control system.

The dynamic study of the spin-down on the emergency bearings, a nonlinear process, is difficult and is performed by numerically integrating the equations of motion in time. The rotor first falls down until it is in contact with the emergency bearings, then a phase of rubbing is initiated, with possible rebounds, until the inner race of the emergency bearings gains the required speed and the rotor spins in a new configuration. This rigid-body behavior is accompanied by deflections and vibrations, and it is very important that the structural integrity of the system is warranted, avoiding undue contact between the rotor and the stator.

Example 27.2 *Consider a rigid, isotropic rotor supported by two magnetic radial bearings and assume that the geometric configuration corresponds with that of Fig. 27.10.*

The main data are $m = 3$ kg; $J_p = 0.02$ kg m^2; $J_t = 0.015$ kg m^2; $a = 200$ mm; $b = 100$ mm. Compute the gain matrix of the control system assuming that each bearing is controlled separately and that a complete axial symmetry is required. Plot the Campbell diagram and compute the control forces at 20,000 rpm, assuming that the rotor has a residual static unbalance following the ISO quality grade G 2.5 (see Chapter 28). The control system should provide uncoupling between translational and rotational modes and locate the first critical speed in the vicinity of 2,500 rpm = 261.8 rad/s. Assume that the time constant of the control system is equal to 1 ms.

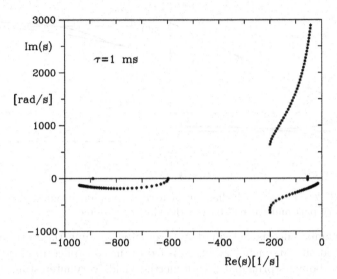

FIGURE 27.12. Roots locus of the same system in Fig. 27.11.

Using complex coordinates, the closed-loop dynamic matrix $A - B_c K_y C$ is

$$
A_{cl} = \begin{bmatrix}
-\dfrac{k_1}{m} & -\dfrac{k_2}{m} & -\dfrac{k_4}{m} & -\dfrac{k_5}{m} \\
-\dfrac{k_2}{J_t} & -\dfrac{k_3}{J_t} & -\dfrac{k_5}{J_t} & -\dfrac{k_6}{J_t} \\
1 & 0 & 0 & 0 \\
0 & 1 & 0 & 0
\end{bmatrix} ,
$$

where

$k_1 = k_{d_1} + k_{d_2}$, $k_2 = -a k_{d_1} + b k_{d_2}$, $k_3 = a^2 k_{d_1} + b^2 k_{d_2} - i\Omega J_p$,

$k_4 = k_{p_1} + k_{p_2}$, $k_5 = -a k_{p_1} + b k_{p_2}$, $k_6 = a^2 k_{p_1} + b^2 k_{p_2}$.

Uncoupling between translational and rotational modes implies that

$$
a k_{d_1} = b k_{d_2} , \qquad a k_{p_1} = b k_{p_2} ,
$$

and the condition on the critical speed of the corresponding undamped system yields

$$
\sqrt{(k_{p_1} + k_{p_2})/m} = \Omega_{cr} = 261.8 \ rad/s .
$$

The values of the stiffness gains thus obtained are

$$
k_{p_1} = 6.85 \times 10^4 \ Nm , \quad k_{p_2} = 1.37 \times 10^5 \ Nm .
$$

The terms to be included in the closed-loop gain matrix are

$$
k_{p_t} = k_{p_1} + k_{p_2} = 2.056 \times 10^5 \ N/m , \quad k_{p_r} = a^2 k_{p_1} + b^2 k_{p_2} = 4,110 \ Nm/rad.
$$

The damping gains can be obtained assuming a given value for the damping ratio of a particular mode, say, the translational mode. Assuming a damping ratio of 1/3, the following values are obtained

$$k_{d_1} = 349.1 \ Ns/m \ , \qquad k_{d_t} = k_{d_1} + k_{d_2} = 523.6 \ Ns/m \ ,$$
$$k_{d_2} = 174.5 \ Ns/m \ , \qquad k_{d_r} = a^2 k_{d_1} + b^2 k_{d_2} = 10.47 \ Nsm/rad \ .$$

The static deflection of the system is immediately obtained

$$x_{st} = -mg/k_{d_t} = \ 0.14 \ mm \ .$$

The control system can be set to compensate for the static deflection and to maintain the rotor in the center of the bearing notwithstanding the weight.
The stiffness gain so computed also includes the negative stiffness linked with the open-loop behavior of the bearing; the time constant of the sensor–controller–actuator subsystems should be applied only to the part of the overall gain linked with K_p and not to that linked with K_u. As a first-order approximation, the contribution of the negative stiffness will be neglected.
The Campbell diagram can be easily computed. As the controller is assumed to have a non-vanishing time constant, the gain matrix K_y is not constant:

$$K_y = \frac{1}{1 + s\tau} \begin{bmatrix} k_{d_1} & 0 & k_{p_1} & 0 \\ 0 & k_{d_2} & 0 & k_{p_2} \end{bmatrix} .$$

By transforming the equations of motion in the Laplace domain and remembering that the translational and rotational modes are uncoupled, the equations allowing the plotting of the Campbell diagram are

$$m\tau s^3 + ms^2 + sk_{d_t} + k_{p_t} = 0$$

for the translational mode and

$$J_t \tau s^3 + (J_t - i\Omega\tau J_p)s^2 + (k_{d_r} - i\Omega J_p)s + k_{p_r} = 0$$

for the rotational mode. The whirl frequency, i.e., the imaginary part of the Laplace variable s, and the decay rate, i.e., the real part of s, are plotted in Fig. 27.11. Also, in the same figure the solution obtained for a vanishingly small time constant of the control system is reported. The roots locus is shown in Fig. 27.12.
Due to uncoupling between translational and rotational rigid-body modes, the response to static unbalance can be plotted by resorting to the equation for translational motion alone

$$\left(-m\Omega^2 + i\frac{\Omega}{1 + i\Omega\tau}k_{v_t} + \frac{1}{1 + i\Omega\tau}k_{d_t} \right)(x + iy) = m\epsilon\Omega^2 e^{i\Omega t} .$$

The amplitude of the response to static unbalance is plotted in nondimensional form in Fig. 27.13. The system is much damped, but the presence of a cut-off frequency, affecting the damping and the stiffness, causes an increase of the amplitude at the critical speed. The rotor self-centers quickly in the supercritical range.

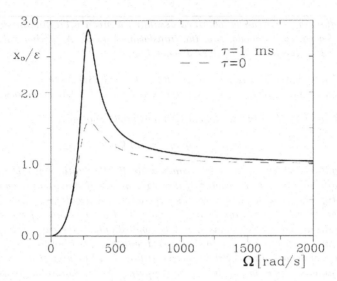

FIGURE 27.13. Unbalance response of a rigid rotor on magnetic bearings.

Unbalance quality grade G 2.5 at 20,000 rpm $=$ 2,094 rad/s corresponds to an eccentricity $\epsilon = 1.19\,\mu m$, i.e., with a mass of 3 kg, to a residual unbalance $m\epsilon = 3.58$ g mm (see Chapter 28). The response of the system at 20,000 rpm is

$$(x + iy)_0 = -1.236 - 0.012i \ \mu m \qquad \text{for } \tau = 1 \text{ ms },$$
$$(x + iy)_0 = -1.203 - 0.102i \ \mu m \qquad \text{for } \tau = 0 \ .$$

The total control force acting on the rotor is simply

$$F_c = \left(i\frac{\Omega}{1 + i\Omega\tau}k_{v_t} + \frac{1}{1 + i\Omega\tau}k_{d_t} \right)(x + iy) =$$
$$= m\Omega^2(x + iy) + m\epsilon\Omega^2 e^{i\Omega t} \ .$$

The modulus of the control force is then

$$|F_c| = 0.594 \ N \qquad \text{for } \tau = 1 \text{ ms },$$
$$|F_c| = 1.347 \ N \qquad \text{for } \tau = 0 \ ,$$

and is shared between the two bearings following their stiffnesses (2/3 on bearing 1 and 1/3 on bearing 2). Note that the control forces are small, even if the balancing quality grade is not high, particularly in the case of the less stiff control, with higher time constant. Actually, smooth running is obtained with even less strict balancing tolerances.

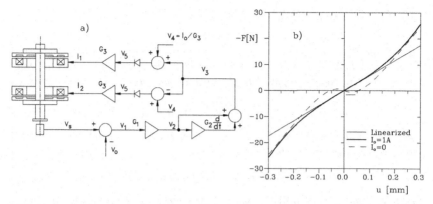

FIGURE 27.14. Active axial magnetic bearing. (a) Scheme of the control system;
(b) force–displacement characteristics.

Example 27.3 *Consider a rotor running on a one-active-axis magnetic suspension. The radial bearings are of the passive type, based on permanent magnets.*

The axial suspension must be active, because the passive radial bearings have a negative (destabilizing) axial stiffness. The mass of the rotor is $m = 0.8$ kg, and the linearized axial stiffness of the radial suspension is $k_z = -5 \times 10^4$ N/m. The axial actuator is made by a pair of electromagnets whose force–current characteristic is

$$F_c = \frac{\mu_0 S N^2 I^2}{4 d^2} ,$$

where I is the current, $\mu_0 = 4\pi \times 10^{-7}$ is the vacuum permeability, $S = 1.2 \times 10^{-3} m^2$ is the surface of the pole pieces, $N = 160$ is the number of turns, and $d = 0.5$ mm is the nominal air gap.

The axial position is measured by a sensor that outputs a voltage proportional to the axial displacement u from the nominal position (air gap of 2.5 mm) through the law

$$V_s = V_0 + au ,$$

where constants V_0 and a take the values 0.85 V and 300 V/m, respectively, when the input voltage to the sensor is 12 V.

The quadratic law linking the magnetic force to the current in the coil compels a choice between two alternatives: to use a control system that supplies the coil with a current proportional to the square root of the displacement in such a way that a restoring force proportional to the displacement is exerted or to add a constant current to the current proportional to the displacement supplied by the control system. In the latter case the coils exert two forces, which in the equilibrium position are equal and opposite; when the shaft is displaced axially the force due to one coil is reduced while the other increases and the restoring effect is obtained. A detailed scheme of the control system, following the second of the two approaches, is shown in Fig. 27.14a.

Note that the voltage signal

$$V_2 = V_1 G_1$$

is differentiated to supply an input to the current generators that is proportional to both the axial displacement and velocity. A damping effect is thus obtained. The voltage supplied to the current generators driving the coils is then

$$V_5 = V_4 \mp G_1 V_1 \mp G_1 G_2 \frac{dV}{dt} = V_4 \mp a G_1 (u + G_2 \dot{u}) \,,$$

where the upper signs hold for the coil nearer the end at which the sensor is located. The current supplied to the coils is then

$$I_i = G_3 V_5 = I_0 \pm a G_1 G_3 (u + G_2 \dot{u}) \,,$$

the upper sign being referred to coil 1. When both coils are operating, the force they exert (positive if directed upward, i.e., in a direction to cause an increase of the displacement u) is

$$F = \frac{\mu_0 S N^2}{4} I_0 \left[\frac{I_2^2}{(d-u)^2} - \frac{I_1^2}{(d+u)^2} \right] .$$

The expression of the force can be linearized, obtaining

$$F = \frac{\mu_0 S N^2}{4} \left[a G_1 G_3 (u + G_2 \dot{u}) - \frac{I_0}{d} u \right] .$$

The total stiffness and damping of the axial suspension are, respectively,

$$K_z = \frac{\mu_0 S N^2}{4} I_0 \left[a G_1 G_3 - \frac{I_0}{d} \right] - k_z \,, \qquad C_z = \frac{\mu_0 S N^2}{4} I_0 a G_1 G_3 G_3 \,.$$

The second term in brackets is the open-loop stiffness, which is negative. It can cause the axial instability of the bearing, which, if the current is not controlled, is unstable.

To achieve a stable behavior in the axial direction, the stiffness must be positive. Assuming the following values for the current I_0 and the gains,

$$I_0 = 1 \ A \,, \quad G_1 = 10 \,, \quad G_2 = 3 \times 10^{-4} \ s \,, \quad G_3 = 0.9 \ A/V \,,$$

the values of the axial stiffness and damping are

$$K_z = 58{,}100 \ N/m \,, \quad C_z = 125 \ Ns/m \,.$$

The system is then axially stable. The axial natural frequency and the damping ratio are

$$\omega_n = 269 \ rad/s \,, \quad \zeta = 0.29 \,.$$

The force–displacement characteristic of the axial suspension is reported in Fig. 27.14b. The curve obtained from the linearized equation is compared with those obtained from the nonlinearized equation and the curve obtained when no constant current is supplied ($I_0 = 0$).

27.3   Flexural vibration dampers

As seen in the previous sections, in many instances it is important to increase nonrotating damping to achieve the required stability. The easiest way is by introducing one or more dampers, which must be nonrotating, i.e., located within the stator or interposed between the stator and the rotor, being careful in the second case that the element that dissipates energy is restrained from rotating. In machines containing rotors working at different speeds, like multi-shaft turbines, it is also possible to use intershaft dampers, i.e., dampers located between two shafts rotating at different speeds; they, however, must be studied carefully because they can cause instability in particular working conditions.

When a damper is nonrotating, the only constraint in its placement is that it must be located between two points of the machine with relative displacement as large as possible in the modes of vibrations that must be controlled. It is possible to use pure dampers, namely, elements that react to a relative velocity between the two attachment points but not to a relative displacement or elements that can also supply a restoring force apart from the damping force.

Generally speaking, flexural vibration dampers are of the dissipative type and use the viscosity of a liquid, usually oil, or the internal damping of some material, usually an elastomer. Active or passive electromagnetic dampers are also entering the application stage.

An example of an elastomeric damper is shown in Fig. 27.15a. It is a damped support, made by a cylindrical elastomeric element into which a rolling element bearing is inserted. Devices of the type shown are simple

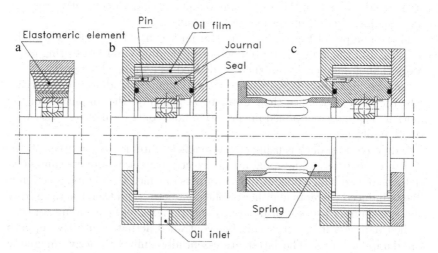

FIGURE 27.15. Flexural vibration dampers: (**a**) elastomeric type; (**b**) squeeze–film damper; (**c**) damped support of the squeeze-film type.

and low cost, but their drawback is that they are subject to overheating. The energy dissipated by the damper is totally converted into heat within the elastomeric mass and, due to the low thermal conductivity of most elastomeric materials, can cause its temperature to increase, even to the point of causing a deterioration of the characteristics of the material. If the damper starts overheating, the internal damping of the material usually decreases, causing an increase of the amplitude of the vibration, which, in turn, increases the heat generation in the damper. An unstable increase of temperature quickly leading to the complete destruction of the element can result. Dampers of this type can be used only after an accurate thermal analysis has demonstrated their applicability.

An oil damper, of the type generally referred to as a squeeze-film damper, is shown in Fig. 27.15b. A journal, which is restrained from rotating by a pin, is connected to the rotor through a rolling element bearing. The journal is located in a bearing and the oil between them is prevented from moving axially by seals at the ends. Radial movements of the journal cause the oil to move circumferentially, and this movement provides the required damping. Using the models seen in the previous section it is easy to state that the stiffness of the oil film is vanishingly small, because no relative rotation between journal and bearing occurs, and that there is no destabilizing effect linked to the circumferential motion of the oil. If the behavior of the device about the centered position can be linearized and the long-bearing model is used because of the axial seals, the damping matrix can be expressed by Eq. (27.7).

Since the clearance between the stator and the rotor is very small, the nonrotating journal can be allowed to touch the stator and to be supported by it. The damper thus becomes a damped support and its rolling bearing supports the weight of the rotor. The damper works in an eccentric position and the linearized solution does not hold any more.

In the damper of Fig. 27.15c, the journal is connected to the bearing by a spring and the device can provide both a damping and a restoring force. More than a damper, it can then be defined as a damped support.

Squeeze-film dampers, which are very common, introduce nonlinearities into the system because their damping depends on the amplitude of the motion. The aforementioned linearization holds only for motions with very small amplitude.

A different approach is using electromagnetic interactions between stator and rotor to generate damping forces. Electromagnetic dampers can essentially be of two kinds: uncontrolled dampers in which the damping action is due to induced currents in either a solid conductor (and then the term eddy currents damper is used) or in a purposely designed electric circuit, or controlled dampers in which a suitably controlled actuator directly supplies the damping forces. The latter are essentially similar to active magnetic bearings, in which the controller has only a derivative branch and thus the actuator is not required to levitate the rotor. They are usually referred to as active magnetic dampers (AMD), with the usual confusion between the terms controlled and active.

Two examples are shown in Fig. 27.16a and b. The general layout is the same as that of the AMB of Fig. 27.8: the cross-sections represented here are through the pole pieces of two opposite electromagnets. While the damper in Fig. 27.16b is identical to a bearing, the only difference being in how the control action is performed, the damped support in Fig. 27.16a has also a nonrotating spring that supports the rotor through a ball bearing. The difference between the two is exactly the same as that between the squeeze film dampers in Fig. 27.15b and c.

Although the layout differences between AMBs and AMDs are minimal, the operating differences are large. The former must supply forces large enough to levitate the rotor and are unstable in open loop, while the second must supply only damping forces and their open-loop instability can be easily overcome by the action of the spring supporting the rotor (Fig. 27.16a) or of the supports that must be at any rate provided (Fig. 27.16b). An AMD can be switched on only when needed, e.g., when crossing the critical speed to limit the amplitude or at high supercritical speeds to fight instability.

Uncontrolled (sometimes referred to as passive) electromagnetic dampers can be subdivided into two classes: motional dampers (MEMDs) and transformer dampers (TEMDs). In the former the damping action is due to induced currents in a conductor moving in a magnetic field (Fig. 27.16c). In the figure the conductor is a solid disc moving (whirling but not rotating) together with rotor, while the permanent magnet and the magnetic circuit

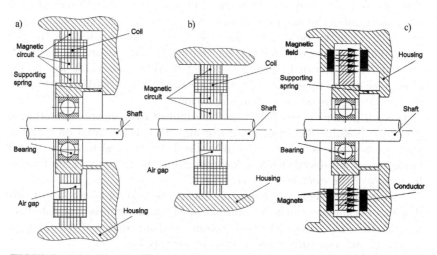

FIGURE 27.16. Electromagnetic dampers. Active magnetic damper (AMD) or transformer electromagnetic damper (TEMD) with (**a**) and without (**b**) supporting spring; (**c**) motional electromagnetic damper (MEMD).

are stationary: The induced currents are therefore eddy currents. A coil could be used instead of a solid disc: In this case it could be shunted on a variable resistor or on an electric circuit of various types, with the result of making it possible to control the damping action by controlling the characteristics of the latter. A MEMD of this type can be considered as a semi-active device, since little power is required for the control action.

As a first approximation a motional electromagnetic damper behaves like a viscous damper, providing a force proportional to the relative velocity. Actually the damping coefficient depends on the frequency and its transfer function is that of a first-order system (Fig. 11.11), with a cut-off frequency

$$\omega_p = \frac{R}{L}, \tag{27.47}$$

where R and L are, respectively, the resistance and the inductance of the electric circuit through which the induced current flows. If a solid disc has a very low inductance, its resistance is very low as well, and the cut-off frequency can fall within the working range of the device. This effect must be carefully considered when designing electromagnetic dampers.

The transfer function (mechanical impedance) of a MEMD is thus

$$G(s) = \frac{F}{\dot{x}} = \frac{c}{1 + \frac{s}{\omega_p}}, \tag{27.48}$$

where c is the damping coefficient at very low frequency. The device is equivalent to a spring and a damper in series (Maxwell's damper, Fig. 3.4c).

In transformer dampers, the induced currents are caused by changes in the reluctance of the magnetic circuit. They have the same physical layout of AMD, with the only difference that the current supplied to the coil to produce the magnetic field is not controlled: Two examples of TEMD are thus those shown in Fig. 27.16a and b. When the shaft whirls, the reluctance of the magnetic circuit changes, inducing a current in the coil and producing the required damping action.

The coil producing the field must be supplied by a source simulating as closely as possible a voltage generator and must be shunted on a circuit whose impedance is as low as possible. The lower the resistance the higher the damping produced, while a low inductance is needed to keep low the frequency of the electric pole (see below).

It is possible to separate the coil producing the field (or even to substitute it with a permanent magnet) from that in which the induced currents flow, but this does not change the basic layout. The impedance of the circuit in which the induced currents flow must be anyway low.

Summarizing, it is possible to say that in MEMD what moves is the copper (the conductor) while in TEMD what moves is the iron (an element of the magnetic circuit).

The mechanical impedance between the electromagnetic force and the relative velocity is

$$G(s) = \frac{F}{\dot{x}} = \frac{1}{s}\frac{k_{em}}{1+\frac{s}{\omega_p}} \, , \qquad (27.49)$$

where constant k_{em} is the negative stiffness of the open-loop magnetic bearing and the frequency of the *electric pole* ω_p has the same expression seen for MEMDs.

By inserting a spring with positive stiffness $k > |k_{em}|$ between stator and rotor, as in Fig. 27.16a, the mechanical impedance becomes

$$G(s) = \frac{F}{\dot{x}} = \frac{1}{s}\left(\frac{k_{em}}{1+\frac{s}{\omega_p}} + k\right) \, . \qquad (27.50)$$

By introducing the frequency

$$\omega_z = \omega_p\frac{k_{em}+k}{k} \, , \qquad (27.51)$$

and the total stiffness of the damper

$$k_t = k_{em} + k \, , \qquad (27.52)$$

the mechanical impedance can be written as

$$G(s) = \frac{F}{\dot{x}} = \frac{k_t}{s}\frac{1+\frac{s}{\omega_z}}{1+\frac{s}{\omega_p}} \, . \qquad (27.53)$$

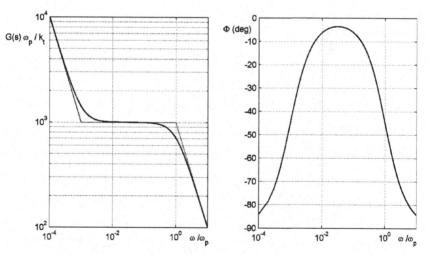

FIGURE 27.17. Nondimensional amplitude and phase of the mechanical impedance of a TEMD with $\omega_p = 1000\omega_z$.

To ensure stability, the frequency ω_z must be smaller than the frequency ω_p of the *electric pole*.

The mechanical impedance of a TEMD is thus equivalent to that of the parallel of a spring and a spring and a damper in series (Kelvin's damper, Fig. 3.4e). It is reported in Fig. 27.17 in nondimensional form for the case with $\omega_p = 1000\omega_z$.

Remark 27.9 *The same device can thus work as a TEMD or as an AMD, depending on whether the controller is switched off or on.*

Remark 27.10 *The damping due to energy dissipations caused by induced currents has the stabilizing effects of nonrotating damping only if the circuit in which the eddy currents flow is stationary. If the solid conductor or the coil rotates, the electromagnetic damping must be considered as a rotating damping, with all the possible stability problems in supercritical conditions.*

27.4   Signature of rotating machinery

The experimental analysis of the vibrations caused by rotating machinery yields much useful information on their working conditions and allows the discovery of possible problems before their consequences become too severe and, in some cases, even predicts their occurrence. The ultimate aim of the analysis is to diagnose the state of the machine to be able to perform preventive maintenance. To monitor the dynamic behavior of the machine,

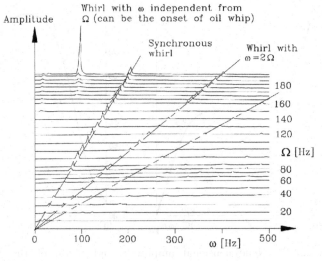

FIGURE 27.18. Cascade plot.

it is often sufficient to attach transducers that can measure the acceleration, velocity, or displacement in selected locations. In some machines, the transducers can be mounted permanently, and the signal they supply can be monitored continuously or at regular intervals or even only when working anomalies become apparent.

Different kinds of sensors are available, but now they are almost always connected to electronic data-acquisition systems that can perform various types of analyses and supply the relevant information in the form the user feels is more useful. In particular, it is expedient to perform a harmonic analysis of their output to obtain the acceleration or displacement spectrum. A common and useful way of representing these spectra is the so-called cascade plot: a sort of tridimensional plot in which the spectra obtained at different spin speeds are reported in the planes with Ω constant in a tridimensional space ω, Ω, amplitude (Fig. 27.18). Often, the frequency is reported in Hertz, and the spin speed is in revolutions per minute or revolutions per second.

Usually, the terms *cascade plot* and *waterfall plot* are used for two different types of diagrams. In the first, the spectra are plotted at different spin speeds; in the second, they are obtained at the same speed but at different times. The waterfall plot is not used to study the behavior of the machine in different working conditions but to follow the evolution in time of its dynamic behavior. The cascade plot has some similarity with the tridimensional plot of Fig. 26.2, except that ω and Ω are exchanged. There is, however, a major difference: In the plot of Fig. 26.2, the amplitude is the total amplitude of a mono-harmonic motion, or better, of circular whirling, while in the cascade plot the amplitude of the various harmonic components is reported as a function of the frequency after performing the harmonic analysis of the non-harmonic waveform. In the plot, it is possible to identify, at each speed, the frequencies of the various motions of the rotor. If different transducers are placed in different radial planes then, by comparing the phases of the relevant signals, it is possible to assess the direction of the various whirl components. From the frequency and phase information, it is possible to study the causes that produce the vibration and to decide the proper correcting actions.

For example, the synchronous component is usually linked with unbalance and can be corrected by performing a more accurate balancing, while a component with frequency equal to twice the rotational speed is generally linked with rotor anisotropy and can be corrected by making it more symmetrical. As a general rule, balancing the rotor has very little effect on all non-synchronous components.

Each machine produces a characteristic vibration spectrum, which is often referred to as the mechanical signature of the machine. Any alteration in time of the signature, as evidenced by a waterfall plot, is the symptom of an anomaly of the working conditions and must be considered very

TABLE 27.1. Characterization of forced and self-excited rotor vibrations.

	Forced or resonant vibrations	Self-excited vibrations
Relationship between frequency and speed	Frequency is equal to (i.e., synchronous with) the spin speed or a whole number or rational fraction of spin speed	Frequency is nearly constant and essentially independent of spin speed or any external excitation or/and is at or near one of the shaft natural frequencies
Relationship between amplitude and speed	Amplitude will peak in a narrow band of spin speed wherein the rotor's natural frequency is equal to the spin speed or to a whole number multiple or a rational fraction of the spin speed on external excitation	Amplitude will suddenly increase at a threshold speed and continues at high or increasing levels as spin speed is increased
Whirl direction	Almost always forward whirling	Generally forward whirling, but backward whirling has been reported
Rotor stressing	Static stressing in case of synchronous whirling	Oscillatory stressing at frequency equal to $\omega - \Omega$
Correcting actions	1. Introduce damping to limit peak amplitudes at critical speeds. 2. Tune the system's critical speeds to be outside the working range. 3. Eliminate all deviations from axial symmetry in the system as built or as induced during operation (e.g., balancing)	1. Increase damping to increase the threshold of instability above the operating speed range. 2. Raise the rotor natural frequencies as high as possible. 3. Identify and eliminate the instability mechanism
Influence of damping	Addition of damping may reduce peak amplitude but does not affect the spin speed at which it occurs	Addition of damping may raise the speed at which instability occurs but usually does not affect the amplitude after onset
Influence of system geometry	Excitation level and hence amplitude are dependent on some lack of axial symmetry in the rotor mass distribution or geometry or external forces applied to the rotor. Amplitudes may be reduced by refining the system to make it more axisymmetric or balanced	Amplitudes are independent of system axial symmetry. Given an infinitesimal deflection to an otherwise axi-symmetric system, the amplitude will self-propagate for whipping speeds above the threshold of instability

carefully. It can actually be linked with a problem that has occurred or is developing. The importance of correctly diagnosing problems before they actually occur, in reducing costs linked with maintenance and with the unavailability of the machine, is clear.

Some tables that can help in the identification of the causes of anomalies in the dynamic behavior of rotating machinery are here reported from a paper by F. Ehrich and D. Childs[10] (Tables 27.1, 27.2, and 27.3).

TABLE 27.2. Diagnostic table of self-excited vibrations of rotating machinery.

Mechanism	Ratio ω/Ω	Direction
Internal rotor damping	$0.2 < \omega/\Omega < 1$ $(\omega/\Omega = 0.5)$	Forward
Hydrodynamic bearings, labyrinth, or liquid seals	$\omega/\Omega < 0.5$ $(0.45 < \omega/\Omega < 0.48)$	Forward
Blade-tip clearance excitation	Dependent on fluid force levels	Forward
Centrifugal pump and compressor whirl	Dependent on fluid force levels	Forward
Propeller and turbomachinery whirl	Dependent on fluid force levels	Backward, if the vertex of the cone described in the whirl motion is after the rotor (referring to the direction of the fluid flow). Forward in the opposite case
Excitation due to fluid trapped in rotors	$0.5 < \omega/\Omega < 1.0$ $(0.7 < \omega/\Omega < 0.9)$	Forward

TABLE 27.3. Design correcting actions to reduce instability.

Mechanism	Correcting action
Internal rotor damping	Minimize number of separate elements in rotor; restrict span of rabbets and shrink-fitted parts; provide secure lock up on assembled elements
Hydrodynamic bearings	Install tilting pad or rolling element bearings
Labyrinth seals	Add swirl brakes at seal inlets to reduce the inlet tangential velocity. Replace rotor-mounted labyrinth vanes with stator-mounted vanes. Replace labyrinth seals with honeycomb seals
Liquid seals for pumps	Introduce swirl webs or brakes at seal inlet; roughen stator elements
Blade-tip clearance	No ready measures that do not affect the unit excitation operating efficiency
Centrifugal pump and compressor whirl	Not well understood
Propeller and turbomachinery whirl	Modify mode shapes to minimize angular motion of the plane of turbomachinery
Excitation due to fluid trapped in rotors	Introduce drain holes to eliminate fluid accumulation or axial webs to inhibit rotation of fluid

---

[10]F. Ehrich, D. Childs, 'Self-excited vibration in high performance turbomachinery', *Mechanical Engineering*, May, (1984).

27.5   Exercises

Exercise 27.1 *Consider the same rigid rotor used for Exercise 26.5, but substitute the ball bearings with a pair of plain journal bearings. Modeling the bearings using the short-bearing assumption, with a fully cavitated film, i.e., with the oil film extending only for 180°, compute the position of the center of the journals at all speeds up to the maximum operating speed, perform the dynamic analysis in the small, and plot the Campbell diagram, studying the stability of the system. Bearing data: diameter of shaft = 60 mm, l = 40 mm, c = 40 μm, μ = 0.02 Ns/m^2.*

Exercise 27.2 *Consider the gas turbine of Example 24.2. Repeat the dynamic analysis substituting the rigid bearings with two journal bearings of the type seen in Exercise 27.1, but with journal diameter 35 mm and length 30 mm.*

Exercise 27.3 *Compute the Campbell diagram of the rotor (Example 24.2) operating on two magnetic bearings. Assume a bearing stiffness of 10^7 N/m and use a simple control in which each bearing is controlled independently, without attempts to specifically control deformation modes.*

Exercise 27.4 *Consider the rotor on active magnetic bearings of the previous exercise, but instead of locating the sensors at the same positions as the actuators (nodes 1 and 11), assume that the sensor for the actuator at node 1 is located at node 2 and that for the actuator at node 11 is located at node 10. Plot again the Campbell diagram and the decay rate plot and compare the results with those of the co-located system. Does non-co-location affect stability?*

28
Rotor Balancing

All rotors, particularly those intended to operate at high rotational speed, must be balanced before starting their service life and sometimes balancing procedures must be repeated from time to time. Balancing is the subject of standards and is dealt with in a wide literature.

28.1 General considerations

The designer must take into account balancing requirements and provide the possibility of removing or adding masses in proper locations since the early design stages. Balancing must be regarded as one of the construction stages, to be performed after assembling the whole rotor or before, on its components, if they must be balanced separately (which usually does not avoid a further balancing process on the assembly), and balancing tolerances must be stated in a way that is not conceptually different from what is done for other types of tolerances, dimensional or geometrical.

The balance conditions of a rotor can change in time, and periodic rebalancing may be needed. In some cases it is not just the case of a slow change of balancing conditions in time, but of continuous variations occurring as consequence of operating and environmental conditions, wear and aging. This phenomenon can be quite severe and is usually referred to as *wandering unbalance*; it can be caused by thermal deformations of the rotor, material inhomogeneity, cracks, loose tolerances in built-up rotors, and the like.

Some rotors must be balanced several times during the first runs at subsequent higher speeds, in order to reach good balancing conditions at operating speed and running temperature. Even if the rotor is correctly balanced in operating conditions, poor balance may be encountered during start-up, until steady-state conditions are reached.

Rotor balancing has been the object of standards and designers must refer to them in stating balancing tolerance at the design stage. Standards are stated for the various types of machines, but it is the duty of the designer to verify that the stresses and deformations caused by the maximum residual unbalance prescribed are not beyond allowable limits. He must also be sure that the prescribed balancing tolerances are strict enough to prevent the rotor from being a source of unwanted vibration and noise for the surrounding environment. As with all tolerances, it must be remembered that it is impossible to reach a perfect balancing and that it is not necessary, and generally not advisable (at least from the economical point of view), to impose too-strict balancing requirements.

From the point of view of balancing, rotors are usually divided into two categories: rigid and deformable rotors. This subdivision, which is accepted by ISO standards, is in a certain way arbitrary, because no rigid body exists in the real world. A rotor can belong to either class, depending on the speed at which it is supposed to operate and, in particular, a speed at which any rigid rotor ceases to behave as such always exists. The balancing of rigid and deformable rotors will be only briefly summarized in the following sections: The reader can find all the required details in specialized monographs, in particular, those published by firms that build balancing machines.[1]

28.2   Rigid rotors

Following the ISO 1925 standard, a rotor can be considered rigid if it can be balanced by adding or removing mass in two arbitrarily chosen planes perpendicular to the rotation axis and if its balance conditions are practically independent from speed up to the maximum allowable speed. If this condition is satisfied, the rotor can be assimilated to a rigid body.

If the inertia of the stator is neglected and its elastic properties, referred to the center of mass of the rotor, are summarized in the stiffness and damping matrices $\mathbf{K}$ and $\mathbf{C}_r$, a model with four real degrees of freedom (two complex ones) is adequate to study its flexural dynamic behavior (See Section 24.1.1). The balance conditions of the machine can then be summarized by two parameters: the eccentricity ϵ (or the static unbalance $m\epsilon$) and angle χ between the principal axis of inertia of the rotor and the rotation axis (or the couple unbalance $(J_p - J_t)\chi$).

---

[1] As an example, see H. Schneider, *Balancing technology*, Schenck, Darmstadt, 1974.

Standards use the peripheral velocity of the center of mass of the rotor as a measure of the static unbalance

$$V_G = \Omega_{max}\epsilon_{max} \tag{28.1}$$

and define a quality grade for balancing, usually referred to as G, as the maximum allowable peripheral velocity of the center of mass, expressed in mm/s. A rotor that has to be balanced in the class $G = 2.5$ at 10,000 rpm, for example, must have an eccentricity smaller than

$$\epsilon_{max} = \frac{V_G}{\Omega_{max}} = 2.38 \ \mu\text{m} \ , \tag{28.2}$$

which results in a peripheral velocity of the mass center of 2.5 mm/s.

The maximum values of the residual eccentricity are plotted as functions of the maximum operating speed for different values of the quality grade in Fig. 28.1. The quality grades suggested by ISO standard 1940 for different types of rotors are reported in Table 28.1.

The mentioned standards consider only eccentricity and then static unbalance. In the case of a rigid rotor running on two bearings, a couple

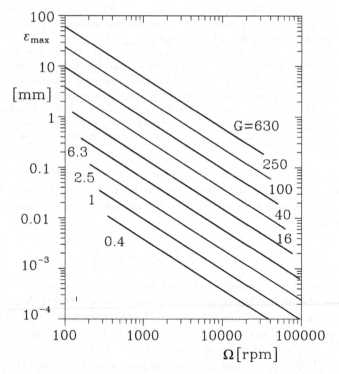

FIGURE 28.1. Maximum residual eccentricity as a function of the maximum operating speed for quality grades between $G = 0.4$ and $G = 630$ mm/s.

TABLE 28.1. Quality grades G suggested for different types of rotors (ISO 1940).

Grade	Examples
G 4000	Crankshaft drives of rigidly mounted slow marine diesel engines with an uneven number of cylinders.
G 1600	Crankshaft drives of rigidly mounted large two-cycle engines.
G 630	Crankshaft drives of rigidly mounted large four-cycle engines, crankshaft drives of elastically mounted marine diesel engines.
G 250	Crankshaft drives of rigidly mounted fast four-cylinder diesel engines.
G 100	Crankshaft drives of fast diesel engines with six or more cylinders, complete engines (gasoline or diesel) for cars, trucks, and locomotives.
G 40	Car wheels, wheel rims, wheel sets, drive shafts, crankshaft drives of elastically mounted fast four-cycle engines (gasoline or diesel) with six or more cylinders, crankshaft drives for engines of cars, trucks, and locomotives.
G 16	Drive shafts (propeller shafts, cardan shafts) with special requirements, parts of crushing machinery, parts of agricultural machinery, individual components of engines (gasoline or diesel) for cars, trucks, and locomotives, crankshaft drives of engines with six or more cylinders under special requirements.
G 6.3	Parts of process plant machinery, marine main turbine gears (merchant service), centrifuge drums, fans, assembled aircraft gas turbine rotors, flywheels, pump impellers, machine tools and general machinery parts, normal electrical armatures, individual components of engines under special requirements.
G 2.5	Gas and steam turbines, including marine main turbines (merchant service), rigid turbogenerator rotors, rotors, turbocompressors, machine-tool drives, medium and large electrical armatures with special requirements, small electrical armatures, turbine-driven pumps.
G 1	Tape recorder and phonograph (gramophone) drives, grinding machine drives, small electrical armatures with special requirements.
G 0.4	Spindles, discs, and armatures of precision grinders, gyroscopes.

unbalance due to two static unbalances, each equal to half the maximum allowable static unbalance placed at the bearing locations and phased at 180°, is considered a limit. Let d be the distance between the bearings; it follows that

$$|\chi(J_p - J_t)|_{max} = m\epsilon_{max}\frac{d}{2}. \qquad (28.3)$$

Rigid rotors are normally balanced using balancing machines. They can be either the high-stiffness or the low-stiffness type. Rigid (high-stiffness) balancing machines are machines in which the rotor can be spun on two very stiff supports provided with force transducers. The unbalance condition of the rotor is obtained from the measurement of the forces it exerts on its supports. Low-stiffness machines are similar to the former but the supports are more or less free to move and the transducers measure a quantity linked with their motion, usually the acceleration but sometimes the displacement or velocity. The unbalance of the rotor is obtained from the displacement measurements. Two planes perpendicular to the rotation axis are chosen, as far from each other as possible, on which masses can be either added or removed.

The maximum values of the residual unbalance on the two balancing planes are computed from the allowable residual unbalance corresponding to the required quality grade. The actual unbalance state of the rotor is then determined, spinning the rotor on a balancing machine.

Modern balancing machines are provided with data-acquisition systems that perform all needed computations and directly supply the values of the unbalance on the two correction planes and all information on the masses to be added or removed in them (amount of mass, radius, and angular position). Once the correction has been performed, a further measurement aimed at checking whether the required tolerance has been achieved usually follows.

The need to add or remove mass in the two planes at the same angular position is a clear symptom of a static unbalance. If the corrections must be phased 180° from each other, the unbalance is purely a couple unbalance. Generally, the phasing is neither at 0° nor at 180°, corresponding to a general state of dynamic unbalance, defined as the sum of static plus couple unbalance.

In some cases, the various parts that constitute a rotor must be balanced separately. The balancing tolerances of the various parts must be stated, remembering that unbalance is a vector quantity and that in the assembly process unbalances usually add to each other in a random way. The absolute value of the unbalance is, in the most unfavorable case, equal to the sum of the absolute values of the unbalances of the parts. Dimensional tolerances of the various parts and their effects on the relative position of the various parts must also be considered.

If possible, the rotor must be balanced after assembly, possibly on its bearings in such a way that the tolerances of the bearings and their seats are also accounted for.

Some rotors (e.g., rotors whose size exceeds the possibilities of available balancing machines, high-precision machinery) are balanced directly on-site. The machine is instrumented and the synchronous component of the vibration of the machine is monitored at different speeds. The amplitude and phase of the synchronous component give all the information needed to identify and correct the unbalance.

28.3   Flexible rotors

Balancing flexible rotors is much more difficult than balancing rigid rotors and, strictly speaking, it cannot be performed on balancing machines. The stiffness of the supports of the machine has, in fact, great influence on the deflected shape of the rotor, and then balancing should be performed directly on the whole machine. The process of balancing the whole machine is generally referred to as *field balancing*. It is, however, still possible, in some cases, to resort to a balancing machine, provided that its stiffness is not too different from that of the stator of the actual machine and proper allowance is taken for the unavoidable differences.

Strictly speaking, a flexible rotor is balanced only if each of its cross-sections is statically and dynamically balanced. In practice, it is not necessary that this condition be met to achieve the required aim, namely maintaining the effects of unbalance (vibrations, stresses, noise, etc.) within tolerable limits in the whole working range of the machine.

The flexible nature of the rotor and the difficulties that can be encountered in the balancing process are directly linked with the ratio between the maximum operating speed and the first flexural critical speed due to a bending mode of the rotor.

If the rotor has to operate at speeds far lower than the first critical speed (below about half of it), it can be assumed to be rigid. When the operating speed is close to the first critical speed, the possibility that the rotor inflects, assuming a shape not far from the first mode shape, must be taken into account. Correspondingly, close to the nth critical speed, the inflected shape is not dissimilar from the nth mode shape (nth eigenvector or eigenfunction, depending on the model used for the dynamic analysis).

The inflected shape can always be considered as a linear combination of the mode shapes[2] and if the rotor operates between two critical speeds, the dominant modes are those related to them. If the rotor has to operate above the nth critical speed, in order to be balanced in the whole working range it must be balanced with reference to the first $n + 1$ mode shapes.

From a practical point of view, ISO standards subdivide rotors into five classes. Rigid rotors, as described in the preceding section, belong to the first class.

Rotors of the second class are defined as semirigid, i.e., rotors that cannot be considered rigid but can be balanced in a low-speed balancing machine. This class is subdivided into eight subclasses, from 2a to 2h, as shown in Table 28.2.

The third class contains the true flexible rotors, which cannot be balanced using a low-speed balancing machine, but requires balancing at high speed. From this class, the rotors belonging to classes 4 and 5 are excluded. Rotors of the fourth class are said to be special flexible rotors, which are defined as rotors that would be rigid or semirigid but to which one or more flexible

---

[2]This property holds also if the gyroscopic effect is taken into account, see G. Genta and F. De Bona, "Unbalance response of rotors: a modal approach with some extensions to damped natural systems", *Journal of Sound and Vibration*, 140 (1), 1990, 129–153.

components have been added. The fifth class contains flexible rotors, which would belong to class 3 or 4, but for which balancing at a single speed (generally at the operating speed) is required.

Two procedures are usually considered for field balancing of flexible rotors: modal balancing and the influence coefficients method.

28.3.1   Modal balancing

The first practice is based on balancing the rotor at the various critical speeds, starting from the lowest. The correction masses must be located in the planes in which the relevant mode shape has large displacements. Generally speaking, the number of the correction planes must be equal to the order of the critical speed.

It is possible to demonstrate that it is possible to correct the unbalance at the general ith critical speed without disturbing the balance conditions of the previous $i-1$ modes, which have already been balanced.

Consider a rotor modeled as a discretized, multi-degree-of-freedom system. The unbalance response can be computed using Eq. (24.46). To balance the system at the first critical speed means to add to the unbalance distribution $\mathbf{f}$ to another unbalance distribution $\mathbf{f}_{b1}$, which causes the modal force due to unbalance related to the first mode to vanish

$$\overline{f}_1 = \mathbf{q}_1^T (\mathbf{f} + \mathbf{f}_{b1}) = 0 . \tag{28.4}$$

If balancing is performed by adding the unbalance $m_1\epsilon_1$ at the jth generalized coordinate (vector $\mathbf{f}_{b1}$ has all elements equal to zero except the jth, whose value is $m_1\epsilon_1\Omega^2$), the modal unbalance that has been added is

$$\mathbf{q}_1^T \mathbf{f}_{b1} = m_1\epsilon_1 q_{j1}\Omega^2 .$$

From Eq. (28.4), the value of the unbalance to be added to balance the first mode shape is immediately obtained

$$m_1\epsilon_1 q_{j1}\Omega^2 = -\mathbf{q}_1^T \mathbf{f} . \tag{28.5}$$

To balance the second mode, the two unbalances $m_{2r}\epsilon_{2r}$ and $m_{2s}\epsilon_{2s}$ are added corresponding to the rth and sth generalized coordinates. The second mode is balanced if

$$(m_{2r}\epsilon_{2r}q_{r2} + m_{2s}\epsilon_{2s}q_{s2})\Omega^2 = -\mathbf{q}_2^T \mathbf{f} + \mathbf{f}_{b1} . \tag{28.6}$$

The second mode can be balanced without disturbing the balancing, already achieved, of the first mode if the modal force corresponding to the first mode remains equal to zero after the addition of the new balancing masses at the generalized coordinates r and s

$$m_{2r}\epsilon_{2r}q_{r1} + m_{2s}\epsilon_{2s}q_{s1} = 0 . \tag{28.7}$$

TABLE 28.2. Rotors of class 2 (semirigid rotors).

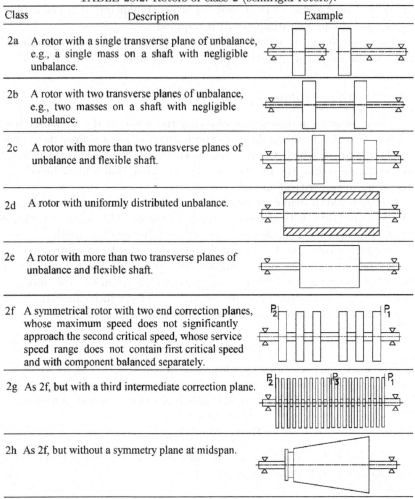

Class	Description	Example
2a	A rotor with a single transverse plane of unbalance, e.g., a single mass on a shaft with negligible unbalance.	
2b	A rotor with two transverse planes of unbalance, e.g., two masses on a shaft with negligible unbalance.	
2c	A rotor with more than two transverse planes of unbalance and flexible shaft.	
2d	A rotor with uniformly distributed unbalance.	
2e	A rotor with more than two transverse planes of unbalance and flexible shaft.	
2f	A symmetrical rotor with two end correction planes, whose maximum speed does not significantly approach the second critical speed, whose service speed range does not contain first critical speed and with component balanced separately.	
2g	As 2f, but with a third intermediate correction plane.	
2h	As 2f, but without a symmetry plane at midspan.	

Equations (28.6) and (28.7) allow computation of the two correction unbalances $m_{2r}\epsilon_{2r}$ and $m_{2s}\epsilon_{2s}$.

The correction of the third mode can be performed in the same way: The three correction unbalances can be computed using an equation of the type of Eq. (28.6), stating that the modal force corresponding to the third mode must vanish, plus two equations of the type of Eq. (28.7) stating that the balancing of the third mode does not affect the balancing conditions at the previously balanced modes. In a similar way, all other modes can be balanced.

What has been shown is actually a demonstration that modal balancing is possible, not the procedure to practically implement the balancing process. The knowledge of the unbalance distribution **f**, which is generally unknown, and of the mode shapes is not required. It is, however, clear that the computation of the mode shapes allows to identify the planes in which the correcting action is most effective, because the balancing masses must be located at the loops of the mode shape, or at least not too near to the nodes, where they would be ineffective.

Before starting modal balancing, the rotor can be balanced as a rigid body on a balancing machine. There is no agreement on the advisability of this practice, but, generally speaking, rigid-body balancing can be omitted if the starting unbalance is not severe enough to prevent operating close the first critical speed, as needed to start the modal balancing process.

It can happen that a rotor, that has been balanced at the first n critical speeds, causes strong vibrations at the maximum operating speed, which is between the nth and the $(n+1)$th critical speeds, without being possible to reach the latter to complete the balancing procedure. The strong vibrations are caused, in this case, by the unbalanced higher modes and, in particular, the $(n+1)$th mode. In this case the last corrections are implemented, taking into account the mode shape that is predicted at the $(n+1)$th critical speed, even if it is not materially possible to reach it.

28.3.2 Influence coefficients method

The influence coefficients method is based on the observation that the vibrations detected in a number of measuring points can be considered the effect of a number of concentrated unbalances in a number of arbitrarily located planes. This addition process implies that the behavior of the system is linear.

Let the number of measuring points be n, the number of speeds at which the balancing process is performed be m, and the number of planes in which the correction masses are to be located be q. Due to the linearity of the system, the $m \times n$ responses z_i obtained in the n measuring points at the m test speeds are linked to the q unknown unbalances $m_i \epsilon_i$ in the q correction planes by the general linear relationship

$$\{z\}_{m \times n} = [A]_{(m \times n) \times q} \{m\epsilon\}_q. \tag{28.8}$$

The responses z_i, which can be displacements, as implicitly assumed in Eq. (28.8), but also accelerations or velocities, and the unbalances $m_i \epsilon_i$ are all vectors and coincide with the complex quantities z and $m\epsilon$ in the preceding sections.

If the influence coefficients in matrix **A**, whose size is $(m \times n) \times q$, were known, Eq. (28.8) could be used directly to compute the unknown unbalances from the vibration measurements, provided that matrix **A** is square, i.e., the number of correction planes q is equal to the product of the number of test speeds m by the number of measuring points n and is non-singular.

The coefficients of influence are easily determined. A known unbalance $m_j r_j$ is introduced in a generic correction plane (the jth), in a known angular position. The tests are repeated, and a set of new $m \times n$ responses z_i^* are measured. They are linked to vector $\{m\epsilon\}^*$ through Eq. (28.8):

$$\mathbf{z}^* = \mathbf{A}\{m\epsilon\}^* ,$$

where vector $\{m\epsilon\}^*$ has all elements that coincide with those of the vector in Eq. (28.8) except for the jth element, which has been incremented by the known unbalance $m_j r_j$.

By subtracting Eq. (28.8) written for unbalances $\{m\epsilon\}$ and $\{m\epsilon\}^*$, it follows that

$$\mathbf{z}^* - \mathbf{z} = \mathbf{A} \begin{bmatrix} 0 & 0 & \dots & m_j r_j & \dots & 0 \end{bmatrix}^T . \qquad (28.9)$$

The jth column of matrix $\mathbf{A}$ is then simply obtained by dividing the difference between vectors $\mathbf{z}^*$ and $\mathbf{z}$ by $m_j r_j$. The procedure can be repeated by removing the previously added unbalance from the jth plane and adding a new known unbalance in another plane. The matrix of the coefficients of influence is thus obtained, column after column, and the unbalance distribution $\{m\epsilon\}$ is obtained:

$$\{m\epsilon\} = \mathbf{A}^{-1}\mathbf{z}. \qquad (28.10)$$

The rotor is then balanced by introducing in the q correction planes suitable unbalances equal and opposite to the ones computed. The balanced conditions of the rotor are thus achieved, at least at the speeds at which the tests have been performed.

The procedure described here is the extension to flexible rotors of the usual procedure for the calibration of the balancing machine for rigid rotors. In the latter case, only one test speed is chosen ($m = 1$), because the balance conditions are independent of the speed. By using two measuring points ($n = 2$), the number of the required correction planes reduces to 2 ($q = 2$).

The quality of the balancing of a flexible rotor can only be stated from the amplitude of the synchronous component of the vibrations measured on the running machine. Table 28.3, from an ISO recommendation, can supply detailed indications.

TABLE 28.3. Balance-quality criteria for flexible rotors.

	Range of effective pedestal synchronous vibration velocity (mm/s), r.m.s.				Correction factor		
	.28|.45|.71|1.12|1.8|2.8|4.5|7.1|11.2|18|28|45| 71				C_1	C_2	C_3
	A B C D						
I	Small electric motors (up to 15 kW)				.63		
	Superchargers				.63	2	
	Gyroscopes				.63	2	
	A B C D						
II	Paper making machines				.63		
	Med. size elec. motors and gener. (17–75 kW), norm. foundations				.63	4	
	Electric motors and gener. up to 3000 kW, special foundations				.63	4	20
	Pumps and compressors				.63	8	15
	Small turbines				.63	4	8
	A B C D						
III	Large electric motors				.63	5	
	Turbines and generators, rigid and heavy foundations				.63	5	20
	A B C D						
IV	Large electric motors, turbines and generators, light foundations				.63	3	10
	Small jet engines				.63		
	A B C D						
V	Jet engines larger than category IV				.63	2	10

Balance quality
A = Precision quality
B = Commercially acceptable
C = In need of attention at next overhaul
D = In need of immediate attention

Correction factors
C_1 = Measurement in high-speed balancing machine at service speed where bearing conditions are different from service conditions
C_2 = Shaft vibrations measured in or adjacent to the bearings
C_3 = Shaft vibrations measured at location of maximum shaft lateral deflection.

Example 28.1 *Consider the turbomolecular pump studied in Example 24.1. Compute the unbalance response at operating speed, knowing that the balancing quality grade required is $G = 2.5$.*

The peripheral velocity of the center of mass corresponding to a quality grade $G = 2.5$ is of 2.5 mm/s. The maximum value of the eccentricity is then $\epsilon_{max} = V_G/\Omega_{max} = 0.00714$ mm.

Because the distance between the two balancing planes is 260 mm, the maximum value of the couple unbalance is $|\chi(J_p - J_t)|_{max} = m\epsilon_{max}d/2 = 5.57 \times 10^{-6}$ kgm^2 .

The responses to static and couple unbalance at 30,000 rpm can be computed from the equations

$$\begin{bmatrix} 5.715 & -5.591 \\ -5.591 & 129 \end{bmatrix} \times 10^7 \mathbf{q} = \left\{ \begin{array}{c} 422.9 \\ 0 \end{array} \right\},$$

$$\begin{bmatrix} 5.715 & -5.591 \\ -5.591 & 129 \end{bmatrix} \times 10^7 \mathbf{q} = \left\{ \begin{array}{c} 0 \\ 54.99 \end{array} \right\},$$

which yield the following values of the displacements and rotations:

$$\mathbf{q} = \left\{ \begin{array}{c} -4.0990 \\ -0.3088 \end{array} \right\} \times 10^{-6} \qquad \textit{(static unbalance)},$$

$$\mathbf{q} = \left\{ \begin{array}{c} -4.001 \\ 4.089 \end{array} \right\} \times 10^{-8} \qquad \textit{(couple unbalance)}.$$

From the results obtained, it is clear that at 30,000 rpm the rotor is almost completely self-centered, where static unbalance is concerned, and couple unbalance causes very small deformations of the shaft.

28.4   Exercises

Exercise 28.1 *Consider the same system studied in Exercise 24.1. Assuming that the central joint is balanced to a quality grade $G = 16$ for a maximum speed of 10,000 rpm and that the central support has a hysteretic damping with loss factor $\eta = 0.1$, compute the unbalance response up to a speed equal to 1.5 times the first critical speed.*

Exercise 28.2 *Compute the driving torque and power needed to overcome the drag due to nonrotating damping in the case of the shaft of Exercise 24.1 balanced to a quality grade $G = 16$. Compute the minimum torque and power needed to avoid stalling when crossing the critical speed.*

Exercise 28.3 *Consider the turbomolecular pump of Example 24.1. Design a magnetic suspension system to achieve the same goals with a shaft stiff enough to allow the rotor to be considered rigid in the whole working range. Compute the response to a residual static unbalance corresponding to a quality grade $G = 6.3$. Assume a time constant for the control system equal to 1 ms.*

Exercise 28.4 *Repeat the computations of the previous exercise without neglecting the compliance of the shaft. Assume that the shaft has a diameter of 8 mm.*

29

Torsional Vibration of Crankshafts

Reciprocating machines usually are provided with crank mechanisms to transform the reciprocating motion of the pistons into the rotating motion of the shaft. The crankshaft is subject to strong dynamic problems, among which torsional vibration is usually the most important. Owing to the geometrical and mechanical complexity of the system, it is impossible to study this dynamics with the same precision that is common in the modeling of rotating machinery.

29.1 Specific problems of reciprocating machines

Machines containing reciprocating elements have some characteristic dynamic problems. In many cases, their solution is difficult and designers cannot ensure working conditions as smooth as those typical of machines containing only rotating elements. In many cases it is already a difficult task to avoid strong vibrations, which are uncomfortable for users and dangerous to structural safety. Often extensive experimentation is required even to reach this limited goal.

Most reciprocating machines are based on the crank mechanism, often in the form of a crankshaft with several connecting rods and reciprocating elements. Such devices cannot, in general, be exactly balanced: The inertial forces they exert on the structure constitute a system of forces whose resultant is not vanishingly small and is variable in time. The geometric configuration of the system made by a crankshaft, connecting rods, and reciprocating elements is complex, and crankshafts not only do not possess axial symmetry but often do not have symmetry planes. In these conditions, the uncoupling among axial, torsional, and flexural behavior, seen for beam-like elements, is no longer possible, if not as a rough approximation, and vibration modes become complicated.

The external forces acting on the elements of reciprocating machines are usually variable in time, often with periodic law. Also the forces exerted by hot gases on the pistons of reciprocating internal-combustion engines and other forces typical of reciprocating machines are periodic. Their time histories are not harmonic but, once harmonic analysis has been performed, they can be considered as the sum of many harmonic components whose frequencies are usually multiples, by a whole number or a rational fraction, of the rotational speed of the machine. The possibilities of resonance between these forcing functions and the natural frequencies of the system are many.

The most dangerous vibrations are usually linked to modes that are essentially torsional. In this chapter the attention will be concentrated on the torsional vibrations of reciprocating machines, dealing with axial and composite vibrations only marginally. It must, however, be kept in mind that all vibrations of crankshafts are composite vibrations, and no uncoupling is strictly possible.

29.2 Equivalent system for the study of torsional vibrations

29.2.1 Equivalent inertia

The traditional approach to the study of torsional vibrations of reciprocating machines is based on the reduction of the actual system made of crankshafts, connecting rods, and reciprocating elements to an equivalent system, usually modeled as a lumped-parameters system, whose torsional behavior can be studied separately from axial and bending modes (Fig. 29.1). The model of the reciprocating machine so obtained is then coupled to a model, usually based on the lumped-parameters approach, including all the rotating elements connected to the crankshaft, e.g., the driven machine in the case of an engine or the motor operating the compressor.

The inertia of the cranks and the reciprocating elements is lumped into a number (one for each crank) of moments of inertia (flywheels). The moments of inertia are then connected to each other by straight shafts, whose diameter is equal to the diameter of the relevant part of the actual shaft, or, more often, has a standard conventional value, and whose length is computed in such a way that the torsional stiffness of the equivalent shaft is as close as possible to the actual stiffness of the shaft.

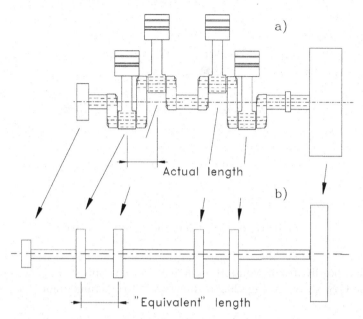

FIGURE 29.1. Sketch of the crankshaft: (**a**) actual system; (**b**) equivalent system, lumped-parameters model.

Remark 29.1 *The equivalent system is made by a straight shaft on which moments of inertia are located and thus the uncoupling among torsional, axial, and flexural behavior is possible. The construction of the equivalent system thus reduces to the computation of the moment of inertia of the flywheels simulating the inertia of the cranks and of the equivalent lengths of the shafts and its dynamic study is straightforward.*

Consider the crank mechanism sketched in Fig. 29.2. It is made of a disc, with a crankpin in B on which the connecting rod PB, whose center of mass is G, is articulated. The reciprocating parts of the machine are articulated to the connecting rod in P. The actual position of the center of mass of the reciprocating elements, which can include the piston as well as the crosshead and other parts, is not important in the analysis; in the following study this point will be assumed to be located directly in P. Note that the axis of cylinder, i.e., the line of motion of point P, does not necessarily pass through the axis of the shaft; the offset d will, however, be assumed to be small. Let J_d, J_b, m_b, and m_p be the moments of inertia of the disc that constitutes the crank and of the connecting rod (about its center of mass G) and the masses of the connecting rod and of the reciprocating parts, respectively.

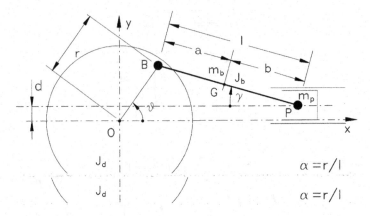

FIGURE 29.2. Sketch of the crank mechanism.

The coordinates of points B, G, and P can be expressed in the reference frame Oxy shown in Fig. 29.2 as functions of the crank angle θ as

$$(\overline{B-O}) = \left\{ \begin{array}{c} r\cos(\theta) \\ r\sin(\theta) \end{array} \right\}, \qquad (\overline{G-O}) = \left\{ \begin{array}{c} r\cos(\theta) + a\cos(\gamma) \\ r\sin(\theta) - a\sin(\gamma) \end{array} \right\},$$

$$\text{(29.1)}$$

$$(\overline{P-O}) = \left\{ \begin{array}{c} r\cos(\theta) + l\cos(\gamma) \\ d \end{array} \right\}.$$

Angle γ is linked to angle θ by the equation

$$r\sin(\theta) = d + l\sin(\gamma), \qquad (29.2)$$

i.e.,

$$\sin(\gamma) = \alpha\sin(\theta) - \beta,$$

where

$$\alpha = \frac{r}{l}, \qquad \beta = \frac{d}{l}.$$

Ratios α and β are expressed by numbers smaller than 1, and in practice they are quite small; usually $\alpha \leq 0.3$ and $\beta = 0$. In the case of an ideal crank mechanism, in which the connecting rod is infinitely long ($\alpha = 0$) and the axis of the cylinder passes through the axis of the crank ($\beta = 0$), the motion of the reciprocating masses is harmonic when the crank speed is constant.

Because $\dot\theta$ is the angular velocity of the crank, its kinetic energy is simply

$$T_d = \frac{1}{2} J_d \dot\theta^2 \,. \tag{29.3}$$

The speed of the reciprocating masses can be easily obtained by differentiating the third equation (29.1) with respect to time and obtaining the expression for $\dot\gamma$ from Eq. (29.2):

$$V_p = -r\dot\theta \sin(\theta) - l\dot\gamma \sin(\gamma) = -r\dot\theta \left[\left(1 + \alpha \frac{\cos(\theta)}{\cos(\gamma)} \right) \sin(\theta) - \beta \frac{\cos(\theta)}{\cos(\gamma)} \right].$$
$$\tag{29.4}$$

The kinetic energy of the reciprocating masses is

$$T_p = \frac{1}{2} m_p r^2 \dot\theta^2 f_1(\theta) \,, \tag{29.5}$$

where

$$f_1(\theta) = \left[\sin(\theta) + \alpha \frac{\sin(2\theta)}{2\cos(\gamma)} - \beta \frac{\cos(\theta)}{\cos(\gamma)} \right]^2 .$$

Instead of computing the kinetic energy of the connecting rod by writing the velocity of its center of mass G, it is customary to substitute the rod with a system made of two masses m_1 and m_2, located at the crankpin B and the wrist pin P, respectively, and a moment of inertia J_0. To correctly simulate the connecting rod, such a system must have the same total mass, moment of inertia, and position of the center of mass. These three conditions allow for three equations yielding the following values for m_1, m_2, and J_0:

$$m_1 = m_b \frac{b}{l} \,, \qquad\qquad m_2 = m_b \frac{a}{l} \,,$$
$$\tag{29.6}$$
$$J_0 = J_b - (m_1 a^2 + m_2 b^2) = J_b - m_b ab \,.$$

Generally speaking, the moment of inertia of masses m_1 and m_2 is greater than the actual moment of inertia of the connecting rod and, consequently, the term J_0 is negative.

Remark 29.2 *The negative moment of inertia has no physical meaning in itself: the minus sign indicates that this is just a term which must be subtracted in the expression of the kinetic energy.*

The kinetic energy of mass m_1 can be computed simply by adding a moment of inertia $m_1 r^2$ to that of the crank.

Similarly, the kinetic energy of mass m_2 can be accounted for by adding m_2 to the reciprocating masses. The effect of the moment of inertia J_0 can be easily computed

$$T_{J_0} = \frac{1}{2} J_0 \dot\gamma^2 = \frac{1}{2} J_0 \dot\theta^2 f_2(\theta) \,, \tag{29.7}$$

where

$$f_2(\theta) = \alpha^2 \left[\frac{\cos(\theta)}{\cos(\gamma)} \right]^2 .$$

The total kinetic energy of the system shown in Fig. 29.2 is, consequently,

$$\mathcal{T} = \frac{1}{2}\dot{\theta}^2 \left[J_d + m_1 r^2 + (m_2 + m_p)r^2 f_1(\theta) + J_0 f_2(\theta) \right] = \frac{1}{2} J_{eq}(\theta)\dot{\theta}^2 .$$

$$(29.8)$$

The whole system can be modeled, from the viewpoint of its kinetic energy, by a single moment of inertia variable with the crank angle $J_{eq}(\theta)$, rotating at the angular velocity $\dot{\theta}$.

In the study of the torsional vibrations, the motion of each section of the crankshaft can be expressed as the superimposition of a rigid-body rotation with angular velocity Ω and a vibrational motion expressed by the torsional rotation $\phi_z(t)$. Angle θ can be expressed as

$$\theta(t) = \Omega t + \phi_z(t) , \qquad (29.9)$$

and the kinetic energy (Eq. (29.8)) is

$$\mathcal{T} = \frac{1}{2} J_{eq}(\theta)(\Omega + \dot{\phi}_z)^2 . \qquad (29.10)$$

Equation (29.10) holds in general, even if the average velocity Ω of the shaft varies in time; in all subsequent computations, however, only constant-rate rotation will be assumed and derivatives of Ω with respect to time will be neglected. To study the vibrations in a shaft rotating at variable speed, a law $\Omega(t)$ can be assumed, and the equations that follow can be accordingly modified.

The equation of motion of the crank in terms of the generalized coordinate $\phi_z(t)$ can be obtained from the Lagrange equation. By performing all the relevant derivatives, it follows that

$$J_{eq}\ddot{\phi}_z + \frac{1}{2}(\Omega + \dot{\phi})^2 \frac{\partial J_{eq}(\theta)}{\partial \theta} = M , \qquad (29.11)$$

where M is the moment due to all external actions on the crank (elastic and damping reaction of the shaft, forces applied on the piston, etc.).

Equation (29.11) is a second-order differential equation in ϕ_z. It is nonlinear even if the moment M is a linear function of the generalized coordinate ϕ_z (as when the shaft has a linear elastic behavior) or of its derivative (as when linear damping is present), due to the fact that its coefficients, apart from being variable in time, are functions of the coordinate ϕ_z.

The conventional approach to the study of reciprocating machines is based on a number of simplifications of the equation of motion (29.11) of the crank mechanism aimed at obtaining a linear differential equation with constant coefficients.

The first simplification is to replace a constant moment of inertia $\overline{J}_{eq}$ in the term $J_{eq}\ddot{\phi}_z$ for the actual value that is a function of θ. A second simplification is to assume that angle ϕ_z is small enough to be neglected in the expression of θ in the second term of Eq. (29.11), that consequently reduces to

$$\overline{J}_{eq}\ddot{\phi}_z = M - \frac{1}{2}\Omega^2 \left[\frac{\partial J_{eq}(\Omega t)}{\partial(\Omega t)}\right]. \qquad (29.12)$$

The last term of Eq. (29.12) is independent from angle ϕ_z and its derivatives. It is a known function of angle Ωt and hence of time and can be dealt with as an external excitation applied to the system. The aforementioned approach clarifies why and under which assumptions it is possible to substitute each crank with a constant moment of inertia on which suitable inertia forces (inertia torques), variable in time with known time history, act together with the actual external forces due to the working fluid and the other parts of the system.

Remark 29.3 *The study of the free oscillations of the system, performed by resorting to the homogeneous equation associated with Eq. (29.12), does not take into account such inertia forces and considers only the mean moment of inertia* $\overline{J}_{eq}$.

For the computation of both $\overline{J}_{eq}$ and the inertia forces, it is possible to resort to series expansions of functions $f_1(\theta)$ and $f_2(\theta)$, which can be expressed as trigonometric series of angle θ, whose coefficients are functions of nondimensional parameters α and β.

In the limiting case of $\alpha = \beta = 0$, corresponding to an infinitely long connecting rod (piston moving with harmonic time history), the expressions for $f_1(\theta)$ and $f_2(\theta)$ are particularly simple

$$f_1(\theta) = \sin^2(\theta) = \frac{1 - \cos(2\theta)}{2}, \qquad f_2(\theta) = 0. \qquad (29.13)$$

In practice, it is impossible to neglect the fact that the length of the connecting rod is finite, even if α is usually not greater than 0.3. By expressing $\cos(\gamma)$ as

$$\sqrt{1 - [\alpha\sin(\theta) - \beta]^2},$$

computing the Taylor series in α and β for functions $f_1(\theta)$ and $f_2(\theta)$, and truncating it retaining all terms containing powers up to the fifth, the following trigonometric polynomials can be obtained:

$$f_1(\theta) = a_0 + \sum_{i=1}^{7} a_i \cos(i\theta) + \sum_{i=1}^{6} b_i \sin(i\theta),$$

$$f_2(\theta) = c_0 + \sum_{i=1}^{4} c_i \cos(i\theta) + \sum_{i=1}^{3} d_i \sin(i\theta). \qquad (29.14)$$

The coefficients of the trigonometric series are

$$a_0 = \frac{8+2\alpha^2(1+6\beta^2)+8\beta^2(1+\beta^2)+\alpha^4}{16},$$

$$a_1 = \alpha\frac{128+16\alpha^2(2+15\beta^2)+48\beta^2(4+5\beta^2)+15\alpha^4}{256},$$

$$a_2 = -\frac{16+\alpha^4-16\beta^2(1+\beta^2)}{32},$$

$$a_3 = -\alpha\frac{128+24\alpha^2(2+15\beta^2)+48\beta^2(4+5\beta^2)+27\alpha^4}{256},$$

$$a_4 = -\alpha^2\frac{2+\alpha^2+12\beta^2}{16},$$

$$a_5 = \alpha^3\frac{16+15\alpha^2+120\beta^2}{256},$$

$$a_6 = \frac{\alpha^4}{32},$$

$$a_7 = -\frac{3\alpha^5}{256},$$

$$b_1 = -\beta\alpha\frac{1+\alpha^2+2\beta^2}{2},$$

$$b_2 = -\beta\frac{128+48\alpha^2(2+5\beta^2)+16\beta^2(4+3\beta^2)+75\alpha^4}{128}.$$

$$b_3 = -\alpha\beta\frac{2+\alpha^2+4\beta^2}{2},$$

$$b_4 = \alpha^2\beta\frac{12+15\alpha^2+30\beta^2}{32},$$

$$b_5 = \beta\frac{\alpha^3}{4},$$

$$b_6 = -\beta\frac{15\alpha^4}{128},$$

$$c_0 = \alpha^2\frac{4+\alpha^2(1+6\beta^2)+4\beta^2}{8},$$

$$c_2 = \alpha^2\frac{1+\beta^2}{2},$$

$$c_4 = -\frac{\alpha^4}{8},$$

$$d_1 = d_3 = -\beta\frac{\alpha^3}{2},$$

$$c_1 = c_3 = d_2 = 0,$$

If the axis of the cylinder passes through the center of the crank ($\beta = 0$), as is usually the case, all coefficients b_i and c_i vanish, as was easily predictable, because for symmetry reasons $f_1(\theta)$ and $f_2(\theta)$ are even functions of θ. Functions $f_1(\theta)$ and $f_2(\theta)$ are reported in Fig. 29.3 for two different values of α with $\beta = 0$. The results obtained using the series (29.14) with simplified expressions of the coefficients (considering only the first three non-vanishing harmonic terms for $f_1(\theta)$ and the first two for $f_2(\theta)$ and only the powers of α up to the third) are compared with the exact solutions. The lines are completely superimposed, and the approximation obtainable in this way is clearly very good. By retaining the first seven harmonic terms in $\cos(\theta)$ and the first six terms in $\sin(\theta)$, the equivalent moment of inertia can be expressed as

$$J_{eq} = \overline{J}_{eq} + \sum_{i=1}^{7} J_{c_i}\cos(i\theta) + \sum_{i=1}^{6} J_{s_i}\sin(i\theta), \tag{29.15}$$

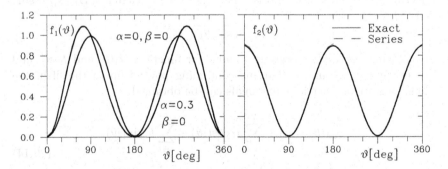

FIGURE 29.3. Functions $f_1(\theta)$ and $f_2(\theta)$.

where

$$\overline{J}_{eq} = J_d + m_1 r^2 + a_0 (m_2 + m_p) r^2 + J_0 c_0 ,$$
$$J_{c_i} = a_i (m_2 + m_p) r^2 + J_0 c_i ,$$
$$J_{s_i} = b_i (m_2 + m_p) r^2 + J_0 d_i .$$

Also in this case, all terms in sine vanish if the axis of the cylinder passes through the center of the crank ($\beta = 0$).

If $\alpha = 0$, the expression for the mean equivalent moment of inertia reduces to

$$\overline{J}_{eq} = J_d + m_1 r^2 + r^2 \frac{m_2 + m_p}{2} . \qquad (29.16)$$

Some authors[1] suggest using for $\overline{J}_{eq}$ the reciprocal of the average value of function $1/\overline{J}_{eq}$ on a complete revolution of the shaft. This procedure, once implemented by plotting function $1/\overline{J}_{eq}$ and then graphically computing the mean, is based on the observation that the square of the torsional natural frequency is proportional to $1/\overline{J}_{eq}$. The equivalent system so computed has a natural frequency equal to the mean square value of the natural frequencies of the actual system, computed in the various positions.

By introducing the series computed earlier for functions $f_1(\theta)$ and $f_2(\theta)$ in the last term of the equation of motion (29.11), approximate expressions for the inertia forces are obtained. This part of the analysis of the system will be included in the study of the forced oscillations, because the inertia forces are introduced into the computation as external excitations.

Instead of resorting to expressions of the type of Eq. (29.14), it is simpler to use the expressions of $f_1(\theta)$ and $f_2(\theta)$ in which $\cos(\gamma)$ has been substituted by its expression as a function of angle θ

$$f_1(\theta) = \left[\sin(\theta) + \frac{\alpha \sin(\theta) \cos(\theta) - \beta \cos(\theta)}{\sqrt{1 - [\alpha \sin(\theta) - \beta]^2}} \right]^2 ,$$

$$f_2(\theta) = \frac{\alpha^2 \cos^2(\theta)}{1 - [\alpha \sin(\theta) - \beta]^2} , \qquad (29.17)$$

and to numerically compute the values of $\overline{J}_{eq}$, J_{c_i}, and J_{s_i} through a fast Fourier transform of function $J_{eq}(\theta)$ computed using Eq. (29.15).

The function usually needs to be computed in a small number of points (32 or 64) but if many harmonics are required like in the present case, more points can be needed. However, even if 1,024 or 2,048 points are needed, the computation is very fast on any computer.

The approximate procedure outlined here leads to results that can be very rough, particularly in those cases in which the inertia forces due to reciprocating masses are high. However, one of the aims of designers of

---

[1] See, for example, C.B. Biezeno and R. Grammel, *Technische Dynamik*, Springer, Berlin, 1953.

reciprocating machinery is reducing the mass of reciprocating elements, particularly in the case of fast machines, and this approach can be used in many cases with confidence.

Remark 29.4 *Even with the mentioned limitations, the approach described here is the only one that allows the behavior of a complex machine to be predicted without resorting to numerical experiments.*

Example 29.1 *Compute the values of $\overline{J}_{eq}$, J_{c_i}, and J_{s_i} up to the seventh harmonic for a crank and piston unit with the following data: stroke: $2r = 85$ mm; mass of complete piston: $m_p = 0.540$ kg; mass of connecting rod: $m_b = 0.420$ kg; length of connecting rod: $l_b = 181$ mm; distance a (Fig. 29.2): $a = 78$ mm; moment of inertia of connecting rod: $J_b = 0.0015$ kg m^2; moment of inertia of crank: $J_d = 0.00447$ kg m^2.*

Compare the results obtained from Eq. (29.14) with those obtained through fast Fourier transform of function $J_{eq}(\theta)$ computed using Eq. (29.15).

The value of ratio α is $\alpha = r/l = 0.235$ and $\beta = 0$. From Eq. (29.6) it follows that $m_1 = 0.239$ kg, $m_2 = 0.181$ kg, and $J_0 = -0.0019$ kg m^2. Because $\beta = 0$, all coefficients b_i and d_i vanish; coefficients a_i and c_i are

$$a_0 = 0.5071, \qquad\qquad a_1 = 0.1191,$$
$$a_2 = -0.5001, \qquad\qquad a_3 = -0.1199,$$
$$a_4 = -007082, \qquad\qquad a_5 = 0.0008509,$$
$$a_6 = 0.00009499, \qquad\qquad a_7 = -0.000008364,$$
$$c_0 = 0.0279, \qquad\qquad c_1 = 0.0001551,$$
$$c_2 = -0.0007029, \qquad\qquad c_3 = 0.0001562,$$
$$c_4 = -0.000008510, \qquad\qquad c_5 = 0.000001108,$$
$$c_6 = 0.0000001237, \qquad\qquad c_7 = -0.00000001089.$$

All J_{s_i} vanish, and the results for $\overline{J}_{eq}$ and J_{c_i} are reported in the following table

	Eq. (29.14)	FFT 32 points	FFT 128 or more points
$\overline{J}_{eq}$	5.50969×10^{-3}	5.50968×10^{-3}	5.50968×10^{-3}
J_{c_1}	1.55056×10^{-4}	1.55056×10^{-4}	1.55056×10^{-4}
J_{c_2}	-7.02940×10^{-4}	-7.02937×10^{-4}	-7.02937×10^{-4}
J_{c_3}	-1.56153×10^{-4}	-1.56157×10^{-4}	-1.56157×10^{-4}
J_{c_4}	-8.51037×10^{-6}	-8.49719×10^{-6}	-8.49719×10^{-6}
J_{c_5}	1.10817×10^{-6}	1.11079×10^{-6}	1.11079×10^{-6}
J_{c_6}	1.23709×10^{-7}	1.20490×10^{-7}	1.20466×10^{-7}
J_{c_7}	-1.08929×10^{-8}	-1.16848×10^{-8}	-1.18246×10^{-8}

Remark 29.5 *The results obtained from the FFT are more correct than those computed through Eq. (29.14) because the latter are obtained by truncating the series in the powers of α and β.*

29.2.2 Equivalent length

To completely define the equivalent system, the stiffness of the portions of shaft connecting the flywheels must be computed. The equivalent system is modeled as a lumped-parameters system and, consequently, the various flywheels have a vanishing axial length: The portions of the actual shaft modeled by the equivalent stiffnesses must be contiguous, each starting in the same position where the previous one ends.

It is not possible to compute their stiffness by modeling each part or the crank as a simple body (beams loaded in torsion for the journals, beams loaded in bending and torsion for the crankpins, beams loaded in bending for the crank webs, etc.). The geometric complexity, the presence of fillets, and the very small slenderness of the beams make it impossible to resort to such approach.

There are three ways to evaluate the equivalent stiffness, namely

- experimental evaluation,

- use of semi-empirical methods, and

- numerical modeling, mainly using the FEM.

The experimental evaluation clearly yields the most reliable results, but it obviously cannot be performed at the design stage without building models or prototypes with additional cost and time required for the dynamic analysis.

Empirical and semi-empirical formulas, which allow at least approximate evaluations to be obtained, have been suggested by many authors and can be found in several handbooks.[2]

They are mainly based on simple mathematical models and modified on the basis of experimental results. They yield the equivalent stiffness or the equivalent length of a shaft with a given cross-section, which is the same. They yield fairly accurate results, provided they are used with care, without attempting to apply them outside the field for which they are suggested. One of the mentioned formulas, usually referred to as Carter's formula, is

$$l_{eq} = (2c + 0,8b) + \frac{3}{4}a\frac{D^4 - d^4}{D'^4 - d'^4} + \frac{3}{2}r\frac{D^4 - d^4}{bs^3}, \qquad (29.18)$$

---

[2]See, for example, E.J. Nestorides, *A Handbook on Torsional Vibration*, Cambridge Univ. Press, 1958; or W. Ker Wilson, *Torsional Vibration Problems*, Chapman & Hall, 1963.

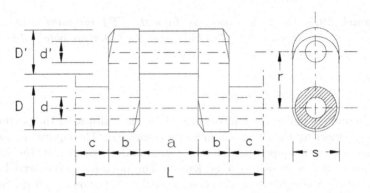

FIGURE 29.4. Sketch of the crankshaft for the computation of the equivalent length.

where the notation refers to Fig. 29.4. It is particularly suited for fast internal-combustion engines. Another common formula is the so-called Tuplin's formula, particularly suited for large diesel engines (same notation as Fig. 29.4)

$$l_{eq} = \frac{2c + 0,15D}{1 - \left(\frac{d}{D}\right)^4} + \frac{(a + 0,15D)(D^4 - d^4)}{\left[1 - \left(\frac{d'}{D'}\right)^4\right](D'^4 - d'^4)} + \left[2b - 0,15(D + D')\right] \times$$

$$\times \frac{D^4 - d^4}{s^4 - d^4} + r\left(0,065\frac{D}{b} + 0,58\right)\frac{D^4 - d^4}{bs^3} + 0,016\frac{D^4 - d^4}{b^2 s}.$$

$$(29.19)$$

It is possible to resort to a third approach based on the construction of numerical models of a single crank and to evaluate their static stiffness by numerical methods, mainly the FEM. This is much simpler than the complete simulation of the crankshaft using the same numerical approach. Only one crank (or half, for symmetry) needs to be modeled, if all cranks are equal, and the computation reduces to a static evaluation.

Nevertheless, the geometric complexity and the uncertainties on how to constrain the mathematical model can make this computation more difficult than it appears.

Strictly speaking, the lack of symmetry couples torsional and flexural deformations, and the stiffness of the basement of the engine and the presence of the oil films in the bearings can affect the results.

The equivalent stiffness and length, computed through any of the mentioned approaches, are linked to each other through the obvious formula

$$k = G\frac{I_p}{l_{eq}}.$$

$$(29.20)$$

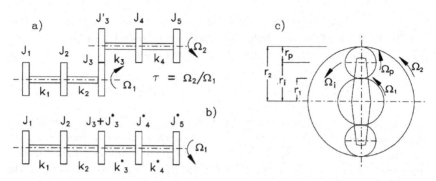

FIGURE 29.5. Geared system: Sketch of the (**a**) actual system and (**b**) equivalent system; (**c**) planetary gear train. Sketch of the system and notation.

29.2.3   Geared systems

Consider the system sketched in Fig. 29.5a in which the two shafts are linked by a pair of gear wheels, with transmission ratio τ. For the study of the torsional vibrations of the system, it is possible to replace the system with a suitable equivalent system, in which one of the two shafts is substituted by an expansion of the other (Fig. 29.5b). This substitution can be performed only if no allowance is taken for backlash, which would introduce nonlinearities. Assuming also that the deformation of gear wheels is negligible, the equivalent rotations ϕ_i^* can be obtained from the actual rotations ϕ_i simply by dividing the latter by the transmission ratio $\tau = \Omega_2/\Omega_1$:

$$\phi_i^* = \frac{\phi_i}{\tau} \, . \tag{29.21}$$

The kinetic energy of the ith flywheel, whose moment of inertia is J_i, and the elastic potential energy of the ith span of the shaft are, respectively,

$$\mathcal{T} = \tfrac{1}{2} J_i \dot{\phi}_i^2 = \tfrac{1}{2} J_i^* \dot{\phi}_i^{*2} \, ,$$

$$\mathcal{U} = \tfrac{1}{2} k_i \left(\phi_{i+1}^2 - \phi_i^2 \right) = \tfrac{1}{2} k_i^* \left(\phi_{i+1}^{*2} - \phi_i^{*2} \right) , \tag{29.22}$$

where the equivalent moment of inertia and stiffness are, respectively,

$$J_i^* = \tau^2 J_i \, , \qquad k_i^* = \tau^2 k_i \, . \tag{29.23}$$

If the system includes a planetary gear train, the computation can be performed without difficulties. The equivalent stiffness can be computed simply from the overall transmission ratio, and in the computation of the equivalent inertia the total kinetic energy of the rotating parts must be taken into account.

The angular velocities of the central gear Ω_1, of the ring gear Ω_2, of the revolving carrier Ω_i, and of the intermediate pinions Ω_p of the planetary gear shown in Fig. 29.5c are linked by the equation

$$\frac{\Omega_1 - \Omega_i}{\Omega_2 - \Omega_i} = -\frac{r_2}{r_1}, \qquad \Omega_p = (\Omega_1 - \Omega_i)\frac{r_1}{r_p} - \Omega_i. \qquad (29.24)$$

The equivalent moment of inertia of the system made of the internal gear, with moment of inertia J_1, the ring gear, with moment of inertia J_2, the revolving carrier, with moment of inertia J_i, and n intermediate pinions, each with mass m_p and moment of inertia J_p, referred to the shaft of the internal gear is

$$J_{eq} = J_1 + J_2\left(\frac{\Omega_2}{\Omega_1}\right)^2 + (J_i + nm_p r_i^2)\left(\frac{\Omega_i}{\Omega_1}\right)^2 + nJ_p\left(\frac{\Omega_p}{\Omega_1}\right)^2. \qquad (29.25)$$

If the deformation of the meshing teeth must be accounted for, it is possible to introduce into the model two separate degrees of freedom for the two meshing gear wheels, modeled as two different inertias, and to introduce between them a shaft whose compliance simulates the compliance of the transmission. This is particularly important when a belt or flexible transmission of some kind is used instead of gear wheels.

In a machine there may be several shafts connected to each other, in series or in parallel, by gear wheels with different transmission ratios. The equivalent system is referred to one of the shafts and the equivalent inertias and stiffnesses of the elements of the others are all computed using the ratios between the speeds of the relevant element and the reference one. The equivalent system will thus be made of a set of elements, in series or in parallel, following the scheme of the actual system, but with rotations that are all consistent. If the compliance of the gears is to be taken into account in detail, the nonlinearities due to the contacts between the meshing teeth and backlash must be considered. The study becomes far more complex and the methods seen in Part II for nonlinear systems must be used.

29.3   Computation of the natural frequencies

In many cases, the equivalent system reduces to a single shaft on which a number of moments of inertia are located. The system is thus in-line, and it is possible to use the transfer-matrices method. The Holzer method, described in detail in Section 14.5, for decades was the most common tool for dealing with this problem. It is, however, possible and more computationally efficient to resort to a stiffness approach. There is no difficulty writing the stiffness matrix of each span of the shaft, modeled as a beam element and assembling them into a global stiffness matrix. In the case of in-line systems, a tridiagonal stiffness matrix is obtained.

Remark 29.6 *The stiffness matrix is singular because there is no con-straint to prevent the system from performing rigid rotations: It cannot be inverted, and the compliance matrix is not available.*

The mass matrix is even simpler: Because the inertia properties are lumped, the mass matrix is diagonal.

The equation of motion for the study of the free behavior of the system of Fig. 29.5b is, for example,

$$
\begin{bmatrix}
J_1 & 0 & 0 & 0 & 0 \\
0 & J_2 & 0 & 0 & 0 \\
0 & 0 & J_3 + J_3^* & 0 & 0 \\
0 & 0 & 0 & J_4^* & 0 \\
0 & 0 & 0 & 0 & J_5^*
\end{bmatrix}
\begin{Bmatrix}
\ddot{\phi}_1 \\
\ddot{\phi}_2 \\
\ddot{\phi}_3 \\
\ddot{\phi}_4^* \\
\ddot{\phi}_5^*
\end{Bmatrix}
+ \qquad (29.26)
$$

$$
+
\begin{bmatrix}
k_1 & -k_1 & 0 & 0 & 0 \\
-k_1 & k_1 + k_2 & -k_2 & 0 & 0 \\
0 & -k_2 & k_2 + k_3^* & -k_3^* & 0 \\
0 & 0 & -k_3^* & k_3^* + k_4^* & -k_4^* \\
0 & 0 & 0 & -k_4^* & k_4^*
\end{bmatrix}
\begin{Bmatrix}
\phi_1 \\
\phi_2 \\
\phi_3 \\
\phi_4^* \\
\phi_5^*
\end{Bmatrix}
=
\begin{Bmatrix}
0 \\
0 \\
0 \\
0 \\
0
\end{Bmatrix}.
$$

The system is underconstrained, the first natural frequency is equal to zero, and the corresponding mode is a rigid-body rotation about the axis of the shaft. Usually only the non-vanishing natural frequencies are consid-ered: They are $n - 1$ if the system has n degrees of freedom. If the system is damped the study can be performed without difficulty using the FEM, but the presence of damping can introduce some complexities when using transfer matrices.

In the case of in-line systems, it is possible to resort to the transfer-matrices approach, but when the geometry is more complicated, it is nec-essary to resort to the stiffness approach. However, in the case of branched systems, when the secondary branches stemming from the main system have little influence on the dynamic behavior of the latter, it is still possi-ble to use the Holzer method by approximating each branch with a single inertia located at the connection point. The simplifications that can be introduced in this way are, however, very small, if modern computing fa-cilities are used, and they do not justify the approximations so introduced.

The stiffness matrix of a branched or multiply connected system can easily be built using the assembly procedures seen for the FEM. It is no longer tridiagonal, even if it usually still has a band structure, and it is possible to resort to the usual algorithms to reorder the list of the general-ized coordinates, reducing the bandwidth to a minimum. The mass matrix, on the contrary, is always a diagonal matrix if the model is based on a lumped-parameters approach. The damping matrix usually has the same type of structure of the stiffness matrix, except that a number of elements can give vanishingly small contributions to damping.

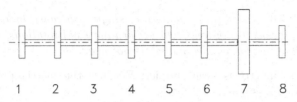

FIGURE 29.6. Equivalent system.

Example 29.2 *Compute the torsional natural frequencies of a diesel six-cylinder, four-stroke-cycle engine driving an electric generator unit.*
The main characteristics of the system are bore: $\phi = 300$ mm; area of the pistons: $A = 706.8$ cm^2; stroke: $2r = 450$ mm; firing order: 1-5-3-6-2-4; speed: $\Omega = 500$ rpm (52.33 rad/s); mass of complete piston: $m_p = 70.5$ kg; mass of connecting rod: $m_b = 100.6$ kg; length of connecting rod: $l_b = 950$ mm; distance a (Fig. 29.2): $a = 270$ mm; moment of inertia of connecting rod: $J_b = 16.383$ kg m^2. Geometric data of crankshaft (Fig. 29.4): $a = 136$ mm, $b = 92$ mm, $c = 73$ mm, $D = 240$ mm, $d = 80$ mm, $D' = 230$ mm, $d' = 80$ mm, $s = 400$ mm. Moment of inertia of crank web: $J_d = 14.880$ kg m^2; moment of inertia of flywheel: $J_6 = 98$ kg m^2; stiffness engine-flywheel shaft: $k_6 = 66 \times 10^6$ Nm/rad; moment of inertia of generator: $J_7 = 49$ kg m^2; stiffness flywheel-generator shaft: $k_7 = 15 \times 10^6$ Nm/rad.
Equivalent system: Moments of inertia. The value of ratio α is $\alpha = r/l = 0.2368$. From Eq. (29.6) it follows that $m_1 = 72.008$ kg, $m_2 = 28.591$ kg, $J_0 = -2.087$ kg m^2. The mean equivalent moment of inertia is thus $\overline{J}_{eq} = 21$ kg m^2.
Equivalent system: Stiffness of cranks. From Eq. (29.19) (Tuplin's formula), it follows that $l_{eq} = 513.79$. The stiffness of the shaft that separates each pair of contiguous equivalent inertias is then

$$k = \frac{GJ_p}{l_{eq}} = 51 \times 10^6 \text{ Nm/rad.}$$

The scheme of the equivalent system is reported in Fig. 29.6.
Natural frequencies and mode shapes. The mass matrix of the system is the following diagonal matrix:

$$M = diag\{\ 21 \quad 21 \quad 21 \quad 21 \quad 21 \quad 21 \quad 98 \quad 49\ \}.$$

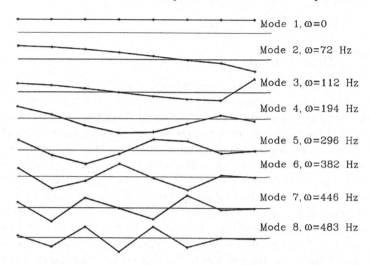

Mode 1, ω=0

Mode 2, ω=72 Hz

Mode 3, ω=112 Hz

Mode 4, ω=194 Hz

Mode 5, ω=296 Hz

Mode 6, ω=382 Hz

Mode 7, ω=446 Hz

Mode 8, ω=483 Hz

FIGURE 29.7. Torsional modes of the system in Fig. 29.6.

The stiffness matrix is a tridiagonal matrix:

$$K = \begin{bmatrix} 51 & -51 & 0 & 0 & 0 & 0 & 0 & 0 \\ -51 & 102 & -51 & 0 & 0 & 0 & 0 & 0 \\ 0 & -51 & 102 & -51 & 0 & 0 & 0 & 0 \\ 0 & 0 & -51 & 102 & -51 & 0 & 0 & 0 \\ 0 & 0 & 0 & -51 & 102 & -51 & 0 & 0 \\ 0 & 0 & 0 & 0 & -51 & 117 & -66 & 0 \\ 0 & 0 & 0 & 0 & 0 & -66 & 81 & -15 \\ 0 & 0 & 0 & 0 & 0 & 0 & -15 & 15 \end{bmatrix} \times 10^6 .$$

The natural frequencies and mode shapes can easily be obtained by solving the eigenproblem, $\det(K - \omega^2 M) = 0$, *which yields the following matrix of the eigenvalues:*

$$[\omega^2]$$

$$= diag\big\{ \ 0 \quad 0.210 \quad 0.499 \quad 1.486 \quad 3.468 \quad 5.789 \quad 7.867 \quad 9.245 \ \big\} \times 10^6 .$$

The natural frequencies of the system are thus

$\omega_0 = 0$,

$\omega_1 = 458 \ rad/s \ = \ 72 \ Hz$,

$\omega_2 = 704 \ rad/s \ = \ 112 \ Hz$,

$\omega_3 = 1,219 \ rad/s \ = \ 194 \ Hz$,

$\omega_4 = 1,862 \ rad/s \ = \ 296 \ Hz$,

$\omega_5 = 2,406 \ rad/s \ = \ 382 \ Hz$,

$\omega_6 = 2,804 \ rad/s \ = \ 446 \ Hz$,

$\omega_7 = 3,040 \ rad/s \ = \ 483 \ Hz$.

The mode shapes are reported in Fig. 29.7.

29.4   Forced vibrations

29.4.1   Driving torque

The elements of reciprocating machines are loaded by forces variable in time, which in many instances can excite torsional vibrations of the shaft. In this section, only the forces originating the driving torque in reciprocating engines will be studied in detail but the same considerations can be referred to reciprocating compressors and similar machines. The fact that a term of the equation of motion coming from the derivatives of the kinetic energy, i.e., from the inertia of the system, can, under widely used assumptions, be considered as due to inertia forces applied to the system with known time history was explained in detail in Section 29.2.1. The corresponding inertia forces will then be included in the external forcing functions. A moment, varying in time with known time history, equal to the sum of a driving torque and an inertia torque, will be considered as acting on each flywheel of the equivalent system.

The pressure of the gases contained in the cylinder $p(t)$ varies in time during the working cycle of the engine. The virtual displacement δs of the piston corresponding to a virtual displacement $\delta \theta$ of the crank and the corresponding virtual work $\delta \mathcal{L}$ performed by the pressure can be expressed, respectively, as

$$\delta s = \frac{V_p}{\theta} \delta \theta \,, \qquad \delta \mathcal{L} = p(t) A \delta s = p(t) r A \sqrt{f_1(\theta)} \delta \theta \,, \qquad (29.27)$$

where function $f_1(\theta)$ is given by Eq. (29.5) and A is the area of the piston. The generalized force M_m due to the pressure $p(t)$ is, consequently,

$$M_m = \frac{d(\delta \mathcal{L})}{d(\delta \theta)} = p(t) r A \sqrt{f_1(\theta)} \,. \qquad (29.28)$$

It must be noted that $M_m(t)$ is a function of the generalized coordinate $\phi_z(t)$ through function $f_1(\theta)$. Usually this dependence is neglected, introducing into Eq. (29.28) a simplified expression of the crank angle $\theta = \Omega t$. This assumption introduces negligible errors, at least in the case of vibrations with small amplitude, and is usually considered acceptable; at any rate, it is consistent with all the other simplifications seen earlier.

In the case of gas compressors, steam engines, or two-stroke-cycle internal-combustion engines, function $p(t)$ is periodical with fundamental frequency equal to the rotational velocity Ω of the shaft. In the case of four-stroke-cycle internal-combustion engines, the period of function $p(t)$ is doubled, i.e., its fundamental frequency is equal to $\Omega/2$ (see, for instance, Fig. 29.8).

Because the generalized force (moment) $M_m(t)$ is periodical, with the same frequency of law $p(t)$, it can be expressed by a trigonometric polynomial, truncated after m harmonic terms

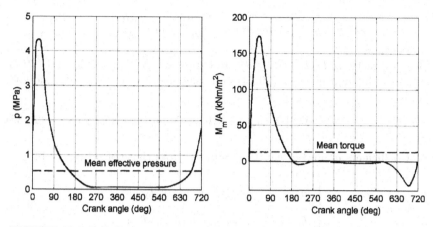

FIGURE 29.8. Pressure acting on the piston and diving torque (divided by the area of the piston) in a four-stroke-cycle engine as functions of the crank angle.

$$M_m = A_0 + \sum_{k=1}^{m} A_k \cos(k\Omega't) + \sum_{k=1}^{m} B_k \sin(k\Omega't), \qquad (29.29)$$

where the frequency Ω' of the fundamental harmonic is equal to Ω, except for the case of four-stroke-cycle internal-combustion engines in which

$$\Omega' = \frac{\Omega}{2},$$

and

$$\begin{cases} A_0 = \dfrac{1}{2\pi} \displaystyle\int_0^{2\pi} M_m d(\Omega't), \\[2mm] A_k = \dfrac{1}{\pi} \displaystyle\int_0^{2\pi} M_m \cos(k\Omega't) d(\Omega't), \qquad k = 1, 2, \ldots, \\[2mm] B_k = \dfrac{1}{\pi} \displaystyle\int_0^{2\pi} M_m \sin(k\Omega't) d(\Omega't), \qquad k = 1, 2, \ldots. \end{cases} \qquad (29.30)$$

The coefficients of the polynomial can be computed from the theoretical or experimental law $p(t)$, and empirical expressions can be found in the literature. The driving torque depends at any rate on the working conditions of the machine; the coefficients A_i and B_i can be assumed to be proportional to the average driving torque A_0 or to the product of half the displacement (the area of the piston times crank radius) by the mean indicated pressure.

29.4.2   Inertia torque

The inertia torque, i.e., the moment acting on the flywheel simulating the crank as a result of the variability of the equivalent moment of inertia, can be obtained by introducing Eq. (29.8) into Eq. (29.12)

$$M_i = -\frac{1}{2}\Omega^2 \left[r^2(m_2 + m_p)\frac{df_1(\theta)}{d\theta} + J_0\frac{df_2(\theta)}{d\theta} \right] , \qquad (29.31)$$

where functions $f_1(\theta)$ and $f_2(\theta)$ are expressed by Eqs. (29.5) and (29.7). By introducing the series (29.14) and remembering that $\theta \approx \Omega t$, it follows that

$$M_i = \Omega^2 \left[\sum_{k=1}^{\infty} a_k^* \cos(k\Omega t) + \sum_{k=1}^{\infty} b_k^* \sin(k\Omega t) \right] . \qquad (29.32)$$

Coefficients a_k^* and b_k^* can be computed from those of series (29.14)

$$\begin{cases} a_k^* = \dfrac{k}{2}\left[r^2(m_2 + m_p)a_k + J_0 c_k \right] , \\[2mm] b_k^* = -\dfrac{k}{2}\left[r^2(m_2 + m_p)b_k + J_0 d_k \right] . \end{cases} \qquad (29.33)$$

Remark 29.7 *The fundamental harmonic of the inertia torque M_i is always Ω. Moreover, if the axis of the cylinder passes through the center of the crank ($\beta = 0$), there are only harmonics in sine (all $b_k^* = 0$). The harmonics in the series for M_i coincide with those in the series for the driving torque M_m, although being less in number.*

The series for the total forcing torque acting on the equivalent flywheels simulating the cranks is then obtained by adding the relevant terms of the series for M_m and M_i.

29.4.3 Torsional critical speeds

The frequency of each harmonic of the forcing function acting on the cranks is proportional to the rotational speed of the machine. In the case of two-stroke-cycle internal-combustion engines or other reciprocating machines in which the duration of the working cycle corresponds to a single revolution of the crankshaft, the frequency of the various harmonics is equal to whole multiples of the rotational speed

$$\omega_i = k\Omega .$$

In the case of four-stroke-cycle internal-combustion engines the frequency of the fundamental harmonic is equal to half of the rotational speed Ω and the frequency of the kth harmonic is

$$\omega_i = \frac{k}{2}\Omega .$$

The torsional natural frequencies of the equivalent system are independent of the rotational speed. The resonance conditions, defining the torsional critical speeds, can then be studied using a Campbell diagram of the type shown in Fig. 29.8. There are many resonance conditions, and it

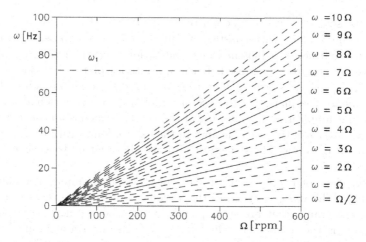

FIGURE 29.9. Torsional Campbell diagram of a six-cylinder four-stroke-cycle internal-combustion engine.

is very difficult, usually impossible, to avoid some of them being located within the working range of the machine. Not all resonance conditions are equally dangerous, and the dynamic stressing of the shaft at the various critical speeds must be evaluated to understand their severity.

Example 29.3 *Compute the torsional critical speeds of the diesel engine electric generator unit studied in Example 29.2.*
The Campbell diagram of the system is plotted in Fig. 29.9. In the speed range from 0 to 600 rpm (overspeed of 20%) there are six critical speeds linked with the first 20 harmonics of the forcing function. The ones nearest to the maximum speed are those related to the 14th and 15th harmonic ($\omega = 7\Omega$ and $\omega = 15\Omega$ /2). They are $\Omega_{cr_{14}} = 624.8$ rpm and $\Omega_{cr_{15}} = 583.1$ rpm.

29.4.4 Forcing functions on the cranks of multi-cylinder machines

The total torque acting on the ith crank, obtainable as a sum of moments M_m and M_i, can be expressed as a trigonometric polynomial, whose terms are characterized by different amplitudes and phases. If the various cranks, reciprocating parts, and working cycles of a multi-cylinder machine are all equal, the time histories of the moments acting on the various nodes of the equivalent system are all equal but are timed in a different way. Because each harmonic component of the moment acting on the cranks can be represented as the projection on the real axis of a vector rotating in the complex plane with constant angular velocity, it is possible to draw, for each harmonic, a plot in which the various vectors acting on the different cranks are represented. If the amplitudes of these vectors are all equal, as in the case when all cranks are equal, the diagram is useful only to compare the phases of the vectors, and they are traditionally plotted with unit amplitude. The phasing of the vectors depends on the geometric characteristics of the machine and, in the case of four-stroke-cycle engines, on the firing order. These diagrams are usually referred to as *phase-angle diagrams*.

Consider, for example, an in-line four-stroke-cycle internal-combustion engine. If the working cycles of the various cylinders are evenly spaced in time, the cranks that fire subsequently must make an angle of $4\pi/n$ rad, where n is the number of cylinders. In the case of a four-in-line engine, this angle is 180°, and the most common geometric configuration of the crankshaft is that shown in Fig. 29.10a, chosen because it allows the best balancing of inertia forces. In the same figure, the configuration of the crankshaft of a six-in-line engine is also shown.

In the case of the four-cylinder engine, the possible firing orders are two: 1-2-3-4 and 1-3-4-2. In both cases, it is impossible to prevent two contiguous cylinders from firing immediately one after the other. The phase-angle diagrams for the first four harmonics are plotted in Fig. 29.10b, for the second of the two firing orders.

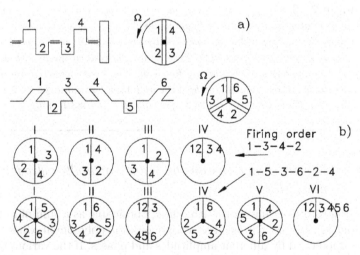

FIGURE 29.10. (a) Configuration of the crankshaft and crank angle diagrams for in-line four-stroke-cycle four- and six-cylinder internal-combustion engines. In the latter case the configuration shown is just one of the possible choices; (b) phase-angle diagrams for the same engines.

If the order of the harmonic is a whole multiple of the number of cylinders, all rotating vectors are superimposed, i.e., the forcing functions acting on all cranks are all in phase. These harmonics are usually the most dangerous and are often referred to as major harmonics. The phase-angle diagrams for the $(n + i)$th harmonic coincide with that related to the ith harmonic and, consequently, only the first n phase-angle diagrams are usually plotted.

The phase-angle diagrams are plotted so that they directly supply the phasing of the excitation on the various cranks with respect to that acting on a crank chosen as reference, usually the first one. The phase of each harmonic with respect to the fundamental one must be taken into account when adding their effects. The forcing function acting on the jth crank can be expressed by the following series, truncated at the mth harmonic

$$M_{tot_j} = M_{m_j} + M_{i_j} = \sum_{k=0}^{m} M_{m_k} e^{i\left(k\Omega' i + \Phi_{m_k} + \delta_{j_k}\right)} + \sum_{i=0}^{m} M_{i_k} e^{i\left(k\Omega' i + \Phi_{i_k} + \delta_{j_k}\right)},$$

(29.34)

where

- M_{m_k} and Φ_{m_k} are the amplitude and phase of the kth harmonic of the driving torque, respectively. With reference to the series (29.29) approximating the driving torque, their values are $M_{m_k} = \sqrt{A_k^2 + B_k^2}$ and $\Phi_{m_k} = \arctan(A_k/B_k)$, respectively.

- δ_{j_k} is the phase of the kth harmonic acting on the jth crank, as obtained from the phase-angle diagram. If the diagram is referred to the first crank, $\delta_{j_k} = 0$ for $j = 1$.

- M_{i_k} and Φ_{i_k} are the amplitude and phase, respectively, of the kth harmonic (referred to the fundamental harmonic Ω' and not to the rotational speed Ω) of the inertia torque. If

$$\Omega' = \Omega,$$

the order of the harmonic to be introduced into Eq. (29.34) coincides with that in Eq. (29.32); if, on the contrary,

$$\Omega' = \frac{\Omega}{2},$$

the amplitude of the odd harmonics vanishes while even harmonics correspond with those whose order is halved in Eq. (29.32). If the axis of the cylinder passes through the center of the shaft, the phases ϕ_{i_k} are all equal to $\pi/2$, as all terms in $\cos(k\Omega t)$ vanish.

Phases δ_{j_k} of the harmonics of the driving torque M_{m_k} coincide with those of the harmonics of the inertia torque M_{i_k}.

29.4.5 Evaluation of the damping of the system

Although the response of the system in conditions far from resonance can be computed from an undamped model, the response at resonance can be obtained only after the damping of the system has been evaluated, and the precision of the results is strictly dependent on the precision with which damping is known. Because the resonant conditions are usually the most dangerous, this part of the dynamic analysis is important, and the difficulty in estimating the damping is one of the factors limiting the usefulness of the dynamic analysis and the need of resorting to extensive experimentation. If damping were mostly due to the internal damping of the material, there would be no difficulty introducing a proportional damping with modal-damping ratio equal for all modes: $\zeta_j = \eta/2$, where η is the loss factor of the material of the crankshaft.

Actually, damping is due to many causes, like friction between moving parts (e.g., the piston and the cylinder wall), electromagnetic forces, and the presence of fluid in which some rotating parts move. Neglecting them would lead to a large underestimate of damping, and it is usually necessary to resort to experimental results, obtained from machines similar to the one under study and to empirical or semi-empirical formulas and numerical values reported in the literature.

Usually the damping due to the crank mechanism is evaluated by introducing a damping force acting on the crankpin proportional to the area of the piston and the velocity of the crankpin. The damping moment acting on the jth crank is

$$M_d(t) = k' A r^2 \,, \tag{29.35}$$

where k' is a coefficient whose dimension is a force multiplied by time and divided by the third power of a length. In SI units, it is expressed in $\mathrm{Ns/m^3}$. Values of k' included in the range between 3,500 and 10,000 $\mathrm{Ns/m^3}$ for in-line aircraft engines and between 15,000 and 1.5×10^6 $\mathrm{Ns/m^3}$ for large internal-combustion engines can be found in the literature. The span of these ranges is large, and these values must be regarded only as indicative; only experimental results on machines similar to the one under study can be reliable. The use of Eq. (29.35) leads to the assumption that in each crank there is a viscous damper with damping coefficient equal to

$$c_j = k' A r^2 \,.$$

The damping of the other elements connected to the shaft (propellers, brakes, electrical machines) can be evaluated by assuming that they provide a braking torque on the relevant node that is proportional to the instantaneous angular velocity at the power p through coefficient k''

$$M_{d_j}(t) = k'' \left[\Omega + \sum_{k=1}^{m} \dot{\phi}_{z_{j_k}}(t) \right]^p \,, \tag{29.36}$$

where subscript j refers to the jth node of the model and k denotes the kth harmonic. All harmonics are assumed to act in phase, an approximation that does not affect the results obtained later. This type of damping is clearly nonlinear, but can be linearized by introducing in the computation a viscous damping, equivalent from the viewpoint of energy dissipation, for each harmonic. By introducing the harmonic time history of each component of the velocity $\dot{\phi}_{z_{j_k}}(t)$, the energy dissipated in a cycle is

$$\mathcal{L} = \int_0^T k'' \left[\Omega + \sum_{k=1}^m k\Omega' \phi_{z_{j_{k_0}}} \cos(k\Omega't) \right]^{p+1} dt . \tag{29.37}$$

By computing the $(p+1)$th power of the binomial within the integral sign, it follows that

$$\mathcal{L} = \int_0^T k'' \Omega^{p+1} dt + \binom{p+1}{1} \int_0^T k'' \Omega^p \Omega' \sum_{k=1}^m k\phi_{z_{j_{k_0}}} \cos(k\Omega't) dt +$$

$$+ \binom{p+1}{2} \int_0^T k'' \Omega^{p-1} \Omega'^2 \left[\sum_{k=1}^m k\phi_{z_{j_{k_0}}} \cos(k\Omega't) \right]^2 dt + \cdots . \tag{29.38}$$

The first integral of Eq. (29.38) is constant and corresponds to the average braking power

$$P = k'' \Omega^{p+1}$$

applied in the relevant node. It can be used to compute coefficient k'', as the average power is generally known.

The second integral vanishes, while the third contains terms in the squares of cosine function and terms in which there are products of cosines with different arguments. The latter vanish once they are integrated over a whole period. The fourth term also vanishes. If the series (29.38) is truncated after the fourth term, the braking torque, excluding the contribution due to the constant term, is

$$\mathcal{L} = \binom{p+1}{2} \pi k'' \Omega^{p-1} \Omega' \sum_{k=1}^m k\phi_{z_{j_{k_0}}}^2 , \tag{29.39}$$

corresponding to that due to a viscous damper with equivalent damping coefficient

$$c_{eq} = \binom{p+1}{2} k'' \Omega^{p-1} . \tag{29.40}$$

If $p = 1$, it is a true viscous damper and c_{eq} coincides with k''.

If $p = 2$, as is often assumed in the case of propellers, k'' is expressed in SI units as Nms^2 and

$$c_{eq} = 3k'' \Omega .$$

If $p = 3$, k'' is expressed in SI units as Nms^3 and

$$c_{eq} = 6k''\Omega^2 \ .$$

The damping matrix of the system can be obtained by applying in the various nodes the viscous dampers whose damping coefficients have been computed above and is diagonal. This way of evaluating the damping of the system is an approximation that can be rough. However, it is often used in practice because there are no simple alternatives.

29.4.6 Forced response of the system

Once the forcing functions have been obtained, there is no difficulty, at least from a theoretical viewpoint, to compute the response of the system at the various speeds. Usually the problem of computing the forced response of the system is subdivided into two parts: the evaluation of the response to the harmonics of the forcing function that are not in resonance and that related to the resonant harmonics. The first problem can be solved by resorting to the undamped system; for the second it is necessary to resort to the damped model.

A vector of the driving functions, whose terms are expressed by the series (29.34), can be added on the right-hand side of the equation of motion (29.26), to which the damping matrix obtained in Section 29.4.5 has also been added

$$\mathbf{J}\{\ddot{\phi}_z\} + \mathbf{C}\{\dot{\phi}_z\} + \mathbf{K}\{\phi_z\} = \sum_{(k=1)}^{m} \mathbf{m}_{s_k} \sin(k\Omega't) + \sum_{(k=1)}^{m} \mathbf{m}_{c_k} \cos(k\Omega't),$$

$$(29.41)$$

where, to avoid confusion with the moments M, the symbol $\mathbf{J}$ (not to be confused with the identity matrix $\mathbf{I}$) has been used for the mass matrix because all its elements are mass moments of inertia; $\mathbf{m}$ indicates the vectors in which the moments are listed. The coefficients of the terms in cosine and sine, respectively, expressed by Eq. (29.34) have been collected in vectors:

$$\begin{aligned}
\mathbf{m}_{c_k} &= \{m_{m_k} \cos(\Phi_{m_k} + \delta_k)\} + \Omega^2\{m_{i_k} \cos(\Phi_{i_k} + \delta_k)\}, \\
\mathbf{m}_{s_k} &= \{m_{m_k} \sin(\Phi_{m_k} + \delta_k)\} + \Omega^2\{m_{i_k} \sin(\Phi_{i_k} + \delta_k)\}.
\end{aligned} \qquad (29.42)$$

They are clearly functions of Ω^2, due to the presence of inertia torques. The series are truncated at the mth harmonic, and the static component of the torque has been neglected. In the case of major harmonics of a machine in which all elements are equal and distance d is equal to zero, all δ_{j_k} vanish and Φ_{i_k} are all equal to $\pi/2$.

The solution of Eq. (29.41) can be obtained directly by adding the responses to the various harmonic components

$$\phi = \sum_{k=1}^{m} \phi_{z_k} = \sum_{k=1}^{m} \phi_{z_{s_k}} \sin(k\Omega't) + \sum_{k=1}^{m} \phi_{z_{c_k}} \cos(k\Omega't). \qquad (29.43)$$

The amplitudes of the components in sine and cosine of the response can be obtained by solving the following linear sets of equations

$$
\begin{bmatrix} \mathbf{K} - k^2\Omega'^2\mathbf{J} & -k\Omega'\mathbf{C} \\ k\Omega'\mathbf{C} & \mathbf{K} - k^2\Omega'^2\mathbf{J} \end{bmatrix} \left\{ \begin{matrix} \boldsymbol{\phi}_{z_{s_k}} \\ \boldsymbol{\phi}_{z_{c_k}} \end{matrix} \right\} = \left\{ \begin{matrix} \mathbf{m}_{s_k} \\ \mathbf{m}_{c_k} \end{matrix} \right\}. \qquad (29.44)
$$

If damping is neglected, as is customary when the response in conditions far from resonance is obtained, Eq. (29.44) uncouples into two separate linear sets of equations. Equation (29.44) must be solved at each rotational speed and for each harmonic of the forcing function; however, the computation is usually not heavy, because the order of the matrices, equal to the number of the degrees of freedom of the equivalent system, is small.

The amplitude of the oscillations at each node of the system, due to each harmonic of the forcing function, can be easily computed. Once the amplitudes are known, there is no difficulty obtaining the dynamic stressing of each span of the shaft. The maximum value of the shear stress in the shaft spanning from the jth to the $(j+1)$th node is

$$
\tau_{max_j}(t) = \frac{k_j(\phi_{z_{j+1}} - \phi_{z_j})}{W_j}, \qquad (29.45)
$$

where W_j is the torsional section modulus of the relevant shaft element. The torsional displacements at the nodes are not in phase, and then it is impossible to obtain the twist angle of each span as difference of amplitude between the end sections. The components in phase and in quadrature must be accounted for separately, and the amplitude of the twist angle, needed to compute the stress, must be obtained from the two components. The shear stress so computed is variable in time with poly-harmonic time history; the time histories of the stresses due to the various harmonics, each with its amplitude, phase, and frequency, should then be added to each other and to the static stressing, and the fact that their consequences on the overall fatigue of the shaft are different should be considered.

In practice, a much simpler approach is followed, not because of the long computations involved, which with a modern computer could be dealt without problems, but because the phasing of the harmonics can be difficult to evaluate (near the resonance the phase is quickly variable) and predicting the fatigue life of a machine element subject to poly-harmonic stressing is still difficult. The amplitudes of the stress cycles due to the various harmonics are computed and added together, often limiting the sum to the few most important harmonics. This procedure leads to overestimating the amplitude of the variable component of the stress and then is conservative.

The shear stresses so computed must be added to those due to other causes and, using a suitable failure criterion, to stresses due to bending, axial forces, shrink fitting, surface forces, etc. The stress concentrations due to the complex geometric shape of crankshafts can be quite high, and the analyst cannot neglect them.

29.4.7 Modal computation of the response

The computation of the response of the system can also be performed following a modal approach, which is essentially equivalent to uncoupling the equations of motion by neglecting the elements of the modal-damping matrix outside the main diagonal. The equation yielding the jth modal amplitude is then

$$\ddot{\eta}_j + 2\zeta_j\omega_j\dot{\eta}_j + \omega_j^2\eta_j = \sum_{(k=1)}^{m} \left[\mathbf{q}_j^T \mathbf{m}_{s_k} \sin(k\Omega't) + \mathbf{q}_j^T \mathbf{m}_{c_k} \cos(k\Omega't) \right] ,$$

(29.46)

where $\mathbf{q}_j$ is the jth eigenvector of the undamped system, normalized in such a way that the corresponding modal mass has a unit value, and the modal-damping ratio ζ_j is

$$\zeta_j = \frac{\mathbf{q}_j^T \mathbf{C} \mathbf{q}_j}{2\omega_j} .$$

(29.47)

The number of modal equations (29.46) is equal to the number of modes with non-vanishing eigenfrequency; however, only a few modes need to be considered, and in most cases only the response corresponding to the first mode is computed.

Usually further approximations are introduced and the computation of the response follows the approach seen in Fig. 7.2d. On the basis of the considerations seen in Section 7.1, a harmonic of the forcing function with frequency $k\Omega'$ is resonant with the sth mode if

$$\omega_s\sqrt{1 - 2\zeta_s} \leq k\Omega' \leq \omega_s\sqrt{1 + 2\zeta_s} .$$

Outside the mentioned frequency range, the undamped system can be used to compute the response of the relevant (jth) mode:

$$\eta_j = \sum_{k=1}^{m} \left[\frac{\mathbf{q}_j^T \mathbf{m}_{s_k} \sin(k\Omega't) + \mathbf{q}_j^T \mathbf{m}_{c_k} \cos(k\Omega't)}{\omega_j^2 + k^2\Omega'^2} \right] .$$

(29.48)

Once the modal responses have been recombined, the amplitudes

$$\boldsymbol{\phi} = \boldsymbol{\Phi}\boldsymbol{\eta}$$

and the shear stresses are obtained.

If one of the harmonics of the forcing function is close to one of the natural frequency, the relevant amplitude must be obtained from the damped model. The deformed shape in resonance is assumed to be equal to the corresponding mode shape, which amounts to the assumption that the modal responses are uncoupled even if the system is damped, i.e., to neglect the out-of-diagonal elements of the modal-damping matrix, and to neglect the contribution of the other modes. The displacements are then in phase with each other because the eigenvector is a real eigenvector of an undamped system.

The oscillations in the various points of the shaft are in phase with each other, but they are neither in phase nor in quadrature with the forcing functions, which are not in phase with each other. Only in the case of the major harmonics are the forcing functions in phase with each other and, in resonance, in quadrature with the response.

Consider the case in which the kth harmonic is in resonance with the sth mode, i.e.,

$$k\Omega' = \omega_s .$$

Because the deformed shape is assumed to be proportional to the sth mode shape, it follows that

$$\phi(t) = \alpha\mathbf{q}_s \sin(k\Omega't), \qquad (29.49)$$

where α is a proportionality coefficient that depends on the criterion used for normalizing the eigenvectors. The phasing of the response was assumed to be equal to zero, i.e., the shaft is assumed to be in the undeformed configuration at time $t = 0$. If the largest value of the eigenvector is assumed to be equal to unity, α is the maximum amplitude of the deflected shape: In modal terms, α is nothing other than the amplitude η_{s_0} of the sth modal coordinate. It can be computed through Eq. (7.7), written using the modal quantities

$$\alpha = \frac{\overline{F}_s}{\overline{C}_s\omega_s} = \frac{\overline{F}_s}{\overline{C}_s k\Omega'}, \qquad (29.50)$$

where

$$\overline{C}_s = \mathbf{q}_s^T\mathbf{C}\mathbf{q}_s = \sum_{j=1}^{n} c_{eq_j}q_{s_j}^2 . \qquad (29.51)$$

The computation of the modal force is more complex. If M_{o_j} is the amplitude of the kth harmonic of the forcing function applied to the jth node and δ_j is the phase lag between the moment applied to the jth node and that applied to the first node obtainable from the phase-angle diagram ($\delta_1 = 0$), it follows that[3]

$$M_j(t) = M_{0_j} \sin(k\Omega't + \delta_j) = \qquad (29.52)$$

$$= \left[M_{0_j}\cos(\delta_j)\right]\sin(k\Omega't) + \left[M_{0_j}\sin(\delta_j)\right]\cos(k\Omega't) .$$

---

[3]In the remainder of this section modal forces are written in phase with the forcing function on the first node, while rotations are written in phase with the rotation of the same node. Because only the amplitude is required, the choice of the instant in which $t = 0$ is immaterial.

The in-phase and in-quadrature (with reference to the first node) components of the modal force must be computed separately. They are

$$\overline{F}_{s_{phase}} = \sum_{j=1}^{n} q_{s_j} M_{0_j} \cos(\delta_j), \qquad \overline{F}_{s_{quad}} = \sum_{j=1}^{n} q_{s_j} M_{0_j} \sin(\delta_j). \quad (29.53)$$

The amplitude of the modal response and the phase delay between the response and the kth harmonic of the forcing function applied at the first node are, respectively,

$$\alpha = \frac{\sqrt{\left[\sum_{j=1}^{n} q_{s_j} M_{0_j} \cos(\delta_j) \right]^2 + \left[\sum_{j=1}^{n} q_{s_j} M_{0_j} \sin(\delta_j) \right]^2}}{k\Omega' \sum_{j=1}^{n} c_{eq_j} q_{s_j}^2}, \quad (29.54)$$

$$\beta = \arctan\left[\frac{\overline{F}_{s_{fase}}}{\overline{F}_{s_{quad}}} \right] = \arctan\left[\frac{\sum_{j=1}^{n} q_{s_j} M_{0_j} \cos(\delta_j)}{\sum_{j=1}^{n} q_{s_j} M_{0_j} \sin(\delta_j)} \right].$$

In the literature, the same result is often obtained using energetic reasoning: The energy dissipated by damping is equated to the work of the forcing functions. The amplitude is often obtained using the graphical construction shown in Fig. 29.11, where the modal force $\overline{F}_s$ for the resonance of the fifth harmonic of the forcing function with the first natural frequency of a four-stroke-cycle in-line six-cylinder engine is computed graphically.

Because the rotations at the various nodes are assumed to be in phase at resonance, the computation of the stresses is particularly simple; the amplitude of the twist angle of each span is just the difference between the amplitudes of the rotations at the ends.

Remark 29.8 *In many cases the approximations linked with the whole modeling of the system, in particular where the equivalent inertia and damping are concerned, justify the use of the approximations seen earlier, particularly those linked with modal uncoupling, and make it useless to compute the response with greater precision. However, there are cases where the contribution of the nonresonant harmonics to the total stressing is not small, even for harmonics that are very far from resonance.*

The modal approximation becomes rough in this instance because the overall shape of the response can be far from the first mode shape, and the response is far better computed using a non-modal approach (see Example 29.4).

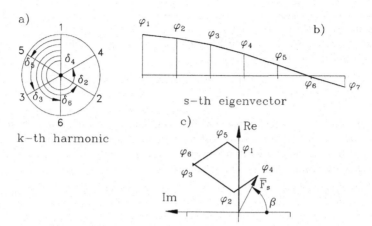

FIGURE 29.11. Graphical computation of the modal force for the resonance between the fifth harmonic and the first natural frequency ($k = 5$, $s = 1$) in an in-line four-stroke-cycle six-cylinder internal-combustion engine: (**a**) phase-angle diagrams for the fifth harmonic; (**b**) mode shape of the first natural frequency; (**c**) computation of the modal force.

Example 29.4 *Consider the diesel engine generator unit of Example 29.2. Compute the stresses due to torsional vibrations in the most critical part of the crankshaft using both modal and non-modal approaches.*
When damping is required, assume a coefficient $k' = 2 \times 10^5$ Ns/m^3 for the cranks and use Eq. (29.40) with $p = 3$ for the generator (compute coefficient k'' from the steady-state power).
Assume that the diameters of shafts number 6 and 7 are 200 and 130 mm, respectively. The harmonic components of the driving torque at full power are

Harmonic	A_k [Nm]	B_k [Nm]	Harmonic	A_k [Nm]	B_k [Nm]
0	3,448.4	–	11	−839.7	1,114.9
1	5,763.3	5,267.8	12	−808.8	853.3
2	3,176.8	8,119.6	13	−727.1	655.6
3	779.8	8,050.4	14	−662.8	541.7
4	−446.4	6,435.8	15	−635.9	420.1
5	−743.0	4,853.6	16	−592.2	274.1
6	−820.8	3,795.8	17	−518.9	170.2
7	−963.5	2,962.4	18	−453.4	111.1
8	−1,024.0	2,193.0	19	−396.9	46.6
9	−939.4	1,645.6	20	−324.0	−13.8
10	−855.0	1,350.3			

The model of the system and the mass and stiffness matrices are the same as in Example 29.2. The damping matrix is a diagonal matrix; the terms corresponding to the cranks are

$$c_{ii} = k'Ar^2 = 2 \times 10^5 \times 706.8 \times 10^{-4} \times 0.225^2 \ Nms/rad \, .$$

For the damping in node 8 (generator), coefficient k'' can be easily computed from the steady-state component of the torque. At 500 rpm = 52.4 rad/s it leads to a power of 180.6 kW per cylinder, i.e., to a total power of 1,084 kW; as $p = 3$, the values of k'' and c_{88} are

$$k'' = P/\Omega^{p+1} = 0.144 \ Nms^3 \, , \qquad c_{88} = 6k''\Omega^2 = 2.364 \ Nms/rad \, .$$

However, this value of the damping is linked with the load on the generator. Because in some cases the most dangerous conditions can be those in which the engine works with no load (only inertia torques act on the cranks), it is conservative to neglect the damping of the generator, which leads in most cases to an overestimate of the dynamic stresses.

The static stresses can be easily computed, obtaining the following values for the shafts from the first to the last:

$$\tau_{s_1} = 1.46 \ MN/m^2, \quad \tau_{s_2} = 2.93 \ MN/m^2, \quad \tau_{s_3} = 4.39 \ MN/m^2,$$
$$\tau_{s_4} = 5.86 \ MN/m^2, \quad \tau_{s_5} = 7.32 \ MN/m^2, \quad \tau_{s_6} = 13.17 \ MN/m^2,$$
$$\tau_{s_7} = 47.96 \ MN/m^2 \, .$$

The most stressed shaft is that linking the engine to the generator. The dynamic stresses are computed using two different procedures, modal and nonmodal. For the modal approach, the modal mass, stiffness, and damping for the first mode are computed, obtaining

$$\overline{M}_1 = \ 103.35 \, , \quad \overline{K}_1 = 2.171 \times 10^7 \, , \quad \overline{C}_1 \ = \ 1,949 \quad (\overline{\zeta}_1 \ = \ 0.021) \, .$$

The first modal system is much underdamped. The maximum dynamic stresses in the first mode occur in element number 7. The dynamic stresses in this element corresponding to the various modes and the total dynamic stress are plotted in Fig. 29.12a as functions of the speed.
When performing the non-modal computation, the element in which the maximum stress takes place depends on the harmonic considered: It is then advisable to plot the total stress due to all harmonics in each element (Fig. 29.12b). From the comparison of the results obtained through the two procedures, it is clear that in this case the approximations due to the modal approach are not acceptable, because of the large contribution of the nonresonant harmonics to the overall deformation, resulting in a response shape that is very different from the first mode shape. For example, at a speed of 50 rad/s = 477 rpm, the stresses in MN/m^2 due to the 18th harmonic (very close to resonance) in the various elements are

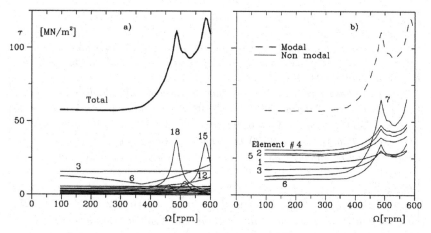

FIGURE 29.12. Dynamic stresses computed through the modal approximation with a single mode (stresses in element number 7 due to the various harmonics, separately and jointly, **a**) and the non-modal approach (stresses in all elements due to all harmonics, **b**).

Element	1	2	3	4	5	6	7
Modal	2.509	4.800	6.677	7.976	8.585	12.666	28.336
Non-modal	2.512	4.815	6.716	8.058	8.728	12.998	28.223

The two approaches lead to close results. The same results, computed for the first harmonic, which is far from resonance, are

Element	1	2	3	4	5	6	7
Modal	0.4822	0.9226	1.283	1.533	1.650	2.434	5.446
Non-modal	3.3200	3.3194	0.007	3.321	3.325	0.010	0.013

The results in this case are different, and the modal computation gives unreliable results.

The maximum value of the dynamic stress within the working range occurs in element number 7 at 487 rpm (resonance of the 15th harmonic) and takes a value of 65.04 MN/m² (the modal approach would lead to a value of 111.075 MN/m²). The total stresses are thus

$$\tau = 47.09 \ \pm \ 65.04 \ MN/m^2 \ .$$

Example 29.5 *Consider a four-stroke-cycle six-cylinder in-line engine with a total capacity of 3,000 cm³.*

Assume that the engine is coupled with the transmission through a very soft coupling, which causes a complete dynamic uncoupling of the engine from the transmission in the whole frequency field of interest. The equivalent system is then made of seven equivalent inertias (the six cranks and the flywheel) and six equivalent shafts (Fig. 29.13a).

The main characteristics of the system are number of cylinders: 6; bore: ϕ = 86 mm; area of the pistons: A = 58.09 cm^2; stroke: $2r$ = 85 mm; firing order: 1-5-3-6-2-4; speed: Ω = 6,000 rpm (638.3 rad/s); reciprocating masses (1 cylinder): $m_p + m_2$ = 0.720 kg; length of connecting rod: l_p = 181 mm; crankpin: outer diameter d_0 = 45 mm, inner diameter d_i = 10 mm; ratio α = 0.235.

Equivalent moments of inertia of the cranks and the flywheel [kg m^2]: $\overline{J}_{eq_1}$ = 0.00551; $\overline{J}_{eq_2}$ = $\overline{J}_{eq_5}$ = 0.00489; $\overline{J}_{eq_3}$ = $\overline{J}_{eq_4}$ = 0.00602; $\overline{J}_{eq_6}$ = 0.00625; $\overline{J}_{eq_7}$ = 0.06753. Equivalent stiffness of the shafts [MNm/rad]: $k_1 = k_2 = k_3 = k_4 = k_4 = 0.392$; $k_6 = 0.455$.

Computation of the natural frequencies. The mass and stiffness matrices of the system are

$$M = diag\left\{ 551 \quad 489 \quad 602 \quad 602 \quad 489 \quad 625 \quad 6753 \right\} \times 10^{-5},$$

$$K = \begin{bmatrix} 392 & -392 & 0 & 0 & 0 & 0 & 0 \\ -392 & 784 & -392 & 0 & 0 & 0 & 0 \\ 0 & -392 & 784 & -392 & 0 & 0 & 0 \\ 0 & 0 & -392 & 784 & -392 & 0 & 0 \\ 0 & 0 & 0 & -392 & 784 & -392 & 0 \\ 0 & 0 & 0 & 0 & -392 & 847 & -455 \\ 0 & 0 & 0 & 0 & 0 & -455 & 455 \end{bmatrix} \times 10^3.$$

The natural frequencies and mode shapes can be easily obtained by solving the usual eigenproblem, which yields the following natural frequencies and eigenvector for the first mode with natural frequency different from zero, normalized in such a way that the largest element has a unit value

$$\begin{array}{ll} \omega_0 = 0, \\ \omega_1 = 2,445 \ rad/s = 391 \ Hz, \\ \omega_2 = 6,242 \ rad/s = 993 \ Hz, \\ \omega_3 = 9,607 \ rad/s = 1,529 \ Hz, \qquad \{q_1\} = \\ \omega_4 = 12,485 \ rad/s = 1,987 \ Hz, \\ \omega_5 = 15,300 \ rad/s = 2,435 \ Hz, \\ \omega_6 = 16,349 \ rad/s = 2,602 \ Hz, \end{array} \left\{ \begin{array}{c} 1.000000 \\ 0.915283 \\ 0.761752 \\ 0.537715 \\ 0.263907 \\ -0.029741 \\ -0.280269 \end{array} \right\}.$$

The corresponding mode shape is plotted in Fig. 29.13b. The modal mass and stiffness for the first mode are, respectively,

$$\overline{M}_1 = q_1^T M q_1 = 0.0205, \qquad \overline{K}_1 = q_1^T K q_1 = 123,450.$$

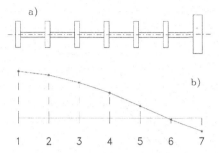

FIGURE 29.13. (a) Equivalent system; (b) first torsional mode.

Relationship between shaft twisting and shear stresses
If one of the shafts is twisted by angle θ_i, the corresponding torsional moment is $M_t = \theta_i k_i$. If the inner and outer diameters of the most loaded part, usually the crankpin, are d_i and d_0, respectively, for all cranks, the maximum value of the shear stress is

$$\tau_{max} = \frac{M_t}{W_t} = 2.191 \times 10^{10}\theta_i \ N/m^2 \ .$$

When the shaft vibrates with the first mode shape, the most stressed crank is the fifth. The corresponding twist angle is $\theta_5 = 0.294q_1$. The proportionality relationship between the maximum value of the shear stress and the amplitude of the first eigenvector is then

$$\frac{\tau_{max}}{q_1} = 0.6434 \times 10^{10} \ N/m^2 rad \ = \ 6,434 \ MN/m^2 rad \ .$$

Computation of the harmonics of the forcing function and of the modal forces *The curves of the power P and the mean indicated pressure p_{mi} as functions of the engine speed are plotted in Fig. 29.14. The curve expressing the mean indicated pressure can be approximated by the expression*

$$p_{mi} = -0.960 \times 10^{-5}\Omega^2 + 8.26 \times 10^{-3}\Omega - 0.475 \ ,$$

where the angular velocity is in rad/s and the pressure is in MPa.
To evaluate the harmonics of the driving torque, the empirical formula for the four-stroke-cycle engines reported on the already-mentioned book by Ker Wilson (Practical Solution of Torsional Vibration Problems, Vol. 2, Chapman & Hall, London, 1963, 218) will be used. Using the notation in this book, and particularly using the symbol k for the order of the harmonic, the formula can be written in the form

$$M_{m_0} = p_{mi}rA\frac{1}{2\pi} \ , \qquad M_{m_k} = p_{mi}rA\frac{25}{50\sqrt{\frac{2}{k} + 5k^2}} \ .$$

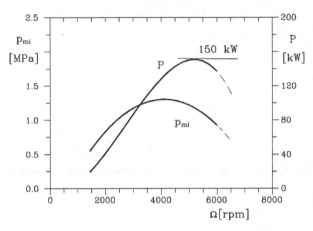

FIGURE 29.14. Power and the mean indicated pressure as functions of the engine speed.

Only the amplitude of each harmonic can be computed from the formula; the phase remains unknown. In practice, the phase is important only to add the inertia torque to the driving torque, and this is important only for harmonics 2, 4, 6, and 8. In the mentioned book, these harmonics are reported separately for the components in phase and in quadrature. The nondimensional amplitudes $M_{m_k}^* = M_{m_k}/p_{mi}rA$ *are reported in the following table:*

k	A_k^*	B_k^*	$M_{m_k}^*$	k	$M_{m_k}^*$
0	–	–	0.160	11	0.040
1	–	–	0.330	12	0.034
2	0.330	0.140	0.360	13	0.029
3	–	–	0.290	14	0.025
4	0.220	−0.040	0.220	15	0.022
5	–	–	0.160	16	0.019
6	0.110	−0.048	0.120	17	0.017
7	–	–	0.092	18	0.015
8	0.051	−0.051	0.072	19	0.014
9	–	–	0.058	20	0.012
10	–	–	0.048		

Note that the fundamental frequency is equal to $\Omega/2$. If the moment of inertia J_0 and all terms in α with powers higher than 2 are neglected, the four non-vanishing harmonics of the inertia torque can be expressed as

$$M_{i_k} = m_a r^2 \Omega^2 b_k \ ,$$

where m_a is the total mass of the reciprocating elements and nondimensional coefficients b_k are

$$b_2 = 0.0588, \quad b_4 = -0.5, \quad b_6 = -0.176, \quad b_8 = -0.0138.$$

The modal forces can be computed by considering the six moments acting on the cranks, which are equal in amplitude but out of phase from each other, following the phase-angle diagram. The amplitude of the force due to the kth harmonic is

$$\overline{F}_{1_k} = \{q_1\}^T \{M\}_k, \quad i.e., \quad |F_{1_k}| = \left| \sum_{j=1}^{6} q_{1_j} M_{k_j} \right|.$$

The sum must be considered a vector sum. The phase-angle diagrams are those shown in Fig. 29.10 for the six-cylinder engine. From the figure, it is clear that the diagrams for harmonics 2 and 4 are the mirror image of each other, and then the relevant modal forces are equal. The same holds for harmonics 1 and 5. Remembering that the diagrams for harmonics beyond the sixth are equal to those of the first six, only four types of phase-angle diagrams exist. Considering the first 24 harmonics, they are as follows:

Group 1: Harmonics 1, 7, 13, and 19 and their mirror images 5, 11, 17, and 23. The phase angles are those shown in the first diagram. The in-phase and in-quadrature components of the modal force are, from Eq. (29.53),

$$\overline{F}_{1_{k_{phase}}} = \mathbf{q}_1^T \{M_j \cos(\delta_j)\} = M_k \big[q_{11} \cos(0) + q_{12} \cos(120°) +$$

$$+ q_{13} \cos(240°) + q_{14} \cos(60°) + q_{15} \cos(300°) + q_{16} \cos(180°) \big] = 0.592 M_k,$$

$$\overline{F}_{1_{k_{quad}}} = \mathbf{q}_1^T \{M_j \sin(\delta_j)\} = M_k \big[q_{11} \sin(0) + q_{12} \sin(120°) +$$

$$+ q_{13} \sin(240°) + q_{14} \sin(60°) + q_{15} \sin(300°) + q_{16} \sin(180°) \big] = 0.370 M_k,$$

The amplitude of the modal force is then $|\overline{F}_{1_k}| = 0.698 M_k$.

Group 2: Harmonics 2, 8, 14, and 20 and their mirror images 4, 10, 16, and 22. By operating as for the first group, it follows that $|\overline{F}_{1_k}| = 0.290 M_k$.

Group 3: Harmonics 3, 9, 15, and 21. $|\overline{F}_{1_k}| = 1.90 M_k$.

Group 4: Major harmonics 6, 12, 18, and 24. $|\overline{F}_{1_k}| = 3.45 M_k$.

Response of nonresonant harmonics.

Computing of the response in conditions far from resonance is straightforward. The modal amplitude of the response is

$$|\eta_{1_k}| = \frac{|\overline{F}_{1_k}|}{K_1 - \left(k \frac{\Omega}{2}\right)^2 \overline{M}_1}.$$

Compute, for example, the response to the sixth harmonic at 6,000 rpm. The mean indicated pressure at that speed is 0.924 MPa. The sixth harmonic belongs to group four (major harmonics). The forcing moment due to the driving torque has components A_k and B_k, which can be computed using the data reported in the previous table:

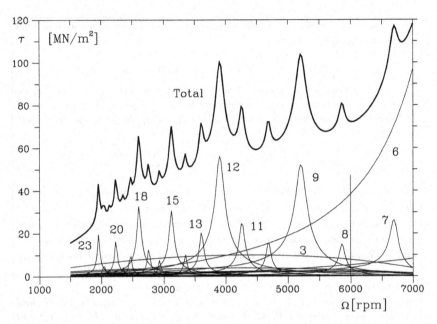

FIGURE 29.15. Dynamic shear stresses in the most loaded part as functions of the speed: the first 24 harmonics of the forcing function and first torsional natural frequency.

$$A_6 = 0.11 p_{mi} Ar = 25.11 \ , \qquad B_6 = -0.048 p_{mi} Ar = -10.96 \ .$$

The moment due to inertia torques is

$$M_{i_6} = -0.176 m_a \Omega^2 r^2 = -90.36,$$

and the total moment is

$$M_6 = \sqrt{(A_6 + M_{i_6})^2 + B_6^2} = 66.16 \ .$$

The total moment, due to gas pressure and inertia forces, is smaller than the moment due to inertia forces alone. The most critical condition for the harmonic considered at the speed of 6,000 rpm occurs when the engine is driven from the outside, without any working fluid in it. The stress analysis will then be performed in the corresponding condition. The value of the modal force is

$$|\overline{F}_{1_6}| = 3.45 M_{i_6} = 311.74 \ .$$

The amplitude of the modal response at 6,000 rpm is then

$$\eta_{1_6} = \frac{|\overline{F}_{1_k}|}{\overline{K}_1 - \left(k\frac{\Omega}{2}\right)^2 \overline{M}_1} = 6.159 \times 10^{-3} \ ,$$

and the corresponding maximum value of the stress is

$$\tau_{max} = 6,434\, q_1 = 39.63\ MN/m^2\ .$$

Response of resonant harmonics. To compute the response in resonant conditions, it is necessary to assume a value of the damping of the system. An empirical procedure reported in the already-mentioned book by Ker Wilson (page 706 and following) will be followed: The amplitude at resonance is obtained as the product of the static deformation of the modal system by a suitable amplification factor

$$\eta_{imax} = H_{imax} \frac{\overline{F}_i}{K_i}\ .$$

The amplification factor is obtained through the empirical formula

$$H_{imax} = \frac{500}{\sqrt{16 + 145\tau^*}}.$$

Shear stress τ^* is the maximum stress, in MN/m^2, that would be in the shaft if it were deformed following the first mode, in static conditions, by the relevant modal force.
Consider the resonance between the seventh harmonic and the first natural frequency. The critical speed is

$$\Omega_{cr7} = \frac{2\omega_1}{7} = 701.43\ rad/\ s\ =\ 6,698\ rpm\ ;$$

the mean indicated pressure at that speed is $p_{mi} = 0.597\ MPa$.
The amplitude of the driving torque and the corresponding modal force are

$$M_{m7} = 0.092 P_{mi} Ar = 13.56\,,\qquad |\overline{F}_{17}| = 0.70\,,\qquad M_7 = 9.46\,.$$

The amplitude of the static deformation and the static maximum value of the shear stress corresponding to the modal force are, respectively,

$$\eta_{1st} = \frac{|\overline{F}_{17}|}{K_1} = 7.669 \times 10^{-5}\,,\qquad \tau^* = 6,434 q_{1st} = 0.494\ MN/m^2\,.$$

The amplification factor, computed using the aforementioned formula, and the damping ratio are, respectively,

$$H_{7max} = \frac{500}{\sqrt{16 + 145\tau^*}} = 53.43\,,\qquad \zeta_7 = \frac{1}{2H_{7max}} = 0.0094.$$

The amplitude of the modal response and the corresponding value of the shear stress are

$$\eta_{17} = H_{max}\eta_{1st} = 4.09 \times 10^{-3}\,,\qquad \tau_{max} = 6434 q_1 = 26.3\ MN/m^2.$$

The responses in terms of maximum shear stress for the first 24 harmonics are shown in Fig. 29.15. They were computed considering the modal system as a damped system with a single degree of freedom whose damping is obtained, for each harmonic, following the same procedure seen for the seventh. The shear stresses due to the various harmonics are then combined by simply adding their amplitudes. This procedure, which neglects the phase, is clearly conservative. Total stresses. The maximum value of the total stress due to torsional vibration in the working speed range of the machine occurs at a speed of 5,200 rpm in correspondence to the resonance peak of the ninth harmonic. The value of the total shear stress is $\tau_{max} = 104$ MN/m^2. At the speed of 5,200 rpm (544.5 rad/s), the engine can produce a power of 150 kW. The more loaded crank is the fifth, where the static stress is equal to 5/6 of the stress corresponding to the maximum power

$$\tau_{max_{st}} = \frac{5}{6} \frac{P_{max}}{\Omega} \frac{16 d_e}{\pi(d_e^4 - d_i^4)} = 12.83 \ MN/m^2 \ .$$

The shaft undergoes fatigue loading with cycles included between the limits

$$\tau = 12.83 \pm 104 \ MN/m^2 \ .$$

Because the stressing is mostly due to the resonant ninth harmonic, the frequency of the fatigue loading can be assumed to be equal to the first natural frequency of the shaft, 391 Hz. Note that in the current case the non-modal computation would have yielded very similar results because most of the stressing is due to resonating harmonics.

29.5   Torsional instability of crank mechanisms

A second-approximation formulation of the equation of motion of the system consisting of the crank, the connecting rod, and the reciprocating parts can be obtained by introducing the trigonometric polynomial (29.15) expressing the equivalent moment of inertia into the equation of motion (29.11)

$$\overline{J}_{eq}\ddot{\phi}_z + \left\{ \sum_{j=1}^{n} J_{c_j} \cos[j(\phi_z + \Omega t)] + \sum_{j=1}^{n} J_{s_j} \sin[j(\phi_z + \Omega t)] \right\} \ddot{\phi}_z +$$

$$+ \frac{1}{2}(\Omega + \dot{\phi}_z) \left\{ -\sum_{j=1}^{n} j J_{c_j} \sin[j(\phi_z + \Omega t)] + \sum_{j=1}^{n} j J_{s_j} \cos[j(\phi_z + \Omega t)] \right\} = M \ .$$

$$(29.55)$$

Equation (29.55) is the exact equation of motion of the system if an infinity of harmonic terms is considered. Actually, a good approximation is obtained by resorting to a small value of n. If the amplitude of the oscillations is small enough, the trigonometric functions of angle $j\phi_z$ can be

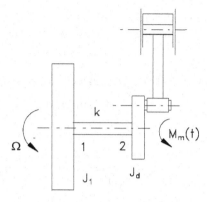

FIGURE 29.16. Crank mechanism: model used for the study of the torsional instability.

linearized. By remembering a few trigonometric identities and linearizing the resulting equation, neglecting the products of small quantities, Eq. (29.55) reduces to

$$\bar{J}_{eq}\ddot{\phi}_z + \ddot{\phi}_z \sum_{j=1}^{n} \left[J_{c_j} \cos(j\Omega_t) + J_{s_j} \sin(j\Omega_t) \right] - \dot{\phi}_z \Omega \sum_{j=1}^{n} j \left[J_{c_j} \sin(j\Omega_t) + \right.$$

$$\left. - J_{s_j} \cos(j\Omega_t) \right] - \frac{1}{2}\phi_z\Omega^2 \sum_{j=1}^{n} j^2 \left[J_{c_j} \cos(j\Omega_t) + J_{s_j} \sin(j\Omega_t) \right] +$$

$$- \frac{1}{2}\Omega^2 \sum_{j=1}^{n} j \left[J_{c_j} \sin(j\Omega_t) + J_{s_j} \cos(j\Omega_t) \right] = M .$$

$$(29.56)$$

Equation (29.56) is a linear differential equation with coefficients that are functions of time. It can be used as a starting point to build a second-approximation model of the reciprocating machine. This approach is, however, quite complex and only a simplified case is reported here.[4] Consider a single-cylinder engine with a simplified geometry such that $\alpha = 0$ and $\beta = 0$. Assume that the crank is connected to a large flywheel, so large that its motion is not affected by torsional vibrations (Fig. 29.16). Assuming that the flywheel rotates at constant speed Ω, the system has a single degree of freedom, namely, the angle of torsion ϕ_z of the shaft. The behavior of the shaft is linear, with stiffness k, a viscous damper acts on the

---

[4]For a detailed study, see E. Brusa, C. Delprete, G. Genta, "Torsional vibration of crankshafts; effects of nonconstant moments of inertia", *Journal of Sound and Vibration*, 205, 2, (1997), 135–150.

crank with damping coefficient c, producing a moment proportional to the speed $\dot{\phi}_z + \Omega$, and the driving torque is $M_m(t)$. The equation of motion (29.56) reduces to

$$
\begin{aligned}
&\ddot{\phi}_z[1 - \epsilon \cos(2\Omega t)] + 2\epsilon\Omega^2\phi_z \cos(2\Omega t) + 2\epsilon\Omega\dot{\phi}_z \sin(2\Omega t) + \\
&+ 2\zeta\omega_0(\Omega + \dot{\phi}_z) + \omega_0\phi_z = -\epsilon\Omega^2 \sin(2\Omega t) + \omega_0^2\frac{M_m(t)}{k},
\end{aligned}
\tag{29.57}
$$

where

$$
\omega_0 = \sqrt{\frac{k}{\overline{J}_{eq}}}, \quad \zeta = \frac{c}{2\sqrt{k\overline{J}_{eq}}}, \quad \epsilon = \frac{(m_2 + m_p)r^2}{2\overline{J}_{eq}}.
$$

The inertial term on the right-hand side of Eq. (29.57) coincides with the inertia torque acting on the equivalent system. If the small terms are neglected, Eq. (29.57) reduces to the equation of motion of the equivalent system studied in the preceding section, with the added conditions that the connecting rod is infinitely long and damping is included in the form studied in Section 29.4.5.

Equation (29.57) is neither a Mathieu nor a Hill equation but can be solved by using the same methods in Section 22.3 and particularly by resorting to a series expansion of the solution and finding the coefficients using an infinite determinant. Due to the complexity of the study, only some results obtained by M.S. Parisha and W.D. Carnegie[5] through numerical experimentation will be reported here. The equation for the free oscillations of the system obtained from Eq. (29.57) by neglecting both the driving torque and the drag torque linked with the constant angular velocity Ω can be written in the nondimensional form as

$$
\begin{aligned}
&\frac{d^2\phi_z}{d\tau^2}[1 - \epsilon \cos(2\tau)] + 2\frac{d\phi_z}{d\tau}[\epsilon \sin(2\tau) + \zeta\frac{\omega_0}{\Omega}] + \\
&+ \phi_z\left[2\epsilon \cos(2\tau) + \left(\frac{\omega_0}{\Omega}\right)^2\right] = 2\zeta\frac{\omega_0}{\Omega} - \epsilon \sin(2\tau),
\end{aligned}
\tag{29.58}
$$

where $\tau = \Omega t$ is the nondimensional time. Because the behavior of the system depends only on three nondimensional parameters – ϵ, ζ, and ω_0/Ω – the study of the stability of the system can be easily performed by numerically integrating the equation of motion with different values of the parameters and checking whether it develops an unstable behavior.

Some of the plots that define the zones in which the system is unstable are summarized in Fig. 29.16. The undamped system has two instability ranges, one at a rotational speed equal to half the natural frequency of the equivalent system and the other at a speed equal to ω_0 (Fig. 29.17).

---

[5]M.S. Parisha, W.D. Carnegie, "Effect of the variable inertia on the damped torsional vibrations of diesel engine systems", *Journal of Sound and Vibration*, 46, 3, (1976).

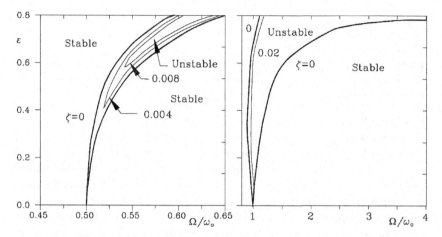

FIGURE 29.17. Instability ranges of the system sketched in Fig. 29.16 with different values of damping.

If ϵ tends to zero, the same results seen in the previous sections are obtained: An infinitely small instability range is found at the torsional critical speeds found on the Campbell diagram by intersecting the horizontal straight line $\omega = \omega_0$ with the lines $\omega = 2\,\Omega$ and $\omega = \Omega$. The instability range reduces to the torsional critical speeds due to the first two harmonics of moment M_i. With increasing ϵ, the two instability ranges grow larger and, if ϵ is large enough, it is impossible to operate at speeds in excess of the natural frequency.

The presence of damping reduces the fields of instability and makes it possible to operate at any speed, provided that ϵ, i.e., the mass of the reciprocating elements, is not too large.

The instability range located close to the critical speed due to the harmonic with $\omega = \Omega$ is less affected by damping than that corresponding to the condition $\omega = 2\Omega$ and thus is the most dangerous. The effect of damping reduces with increasing values of ϵ. These results confirm that there is no instability range if ϵ is lower than 4ζ in the zone close to $\omega/\Omega = 1$ and lower than $4\sqrt{2}\zeta$ in the zone close to $\omega/\Omega = 1/2$. They are applicable, at least qualitatively, well beyond the simplified model used to obtain them. They also hold for the cranks of multi-cylinder engines for the evaluation of the effects of the mass of the reciprocating elements on the overall stability of the system.

29.6   Exercises

Exercise 29.1 *Consider the engine of Example 29.5 and repeat the dynamic analysis using a non-modal approach. Because the way of taking into account damping used in the example is strictly linked with the modal approach, use the* *procedure described in Section 29.4.5 with a coefficient $k' = 56,000$. Assume that the diameter of element 6 is 45 mm.*

Exercise 29.2 *Repeat the computation of Exercise 29.1 using the modal approach, with the damping defined earlier and compare the results with those obtained in Example 29.5 and in Exercise 29.1.*

Exercise 29.3 *The engine of Exercise 29.1 is connected to a blower with a moment of inertia of 0.6 kg m² through a shaft of length 3 m and a joint whose stiffness is 20 kNm/rad. The moment of inertia of the part of the joint on the blower shaft is 0.05 kg m², and the inertia of the other part of the joint is already included in the moment of inertia of the flywheel. Compute the diameter to the drive shaft in order not to exceed a static stressing of 40 MN/m² in steady-state working and repeat the eigenanalysis of the system. Assume that the damping of the blower can be computed through Eq. (29.40) with $p = 2$ and compute the dynamic stressing of the system. Does the presence of the blower significantly affect the behavior of the engine?*

Exercise 29.4 *Just after the elastic joint of the system of Exercise 29.3, connect a water pump, whose moment of inertia is 0.1 kg m², through two gear wheels with transmission ratio 1.3. The pinion gear has a diameter of 200 mm, and both gears are 45 mm thick. Assume that the shaft leading to the pump is 4 m long and has a diameter of 50 mm, and estimate the moments of inertia of the gear wheels, assuming they are made of steel. Neglect the compliance of the meshing gears. Repeat the dynamic analysis, with the main aim of evaluating the dynamic torque on the gear wheels and the dynamic stressing on the shaft of the pump.*

Exercise 29.5 *A marine unit is made by four engines of the type described in Example 10.5 driving two propellers through gear wheels. The engines work in pairs, the two engines being assembled with the flywheels one close to the other, connected through a shaft 600 mm long, whose diameter is computed in order not to exceed a shear stress of 30 MN/m² in static conditions. The two pairs are located side by side, driving a common shaft through gear wheels. The transmission ratio is 0.9, the diameter of the pinion gear is 208 mm, and the thickness of the gear wheels is 30 mm. The common shaft is 4 m long, and its diameter is chosen for the same value of the static shear stress as the other shafts connecting the engines. Two propeller shafts, 6 m long, are then driven with a transmission ratio of 0.4. The diameter of the pinion gear is 200 mm, and the thickness of the wheels is 30 mm.*

Draw a detailed sketch of the system. Knowing that the moments of inertia of the propellers are 12 kg m², estimating the moments of inertia and the geometric properties of all elements, and neglecting damping due to the propellers, compute the natural frequencies and the dynamic stressing of the system.

30

Vibration Control in Reciprocating Machines

Many reciprocating machines, in particular internal-combustion engines, are provided with suitable devices to control torsional vibration of the crankshaft. There are many different types of torsional vibration dampers and, after more than a century of development, research in this field is still very active.

30.1 Dissipative dampers

Various kinds of damping devices are used to control torsional vibration of reciprocating machines when it is impossible to prevent one of the critical speeds yielding severe dynamic stressing from falling within the working range or, more generally, when the amplitude of the torsional vibrations is incompatible with the safe operation of the machine. Many different types of such devices were developed, each with its field of application due their different mechanical, thermo-mechanical, and cost characteristics.

Often torsional vibration dampers are applied at one end of the crankshaft and are made of a flywheel (usually referred to as a *seismic mass*), whose geometric configuration can be of a wide variety of types, connected to the shaft by suitable elastic and damping elements. In many applications there may be only one of these elements, and sometimes the restoring force can be supplied by the centrifugal field due to rotation and the seismic mass may have the shape of a counterbalance of the crankshaft. The conceptual layout of almost all torsional vibration dampers can be reduced to the damped vibration absorber of Fig. 7.12 or to the Lanchester damper, with the obvious differences that now it is applied to a multi-degree-of-freedom system and that instead of masses, stiffnesses, and dampers, there are moments of inertia and torsional stiffnesses and dampers. Without including all the possible types of torsional vibration dampers, they can be subdivided into three types:

- dissipative dampers,

- damped vibration absorbers, and

- rotating-pendulum vibration absorbers.

A viscous torsional vibration damper is the Lanchester damper applied to one of the ends of the crankshaft and consists of a flywheel, generally shaped as a ring free to rotate within a casing filled with a fluid with high viscosity, for example, a silicon-based oil (Fig. 30.1a).

Damping in this case is of the viscous type, i.e., the drag torque is proportional to the angular velocity, and the damping coefficient depends on the clearance between the ring and the housing and on the characteristics of the fluid. It is greatly influenced by the temperature of the fluid. The model for the dynamic study of the system must be modified by adding the moment of inertia of the casing, in the node in which the damper is applied, and adding a new node in which the inertia of the ring is lumped. The two nodes are connected by a viscous damper and a spring with zero stiffness.

For a first-approximation evaluation of the optimum damping it is possible to assume that the presence of the damper does not significantly affect

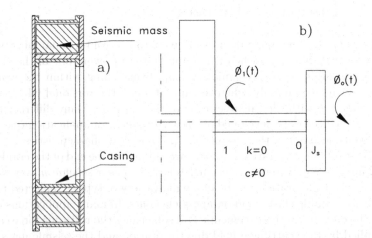

FIGURE 30.1. Dissipative torsional vibration damper: (a) sketch of the system and (b) model for the dynamic study.

the natural frequencies. Under this assumption, the behavior of the damper can be studied separately, assuming the time history of the motion of the node where it is applied. With reference to Fig. 30.1b, assume that the time history at node 1 where the damper is applied (the seismic mass is lumped in node 0) is harmonic,

$$\phi_1(t) = \phi_{1_0} \sin(\omega t) \, .$$

Also, the time history of node 0 is harmonic but with by a phase lag Φ :

$$\phi_0(t) = \phi_{0_0} \sin(\omega t + \Phi) \, .$$

The model so obtained is formally identical to that of a system with a single degree of freedom excited by the displacement of the supporting point. The response can be computed using Eq. (7.6), where J_s replaces m and $k = 0$. Remembering that

$$2\zeta \omega_n = \frac{c}{m}$$

and introducing the ratio

$$\alpha = \frac{J_s}{c} \, ,$$

it follows that

$$\Re(H) = \frac{1}{1 + (\alpha\omega)^2} \, , \qquad \Im(H) = \frac{(\alpha\omega)}{1 + (\alpha\omega)^2} \, . \qquad (30.1)$$

At time $t = 0$ the angular displacement between the seismic mass and its housing reaches its maximum, and the relative displacement can be expressed as

$$[\phi_0(t) - \phi_1(t)] = |\phi_0 - \phi_1| \cos(\omega t) = \phi_{1_0} \frac{\alpha\omega}{\sqrt{1 + (\alpha\omega)^2}} \cos(\omega t) \, . \qquad (30.2)$$

The energy dissipated in a period by the damper is

$$E_d = \int_0^T c \left[\dot{\phi}_0(t) - \dot{\phi}_1(t) \right]^2 dt = c\phi_{1_0}^2 \pi \frac{J_s^2 \omega^3}{c^2 + J_s^2 \omega^2} \, . \qquad (30.3)$$

It is easy to verify that both conditions $c = 0$ and $c \to \infty$ lead to a vanishingly small energy dissipation: in the first case because the seismic mass does not interact with the system and in the second because nodes 1 and 0 are rigidly connected. The value of the damping coefficient leading to a maximum energy dissipation is obtained by differentiating Eq. (30.3) and equating the derivative to zero. The value of the optimum damping so obtained is

$$c_{opt} = J_s \omega \, . \qquad (30.4)$$

Even if the value of the optimum damping depends on the frequency, dampers of this type allow a substantial reduction of the amplitude of vibration in a wide frequency range.

The type shown in Fig. 30.1 is quite common, but other types are also used. In that shown in Fig. 30.2a the seismic mass is made by a hollow flywheel inside which a sort of lever or balance wheel is located. Some cavities inside the flywheel are filled with a high-viscosity fluid that is pumped by the balance wheel from one cavity to the other through calibrated passages during torsional vibration. Tuning screws allow the passages to be restricted to tune the value of the damping coefficient.

Another type of damper is shown in Fig.30.2b. The seismic mass is made of two disc flywheels pressed against a disc rigidly connected to the shaft by spring-loaded set screws. The mating surfaces of the flywheels and the disc are lined by friction material, constituting the damping element. In this case damping is due to dry friction, and the behavior of the system is nonlinear. The term *Lanchester damper* should be used, strictly speaking, only for dampers of this type, not for all dissipative dampers with zero stiffness.

All dissipative dampers are subject to heating, even overheating, because they convert mechanical energy into heat. It is thus necessary to check that they can dissipate all the thermal energy they produce, which can be done using formulas of the type of Eq. (30.3), at least in terms of average power in a given period of time. Usually a limit to the ratio between the thermal power and the external surface of the damper is assumed. For the type in

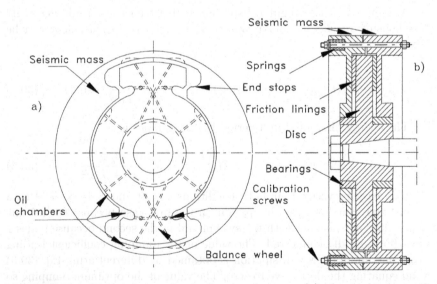

FIGURE 30.2. Dissipative torsional vibration dampers: (a) viscous damper and (b) dry friction damper (Lanchester damper).

Fig. 30.1a, for example, it is suggested not to exceed 1.9×10^4 kW/m^2 in continuous working and 5×10^4 kW/m^2 for short periods of time.

30.2   Damped vibration absorbers

Vibration absorbers were dealt with in Section 7.8. Also in the case of torsional vibration absorbers, they introduce a new natural frequency and change the natural frequency on which they are tuned. The distance between the two resonance peaks increases with increasing moment of inertia of the seismic mass. Because an undamped vibration absorber is effective in a very narrow frequency range, outside which it is not only ineffective but also causes new resonances, the seismic mass is connected to the shaft through a system that has a certain amount of damping. In this case it is possible to obtain a response that is fairly flat in a wide range of frequencies, as clearly shown in Fig. 7.12d.

From a practical viewpoint, all dampers shown in the previous section can be converted into damped vibration absorbers simply by adding an elastic element between the shaft and the seismic mass, which allows tuning the damper on the required frequency. The seismic mass can be shaped as a disc or a balance wheel, as in the case of the damper of Fig. 30.3a, which is very similar to that shown in Fig. 30.2a, except that in the former the balance wheel is the seismic mass and the casing is attached to the shaft.

The elastomeric dampers widely used on small automotive diesel engines (Fig. 30.3b) can be considered dissipative vibration absorbers. The elastomeric elements provide both elasticity and damping and can be shaped

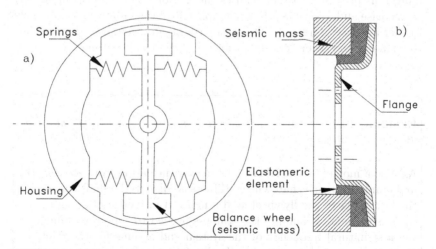

FIGURE 30.3. Damped torsional vibration absorbers: (a) viscous damper with balance wheel and (b) elastomeric damper.

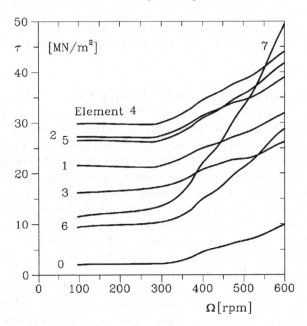

FIGURE 30.4. Dynamic stresses computed through a non-modal approach.

in such a way as to obtain the required values for both. Care must be exerted to insure that the amount of heat generated within the damper is not too large, particularly considering the low thermal conductivity and poor high-temperature characteristics of elastomeric materials. Overheating is dangerous because it usually causes the damping capacity of the material to decrease and, consequently, the amplitude of the vibration to grow, with further increase of heating. This easily results in a complete destruction of the damper and severe fatigue problems of the whole system.

Example 30.1 *Add a damped vibration absorber to the front end of the diesel engine of Example* 29.4.

Assume a moment of inertia of 6 kg m^2 for the casing, connected to the first crank through a shaft whose stiffness is equal to the shaft connecting the last crank to the flywheel, and a moment of inertia of 26 kg m^2 for the seismic mass. Compute the stiffness needed to tune the damper on the first natural frequency of the system and assume for the damping the optimum value computed in Section 30.1 for the springless damper. Repeat the computation of the stresses and compare the results.

The equivalent system now includes 10 *nodes, with node* 1 *located at the seismic mass and node* 2 *at the damper case. The diameter of element* 2 *is assumed to be* 200 *mm. A first computation is run with a zero value of the stiffness of the damper, obtaining the first natural frequency of the system without seismic mass of* 445.2 *rad/s* = 70.86 *Hz; the casing of the damper has little effect on the first natural frequency. To tune the damper on the first natural frequency, its stiffness must be*

$$k_{damp} = J_{damp}\omega_1^2 = 5.53 \times 10^6 \ Nm/rad.$$

The nonvanishing natural frequencies of the undamped system are then computed:

$$\omega_1 = 55 \ Hz \,, \qquad\qquad \omega_6 = 376 \ Hz \,,$$
$$\omega_2 = 67 \ Hz \,, \qquad\qquad \omega_7 = 442 \ Hz \,,$$
$$\omega_3 = 114 \ Hz \,, \qquad\qquad \omega_8 = 482 \ Hz \,,$$
$$\omega_4 = 192 \ Hz \,, \qquad\qquad \omega_9 = 627 \ Hz \,,$$
$$\omega_5 = 290 \ Hz \,.$$

The presence of the damper introduces two new values in place of the old value for the first mode and a high natural frequency linked mostly to the presence of the damper casing. The remaining values are little affected. The value of the damping can be assumed to be

$$c_{damp} = J_{damp}\omega_1 = 1.16 \times 10^4 Nms/rad :$$

The computation of the dynamic response is only performed using a non-modal approach, because the presence of the damper makes the assumption of modal uncoupling even less realistic than in Example 10.4 (Fig. 30.4). The maximum value of the dynamic stress within the working range occurs in element number 4 (not taking into account the shaft between damper and first crank, which will be indicated as element number 0) at 500 rpm and takes a value of 38.70 MN/m². The total stress is thus

$$47.09 \pm 38.70 \ \mathrm{MN/m}^2 \,.$$

The damper is quite effective and reduces dynamic stressing by about 40%. Element number 7 is the most stressed only at speeds above the working range and that no well-defined resonance peak occurs.

Also torsional vibration dampers can exploit eddy currents produced in a conductor moving in a magnetic field as a means to dissipate energy. The functional layout can be that of the viscous damper of Fig. 7.12a, with or without a spring to tune the seismic mass on a given frequency. An example is shown in Fig. 30.5: a solid conductor disc is rigidly connected to the shaft, while the housing acts as a seismic mass. A number of permanent magnets are attached to the inside of the housing, so that the magnetic field crosses the disc: Since the magnets have the shape of sectors and are magnetized

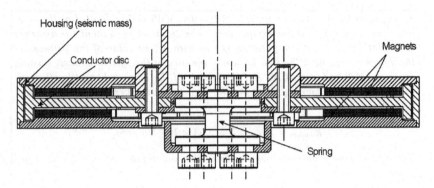

Housing (seismic mass)

Conductor disc

Magnets

Spring

FIGURE 30.5. Torsional damped vibration absorber based on a motional electromagnetic damper (MEMD).

in opposite way, the field has no axial symmetry, and any relative rotation of the housing with respect to the disc causes eddy currents to flow in the latter.

Following the classification seen in Chapter 27, it is a motional electromagnetic damper (MEMD). In the case of the figure there is also a spring, so that the device can be considered as a damped vibration absorber.

30.3   Rotating-pendulum vibration absorbers

Rotating-pendulum vibration absorbers are among the few non-dissipative dampers that have been widely used. The conceptual scheme of such devices was shown in Fig. 23.1, the only difference being that in this case the pendulum is constrained to move in the xy-plane. The natural frequency of the free oscillations of the pendulum is proportional to the rotational speed

$$\omega = \Omega\sqrt{\frac{r}{l}}.$$

Because the various harmonics of the forcing function are characterized by a frequency proportional to the rotational speed, once the vibration absorber is tuned at a certain speed on a given harmonic, it remains locked on that in the whole working range of the machine.

To analyze the behavior of rotating-pendulum vibration absorbers, a more complex model than that shown in Fig. 23.1 is needed, as the torsional vibration of the disc at which the pendulum is attached cannot be neglected. Consider the scheme of Fig. 30.6, in which a pendulum of mass m and length l is hinged to a disc whose moment of inertia is J. Let the disc be in node 1 of the discretized model and the node where the system is connected, through a shaft with stiffness k, be node 0. The aim of the

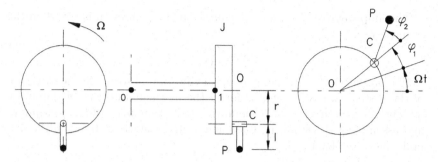

FIGURE 30.6. Rotating-pendulum vibration absorber; model with two degrees of freedom.

analysis is building a rotating-pendulum vibration absorber element to be assembled to the finite element model of the system. The velocity of point P can be computed as in Example 23.1, where ϕ_2 and $\Omega + \dot{\phi}_1$ are substituted for θ and Ω, respectively, and angle ϕ is assumed to be zero, because the pendulum is constrained to remain in the xy-plane. The kinetic and potential energies of the system sketched in Fig. 30.3 are, respectively,

$$2\mathcal{T} = J(\Omega + \dot{\phi}_1)^2 + m|V_P|^2 = (J + mr^2)(\Omega + \dot{\phi}_1)^2 +$$
$$+ml^2(\Omega + \dot{\phi}_1 + \dot{\phi}_2)^2 + mrl(\Omega + \dot{\phi}_1)(\Omega + \dot{\phi}_1 + \dot{\phi}_2)\cos(\phi_2),$$
$$2\mathcal{U} = k\phi_1^2 .$$

$$(30.5)$$

By performing the required derivatives, linearizing the trigonometric functions of angle ϕ_2, and neglecting the terms containing products of generalized coordinates and their derivatives, the equation of motion of the free oscillations of the system sketched in Fig. 30.6 is

$$\begin{bmatrix} J + m(r+l)^2 & ml(r+l) \\ ml(r+l) & ml^2 \end{bmatrix} \begin{Bmatrix} \ddot{\phi}_1 \\ \ddot{\phi}_2 \end{Bmatrix} + \begin{bmatrix} k & 0 \\ 0 & mrl\Omega^2 \end{bmatrix} \begin{Bmatrix} \phi_1 \\ \phi_2 \end{Bmatrix} = \begin{Bmatrix} 0 \\ 0 \end{Bmatrix} .$$

$$(30.6)$$

It is easy to verify that if a moment with harmonic time history with frequency ω is applied to node 1 or the same node is excited by a harmonic motion of the constraint, the ratio between the amplitude of the oscillation of the disc and that of the pendulum is

$$\frac{\phi_{1_0}}{\phi_{2_0}} = \frac{\dfrac{l}{r} - \left(\dfrac{\Omega}{\omega}\right)^2}{1 + \dfrac{l}{r}} .$$

$$(30.7)$$

The amplitude at node 1 then vanishes if the pendulum is tuned on the frequency ω, i.e., if

$$\frac{r}{l} = \left(\frac{\omega}{\Omega}\right)^2.$$

The presence of a rotating-pendulum vibration absorber in the nth node of the system can be accounted for in the mathematical model simply by adding a further degree of freedom, the oscillation angle of the pendulum, and assembling the following mass and stiffness matrices between the nth node and the added one:

$$\mathbf{M} = m \begin{bmatrix} (r+l)^2 & l(r+l) \\ l(r+l) & l^2 \end{bmatrix}, \qquad \mathbf{K} = \Omega^2 \begin{bmatrix} 0 & 0 \\ 0 & mrl \end{bmatrix}. \qquad (30.8)$$

The length of the pendulum must, in general, be very small, particularly when the vibration absorber is to be tuned on a high-order harmonic. If tuning has to be performed on the sixth harmonic of a four-stroke-cycle internal-combustion engine, for example, characterized by ratio $\omega/\Omega = 3$, the length of the pendulum must be equal to $1/9$ of the radius at which it is hinged. Very short pendulums result from the need of tuning at

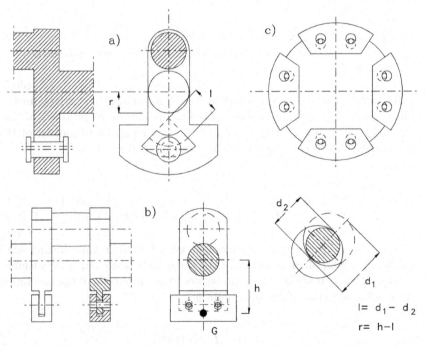

FIGURE 30.7. Rotating-pendulum vibration absorbers: (a) roller type, dimensions refer to the case in which the rotational inertia of the roller is neglected, and (b) and (c) bi-filar pendulum type.

higher-order frequencies. Different solutions to the problem of building very short pendulums with sufficient mass have been forwarded. Two of them are sketched in Fig. 30.7. The rotating pendulum of Fig. 30.7a is made by a roller, free to move in a circular slot in one of the crank webs. The solution of Fig. 30.7b is equivalent to a bi-filar pendulum and allows very small values of length l to be achieved. The system of Fig. 30.7c is made by four rotating pendulums of the type shown in Fig. 30.7b attached to a single disc. They can have different values of the length l, in order to be tuned on different harmonics of the forcing function.

30.4   Experimental measurement of torsional vibrations

The torsional vibrations of reciprocating machines are particularly dangerous because they have little effect on the overall motion of the machine and, hence, are very difficult to detect without suitable instrumentation. This is because the shaft is not constrained against rotation and the inertia forces due to torsional vibration are completely balanced, at least if the relevant modes are completely uncoupled from other modes. Their effects on the nonrotating parts of the machine, which would be completely nil in the case of the equivalent system, are, at any rate, small. Those warning signs, such as noise and vibrations transferred to other parts of the machine that usually allow detection of dynamic problems, are absent or at least small. An engine or a compressor can thus work with high levels of dynamic stresses without the operator realizing it until failure occurs. Only the use of suitable instruments allows verification that the level of torsional vibrations is low enough not to cause problems.

The preceding statement is not completely true for modern car engines, which drive a number of accessories (air conditioner, a large alternator, etc.) through belts. When torsional vibrations are excited, they exert forces that are variable in time on the supports of the accessories, and the resulting vibration can be detrimental to the acoustic and vibrational comfort of the vehicle and to structural safety. Recently, torsional vibration dampers, which were limited to diesel engines, started to be used on many spark-ignition engines and complicated types aimed to reduce the transmission of vibrations to the accessories appeared.

Torsional vibration can be detected from cyclic variations of the instant rotational velocity or directly from measurements of the dynamic stressing of the rotating parts of the machine. Speed variations can be detected using the devices usually included in all rotating machinery to measure the rotational speed, provided they are sensitive enough.

A common device is a magnetic pick-up that produces a pulse each time a tooth of a ferromagnetic gear wheel passes in front of it. If the speed varies periodically in time, the frequencies of the torsional oscillations can be obtained from a Fourier analysis of the signal.

However, this method is limited where the maximum vibration frequency detectable is concerned, because the period of time between the passages of two subsequent teeth must be much smaller than the period of the frequency to be detected, particularly if the amplitude also has to be measured. Laser velocity transducers can detect the peripheral velocity of the rotor, allowing the torsional vibrations of the machine to be accurately assessed. An advantage of these methods is the absence of the need to transfer signals from the rotor to the stationary parts of the machine.

Alternatively, it is possible to use electric strain gauges on the shaft to directly measure the variable components of the stresses. The strain gauges must be located close to the nodes of the mode shape of the harmonic under study, because the stresses reach their maximum value in these locations. When using strain gauges it is necessary to transfer the signal from rotating to stationary parts of the machine. This can be done either by using slip rings and brushes, which are similar to those of DC motors, or by more modern magnetic, radio, or optical contactless systems. Instead of locating the strain gauges directly on the shaft, it is possible to add a seismic mass connected to the shaft through a low-stiffness spring system and to locate the strain gauges on these springs.

If the natural frequency of the seismic mass is lower than the first torsional natural frequency of the system, all the harmonics of the oscillation can be detected. Also in this case, a device that can transfer the signal to the stationary parts of the machine must be provided. Many types of instruments, mostly of the mechanical type, which were used in the past, are now obsolete except for particular applications.

30.5 Axial vibration of crankshafts

Crankshafts of reciprocating machines can be subject to axial vibration, or better, since the modes of the various types are all coupled, to vibration modes that are mostly axial. They can be dangerous because of the possibility of excitation by gas pressures and inertia forces of reciprocating parts due to the coupling of the various modes. In the past, axial vibrations were an actual danger only in a few cases, mainly linked with slow large internal-combustion engines, but the modern tendency toward higher stress levels, speed, and power/mass ratio caused them to become an important factor in the design of a wider class of reciprocating machines.

The simplified models used for the study of axial vibration are similar to those seen in the context of torsional vibration. An equivalent system is obtained by lumping the masses in some nodes and evaluating the axial stiffness of the various parts of the crankshaft. Usually, the largest compliance is due to the cranks, because the crankwebs and crankpins bend during axial vibration, and the straight parts of the shaft are much stiffer and can often be considered rigid in the study of the lower modes.

The stiffness can be computed using the semi-empirical or empirical formulas reported in the literature, numerical methods, or experimental tests. The equivalent masses are, in this case, constant and no inertia forces due to the reciprocating elements must be taken into account. The masses are only those of the shaft because, due to the axial clearance in the bearings, the mass of the connecting rods is not involved in axial vibration.

The equivalent system for the study of torsional vibration is underconstrained, but that for axial vibrations is axially kept in position by the thrust bearing, usually modeled as an elastic constraint. All other bearings are usually not considered: They allow axial displacements of the order of magnitude of the amplitude of axial vibration. Once the equivalent system has been obtained, there is no difficulty studying free vibration. Usually, an in-line system is obtained and transfer-matrices procedures, like the Holzer method, can be used even if more modern approaches based on the FEM are now common. Because the equivalent system is a lumped-parameters system, there is no difficulty obtaining the modal forces due to the connecting rods by measuring or computing, using empirical formulas or numerical methods, the axial displacements due to unit forces applied by the connecting rods. The study of forced vibrations can thus be performed.

Also in this case there can be dangerous resonance conditions, even if the excitation is much smaller for axial than for torsional vibration, due to the lower damping of the system. The same procedures seen for torsional vibration allow the amplitude and the stresses to be obtained. If they exceed the allowable limits, it is possible to resort to suitable dampers, possibly made of a seismic mass, free to move in the axial direction, or a piston that can move in axial direction (and to rotate, because it is connected to the shaft) together with the shaft, in a cylinder full of viscous fluid, rigidly or elastically mounted to the basement of the machine.

Note that the elastomeric damper shown in Fig. 30.7b also acts as an axial vibration damper. It is possible to tune the damper on two different frequencies, for torsional and axial vibrations, by tailoring the shape of the elastomeric element and thus the ratio between the axial and the torsional stiffness. The ratio between the mass and the moment of inertia can also be changed, at least within some limits, to achieve the required tuning.

30.6   Short outline on balancing of reciprocating machines

Reciprocating machines are a source of vibration, not only as a result of the variation in time of the driving torque, causing torsional vibration of the crankshaft and a reaction torque on the basement that varies in time, but also for the inertia forces due to reciprocating parts. The crank mechanism itself is a source of vibration even when it produces no driving or braking torque. Crank mechanisms are included in a wide variety of common machines, and the problem of balancing their inertia forces has been widely studied: Whole books have been devoted to this subject.[1] Many devices, sometimes quite complicated, have been suggested, and, in many cases, used with the aim of improving the smoothness of reciprocating engines and compressors. Due to the complexity of the problem, only a short outline of the subject is presented here.

The inertia forces due to the crank mechanism can be studied using the same scheme seen for the study of the effects of the static unbalance of a rotor. Consider a single-cylinder machine supported on elastic mountings that allow the whole system to be displaced in the x- and y-directions (Fig. 30.8a). The crank is assumed to be made of a statically balanced disc D, the crankpin B, whose mass is included in the mass m_1 of the connecting rod, and a counterweight C, that is considered a point mass. If the rotational speed Ω of the disc is assumed to be constant, remembering

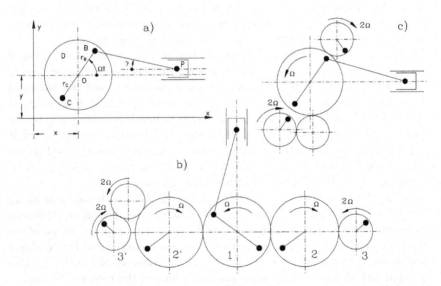

FIGURE 30.8. (a) Model for the study of the inertia forces acting in the x- and y-directions in a crank system; (b) scheme of the arrangement for balancing first- and second-order forces in a single-cylinder machine. First-order moments of the counterweights: shaft 1: $m_1 r_B + F_1/2\Omega^2$; shafts 2 and 2': $F_1/4\Omega^2$; shafts 3 and 3': $F_2/8\Omega^2$; and (c) balancing second-order forces.

---

[1]See, for example, W.Thomson, *Fundamentals of Automobile Engines Balancing*, Mech. Eng. Publ. Ltd., 1978).

the definition of function $f_1(\theta)$ (Eq. 29.5), the velocities of points O, C, B, and P are

$$V_O = \left\{ \begin{matrix} \dot{x} \\ \dot{y} \end{matrix} \right\}, \qquad V_B = \left\{ \begin{matrix} \dot{x} - r_B\Omega\sin(\Omega t) \\ \dot{y} + r_B\Omega\cos(\Omega t) \end{matrix} \right\},$$

$$V_C = \left\{ \begin{matrix} \dot{x} + r_C\Omega\sin(\Omega t) \\ \dot{y} - r_B\Omega\cos(\Omega t) \end{matrix} \right\}, \qquad V_P = \left\{ \begin{matrix} \dot{x} + r_B\Omega\sqrt{f_1(\Omega t)} \\ \dot{y} \end{matrix} \right\}.$$

$$(30.9)$$

If the connecting rod is modeled in the usual way (two point masses m_1 in B and m_2 in P and a massless moment of inertia J_0), the kinetic energy of the whole system made by the crank, the connecting rod, and the reciprocating masses is

$$\mathcal{T} = \frac{1}{2}m_t(\dot{x}^2 + \dot{y}^2) + \frac{1}{2}J_{eq}\Omega^2 + \Omega(m_C r_C - m_1 r_B)\times \qquad (30.10)$$

$$\times\,[\dot{x}\sin(\Omega t) - \dot{y}\cos(\Omega t)] - (m_2 + m_p)r_B\dot{x}\Omega\sqrt{f_1(\Omega t)}\,,$$

where

$$m_t = m_d + m_C + m_1 + m_2 + m_P\,,$$

$$J_{eq} = J_d + m_C r_c^2 + m_1 r_B^2 + (m_2 + m_P)r_B^2 f_1(\Omega t) + J_0 f_2(\Omega t)\,.$$

In computing the derivatives of the kinetic energy that enter the equation of motion, the rotational velocity Ω will be assumed as a constant. If this assumption is dropped, a further generalized coordinate, for example, the rotation angle θ, must be introduced, and the torsional behavior of the shaft can be studied with the effects of unbalance. By introducing the series for $1/\cos(\gamma)$ truncated at the second term into the expression for $\sqrt{f_1(\theta)}$, it follows that

$$\sqrt{f_1(\theta)} = \sin(\theta) + \frac{\alpha}{4}(4 + \alpha^2 + 2\beta^2)\sin(2\theta) + \qquad (30.11)$$

$$-\frac{\alpha^3}{8}\sin(4\theta) - \frac{\alpha^2\beta}{2}\cos(\theta) + \frac{\alpha^2\beta}{2}\cos(3\theta)\,.$$

By performing the relevant derivatives, using Eq. (30.11) for $\sqrt{f_1(\theta)}$ and assuming that the axis of the cylinder passes through the center of the crank ($\beta = 0$), the following equations of motion are obtained:

$$\left\{ \begin{array}{l} m_t\ddot{x} = Q_x - \Omega^2(m_C r_C - m_1 r_B)\cos(\Omega t) + \Omega^2(m_2 + m_P)r_B^2\times \\ \qquad \times\left[\cos(\Omega t) + \dfrac{\alpha}{2}(4 + \alpha^2)\cos(2\Omega t) - \dfrac{\alpha^3}{2}\cos(4\Omega t)\right], \\ m_t\ddot{y} = Q_y - \Omega^2(m_C r_C - m_1 r_B)\sin(\Omega t)\,. \end{array} \right.$$

$$(30.12)$$

The last terms on the right-hand side of Eq. (30.12) are the unbalance forces acting on the crank. If the first-order moment $m_C r_C$ is equal to that of mass m_1, the inertia forces due to rotating masses are balanced and no force acts in the y-direction. Nevertheless, reciprocating masses m_2 and m_P cause inertia forces in the x-direction to be produced. They are periodic in time with fundamental frequency equal to Ω and cannot be balanced by simple means such as adding counterweights. The unbalance forces in the x-direction consist of a first-order component with amplitude F_1 and frequency Ω, a second-order force with amplitude F_2 and frequency 2Ω, and a third-order component with amplitude F_3 and frequency 4Ω. Their amplitudes are

$$F_1 = \Omega^2 r_B(m_2 + m_P),$$

$$F_2 = \Omega^2 \frac{\alpha(4+\alpha^2)}{2} r_B(m_2 + m_P), \tag{30.13}$$

$$F_3 = \Omega^2 \frac{\alpha^3}{2} r_B(m_2 + m_P).$$

If more terms in the series for $1/\cos(\gamma)$ were considered, slightly different expressions of the forces would have been obtained together with higher-order terms. From a practical viewpoint, however, all forces of third and higher order are considered negligible.

A first action aimed at reducing unbalance forces is that of overbalancing the shaft, i.e., of using a counterbalance greater than that needed to balance the rotating masses alone. If half of the reciprocating masses are balanced, i.e., if

$$m_C r_C = m_1 r_B + \frac{1}{2} r_B(m_2 + m_p), \tag{30.14}$$

the unbalance forces acting on the system in the x- and y-directions are, respectively,

$$\begin{cases} F_x = \frac{1}{2} F_1 \cos(\Omega t) + F_2 \cos(2\Omega t) + F_3 \cos(4\Omega t), \\ \\ F_y = -\frac{1}{2} F_1 \sin(\Omega t). \end{cases} \tag{30.15}$$

The resultant of the unbalanced first-order forces is a force with amplitude $F_1/2$, rotating with angular velocity Ω in a direction opposite the direction of rotation. The first-order force is so halved and transformed from a force pulsating in the x-direction to a force rotating in the backward direction. It can be completely balanced by a counterweight with first-order moment $(m_2 + m_P)r_B/2$ rotating in the backward direction.

To avoid inertia torques, such a counterweight should rotate in the same plane of disc D and have the same axis of rotation. In practice, two counterweights located at the sides of the crank, each with a first-order moment equal to half that computed, must be used. In a similar way, second-order forces can be balanced by using two counterbalances, each having a first-order moment equal to $F_2/8\Omega^2$, rotating at a speed that is double the rotational speed in opposite directions. The scheme of Fig. 30.8b shows a possible arrangement in which first- and second-order forces are completely balanced. However, this scheme is very complicated, and in practice single-cylinder machines are seldom balanced.

In the case of multi-cylinder machines, first-order forces can be balanced simply by choosing a suitable phasing of the cranks. Because the cranks are located in different planes, care must be taken to balance both the forces and the moment due to unbalance. All first-order forces and moments are balanced in the case of the schemes of Fig. 29.10. To balance second-order forces it is possible to use a scheme of the type shown in Fig. 30.8c. Arrangements of this type are used to improve the smoothness of the engine in high-class motor cars, for example. In this case, often the axes of the two balancing shafts are not located in a plane perpendicular to the axis of the cylinder to produce a torque with a frequency equal to 2Ω, which can balance the harmonic with the same frequency in the driving torque of the engine (Fig. 30.8c). This can be done only approximately, because the amplitude of the inertia torque is proportional to the square of the rotational speed, while that of the driving torque varies in time in an arbitrary way, being controlled by the driver.

30.7 Exercises

Exercise 30.1 *A large marine drive system consists of a four-cylinder diesel engine with a detuner and a long drive shaft with variable diameter.*

The geometry of the system can be summarized by subdividing it into 17 elements connecting 18 stations. In station 1 is the seismic mass of the detuner, in station 2 the part of the detuner connected to the crankshaft, in stations 3, 4, 5, and 6 the cranks, in station 7 the flywheel, and in station 18 the propeller.

The inertial and geometric properties of the fields from 2 to 17, which can be either cylindrical or conical, are reported in Table 30.1.

The stiffness of the detuner (field 1) is equal to 7.8×10^6 Nm/rad and the Young's modulus of the material is $E = 2.1 \times 10^{11}$ N/m^2. Draw a detailed sketch of the system. Compute the natural torsional frequencies of the system, with and without the detuner, by using a finite element model with more than 18 nodes to model the conical parts of the propeller shaft with prismatic elements. For the eigensolution, the model can be condensed using only eight master degrees of freedom. Plot the first four mode shapes, computing the rotations in all nodes, so as to obtain a detailed deformed shape of the propeller shaft.

Exercise 30.2 *Consider the marine drive of the previous exercise. In node 8 the engine shaft ends with a gear wheel, which meshes with two gear wheels*

TABLE 30.1. Exercise 30.1; inertial and geometrical properties.

Node	J [kg m^2]	Node	J [kg m^2]
1	280	5	2,580
2	111	6	2,218
3	2,260	7	450
4	2,620	18	4,180

Field	Type	D [mm]	l [mm]	Field	Type	D [mm]	l [mm]
2	cyl.	500	1,584	10	con.	585–495	145
3	cyl.	500	2,592	11	cyl.	495	276
4	cyl.	500	3,264	12	con.	495–585	145
5	cyl.	500	2,592	13	cyl.	585	5,532
6	con.	500–450	2,520	14	con.	585–495	145
7	cyl.	450	1,200	15	cyl.	495	276
8	con.	450–585	145	16	con.	495–585	145
9	cyl.	585	3,276	17	cyl.	585	5,040

located on propeller shafts identical to the one in the previous exercise. Neglect the compliance of the meshing gears.

Compute the torsional eigenfrequencies and the mode shapes of the system, with the following data: moment of inertia of the propellers, the gear wheel on the engine shaft, and the propeller shafts are 3,200 kg m^2, 150 kg m^2, and 220 kg m^2, respectively; gear ratio is 0.86.

Exercise 30.3 *Repeat the computations of the previous exercise, assuming that the stiffness of the meshing gears, relative to the driven systems, is 120×10^6 Nm/rad.*

Exercise 30.4 *Repeat the study of Example 30.1 using a springless dissipative damper.*

Exercise 30.5 *Study a dissipative damper to be added at the free end of each crankshaft of the engines of Exercise 29.5 to reduce dynamic stressing.*

Appendix A
Solution Methods

A.1 General considerations

The computations needed to perform the dynamic analysis of systems with many degrees of freedom are, in many cases, very involving. Many numerical methods have been developed to efficiently compute the natural frequencies, the mode shapes, and the forced response dynamic systems, or better, of their discretized models. Their formulation and implementation have been, and still are, the object of intensive research work. The detailed study of these methods is well beyond the scope of this text, but an engineer using structural analysis codes needs to have at least an approximate knowledge of the relevant solution methods, particularly when he has to choose among various possibilities offered by computer programs. He does not need to be an applied mathematician, but it is important that he has at least an idea of how the machine 'crunches the number'.

Four mathematical problems will be dealt with here, namely, the solution of a set of linear equations, the solution of eigenproblems, the solution of sets of nonlinear equations, and the numerical integration of sets of differential equations, both linear and nonlinear. Many books[1] and countless

---

[1]See, for example, F.G. Curtis, *Applied Numerical Analysis*, Addison Wesley, Reading, 1978; J.H. Wilkinson, C. Reinsh, *Linear Algebra: Handbook for Automatic Computing*, Springer, New York, 1971; J. Stoer, R. Burlish, *Introduction to Numerical Analysis*, Springer, New York, 1980; W.H. Press, B.P. Flannery, S.A. Teukolsky, W.T. Vetterling, *Numerical Recipes, the Art of Scientific Computing*, Cambridge Univ. Press, Cambridge, 1986.

papers were devoted to these four problems; the short outline shown here has the aim of supplying general information to the structural analyst who has to use the relevant computer codes. No detailed formulas or practical details on coding are included because many subroutines are available, and the author does not advise the preparation and use of home-brewed codes, particularly in this case.

A.2   Solution of linear sets of equations

The basic mathematical problem encountered in the static analysis of linear systems with many degrees of freedom is the solution of a set of linear equations, whose matrix of the coefficients, the stiffness matrix, is generally real, positive defined, and symmetrical and often has a narrow-band structure. The response of the system to a harmonic excitation can be computed by solving the frequency domain Eq. (7.4) where the dynamic stiffness matrix, while being symmetrical and usually retaining a band structure, can be nonpositive definite. If the model includes damping, the dynamic stiffness matrix is complex.

The set of linear equations

$$\mathbf{A}\mathbf{x} = \mathbf{b}, \tag{A.1}$$

where both $\mathbf{A}$ and $\mathbf{b}$ are complex, can be transformed into the set of real equations, at the expense of doubling the size of the problem

$$\begin{bmatrix} \Re(\mathbf{A}) & -\Im(\mathbf{A}) \\ \Im(\mathbf{A}) & \Re(\mathbf{A}) \end{bmatrix} \left\{ \begin{array}{c} \Re(\mathbf{x}) \\ \Im(\mathbf{x}) \end{array} \right\} = \left\{ \begin{array}{c} \Re(\mathbf{b}) \\ \Im(\mathbf{b}) \end{array} \right\}. \tag{A.2}$$

The matrix of the coefficients of Eq. (A.2) is not symmetrical even if that of Eq. (A.1) is such. Also, its band structure can be far less prominent than that of the original matrix.

When the conjugates $\bar{\mathbf{x}}$ of the unknowns $\mathbf{x}$ explicitly enter the equations

$$\mathbf{A}\mathbf{x} + \mathbf{B}\bar{\mathbf{x}} = \mathbf{c}, \tag{A.3}$$

the corresponding real set of equations is

$$\begin{bmatrix} \Re(\mathbf{A}) + \Re(\mathbf{A}) & -\Im(\mathbf{A}) + \Im(\mathbf{B}) \\ \Im(\mathbf{A}) + \Im(\mathbf{B}) & \Re(\mathbf{A}) - \Re(\mathbf{B}) \end{bmatrix} \left\{ \begin{array}{c} \Re(\mathbf{x}) \\ \Im(\mathbf{x}) \end{array} \right\} = \left\{ \begin{array}{c} \Re(\mathbf{c}) \\ \Im(\mathbf{c}) \end{array} \right\}. \tag{A.4}$$

The solution of a linear set of equations is usually indicated by writing the inverse of the matrix of the coefficients, i.e.,

$$\mathbf{x} = \mathbf{A}^{-1}\mathbf{b}. \tag{A.5}$$

The inversion of the matrix of coefficients is, however, a most inefficient way of solving linear sets of equations, and notations involving it must not be considered as an indication of how to perform the computations.

When the solution to Eq. (A.1) is actually sought, two different types of techniques can be used, namely, direct and iterative algorithms. While in the early developments of the FEM iterative algorithms were widely used, there is now a general agreement on the application of direct techniques. They are all more or less related to the Gauss elimination technique, introduced more than a century ago. It is based on the transformation of the set of equations to eliminate the unknowns one by one until a single equation with one unknown is obtained. The unknowns can thus be computed one by one. The procedures of this type can be subdivided into two steps: The triangularization of the matrix of the coefficients, in which the equation yielding one of the unknowns is obtained, and the subsequent backsubstitution in which all other unknowns are subsequently found. The first part is by far the longest, where the computer time is concerned. The total number of elementary operations to be performed is of the order of $n^3/3$.

The Gauss method can be interpreted as a series of $n-1$ transformations of the matrix of coefficients $\mathbf{A}$ and vector $\mathbf{b}$ of Eq. (A.1), yielding an equation in which the matrix of coefficients is an upper triangular matrix. Such a transformation can be summarized as the multiplication of both sides of Eq. (A.1) by a non-singular matrix $\mathbf{G}$, such that matrix

$$\mathbf{U} = \mathbf{GA}$$

is an upper triangular matrix. The first of the two steps of the Gauss method is then the construction of matrix $\mathbf{G}$ and the computation of the products

$$\mathbf{U} = \mathbf{GA} \quad \text{and} \quad \mathbf{b}^* = \mathbf{Gb} ,$$

while the second is the solution of the equation

$$\mathbf{Ux} = \mathbf{b}^* .$$

It is then clear that the products of matrix $\mathbf{A}$ and vector $\mathbf{b}$ by the transformation matrix $\mathbf{G}$ can be performed separately and, in particular, when several sets of equations with the same matrix of coefficients but different vectors $\mathbf{b}$ have to be solved, matrix $\mathbf{U}$ can be computed only once, and the parts of the computation to be performed several times are only those related to products $\mathbf{Gb}$ and the backsubstitution, which are less costly from a computational viewpoint.

When no exchange of lines is required, matrix $\mathbf{A}$ can be decomposed in the form

$$\mathbf{A} = \mathbf{LU} ,$$

where $\mathbf{L}$ is a lower triangular matrix and $\mathbf{U}$ is the aforementioned upper triangular matrix. Such decomposition is often referred to as LU factorization or, in two forms slightly different from each other, Doolittle and Crout factorizations.

When matrix $\mathbf{A}$ is symmetrical and positive definite the two triangular matrices $\mathbf{L}$ and $\mathbf{U}$ are the transposes of each other and the transformation takes the form

$$\mathbf{A} = \mathbf{L}\mathbf{L}^T .$$

This form is referred to as Choleski factorization. Because the solution of the set of equations through Choleski factorization is faster than the use of the regular Gauss method, involving only about $n^3/6$ operations, most finite element codes use, for the static solution, this algorithm. The presence of a band structure can further simplify the computation and many algorithms that take into account this feature have been developed.

When solving several sets of equations with the same matrix of coefficients, it is customary to write the equation in the form

$$\mathbf{A}\mathbf{X} = \mathbf{B}, \tag{A.6}$$

where matrix $\mathbf{B}$ is a rectangular matrix whose columns are the various vectors $\mathbf{b}$ of the different sets of equations and, similarly, the columns of matrix $\mathbf{X}$ are the unknown vectors $\mathbf{x}$. Equation (A.6) is not only a notational shortcut, $\mathbf{X}$ is an unknown matrix that when multiplied by matrix $\mathbf{A}$ yields matrix $\mathbf{B}$. If the number of sets of equations is m, Eq. (A.6) is actually a set of $m \times n$ equations yielding the $m \times n$ unknown elements of matrix $\mathbf{X}$.

The precision obtainable with the aforementioned techniques depends on the structure of the matrix of the coefficients. If it is well conditioned, the result is usually good; however, it is possible to apply iterative procedures to refine the results obtained through direct techniques. Consider the set of equations

$$\mathbf{A}\mathbf{x} = \mathbf{b}$$

and the approximate solution $\mathbf{x}^{(1)}$. The exact solution $\mathbf{x}$ can be written in the form

$$\mathbf{x} = \mathbf{x}^{(1)} + \delta\mathbf{x}^{(1)},$$

where the last term expresses the errors introduced by the approximate solution technique. Equation (A.1) can be written in the form

$$\mathbf{A}\delta\mathbf{x}^{(1)} = \mathbf{r}^{(1)} \quad \text{where} \quad \mathbf{r}^{(1)} = \mathbf{b} - \mathbf{A}\mathbf{x}^{(1)}, \tag{A.7}$$

which can be used to compute the error $\delta\mathbf{x}^{(1)}$. The solution of Eq. (A.7) is straightforward, because it requires only the factorization of matrix $\mathbf{A}$, which has already been performed. The procedure can be repeated several times, each time getting closer to the correct solution. In most cases, however, the precision of the result directly obtained is sufficient, and no iterative refinement is required.

The solution of a set of equations with a diagonal matrix of the coefficients, is straightforward because each equation directly yields one unknown. The elements of the main diagonal of the matrix of coefficients can be separated from the others

$$\mathbf{A} = \mathbf{A}^* + \mathbf{A}^{**}, \tag{A.8}$$

where $\mathbf{A}^*$ is diagonal and $\mathbf{A}^{**}$ is a matrix with zero element on the main diagonal. Equations (A.1) can be written in the form

$$\mathbf{A}^*\mathbf{x} = \mathbf{b} - \mathbf{A}^{**}\mathbf{x}, \tag{A.9}$$

which can easily be solved iteratively. A trial vector $\mathbf{x}^{(0)}$, usually with all elements equal to 0, is introduced on the right-hand side and a new value $\mathbf{x}^{(1)}$ is computed by solving a set of uncoupled equations. The procedure is then repeated until convergence is obtained. This iterative solution scheme is referred to as the Jacobi method. When solving the ith equation, the new values of the first $(i - 1)$ unknowns have already been obtained and the new values can be used directly. The latter scheme is known as the Gauss–Seidel method.

A condition that is sufficient, although not necessary, for ensuring convergence of the Jacobi method is that matrix $\mathbf{A}$ be diagonally dominant, i.e., in each row the element on the diagonal is greater than the sum of the other elements.

It is possible to demonstrate that when the matrix of the coefficients is symmetrical and positive defined the Gauss–Seidel method converges. When the Gauss–Seidel and Jacobi methods both converge, the first is faster than the latter. To maximize the chances of obtaining convergence and to make it faster, the order of the equations should be rearranged in such a way that the largest elements lie on the main diagonal.

In some cases it is necessary to explicitly obtain the inverse of a matrix, as when performing matrix condensation. Remembering Eq. (A.6), if matrix $\mathbf{B}$ is the identity matrix $\mathbf{I}$, the unknown matrix $\mathbf{X}$ is nothing other than the inverse $\mathbf{A}^{-1}$. A simple way to compute the inverse of a matrix is by factorizing it and then obtaining the various columns by solving n sets of equations in which the various vectors $\mathbf{b}$ have all terms equal to zero except the term corresponding to the number of the column to be found, which has a unit value.

If matrix $\mathbf{A}$ is complex, the real and imaginary parts of its inverse $\mathbf{A}^{-1}$ can be computed from the following real equation:

$$\begin{bmatrix} \Re(\mathbf{A}) & -\Im(\mathbf{A}) \\ \Im(\mathbf{A}) & \Re(\mathbf{A}) \end{bmatrix} \left\{ \begin{array}{c} \Re(\mathbf{A}^{-1}) \\ \Im(\mathbf{A}^{-1}) \end{array} \right\} = \left\{ \begin{array}{c} \mathbf{I} \\ \mathbf{0} \end{array} \right\}. \tag{A.10}$$

When matrix $\mathbf{A}$ has a band structure its inverse usually does not have the same type of structure. If the matrix is stored in the memory of the computer taking advantage of its band structure, the memory required for storage of the inverse can be much greater than that needed for the original matrix.

A.3   Computation of eigenfrequencies

A.3.1   General considerations

The first and most important step for the study of the dynamic behavior of a linear system is the evaluation of its eigenfrequencies and mode shapes. When using discretized models, this basic step reduces to the mathematical problem of finding the eigenvalues and eigenvectors of the dynamic matrix of the system. Even if the size of the matrices can be reduced by applying the condensation and substructuring techniques seen in Chapter 10, the solution of an eigenproblem whose size is only a few hundred can still require long and costly computations.

The complexity of the problem depends not only on its size but also on the characteristics of the relevant matrices and the requirements of the particular problem. First, the user can be interested in obtaining only the eigenvalues or both eigenvalues and eigenvectors. Generally speaking, the problem can then be attacked at three different levels, namely, it can reduce to the search of

- a single, usually the lowest, eigenfrequency;

- a selected number of eigenfrequencies, usually the lowest or those included in a given range;

- all eigenfrequencies.

The first alternative was very popular when automatic computation was not available or very costly. The first natural frequency could be rapidly evaluated with limited costs, but there was no chance of performing any modal analysis. Nowadays this approach is only used in the first steps of the design procedures, to be sure that no natural frequency is lower than a given value, usually stated in the design specifications. Approximate techniques, yielding a value lower than the correct one, are sufficient in this case, and more detailed computations can be postponed to a subsequent stage of the analysis, when the design is better defined.

Usually the dynamic analysis of models with many degrees of freedom follows the second approach. The knowledge of a certain number of eigenvalues and eigenvectors allows the performance of an approximate modal analysis and the computation of all the required dynamic characteristics of the system. In particular, the FEM yields a large number of vibration modes, due to the large number of degrees of freedom of the mathematical model, but many of them, usually those with the highest eigenfrequencies, have little physical relevance and are strongly influenced by the discretization scheme used. If they are discarded, no relevant information on the dynamic behavior of the system is lost.

The last approach is so demanding where the complexity of the computations is concerned that it is used only when dealing with systems with a small number of degrees of freedom, perhaps obtained through a large-scale condensation of a more complex model. However, the algorithms yielding all the eigenvalues are more efficient than the selective ones, for a given number of eigenvalues found, and it is a common opinion that, when more than about 20% of the eigenvalues are required, it is more convenient to find all of them. Note that the algorithms that search all eigenvalues do not usually find them in any prescribed order; as a consequence it is not possible to start the search and stop the algorithm after a given number of solutions has been found, because there is always the possibility that a solution lying within the field of interest has been lost.

As already stated for the solution of linear sets of equations, but to a greater extent, the choice of the most convenient method depends on the structure of the relevant matrices and the aims of the search, and it is not possible to state which is the best method, in general. No attempt to deal in detail with the various mathematical aspects of the problem will be done here because the aim of this section is only to supply some general information on the more common algorithms. The user can find more details in any good textbook on numerical analysis.

If the matrix whose eigenvalues and eigenvectors are required is real, the eigenanalysis can yield either real or complex-conjugate results. If, on the contrary, the starting matrix is complex, the complex results are not conjugate. Consider the general eigenproblem

$$(\mathbf{A} - \lambda \mathbf{I})\mathbf{x}_0 = \mathbf{0}, \qquad (A.11)$$

already written in standard form, where matrix $\mathbf{A}$ is complex. It can be transformed into the real eigenproblem

$$\left(\begin{bmatrix} \Re(\mathbf{A}) & -\Im(\mathbf{A}) \\ \Im(\mathbf{A}) & \Re(\mathbf{A}) \end{bmatrix} - \lambda \mathbf{I} \right) \left\{ \begin{matrix} \Re(\mathbf{x}_0) + i\Im(\mathbf{x}_0) \\ \Re(\mathbf{x}_0) - i\Im(\mathbf{x}_0) \end{matrix} \right\} = \mathbf{0}, \qquad (A.12)$$

whose size is double that of the original problem. Equation (A.12) can be easily solved using the standard algorithms for non-symmetrical matrices and yields $2n$ solutions: the n eigenvalues and eigenvectors of Eq. (A.11) and their conjugates. In all those cases in which the sign of the imaginary part of the eigenvalues is important, a procedure that can distinguish between the actual eigenvalues of the original problem and those added when doubling the size of the matrices must be devised. This can easily be done by checking the structure of the eigenvector corresponding to each eigenvalue: If the real part of the first n elements is equal to the imaginary part of the remaining ones, a solution of the original problem was found; otherwise the solution is discarded.

The Routh–Hurwitz criterion, which allows the assessing of whether some of the eigenvalues have a positive real part, is sometimes used as an alternative to the actual solution of the eigenproblem for the study of the stability of the system. The Routh–Hurwitz criterion is based on computations that are much simpler than those required to actually solve the eigenproblem; however, when the size of the matrix is not very small, the application of the criterion itself can lead to long computations and, now that the direct solution of the eigenproblem is possible, it can be questionable whether it is worthwhile to resort to an approach that, at any rate, yields results of lesser interest. The main disadvantage of the Routh–Hurwitz criterion is actually that it shows whether a system is stable but not how much stable it is. This can be circumvented by performing an eigenvalue shift, i.e., by substituting s with $(s_1 - \sigma)$, modifying accordingly the eigenproblem, and repeating the stability study. It is possible to state whether the real part of some of the eigenvalues is greater than $-\sigma$, i.e., whether in the complex plane some eigenvalues lie on the right of the line of equation $\Re(s) = -\sigma$. Note that the computation of the new eigenproblem can involve long computations, particularly if n is high.

Due to the mentioned drawbacks, the Routh–Hurwitz criterion will not be dealt with more here; the relevant equations can be found in many texts of applied mathematics and dynamics, as in, for example, A.F. D'Souza, *Design of Control Systems*, Prentice-Hall, Englewood Cliffs, 1988, 199.

A.3.2   The Raleigh quotient

Assume an arbitrary n-dimensional vector $\mathbf{x}$. The ratio

$$R = \frac{\mathbf{x}^T \mathbf{K} \mathbf{x}}{\mathbf{x}^T \mathbf{M} \mathbf{x}} \tag{A.13}$$

is a number that lies between the smallest and largest eigenvalues. If $\mathbf{x}$ is an eigenvector, the Raleigh quotient expressed by Eq. (A.13) coincides with the corresponding eigenvalue. Moreover, if the arbitrary vector $\mathbf{x}$ is a linear combination of a reduced set of eigenvectors, the Raleigh quotient is included in a field spanning the minimum and maximum of the eigenvalues corresponding to the given eigenvectors. If vector $\mathbf{x}$ is close to a generic eigenvector with an error ϵ, the Raleigh quotient is close to the corresponding eigenvalue with an error of the order of the square of ϵ. This means that if the Raleigh quotient is considered a function of vector $\mathbf{x}$, it is stationary in the neighborhood of any eigenvector.

A.3.3   The Dunkerley formula

The so-called Dunkerley formula was a common tool for the computation of an approximate value of the lowest eigenfrequency of undamped systems and is still reported in many handbooks, even if often with other names and in modified forms. Its usefulness lies mostly in the feature of supplying an approximation of the lowest natural frequency that is surely lower than the exact value. It can be used with confidence when checking that the first natural frequency is higher than a given value.

It is based on the compliance formulation (first Eq. (4.22)), in which the highest eigenvalue corresponds to the lowest eigenfrequency. It is well known that the sum of the eigenvalues of a matrix is equal to the sum of the elements on its main diagonal. If, as is usually the case, the first natural frequency is much lower than the others, the square of its reciprocal is very close to the sum of the eigenvalues and then to the trace of the dynamic matrix in the compliance formulation.

Because the Dunkerley formula is mostly used for lumped-parameters models, the mass matrix is diagonal and the elements on the main diagonal of the dynamic matrix

$$\mathbf{D} = \mathbf{K}^{-1}\mathbf{M}$$

can be computed simply as

$$d_{ii} = \beta_{ii} m_{ii} \ .$$

It thus follows that

$$\frac{1}{\omega_1^2} \approx \sum_{\forall i} \frac{1}{\omega_i^2} = \sum_{\forall i} \beta_{ii} m_{ii} \ . \tag{A.14}$$

The use of Eq. (A.14) requires computation of the compliance matrix; its use is therefore simple when the elastic behavior of the system is expressed in terms of coefficients of influence. In such a case it is not even necessary to compute all coefficients, being sufficient to know those that are on the main diagonal. If, however, the stiffness approach is followed, as when using the FEM, the compliance matrix must be computed by inverting the stiffness matrix and the use of the Dunkerley formula may be inconvenient. An exception is when the stiffness matrix has already been factorized, e.g., for the solution of the static problem.

A.3.4   Vector iteration method

The lowest natural frequency can be easily computed using an iterative procedure, known in general as the *vector iteration method*, and, sometimes, in structural dynamics, as the *Stodola method*. It allows the computation of the highest eigenvalue of the dynamic matrix and, when used to obtain the lowest eigenfrequency, the compliance formulation must be used.

By introducing the dynamic matrix

$$\mathbf{D} = \mathbf{K}^{-1}\mathbf{M}$$

into the first Eq. (4.22), it can be rewritten as

$$\mathbf{D}\mathbf{x}_0 = \frac{1}{\omega^2}\mathbf{x}_0 \ . \tag{A.15}$$

If vector $\mathbf{x}_0$ coincides with one of the eigenvectors, the result obtained by premultiplying it by the dynamic matrix is a vector proportional to $\mathbf{x}_0$. The constant of proportionality is the relevant eigenvalue, i.e., the reciprocal of the square of the corresponding eigenfrequency ω_i.

A fast-converging iterative procedure can be devised. Assume a trial vector $\mathbf{x}'$: Because any vector in the space of the configurations of the system can be expressed as a linear combination of the eigenvectors through the modal coordinates, it can be written in the form

$$\mathbf{x}' = \mathbf{\Phi}\boldsymbol{\eta}'. \tag{A.16}$$

Equation (A.16) is, at this stage of the computation, just a formal statement, because the matrix of the eigenvectors $\mathbf{\Phi}$ is still unknown and the modal coordinates corresponding to $\mathbf{x}'$ cannot be computed.

Premultiplying vector $\mathbf{x}'$ by the dynamic matrix, a second vector $\mathbf{x}''$ is readily obtained:

$$\mathbf{x}'' = \mathbf{D}\mathbf{x}' = \mathbf{D}\mathbf{\Phi}\boldsymbol{\eta}' = \sum_{\forall i} \mathbf{D}\mathbf{q}_i \eta'_i. \tag{A.17}$$

Remembering Eq. (A.17), it follows that

$$\mathbf{x}'' = \mathbf{D}\mathbf{\Phi}\boldsymbol{\eta}' = \sum_{\forall i} \frac{1}{\omega_i^2} \mathbf{D}\mathbf{q}_i \eta'_i. \tag{A.18}$$

Because the first vector is not an eigenvector, $\mathbf{x}''$ is not proportional to $\mathbf{x}'$. It can, however, be expressed as a linear combination of the eigenvectors of the system:

$$\mathbf{x}'' = \mathbf{\Phi}\boldsymbol{\eta}''.$$

The modal coordinates of $\mathbf{x}''$ and $\mathbf{x}'$ are then linked by the relationship

$$\eta''_i = \frac{1}{\omega_i^2} \eta'_i. \tag{A.19}$$

Equation (A.19) states that by premultiplying vector $\mathbf{x}'$, corresponding to the modal coordinates $\boldsymbol{\eta}'$, by the dynamic matrix, a second vector $\mathbf{x}''$ is obtained whose modal coordinates $\boldsymbol{\eta}''$ can be obtained from those of the former simply by multiplying them by $1/\omega_i^2$. Because ratio $1/\omega_i^2$ for the first mode is greater (usually much greater, but here it is not strictly needed) than the same ratio for the other modes, it is clear that the first modal coordinate of vector $\mathbf{x}''$ is greater, in relative terms, than that of vector $\mathbf{x}'$. This physically means that the shape of vector $\mathbf{x}''$ is more similar to the first mode shape than that of vector $\mathbf{x}'$.

By repeating the procedure, this similarity increases, iteration after iteration, because the modal coordinates of the nth iterate $\mathbf{x}^{(n)}$ are

$$\eta_i^{(n)} = \frac{1}{\omega_i^{2n}} \eta'_i. \tag{A.20}$$

After a certain number of iterations, it is possible to obtain a vector that coincides, apart from an error that can be arbitrarily small, with the first eigenvector. The first eigenfrequency is then computed through Eq. (A.15). Practically, the starting vector can be chosen arbitrarily: It can be coincident with the static deflected shape but this is not really important. The choice of a vector that is not too different from the first eigenvector allows convergence in a smaller number of iterations to be obtained, but the method converges so fast that the number of iterations is usually very low, even if the starting vector is chosen randomly. At each iteration, the vector is normalized and premultiplied by the dynamic matrix, until the normalization factor at the ith iteration is different from that at the $(i - 1)$th iteration by a quantity smaller than a given tolerance. The last normalization factor so obtained is the reciprocal of the square of the lowest natural frequency.

If the starting vector $\mathbf{x}'$ has a first modal coordinate η_1' that is exactly 0, the procedure should theoretically converge to the second eigenvector. Actually, it is sufficient that the modal coordinate η_1', although very small, is not exactly zero, as happens as a consequence of computational approximation, and convergence to the first mode is at any rate obtained.

The vector iteration method is similar in aim to the Dunkerley formula, with important differences. In the former case, the result can be refined to obtain an error that is arbitrarily small, independent of how much smaller the first natural frequency is with respect to the others (if the first two eigenfrequencies are very close, convergence can be slow but is, at any rate, ensured). In the latter case, the error cannot be corrected and depends on the relative magnitude of the eigenvalues. While the former also allows computation of the mode shape, the latter yields only the value of the frequency.

To obtain the second eigenfrequency, i.e., to make the vector iteration method converge to the second eigenvector, it is necessary to use a starting vector whose first modal coordinate is exactly equal to zero and to verify at each iteration that this feature is maintained. The last statement means that at each iteration the vector obtained must be modified to remove the small component of the first mode that creeps in due to computational errors. If this component is not removed, it would, in subsequent iterations, outgrow all other components.

Consider a generic vector $\mathbf{x}'$, whose modal coordinates are $\boldsymbol{\eta}'$. It is possible to demonstrate that if the first modal coordinate is equal to zero then

$$\mathbf{x}'^{T}\mathbf{M}\mathbf{q}_1 = 0. \tag{A.21}$$

Equation (A.21) can be written in the form

$$\sum_{i=1}^{n} \eta_i' \mathbf{q}_i^T \mathbf{M}\mathbf{q}_1 = 0. \tag{A.22}$$

All terms of the sum on the right-hand side are equal to zero, the first because the first modal coordinate is equal to zero, all other terms owing to the fact that the eigenvectors are m-orthogonal. To ensure that the first modal coordinate of vector x' is vanishingly small, it is then enough to verify that it satisfies Eq. (A.21).

Because, in general, Eq. (A.21) is not satisfied, it is possible to use it to modify one of the elements of vector x' in order to transform it into a new vector x'^* with the required characteristics. Equation (A.21) can be written in the form

$$\sum_{i=1}^{n} \left(x_i' \sum_{j=1}^{n} m_{ij} q_{j1} \right) = 0, \tag{A.23}$$

which can readily be solved in the first element x_1':

$$x_1' = \frac{\sum_{i=2}^{n} \left(x_i' \sum_{i=1}^{n} m_{ij} q_{j1} \right)}{\sum_{j=1}^{n} m_{1j} q_{j1}}. \tag{A.24}$$

This transformation can be implemented by premultiplying vector x' by a matrix S, which is usually referred to as the sweeping matrix

$$x'^* = Sx', \tag{A.25}$$

where

$$S = \begin{bmatrix} 0 & \alpha_1 & \alpha_2 & \alpha_3 & \cdots & \alpha_n \\ 0 & 1 & 0 & 0 & \cdots & 0 \\ 0 & 0 & 1 & 0 & \cdots & 0 \\ \cdots & \cdots & \cdots & \cdots & \cdots & \cdots \\ 0 & 0 & 0 & 0 & \cdots & 1 \end{bmatrix} \tag{A.26}$$

and

$$\alpha_i = -\frac{\sum_{j=1}^{n} m_{(i+1)j} q_{j1}}{\sum_{j=1}^{n} m_{1j} q_{j1}}.$$

Instead of premultiplying the vector obtained at each iteration by the sweeping matrix, it is computationally more efficient to postmultiply the dynamic matrix by the sweeping matrix and to perform the iterative computation using the modified dynamic matrix

$$D^{(2)} = DS. \tag{A.27}$$

Once the second eigenvector has also been computed, the computation can proceed by computing a new sweeping matrix, which also ensures that the second modal coordinate of the relevant vector vanishes, postmultiplying the original dynamic matrix by the sweeping matrix, and repeating the iterative computation. Generally speaking, to obtain the sweeping matrix for the computation of the $(m+1)$th eigenvector, a set of m coupled linear equations must be solved. The computation of the sweeping matrix gets more complex while increasing the order of the eigenvector to be computed.

Alternatively, instead of using the sweeping matrix it is possible to resort to the so-called *deflated dynamic matrices*. The deflated matrix for the computation of the second eigenvector can be computed using the formula

$$\mathbf{D}^{(2)} = \mathbf{D} - \frac{1}{\omega_1^2}\mathbf{q}_1\mathbf{q}_1^T\mathbf{M}, \tag{A.28}$$

where the first eigenvector has been normalized in such a way that the first modal mass has a unit value.

Equation (A.28) can be proved by simply noting that when a vector $\mathbf{x}'$ is premultiplied by matrix $\mathbf{D}^{(2)}$, it follows that

$$\mathbf{D}^{(2)}\mathbf{x}' = \mathbf{D}\mathbf{x}' - \frac{1}{\omega_1^2}\mathbf{q}_1\mathbf{q}_1^T\mathbf{M}\mathbf{x}', \tag{A.29}$$

i.e., writing the equation in terms of modal coordinates,

$$\mathbf{D}^{(2)}\mathbf{x}' = \sum_{i=1}^{n}\eta_i\mathbf{D}\mathbf{q}_i - \frac{1}{\omega_1^2}\sum_{i=1}^{n}\eta_i\mathbf{q}_1\mathbf{q}_1^T\mathbf{M}\mathbf{q}_i. \tag{A.30}$$

Remembering that the eigenvectors are m-orthogonal, only one of the terms of the last sum in Eq. (A.30) is different from zero, the term in which the first modal mass, which has a unit value, is present. Equation (A.30) reduces to

$$\mathbf{D}^{(2)}\mathbf{x}' = \eta_1\left[\mathbf{D}\mathbf{q}_1 - \frac{1}{\omega_1^2}\mathbf{q}_1\right] + \sum_{2=1}^{n}\eta_i\mathbf{D}\mathbf{q}_i = \sum_{i=2}^{n}\eta_i\mathbf{D}\mathbf{q}_i. \tag{A.31}$$

The right-hand side of Eq. (A.31) does not contain the first eigenvector, and the deflated matrix expressed by Eq. (A.28) allows performance of the iterations without the danger that the computation again converges to the first mode. In a similar way, it is possible to show that the deflated matrix for the computation of the third eigenfrequency can be obtained from $\mathbf{D}^{(2)}$ using the formula

$$\mathbf{D}^{(3)} = \mathbf{D}^{(2)} - \frac{1}{\omega_2^2}\mathbf{q}_2\mathbf{q}_2^T\mathbf{M}. \tag{A.32}$$

Similar formulas hold for all subsequent modes. Note that in this case the computation of each mode is not more difficult than the computation of the previous ones. However, the approximation with which the results are obtained gets worse and the use of the vector iteration method, using both sweeping matrices and deflated matrices, is advisable only when a very small number of eigenfrequencies are to be obtained.

A.3.5 *Transformation of the matrices of the eigenproblem*

Many techniques aimed at solving the eigenproblem take advantage of various transformations of the relevant matrices, which, while leaving unmodified the eigenvalues and eigenvectors or modifying them in a predetermined

way, allow the solution to be obtained in a much simpler way. The first is
usually referred to as *eigenvalue shifting*. If the stiffness matrix is substituted by

$$\mathbf{K}^* = \mathbf{K} - a\mathbf{M}\,, \tag{A.33}$$

the eigenvalues of the modified problem $(\mathbf{K}^* - \omega^{*2}\mathbf{M}^*)\mathbf{x} = 0$ are related to
those of the original eigenproblem by the simple relationship

$$\omega^{*2} = \omega^2 - a\,. \tag{A.34}$$

Transformation (A.33) can be useful when the original stiffness matrix
is singular since, with an appropriate choice of the eigenvalue shift a, it
is possible to obtain a matrix $\mathbf{K}^*$ that is positive definite. Another use
of the eigenvalue shifting is that of hastening the convergence of iterative
techniques: Because the speed of convergence depends on the ratio between
the second eigenvalue and the first one, an appropriate shift that increases
this ratio can allow faster computations.

Consider a transformation of the type

$$\mathbf{K}^* = \mathbf{Q}\mathbf{K}\mathbf{Q}^T\,; \qquad \mathbf{M}^* = \mathbf{Q}\mathbf{M}\mathbf{Q}^T\,. \tag{A.35}$$

Under wide assumptions on the transformation matrix $\mathbf{Q}$, the eigenvalues
of the transformed problem are the same as those of the original one. If
the transformed matrices are diagonal, the eigenproblem is immediately
solved.

Many methods have been developed with the aim of determining a transformation matrix that can diagonalize the mass and stiffness matrices, usually working in subsequent steps. A particular case is that of Jacobi method,
devised to deal with the case in which the eigenproblem, reduced to standard form, has a symmetrical dynamic matrix, i.e., matrix $\mathbf{M}$ is an identity
matrix, possibly multiplied by a constant, and then the eigenvectors are orthogonal. Because matrix $\mathbf{M}$ is already diagonal, the transformation matrix
must be orthogonal and can be assumed to be a rotation matrix. The transformation of the stiffness matrix can thus be thought of as a sequence of
rotations of the dynamic matrix, until a diagonal matrix is obtained. Note
that an infinity of rotations is theoretically needed to obtain exactly a diagonal matrix, but, in practice, with a finite number of steps a matrix that
is diagonal within the required accuracy is obtained.

A set of n^2 rotations applied to all combinations of rows and columns, or
better, of $(n^2 - n)/2$ rotations, is referred to as a *Jacobi sweep*. The number
of sweeps needed to achieve the required precision is in most cases between
6 and 10, if particular strategies are followed in the procedure. The total
number of matrix rotations is then between $3n^2$ and $6n^2$. Many computer
programs based on the Jacobi procedure, with different modifications to
hasten convergence and extend it to cases that cannot be reduced to a
symmetric dynamic matrix, are in common use and have been included in
dynamic analysis codes.

Other methods use similar iterative sequences of matrix transformations, such as the LR algorithm and the QR algorithm. The latter is often considered the most efficient general-purpose algorithm to find all eigenvalues and eigenvectors, real or complex, of a general matrix. The Lanczos method is based on the transformation of the relevant matrices into tridiagonal matrices and in the subsequent computation of selected eigenvalues. The various factorization techniques mentioned in Section A.2, such as LU or Choleski factorization, are often used before starting the eigenvalue solution procedure. Also balancing procedures, aimed at avoiding large differences between the elements of the matrices can be very useful. An eigensolution code is actually a sequence of many procedures that transform the relevant matrices, find the eigenvalues and the eigenvectors, and backtransform the results.

A.3.6   Subspace iteration technique

The subspace iteration method is one of the most popular approaches to the computation of the first m eigenvalues and eigenvectors, where $m < n$. The method starts by stating that the m eigenvectors of interest are a linear combination of p (with $p > m$) vectors $\mathbf{r}$ arbitrarily chosen:

$$\mathbf{q}_i = \mathbf{Q}\mathbf{a}_i \,, \tag{A.36}$$

where

$$\mathbf{Q} = [\mathbf{r}_1 \mathbf{r}_2 \ldots \mathbf{r}_p] \qquad \text{and} \qquad i = 1, 2, \ldots, m.$$

Vector $\mathbf{a}_i$ contains the p coefficients of the linear combination yielding the ith eigenvector. The size of the vectors and matrices is n for $\mathbf{q}_i$ and $\mathbf{r}$, $n \times p$ for $\mathbf{Q}$, and p for $\mathbf{a}_i$. This procedure has an immediate geometrical meaning: It states that the m eigenvectors being sought lie in a p-dimensional subspace of the configurations space, which is identified by vectors $\mathbf{r}_i$.

The Raleigh quotient

$$R = \frac{\mathbf{q}_i^T \mathbf{K} \mathbf{q}_i}{\mathbf{q}_i^T \mathbf{M} \mathbf{q}_i} = \frac{\mathbf{a}_i^T \mathbf{K}^* \mathbf{a}_i}{\mathbf{a}_i^T \mathbf{M}^* \mathbf{a}_i}, \tag{A.37}$$

where matrices $\mathbf{M}^*$ and $\mathbf{K}^*$ are obtained through transformation (A.35), coincides with the ith eigenvalue if $\mathbf{q}_i$ coincides exactly with the eigenvector. Moreover, it is possible to state that the linear combination coefficients $\mathbf{a}_i$ leading to an eigenvector can be obtained by imposing a stationarity condition of the Raleigh quotient (A.37). This stationarity condition can be expressed by the equation

$$(\mathbf{K}^* - R\mathbf{M}^*)\mathbf{a} = 0. \tag{A.38}$$

Equation (A.38) defines an eigenproblem yielding the Raleigh quotients, i.e., the eigenvalues and the corresponding vectors $\mathbf{a}_i$, which allow the eigenvectors of the original problem to be found. Obviously, because the size of the eigenproblem is p, only p eigenvalues can be found. Due to the reduced size of the eigenproblem, any standard technique, such as the Jacobi method, can be used without further problems. The eigenvectors so obtained can be transformed back to the original n-dimensional space by premultiplying them by matrix $\mathbf{Q}$. If the first p eigenvectors of the original problem lie exactly in the subspace identified by matrix $\mathbf{Q}$, the solution so obtained would be exact. Because vectors $\mathbf{r}_i$ have been chosen more or less arbitrarily, the solution is only an approximation. Ritz vectors, as defined in Section 6.4, can be used and the approach outlined earlier is usually referred to as the *Ritz method* for the computation of eigenvalues and eigenvectors.

An iterative technique aimed at refining the result obtained through the Ritz method is the essence of the subspace iteration technique. The computation starts by choosing a set of p initial trial vectors, where the number p of dimensions of the subspace is greater than the number m of eigenvalues to be computed. A rule generally followed is to choose p as the minimum value between $2m$ and $m + 8$. From the p trial vectors $\mathbf{x}_i$ a set of Ritz vectors $\mathbf{r}_i{}^*$ are computed through the equation

$$\mathbf{KQ} = \mathbf{MX}, \tag{A.39}$$

where matrix $\mathbf{X}$ contains vectors $\mathbf{x}_i$. The matrices are then transformed to the subspace defined by the Ritz vectors and the eigenproblem is solved using the Jacobi method, as outlined earlier.

A convergence test, aimed at verifying whether the first m eigenvalues obtained are close enough to the true eigenvalues, is performed. If this is not the case, the eigenvectors so obtained are assumed to be new trial vectors $\mathbf{x}_i$, and the procedure is repeated. The procedure converges to the first m eigenvalues unless one of the trial vectors is m-orthogonal to one of the eigenvectors to be found. It is possible to devise a procedure to verify this occurrence and modify the initial choice accordingly. The first trial vector can be assumed arbitrarily, for example, a vector with all unit elements, the second as a vector with all zero terms except the one in the position in which the original matrices have the smallest ratio k_{ii}/m_{ii}, which is assumed to have unit value. The subsequent vectors are similar to the second one, with the second, third, and so on smallest value of k_{ii}/m_{ii}. Because at each iteration a set of equations whose coefficient matrix is $\mathbf{K}$ have to be solved, the factorization of this matrix can be performed only once at the beginning of the computation and need not be repeated anymore.

A.4   Solution of nonlinear sets of equations

The solution of nonlinear sets of equations is still a difficult problem for which a completely satisfactory general solution does not exist. If the set can be reduced to a single nonlinear equation, the bisection method, although usually not very efficient, ensures that all real solutions are found. Many other methods are applicable to this case. In the case of a set with many equations, two approaches, both iterative, are usually possible.

The simplest one is the use of a Jacobi or Gauss–Seidel iterative procedure, already seen for linear sets of equations. If the set of nonlinear equations is written separating the diagonal part of the matrix of the coefficients of the linear part from the out-of-diagonal part as in Eq. (A.8), the equation allowing the obtaining of $\mathbf{x}^{(i+1)}$, at the $(i+1)$th iteration, from vector $\mathbf{x}^{(i)}$ at the ith iteration is

$$\mathbf{A}^*\mathbf{x}^{(i+1)} = \mathbf{b} - \mathbf{A}^{**}\mathbf{x}^{(i)} + \mathbf{g}^{(i)} , \tag{A.40}$$

where $\mathbf{g}(x)$ is the nonlinear part of the set of equations.

The convergence of the procedure is, in general, not sure, depending on the choice of the trial vector used to start the computation. If multiple solutions exist, their domains of attraction may have complex shapes.

The Newton–Raphson algorithm is often regarded as the best choice for the solution of sets of nonlinear equations. It is based on an iterative solution of linear equations obtained through a series expansion of the nonlinear original equations truncated after the first term. It is performed by writing the equations to be solved in the form

$$\mathbf{p}(x) = \mathbf{0} \tag{A.41}$$

and expanding the nonlinear functions $\mathbf{p}(x)$ in the neighborhood of the solution $\mathbf{x}^{(o)}$ in the form

$$\mathbf{p}(x) = \mathbf{p}(x^{(o)}) + \mathbf{S}(x^{(o)}) \left(\mathbf{x} - \mathbf{x}^{(o)}\right) , \tag{A.42}$$

where the elements of the Jacobian matrix $\mathbf{S}$ are

$$S_{ij} = \frac{\partial p_i(x)}{\partial x_j}. \tag{A.43}$$

The equation allowing $\mathbf{x}^{(i+1)}$ at the $(i+1)$th iteration to be obtained from vector $\mathbf{x}^{(i)}$ at the ith iteration is

$$\mathbf{x}^{(i+1)} = \mathbf{x}^{(i)} - h\mathbf{S}^{-1}\{p(x^{(i)})\} , \tag{A.44}$$

where h is a relaxation factor that can be used to hasten convergence but is usually taken equal to unity.

Usually the method converges to one of the solutions of the equation, but the convergence characteristics are strongly influenced by the initial assumption of the trial vector $\mathbf{x}^{(0)}$. Often, for selected values of $\mathbf{x}^{(0)}$, the iterative procedure does not lead to convergence but locks itself in a cycle in which a number of vectors $\mathbf{x}$ are cyclically found. When multiple solutions exist, the domains of attraction of the various solutions can have very complicated shapes, with fractal geometries often found. Moreover, solutions that are physically unstable also have their own domain of attraction: Solutions that are physically impossible can be found when starting from selected initial values. The numerical stability of the solution obtained through the Newton–Raphson method has nothing to do with the physical stability of the same solution. Much research work has been devoted to the Newton–Raphson method.[2]

A.5   Numerical integration in time of the equation of motion

An increasingly popular approach to the computation of the time history of the response from the time history of the excitation is the numerical integration of the equation of motion. However, any solution obtained through this numerical approach must be considered as the result of a numerical experiment and usually gives little general insight into the relevant phenomena. It does not substitute other analytical methods but rather provides a powerful tool to investigate cases that cannot be dealt with in other ways.

The integration of the equation of motion can be performed using a variety of methods. All of them operate following the same guidelines: The state of the system at time $t + \Delta t$ is computed from the known conditions characterizing the state of the system at time t. The finite time interval Δt must be small enough to allow the use of simplified expressions of the equation of motion without introducing large errors. The mathematical simulation of the motion of the system is thus performed step by step, increasing the independent variable t with subsequent finite increments Δt.

The various methods use different simplified forms of the equation of motion, and consequently, the precision with which the conditions at the end of each step are obtained depends on the particular method used. All the simplified forms must tend to the exact equation of motion when the time interval Δt tends to zero. It is therefore obvious that the higher the precision obtainable at each step from a given method is, the longer the time interval that allows to obtaining the required overall precision. The choice of the method is therefore a trade-off between the simplicity of the expression used to perform the computation in each step, which influences the computation time required, and the total number of steps needed to follow the motion of the system for the required time.

The simplest methods are those based on the substitution of the differentials in the equation of motion with the corresponding finite differences

---

[2]The books by H.O. Peitgen: *Newton's Method and Dynamical Systems*, Kluwer Academic Publishers, Dordrecht, 1988, and *The Beauty of Fractals*, Springer, New York, 1986, can be very useful.

and, among them, the simplest is the so-called Euler method, which operates in the phase space, i.e., requires the transformation of the second-order differential equations of motion into a set of first-order equations. In the case of linear systems, it deals with Eq. (1.30).

The ratio between the finite differences

$$\frac{\mathbf{z}_2 - \mathbf{z}_1}{\Delta t}$$

computed between instants t_2 and t_1, separated by the time interval Δt, is replaced for the derivative $\dot{\mathbf{z}}$, and the approximate average value

$$\frac{\mathbf{z}_2 + \mathbf{z}_1}{2}$$

is replaced for the instantaneous value of the same variables. The equation that allows the computation of the state variables at time t_2 is thus

$$(2\mathbf{I} - \Delta t \mathcal{A})\, \mathbf{z}_2 = (2\mathbf{I} + \Delta t \mathcal{A})\, \mathbf{z}_1 + \Delta t \mathcal{B}\,[\mathbf{u}(t_2) + \mathbf{u}(t_1)] \ . \tag{A.45}$$

The matrix of the coefficients of this set of linear equations is constant if the step of integration is not changed during the simulations and needs to be factorized only once.

In the case of systems with parameters that are variable with time, this does not occur, and the relevant matrix must be factorized at each step. In the case of nonlinear systems, the nonlinear part of the equation can be introduced into functions $\mathbf{u}$, and the discretization in time is usually performed by replacing the differentials with the finite differences and using the values of the state variables at time t_1:

$$\mathbf{z}_2 = \mathbf{z}_1 + \Delta t \, (\mathcal{A}\mathbf{z}_1 + \mathcal{B}\mathbf{u}(\mathbf{z}_1, t_1)) \ . \tag{A.46}$$

Another method based on the direct replacement of the finite differences for the differentials but operates directly with the second-order equations is the central differences method. The first and second derivatives of the displacement at time t_i can be expressed as functions of the positions assumed by the system in the same instant and in those that precede and follow it by the time interval Δt:

$$
\begin{aligned}
\dot{\mathbf{x}}_i &= (\mathbf{x}_{i+1} - \mathbf{x}_{i-1})\frac{1}{2\Delta t} \ , \\
\ddot{\mathbf{x}}_i &= (\mathbf{x}_{i+1} - 2\mathbf{x}_i + \mathbf{x}_{i-1})\frac{1}{(\Delta t)^2} \ .
\end{aligned}
\tag{A.47}
$$

By writing the dynamic equilibrium equation of the system at time t_i using expression (A.47) for the derivatives and solving it in $\mathbf{x}_{i+1}$, it follows that

$$
\begin{aligned}
\left(2\mathbf{M} + \Delta t \mathbf{C}\right)\mathbf{x}_{i+1} &= \left(4\mathbf{M} - 2(\Delta t)^2\mathbf{K}\right)\mathbf{x}_i + \\
&+ \left(-2\mathbf{M} + \Delta t \mathbf{C}\right)\mathbf{x}_{i-1} + 2(\Delta t)^2\mathbf{f}(t_i) \ .
\end{aligned}
\tag{A.48}
$$

Equation (A.48) allows computation of the position at time t_{i+1} once the positions at times t_{i-1} and t_i are known. If the time increment Δt is kept constant, the factorization of the matrix of coefficients of Eq. (A.48) can be performed only once.

The computation must start from the initial conditions, which are usually expressed in terms of positions and velocities at time t_o. The central differences method, however, requires knowledge of the positions at two instants preceding the relevant one and not of positions and velocities at a single instant. The positions at the instant before the initial time are easily extrapolated:

$$\mathbf{x}_{-1} = \mathbf{x}_0 - \mathbf{v}_0\Delta t + \ddot{\mathbf{x}}_0 \frac{(\Delta t)^2}{2}\,, \qquad (A.49)$$

where the acceleration $\ddot{\mathbf{x}}_0$ at time t_o is easily computed from the equation of motion. The iterative computation can thus start.

The central differences method is said to be *explicit*, because the dynamic equilibrium equation is written at time t_i, at which the position is known. On the contrary, the dynamic equilibrium equation can be written at time t_{i+1}, at which the position is unknown. In this case, the method is said to be *implicit*.

One of the most common implicit methods is the Newmark method. It is based on the extrapolation of the positions and velocities at time t_{i+1} as functions of the unknown accelerations at the same time:

$$\begin{aligned}
\dot{\mathbf{x}}_{i+1} &= \dot{\mathbf{x}}_i + \Delta t \left[(1+\gamma)\ddot{\mathbf{x}}_i + \gamma\ddot{\mathbf{x}}_{i+1} \right], \\
\mathbf{x}_{i+1} &= \mathbf{x}_i + \Delta t\, \dot{\mathbf{x}}_i + (\Delta t)^2 \left[\left(\tfrac{1}{2} - \beta\right)\ddot{\mathbf{x}}_i + \beta\ddot{\mathbf{x}}_{i+1} \right].
\end{aligned} \qquad (A.50)$$

Parameter β is linked to the assumed time history of the acceleration in the time interval from t_i to t_{i+1}. If, for example, the acceleration is constant, $\beta = 1/4$; a linearly variable acceleration leads to $\beta = 1/6$. Parameter γ, usually stated at the value $1/2$, controls the stability of the method.

The equation of motion written at time t_{i+1} can be used to obtain the displacements $\mathbf{x}_{i+1}$ at the end of the time step:

$$\begin{aligned}
\left(2\mathbf{M} + \Delta t\mathbf{C} + 2\beta(\Delta t)^2\mathbf{K} \right)\mathbf{x}_{i+1} = {}& \left(4\mathbf{M} - 2(1 - 2\beta)(\Delta t)^2\mathbf{K} \right)\mathbf{x}_i + \\
& + 2\beta(\Delta t)^2 \left[\mathbf{f}(t_{i+1}) + \left(\tfrac{1}{\beta} - 2\right)\mathbf{f}(t_i) - \mathbf{f}(t_{i-1}) \right] + \\
& - \left(2\mathbf{M} + \Delta t\mathbf{C} + 2\beta(\Delta t)^2\mathbf{K} \right)\mathbf{x}_{i-1}.
\end{aligned}$$

$$(A.51)$$

Also in this case the matrix of the coefficients can be factorized only once.

Another implicit method is the Houbolt method, in which the response of the system is approximated by a third-power law spanning for three time intervals of amplitude Δt. The acceleration and velocity at time $t + \Delta t$ are expressed by the following functions of the position at time $t - 2\Delta t$, $t - \Delta t$, t, and $t + \Delta t$:

$$\ddot{\mathbf{x}}_{i+1} = (2\mathbf{x}_{i+1} - 5\mathbf{x}_i + 4\mathbf{x}_{i-1} - \mathbf{x}_{i-2})\frac{1}{(\Delta t)^2} \, ,$$

$$\dot{\mathbf{x}}_{i+1} = (11\mathbf{x}_{i+1} - 18\mathbf{x}_i + 9\mathbf{x}_{i-1} - 2\mathbf{x}_{i-2})\frac{1}{6\Delta t} \, . \tag{A.52}$$

By introducing the values of the velocity and acceleration computed earlier into the equation of motion at time $t + \Delta t$, the following expression, which can be solved in the unknown values of the displacements, is readily obtained:

$$\left(2\mathbf{M} + \tfrac{11}{6}\Delta t\mathbf{C} + (\Delta t)^2\mathbf{K} \right)\mathbf{x}_{i+1} = \left(5\mathbf{M} + 3\Delta t\mathbf{C} + (\Delta t)^2\mathbf{K} \right)\mathbf{x}_i +$$

$$- \left(4\mathbf{M} + \tfrac{3}{2}\Delta t\mathbf{C} \right)\mathbf{x}_{i-1} + \left(\mathbf{M} + \tfrac{1}{3}\Delta t\mathbf{C} \right)\mathbf{x}_{i-2} + (\Delta t)^2\mathbf{f}(t_{i+1}) \, . \tag{A.53}$$

To start the computation, the positions at times 0, $-\Delta t$, and $-2\Delta t$ must be known. Because this step of the computation is not critical, a rough approximation can be obtained by simply assuming constant speed before time $t = 0$. A better approximation can be obtained using the second equation (A.47) for the acceleration at time 0 (which can be computed directly from the equation of motion in which the initial conditions for the position and speed have been introduced), the following equation expressing the velocity

$$\dot{\mathbf{x}}_0 = (2\mathbf{x}_1 + 3\mathbf{x}_0 - 6\mathbf{x}_{-1} + \mathbf{x}_{-2})\frac{1}{6\Delta t} \tag{A.54}$$

and Eq. (A.53) with $i = 0$. There are three linear equations that yield the unknown positions at times Δt, $-\Delta t$, and $-2\Delta t$.

Other methods, such as the Wilson method, are based on the incremental formulation of the equation of motion. The dynamic equilibrium equation at time t_{i+1} can be written in the form

$$\mathbf{M}(\ddot{\mathbf{x}}_i + \Delta\ddot{\mathbf{x}}) + \mathbf{C}(\dot{\mathbf{x}}_i + \Delta\dot{\mathbf{x}}) + \mathbf{K}(\mathbf{x}_i + \Delta\mathbf{x}) = \mathbf{f}(t_i) + \Delta\mathbf{f}, \tag{A.55}$$

which can be solved in $\Delta\ddot{\mathbf{x}}$ and $\Delta\dot{\mathbf{x}}$, yielding

$$\Delta\ddot{\mathbf{x}} = \frac{6}{(\Delta t)^2}\Delta\mathbf{x} - \frac{6}{\Delta t}\dot{\mathbf{x}}_i - 3\ddot{\mathbf{x}}_i \, ,$$

$$\Delta\dot{\mathbf{x}} = \frac{3}{\Delta t}\Delta\mathbf{x} - 3\dot{\mathbf{x}}_i - \frac{\Delta t}{2}\ddot{\mathbf{x}}_i \, . \tag{A.56}$$

Equations (A.56) can be introduced into Eq. (A.55), obtaining

$$\left(6\mathbf{M} + 3\Delta t\mathbf{C} + (\Delta t)^2\mathbf{K} \right)\Delta\mathbf{x} = \mathbf{M}\left(6\Delta t\dot{\mathbf{x}}_i + 3(\Delta t)^2\ddot{\mathbf{x}}_i \right) +$$

$$+ \mathbf{C}\left(3(\Delta t)^2\dot{\mathbf{x}}_i - \frac{(\Delta t)^3}{2}\ddot{\mathbf{x}}_i \right) + (\Delta t)^2\Delta\mathbf{f} \, . \tag{A.57}$$

Equation (A.57) thus allows the position of the system at time t_{i+1} to be computed.

Step-by-step integration methods can be unconditionally stable, when errors remain bounded at increasing values of the time step Δt. If the method is stable only for small enough values of Δt, beyond which it becomes unstable, it is said to be conditionally stable.

The central differences method, for example, is only conditionally stable, and a value of Δt smaller than the period of the free oscillations T divided by π has to be chosen in single-degree-of-freedom systems:

$$\Delta t < \frac{T}{\pi} = 2\sqrt{\frac{m}{k}} \,. \qquad (A.58)$$

The aforementioned formulations of the Wilson and Houbolt methods are unconditionally stable while the Newmark method is unconditionally stable only if $\gamma = 1/2, \beta \geq 1/4$. Values of the time interval smaller than those needed to achieve stability must be used to obtain accurate results. In the case of systems with a single degree of freedom, it is, consequently, not very important to use unconditionally stable methods.

When studying systems with many degrees of freedom, however, the choice can become quite important: The higher-order modes, although physically of little importance, can drive the solution to instability and compel to using a value of the time increment much smaller than that needed to achieve the required precision. In this case there are two viable choices: To use unconditionally stable methods or to resort to the modal approach, which allows high-frequency modes to be discarded and has the added advantage of accepting different values of the step of integration for the various modes.

This is particularly true in the case of *stiff* systems. Although it is difficult to give an exact definition of a stiff system, it may be said that it is not only a matter of having high natural frequencies (and then requiring small time increments for numerical integration) but mainly in being characterized by widely different time scales of their time histories. In other words, a stiff system has natural frequencies spanning a wide range. They are difficult to integrate, and purposely developed (possibly implicit) algorithms are required.

To evaluate the errors introduced by the numerical integration, it is possible to simulate the motion of a system with a single degree of freedom excited by a harmonic forcing function. In this case, two types of errors are readily seen: a decrease of the amplitude in time, which is greater than that due to the actual damping of the system, and an increase of the period. The first effect is equivalent to the introduction of a numerical damping which, in some cases, can be exploited to reduce stability problems. The second effect is well shown by the curves reported in Fig. A.1, obtained by integrating the equation of motion of an undamped system with a single degree of freedom using the Newmark algorithm. No forcing function has

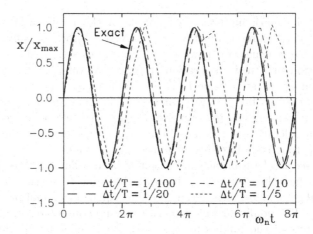

FIGURE A.1. Numerical integration of the equation of motion of a system with a single degree of freedom and comparison with the analytical solution; Newmark method with different values of the time step.

been introduced, and the initial conditions corresponding to an impulse excitation have been used.

In any case, a correct choice of the time step Δt ensures that both effects are very small and that the required accuracy is obtained.

All numerical integration methods can be used with small modifications to study nonlinear systems, provided the nonlinear forces are included in the forcing vector $\mathbf{f}$. In the case of explicit methods there is no difficulty, because the equilibrium equations are written at time t_i in which the displacements are known and the nonlinear functions can be easily computed. In the case of the central differences method, the equation yielding the values of the unknowns at time x_{i+1} is

$$\left(2\mathbf{M} + \Delta t\mathbf{C}\right)\mathbf{x}_{i+1} = \left(4\mathbf{M} - 2(\Delta t)^2\mathbf{K}\right)\mathbf{x}_i + $$
$$+\left(-2\mathbf{M} + \Delta t\mathbf{C}\right)\mathbf{x}_{i-1} - 2(\Delta t)^2\mathbf{g}_i + 2(\Delta t)^2\mathbf{f}(t_i)\,, \quad \text{(A.59)}$$

where vector function $\mathbf{g}$ expresses the nonlinear part of the model. The matrix of the coefficients is also, in this case, a constant and can be factorized only once at the beginning of the computation.

Other methods, particularly implicit ones, require the solution of a set of nonlinear equations at each step, or, more often, the inclusion of the derivatives of the nonlinear functions in the matrix of the coefficients of the set of linear equations. This has the advantage of obtaining greater precision, given the time step Δt, at the cost of performing the factorization of the matrix of the coefficients at each step.

In the Newmark method, for example, if the nonlinear function $\mathbf{g}$ is included in the forcing functions, Eq. (A.51) cannot be solved directly. An iterative scheme can, however, be easily devised. When starting the computation at a new integration step, a trial value of the acceleration at the end of the step (time t_{i+1}) is assumed: $\ddot{\mathbf{x}}_{i+1}^*$. It can be chosen in an arbitrary way, but if a value not too different from the actual one is chosen, the convergence is faster. At the first trial, it is possible to assume that the acceleration at the end of the step is the same as at the beginning or, better, to extrapolate the value from the previous ones.

Once the acceleration at time t_{i+1} has been assumed, it is possible to compute the velocities and position at the same time using Eq. (A.50). These values can be introduced into the equation of motion written at time t_{i+1}. The nonlinear functions can now be computed, obtaining a set of equations in the unknowns $\ddot{\mathbf{x}}_{i+1}$, which is, generally speaking, linear because the equations of motion are usually linear in the accelerations.

The values $\ddot{\mathbf{x}}_{i+1}$ so obtained are compared with the trial values $\ddot{\mathbf{x}}_{i+1}^*$. If they are close enough, accelerations $\ddot{\mathbf{x}}_{i+1}$ can be assumed to be correct; the state of the system obtained at the end of the step is retained and the computation proceeds to the following step.

If, on the contrary, they are not close enough, a new set of accelerations $\ddot{\mathbf{x}}_{i+1}^{**}$ is assumed and the computation is repeated. The choice of $\ddot{\mathbf{x}}_{i+1}^{**}$ is delicate because it strongly influences the number of iterations. If the values computed at the end of the step are directly assumed for the new trial, the correction is generally too large, and oscillations about the correct value may be caused. To avoid these mathematical oscillations, some sort of numerical damping can be introduced, assuming that

$$\ddot{\mathbf{x}}_{i+1}^{**} = \ddot{\mathbf{x}}_{i+1}^* + \frac{1}{\beta'}(\ddot{\mathbf{x}}_{i+1} - \ddot{\mathbf{x}}_{i+1}^*). \tag{A.60}$$

If $\beta' = 1$, no damping is introduced. With growing β' the oscillations are reduced and, for a 'critical value' of β', the maximum speed of convergence is reached. If β' is further increased, the number of iterations needed increases again. The optimum value of β' must be obtained by trials for each case.

The Newmark method allows a good precision to be achieved with relatively long steps of integration. Other choices are various formulations of the Runge–Kutta method, of different order, and with or without adaptive time steps, various predictor–corrector methods, and Nystrom and Bulirsh–Stoer methods.[3]

---

[3]They are described in many books, such as J. Stoer, R. Burlish, *Introduction to Numerical Analysis*, Springer-Verlag, New York, 1980, and W.H. Press, B.P. Flannery, S.A. Teukolsky, W.T. Vetterling, *Numerical Recipes, the Art of Scientific Computing*, Cambridge Univ. Press, Cambridge, England, 1986.

Appendix B
Laplace Transform Pairs

$f(t)$	$\tilde{f}(s)$
1	$\dfrac{1}{s}$
t	$\dfrac{1}{s^2}$
$t^n \quad (n = 1,\, 2,\, ...)$	$\dfrac{n!}{s^{n+1}}$
$\dfrac{t^{n-1}}{(n-1)!}$ (with $0!=1$)	$\dfrac{1}{s^n}$
$e^{\omega t}$	$\dfrac{1}{s-\omega}$
$te^{\omega t}$	$\dfrac{1}{(s-\omega)^2}$
$e^{\omega t}\,\dfrac{t^{n-1}}{(n-1)!}$ (with $0!=1$)	$\dfrac{1}{(s-\omega)^n}$
$1 - e^{\omega t}$	$\dfrac{-\omega}{s(s-\omega)}$
$\sin(\omega t)$	$\dfrac{\omega}{s^2+\omega^2}$
$\cos(\omega t)$	$\dfrac{s}{s^2+\omega^2}$
$\sin(\omega t - \phi)$	$\dfrac{\omega\cos(\phi) - s\sin(\phi)}{s^2+\omega^2}$
$\sinh(\omega t)$	$\dfrac{\omega}{s^2-\omega^2}$
$\cosh(\omega t)$	$\dfrac{s}{s^2-\omega^2}$

$f(t)$	$\tilde{f}(s)$
$1 - \cos{(\omega t)}$	$\dfrac{\omega^2}{s\left(s^2 + \omega^2\right)}$
$\omega t - \sin{(\omega t)}$	$\dfrac{\omega^3}{s^2\left(s^2 + \omega^2\right)}$
$\omega t \cos{(\omega t)}$	$\dfrac{\omega\left(s^2 - \omega^2\right)}{\left(s^2 + \omega^2\right)^2}$
$\omega t \sin{(\omega t)}$	$\dfrac{2\omega^2 s}{\left(s^2 + \omega^2\right)^2}$
$\delta(t)$ (Dirac function)	1
$u(t)$ (step function)	$\dfrac{1}{s}$
$\dfrac{1}{\sqrt{1-\zeta^2}}e^{-\zeta\omega t}\sin\left(\sqrt{1-\zeta^2}\omega t\right)$ $(\zeta < 1)$	$\dfrac{1}{s^2 + 2\zeta\omega s + \omega^2}$
$e^{-\zeta\omega t}\left[\cos\left(\sqrt{1-\zeta^2}\omega t\right) + \dfrac{\zeta}{\sqrt{1-\zeta^2}}\right.$ $\left. \sin\left(\sqrt{1-\zeta^2}\omega t\right)\right]$	$\dfrac{s + 2\zeta\omega}{s^2 + 2\zeta\omega s + \omega^2}$
$\dfrac{e^{bt} - e^{at}}{b - a}$ if $a \neq b$	$\dfrac{1}{(s - a)(s - b)}$
$\dfrac{be^{bt} - ae^{at}}{b - a}$ if $a \neq b$	$\dfrac{s}{(s - a)(s - b)}$
$\dfrac{1}{2\omega^2}\left[\cosh{(\omega t)} - \cos{(\omega t)}\right]$	$\dfrac{s}{s^4 - \omega^4}$
$\dfrac{1}{2\omega}\left[\sinh{(\omega t)} + \sin{(\omega t)}\right]$	$\dfrac{s^2}{s^4 - \omega^4}$
$\dfrac{1}{2}\left[\cosh{(\omega t)} + \cos{(\omega t)}\right]$	$\dfrac{s^3}{s^4 - \omega^4}$
Triangular wave function (Fig. B.1a)	$\dfrac{1}{as^2}\tanh\left(\dfrac{as}{2}\right)$
Square wave function (Fig. B.1b)	$\dfrac{1}{s}\tanh\left(\dfrac{as}{2}\right)$
Rectified sine wave function (Fig. B.1c)	$\dfrac{\pi a}{a^2 s^2 + \pi^2}\coth\left(\dfrac{as}{2}\right)$
Half rectified sine wave function (Fig. B.1d)	$\dfrac{\pi a}{\left(a^2 s^2 + \pi^2\right)\left(1 - e^{-as}\right)}$
Saw tooth wave function (Fig. B.1e)	$\dfrac{1}{as^2} - \dfrac{e^{-as}}{s\left(1 - e^{-as}\right)}$
Triangular function (Fig. B.1f)	$\dfrac{F_0}{as^2}\left(1 + e^{-as}\right)$
Pulse function (Fig. B.1g)	$\dfrac{e^{-as}\left(1 - e^{-as}\right)}{s}$
Multistep function (Fig. B.1h)	$\dfrac{1}{s\left(1 - e^{-as}\right)}$
Time-limited ramp (Fig. B.1i)	$\dfrac{F_0}{a}\left[\dfrac{1}{s^2} - e^{-as}\left(\dfrac{a}{s} - \dfrac{1}{s^2}\right)\right]$
$\sin\left(\dfrac{\pi t}{a}\right)$ for $0 \leq t < a$; 0 for $t > a$	$\dfrac{a\left(1 + e^{-as}\right)}{a^2 s^2 + \pi^2}$

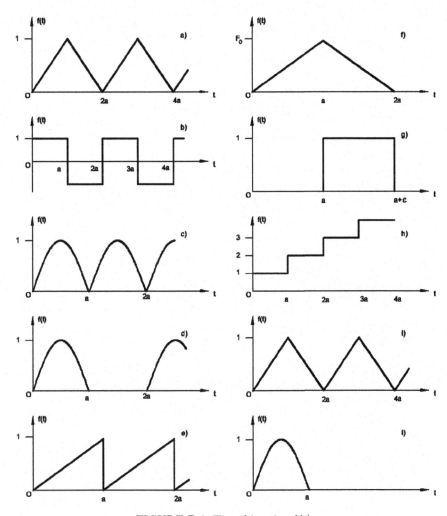

FIGURE B.1. Time histories $f(t)$.

Bibliography

So many papers and books have been published on mechanical vibrations that any attempt to supply a complete or even representative bibliography is bound to fail. Only on the subject of rotor dynamics, for example, Loewi and Piarulli (1969) list 554 titles. Every year, there are many conferences on rotor dynamics, and tens of papers are presented at each.

Also, many papers and books that deal with other subjects contain parts related to mechanics of vibration. For example, books on springs, gears, bearings, and so on, contain chapters on the dynamic problems of these machine elements, which are often of interest to those who study the general problem of mechanical vibrations.

The author chose to give a list of books he considers oriented toward the subject of this book. It is not aimed at being comprehensive; the author is sure that many books that should have been included were left out. Proceedings of conferences or collections of papers are also not included, or the list would have been too long. Where translations in different languages exist and an English edition was found, the latter was listed.

The bibliography is subdivided into three parts, one for each part of this book, and the titles are listed in each part by order of the year of publication. Books published in the same year are listed in alphabetical order by last name of the first author.

As a personal suggestion of the author to all who start the study of the subject of mechanical vibrations, the short book by Bishop, *Vibration*, (Cambridge University Press, 1979), is well worth reading for its nonmathematical approach, which makes the reading easy, pleasant, and involving. For a more mathematical approach to the subject, the classical book by

Lord Rayleigh (1877) is still worthwhile reading. Its title is somewhat misleading, because it deals with mechanical vibrations in general.

General mechanical vibrations

[1] J.S. Anderson, M. Bratos Anderson, *Solving Problems in Vibrations*, Longman, Singapore, 1987.

[2] A.A. Andronov, C.E. Chaikin, *Theory of Oscillations*, Princeton, 1948.

[3] A.A. Andronov, A.A. Vitt, S.E. Khaikin, *Theory of Oscillators*, Dover, New York, 1987.

[4] N.K. Bajaj, *The Physics of Waves and Oscillations*, McGraw-Hill, New Delhi, 1988.

[5] R. Baldacci, C. Ceradini, E. Giangreco, *Dinamica e Stabilità*, CISIA, Milano, 1971.

[6] J.R. Barker, *Mechanical and Electrical Vibrations*, Wiley, New York, 1964.

[7] C.F. Beards, *Vibrations and Control Systems*, Ellis Horwood, Chichester, 1988.

[8] C.F. Beards, *Engineering Vibration Analysis with Applications to Control Systems*, Arnold, London, 1995.

[9] C.F. Beards, *Structural Vibration, Analysis and Damping*, Arnold, London, 1996.

[10] J.S. Bendat, A.G. Piersol, *Random Data: Analysis and Measurement Procedures*, Wiley, New York, 1971.

[11] J.S. Bendat, A.G. Piersol, *Engineering Applications of Correlation and Spectral Analysis*, Wiley, New York, 1980.

[12] J.S. Bendat, A.G. Piersol, *Random Data*, Wiley, New York, 1986.

[13] L.L. Beranek, *Noise and Vibration Control*, McGraw-Hill, New York, 1971.

[14] G.V. Berg, *Elements of Structural Dynamics*, Prentice Hall, Englewood Cliffs, NJ, 1989.

[15] W.G. Bickley, A. Talbot, *An Introduction to the Theory of Vibrating Systems*, Clarendon Press, Oxford, 1961.

[16] J.M. Biggs, *Introduction to Structural Dynamics*, McGraw-Hill, New York, 1964.

[17] R.E.D. Bishop, *Vibration*, Cambridge Univ. Press, Cambridge, 1979.

[18] R.E.D. Bishop, G.M.L. Gladwell, *The Matrix Analysis of Vibration*, Cambridge Univ. Press, Cambridge, 1965.

[19] R.E.D. Bishop, D.C. Johnson, *Vibration Analysis Tables*, Cambridge Univ. Press, Cambridge, 1956.

[20] W.K. Blake, *Mechanics of Flow-Induced Sound and Vibration*, Academic Press, New York, 1986.

[21] R.D. Blevins, *Flow Induced Vibration*, Van Nostrand, New York, 1977.

[22] R.D. Blevins, *Formulas for Natural Frequency and Mode Shape*, Van Nostrand, New York, 1979.

[23] V.V. Bolotin, *Random Vibration of Elastic Systems*, Martinus Nijoff Publ., The Hague, 1984.

[24] A. Bolton, *Structural Dynamics in Practice, a Guide for Professional Engineers*, McGraw-Hill, New York, 1994.

[25] W.J. Bottega, *Engineering Vibrations*, Taylor & Francis, Boca Raton, FL, 2006.

[26] J.T. Broch, *Mechanical Vibration and Shock Measurements*, Brüel & Kjaer, Naerum, 1972.

[27] R. Buckley, *Oscillations and Waves*, Adam Hilger, Bristol, 1985.

[28] R. Burton, *Vibration and Impact*, Dover, New York, 1958.

[29] G. Buzdugan, *La Mesure des Vibrations m écaniques*, Eyeralles, Parigi, 1964.

[30] G. Buzdugan, E. Mihailescu, M. Rades, *Vibration Measurement*, Martinus Nijoff Publication, Dordrecht, 1986.

[31] R.H. Cannon, *Dynamics of Physical Systems*, McGraw-Hill, New York, 1967.

[32] C. Carmignani, *Fondamenti di Dinamica Strutturale*, ETS, Pisa, 1990.

[33] F. Cesari, *Metodi di Calcolo nella Dinamica delle Strutture*, Pitagora, Bologna, 1983.

[34] W. Chan-Xie, *Dynamic Stability of Structures*, Cambridge University Press, Cambridge, 2006.

838   Bibliography

[35] A.H. Church, *Mechanical Vibrations*, Wiley, New York, 1957.

[36] R.W. Clough, J. Penzien, *Dynamics of Structures*, McGraw-Hill, New York, 1975.

[37] I. Cochin, *Analysis and Design of Dynamic Systems*, Harper & Row, New York, 1980.

[38] R.R. Craig, *Structural Dynamics*, Wiley, New York, 1981.

[39] R.R. Craig Jr., A.J. Kurdila, *Fundamental of Structural Dynamics* (2nd Ed.), Wiley, New York, 2006.

[40] S.H. Crandall, *Engineering Analysis*, McGraw-Hill, New York, 1956.

[41] S.H. Crandall, *Random Vibration*, MIT Press, Cambridge, MA, 1958.

[42] S.H. Crandall, *Random Vibration*, vol. 2, MIT Press, Cambridge, MA, 1963.

[43] S.H. Crandall, W.D. Mark, *Random Vibrations in Mechanical Systems*, Academic Press, New York, 1963.

[44] S. H. Crandall, W.D. Mark, *Random Vibrations in Mechanical Systems*, Academic Press, New York, 1973.

[45] C.E. Crede, *Vibration and Shock Isolation*, Wiley, New York, 1951.

[46] C.W. Da Silva, *Vibration and Shock Handbook*, Taylor & Francis, Boca Raton, FL, 2005.

[47] C.W. Da Silva, *Vibration* (2nd Ed.), Taylor & Francis, Boca Raton, FL, 2007.

[48] O. Danek, L. Spacek, *Selbsterregte Schwingungen an Werkzengmaschinen*, V.E.B., Berlin, 1962.

[49] M.T. De Almeida, *Vibrações Mech ânicas para Engenheiros*, Blücher, San Paulo, 1987.

[50] M. Del Pedro, P. Pahud, *Vibration Mechanics*, Kluwer Academic Publishers, Dordrecht, 1989.

[51] M. Del Pedro, P. Pahud, *Vibration Mechanics, Linear Discrete Systems*, Kluver Academic Publishers, Dordrecht, 1991.

[52] J.P. Den Hartog, *Mechanical Vibrations*, McGraw-Hill, New York, 1956.

[53] A.D. Dimarogonas, S. Haddad, *Vibration for Engineers*, Prentice Hall, Englewood Cliffs, NJ, 1992.

[54] J. Donéa (Editor), *Advanced Structural Dynamics*, Applied Science Publication, London, 1980.

[55] J.F. Doyle, *Static and Dynamic Analysis of Structures*, Kluwer Academic Publishers, Dordrecht, 1991.

[56] J.F. Doyle, *Wave Propagation in Structures*, Springer, New York, 1997.

[57] T.V. Duggan, *Power Transmission and Vibration Considerations in Design*, Iliffe Books, London, 1971.

[58] F.R. Erskine Crossley, *Dynamics in Machines*, Ronald Press Co., New York, 1954.

[59] D.J. Ewins, *Machinery Noise and Diagnostics*, Butterworth, London, 1986.

[60] F.J. Fahi, P. Gardonio, *Sound and Structural Vibration, Second Edition: Radiation, Transmission and Response*, Academic Press, Oxford, 2007.

[61] F. Fahy, *Sound and Structural Vibration* (6th printing), Academic Press, New York, 2000.

[62] D.G. Fertis, *Dynamics and Vibration of Structures*, Wiley, New York, 1973.

[63] O. Foppl, *Grundzüge der Technischen Schwingungslehere*, Springer, Berlin, 1923.

[64] L. Fryba, *Vibration of Solids and Structures Under Moving Loads*, Noordhoff, Groningen, 1972.

[65] M. Géradin, A. Cardona, *Flexible Multibody Dynamics*, Wiley, New York, 2001.

[66] M. Géradin, D. Rixen, *Mechanical Vibrations, Theory and Application to Structural Dynamics*, Wiley, New York, 1997.

[67] M.F. Gardner, J.L. Barnes, *Transients in Linear Systems*, Wiley, New York, 1942.

[68] G. Genta, *Vibration of Structures and Machines*, Springer, New York, 1999.

[69] G.M.L. Gladwell, *Inverse Problems in Vibration*, Martinus Nijoff Publication, Dordrecht, 1986.

[70] S. Goldman, *Vibration Spectral Analysis*, Industrial Press, New York, 1999.

[71] D.J. Gorman, *Free Vibration Analysis of Rectangular Plates*, Elsevier, New York, 1982.

[72] W. Gough, J.P.G. Richards, R.P. Williams, *Vibrations and Waves*, Wiley, New York, 1983.

[73] D. Guicking, *Active Noise and Vibration Control, Reference Bibliography*, University of Göttingen, 1992.

[74] A.R. Guido, S. Della Valle, *Meccanica delle Vibrazioni, Dinamica dei Sistemi a Molti Gradi di Libertà*, Cooperativa Universitaria Editrice Napoletana, Napoli, 1988.

[75] R. Gutowski, V.A. Swietlicki, *Dynamika Idragania Ukladow Mechanicznych, Panstwowa Wydawnictwo Naulowe*, Warsaw, 1986.

[76] C.M. Harris, C.E. Crede, *Shock and Vibration Handbook*, McGraw-Hill, New York, 1961.

[77] C.M. Harris, A.G. Piersol, *Harris' Shock and Vibration Handbook*, McGraw-Hill, New York, 2001.

[78] J.B. Hartman, *Dynamics of Machinery*, McGraw-Hill, New York, 1956.

[79] H.M. Hausen, P.F. Chenea, *Mechanics of Vibration*, Wiley, New York, 1952.

[80] A.R. Holowenko, *Dynamics of Machines*, Wiley, New York, 1955.

[81] J.B. Hunt, *Dynamic Vibration Absorbers*, Mech. Eng. Publications, London, 1979.

[82] W.C. Hurty, M.F. Rubinstein, *Dynamics of Structures*, Prentice Hall, Englewood Cliffs, 1964.

[83] R.A. Ibrahim, *Parametric Random Vibration*, R.S.P., Wiley, New York, 1985.

[84] D.J. Inman, *Engineering Vibration*(3rd Ed.), Prentice Hall, Englewood Cliffs, NJ, 2007.

[85] L.S. Jacobsen, R.S. Ayre, *Engineering Vibrations*, McGraw-Hill, New York, 1958.

[86] D.S. Jones, *Electrical and Mechanical Oscillations*, Routledge & Kegan, London, 1961.

[87] J.L. Junkins, Y. Kim, *Introduction to Dynamics and Control of Flexible Structures*, AIAA, Washington, 1993.

[88] S. Kaliski (Editor), *Vibrations and Waves*, Elsevier, Amsterdam, 1992.

[89] A. Kalnins, C.L. Dym (Editors), *Vibration. Beams, Plates & Shells*, Dowden, Hutchinson & Ross, Stroudsburg, 1976.

[90] S.G. Kelly, *Fundamental of Mechanical Vibrations*, McGraw-Hill, New York, 1993.

[91] S.G. Kelly, *Schaum's Outline of Mechanical Vibrations*, McGraw-Hill, New York, 1996.

[92] J.D. Koplunov, L.Y. Kossovich, F.V. Nolte, *Dynamics of Thin Walled Elastic Bodies*, Academic Press, New York, 1998.

[93] B.G. Koronev, L.M. Reznikov, *Dynamic Vibration Absorbers, Theory and Technical Applications*, Wiley, Chichester, 1993.

[94] J. Kozesnik, *Dynamics of Machines*, SNTL, Praga, 1962.

[95] G. Krall, *Meccanica Tecnica delle Vibrazioni*, Zanichelli, Bologna, 1940.

[96] B.T. Kulakowski, J.E. Gardner, J.L. Shearer, *Dynamic Modeling and Control of Engineering Systems*, Cambridge University Press, Cambridge, 2007.

[97] F. Küçükay, *Dynamic der Zahnradgetriebe*, Springer, Berlin, 1987.

[98] M. Lalanne, P. Berthier, J. Der Hagopian, *Mechanical Vibrations for Engineers*, Wiley, New York, 1983.

[99] B.J. Lazan, *Damping of Materials and Members in Structural Mechanics*, Pergamon Press, Oxford, 1968.

[100] S. Levy, J.P.D. Wilkinson, *The Component Element Method in Dynamics*, McGraw-Hill, New York, 1976.

[101] R.H. Lyon, *Modal Analysis of Large Structures*, Brï & Kjaer, Naerum, 1987.

[102] R. Mathey, *Physique des Vibrations M écaniques*, Dunod, Parigi, 1963.

[103] R. Mazet, *Mécanique Vibratoire*, Librairie Polytechnique Béranger, Paris, 1955.

[104] K.G. McConnel, *Vibration Testing, Theory and Practice*, Wiley, New York, 1995.

[105] L. Meirovitch, *Analytical Methods in Vibrations*, Macmillan, New York, 1967.

[106] L. Meirovitch, *Elements of Vibration Analysis*, McGraw-Hill, New York, 1975.

[107] L. Meirovitch, *Introduction to Dynamics and Control*, Wiley, New York, 1985.

[108] L. Meirovitch, *Elements of Vibrations Analysis*, McGraw-Hill, New York, 1986.

[109] L. Meirovitch, *Dynamics and Control of Structures*, Wiley, New York, 1990.

[110] L. Meirovitch, *Principles and Technology of Vibration*, Prentice Hall, Englewood Cliffs, NJ, 2007.

[111] V. Migulin, *Basic Theory of Oscillations*, Mir, Moscow, Russia, 1983.

[112] D.K. Miu, *Mechatronics, Electromechanics and Contromechanics*, Springer, New York, 1993.

[113] R.K. Mobley, *Vibration Fundamentals*, Butterworth-Heinemann, Woburn, 1999.

[114] P.C. Müller, W.O. Schiehlen, *Linear Vibrations*, Martinus Nijoff Publication, Dordrecht, 1985.

[115] N.O. Myklestad, *Fundamentals of Vibration Analysis*, McGraw-Hill, New York, 1956.

[116] A.D. Nashif, D.I.G. Jones, J.P. Henderson, *Vibration Damping*, Wiley, New York, 1985.

[117] D.E. Newland, *Random Vibrations and Spectral Analysis*, Longman, London, 1975.

[118] D.E. Newland, *An Introduction to Random Vibration and Spectral Analysis* (II Ed.), Longman, London, 1984.

[119] D.E. Newland, *Mechanical Vibration Analysis and Computations*, Longman, Singapore, 1989.

[120] D.E. Newland, *An Introduction to Random Vibrations, Spectral and Wavelet Analysis*, Longman, Singapore, 1993.

[121] D.E. Newland, *An Introduction to Random Vibrations, Spectral & Wavelet Analysis*,(3rd Ed.), Longman's, New York, 2005.

[122] C.H. Norris, R.J. Hansen, M.J. Holley, J.M. Biggs, S. Namyet, J.K. Minami, *Structural Design for Dynamic Loads*, Mc Graw-Hill, New York, 1959.

[123] M.P. Norton, *Fundamentals of Noise and Vibration Analysis for Engineers*, Cambridge Univ. Press, Cambridge, 1989.

[124] H.J. Pain, *The Physics of Vibrations and Waves*, Wiley, New York, 1983.

[125] B. Parker, R. Crossley, *Modal Control, Theory and Applications*, Taylor & Francis, London, 1972.

[126] M. Petyt, *Introduction to Finite Element Vibration Analysis*, Cambridge University Press, Cambridge, 1990.

[127] A.B. Pippard, *The Physics of Vibration*, Cambridge University Press, Cambridge, 1978.

[128] K. Piszczek, J. Niziot, *Random Vibration of Mechanical Systems*, Ellis Horwood, Chichester, 1986.

[129] J.M. Prentis, *Dynamics of Mechanical Systems*, Longman, London, 1970.

[130] A. Preumont, *Random Vibration and Structural Analysis*, Kluwer Academic Publishers, Dordrecht, 1990.

[131] A. Preumont, *Vibration Control of Active Structures*, Kluwer Academic Publishers, Dordrecht, 1997.

[132] V. Prodonoff, *Vibrações Mech ânicas, Simulaçao e Anàlise*, Maity Comunicaçao, Rio de Janeiro, 1990.

[133] J.S. Rao, K. Gupta, *Theory and Practice of Mechanical Vibrations*, Wiley Eastern, Delhi, 1984.

[134] S.S. Rao, *Mechanical Vibrations*, Addison Wesley, Reading, MA, 1986.

[135] S.S. Rao, *Mechanical Vibrations*, Addison Wesley, Reading, MA, 1995.

[136] S.S. Rao, *Mechanical Vibrations* (4th Ed.), Prentice Hall, Englewood Cliffs, NJ, 2003.

[137] L. Rayleigh, *The Theory of Sound*, McMillan, London, 1877 (Last edition 1944).

[138] E.I. Rivin, *Passive Vibration Isolation*, ASME Press, New York, 2003.

[139] J.D. Robrun, C.J. Dodds, D.B. Macvean, V.R. Paling, *Random Vibration*, Springer, Vienna, 1971.

[140] Y. Rocard, *Dynamique Générale des Vibrations*, Masson, Paris, 1948.

[141] V. Rocard, *General Dynamics of Vibrations*, Ungar, New York, 1960.

[142] M. Roseau, *Vibrations in Mechanical Systems*, Springer, Berlin, 1987.

[143] D. Schiff, *Dynamic Analysis and Failure Modes of Simple Structures*, Wiley, New York, 1990.

[144] G.I. Schuëller (Editor), *Structural Dynamics*, Springer, Berlin, 1991.

[145] R.H. Scxanlan, R. Rosenbaum, *Aircraft Vibration and Flutter*, Dover, New York, 1968.

[146] W.W. Seto, *Theory and Problems of Mechanical Vibrations*, Schaum, New York, 1964.

[147] E. Sevin, W.D. Pilken, *Optimum Shock and Vibration Isolation*, The Shock and Vibration Information Center, Washington, DC, 1971.

[148] A.A. Shabana, *Theory of Vibration. Vol. I: An Introduction; Vol. II: Discrete and Continuous Systems*, Springer, New York, 1991.

[149] J.L. Shearer, A.T. Murphy, H.H. Richardson, *Introduction to System Dynamics*, Addison Wesley, Reading, MA, 1971.

[150] J.C. Snowdon, *Vibration and Shock in Damped Mechanical Systems*, Wiley, New York, 1968.

[151] W. Soedel, *Vibrations of Shells and Plates*, Dekker, New York, 1981.

[152] V.A. Svetlickij, *Vibrations Aleatoires des Systèmes Mécaniques*, Technique et Documentation, Parigi, 1980.

[153] J.I. Taylor, *The Vibration Analysis Handbook*, Vibration Consultants, Bayonet Point, 2003.

[154] A. Tenot, *Mesure des Vibrations et Isolation des Assises de Machines*, Dunod, Paris, 1953.

[155] W.T. Thompson, *Vibration Theory and Applications*, Prentice Hall, Englewood Cliffs, NJ, 1965.

[156] W.T. Thompson, M.D. Dahleh, *Theory of Vibrations with Applications* (5th Ed.), Prentice Hall Englewood Cliffs, 1997.

[157] W.T. Thomson, *Theory of Vibration with Applications*, Chapman & Hall, London, 1993.

[158] D. Thorby, *Structural Dynamics and Vibration in Practice: An Engineering Handbook*, Butterworth-Heinemann, Woburn, 2008.

[159] S. Timoshenko, D.H. Young, *Advanced Dynamics*, McGraw-Hill, New York, 1948.

[160] S. Timoshenko, D.H. Younger, W. Weaver, *Vibration Problems in Engineering*, Wiley, New York, 1974.

[161] S.A. Tobias, *Machine-tool Vibrations*, Blackie, Glasgow, 1965.

[162] F.S. Tse, I.E. Morse, R.T. Hinkle, *Mechanical Vibrations*, Allyn & Bacon, Boston, 1978.

[163] G.W. Van Santen, *Vibrations Mé chaniques*, Bibliothèque Technique Philips, Eindhoven, 1957.

[164] I.L. Vér, L.L. Beranek, *Noise and Vibration Control Engineering: Principles and Applications*, Wiley, New York, 2006.

[165] M.A. Wahab, *Dynamics and Vibration: An Introduction*, Wiley, New York, 2008.

[166] R.H. Wallace, *Understanding and Measuring Vibrations*, Springer, New York, 1970.

[167] G.B. Warburton, *The Dynamical Behaviour of Structures*, Pergamon Press, 1964.

[168] G.B. Warburton, *Reduction of Vibrations*, Wiley, New York, 1991.

[169] E.A. Wedemeyer, *Automobilschwingungslehere*, Friedr. Vieweg & Sohn, Braunschweig, 1930.

[170] V. Wowk, *Machinery Vibration, Measurement and Analysis*, McGraw Hill, New York, 1991.

[171] V. Wowk, *Machinery Vibration: Balancing, Special Reprint Edition*, McGraw-Hill, New York, 1998.

[172] C.Y. Young, *Random Vibration of Structures*, Wiley, New York, 1986.

[173] K. Zaveri, M. Phil, *Modal Analysis of Large Structures*, Brüel & Kjaer, Naerum, 1984.

Nonlinear and chaotic vibrations

[1] S.R. Bishop, *Nonlinearity and Chaos in Engineering Dynamics*, Wiley, Chichester, 1994.

[2] N. Bogoliubov, I. Mitropolski, *Les M éthodes Asymptotiques en Théorie des Oscillations Non Linéaires*, Mir, Mosca, 1962.

[3] N.V. Butenin, *Elements of the Theory of Nonlinear Oscillations*, Blaisdell, New York, 1965.

[4] F. Dinca, C. Teodosiu, *Nonlinear and Random Vibrations*, Editura Academiei Rep. Soc. Romania, Bucarest, 1973.

[5] J. Guckenheimer, P. Holmes, *Nonlinear Oscillations, Dynamical Systems, and Bifurcations of Vector Fields*, Springer, New York, 1983.

[6] P. Hagedorn, *Nonlinear Oscillations*, Clarendon Press, Oxford, 1981.

[7] C. Hayashi, *Forced Oscillations in Non-linear Systems*, Nippon Printing and Publ. Comp., Osaka, 1953.

[8] C. Hayashi, *Nonlinear Oscillations in Physical Systems*, McGraw-Hill, New York, 1964.

[9] C. Hayashi, *Nonlinear Oscillations in Physical Systems*, Princeton University Press, Princeton, NJ, 1985.

[10] R.C. Hillborn, *Chaos and Nonlinear Dynamics*, Oxford University Press, Oxford, 1994.

[11] T. Kapitaniak, *Chaos in Systems with Noise*, World Scientific, Singapore, 1988.

[12] H. Kauderer, *Nichtlineare Mechanik*, Springer, Berlin, 1958.

[13] M. Minorsky, *Nonlinear Oscillations*, Van Nostrand, Princeton, NJ, 1962.

[14] R.R. Mohler, *Nonlinear Systems*, Prentice Hall, Englewood Cliffs, NJ, 1991.

[15] F.C. Moon, *Chaotic Vibrations*, Wiley, New York, 1987.

[16] A.H. Nayfeh, D.T. Mook, *Nonlinear Oscillations*, Wiley, New York, 1979.

[17] M. Roseau, *Vibrations Non Liné aires et Théorie de la Stabilité*, Springer, Berlin, 1966.

[18] G. Schmidt, A. Tondl, *Nonlinear Vibration*, Cambridge University Press, Cambridge, 1986.

[19] C. Sparrow, *The Lorentz Equations: Bifurcations, Chaos, and Strange Attractors*, Springer, New York, 1982.

[20] V.M. Starzhinskii, *Applied Methods in the Theory of Nonlinear Oscillations*, Mir, Moscow, 1980.

[21] J.J. Stoker, *Nonlinear Vibrations*, Interscience, New York, 1950.

[22] W. Szemplinska-Stupnicka, *The behaviour of Nonlinear Vibrating Systems*, Kluwer Academic Publishers, Dordrecht, 1990.

[23] J.M.T. Thompson, H.B. Stewart, *Nonlinear Dynamics and Chaos*, Wiley, New York, 1986.

[24] A. Tondl, *Quenching of Self-Excited Vibrations*, Academia, Prague, Czechoslovakia, 1991.

[25] A. Tondl, T. Ruijgrok, F. Verhulst, R. Nabergoj, *Autoparametric Resonance in Mechanical Systems*, Cambridge University Press, Cambridge, 2000.

[26] M. Urabe, *Nonlinear Autonomous Oscillations*, Academic Press, New York, 1967.

Dynamics of rotating and reciprocating machines

[1] L. Buzzi, *Equilibratura*, CEMB, Mandello del Lario, 1971.

[2] D. Childs, *Turbomachinery Rotordynamics*, Wiley, New York, 1993.

[3] M.S. Darlow, *Balancing of High Speed Machinery*, Springer, New York, 1989.

[4] A.D. Dimarogonas, S.A. Paipetis, *Analytical Methods in Rotor Dynamics*, Applied Science Publishers, London, 1983.

[5] F.M. Dimentberg, *Flexural Vibrations of Rotating Shafts*, Butterworth, London, 1961.

[6] G. Genta, *Dynamics of Rotating Systems*, Springer, New York, 2005.

[7] M.J. Goodwin, *Dynamics of Rotor-Bearing Systems*, Unwin Hyman, London, 1989.

[8] W. Ker Wilson, *Torsional Vibration Problems*, Chapman & Hall, London, 1963.

[9] E. Krämer, *Dynamics of Rotors and Foundations*, Springer, Berlin, 1993.

[10] M. Lalanne, G. Ferraris, *Rotordynamics Predictions in Engineering* (2nd Ed.) Wiley, New York, 1990.

[11] M. Lalanne, G. Ferraris, *Rotordynamics Predictions in Engineering* (2nd Ed.), Wiley, New York, 1998.

[12] C.W. Lee, *Vibration Analysis of Rotors*, Kluwer Academic Publishers, Dordrecht, 1993.

[13] R.G. Loewi, V.J. Piarulli, *Dynamics of Rotating Shafts*, The Shock and Vibration Information Center, Naval Res. Lab., Washington, DC, 1969.

[14] O. Marenholtz, *Dynamics of Rotors*, Springer, Vienna, 1984.

[15] A. Muszyńska, *Rotordynamics*, Springer, CRC Press, Boca Raton, FC, 2005.

[16] E.J. Nestorides, *A Handbook on Torsional Vibrations*, Cambridge University Press, Cambridge, 1958.

[17] T. O'Callaghan, *Berechnung von Torsionsschwingungen an Hand der Theorie der Effektiven Massen*, Fachverlag Schiele & schön, Berlin, 1958.

[18] J. Rao, *Rotor Dynamics*, Wiley Eastern, Delhi, 1985.

[19] N.F. Rieger, *Balancing of Rigid and Flexible Rotors*, The Shock and Vibration Information Center, U.S. DoD, Washington, DC, 1986.

[20] H. Schneider, *Balancing Technology*, Schenck, Darmstadt, 1977.

[21] G. Schweitzer, *Critical Speeds of Gyroscopes*, Springer, Vienna, 1972.

[22] J.I. Taylor, D.W. Kirkland, *The Bearing Analysis Handbook: A Practical Guide for Solving Vibration Problems in Bearings*, Vibration Consultants, Bayonet Point, 2004.

[23] A. Tondl, *Some Problems of Rotor Dynamics*, Chapman & Hall, London, 1965.

[24] W.A. Tuplin, *Torsional Vibration, Elementary Theory and Design Calculations*, Chapman & Hall, London, 1934.

[25] J.M. Vance, *Rotordynamics of Turbomachinery*, Wiley, New York, 1988.

[26] T. Yamamoto, J. Ishida, *Linear and Nonlinear Rotordynamics*, Wiley, New York, 2001.

Index

Mechanical Engineering Series

(continued from page ii)

Printed in the United States
By Bookmasters